Strength of Materials

While designing engineering structures, whether they are supporting girders, shock absorbers, or wings of an aircraft, an understanding of structural behavior and the influence of stresses is necessary. Written with a distinct approach of explaining concepts through solved problems this text discusses all fundamental concepts of the strength of materials including stress, strain, elastic constants, shear force, bending moment, and bending stress.

The study of flexural shear stress, conjugate beam method, method of sections and joints, statically determinate trusses, and thin cylinders is presented in detail with the help of solved numerical exercises. The text also discusses advanced concepts of strength of materials such as shear center, rotating discs, unsymmetrical bending, and deflection of trusses. Designed as a foundation text for undergraduates pursuing courses in civil engineering, mechanical engineering, and metallurgical engineering, this book also has use value for candidates appearing for competitive examinations.

T. D. Gunneswara Rao is Professor in the Department of Civil Engineering, National Institute of Technology, Warangal. He has more than 20 years of teaching and research experience. He has been teaching courses on strength of materials, theory of structures, and theory of plates and shells at undergraduate and graduate levels. His research interests include fracture mechanics of concrete structures and fiber reinforced structures.

Mudimby Andal is Associate Professor in the Department of Civil Engineering, Kakatiya Institute of Technology and Science, Warangal. She teaches courses on strength of materials, engineering mechanics, highway and railway engineering, and geotechnical engineering at undergraduate and graduate levels. Her research interests include applied mechanics, strength of materials, and geo-environmental engineering.

Strength of Materials

Strength of Materials

Fundamentals and Applications

T. D. Gunneswara Rao
Mudimby Andal

CAMBRIDGE
UNIVERSITY PRESS

University Printing House, Cambridge CB2 8BS, United Kingdom

One Liberty Plaza, 20th Floor, New York, NY 10006, USA

477 Williamstown Road, Port Melbourne, vic 3207, Australia

314 to 321, 3rd Floor, Plot No.3, Splendor Forum, Jasola District Centre, New Delhi 110025, India

79 Anson Road, #06–04/06, Singapore 079906

Cambridge University Press is part of the University of Cambridge.

It furthers the University's mission by disseminating knowledge in the pursuit of education, learning and research at the highest international levels of excellence.

www.cambridge.org
Information on this title: www.cambridge.org/9781108454285

First published 2018

Printed in India by Rajkamal Electric Press

A catalogue record for this publication is available from the British Library

ISBN 978-1-108-45428-5 Paperback

Additional resources for this publication at www.cambridge.org/9781108454285

To

Late T. Venkateswara Sarma and Late T. Anasuyamma

(Parents of T. D. Gunneswara Rao)

and

Late M. S. Krishnamacharyulu and Late M. Varalakshmi

(Parents of Mudimby Andal)

Contents

PREFACE

This book *Strength of Materials: Fundamentals and Applications* is brought to the readers with an aim to provide sufficient information and point out ways this information can be applied to practical problems in the domain of mechanics, as is done by design engineers in real life situations. It is structured in the form of a text book for undergraduates pursuing civil engineering, mechanical engineering and metallurgical engineering. Our experience in teaching 'strength of materials' over the past 30 years finds its place in this book. Many questions frequently raised by our students, and are also common problems faced by a lot of other students, while attempting to understand the subjects 'strength of materials' or 'mechanics of solids' are addressed in this book in the form of worked out examples and in the detailed treatment of the theoretical aspects.

Each chapter is provided with objectives and these objectives are mapped to the worked-out example problems and the exercise problems. This mapping strategy will also help the teaching faculty in deciding the course objectives and in evaluating the course objectives.

At the end of every chapter, previous GATE examination and UPSC competitive examination objective-type questions are provided with solutions. This book is useful for students preparing for competitive examinations.

The first ten chapters are devoted to understanding the effects of basic structural actions, which would be sufficient for an elementary treatment of the 'strength of materials' course. The remaining seven chapters focus on advanced topics wherein combined structural actions are considered.

Care is exercised so that mistakes or typographical errors are minimized. All the same we request the readers to comment and provide suggestions for improving the next edition of this book.

ACKNOWLEDGEMENTS

We thank God Almighty for giving us the moral support and good health required for bringing this book into existence. We would like to thank each and every person who helped us, directly or indirectly, in the successful completion of this project.

We thank the reviewers for their patience in going through each chapter of the book and providing valuable suggestions for improving the technical aspects as well as readability of the book. We sincerely thank Gauravjeet Singh Reen, Commissioning Editor at Cambridge University Press, who spent a good amount of time making suggestions for editorial improvements and for better appearance of this book.

We are in debt to our research scholars Mallikarjuna Rao, T. Chaitanya, M. Venu, K. V. Ramana and Vinay Kumar who helped us at various stages during the writing of this book.

We cannot find words to describe the blessings and wishes of our family members, without whose support we would not have completed this book.

Last but not least, thanks are due to T. A. Kamakshi, a student of final year B. Tech, Civil Engineering, who acted as a critical reader for each and every worked-out example and unsolved problem, suggesting modifications wherever required.

CHAPTER 1

STRESS–STRAIN

UNIT OBJECTIVE

This chapter provides information about the theory and derivation of formulae for stresses, strains, and deformations. The presentation attempts to help the student achieve the following:

Objective 1: Determine the normal stress and shear stresses.

Objective 2: Determine normal strain and shear strains.

Objective 3: Determine deformations of different structural elements under axial loads.

Objective 4: Calculate the variation in the dimensions caused due to loads.

1.1 INTRODUCTION

The study of strength of materials includes the understanding of internal stresses and deformations of members subjected to external loading. It also includes the study of failure criterion applicable for the solids subjected to loads. The major actions on the bodies subjected to external loading can be considered as axial force, which include axial compression and axial tension, shear force, bending moment, and torsion. Often members are subjected to the mentioned actions either individually or in combined state such as combined bending, torsion, and axial thrust. The internal reactions due to the external forces cannot be visualized, whereas the deformations can be observed, thus can be measured. Hence generally, the failure criterion of a body subjected to external loads depends not only on the internal actions but also on the deformations. To quantify the internal actions due to the above-said forces, the action of the forces and the corresponding deformations are to be studied in detail in the subsequent chapters. The different individual actions on members were presented in Figure 1.1(a)–(e). However, in this chapter concept of internal reactions and their effects will be discussed. In the subsequent chapters, the effects of mentioned individual actions and combined actions will be discussed in detail.

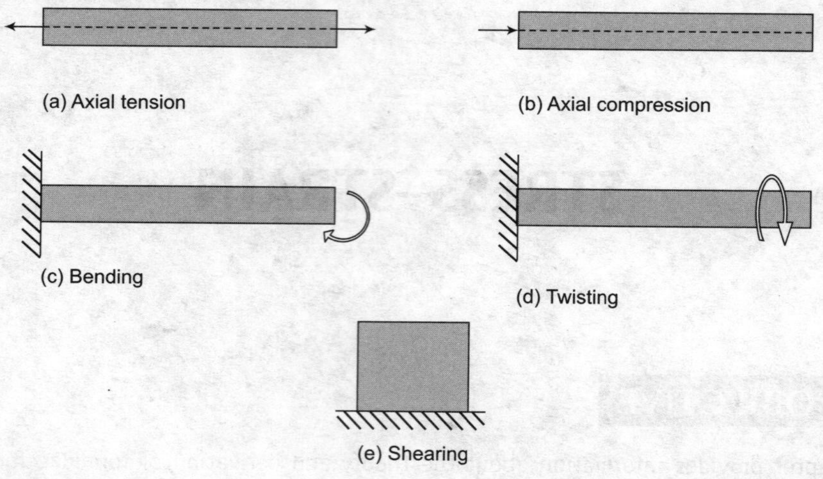

FIGURE 1.1

1.2 NORMAL STRESS AND SHEAR STRESS

Consider a body subjected to several forces and surface tractions as shown in Figure 1.2. Take a section 1-1, to observe the effect of all forces on the section considered. Let 'P' be the net resultant of the forces. The resistance developed by the body to this resultant at any point within the domain of the body is referred as stress.

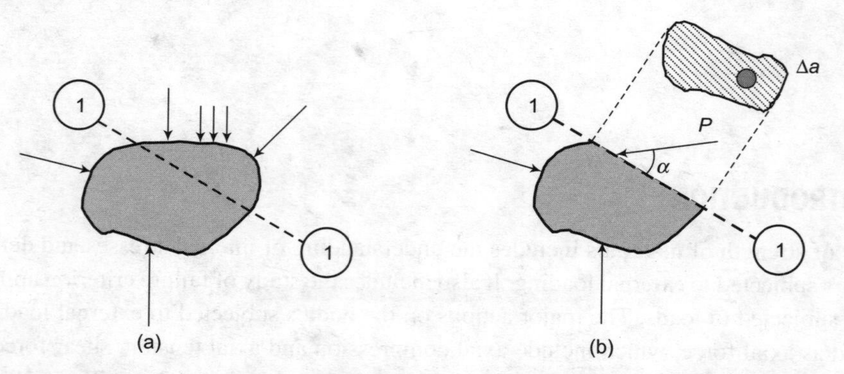

FIGURE 1.2 (a) A body acted upon by external forces; (b) resultant force 'P' acting on section 1-1.

The resultant of forces acting on one side of the section may be resolved into two components. One component is along the plane $P \cos \alpha$, whereas the other component is perpendicular to the plane $P \sin \alpha$.

Consider an elemental area Δa in the plane 1-1, the internal resistance offered by this elemental area for the normal force $P \sin \alpha$ may be written as $Lt_{\Delta a \to 0} \dfrac{\Delta(P \sin \alpha)}{\Delta a}$. This quantity reduces a particular value called normal stress, as the direction of the component is normal to the plane. The letter σ generally denotes this normal stress.

$$Lt_{\Delta a \to 0} \frac{\Delta(P \sin \alpha)}{\Delta a} = \sigma.$$

Similarly there exists internal resistance in tune of the tangential force $P\cos\alpha$. The resistance to tangential force offered by the elemental area can be written as $Lt_{\Delta a \to 0} \frac{\Delta(P \cos \alpha)}{\Delta a}$. This quantity also reduces to a particular value called shear stress or tangential stress. The letter τ generally denotes this shear stress.

$$Lt_{\Delta a \to 0} \frac{\Delta(P \cos \alpha)}{\Delta a} = \tau.$$

In simple terms, the stress may be defined as the internal resistance offered by a body per unit area.

PROBLEM 1.1 **Objective 1**

Referring to Figure 1.3, determine the normal stress and shear stress induced along the sections 1-1 and 2-2 inclined 30° to the longitudinal axis of the bar of square section 40 mm × 40 mm and length 0.5 m. The axial force acting on the bar is 120 kN.

SOLUTION

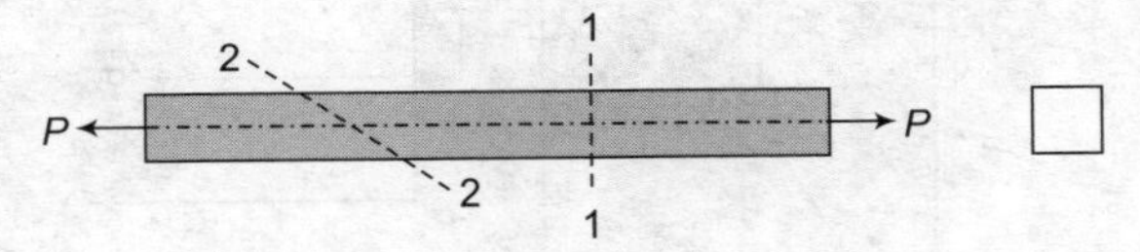

FIGURE 1.3 Bar subjected to axial tension *P*.

Along section 1-1:

The resultant force normal to the section is = P = 120 kN.

Normal stress at this section $\sigma = \frac{P}{A} = \frac{120 \times 1000}{40 \times 40} = 75$ MPa (tensile stress).

Shear force along the section = 0.

Hence the shear stress at this section is zero.

Along section 2-2:

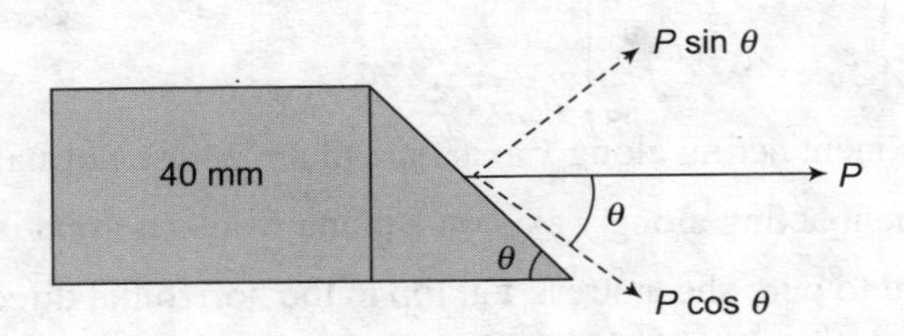

FIGURE 1.4 Internal forces at section 2-2.

The resultant force normal to the section is = 120 sin θ = 120 × sin 30 = 60 kN.

Cross-sectional area of the normal to section 2-2 is 40 × 40/sin 30 = 3200 mm^2.

Normal stress at this section $\sigma = \frac{P}{A} = \frac{60 \times 1000}{3200} = 18.75$ MPa (tensile stress).

Shear force along the section = $P = 120 \cos \theta = 120 \times \cos 30 = 103.92$ kN.

Hence the shear stress at this section $\tau = \frac{103.92 \times 1000}{3200} = 32.48$ MPa.

For a general force system on a body, any plane will carry three stress components due to external loading. Of these three stress components, one is normal stress and the other two are shear stresses. To get the state of stress at a point, we shall represent the stresses over a cube, when this cube reduces to a point, the resulting stresses would be the stresses at point. The possible stresses over such a cube were shown in Figure 1.5.

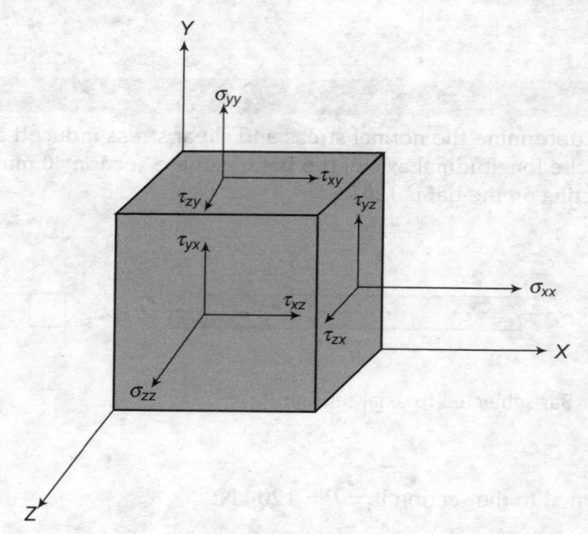

FIGURE 1.5 Possible stresses on a body at a point

The nine stress components are generally represented in a tensor form.

$$[(\sigma)] = \begin{bmatrix} \sigma_{xx} & \tau_{yx} & \tau_{zx} \\ \tau_{xy} & \sigma_{yy} & \tau_{zy} \\ \tau_{xz} & \tau_{yz} & \tau_{zz} \end{bmatrix}$$

σ_{xx} = normal stress component acting along x axis on a plane whose normal is along x axis.

τ_{yx} = shear stress component acting along y axis on a plane whose normal is along x axis.

Consider a body subjected to pure shear stress τ at top in the horizontal direction of the body shown in Figure 1.6. Let 'a' be the length, 'h' be the height and width of the body perpendicular to the plane of the paper be unit.

Net force $\vec{F}$ acting at the top of the block = $\tau \times a \times 1$.

The resisting force that develops at the base = F (←).

Now the tangential force at top and bottom are same and hence produce a clockwise couple equal to $\tau \times a \times 1 \times h$.

To maintain equilibrium, an anticlockwise couple of same intensity should develop.

The shear stress components (τ_*) developed on orthogonal planes gives anticlockwise couple of $\tau_* \times h \times 1 \times a$.

Equating the clockwise couple to the anticlockwise couple

$$\tau \times a \times 1 \times h = \tau_* \times h \times 1 \times a$$

$$\Rightarrow \qquad \tau = \tau_*$$

in which τ_* is referred as complimentary shear stress.

Thus, complimentary shear stress is always equal to the applied shear stress and acts on a plane orthogonal to the plane in which the applied shear stress acts.

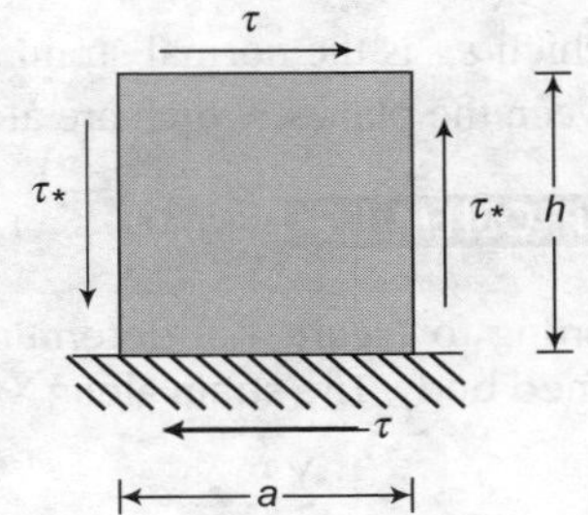

FIGURE 1.6 Shear stress and complimentary shear stress.

1.3 NORMAL STRAIN AND SHEAR STRAIN

Normal stress and shear stress are the internal resistances thus cannot be visualized. The deformations are only the measurable quantities. Thus, normal strain is a measurable deformation parameter corresponding to normal stress and shear strain corresponds to shear stress. Consider the deformations of a block in a plane $ABCD$ shown in Figure 1.7. After the load application, let the deformed shape be $A'B'C'D'$.

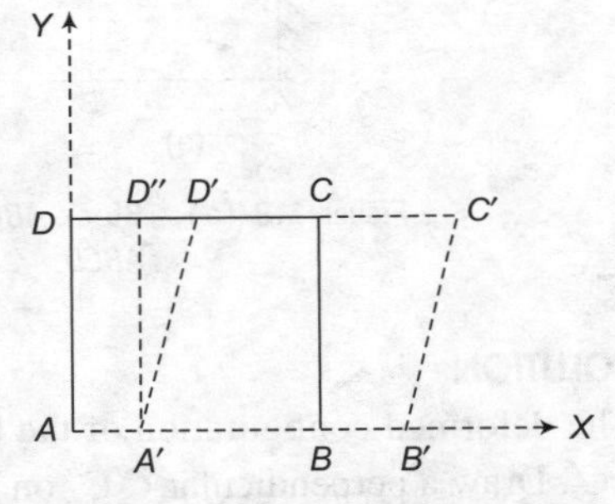

FIGURE 1.7 Deformed shape of a block in a plane.

Change in the length of the part $AB = A'B' - AB$.

Normal strain is defined as the ratio of change in length of a segment to the original length of the segment.

Hence, normal strain along AB is given by $\dfrac{\text{Change in } AB}{\text{Original length of } AB} = \dfrac{A'B' - AB}{AB}$.

The normal strain is denoted by the letter ε. As this quantity is a ratio, strain does not have any units.

Shear strain is defined as the change of angle between two planes due to loading. In Figure 1.7, consider the plane AB and AD. Before the loading, the angle between AB and AD is 90°. After the loading, the included angle between AB and AD reduced by $\angle D''AD'$. This angular change ($\angle D''AD'$) is referred as shear strain, generally denoted by the letter γ. Shear strain also does not have any units like normal strain.

To represent the strain at a point, we shall represent the strains over a cube, as it was done in the case of stresses, when this cube reduces to a point, the resulting strains would be the stain at point.

Thus, the strain tensor at a point can be presented as

$$[\varepsilon] = \begin{bmatrix} \varepsilon_{xx} & \gamma_{yx} & \gamma_{yz} \\ \gamma_{xy} & \varepsilon_{yy} & \gamma_{yz} \\ \gamma_{xz} & \gamma_{yz} & \varepsilon_{zz} \end{bmatrix}$$

in which ε_{xx} is the normal strain along x axis. γ_{yx} is the shear strain or change in the included angle between the planes, which are along y and x axes.

PROBLEM 1.2

Objective 2

Referring to Figure 1.8, determine the normal strain and shear strain along the diagonal AC in a strained body. The strain along X axis is 0.2×10^{-3}. AB = 40 mm and AD = 30 mm. Face AD is fixed.

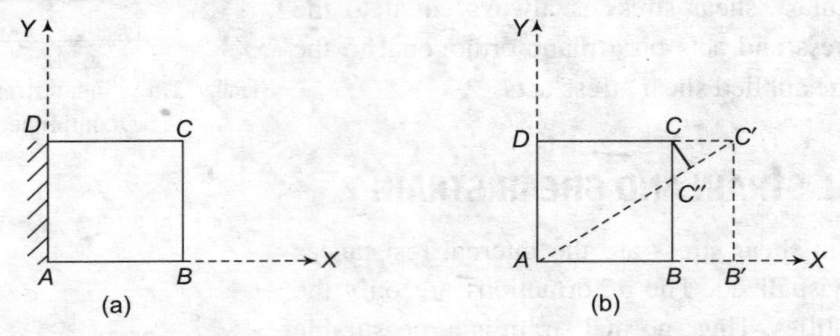

FIGURE 1.8 (a) Block *ABCD* before deformations; (b) deformed configuration of block *ABCD*.

SOLUTION

The deformed configuration of the block *ABCD* is shown as *AB′C′D* to an exaggerated view. Join *AC′*. Draw a perpendicular *C′C″* on to *AC′* from *C*. *AC* is approximately equal to *AC″*.

When the deformations are very small, the following approximation holds good.

$$\angle CAB \approx \angle C'AB = \angle CC'C' = \tan^{-1}\left(\frac{BC}{AB}\right) = 36.87°$$

$$\text{Normal strain along AC} = \frac{\text{Increase in the length } AC}{AC} = \frac{C''C'}{AC} = \frac{C''C'}{AC''}$$

$C''C' = CC' \cos 36.87°$.

$CC' = BB' = \varepsilon_x \times AB = 40 \times 0.2 \times 10^{-3} = 0.008$ mm.

$C''C' = 0.008 \times \cos 36.87° = 0.0064$ mm.

$$\text{Normal strain along } AC = \frac{0.0064}{50} = 0.000128.$$

Shear strain between the planes *AC* and *AB* is $\gamma = \angle CAC''$.

$$\text{Shear strain along } AC = \gamma = \frac{CC''}{AC} = \frac{CC'' \sin 36.87°}{AC} = \frac{0.008 \times 0.75}{50} = 0.00012.$$

1.4 RELATIONSHIP BETWEEN STRESS AND STRAIN

The stress (may be normal stress or shear stress) is related with the corresponding strain (normal strain or shear strain) in terms of elastic constants called modulus of elasticity and rigidity modulus.

The modulus of elasticity is the ratio of normal stress to normal strain, whereas the rigidity modulus is the ratio of shear stress to shear strain. The units for modulus of elasticity or shear modulus are gigapascal (GPa).

$$\text{Modulus of elasticity } (E) = \frac{\text{Normal stress } (\sigma)}{\text{Normal strain } (\varepsilon)}$$

$$\text{Rigidity modulus } (G) = \frac{\text{Shear stress } (\tau)}{\text{Shear strain } (\gamma)}.$$

These elastic constants are constant for individual materials and largely depend on the crystalline structure, orientation, and bond energies. Table 1.1 gives the details of modulus of elasticity and rigidity modulus of different materials within the elastic limit.

Table 1.1 Modulus of elasticity of different materials

Sl. No.	Material	Modulus of Elasticity (GPa)
1	Steel	200
2	Copper	100–80
3	Aluminum	60–80
4	Concrete	25–35
5	Timber	10–15

1.5 ADDITIONAL PROBLEMS ON DIRECT OR AXIAL STRESSES

PROBLEM 1.3 **Objective** *

Derive an expression for the extension of a prismatic bar subjected to axial tension.

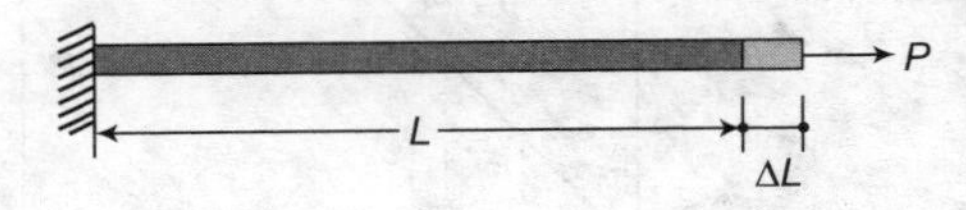

SOLUTION

Normal stress in the bar $\sigma = \frac{P}{A}$.

Let ΔL be the extension of the bar.

Then, normal strain in the bar = $\varepsilon = \frac{\Delta L}{L}$.

Let E be the modulus of elasticity of the bar.

Then, $E = \dfrac{\text{Normal stress}}{\text{Normal strain}} = \dfrac{P/A}{\Delta L/L}$

$$\Rightarrow \qquad \Delta L = \frac{PL}{AE}$$

Stiffness of the bar is defined as the load required for unit extension. Hence, axial stiffness of the bar generally denoted as 'k' is given by

$$k = \frac{P}{\Delta} = \frac{AE}{L}$$

In the above expression of stiffness, the term 'AE' is referred as axial rigidity. It depends on the cross-section as well as material of the member.

A possible doubt to the reader: If the extension of the bar is ΔL due to P, then for additional load say P' in the expression to determine additional extension $\dfrac{P'L}{\text{AE}}$, should L be used or $L + \Delta L$?

This possible doubt makes the reader to understand many important assumptions to be followed in solid mechanics.

1. **Order of loading should not have any effect on the deformations or internal stresses.** This means that whether P is applied first then P', or P' first then P or P and P' be applied simultaneously should not have any effect on the deformation. This is true for the materials, which follow linear force–displacement relationship. This law of superposition does not hold well in case materials which exhibit nonlinear force–displacement relationship. This can be observed from the figures shown below.

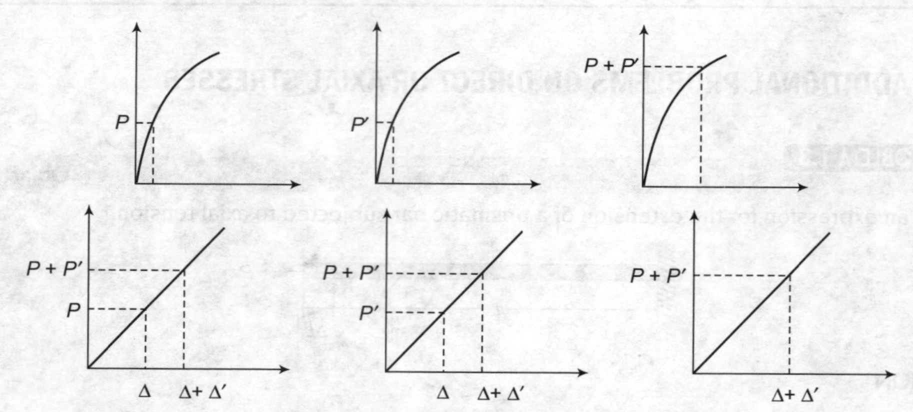

FIGURE 1.9 A case, where law of superposition is valid.

2. **Higher order deformations are neglected.** This means that deformations due to deformation are very small and can be neglected. That is in the axial extension, if $L + \Delta L$ is used in place of L, then $\Delta L + \Delta\Delta L = \dfrac{PL}{AE} + \dfrac{P(\Delta L)}{AE}$. If this is accepted, the next question that may arise is should we use $L + \Delta L + \Delta(\Delta L)$ in place

of L? If we continue like this there will be no end for it. For most of the materials within the working range of loads, $\Delta(\Delta L)$ is very small compared to ΔL, and hence $\Delta(\Delta L)$ can be ignored. This $\Delta(\Delta L)$ is referred as deformation due to deformation or second-order deformation. Thus, in strength of materials the effect of deformations due to deformations or second-order deformations is neglected.

PROBLEM 1.4 **Objective 3**

Estimate the deformation of points *B*, *C*, and *D* of the compound bar *ABCD* subjected to loading as shown in Figure 1.10. Take modulus of elasticity of the material as 200 GPa. P = 10 kN, section at 1-1 is 50 mm × 50 mm solid, section at 2-2 is hollow section of external dimensions 50 mm × 50 mm and inner dimensions 25 mm × 25 mm, and section at 3-3 is circular section of 40 mm diameter. AB = 1 m, BC = 1.2 m, and CD = 1.1 m.

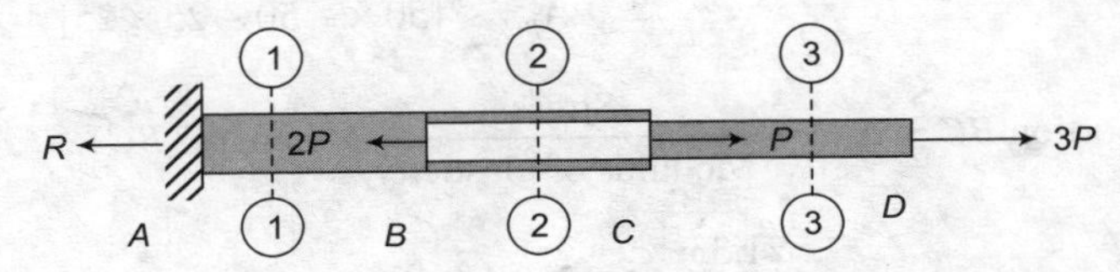

FIGURE 1.10

SOLUTION

Structural systems or components of structural systems must be in equilibrium. This is an essential condition.

Consider the equilibrium of the system.

Thus, sum of the forces in X direction must be equal to zero.

$$\Rightarrow \quad \Sigma F_x = 0$$

$$\Rightarrow \quad 3P + P - 2P - R = 0$$

$$\Rightarrow \quad R = 2P.$$

Consider the free body diagram of the three parts of the *ABCD* as shown in Figure 1.11.

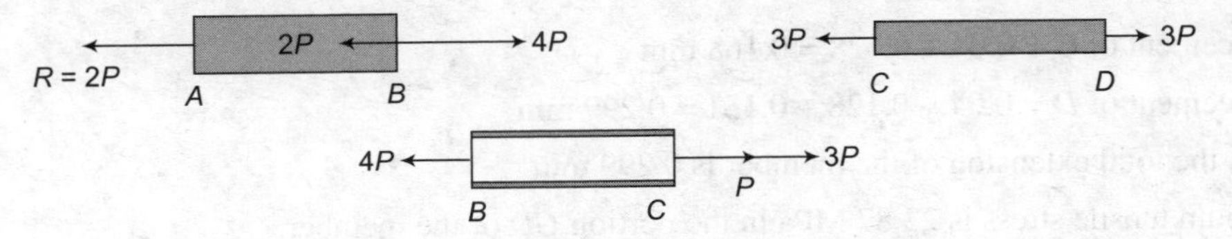

FIGURE 1.11

It is always convenient to draw the free body diagrams from free end of the member. In the portion *CD*, axial tensile load of $3P$ is acting at the free end. To keep the member *CD* in equilibrium, there should an axial force of $3P$ acting at *C* in the opposite direction, that is, toward left. Thus, the portion *CD* of the member is acted upon by a tensile force of $3P$.

$$\text{Tensile stress in the region } CD \text{ of the member} = \sigma_{CD} = \frac{3P}{A_{CD}} = \frac{3 \times 10 \times 1000}{\frac{\pi}{4} \times 40 \times 40} = 23.87 \text{ MPa}$$

Extension in the portion $CD = \Delta_{CD} = \dfrac{\text{Stress}}{\text{Modulus of elasticity}} \times \text{Length } (CD)$

$$\Delta_{CD} = \frac{23.87}{200 \times 10^3} \times 1100 = 0.131 \text{ mm}$$

In the portion *CD*, at *C* a tensile load of 3*P* was applied to maintain equilibrium. Hence, a tensile load of 3*P* at *C* in the portion *BC* should be applied. Then, apply the load *P* acting at *C* in the portion *BC*. Thus, apply a tensile load of 4*P* at *C*. Apply a tensile load of 4*P* at *B* to maintain equilibrium. Therefore, the portion BC of the member is subjected to a tensile load of 4*P*.

Hence, tensile stress in the portion $BC = \sigma_{BC} = \dfrac{4P}{A_{BC}} = \dfrac{4 \times 10 \times 1000}{[50 \times - 50 - 25 \times 25]} = 21.33 \text{ MPa}$.

Extension in the portion $BC = \Delta_{BC} = \dfrac{\text{Stress}}{\text{Modulus of elasticity}} \times \text{Length } (BC)$

$$\Delta_{BC} = \frac{21.33}{200 \times 10^3} \times 1200 = 0.128 \text{ mm}$$

Similarly applying equilibrium for the portion *AB*, the tensile load acting = 2*P*.

Hence, tensile stress in the portion $AB = \sigma_{AB} = \dfrac{2P}{A_{AB}} = \dfrac{2 \times 10 \times 1000}{[50 \times 50]} = 8.00 \text{ MPa}$

Extension in the portion $AB = \Delta_{AB} = \dfrac{\text{Stress}}{\text{Modulus of elasticity}} \times \text{Length } (AB)$

$$\Delta_{AB} = \frac{8.00}{200 \times 10^3} \times 1000 = 0.04 \text{ mm}$$

Finally, the displacement at $A = 0.0$

Displacement of $B = 0.04$ mm

Displacement of $C = 0.04 + 0.128 = 0.168$ mm

Displacement of $D = 0.04 + 0.128 + 0.131 = 0.299$ mm

Hence, the total extension of the member is 0.299 mm

Maximum tensile stress is 23.87 MPa in the portion *CD* of the member.

PROBLEM 1.5

Objective 3

Derive an expression for the extension of a conical bar of length *L* fixed at the base and hanging due to its own weight. Specific weight of the material of the bar is γ and modulus of elasticity is *E*.

SOLUTION

Let d_0 be the base diameter of the conical bar.

Consider a fiber located at a distant *x* from the bottom of the bar of thickness Δx.

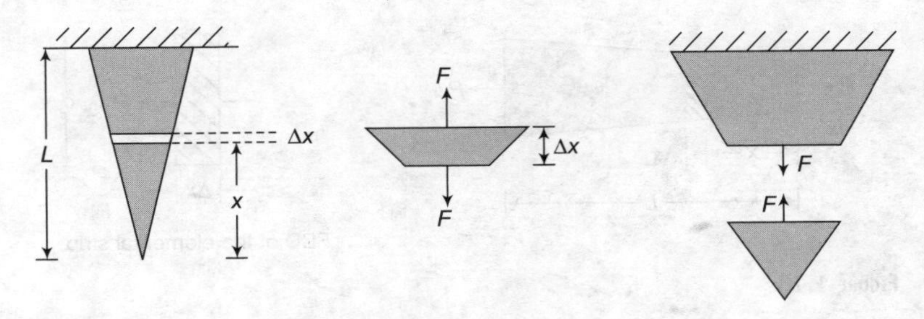

FIGURE 1.12

The free body diagram of the elemental strip of length Δx is shown in Figure 1.12. The force acting on the elemental strip is nothing but the weight of the portion clinging to the section under consideration.

Diameter of the conical bar at the section under consideration $d_x = \dfrac{d_0}{L}x$.

Therefore, weight of the portion of the conical bar up to the section $= F = \dfrac{1}{12}\pi d_x^2 x\gamma$

$$\Rightarrow \qquad F = \frac{1}{12}\pi\gamma\frac{d_0^2}{L^2}x^3.$$

Extension of this elemental strip of length Δx is $\Delta(\Delta L) = \dfrac{F\Delta x}{A_x E}$

$$\Rightarrow \qquad \Delta(\Delta L) = \frac{1}{12E}\pi\gamma\frac{d_0^2}{L^2}x^3 \times \frac{1}{\frac{\pi}{4}\frac{d_0^2}{L^2}x^2} \times \Delta x$$

$$\Rightarrow \qquad \Delta L = \int_0^L \frac{1}{3E}\gamma\, x\, dx = \frac{\gamma L^2}{6E}$$

Thus, the extension of conical bar due to its own weight is $\dfrac{\gamma L^2}{6E}$.

Similarly, it can be shown that the extension of a prismatic member (same cross-section throughout the length) hanging on its own weight is $\dfrac{\gamma L^2}{2E}$.

PROBLEM 1.6 **Objective 3**

Derive an expression for the extension of a tapering bar of length L subjected to an axial pull of intensity P. Modulus of elasticity of the material of the bar is E. The diameter at one end is d_1 and at the other end it is d_2.

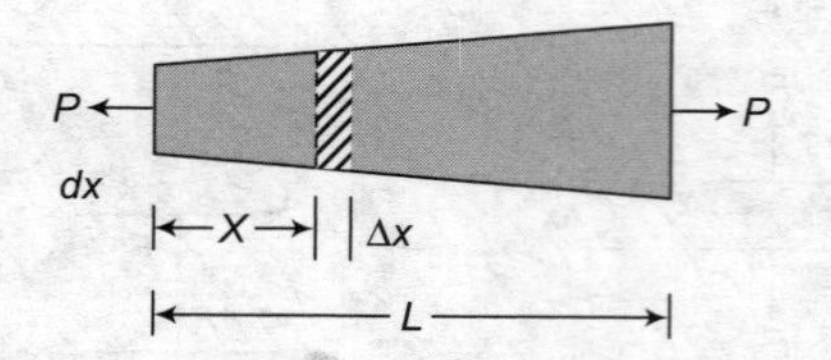

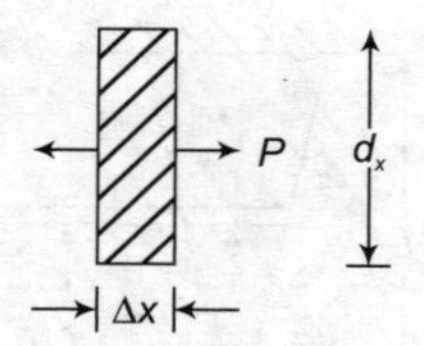

FIGURE 1.13

Consider an elemental strip of length Δx, located at a distant X from the one end of the member. Let d_x be the diameter of the bar located at a distant x from one end of the bar, where the diameter is d_1.

Then, $$d_x = d_1 + \frac{d_2 - d_1}{L}x$$

Consider an elemental strip of length Δx at the section under consideration.
The free body diagram of the elemental strip of length Δx is as shown in Figure 1.13.

Extension of this elemental strip of length Δx is $\Delta(\Delta L) = \dfrac{P\Delta x}{A_x E}$

$$\Rightarrow \quad \Delta(\Delta L) = \frac{P}{E} \times \frac{1}{\frac{\pi}{4} d_x^2} \times \Delta x$$

$$\Rightarrow \quad \Delta L = \int_0^L \frac{P}{E} \times \frac{1}{\frac{\pi}{4}\left[d_1 + \frac{d_2 - d_1}{L}x\right]^2} \times dx$$

$$= \frac{P}{E} \times \left[-\frac{\left(d_1 + \frac{d_2 - d_1}{L}x\right)^{-1}}{\frac{\pi}{4}\left(\frac{d_2 - d_1}{L}\right)} \right]_0^L$$

$$= \frac{PL}{E}\left[\frac{\frac{1}{d_1} - \frac{1}{d_2}}{\frac{\pi}{4}(d_2 - d_1)}\right]$$

$$= \frac{PL}{E\left(\frac{\pi}{4} d_1 d_2\right)}$$

In the above expression if $d_1 = d_2 = d$ then, the expression for extension reduces to the expression for the extension of a solid circular uniform bar $\dfrac{PL}{AE}$.

PROBLEM 1.7

Objective 4

A prismatic solid circular bar of length L is subjected to axial load. This solid bar is bored for a length of $0.5L$, such that the inner diameter of the bored portion is 0.6 times the diameter of the solid portion. Estimate the percentage increase in the extension of the bored bar under same load, when compared with that of the solid prismatic bar.

SOLUTION

Let P be the axial load acting on the bar. Let d be the diameter of the bar.

The extension of the prismatic bar $= \dfrac{4PL}{\pi d^2 E}$.

If half portion of the bar is bored then,

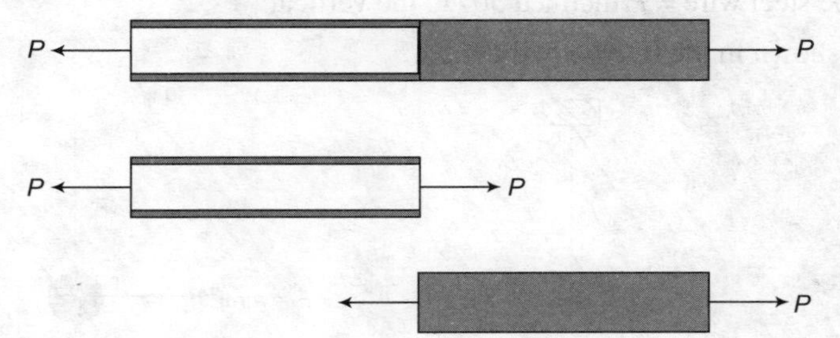

FIGURE 1.14

Free body diagram of the two parts of the bored bar is shown in Figure 1.14.

The extension of the bored half portion of the bar $= \Delta_1 = \dfrac{P(0.5L)}{\frac{\pi}{4}[d^2 - (0.6d)^2 E]}$

$$\Rightarrow \qquad \Delta_1 = \frac{2PL}{0.64\pi d^2 E}.$$

Extension of the remaining half portion $= \Delta_2 = \dfrac{P(0.5L)}{\frac{\pi}{4}d^2 E} = \dfrac{2PL}{\pi d^2 E}$.

The total extension of the bored bar is $\Delta_* = \Delta_1 + \Delta_2 = (1.281)\dfrac{4PL}{\pi d^2 E}$.

Percentage increase in the extension $= \dfrac{\Delta_* - \Delta}{\Delta} \times 100 = 28.1\%$.

PROBLEM 1.8

Objective 4

A 5-kg mass rotates in a horizontal circle with constant angular speed at the end of 1.5 m steel wire such that the steel wire makes 30° with the vertical. Determine the speed, stress in the steel wire due to the rotation, and also evaluate the extension of the steel wire taking $E = 200$ GPa. Diameter of the steel wire is 0.5 mm.

SOLUTION

Let P be the tension in the steel wire and ω be the angular speed of the steel ball. As the ball is rotating about a fixed vertical axis, the ball is subjected to normal force $\left(m\dfrac{v^2}{r}\right)$ and tangential force $\left(m\dfrac{dv}{dt}\right)$, in which v is the speed of the ball, in the present case this v is constant. Hence, the tangential force acting on the ball vanishes.

Velocity of the ball = $v = r\omega$

If we draw the free body diagram of the ball, the forces acting on that will be

(a) Self-weight of the ball = mg = 5 × 9.81 = 49.5 ≈ 50 N.

(b) Normal force (acting normal to the path) = $mr\omega^2$.

(c) Tension in the steel wire = P inclined 30° to the vertical.

These forces were shown in the free body diagram.

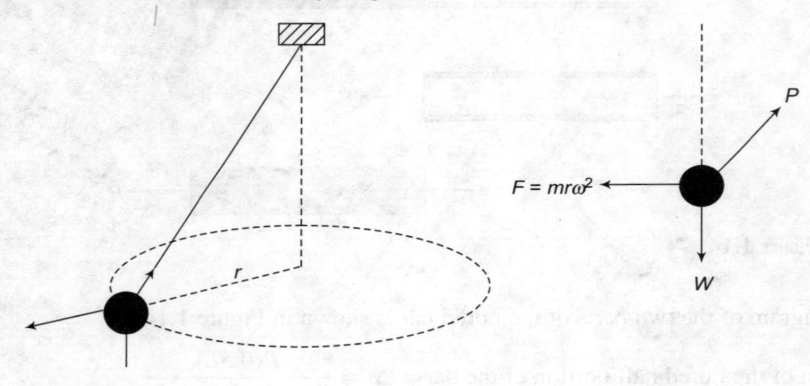

FIGURE 1.15

Resolving the forces vertically,

$$W = P\cos\theta$$

$$\Rightarrow \quad P\cos 30 = 50$$

$$\Rightarrow \quad P = \frac{50}{\cos 30} = 57.735 \text{ N.}$$

$$\Rightarrow \quad \text{Normal stress in the wire } \sigma = \frac{57.735}{\frac{\pi}{4}(0.5)^2} = 294 \text{ MPa.}$$

$$\text{Extension of the steel wire} = \Delta = \frac{\sigma L}{E} = \frac{294 \times 1500}{200 \times 1000} = 2.205 \text{ mm.}$$

To determine the angular speed of the steel ball, resolve the forces horizontally.

$$\Rightarrow \quad P\sin\theta = F = mr\omega^2$$

$$\Rightarrow \qquad \omega^2 = \frac{57.735 \sin 30}{5 \times 1.5 \sin 30}$$

$$\Rightarrow \qquad \omega = 2.775 \text{ radians/s}$$

Angular speed of the ball about the vertical axis is $\dfrac{60\omega}{2\pi} = 26.5$ RPM.

PROBLEM 1.9 **Objective 1**

A rigid bar of length 0.5 m and negligible weight hangs by means of two wires of length 1.2 m each, as shown in Figure 1.16. If a gravity load of 100 kN is applied at the left middle third point of the rigid bar, determine the stresses in the two wires and the inclination of the rigid bar with the horizontal. The cross-sectional area of aluminum bar is 1000 mm^2 and that of steel bar is 500 mm^2. Modulus of elasticity of steel bar is 202 GPa while that of aluminum is 65 GPa.

SOLUTION

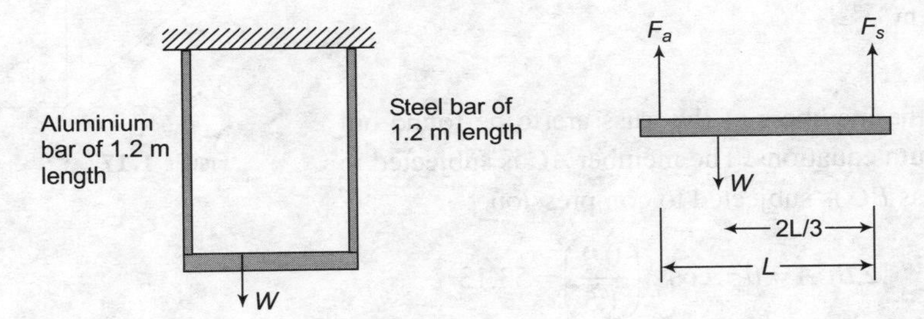

FIGURE **1.16**

Sketch the free body diagram of the rigid bar.
Let F_a be the force the aluminum bar and F_s be the force in the steel bar.
Use equilibrium equations to determine the forces F_a and F_s.
That is, sum of the horizontal forces is equal to zero. (Equilibrium equation (1))

$$\Rightarrow \qquad F_a + F_s = W = 100 \text{ kN}.$$

Sum of the moments about any arbitrary point is zero. (Equilibrium equation (2))
Taking moments about a point through which F_a is passing,

$$\Rightarrow \qquad F_s \times L - 100 \times \frac{2L}{3} = 0$$

$$\Rightarrow \qquad F_s = 66.67 \text{ kN}$$

$$\Rightarrow \qquad F_a = 33.33 \text{ kN}.$$

As the bar is rigid, the bar itself will not undergo any deformations, but aluminum and steel bars undergo extensions.

Extension in the aluminum bar $\Delta_a = \dfrac{F_a \times L_a}{A_a E_a} = \dfrac{66.67 \times 1000 \times 1200}{1000 \times 65 \times 1000} = 1.231$ mm

Extension in the steel $\Delta_s = \dfrac{F_s \times L_s}{A_s E_s} = \dfrac{33.33 \times 1000 \times 1200}{500 \times 202 \times 1000} = 0.396$ mm.

From the extensions of steel bar and aluminum bar, it is clear that the left end of rigid bar moves down by 1.231 mm and the right end of the same moves down by 0.369 mm. Thus, rigid bar rotates in the anticlockwise direction.

Let θ be the rotation of the rigid bar.

Then, $\theta = \dfrac{\Delta_a - \Delta_s}{L} = \dfrac{1.231 - 0.396}{500} = 0.00167$ radians.

Inclination of the rigid bar in anticlockwise direction is 0.00167 radians.

PROBLEM 1.10

Objective 3

Estimate the vertical and horizontal deflection at the point C of the two-member truss shown in Figure 1.17. Take $E = 200$ GPa. Cross-sectional area of each member is 2000 mm^2. $AC = 1.5$ m; $BC = 0.9$ m.

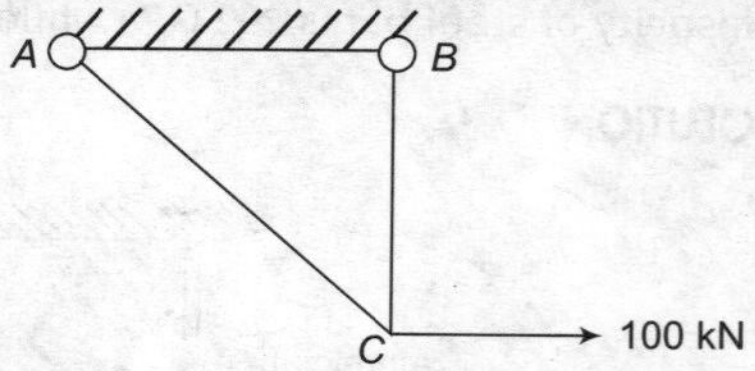

A B C 100 kN

FIGURE 1.17

SOLUTION

The forces in the members of the truss are to be found out using equilibrium equations. The member *AC* is subjected to tension, whereas *BC* is subjected to compression.

$$\angle BCA = \theta = \cos^{-1}\left(\frac{0.9}{1.5}\right) = 53.13$$

$$\Rightarrow \qquad \cos\theta = 0.6$$

$$\Rightarrow \qquad \sin\theta = 0.8$$

Resolving the forces horizontally, $F_{AC} \sin\theta = 100$ kN

$$\Rightarrow \qquad F_{AC} = 125 \text{ kN (tensile).}$$

Resolving the forces vertically, $F_{AC} \cos\theta - F_{BC} = 0$

$$\Rightarrow \qquad F_{BC} = 75 \text{ kN (compressive).}$$

Normal tensile stress in the bar *AC* is $\dfrac{125 \times 1000}{2000} = 62.5$ MPa.

Normal compressive stress in the bar *BC* is $\dfrac{75 \times 1000}{2000} = 37.5$ MPa.

We are required to evaluate the vertical and horizontal deflection of point *C*. The force in the bar *AC* moves the point *C* in the direction of *AC*, CC_1 as shown in Figure 1.18, whereas the compressive force in the bar *BC* tries to move the point *C* in the upward direction along *CB*, CC_2 as shown in Figure 1.18. Because of this, the final position of the point *C* will be the meeting point of arcs drawn to the points C_1 and C_2, taking centers as *A* and *B*, respectively. As the displacements are negligible, rather than drawing arcs, it is convenient to draw perpendiculars. Thus, draw perpendiculars at C_1 to

AC_1 (CC_1) and C_2 to BC_2 (CC_2). These perpendiculars meet at point C_3. Thus, net movement of the point C will be from C to C_3, as shown in Figure 1.18. The horizontal and vertical projection of CC_3 will be the horizontal and vertical displacement of the point C.

Free extension of $AC = CC_1 = \dfrac{62.5 \times 1500}{200 \times 1000} = 0.469$ mm.

Free extension of $BC = CC_2 = \dfrac{37.5 \times 900}{200 \times 1000} = 0.169$ mm.

Angle $C_2CC_1 = 180° - \theta = 126.87°$.

C_1C_3 is perpendicular to CC_1; C_2C_3 is perpendicular to CC_2; and CC_3 is the net displacement of the point C.

It is clear that $CC_3 = CC_1/\cos\theta_1 = CC_2/\cos\theta_2$. (1)

Moreover,

$$\theta_1 + \theta_2 = 126.87° \quad (2)$$

From equation (1)

$$\frac{\cos\theta_1}{\cos\theta_2} = \frac{1.563}{0.169} = 2.775$$

From equation (2)

$$\frac{\cos\theta_1}{\cos(126.87 - \theta_1)} = 2.775$$

$$\Rightarrow \qquad \theta_1 = 50.2°.$$

Thus, $CC_3 = \dfrac{0.169}{\cos(50.2)} = 0.264$ mm .

Hence, the horizontal displacement of C is $CC_3 \sin\theta_1 = 0.264 \sin\theta_1 = 0.768$ mm. Toward right of C, vertical displacement of the point C is $CC_3 \cos\theta_1 = 0.264 \times \cos(50.2) = 0.169$ mm (upward).

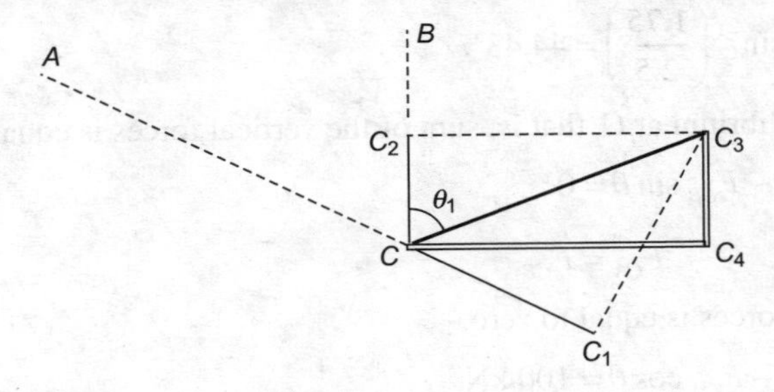

FIGURE 1.18

PROBLEM 1.11

Objective 3

A steel bar *AB* of length 3.5 m and diameter 25 mm are connected by four inextensible cables of length 2.5 m each, forming a rhombus with *AB* as diagonal, as shown in Figure 1.19. A 100 kN forces act at the points *C* and *D*. Determine the decrease in the length of the strut *AB* and increase in the length between the points *C* and *D*. Take modulus of elasticity of steel as 201 GPa.

SOLUTION

The cables *CA*, *CB*, *DA*, and *DB* are inextensible means that they do not undergo any deformation but the points *C* and *D* move due to the deformation of the strut *AB*. A strut is a member subjected to axial compression. The axial deformation of the strut is to be found, to determine the displacement between the points *C* and *DE*. Using equilibrium equations, determine the force in the strut *AB*.

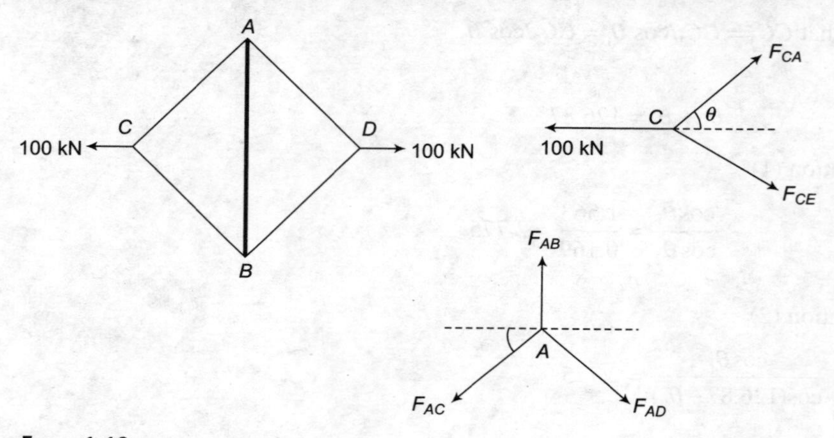

FIGURE 1.19

Let F_{AC} be the force in the cable *AC* and F_{CB} be the force in the cable *CB*. Let θ be the inclination of AC with the horizontal.

$$\sin\theta = \frac{(AB/2)}{AC}$$

$$\Rightarrow \qquad \theta = \sin^{-1}\left(\frac{1.75}{2.5}\right) = 44.43°.$$

Consider the joint equilibrium at *C*, that is, sum of the vertical forces is equal to zero.

$$\Rightarrow \qquad F_{CA}\sin\theta - F_{CB}\sin\theta = 0$$

$$\Rightarrow \qquad F_{CA} = F_{CB}.$$

Sum of the horizontal forces is equal to zero.

$$\Rightarrow \qquad F_{CA}\cos\theta - F_{CB}\cos\theta = 100 \text{ kN}.$$

$$\Rightarrow \qquad F_{CA} = F_{CB} = \frac{100}{2\cos\theta} = 70.01 \text{ kN}.$$

To determine the force in the strut *AB*, consider the equilibrium of the joint *A*.

Sum of the horizontal forces is equal to zero.

$$\Rightarrow \quad F_{CA}\cos\theta - F_{AD}\cos\theta = 0.$$

$$\Rightarrow \quad F_{CA} = F_{AD}\ 70.01\text{ kN}.$$

Sum of the vertical forces is equal to zero.

$$\Rightarrow \quad F_{CA}\sin\theta + F_{AD}\sin\theta = F_{AB}$$

$$\Rightarrow \quad F_{AB} = 2F_{CA}\sin\theta = 98.02\text{ kN}.$$

Axial deformation (compression) of the strut AB is $\dfrac{98.02\times 1000\times 3500}{\frac{\pi}{4}(25)^2\times 201\times 1000} = 3.48$ mm.

To determine the displacement between the points C and D, consider the triangle ACD.

$$L_{CD} = 2\times\left(\frac{AB}{2}\right)\times\tan\theta.$$

Differentiating on both sides

$$\Delta(L_{CD}) = 2\times\tan\theta\times\Delta\left(\frac{AB}{2}\right)$$

$$\Rightarrow \quad \Delta(L_{CD}) = 2\times 0.98\times\frac{3.48}{2} = 3.41\text{ mm}.$$

Displacement between the points C and D is 3.41 mm.

PROBLEM 1.12

Objective 4

Two bars of length 1.5 m each are connected to a rigid bar of length L, as shown in Figure 1.20. The axial stiffness of the bar AB is 2.5 N/mm. Determine the axial stiffness as well as axial rigidity of the bar CD, if the rigid bar has to be horizontal, when a load $W = 600$ N is applied at $L/3$ distant from the bar AB.

SOLUTION

Let F_{AB} and F_{CD} be the forces in the members AB and CD. The condition is that the rigid bar should be horizontal. Thus, the extension of the bar AB and CD should be same.

$$\therefore \quad \left(\frac{F_{AB}L}{AE}\right)_{AB} = \left(\frac{F_{CD}L}{AE}\right)_{CD}$$

$\dfrac{AE}{L}$ of member AB is the stiffness of AB and is given by 2.5 N/mm. Axial stiffness member CD is to be determined.

$$\therefore \quad \frac{F_{AB}}{\left(\frac{AE}{L}\right)_{AB}} = \frac{F_{CD}}{\left(\frac{AE}{L}\right)_{CD}}.$$

F_{AB} and F_{CD} are to be determined from equilibrium equations. Consider the equilibrium of the rigid bar.

That is, sum of the horizontal forces is equal to zero.

$$\Rightarrow \qquad F_{AB} + F_{CD} = W = 600 \text{ N.}$$

Sum of the moments about any arbitrary point is zero.

Taking moments about a point through which F_a is passing

$$\Rightarrow \qquad F_{CD} \times L - 600 \times \frac{L}{3} = 0$$

$$\Rightarrow \qquad F_{CD} = 200 \text{ N}$$

$$\Rightarrow \qquad F_{AB} = 400 \text{ N}$$

$$\therefore \qquad \frac{400}{2.5} = \frac{200}{\left(\frac{AE}{L}\right)_{CD}}$$

$$\Rightarrow \qquad \left(\frac{AE}{L}\right)_{CD} = k_{CD} = 1.25 \text{ N/mm.}$$

Axial stiffness of the member CD required is 1.25 N/mm.

Axial rigidity AE of the member $CD = L_{CD} \times k_{CD} = 1.25 \times 1500 = 1875 \text{ N} \cdot \text{mm}^2$.

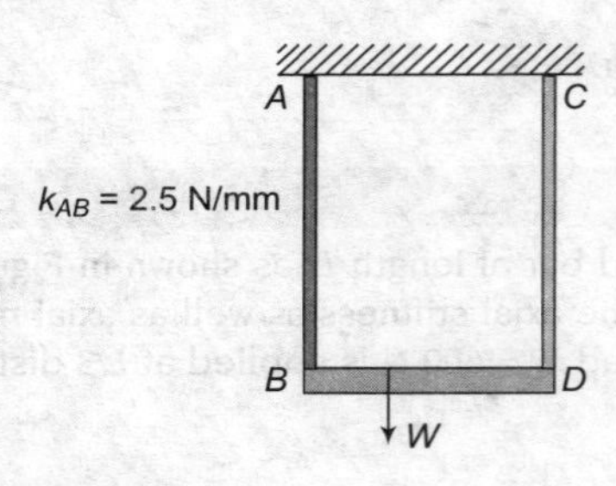

FIGURE 1.20

Statically Indeterminate Structures

Statically indeterminate structures are those which cannot be analyzed with the help of equilibrium equations alone. Most of the structures fall under the category of indeterminate structures. In this section, we consider the analysis of few statically indeterminate structures having axially loaded members. For example, consider a member clamped at both ends subjected to axial forces as shown in Figure 1.21.

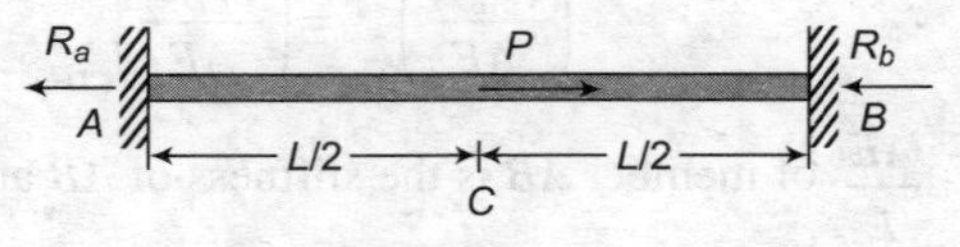

FIGURE 1.21

Let the reactions at A and B, R_a and R_b, respectively, use the condition of equilibrium that sum of the horizontal force is equal to zero.

$$\Rightarrow \qquad R_a + R_b = P. \tag{1.1}$$

The two unknown quantities R_a and R_b cannot be found in equation (1.1), that is, R_a and R_b cannot be determined using equilibrium equation. Thus, this problem falls under the category of statically indeterminate problem. To solve this problem, condition pertaining to deformations shall be used. The conditions pertaining to deformations are called 'compatibility conditions'. In the present example, the compatibility condition is that the total extension of the bar *AB* between the fixed supports is zero. That is, extension in the portion *AC* must be equal to the compressive deformation in the portion *CB*. From this compatibility condition, one more equation can be formed, thus R_a and R_b (two unknowns from two equations) can be determined.

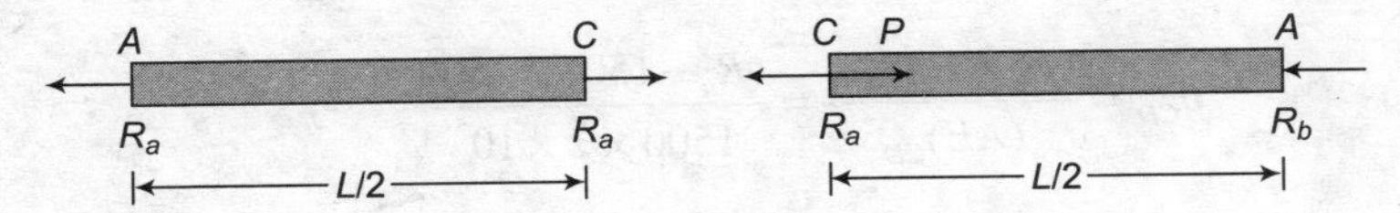

FIGURE 1.22

$$R_b = P - R_a. \tag{1.1}$$

From compatibility condition,

Extension in the bar AC = Compressive deformation in the bar CB

$$\frac{R_a(L/2)}{[AE]_{AC}} = \frac{R_b(L/2)}{[AE]_{CB}} \tag{1.2}$$

From equations (1.1) and (1.2), R_a and R_b can be evaluated.

PROBLEM 1.13

Objective 1

A bar having cross-sectional arcs of 1500 mm^2 is fixed between two rigid walls. Two loads P_1 and P_2 are applied at points *C* and *DE* shown in Figure 1.23. Determine the stresses induced in the portions AC, CD, and DB. Take E = 200 GPa; P_1 = 150 kN; and P_2 = 100 kN.

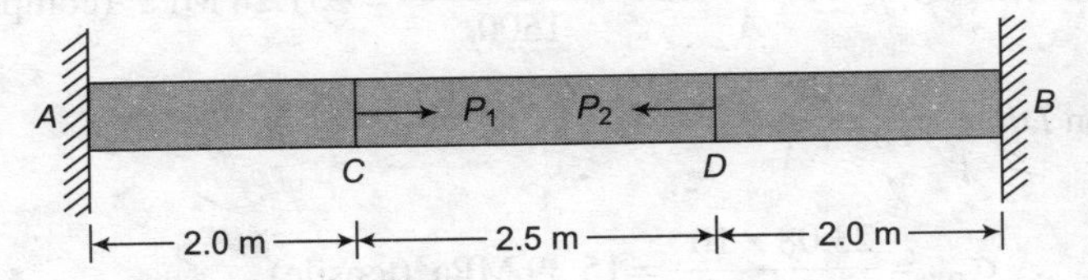

FIGURE 1.23

SOLUTION

Let R_a and R_b be the reactions at *A* and *B*, respectively.

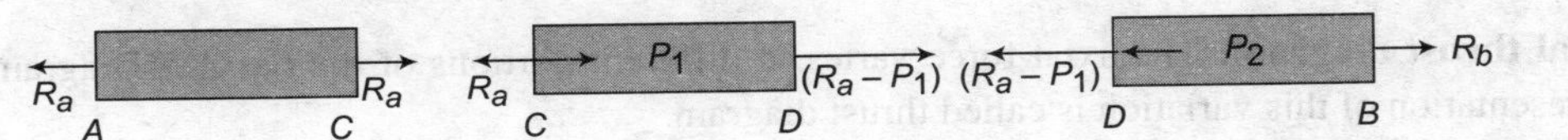

FIGURE 1.24 *FBD of members AC, CD, and DB.*

From the equilibrium condition, the sum of horizontal force is equal to zero.

$$R_a - P_1 + P_2 - R_b = 0 \quad (1.3)$$

$\Rightarrow$ $$R_a - R_b = P_1 - P_2 = 150 - 100.$$

$\therefore$ $$R_a - R_b = 50 \quad (1.4)$$

As per the compatibility condition, total extension between A and B shall be zero.

$\Rightarrow$ $$\delta_{AC} + \delta_{CD} + \delta_{DB} = 0 \quad (1.5)$$

$$\delta_{AC} = \frac{R_a \times L_{AC}}{(AE)_{AC}} = \frac{R_a(2000)}{1500 \times 2 \times 10^5}$$

$$\delta_{CD} = \frac{(R_a - P_1)L_{CD}}{(AE)_{CD}} = \frac{(R_a - 150) \times 2500}{1500 \times 2 \times 10^5}$$

$$\delta_{DB} = \frac{R_b \times L_{DB}}{(AE)_{DB}} = \frac{R_b \times 2000}{1500 \times 2 \times 10^5}.$$

Substituting the above in equation (1.5)

$\therefore$ $$\frac{R_a(2000)}{3 \times 10^8} + \frac{(R_a - 150) \times 2500}{3 \times 10^8} + \frac{R_b \times 2000}{3 \times 10^8} = 0 \quad (1.6)$$

Using R_b value from equation (1.4) into equation (1.6)

$$2R_a + 2.5(R_a - 150) + 2 \times (R_a - 50) = 0$$

$\Rightarrow$ $$R_a = 475/6.5 = 73.08 \text{ kN}.$$

$$R_b = 23.08 \text{ kN}$$

$\therefore$ Stress in the portion $AC = \sigma_{AC} = \dfrac{73.08 \times 10^3}{1500} = 48.72$ MPa

Stress in the portion $CD = \sigma_{CD} = \dfrac{R_a - P_1}{A} = \dfrac{-76.92 \times 10^3}{1500} = -51.28$ MPa (compressive)

Stress in the portion $DB = \sigma_{DB} = \dfrac{R_b}{A}$

$$\sigma_{DB} = \frac{23.08 \times 10^3}{1500} = 15.39 \text{ MPa (tensile)}$$

$$\sigma_{AC} = 48.72 \text{ MPa (tensile)}$$

$$\sigma_{CD} = 51.28 \text{ MPa (compressive)}$$

$$\sigma_{DB} = 15.39 \text{ MPa (tensile).}$$

Axial thrust diagram: The axial force varies in different portions of the bar AB. Diagrammatic representation of this variation is called thrust diagram.

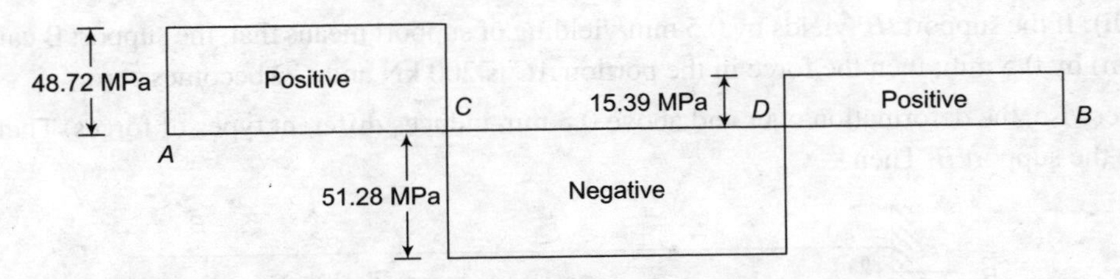

FIGURE 1.25

Axial tension is +ve (positive) and axial compression is –ve (negative).

PROBLEM 1.14 **Objective 4**

A bar *AB* of length 3 m is fixed between the rigid supports. If an axial load of 200 kN at 2 m from support A, determine the axial force in the portion *AC* and *CB*. Also determine the variation in the axial forces, if support *B* yields by 0.5 mm. Take cross-sectional area of the bar *AB* as 1000 mm^2 and E = 200 GPa.

SOLUTION

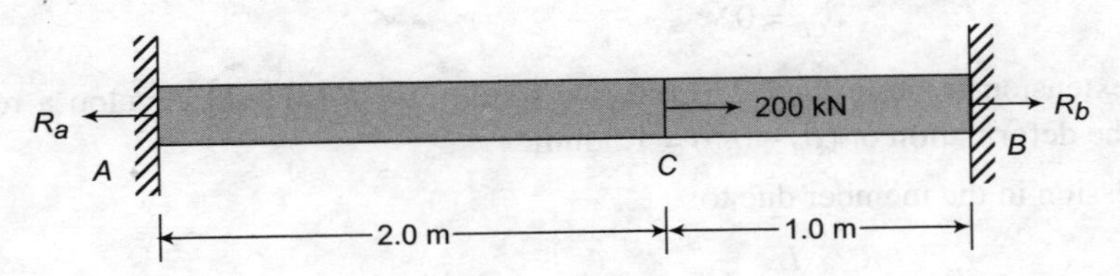

FIGURE 1.26

Case (i): supports do not yield:

$$\sum F_x = 0 \text{ (equilibrium equation)}$$

$$\Rightarrow \quad R_a - R_b = 200 \text{ kN} \tag{1.7}$$

Total extension between *A* and *B* is zero.

$$\Rightarrow \quad \delta_{AC} + \delta_{CB} = 0$$

$$\Rightarrow \quad \frac{R_a \times L_{AC}}{(AE)_{AC}} + \frac{R_b \times L_{CB}}{(AE)_{CB}} = 0$$

$$\Rightarrow \quad \frac{R_a \times 2000}{1000 \times 2 \times 10^5} + \frac{R_b \times 1000}{1000 \times 2 \times 10^5} = 0$$

$$\Rightarrow \quad 2R_a + R_b = 0. \tag{1.8}$$

Solving equations (1.7) and (1.8)

$$R_a = 66.67 \text{ kN}$$

$$R_b = -133.33 \text{ kN}.$$

In this problem, the axial rigidity of the members *AC* and *CB* is not affecting the forces in the portions *AC* and *CB*.

Case (ii): If the support '*B*' yields by 0.5 mm/yielding of support means that, the support B can relax (deform) by 0.5 mm; then the force in the portion *AC* is 200 kN and *CD* becomes zero.

Otherwise, the deformation over and above 0.5 mm induces different types of forces. Therefore, release the support *B*. Then

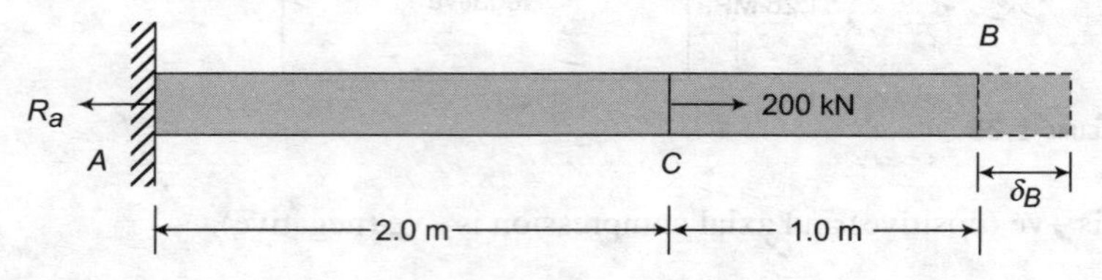

FIGURE 1.27

Free extension $\delta_B = \delta_{AC} + \delta_{CB}$

$$\delta_{AC} = \frac{200 \times 10^3 \times 2000}{1000 \times 2 \times 10^5} = 2 \text{ mm}$$

$$\delta_{CB} = 0$$

As the free extension is more than 0.5 mm, the fixidity at '*B*' should develop a reaction R_b to compensate the deformation of $(\delta_B - 0.5) = 1.5$ mm.

Axial compression in the member due to

$$R_b = \frac{R_b \times L}{AE} = 1.5 \text{ mm}$$

$$\frac{R_b \times 3000}{1000 \times 2 \times 10^5} = 1.5$$

$$R_b = 100 \times 10^3 = 100 \text{ kN.}$$

Force in the portion AC = 100 kN (T)

Force in the portion CB = 100 kN (C)

	Force in the Portion *AC*	Force in the Portion *CB*
Case (i)	66.67 kN (T)	133.33 kN (C)
Case (ii)	100 kN (T)	100 kN (C)

PROBLEM 1.15

Objective 4

Three bars 1, 2, and 3 of length 2, 1.5, and 2.5 m respectively, are connected to a rigid plate, which carries a 200 kN point load as shown in Figure 1.28. The cross-sectional area of each bar is 1000 mm^2. Bars 1 and 3 are of mild steel, while the bar 2 is aluminum. Before placing 200 kN load, the rigid plate is horizontal. Determine the final configuration of the plate due to 200 kN load. Take E_S = 200 GPa and E_a = 60 GPa.

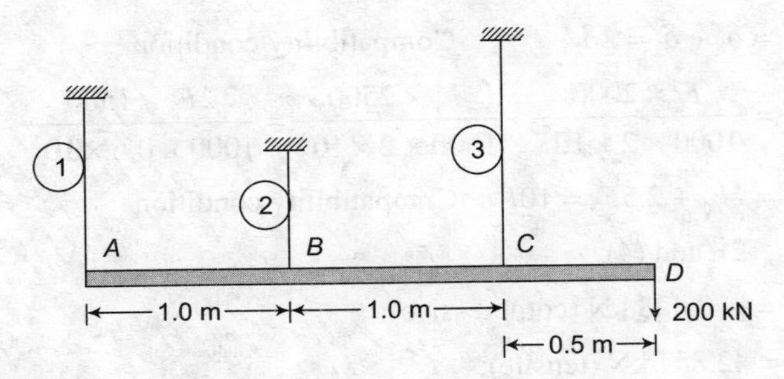

FIGURE 1.28

SOLUTION

Let F_1, F_2, and F_3 be the forces in the members 1, 2, and 3, respectively. Considering the free body diagram of the bar $ABCD$,

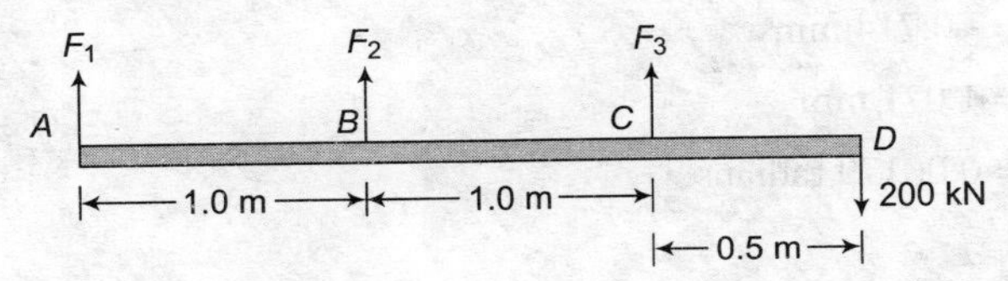

FIGURE 1.29

Sum of the forces in vertical direction is zero.

$$= F_1 + F_2 + F_3 = 200 \text{ kN}.$$

Sum of the moments about point 'A' is zero.

$$= 2F_3 + F_2 = 200 \times 2.5 = 500.$$

From the two equilibrium equations, the three unknown quantities F_1, F_2, and F_3 cannot be found. Hence, compatibility condition is to be used.

The rigid bar $ABCD$ cannot deform, but owing to the extensions of bars 1, 2, and 3 the rigid bar rotates. Let δ_1, δ_2, and δ_3 be the extensions of the bars 1, 2, and 3.

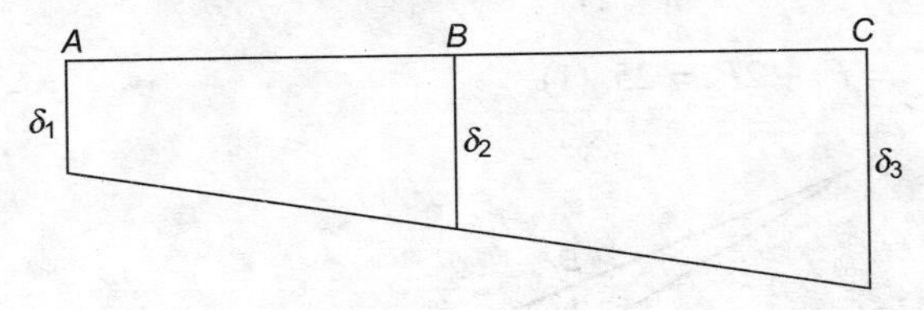

FIGURE 1.30

As the rigid bar just rotates, the deformations δ_1, δ_2, and δ_3 adjust in such way that,

$$\frac{\delta_2 - \delta_1}{1} = \frac{\delta_3 - \delta_1}{2} = \frac{\delta_3 - \delta_2}{1}$$

$$\delta_2 - \delta_1 = \delta_3 - \delta_2$$

$$= \delta_1 + \delta_3 = 2\delta_2 \qquad \text{Compatibility condition} \qquad (1.9)$$

$$= \frac{F_1 \times 2000}{1000 \times 2 \times 10^5} + \frac{F_3 \times 2500}{1000 \times 2 \times 10^5} = \frac{2F_2 \times 1500}{1000 \times 0.6 \times 10^5}$$

$$= 2F_1 + 2.5F_3 = 10F_2 \quad \text{Compatibility condition} \qquad (1.10)$$

Solving equations (1), (2), and (4)

$F_1 = -70.428$ kN (compressive);

$F_2 = 42.857$ kN (tensile);

$F_3 = 228.571$ kN (tensile).

Rotation of the rigid bar *ABCD* is given by

$$\theta = \frac{\delta_2 - \delta_1}{1000} :: \delta_2 = \frac{F_2 L_2}{(AE)_2} :: \delta_1 = \frac{F_1 L_1}{(AE)_1}$$

$\delta_1 = -0.714$ mm

$\delta_2 = 1.071$ mm

$\therefore \qquad \theta = 0.00179$ radians.

PROBLEM 1.16

Objective 1

A rigid bar is supplied by a pin at 'A' and two bars (1) and (2) at points B and C as shown in Figure 1.31. The bar (1) is copper having 100 mm^2 cross-sectional area. Bar (2) is of steel with cross-sectional area of 120 mm^2. Determine the stresses in bars 1 and 2 due to a 10 kN load at 'D'. E_S = 200 GPa and E_C = 80 GPa.

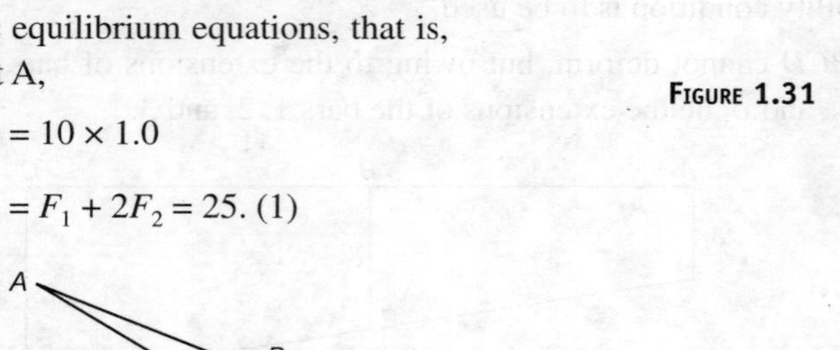

FIGURE 1.31

SOLUTION

Let F_1 and F_2 be the forces in the bars 1 and 2, respectively. Using static equilibrium equations, that is, by taking moments about A,

$$F_1 (0.4) + F_2 (0.8) = 10 \times 1.0$$

$$= F_1 + 2F_2 = 25. \ (1)$$

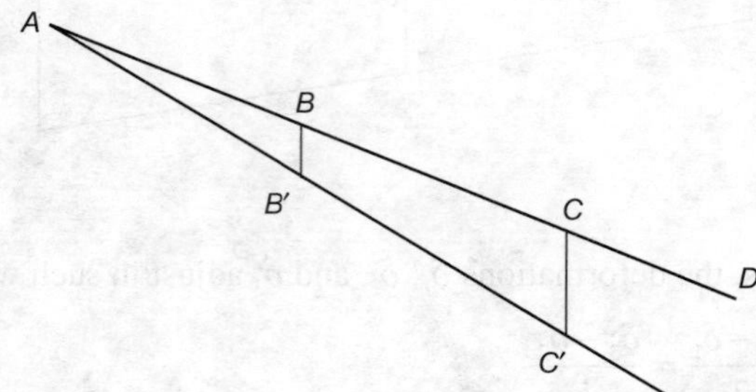

FIGURE 1.32

Bar $ABCD$ is rigid bar. Hence, it takes configurations $AB'C'D$

$$BB' = \delta_B = \frac{F_1 L_1}{A_1 E_1}$$

$$CC' = \delta_C = \frac{F_2 L_2}{A_2 E_2}$$

$$L_1 = 0.4 \text{ m} \qquad \left(\because L_1 = \frac{1000}{400 + 400 + 200} \times 400 \text{ mm}\right)$$

$$L_2 = 0.8 \text{ m}$$

'θ' rotation of bar is equal to

$$\frac{\delta_B}{(400/\cos\alpha)} = \frac{\delta_c}{(800/\cos\alpha)}$$

$$\delta_B = \frac{\delta_C}{2}$$

$$\frac{F_1 \times 400}{100 \times 0.8 \times 10^5} = \frac{F_2 \times 800}{120 \times 2 \times 10^5} \times \frac{1}{2} \tag{1.11}$$

$$F_1 = \frac{1}{3} F_2 \tag{1.12}$$

From equations (1.11) and (1.12)

$$F_1 = 3.571 \text{ kN}; F_2 = 10.714 \text{ kN}$$

Stress in the copper bar (1) = 35.71 N/mm^2 (tensile)

Stress in the steel bar (2) = 89.28 N/mm^2 (tensile).

PROBLEM 1.17 **Objective 1**

A rigid plate form rests on two aluminum rods having 1200 mm^2 cross-sectional area and length 250 mm. A third bar made of steel 249 mm long and cross-sectional area 2400 mm^2 is in between the two aluminum bars as shown in Figure 1.33. A load of 750 kN is applied on the platform. Determine the stresses induced in aluminum and steel bars. Take E_S = 210 GPa and E_{Al} = 70 GPa.

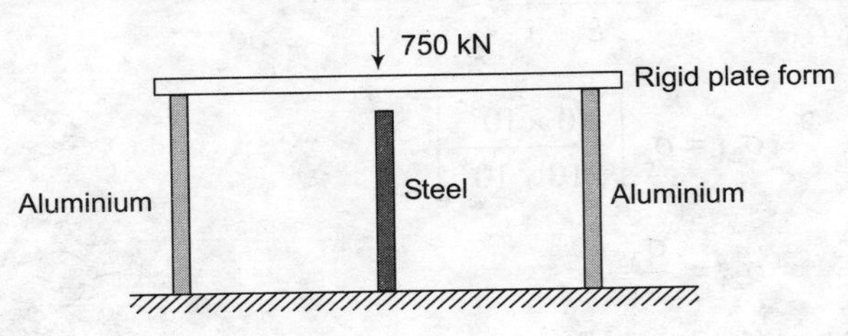

FIGURE 1.33

SOLUTION

In the given system, aluminum bars alone take the load till the gap of 1 mm is closed. Once the gap is closed, then aluminum and steel bars resist the load. Thus, the solution of the problem can be split into two cases. Final stresses will be the algebraic sum of stresses from two cases.

Case (i): Load required by the aluminum bars to deform by 1 mm.

Case (ii): After the gap is closed, sharing the remaining load between aluminum and steel.

Case (i): Let σ_{1A} be the stress in aluminum, which closes the gap between aluminum and steel. In this stage, steel bar is not at all stressed $\sigma_{1S} = 0$.

Then $$\frac{\sigma_{1A} \times L_A}{(AE)_A} = 1 \text{ mm}$$

$$\Rightarrow \quad \frac{\sigma_{1A} \times 250}{70 \times 10^3} = 1 \text{ mm}$$

$$\Rightarrow \quad \sigma_{1A} = 280 \text{ N/mm}^2$$

$\Rightarrow$Load required to close the gap $= \sigma_{1A} \times 2 \times A_{A1}$

$$P_1 = 672 \text{ kN.}$$

$\therefore$ In case (i), stress in aluminum = 280 MPa

$$= 0$$

Case (ii): Let P_2 be the load remaining after closing the gap of 1 mm.

$$P_2 = 750 - 672 = 78 \text{ kN.}$$

Let σ_{2A} and σ_{2S} be the stress developed in aluminum and steel, respectively, in the second stage.

$$\therefore \quad \sigma_{2A} \cdot A_{S1} + \sigma_{2S} \cdot A_S = P_2 \quad \text{(equilibrium condition)}$$

$$\Rightarrow \quad \sigma_{2A}(2400) + \sigma_{2s}(2400) = 78{,}000 \tag{1.13}$$

$$\Rightarrow \quad \sigma_{2A} + \sigma_{2S} = 32.5.$$

Once the gap is closed, the shortcoming of aluminum bar and steel bar should be same for compatibility.

$$\Rightarrow \quad \frac{\sigma_{2A} \times 250}{E_{Al}} = \frac{\sigma_{2S} \times 250}{E_S}$$

$$\therefore \quad \sigma_{2A} = \sigma_{2s}\left[\frac{70 \times 10^3}{210 \times 10^3}\right]$$

$$\sigma_{2A} = \frac{\sigma_{2S}}{3} \tag{1.14}$$

Using equations (1.14) and (1.13)

$$\sigma_{2S} = 24.375 \text{ MPa}$$

$$\sigma_{2A} = 8.125 \text{ MPa}$$

$\therefore$ Final stress in aluminum $= \sigma_A = \sigma_{1A} + \sigma_{2A} = 288.13$ MPa

Final stress in steel $= \sigma_S = \sigma_{1S} + \sigma_{2S} = 24.375$ MPa.

PROBLEM 1.18

Objective 1

A concrete column of cross-section 230 mm × 230 mm carries four numbers of 12 mm diameter steel bars. If an axial load of 200 kN acts on the composite column, determine the stress induced in concrete and steel. E_S/E_C = modular ratio = 10.

SOLUTION

Let σ_S and σ_C be the stress developed in concrete. For static equilibrium condition

$$\sigma_S \times A_S + \sigma_C \times A_C = P$$

in which A_S is cross-sectional area of steel; A_C is cross-sectional area of concrete.

$$A_S = 4 \times \frac{\pi}{4}(12^2) = 452 \text{ mm}^2$$

$$A_C = 52{,}447.6 \text{ mm}^2$$

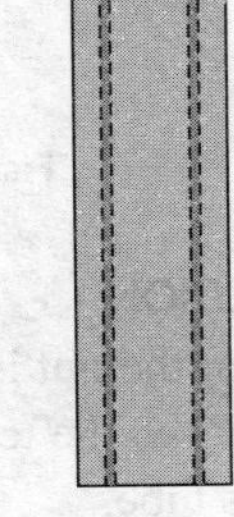

FIGURE 1.34

$$\therefore \quad 452\,\sigma_S + 52{,}447.6\,\sigma_C = 200{,}000 \qquad (1.15)$$

Compatibility condition: Perfect bond exists between concrete and steel.

$$\Rightarrow \quad \delta_C = \delta_S$$

$$\Rightarrow \quad \frac{\sigma_C \times L}{E_C} = \frac{\sigma_S \times L}{E_S}$$

$$\Rightarrow \quad \sigma_S = \frac{E_S}{E_C} \times \sigma_C.$$

Here $\frac{E_S}{E_C}$ is called as modular ratio.

$$\therefore \quad \sigma_S = 10\sigma_C \qquad (1.16)$$

From equations (1.15) and (1.16)

$$\sigma_S = 35.11 \text{ MPa}; \ \sigma_C = 3.511 \text{ N/mm}^2$$

Load shared by concrete = 184.131 kN

Load shared by steel = 15.869 kN

% load shared by steel is 7.9% only.

Thus in reinforced concrete columns, capacity of the member can be better increased by increasing concrete rather than steel.

PROBLEM 1.19

Objective 1

A brass tubes of cross-sectional area 200 mm^2 and length 1 m is clamped between two rigid plates as shown in Figure 1.35. A steel bar of cross-sectional area 150 mm^2 passes centrally through the rigid plates and tightened by ruts. Pitch of the nut is 2 mm. If the nut is rotated by ½ revolutions, estimate the stresses produced in steel and brass. Take E_S = 210 GPa and E_B = 100 GPa.

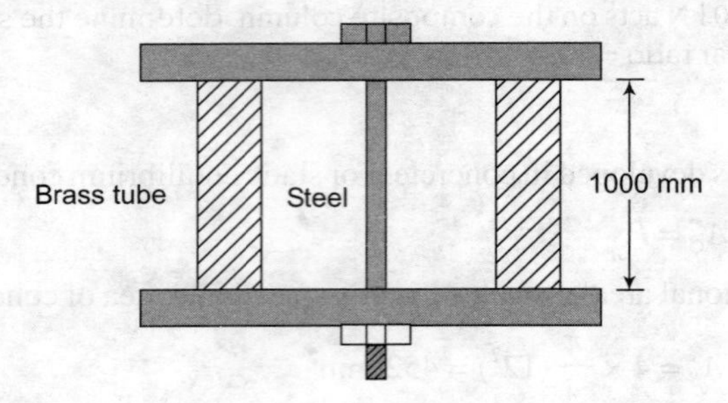

FIGURE 1.35

SOLUTION

When the nut is tightened, the steel both extend and brass bars between the rigid plates compress. Let P_s be the tensile force induced in the steel bolt and P_b be the compressive force induced in the brass tube.

For equilibrium,

$$P_s = P_b = P. \tag{1.17}$$

For compatibility, movement of the nut is equal to the sum of extension in the steel bolt and compression in the brass tube. The above compatibility condition can be explained as below. The steel bar can slide between the rigid plates freely.

Forget tightening of nut for the time being. Now, apply P_s force in steel bolt, the steel bolt extends by δ_s. Then, apply P_b compressive force in the brass tube. Brass tube compress by δ_b when the nut is not tightened, but the force P_b and P_s were applied by external means, then the nut will be $\sigma_s + \delta_b$ distant from the rigid plate. Now, rotate the nut to cover the distance $\Delta = \delta_s + \delta_b$ freely.

This is same as tightening the nut, such that P_b and P_s forces are induced in the brass tube and bolt, simultaneously displacing the nut by δ.

$$\therefore \qquad \Delta = \delta_s + \delta_b \tag{1.18}$$

Δ = distance traveled by the nut in ½ revolution

$$= \text{pitch} \times \frac{1}{2}$$

$$= 2 \times \frac{1}{2} = 1 \text{ mm}$$

$$\delta_S = \frac{P_s \times L_s}{A_S E_S} = \frac{P_s \times 1000}{150 \times 2 \times 10^5}$$

$$\delta_b = \frac{P_b \times L_b}{A_b E_b} = \frac{P_b \times 1000}{200 \times 1 \times 10^5}.$$

Substituting the above in equations (2) and (1)

$$P\left[\frac{1000}{200 \times 1 \times 10^5} + \frac{1000}{150 \times 2 \times 10^5}\right] = 1.0 \text{ mm}$$

$$P = 12{,}000 \text{ N}.$$

Stress in the steel bolt (tensile stress) $\sigma_s = \dfrac{120{,}00}{150} = 80$ MPa (tensile)

Stress in the brass tube $= \sigma_s = \dfrac{12{,}000}{200} = 60$ MPa (compressive stress).

PROBLEM 1.20 **Objective 1**

A rigid bar AB, hinged at C, is connected by two bars (1) and (2) at A and B, respectively, as shown in Figure 1.36. Bar (1) is 2 mm short in length. Forcibly bar (1) is connected to complete the form. Estimate the stresses induced in the bars for the following data:

$A_1 = 200$ mm^2; $A_2 = 400$ mm^2; $E_1 = 100$ GPa; $E_2 = 10$ GPa.

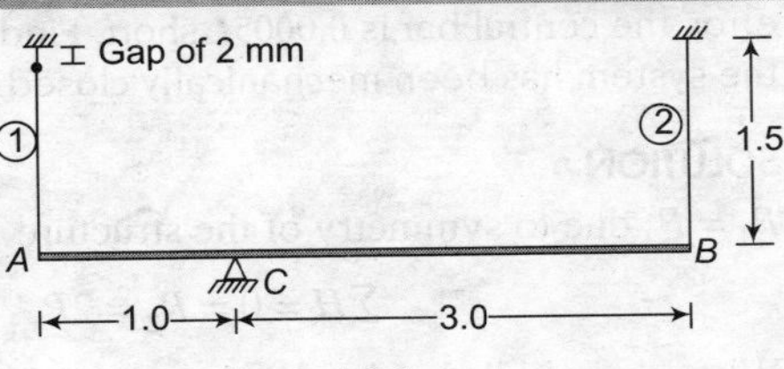

FIGURE 1.36

SOLUTION

To close the gap of 2 mm, say force F_1 is to be applied for the first bar. Because of F_1, let F_2 be the force developed in bar (2). Apply condition of equilibrium,

$$\sum M_C = 0$$

$$= F_1 \times 1 = 3F_2. \tag{1.19}$$

Let δ_1 be the extension in the bar (1) due to F_1. Let δ_2 be the extension in the bar (2) due to F_2. If B moves down by δ_1, point A moves up by $\dfrac{\delta_1}{3}$.

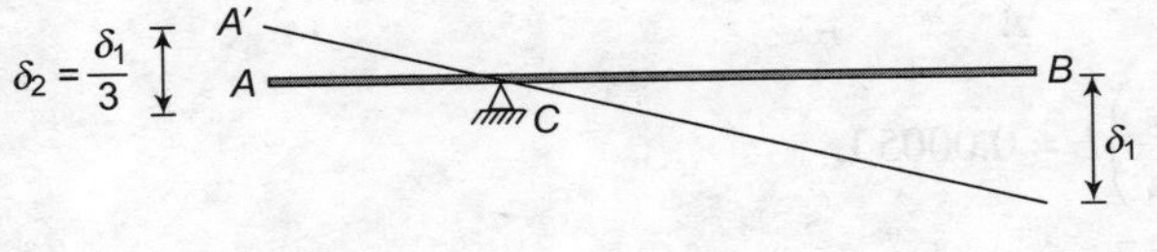

FIGURE 1.37

For closing the gap, $\delta_1 + \dfrac{\delta_2}{3} = \Delta$. (1.20)

This is the compatibility condition.

$$\delta_1 = \frac{F_1 L_1}{A_1 E_1};\ \delta_2 = \frac{F_2 L_2}{A_2 E_2}$$

$$\therefore \quad \frac{F_1 \times 1500}{200 \times 1 \times 10^5} + \frac{1}{3} \times \frac{F_2 \times 1500}{400 \times 0.7 \times 10^5} = 2 \text{ mm.} \quad (1.21)$$

Using the value of F_1 in the above equation

$$F_1 = 8.235 \text{ kN}$$

$$F_2 = 24.706 \text{ kN}$$

$\therefore$ Stress in the bar (1) $\sigma_1 = 123.53$ MPa (tensile stress)

Stress in the bar (2) $\sigma_2 = 20.588$ MPa (tensile stress).

PROBLEM 1.21

Objective 1

Three horizontal bars of same cross-sectional area connected between two rigid plates of length L between them. Because of a fabrication error, the central bar is $0.0005L$ short. Find the stress in each bar after the system has been mechanically closed. $E = 75$ GPa.

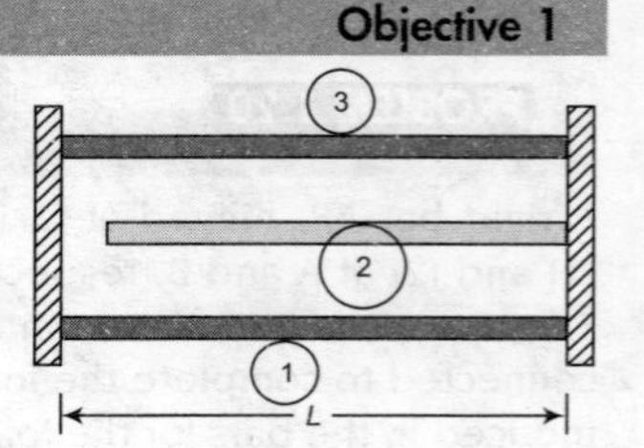

FIGURE 1.38

SOLUTION

$P_1 = P_3$ due to symmetry of the structure

$$\Sigma H = 0 = P_2 = 2P_1 \quad (1.22)$$

When the middle bar is pulled to close the gap, compressive stress is induced in bars (1) and (3).

Hence, as per the compatibility condition

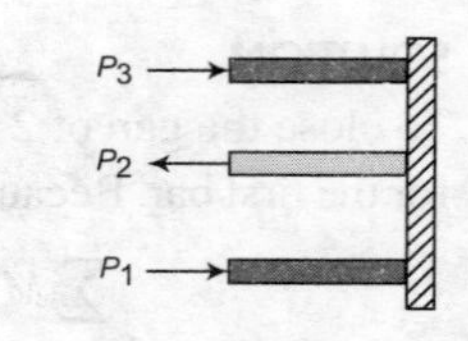

FIGURE 1.39

$$\delta_2 + \delta_1 = \delta_2 + \delta_3 = \Delta.$$

$$\frac{P_2 L}{AE} + \frac{P_1 L}{AE} = \Delta \quad (1.23)$$

$$\frac{P_2 L}{AE} + \frac{P_2 L}{2AE} = 0.0005\, L$$

Stress in the bar 2 is $\sigma_2 = \frac{P_2}{A}$

$$\left(\frac{\sigma_2}{E} + \frac{\sigma_2}{2E}\right) L = 0.0005 \text{ L}$$

$$\sigma_2 = \frac{75 \times 10^3 \times 0.0005}{1.5} = 25 \text{ MPa}$$

Stress in the middle bar $\sigma_2 = 25$ MPa (T)

Stress in the bar (1) $= \sigma_1 = \sigma_2/2 = 12.5$ MPa (C)

Stress in the bar (3) $= \sigma_3 = \sigma_1 = 12.5$ MPa (C).

PROBLEM 1.22

Objective 3

A rigid bar AB is pinned at A and supported by a steel rod at D as shown in Figure 1.40. A linear spring of stiffness 20 kN/mm is located at C and concentrated load of 30 kN at D. Determine the vertical displacement of the point B $A_s = 100$ mm^2, $E_S = 200$ GPa; $L_S = 0.5$ m.

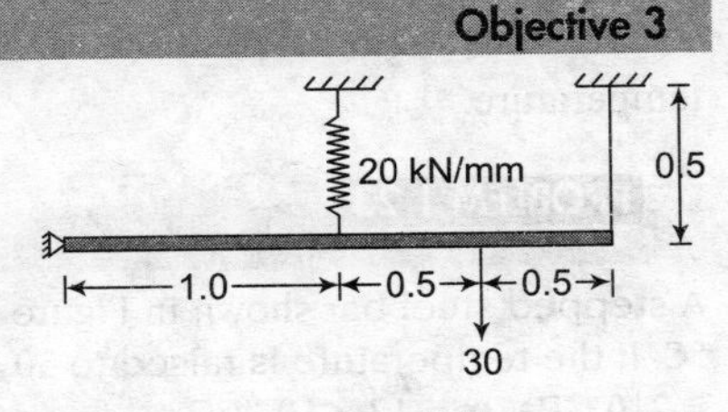

FIGURE 1.40

SOLUTION

A linear spring means force–displacement response of the spring is linear. Let F_C be the force in the spring and F_S be the force in the string.

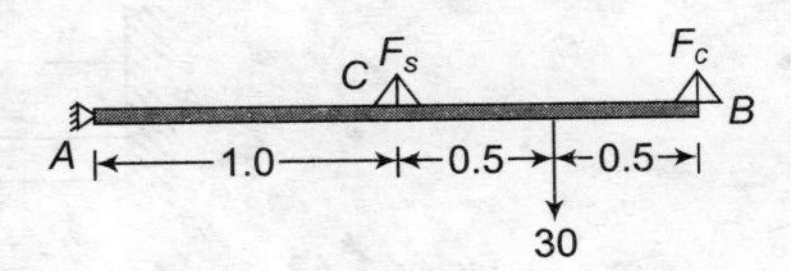

FIGURE 1.41

Taking moments about A,

$$2 \times F_S + 1 \times F_C = 30 \times 1.5$$

$$\Rightarrow \quad 2F_S + F_C = 45$$

Compatibility condition,

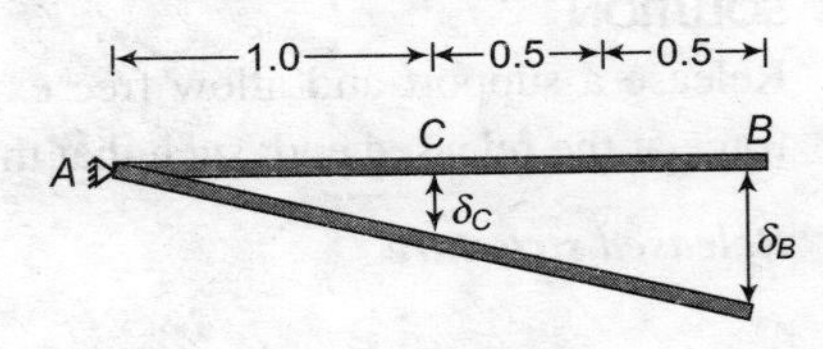

FIGURE 1.42

$$\delta_C/AC = \delta_B/AB$$

$$\delta_C/1 = \delta_B/2$$

$$\Rightarrow \quad 2\delta_C = \delta_B$$

δ_C = extension of the spring = F_C/K

$$\frac{2F_C}{20} = \frac{F_S \times 500 \times 1000}{100 \times 2 \times 10^5} \quad (F_S \text{ and } F_C \text{ are in kN})$$

$$F_C = \frac{F_S}{4}$$

$$\Rightarrow \quad F_S = 20 \text{ kN};\ F_C = 5 \text{ kN}.$$

Displacement at $B = \delta_B = 2 \times \delta_C = 2 \times F_C/K = 2 \times 5/20 = 0.5$ mm.

1.5.1 Thermal Stress

Temperature variations cause change in the dimensions of the body depending on the material properties. If the deformations with change in temperature are restrained, the stress induced in the thermal property of the body is called 'coefficient of thermal expansion'. Coefficient of thermal expansion is the increase in length per unit length of the body due to unit rise in temperature. Generally, temperature is expressed in Kelvin (K) or degree Celsius (°C). Coefficient of thermal expansion is denoted by 'α'.

Increase in length with change in temperature = $\Delta = L\,\alpha\,T$

in which L = length of the body

α = coefficient of thermal expansion

T = rise in temperature.

Thermal strain = $\Delta L/L = \alpha \cdot T$

Thermal strain is positive, if T is rise in temperature and is negative, if T decreases in temperature.

PROBLEM 1.23

Objective 1

A stepped steel bar shown in Figure 1.43 is fixed between two rigid walls at room temperature of 27 °C. If the temperature is raised to 50 °C determine the maximum stress produced in the bar. Take E_S = 210 GPa, $\alpha_s = 12 \times 10^{-6}$/°C.

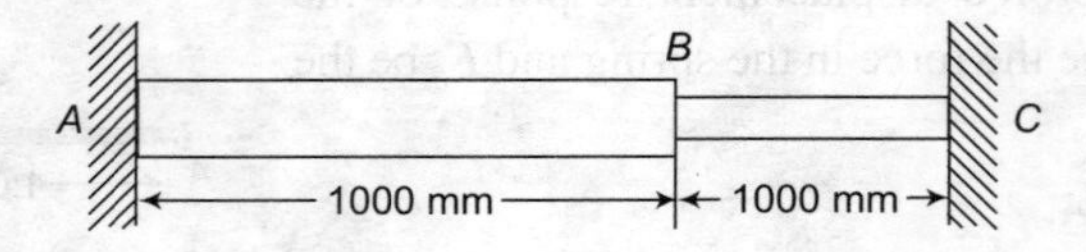

FIGURE 1.43

SOLUTION

Release a support and allow free expansion due to rise in temperature. Then, apply compressive force at the released end, such that the net extension becomes zero.

Released structure

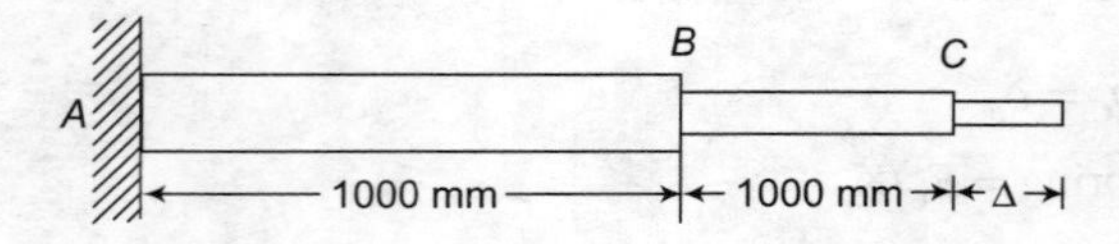

FIGURE 1.44

$\Delta = L\,\alpha\,T$ (It do not depend on cross-sectional area)

$\Delta = 2000 \times 12 \times 10^{-6} \times (50 - 27)$

$= 0.552$ mm.

Apply force 'P' at 'C' such that, the bars compress by 0.552 mm so that net extension at C due to P and rise in temperature is zero.

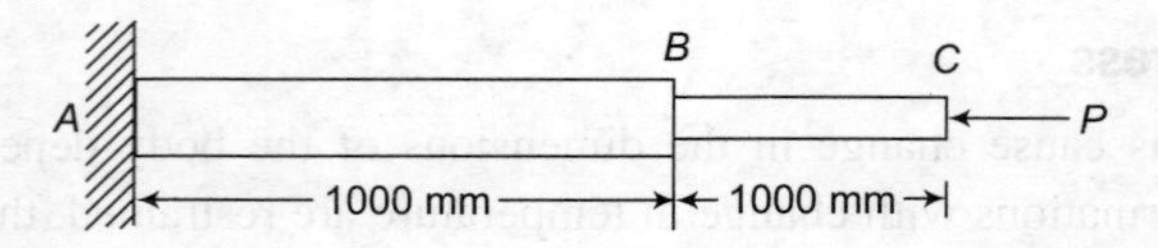

FIGURE 1.45

$\Rightarrow \quad (P \times 1000)/(75 \times 2.10 \times 10^5) + (P \times 1000)/(150 \times 2.1 \times 10^5) = 0.552$

$P = 5796$ N; $\sigma_{max} = 77.28$ MPa (C).

1.5.2 Thermal Stresses in Composite Members

Composite members are more common in structural members. Change in temperature induces stress in the composite members as the thermal properties of the individual materials vary. Consider two bars of different materials say steel and brass; brazed together and subjected to rise in temperature by ΔT. To find the stresses produced due to the temperature variation, release the structure and apply force in the bars to maintain compatibility condition; in this case steel and brass bars undergo same extension or strain.

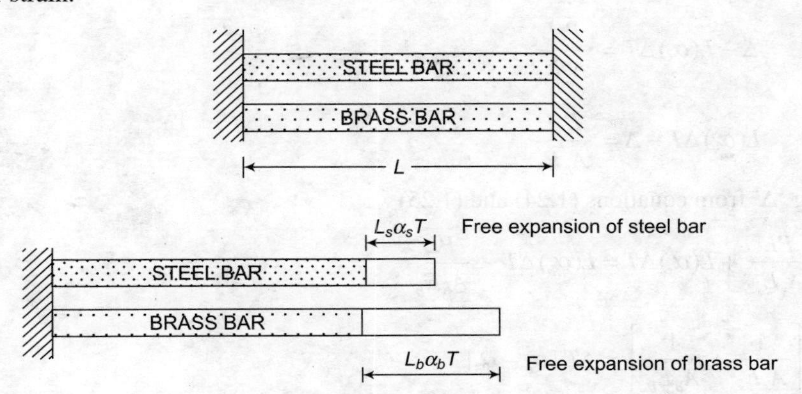

In the absence of rigid plate at the end,

Free expansion of steel = $\alpha_s(L)\,\Delta T$

Free expansion of brass = $\alpha_b(L)\,\Delta T$

in which α_s = coefficient of thermal expansion and α_b = coefficient of thermal expansion of brass.

But as per compatibility condition, both steel and brass bars have to undergo same extension. Thus, steel bar extends little more than the free expansion, whereas brass undergoes extension less than free expansion.

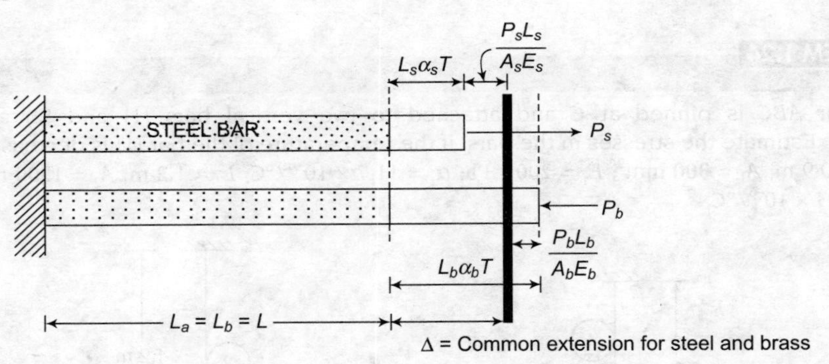

FIGURE 1.46

P_s is the tensile force in the steel bar, responsible for deformation $\Delta - \alpha_s(L)\Delta T$).

P_b is the compressive force in the brass bar, responsible for deformation $(L)\,\alpha_b\Delta T - \Delta$.

Extension due to $P_s = \dfrac{P_S \times L}{A_s E_s}$.

Compression due to $P_b = \dfrac{P_b \times L}{A_b E_b}$.

As per equilibrium condition,

$$P_b = P_s = P.$$

Using the compatibility,

$$\Delta - L(\alpha_s)\Delta T = \frac{P_S L}{A_s E_s} \tag{1.24}$$

$$L(\alpha_b)\Delta T - \Delta = \frac{P_B L}{A_b E_b} \tag{1.25}$$

Equating 'Δ' from equations (1.24) and (1.25)

$$\frac{PL}{A_s E_s} + L(\alpha_s)\Delta T = L(\alpha_b)\Delta T - \frac{PL}{A_b E_b}$$

$$P\left[\frac{1}{A_s E_s} + \frac{1}{A_b E_b}\right] = \Delta T\,[\alpha_b - \alpha_s]$$

$$P = \frac{\Delta T(\alpha_b - \alpha_s) \times A_s E_s A_b E_b}{(A_b E_b + A_s E_s)}$$

Stress in steel (tension) $= \sigma_s = \dfrac{\Delta T(\alpha_b - \alpha_s) \times E_s A_b E_b}{(A_b E_b + A_s E_s)}$

Stress in brass (compressive) $= \sigma_b = \dfrac{\Delta T(\alpha_b - \alpha_s) \times E_b A_s E_s}{(A_b E_b + A_s E_s)}$

PROBLEM 1.24

Objective 1

A rigid bar *ABC* is pinned at *C* and attached to two vertical bars (1) and (2) as shown in Figure 1.47. Estimate the stresses in the bars, if the temperature of the bar (1) is decreased by 40°C. Data: L_1 = 0.9 m; A_1 = 300 mm^2; E_1 = 200 GPa; $\alpha_1 = 11.7 \times 10^{-6}$/°C; L_2 = 1.2 m; A_2 = 1200 mm^2; E_2 = 70 GPa; $\alpha_2 = 23 \times 10^{-6}$/°C;

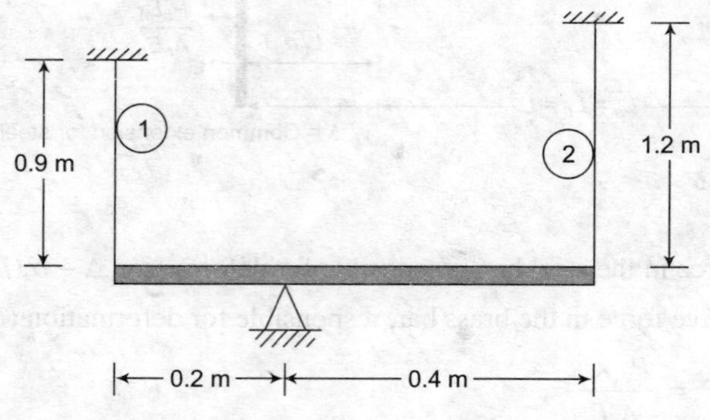

FIGURE 1.47

SOLUTION

Let P_1 and P_2 be the force in the members due to change in temperature.

From the condition of equilibrium

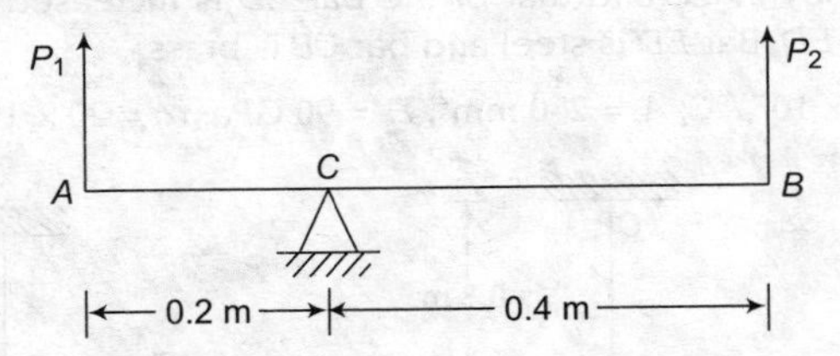

FIGURE 1.48

$$\Sigma M_C = 0$$

$\Rightarrow$ $$P_1 (0.2) = P_2 (0.4) \tag{1.26}$$

$\Rightarrow$ $$P_1 = 2P_2.$$

Let the displacement at A be δ_A and at B be δ_B.

$$\frac{\delta_A}{0.2} = \frac{\delta_B}{0.4} \Rightarrow \delta_A = \frac{\delta_B}{2} \tag{1.27}$$

δ_A is displacement, which is the decrease in length due to thermal variation and increase in length due to P_1.

$\therefore$ $$\delta_A = L_1\alpha_1 T - \frac{P_1 L_1}{A_1 E_1}$$

As the bar (2) is not subjected to any thermal movement

$\therefore$ $$\delta_B = \frac{P_2 L_2}{A_2 E_2}$$

$$\delta_B = 2\delta_A \text{ (compatibility condition)}$$

$\therefore$ $$2\left\{900 \times 11.7 \times 10^{-6} \times 40 - \frac{P_1 \times 900}{300 \times 2 \times 10^5}\right\} = \frac{P_2 \times 1200}{1200 \times 0.7 \times 10^5}$$

$\Rightarrow$ $$0.8424 - 0.3 \times 10^{-4} P_1 = 1.429 \times 10^{-5} P_2 \tag{1.28}$$

Using equation (1.26) in equation (1.28)

$$0.8424 - 0.6 \times 10^{-4} P_1 = 1.429 \times 10^{-5} P_2$$

$\Rightarrow$ $$P_2 = 11{,}339 \text{ N}$$

$\Rightarrow$ $$P_1 = 22678 \text{ kN}.$$

Stress in bar (1) $\sigma_1 = 75.59$ MPa.

Stress in bar (2) $\sigma_2 = 9.45$ MPa.

PROBLEM 1.25

Objective 1

A rigid bar AD pinned at 'A' and attached to the bars BC and ED is shown in Figure 1.49. Temperature of the bar CB is decreased by 25 °C and that of the bar ED is increased by 25 °C. Find the stress induced in the bars CB and ED. Bar ED is steel and bar CB is brass.

$E_S = 200$ GPa; $\alpha_s = 12 \times 10^{-6}$/°C; $A_s = 250$ mm^2; $E_b = 90$ GPa; $\alpha_b = 20 \times 10^{-6}$/°C; $A_b = 500$ mm^2

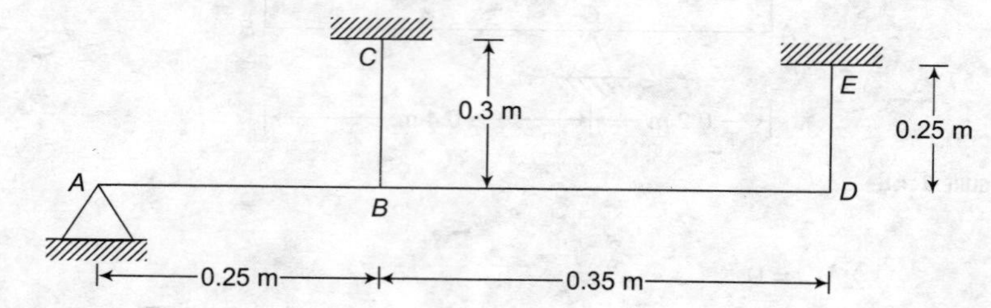

FIGURE 1.49

SOLUTION

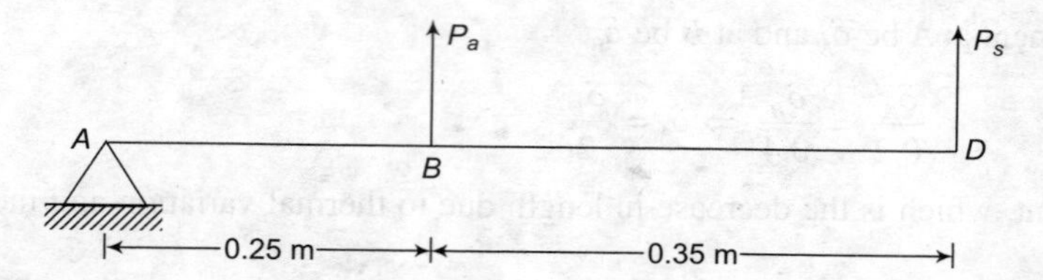

FIGURE 1.50

Let P_s and P_b be the forces developed in the steel and brass bars, respectively. As per equilibrium condition,

Algebraic sum of moments about $A = 0$

$$\Sigma M_A = 0$$

$$\Rightarrow \quad P_s \times 0.6 = P_b \times 0.25$$

$$\Rightarrow \quad P_b = 2.4P_s \tag{1.29}$$

For compatibility condition,

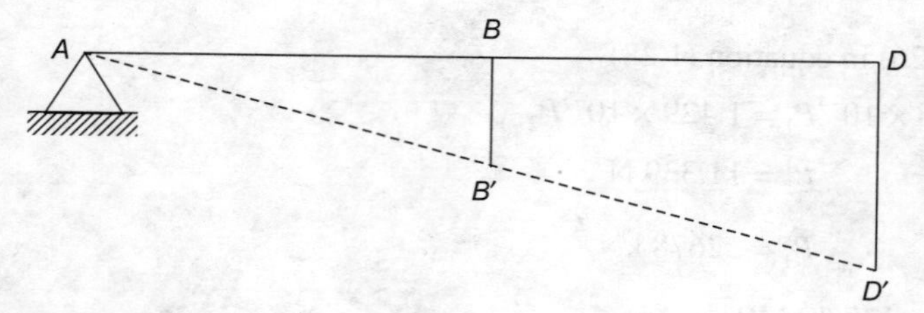

FIGURE 1.51

$BB' = \delta_B$ = displacement of the point $B\downarrow$

= [Extension of the bar BC due to P_b] – [Contraction of BC due to drop in temperature]

$$= \frac{P_b L_b}{A_b E_b} - L_b \alpha_b T$$

$$= \frac{P_b \times 300}{500 \times 0.9 \times 10^5} - 300 \times 20 \times 10^{-6} \times 25$$

$$= 0.667 \times 10^{-5} P_b$$

$DD' = \delta_D$ = displacement of the point $D\downarrow$

= [Extension of the bar ED due to P_s] – [Extension of ED due to rise in temperature]

$$= \frac{P_s L_s}{A_s E_s} + L_s \alpha_s T$$

$$= \frac{P_s \times 250}{250 \times 2 \times 10^5} + 250 \times 12 \times 10^{-6} \times 25$$

$$= 0.5 \times 10^{-5} P_s + 0.075.$$

For compatibility

$$\frac{BB'}{AB} = \frac{DD'}{AD}$$

$\Rightarrow$ $DD' = BB'$

$\therefore$ $\delta_D = 2.4\delta_B$

$\therefore$ $0.5 \times 10^{-5} P_S + 0.075 = 2.4[0.667 \times 10^{-5} P_b - 0.15]$

$\Rightarrow 1.60 \times 10^{-5} P_b - 0.5 \times P_S = 0.435$

$\Rightarrow$ $1.60\ P_b - 0.5 \times P_S = 43{,}500.$

From equation (1), $P_b = 2.4P_s$

$\therefore$ $P_s = 13{,}024$ N

$\therefore$ $P_b = 31{,}257.5$ N

Stress in steel bar = $\sigma_s = 52.10$ MPa (tensile stress)

Stress in steel bar = $\sigma_b = 62.52$ MPa (tensile stress).

PROBLEM 1.26

Objective 1

A rigid bar of negligible weight is supported by two bars (1) and (2) and a change at A as shown in Figure 1.52. Determine the temperature change required to cause a stress of 55 MPa in bar (1) for the following data.

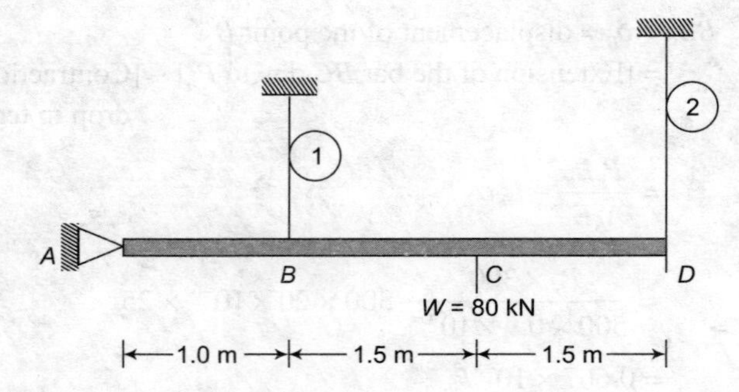

FIGURE 1.52

$$L_1 = 1.5 \text{ m};\ A_1 = 320 \text{ mm}^2;\ E_1 = 200 \text{ GPa};\ \alpha_1 = 11.7 \times 10^{-6}/°\text{C}$$
$$L_2 = 3.0 \text{ m};\ A_2 = 1300 \text{ mm}^2;\ E_2 = 83 \text{ GPa};\ \alpha_2 = 18.9 \times 10^{-6}/°\text{C}$$

SOLUTION

Let σ_1 and σ_2 be the tensile stress induced in bars (1) and (2), respectively.

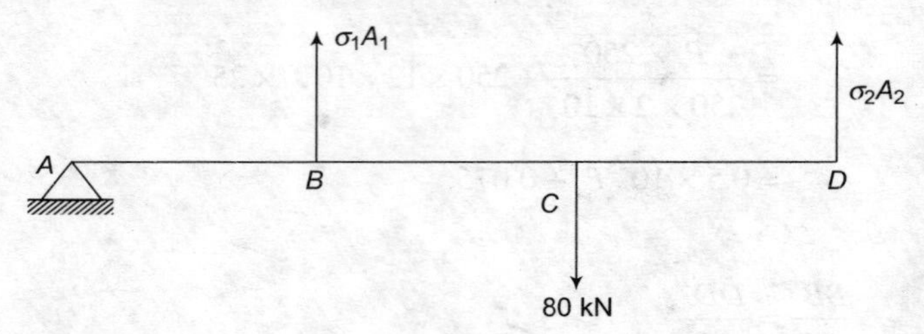

FIGURE 1.53

Algebraic sum of moments about $A = 0$

$$\Sigma M_A = 0$$

$$\Rightarrow \quad \sigma_1 A_1 \times 1 + \sigma_2 A_2 \times 4 = 80 \times 1000 \times 2.5$$

$$\Rightarrow \quad 55 \times 320 \times 1 + 4 \times \sigma_2 \times 1300 = 2 \times 10^5$$

$$\Rightarrow \quad \sigma_2 = 35.08 \text{ MPa}.$$

Let 'T' be the rise in temperature that develops stress in bar (1) by 55 MPa.

From compatibility consideration,

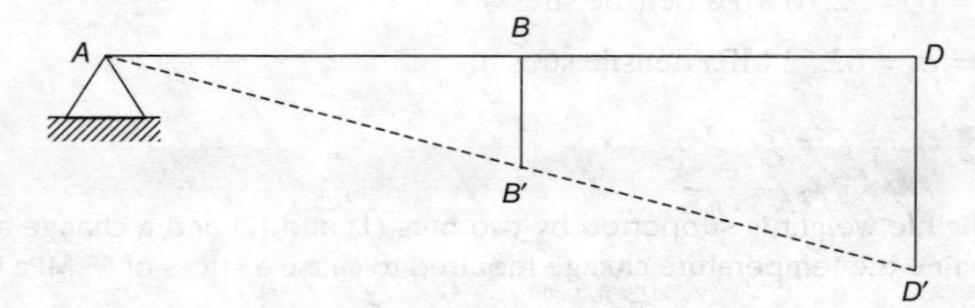

FIGURE 1.54

$$\frac{\delta_B}{1} = \frac{\delta_D}{4}$$

$$\delta_D = 4\delta_B \tag{1.30}$$

δ_B = Displacement of point B, due to rise in temperature and σ_1

$$\delta_B = \frac{\sigma_1 L_1}{E_1} + L_1\alpha_1 T = 0.4124 + 0.0176T$$

δ_D = Displacement of point D due to σ_2 and rise in temperature T

$$= \frac{\sigma_2 L_2}{E_2} + L_2\alpha_2 T$$

$$= \frac{35.08 \times 3000}{0.83 \times 10^5} + 3000 \times 18.9 \times 10^{-6} \times T$$

$$= 1.2678 + 0.0567T.$$

Using δ_D and δ_B values in equation (1.30)

$$1.2678 + 0.0567T = 4 \times [0.4125 + 0.0176T]$$

$$\Rightarrow \qquad T = -27.9\ °C.$$

A drop in temperature about 27.9 °C is required to create tensile stress of 55 MPa in the bar (1).

PROBLEM 1.27

Objective 1

In an assembly of brass tube and steel bolt shown in Figure 1.55, the pitch of the bolt thread is 1 mm. The cross-sectional area of the tube is 1000 mm^2 and that of steel bolt is 500 mm^2. If the nut is turned by 1.5 revolutions and the temperature of the system is raised by 100 °C. Find the stresses in the tube and the bolt. Take E_s = 210 GPa; E_b = 85 GPa; $\alpha_s = 12 \times 10^{-6}$/°C; and $\alpha_b = 20 \times 10^{-6}$/°C.

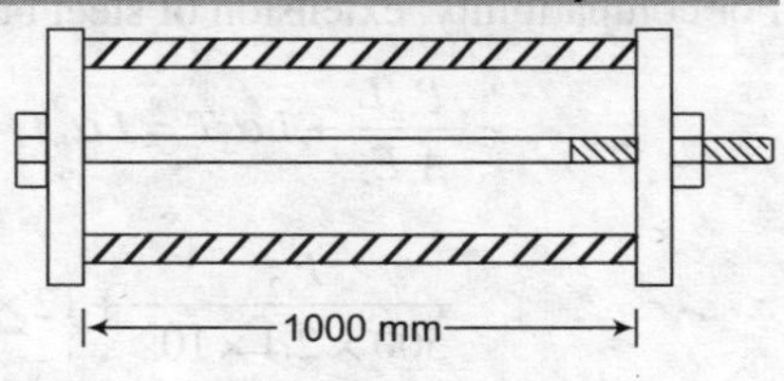

FIGURE 1.55

SOLUTION

This problem can be solved by dividing the problem into two cases.

Case (i): Stresses due problem into only nut tightening. Because of this, steel bolt will be tensioned and brass tube will be compressed.

Case (ii): Stresses tube due to only temperature rise. Because of this, steel bolt will be tensioned and brass tube will be compressed; final stress in the materials will be the algebraic sum of stresses of case (i) and case (ii).

Case (i): Stresses due to nut tightening

Let P_{s1} and P_{b1} be the tensile force in the bolt and compressive force in the tube due to nut tightening by 1.5 revolutions.

For equilibrium,

$$P_{s1} = P_{b1} = P_1 \text{ (say)}.$$

For compatibility

Distance traveled by the nut (Δ) = Extension in the steel bolt + compression in the tube

Distance traveled by the nut in 1.5 evolutions = Δ = 1.5 × pitch

$$= 1.5 \times 1$$

$$= 1.5 \text{ mm.}$$

$$\therefore \quad \frac{P_{s1}L_{s1}}{A_sE_s} + \frac{P_{b1}L}{A_bE_b} = \Delta$$

$$\Rightarrow \quad \frac{P_{s1} \times 1000}{500 \times 210 \times 10^3} + \frac{P_{b1} \times 1000}{1000 \times 85000} = 1.5$$

As $P_{s1} = P_{b1} = P_1$

$$P_1 = 70{,}461 \text{ N.}$$

$$\therefore \quad \sigma_{s1} = 140.92 \text{ MPa (tensile)}$$

$$\therefore \quad \sigma_{b1} = 70.46 \text{ MPa (compressive).}$$

Case (ii): Stresses due to only thermal variation

Let P_{s2} and P_{b2} be the tensile force in the steel bolt and compressive force in the brass tube, respectively.

For equilibrium, $P_{s2} = P_{b2} = P_2$ say. (1.31)

For compatibility, extension of steel bolt and compression of brass tube should be same.

$$\frac{P_{s2}L}{A_sE_s} + L\alpha_sT = L\alpha_bT - \frac{P_{b2}L}{A_bE_b}$$

$$\therefore \quad \frac{P_{s2}}{500 \times 2.1 \times 10^5} + 12 \times 10^{-6} \times 100 = 20 \times 10^{-6} \times 100 - \frac{P_{b2}}{1000 \times 0.85 \times 10^5}$$

$$P_{s2} = P_{b2} = P_2$$

$$P_2 = 37{,}579 \text{ N}$$

$$\text{Stress in bolt } \sigma_{s2} = \frac{37{,}579}{500} = 75.16 \text{ MPa (tensile)}$$

$$\text{Stress in tube } \sigma_{b2} = \frac{37{,}579}{1000} = 37.58 \text{ MPa (compressive)}$$

Final stress in the steel bolt $\sigma_s = \sigma_{s1} + \sigma_{s2}$

$$= 140.92 + 75.16$$

$$= 216.08 \text{ MPa (T)}$$

Final stress in brass tube $\sigma_b = \sigma_{b1} + \sigma_{b2}$

$$= 70.46 + 37.58$$

$$= 108.04 \text{ MPa (C)}$$

1.5.3 Strain Energy due to Axial Loading

Work done by a force is the product of the magnitude of the force and the displacement or deformation of the body in the direction of the force. Thus, forces acting on deformable bodies 'work' as bodies deform, though the bodies are at rest configuration. This work done by the external force must be conserved.

The internal stresses and the corresponding strains due to external force produce internal work (within the body). From low conservation of energy, external work must be equal to internal work. This internal work is defined as strain energy.

$$We = U + I$$

We = External work

U = Strain energy

I = Internal heat energy

In strength of materials, we assume that the system is adiabatic. Thus, no heat is given to the system or taken out of the system. Thus, internal heat energy becomes zero ($I = 0$).

$$\therefore \quad We = U$$

In general, the loads applied are gradual and corresponding deformations are also gradual. Thus, the external work done (We)

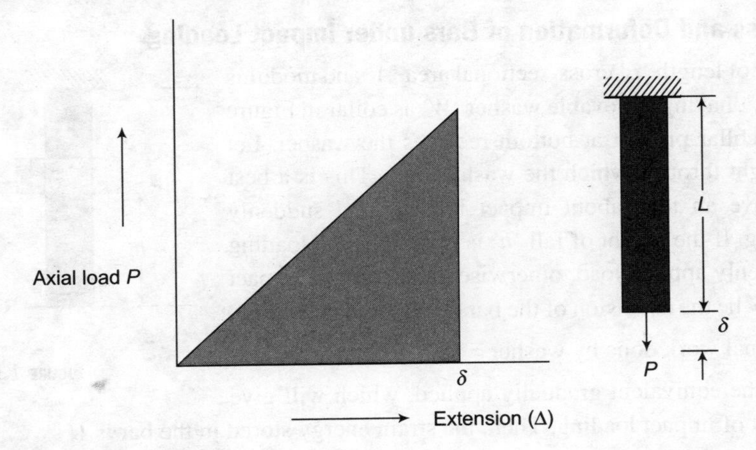

FIGURES 1.56 AND 1.57

$$\therefore \text{ Strain energy } U = \frac{1}{2}P\delta$$

$$U = \frac{1}{2}P\frac{PL}{AE}$$

Rearranging the above expression,

$$U = \frac{1}{2}P\frac{PL}{AE} \times \frac{A}{A}\text{; Taking } P/A \text{ as } \sigma,$$

$$U = \frac{1}{2}\sigma\frac{\sigma L}{E}A = \frac{1}{2}\sigma\frac{\sigma}{E} \times \text{(Volume of the bar)}$$

or strain energy density = U/volume = $\frac{1}{2}\frac{\sigma^2}{E}$

in which 'σ' is stress and E is modulus of elasticity.

Resilience

Strain energy stored in the body, when the stress at its proportionality limit is referred as 'resilience modulus'. This represents the ability of the body to absorb energy. (This value is high for springs.)

$$U_R = \frac{1}{2}\frac{\sigma^2{}_y}{E}$$

Toughness Modulus

Toughness modulus is the strain energy stored in the body up to complete rupture. This can be obtained from the area bounded by the $P - \Delta$ (curve) up to rupture/failure.

Strain energy concept is of great help, especially when we deal with suddenly applied loads or moving loads or impact loads, etc.

1.5.4 Stress and Deformation of Bars under Impact Loading

Consider a bar of length 'L' cross-sectional area 'A' and modulus of elasticity 'E', having a movable washer 'W' as collar in Figure 1.58. A rigid collar present at bottom receives the washer. Let 'h' be the height through which the washer falls. This is a best example to give an idea about impact loading and suddenly applied loading. If the height of fall 'h' is zero, then the loading becomes suddenly applied load, otherwise the loading is impact loading. Let 'δ' be the extension of the bar due to impact loading.

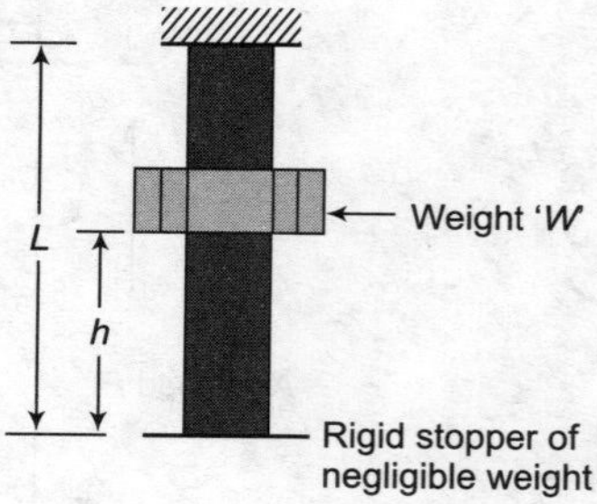

FIGURE 1.58

External work done by washer = $W(h + \delta)$ (1.32)

Let W_* be the equivalent gradually applied, which will give the same effect of impact loading. Then, the strain energy stored in the bar is U

$$U = \frac{1}{2}W_*\delta$$

Strain energy stored in the body due to impact loading of $W_* = U$

As W_* is a gradually applied and gives the same effect of impact loading, 'δ' should be replaced by $\frac{W_*L}{AE}$.

$$U = \frac{1}{2}W_*\frac{W_*L}{AE} \tag{1.33}$$

Equating the external work done to strain energy absorbed, that is, equating equations (1.32) and (1.33) and replacing δ by $\frac{W_* L}{AE}$,

$$\frac{1}{2}W_*\frac{W_* L}{AE} = W\left[h + \frac{W_* L}{AE}\right]$$

$$\Rightarrow \quad W_*^2\left(\frac{L}{AE}\right) - 2W_*\left(\frac{WL}{AE}\right) - 2Wh = 0 \tag{1.34}$$

Solving the above quadratic equation and taking the higher value of W_*

$$W_* = W\left\{1 + \sqrt{1 + \frac{2h}{\left(\frac{WL}{AE}\right)}}\right\}. \tag{1.35}$$

$\frac{WL}{AE}$ is extension for the bar considering the load 'W' as a gradually applied load.

Thus, $\frac{WL}{AE}$ is referred as static deformation

$$W_* = W\left\{1 + \sqrt{1 + \frac{2h}{\delta_{st}}}\right\}. \tag{1.36}$$

δ_{st} = Static deformation due to 'W'.

The instantaneous maximum extension is given by

$$\delta = \frac{W_* L}{AE}$$

Instantaneous maximum stress $\sigma = \frac{W_*}{A}$

For suddenly applied load, $h = 0$ and hence $W_* = 2W$

$$\sigma = \frac{W_*}{A} = \frac{2W}{A}.$$

In the design of energy absorbing structures, such as shock absorbers and springs, it is required that δ_{st} should be high. That means, body must be more flexible. Then, the force attracted by the system W_* is going to be less. Otherwise, if the structure is more rigid, that is, δ_{st} is very less, then W_* force attracted by the system will be very high. Even in the case of earthquake-resistant design of structures, this principle holds good. In case of heavy winds, small plants will not break owing to their flexibility, while very big trees break down/get uprooted because of their high rigidity.

The ratio $\frac{W_*}{W}$ is referred as impact factor.

PROBLEM 1.28

Objective 1

A short steel piece of length 200 mm and cross-sectional area 500 mm^2 receives a falling weight of mass 5 kg on its top; height of the fall is 50 mm. If E = 210 GPa, estimate the instantaneous maximum stress induced in the steel piece. Estimate instantaneous maximum stress, if cross-sectional area of the piece is reduced by 20% in the top half portion of the piece.

SOLUTION

From the first part of the problem, data are

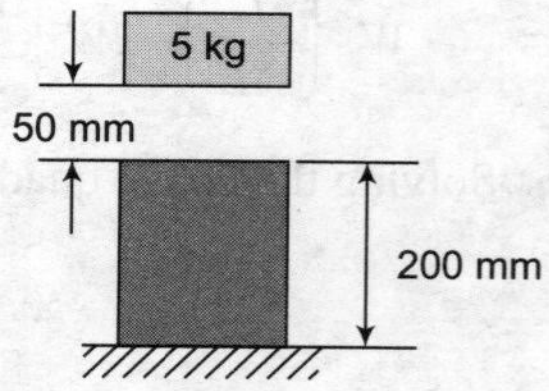

$$W = 5 \times 9.81 = 49.05 \text{ N}$$

$$L = 200 \text{ mm}$$

$$A = 500 \text{ mm}^2$$

$$E = 210 \text{ GPa}$$

$$h = \text{Height of the fall} = 50 \text{ mm}$$

$$\delta_{St} = \text{Static compression due to } W$$

$$= \frac{WL}{AE}$$

$$= \frac{49.05 \times 200}{500 \times 2.1 \times 10^5} = 0.934 \times 10^{-4}.$$

FIGURE 1.59

Equivalent gradually applied load be W_*

$$W_* = W\left\{1 + \sqrt{1 + \frac{2h}{\delta_{st}}}\right\}$$

$$W_* = 49.05\left\{1 + \sqrt{1 + \frac{2 \times 50}{0.934 \times 10^{-4}}}\right\} = 50{,}802.5 \text{ N}$$

Instantaneous maximum stress developed

$$\sigma = \frac{50802.5}{500} = 101.61 \text{ MPa (C)}$$

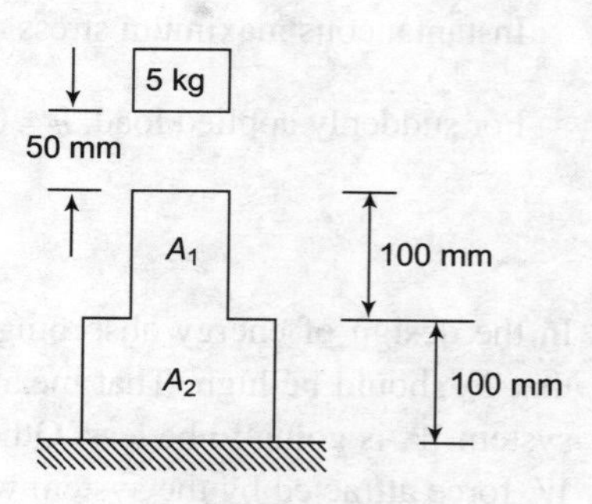

FIGURE 1.60

In second part of the problem, cross-sectional area of the top half of the steel piece is reduced by 20%.

$\therefore$ C/S area of top portion = 400 mm^2.

Let W_{*2} be the statically equivalent load in second case.

δ_{St2} = static compression in stepped column

$$\delta_{st2} = \frac{P_1L_1}{A_1E_1} + \frac{P_2L_2}{A_2E_2}$$

$$= \frac{49.05 \times 100}{400 \times 2.1 \times 10^5} + \frac{49.05 \times 100}{500 \times 2.1 \times 10^5}$$

$$= 1.05 \times 10^{-4} \text{ mm}$$

$$W_{*2} = 49.05 \times \left\{1 + \sqrt{1 + \frac{2h}{\delta_{st2}}}\right\}$$

$$W_{*2} = 47{,}916.97 \text{ N}$$

Instantaneous maximum stress developed = $\sigma = \dfrac{47{,}916.97}{400} = 119.73$ MPa.

Reduction in cross-sectional area is not much beneficial in reducing the impact effect, but reduction in 'E' value will give lot of advantage. This can be verified by decreasing the value of E in this problem.

PROBLEM 1.29 **Objective 1**

A steel piece of height 200 mm receives an impact load of 5 kg mass falling through 50 mm height. Estimate the instantaneous stress produced in the steel piece. To reduce the impact effect, a short rubber pad of same cross-sectional area and thickness of 20 mm is used. Estimate reduction in impact factor with the provision of rubber pad. Take $E_s = 200$ GPa; $E_r = 1.5$ GPa; and $A = 250$ mm^2.

SOLUTION

$$\text{Impact factor} = \frac{\text{Statically equivalent load}}{\text{Falling weight}}$$

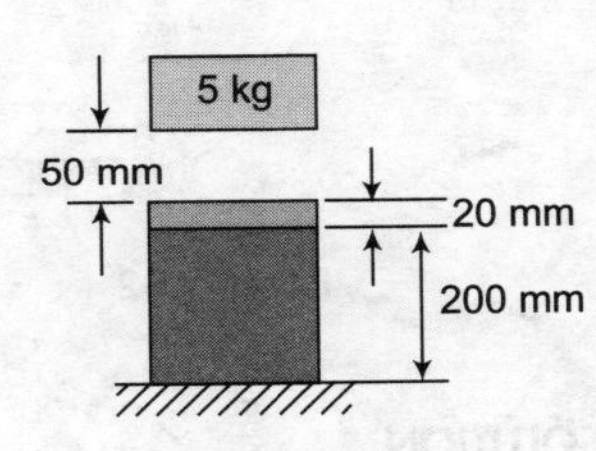

FIGURE 1.61

Let W_* be the statically equivalent load.

$$W = 9.81 \times 5 = 49.05 \text{ N}$$

$$W_* = W\{1 + \sqrt{\frac{2h}{\delta_{st}} + 1}\}$$

$$\delta_{St} = \frac{5 \times 9.81 \times 200}{250 \times 2 \times 10^5} = 1.962 \times 10^{-4} \text{ mm}$$

$$W_* = 49.05\{1 + \sqrt{\frac{2 \times 50}{1.962 \times 10^{-4}} + 1}\}$$

$$= 49.05 \times 714.92 = 35{,}067 \text{ N}$$

$$\therefore \text{Impact factor} = \frac{W_*}{W} = 714.92.$$

Effect of providing rubber pad

Let W_{*2} be the statically applied equivalent load.

$$\delta_{St} = \frac{49.05 \times 200}{250 \times 2 \times 10} + \frac{49.05 \times 20}{250 \times 1.5 \times 1000}$$

$$= 0.00281 \text{ mm}$$

$$W_{*2} = 49.05\left\{\sqrt{1 + \frac{2 \times 50}{0.00281}} + 1\right\}$$

$$= 49.05 \times 189.65 = 9302.25 \text{ N.}$$

Impact factor $\dfrac{W_{*2}}{W} = 189.65.$

From this example, it is clear that provision rubber pad reduces the impact effect.

Impact factor in the absence of rubber pad = 714.92.

Impact factor with the provision of rubber pad is 189.65.

PROBLEM 1.30

Objective 1

A rigid bar ABC, shown in Figure 1.62, is subjected to an impact load factor of 100 N, falling through a height of 0.2 m. A steel bar of cross-sectional area 1200 mm² and 1.5 m long is attached to the rigid bar at B. Determine the instantaneous maximum stress induced in the steel bar. Take E_s = 100 GPa.

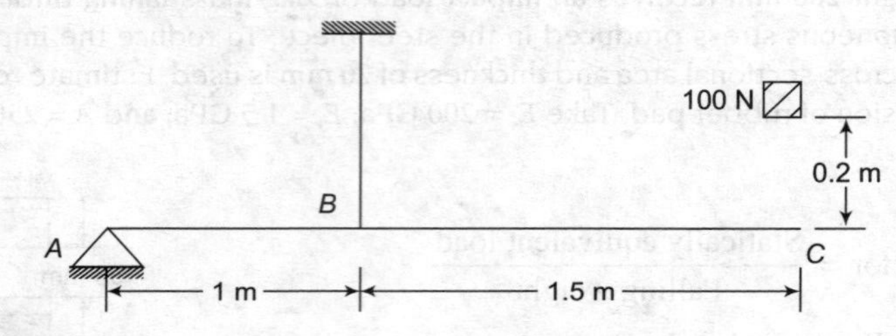

FIGURE 1.62

SOLUTION

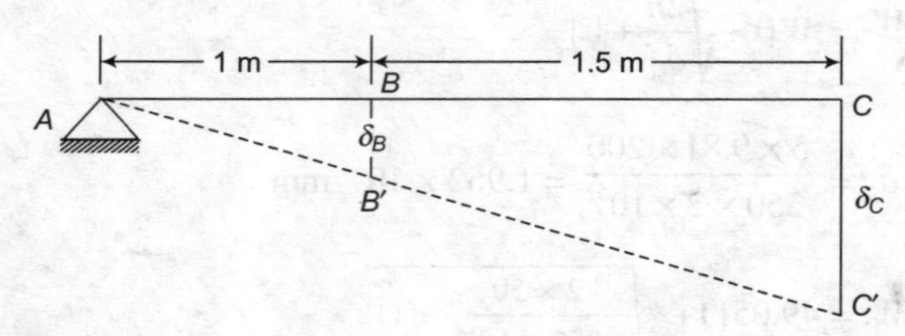

FIGURE 1.63

Let δ_c and δ_b be the deformation at C and B, respectively.

From compatibility,

$$\frac{\delta_C}{2.5} = \frac{\delta_B}{1}$$

Or

$$\delta_c = 2.5\ \delta_B$$

External work done by falling weight

$$= 100\ (h + \delta_C)$$

$$= 100\ (200 + \delta_C)$$

Energy absorbed by the steel bar

Let W_* be the statically applied equivalent tensile load in the steel bar.

$$\text{Energy absorbed} = \frac{1}{2} W_* \delta_B$$

$$= \frac{1}{2} W_* \frac{W_* L}{AE}.$$

Equating the energy absorbed to the external work done,

$$100\,(200 + \delta_C) = \frac{1}{2} \frac{W_*^2 L}{AE}$$

Also $\delta_C = 2.5\ \delta_B$ (from compatibility condition)

$$= 2.5 \times \frac{W_* L}{AE}$$

$$\therefore \quad 100\left\{200 + 2.5\frac{W_* L}{AE}\right\} = \frac{1}{2}\frac{W_*^2 L}{AE}$$

$$\Rightarrow \quad 100\left[200 + \frac{2.5 W_* \times 1500}{1200 \times 2 \times 10^5}\right] = \frac{1}{2} \times \frac{W_*^2 \times 1500}{1200 \times 2 \times 10^5}$$

$$\Rightarrow \quad 0.3125 \times 10^{-5} W_*^2 - 1.5625 \times 10^{-3} W_* - 20000 = 0$$

$$\Rightarrow \quad W_* = 80{,}250.4 \text{ N}.$$

$\therefore$ Instantaneous maximum stress produced in steel = 66.88 MPa (tensile).

PROBLEM 1.31 **Objective 1**

An unknown weight falls 4 cm on to a collar rigidly attached to the lower end of a vertical bar 4 m long and 8 cm^2 in section. If the maximum instantaneous extension is found to be 0.42 cm, find the corresponding stress and the value of the unknown weight. E = 200 kN/mm^2.

SOLUTION

Height of fall h = 40 mm

Length of the bar L = 4 m = 4000 mm

C/S area of the bar A = 8 cm^2 = 800 mm^2

Max. instantaneous extension δ = 0.42 cm = 4.2 mm

Modulus of elasticity E = 200 GPa = 2 × 10^5 MPa.

Let W_* be the equivalent gradually applied load that gives the same effect of impact load.

$$\text{Then} \quad \delta = \frac{W_* L}{AE}$$

$$\Rightarrow \quad 4.2 = \frac{W_* \times 4000}{800 \times 200 \times 10^3}$$

$$\Rightarrow \quad W_* = 168{,}000 \text{ N}.$$

Let W be the falling weight.

Work done, that is, energy associated with falling weight $= W(h + \delta)$

$$= W[40 + 4.2]$$

Energy absorbed by the wire $U = \frac{1}{2} W_* \delta = \frac{1}{2} \times 168{,}000 \times 4.2$

Energy work done to the energy absorbed $W[44.2] = \frac{1}{2} \times 168{,}000 \times 4.2$

$\Rightarrow$ $W = 7981.9$ N.

PROBLEM 1.32

Objective 4

An aluminum bar 60 mm diameter when subjected to an axial tensile load 100 kN elongates 0.20 mm in a gauge length 300 mm and the diameter is decreased by 0.012 mm. Calculate the modulus of elasticity and the Poisson's ratio of the material.

SOLUTION

Diameter of the bar $= 60$ mm

Tensile load $(P) = 100 \text{ kN} = 100 \times 10^3$ N

Extension or elongation $\delta = 0.2$

Gauge length $L = 300$ mm

Decrease in diameter $\delta d = 0.012$ mm

Linear strain $\in = \frac{\delta}{L} = \frac{0.2}{300} = 0.00067$

Linear stress $\sigma = \frac{P}{A} = \frac{100 \times 10^3}{\frac{\Pi}{4} 60^2} = 35.368 \text{ N/mm}^2$

Modulus of elasticity $\frac{\sigma}{\in} = \frac{35.368}{0.00067} = 0.5305 \times 10^5$ MPa

Poisson's ratio $\mu = \frac{\text{Linear strain}}{\text{Lateral strain}}$

Lateral strain $= \frac{\text{Change in diameter}}{\text{Diameter}} = \frac{0.012}{60} = 0.0002$

Poisson's ratio $\mu = \frac{0.0002}{0.00067} = 0.3$

$\mu = 0.3$ and $E = 0.53 \times 10^5$ MPa $= 53$ GPa.

PROBLEM 1.33

Objective 4

A compound bar 1 m long is 40 mm diameter for 300 mm length, 30 mm diameter for the next 350 mm length. Determine the diameter of the remaining length, so that its elongation under an axial load of 100 kN does not exceed 1 mm. Take $E = 2 \times 10^5$ N/mm^2.

SOLUTION

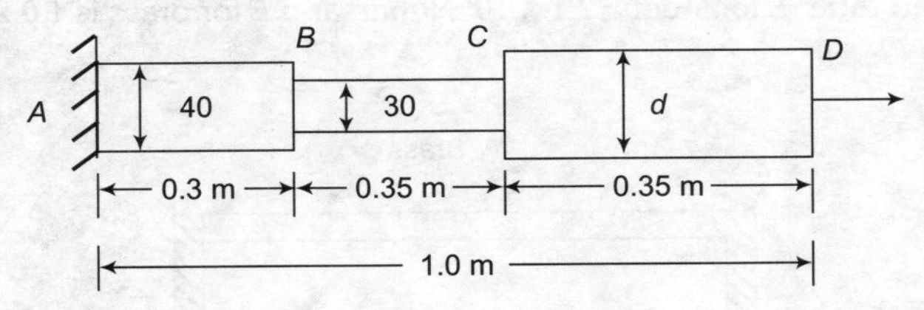

FIGURE 1.64

Length of proportion $CD = 1000 - 300 - 350 = 350$ mm

Total extension: $\delta_{AB} + \delta_{BC} + \delta_{CD} < 1$ mm

Given that, diameter of bar in portion $AB = d_{AB} = 40$ mm

$$E = 200 \text{ GPa and } P = 100 \text{ kN}$$

$$d_{BC} = 30 \text{ mm}$$

$$d_{CD} = d \text{ (to be found)}$$

Length of proportion AB $\quad L_{AB} = 300$ mm

$$L_{BC} = 350 \text{ mm}$$

$$L_{CD} = 350 \text{ mm}$$

$$\delta_{AB} = \frac{PL_{AB}}{A_{AB}E} = \frac{100 \times 10^3 \times 300}{\frac{\Pi}{4}(40)^2 \times 2 \times 10^5} = 0.119 \text{ mm}$$

$$\delta_{BC} = \frac{PL_{BC}}{A_{BC}E} = \frac{100 \times 10^3 \times 350}{\frac{\Pi}{4}(30)^2 \times 2 \times 10^5} = 0.248 \text{ mm}$$

$$\delta_{CD} = \frac{PL_{CD}}{A_{CD}E} = \frac{100 \times 10^3 \times 350}{\frac{\Pi}{4}(d)^2 \times 2 \times 10^5} = \frac{222.82}{d^2}$$

Given that, $\delta_{AB} + \delta_{BC} + \delta_{CD} \leq 1$ mm

$$\Rightarrow \quad 0.119 + 0.248 + \frac{222.82}{d^2} \leq 1$$

$$\Rightarrow \quad \frac{222.82}{d^2} \leq 1$$

$$\Rightarrow \quad d \leq 18.76 \text{ mm.}$$

PROBLEM 1.34

Objective 1

A round steel bar of 25 mm diameter and 360 mm long is placed concentrically within a brass tube which has an outside diameter of 35 mm and an inside diameter of 27.5 mm. The length of the tube exceeds that of the bar by 0.15 mm. Rigid plates are placed on the ends of the tube, through which an axial compressive force of 80 kN is applied on the compound bar. Determine the compressive stresses in the bar and tube. E for steel = 2.1×10^5 N/mm^2 and E for brass is 1.0×10^5 N/mm^2.

SOLUTION

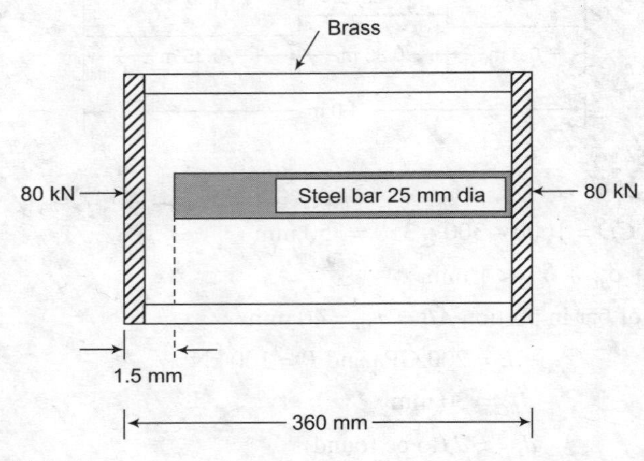

FIGURE 1.65

Cross-sectional area of the brass tube $A_b = \dfrac{\Pi}{4}(35^2 - 27.5^2)$

$$A_b = 368.16 \text{ mm}^2$$

Cross-sectional area of the steel bar $A_s = \dfrac{\Pi}{4}(25)^2 = 490.87 \text{ mm}^2$

Length of the tube = 360 mm

$$E_b = 1 \times 10^5 \text{ N/mm}^2$$

$$E_s = 2.1 \times 10^5 \text{ N/mm}^2$$

A part of 80 kN closes the gap between the steel bar and brass tube. Let this part be P_1

$$\frac{P_1 \times L}{A_b E_b} = 0.15$$

$$0.15 = \frac{P_1 \times 360}{368.16 \times 1 \times 10^5}$$

$\Rightarrow$ $$P_1 = 15.34 \text{ kN}$$

Load $P_2 = 80 - P_1$ is shared by brass tube and steel bar

$$P_2 = 80 - 15.34 = 64.66 \text{ kN}$$

Let P_{2s} and P_{2b} be the load shared by the steel bar and brass tube.

$$P_{2s} + P_{2b} = 64{,}660 \text{ N} \tag{1.37}$$

The compression due to 64.66 kN, steel bar and brass tube undergo same compression.

$$\Delta = \frac{P_{2s}L_s}{A_sE_s} = \frac{P_{2b}L_b}{A_bE_b}$$

$$P_{2s} = 2.8P_{2b}$$

Now, $P_{2b} + 2.8P_{2b} = 64{,}660$

$$P_{2b} = 17{,}015.79 \text{ kN}$$

$$P_{2s} = 47{,}644.21 \text{ kN}$$

Finally due to load 80 kN

Load on steel bar $P_s = P_{2s} = 47{,}644.21$ N

Load on brass tube $P_b = P_1 + P_{2b} = 15{,}340 + 17{,}015.79 = 32{,}355.79$ N

Steel in steel bar = (47,644.21/490.87) = 97.06 N/mm^2

Stress in brass tube = (32,355.79/368.16) = 87.89 N/mm^2.

PROBLEM 1.35 **Objective 4**

A steel rod 28 mm diameter is fixed concentrically in a brass tube of 42 mm outer diameter and 30 mm inner diameter. Both the rod and tube are 450 mm long. The compound rod is held between two stops which are exactly 450 mm apart and the temperature of the bar is raised by 70 °C.

SOLUTION

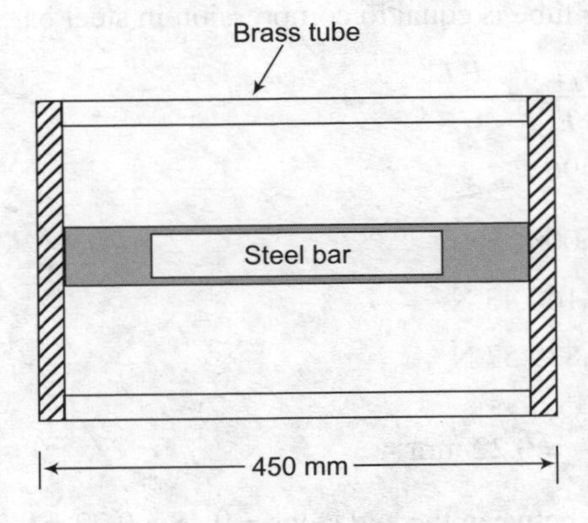

FIGURE 1.66

$$E_S = 200 \text{ GPa}; \ \alpha_s = 11.2 \times 10^{-6}/°\text{C}; \ E_b = 90 \text{ GPa}; \ \alpha_b = 21 \times 10^{-6}/°\text{C}; \ T = 70 \text{ °C}.$$

Diameter of the steel bar = 28 mm

Cross-sectional area of the steel bar $A_s = 615.75$ mm^2

Outer diameter of the brass tube = 42 mm

Inner diameter of the brass tube = 30 mm

Cross-sectional area of the brass tube is A_b = 678.58 mm^2

Stresses due to temperature rise of 70 °C

$$\text{Stress in steel} = \sigma_s = \frac{T(\alpha_b - \alpha_s)E_b E_s A_b}{(E_s A_s + E_b A_b)} = 45.48 \text{ N/mm}^2 \text{ (tensile stress)}$$

$$\text{Stress in brass} = \sigma_b = \frac{T(\alpha_b - \alpha_s)E_b E_s A_s}{(E_s A_s + E_b A_b)} = 41.27 \text{ MPa (compressive stress)}$$

Extension of the tube = 0.66 – 0.18 = 0.48 mm

Case (a): If the total tube is extended by 0.3 mm steel bar and brass tube undergo same extension of 0.3 mm

Stresses due to extension of 0.3 mm, δL = 0.3 mm

Tensile stress in steel bar $\sigma_s = E_s \times \delta L/L$ = 133.33 N/mm^2 (tensile)

Similarly, tensile stress in brass tube $\sigma_b = E_b \times \delta L/L$ = 60 N/mm^2 (tensile)

Because of temperature raise and extension of 0.3 mm,

Stress in brass tube = –35.92 + 60 = 24.08 N/mm^2

Stress in steel bar = 45.48 + 133.33 = 178.81 N/mm^2 (tensile).

Case (b): Because of temperature, the increase in the length of the combined system = 0.48 mm. A compressive force of 90 kN is applied.

Let P_{sb} and P_{bb} be the force in steel bar and brass tube, respectively.

$$P_{sb} + P_{bb} = 90 \text{ kN} = 90{,}000 \text{ N}$$

The compression in the brass tube is equal to compression in steel bar.

$$\Delta = \frac{P_{bb}L}{A_b E_b} = \frac{P_s L}{A_s E_s}$$

$$P_{bb} = 0.496\, P_{sb}$$

$$P_{sb} + 0.496\, P_{sb} = 90{,}000$$

$$P_{sb} = 60{,}160.43 \text{ N}$$

$$P_{bb} = 29{,}839.57 \text{ N}$$

$$\text{Decrease in length } \Delta = \frac{P_{bb}L}{A_b E_b} = 0.22 \text{ mm}$$

Total increase in the distance between the end stops = 0.48 – 0.22 = 0.26 mm.

SUMMARY

- **The resistance per unit area, offered by a body against deformation, is known as stress.**
- **The ratio of change of dimension to the original dimension is known as strain.**
- **The stress induced in a body, which is subjected to equal and opposite pulls, is known as tensile stress and if it is of pushing nature it is known as compressive stress.**
- **Hooke's law states that within elastic limit the stress is proportional to the strain.**
- **In case of composite bar having equal length: strain in each bar is equal and total load on the composite bar is equal to the sum of loads carried by each different materials.**

OBJECTIVE TYPE QUESTIONS

1. A tapered bar of 40 mm diameter at fixed end, 20 mm diameter at free end is 1.5 m long. The free end is subjected to an axial tensile force of 3.14 kN. The maximum stress developed in the bar is
(a) 10 GPa (b) 10 MPa (c) 2.5 GPa (d) 2.5 MPa

2. The analysis of a concrete column with embedded reinforcing steel is
(a) Statically determinate problem (b) Statically indeterminate problem
(c) Both (a) and (b) (d) None of (a) and (b)

3. Two bars A and B of same length are attached to two rigid plates at their ends. The total assembly is subjected to temperature rise: $\alpha_A > \alpha_B$. The bar A is subjected to
(A) Tension (B) Compression
(C) Neither compression nor tension (D) Shear

4. A steel bolt of diameter 20 mm passes through a brass tube of outer diameter 40 mm and inner diameter 30 mm. The bolt is tightened by a nut of pitch 4 mm. If the brass tube compresses by 0.75 mm for 1/4th rotation of the nut, what would be the extension of the steel rod?
(a) Data are insufficient
(b) 0.25 mm
(c) 3.75 mm
(d) 1.0 mm

5. A bar of cross-sectional area 100 mm^2 is subjected to a suddenly applied load of 50 kN. The instantaneous maximum stress induced in the bar is
(a) 10 MPa (b) 250 GPa (c) 1000 MPa (d) 500 MPa

6. A steel rod of length 2 m is subjected to an axial pull of 80 kN. What is the cross-sectional area of the bar required, if the elongation is not to exceed 4 mm. The Young's modulus of material is 200 GPa.
(a) 200 mm^2 (b) 250 mm^2 (c) 1000 mm^2 (d) 500 mm^2

7. A stepped bar ABC is fixed at A and free at C. An axial load of 20 kN is acting at B such that the portion AB of the bar is under tension. If the cross-sectional area of the portion AB of the bar is 200 mm^2 and that of the portion BC is 400 mm^2, what is the maximum stress induced in the bar?
(a) 50 MPa (b) 100 MPa (c) 150 MPa (d) 200 MPa

8. A stepped bar ABC is fixed at A and free at C. An axial load of 40 kN is acting at C such that the portion AB of the bar is under tension. If the cross-sectional area of the portion AB of the bar is 200 mm^2 and that of the portion BC is 400 mm^2, what is the maximum stress induced in the bar?

(a) 50 MPa (b) 100 MPa (c) 150 MPa (d) 200 MPa

9. A stepped bar ABC is fixed at A and free at C. An axial load of 40 kN is acting at B such that the portion AB of the bar is under tension and 40 kN is acting at C, which causes compression in the portion BC. If the cross-sectional area of the portion AB of the bar is 200 mm^2 and that of the portion BC is 400 mm^2, what is the stress induced in the portion AB of the bar?
(a) 50 MPa (b) 100 MPa (c) 0 MPa (d) 200 MPa

10. A stepped bar ABC is fixed at A and free at C. An axial load of 40 kN is acting at B such that the portion AB of the bar is under tension and 40 kN is acting at C, which causes compression in the portion BC. If the cross-sectional area of the portion AB of the bar is 200 mm^2 and that of the portion BC is 400 mm^2, what is the stress induced in the portion BC of the bar?
(a) 50 MPa (b) 100 MPa (c) 0 MPa (d) 200 MPa

11. A stepped bar ABC is fixed at A and free at C. An axial load of 40 kN is acting at B such that the portion AB (2 m) of the bar is under tension and 40 kN is acting at C, which causes compression in the portion BC. If the cross-sectional area of the portion AB of the bar is 200 mm^2 and that of the portion BC (1 m) is 400 mm^2, what is the extension of B, if modulus of elasticity of the material of the bar is 200 GPa?
(a) 5 mm (b) 1 mm (c) 2 mm (d) 0

12. A stepped bar ABC is fixed at A and free at C. An axial load of 20 kN is acting at B such that the portion AB (2 m) of the bar is under tension. If the cross-sectional area of the portion AB of the bar is 200 mm^2 and that of the portion BC (1 M) is 400 mm^2, what is the extension of B, if modulus of elasticity of the material of the bar is 200 GPa?
(a) 5 mm (b) 1 mm (c) 2 mm (d) 0

13. A stepped bar ABC is fixed at A and free at C. An axial load of 20 kN is acting at B such that the portion AB (2 m) of the bar is under tension. If the cross-sectional area of the portion AB of the bar is 200 mm^2 and that of the portion BC (1 M) is 400 mm^2, what is the extension of C, if modulus of elasticity of the material of the bar is 200 GPa?
(a) 5 mm (b) 1 mm (c) 2 mm (d) 0

14. A concrete column of length 3 m, size 400 mm × 400 mm is reinforced with rebars of 800 mm^2. The modulus of elasticity of rebar is 10 times that of concrete. What is the ratio of stress in concrete to stress in steel due to an axial load of 100 kN?
(a) 10 (b) 0.1 (c) 1 (d) Data are not sufficient

15. A concrete column of length 3 m, size 400 mm × 400 mm is reinforced with rebars of 800 mm^2. The modulus of elasticity of rebar is 10 times that of concrete. What is the ratio of strain in concrete to strain in steel due to an axial load of 100 kN?
(a) 10 (b) 0.1 (c) 1 (d) Data are not sufficient

16. A concrete column of length 3 m, size 400 mm × 400 mm is reinforced with rebars of 800 mm^2. The modulus of elasticity of rebar is 10 times that of concrete. What is the stress in concrete due to an axial load of 210 kN?
(a) 10 (b) 0.1 (c) 1 (d) 1.25

17. A concrete column of length 3 m, size 400 mm × 400 mm is reinforced with rebars of 800 mm^2. The modulus of elasticity of rebar is 10 times that of concrete. What is the stress in steel due to an axial load of 210 kN?
(a) 10 (b) 0.1 (c) 12.5 (d) 1.25

18. A concrete column of axial stiffness 10 kN/mm is reinforced with rebars of stiffness 2 kN/mm. What is the axial deformation of the composite column due to an axial load of 120 kN?
(a) 10 mm (b) 200 mm (c) 0.1 mm (d) 120 mm

19. What is the extension of a conical bar of length (L), density (γ), and modulus of elasticity (E) vertically hanging on its own weight?

(a) $\frac{\gamma L^2}{3E}$ (b) $\frac{\gamma L^2}{2E}$ (c) $\frac{\gamma L^2}{6E}$ (d) $\frac{\gamma L^2}{8E}$

20. What is the variation of longitudinal stress along the length of a conical bar vertically hanging on its own weight?
(a) Linear (b) Constant (c) Parabolic (d) Cubic

21. What is the variation of longitudinal stress along the length of a prismatic bar vertically hanging on its own weight?
(a) Linear (b) Constant (c) Parabolic (d) Cubic

22. What is the variation of longitudinal stress along the length of a tapering bar vertically hanging on its own weight?
(a) Linear (b) Constant (c) Parabolic (d) Cubic

23. What is the extension of a prismatic bar of length (L), density (γ), and modulus of elasticity (E) vertically hanging on its own weight?

(a) $\frac{\gamma L^2}{3E}$ (b) $\frac{\gamma L^2}{2E}$ (c) $\frac{\gamma L^2}{6E}$ (d) $\frac{\gamma L^2}{8E}$

24. The strain energy, expressed as kN·mm, stored in a bar of stiffness 2 kN/mm, subjected to a load of 5 kN is
(a) 10 (b) 12.5 (c) 5.00 (d) 6.25

25. The strain energy, expressed as kN·mm, stored in a bar of stiffness 2 kN/mm, extends by 10 mm due to axial load is
(a) 100 (b) 125 (c) 50 (d) 62.5

Solutions for Objective Questions

Sl. No.	1.	2.	3.	4.	5.	6.	7.	8.	9.	10.
Answer	(b)	(b)	(b)	(c)	(c)	(a)	(b)	(d)	(C)	(B)

Sl.No.	11.	12.	13.	14.	15.	16.	17.	18.	19.	20.
Answer	(d)	(b)	(b)	(b)	(c)	(d)	(c)	(a)	(C)	(C)

Sl.No.	21.	22.	23.	24.	25.
Answer	(a)	(c)	(b)	(d)	(a)

EXERCISE PROBLEMS

1. A bar of size 100 mm × 75 mm is subjected to an axial tensile load of 150 kN. What is the shear stress developed along a plane inclined 30° to the longitudinal axis of the beam. Also determine the normal stress in the plane.
2. A bar of size 100 mm × 75 mm is subjected to an axial tensile load of 57 kN. What is the extension of the bar if the length of the bar is 2.0 m? Take E = 200 GPa.
3. A rectangular concrete member is to sustain an axial compressive load of 450 kN. Determine the dimensions of a suitable cross-section for the column with one of the dimension as 230 mm. The

allowable stress in concrete is limited to 4 MPa. What is the allowable load if the deformation of the 3.0-m-high column should not exceed 1.5 mm? Assume Young's modulus for concrete as 22.0 GPa.
[*Ans:* 490 mm; 1239.7 kN.]

4. A block of size 100 mm × 100 mm × 75 mm is subjected to a tangential load of 75 kN on 100 mm × 100 mm face. The same face on the other side is fixed. What is the shear stress developed? Estimate the shear if rigidity modulus of the material is 80 GPa. Determine the deformation at the free end of the block.

5. A 1.5-m-long steel bar is subjected to an axial tensile load of 100 kN. Determine the bar cross-section so that (i) the stress does not exceed 80 MPa and (ii) the extension is not more than 0.8 mm. Assume Young's modulus as 200 GPa.
[*Ans:* 1250 mm2; 937.5 mm^2.]

6. Determine the stresses and deformation in the steel rope of a pulley supporting a load of 20 kN. The rope is made of seven wires of 5 mm diameter. The rope is 10 m on one side of the pulley and 5 m on the other side. E = 150 GPa.

7. Determine the size of the steel rod to sustain a load of 90 kN with a factor of safety of 1.85. What is the maximum permissible length of the rod, if the allowable deformation is 0.25mm? Assume a yield of 280 MPa and E = 200 GPa.

[*Ans:* 27.52 mm; 594 mm.]

8. Calculate the length of the CD(X) component in the combined bars shown in Figure 1, so that the total deformation of the member is zero? Take E of AB and CA = 210 GPa and E of BC as 105 GPa. $A_{AB} = A_{CD} = A$ and $A_{BC} = 2A$. Also find the stress in different parts of the member.

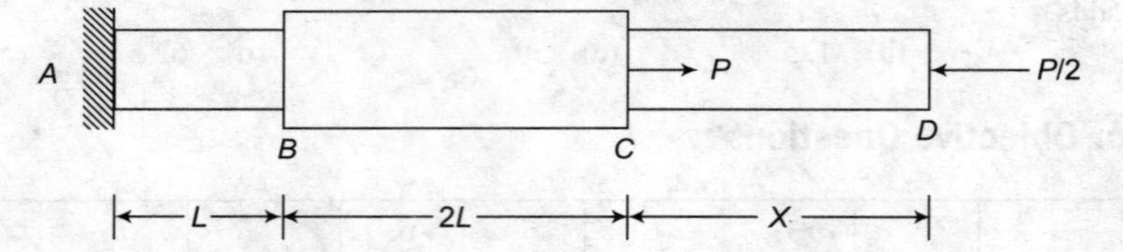

FIGURE 1

9. Determine the distribution of stresses at salient section of the circular bars indicated in Figure 2. Compute total deformation of the members as well. Take E of AB and CA = 210 GPa and that of BC as 105 GPa. $A_{AB} = A_{CD} = A$ and $A_{BC} = 2A$. A = 1200 mm^2; L = 0.5 m; P = 50 kN.

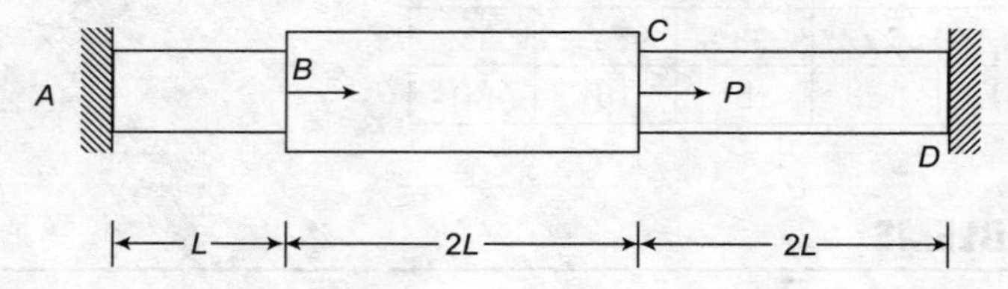

FIGURE 2

10. A bar of length L has its diameter increasing from D at one end to 2D at another end. Determine the deformation of the member subjected to a tensile force of P.

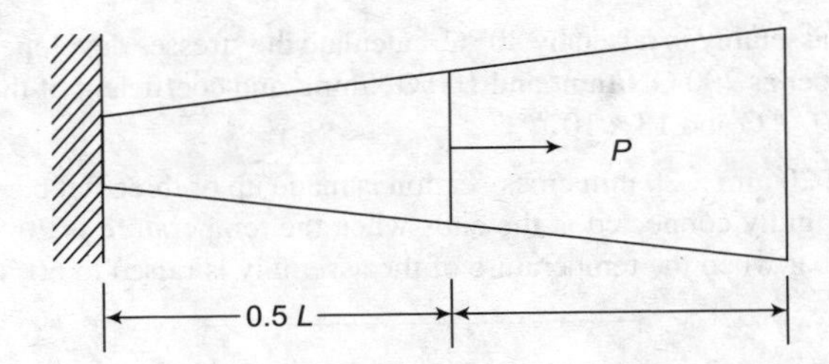

FIGURE 3

11. A steel bar of weight 200 N connected to a steel wire of 10 mm diameter is rotated about one of its end in a horizontal plane with a speed of 2 Hz. Find the maximum stress induced in the bar. If the length of the bar is 1.2 m, determine the extension of the bar. Take E = 200 GPa. Neglect the body force developed in the wire owing to its mass.

12. A load of 500 kN is applied to a Reinforced cement concrete column of diameter 450 mm which has six steel rods of 20 mm diameter embedded in it. Determine the stress in concrete and in the steel. Take E_s = 200 GPa and E_{con} = 12 GPa.

13. A 18 mm steel rod passes centrally through a copper tube of 25 mm external diameter and 20 mm internal diameter and 100 mm long. The tube is closed at each end by rigid washers and nuts screwed to the rod. The nuts are tightened till the compressive force in the copper tube is 6 kN. Determine the stresses in the rod and the tube. E_s = 200 GPa and E_c = 80 GPa.

14. A tensile load of 50 kN is gradually applied to a circular bar of 10 mm diameter and 2 m long. If E = 80 GPa, determine the extension of the rod, stress in the rod, and strain energy absorbed by the rod.

15. Calculate instantaneous stress produced in the bar of cross-sectional area 2000 mm^2 and 2.5 m long by the sudden application of a tensile load of unknown magnitude, if the extension of the bar is due to suddenly applied load is 1.35 mm. Also determine the suddenly applied load. Take E = 210 GPa.

16. A load of 200 N falls through a height of 50 mm on to a collar rigidly attached to the lower end of the vertical bar. Determine the max instantaneous stress induced in the vertical bar, maximum instantaneous elongation and strain energy store in vertical bar. Take E = 200 GPa.

17. An unknown weight fall through a height of 40 mm on a collar rigidly attached to the lower end of a vertical bar 1.5 m long and 800 mm2 in section. If the maximum extension of the rod is to be 3.5 mm, what is the corresponding stress and magnitude of unknown weight? Take E = 200 GPa.

18. A vertical compound tie member fixed rigidly at its upper end. It consists of a steel rod 2 m long and 20 mm diameter placed within a long brass tube of 20 mm internal diameter and 30 mm ext diameter. The rod and the tube are fixed together at the ends. The compound member is suddenly loaded in tension by a weight of 800 N, falling through a height of 10 mm. Estimate the stresses in steel and brass. E_s = 200GPa and E_b = 105 GPa

19. Two vertical rods one of steel and other of brass are fixed at the top and 60 cm apart. Diameters and length of each rod are 20 mm and 30 mm, respectively. A cross (horizontal) rigid bar fixed to the rods at the lower ends caries a load of 2000 N such that, the cross bar remains horizontal even after loading. Find the stress in each rod and the position of the load on the bar. Assume E_s = 210 GPa and E_b = 105 GPa.

20. A rod is 3 m long at a temperature of 10 °C. Find the expansion of the rod, when the temperature is raised to 45 °C. If this expansion is prevented, find the stress induced in the material of the rod. Take E = 105 N/mm^2 and $a = 10 \times 10^{-6}$/°C.

21. A steel rod of 20 mm diameter passes centrally through a copper tube 40 mm external diameter and 30 mm internal diameter. The tube is closed at each end by rigid plates of negligible thickness. If the

temperature of the assembly is raised by 80 °C, calculate the stresses developed in copper and steel. Take E for steel and copper as 200 GN/mm^2 and 100 GN/mm^2 and coefficient of thermal expansion of copper and steel as 12×10^{-6}/°C and 18×10^{-6}/°C.

22. A composite bar of 20 mm × 20 mm cross-section is made up of three flat bars as shown in Figure 4. All the three bars are rigidly connected at the ends when the temperature is 20 °C. Determine the stresses developed in each bar when the temperature of the assembly is raised to 60 °C.

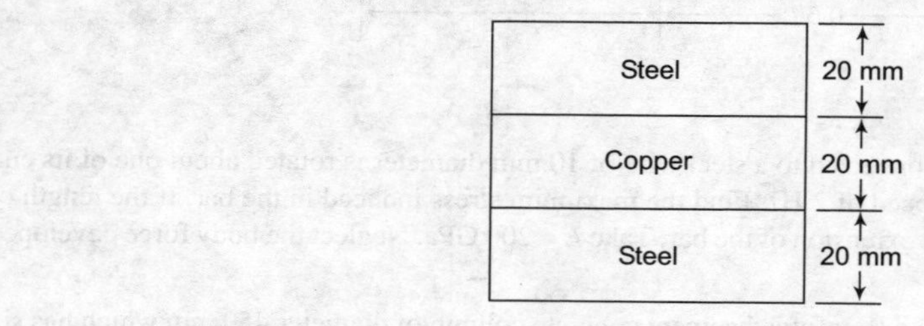

FIGURE 4

CHAPTER 2

ELASTIC CONSTANTS

UNIT OBJECTIVE

This chapter explains in detail about elastic constants and its importance to understand the material behavior. It also provides the formulae to determine the elastic constants for various structural elements. The presentation attempts to help the student achieve the following:

Objective 1: Determine the Young's, shear, and bulk moduli for given loading and deformations.

Objective 2: Quantify the effect of Poisson's ratio of the material on deformations.

Objective 3: Determine relation between elastic constants.

Objective 4: Calculate the variation in the dimensions caused due to loads/pressure.

2.1 INTRODUCTION

Two elastic constants of an elastic body are modulus of elasticity and Poisson's ratio. Modulus of elasticity is the ratio of normal stress to normal strain. It is generally denoted by the letter 'E'. Units are gigapascal (GPa). The other elastic constant is Poisson's ratio.

Poisson's ratio is defined as the ratio of lateral strain to the longitudinal strain. It is denoted by the letter μ.

Consider a bar of length 'L' and diameter 'd' subjected to an axial tensile stress 'σ'. ΔL is the longitudinal extension due to load P. Then, the longitudinal strain

$$\varepsilon = \frac{\Delta L}{L}.$$

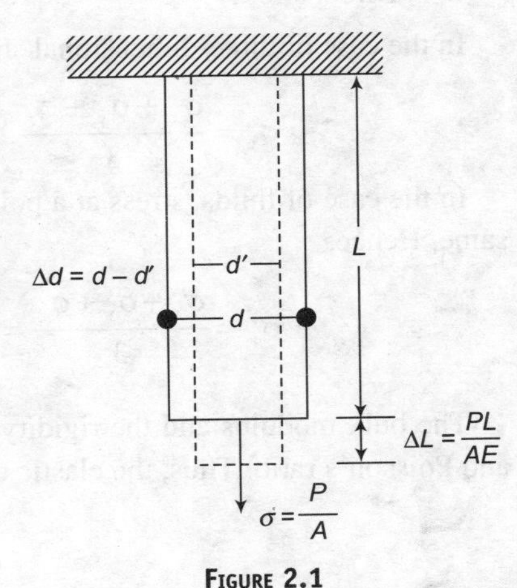

FIGURE 2.1

Let Δd be the decrease in the lateral dimension of the bar.

$$\text{Lateral strain} = \frac{\Delta d}{d}$$

$$\therefore \quad \text{Poisson's ratio} = \mu = -\frac{\left(\frac{\Delta d}{d}\right)}{\left(\frac{\Delta L}{L}\right)}$$

The negative sign indicates that positive strain (increase) in longitudinal strain decreases the lateral dimension, thereby compressive strain in the lateral direction is induced.

$$\mu = \frac{\text{Lateral strain}}{\text{Longitudinal strain}}$$

Rigidity Modulus

Rigidity modulus is the ratio of shear stress to the corresponding shear strain. The unit is gigapascal. This is generally denoted by the letter '*N*'. The shear modulus or rigidity modulus appears in calculation where shape change occurs in the body.

Bulk Modulus

Bulk modulus is the ratio of direct stress or volumetric stress to the volumetric strain. It is denoted by the letter '*K*' and unit is gigapascal. Generally in solids, the change in volume is very small, but this change in volume in case of fluids will be considerable. Thus, bulk modulus appears in fluid mechanics especially in compressible fluid flow problems.

$$\text{Bulk modulus} = K = \frac{\sigma_v}{\frac{\delta v}{V}}$$

δv = Change in the volume

V = Volume of the body

σ_v = Direct stress

In the case of three-dimensional stress system,

$$\sigma_v = \frac{\sigma_x + \sigma_y + \sigma_z}{3}.$$

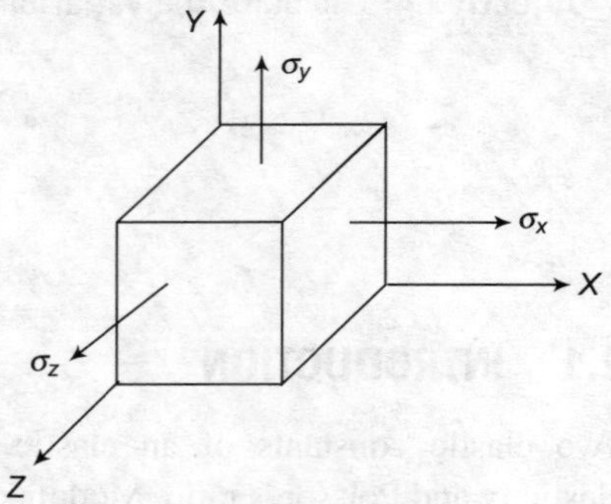

FIGURE 2.2

In the case of fluids, stress at a point in all directions is same. Hence,

$$\sigma_v = \frac{\sigma_x + \sigma_y + \sigma_z}{3} = \sigma.$$

The bulk modulus and the rigidity modulus can be expressed in terms of modulus of elasticity and Poisson's ratio. Thus, the elastic constants for a homogeneous and isotropic body are two (2).

2.2 RELATIONSHIP BETWEEN E AND N

Consider a square block of unit width subjected to shear stress τ as shown in Figure 2.3. The shear stress τ induces vertical shear stress on planes AC and DB of same intensity 'τ'. The shear stress on these planes is referred as complimentary shear stress.

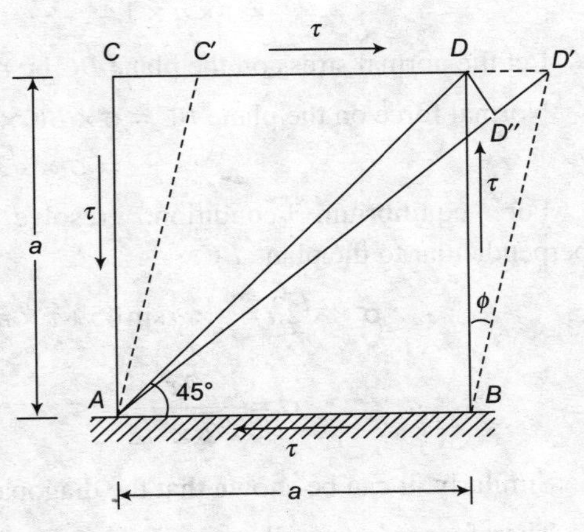

FIGURE 2.3

Let $ABD'C'$ be the deformed configuration of the block. Shear strain is defined as the change of angle between two planes.

$$\therefore \quad \text{Shear strain } \gamma = \tan\phi = \frac{DD'}{a}$$

ϕ is very small and in radians.

$$\therefore \quad \gamma = \phi = \frac{DD'}{a}.$$

Consider the diagonal AD. Because of shear stress, AD has become AD'.

Draw a perpendicular from D on to AD' as D''.

Increase in the length of the diagonal $AD = D''D'$.

Strictly speaking, DD'' is not a perpendicular rather it is an arc taking A as center. But as deformations are very small arc and perpendicular can be treated as same.

$$\therefore \text{ Normal strain along the diagonal } AD = \frac{D'D''}{AD}.$$

Normal strain along the diagonal is called diagonal strain.

$$\text{Diagonal strain } \varepsilon_{AD} = \frac{DD'\cos 45}{AD}$$

$D'D'' = DD' \cos 45$ as the $\angle DAD'$ is very small and hence can be neglected.

$$AD = a \times \sqrt{2}$$

$$\varepsilon_{AD} = \frac{DD'\cos 45}{a \times \sqrt{2}}$$

$$\varepsilon_{AD} = \frac{DD'}{2a} = \frac{\phi}{2} \quad \left(\because \frac{DD'}{a} = \phi\right)$$

$$\varepsilon_{AD} = \frac{\phi}{2}. \tag{2.1}$$

The above condition is obtained using simple geometry principles. Now, we develop an expression for strain along the diagonal AD using the principles of mechanics. Consider the stress along the diagonal AD. Consider the force acting on the block CBD.

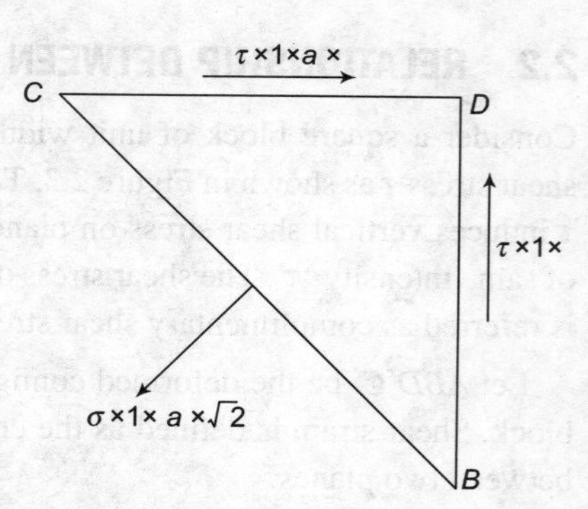

FIGURE 2.4

Shear force on plane CD = Shear force on plane BD

$$= \tau \times a \times 1.$$

Let the normal stress on the plane BC be σ.

Normal force on the plane $BC = \sigma \times BC \times 1$

$$= \sigma \times \sqrt{2}a.$$

For equilibrium condition, resolve the forces perpendicular to the plane BC.

$$\therefore \quad \sigma \times \sqrt{2}a = \tau \times a \sin 45 + \tau \times a \cos 45$$

$$\Rightarrow \quad \sigma = \frac{2\tau}{\sqrt{2} \times \sqrt{2}} = \tau.$$

Similarly, it can be shown that the diagonal stress along $BC = -\tau$(compressive).

Therefore, if a small piece is consider in the block $ABCD$ oriented along the diagonals, the strain will be as follows.

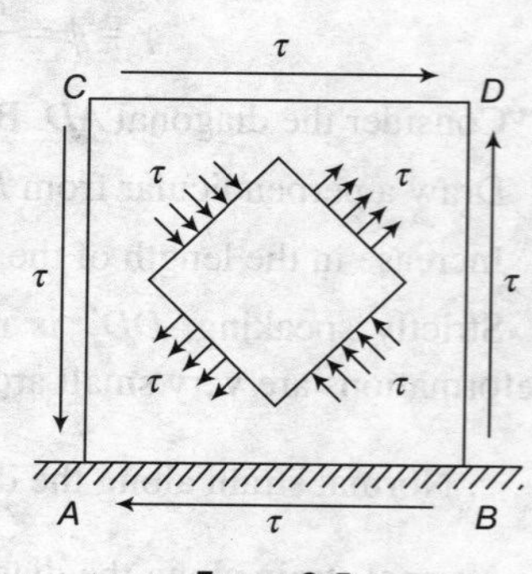

FIGURE 2.5

$$\text{Strain along } AD = \begin{pmatrix}\text{Strain due to} \\ \text{stress acting} \\ \text{along } AD\end{pmatrix} + \begin{pmatrix}\text{Lateral strain} \\ \text{due to} \\ \text{compressive} \\ \text{stress acting} \\ \text{along } BC\end{pmatrix}$$

$$\varepsilon_{AD} = \frac{\sigma_{AD}}{E} - \mu\left(-\frac{\sigma_{BC}}{E}\right)$$

$$\varepsilon_{AD} = \frac{\tau}{E}(1+\mu)(\because \sigma_{BC} = \sigma_{AD} = \tau). \tag{2.2}$$

Equating equations (1) and (2)

$$\frac{\tau}{E}(1+\mu) = \frac{\phi}{2}$$

or

$$\frac{\tau}{\phi} = \frac{E}{2(1+\mu)}$$

But we know that

$$\text{Rigidity modulus } N = \frac{\text{Shear stress}}{\text{Shear strain}} = \frac{\tau}{\phi}$$

$$\therefore \quad N = \frac{E}{2(1+\mu)}$$

or $\quad E = 2\text{N}(1+\mu)$

2.3 RELATIONSHIP BETWEEN E AND K

Consider a cube subjected to stresses in three directions as shown in Figure 2.5.

Let the stresses along the principal directions be σ_x, σ_y, and σ_z.

Let the dimensions of the block be X, Y, and Z along the three axes, respectively.

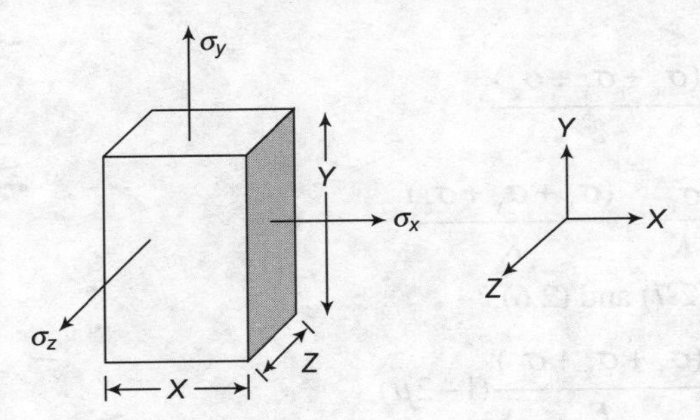

FIGURE 2.6

Volume of the block, $V = XYZ$ (2.3)

There is going to be change in the dimensions of X, Y, and Z due to the stresses σ_x, σ_y, and σ_z.

Let δV be the change in the volume of block.

Differentiating the expression

$$\delta V = \delta X \cdot YZ + \delta Y \cdot ZX + \delta Z \cdot XY$$

$$\frac{\delta V}{V} = \frac{\delta X}{X} + \frac{\delta Y}{Y} + \frac{\delta Z}{Z} \tag{2.4}$$

$$\frac{\delta V}{V} = \frac{\text{Change in volume}}{\text{Original Volume}} = \text{Volumteric strain}$$

$\frac{\delta X}{X}, \frac{\delta Y}{Y}$, and $\frac{\delta Z}{Z}$ are linear strains along X, Y, and Z dimensions. $\frac{\delta X}{X}$, the normal strain depends on the stresses σ_x, σ_y, and σ_z.

$$\therefore \quad \frac{\delta X}{X} = \frac{\sigma_x}{E} - \mu\frac{\sigma_y}{E} - \mu\frac{\sigma_z}{E}$$

$$\Rightarrow \quad \frac{\delta X}{X} = \frac{1}{E}\{\sigma_x - \mu\sigma_y - \mu\sigma_z\}.$$

Similarly, $\frac{\delta Y}{Y} = \frac{1}{E}\{\sigma_y - \mu\sigma_x - \mu\sigma_z\}$ and $\frac{\delta Z}{Z} = \frac{1}{E}\{\sigma_z - \mu\sigma_y - \mu\sigma_x\}$

$$\therefore \quad \frac{\delta V}{V} = \frac{\delta X}{X} + \frac{\delta Y}{Y} + \frac{\delta Z}{Z}$$

$$= \frac{1}{E}\{\sigma_x + \sigma_y + \sigma_z - 2\mu(\sigma_x + \sigma_y + \sigma_z)\}$$

$$\frac{\delta V}{V} = \frac{(\sigma_x + \sigma_y + \sigma_z)}{E}(1 - 2\mu) \tag{2.5}$$

Bulk modulus $K = \dfrac{\sigma_V}{\dfrac{\delta V}{V}}$

$$\delta V = \frac{(\sigma_x + \sigma_y + \sigma_z)}{3} \tag{2.6}$$

$$\frac{\delta V}{V} = \frac{\sigma_V}{K} = \frac{(\sigma_x + \sigma_y + \sigma_z)}{K}. \tag{2.7}$$

Equating expressions (2.7) and (2.6),

$$\frac{(\sigma_x + \sigma_y + \sigma_z)}{3K} = \frac{(\sigma_x + \sigma_y + \sigma_z)}{E}(1 - 2\mu)$$

$$\Rightarrow \qquad E = 3K(1 - 2\mu).$$

2.4 RELATIONSHIP BETWEEN *E*, *N*, AND *K*

We know that,

$$E = 2N\,(1 + \mu) \tag{2.8}$$

$$E = 3K\,(1 - 2\mu) \tag{2.9}$$

From equation (2.8), $\mu = \dfrac{E - 2\text{N}}{2\text{N}}$ (2.10)

From equation (2.9), $\mu = \dfrac{3K - E}{6\,\text{K}}$ (2.11)

Equating equations (2.10) and (2.11)

$$\frac{E - 2\text{N}}{2\text{N}} = \frac{3\text{K} - \text{E}}{6\text{K}}$$

$$\Rightarrow \qquad E = \frac{9\,\text{KN}}{3\,\text{K} + \text{N}}.$$

The modulus of elasticity of a material cannot be negative (–ve) or less than zero.

From equation (2.8),

$$\therefore \qquad 2\,\text{N}\,(1 + \mu) > 0$$

$$\Rightarrow \qquad \mu > -1.$$

From equation (2.9),

$$3\,K\,(1-2\mu) > 0$$

$\Rightarrow$ $$\mu < \frac{1}{2}.$$

$\therefore$ Range of Poisson's ratio μ is (0.5 to -1).

PROBLEM 2.1 **Objective 4**

Determine the change in the volume of the cylinder of height 300 mm and diameter 150 mm, subjected to a compressive stress of 10 N/mm^2 in all direction. Also determine the change in the height and diameter, if $E = 25$ GPa and $\mu = 0.20$.

SOLUTION

Normal stresses in all direction are same. This state of stress is called hydrostatic state of stress.

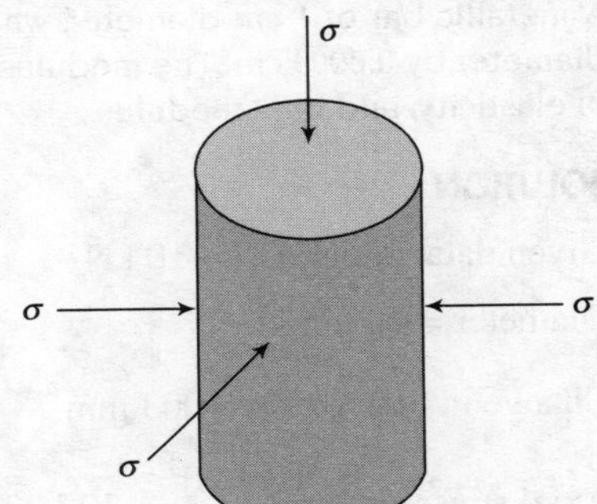

FIGURE 2.7

i.e., $$\sigma_x = \sigma_y = \sigma_z = \sigma$$

$\therefore$ $$\sigma_v = \sigma = 10 \text{ MPa}.$$

Let the change in diameter be δD.

$$\frac{\delta D}{D} = \frac{\sigma_x}{E} - \mu\frac{\sigma_y}{E} - \mu\frac{\sigma_z}{E}$$

$\Rightarrow$ $$\frac{\delta D}{D} = \frac{\sigma}{E}(1-2\mu)$$

$\therefore$ $$\delta D = 150 \times \frac{10}{25\times 10^3}(1-2\times 0.2)$$

Change in diameter = 0.036 mm (decrease)

Let δL be the change in length.

$$\frac{\delta L}{L} = \frac{\sigma}{E}(1-2\mu)$$

$\Rightarrow$ $$\delta L = 300 \times \frac{10}{25\times 10^3}(1-2\times 0.2) = 0.072 \text{ mm}$$

$$K = \frac{E}{3\times(1-2\mu)}$$

$\therefore$ Bulk modulus $$K = \frac{25{,}000}{3(1-2\times 0.2)} = 13{,}889 \text{ MPa}$$

$$\frac{\delta V}{V} = \frac{\sigma}{K} \text{ or } \delta V = \frac{\sigma}{K}\times V.$$

$$\therefore \text{Change in volume} = \frac{10}{13,889} \times \frac{\pi}{4} \times 150^2 \times 300$$

$$= 3817 \text{ mm}^3.$$

Generally, change in volume is expressed as cube centimeters (cc) because mm^3 is a very small unit while m^3 is a very big unit.

$\therefore$ Change in volume = 3.82 cc.

PROBLEM 2.2

Objective 1,2

A metallic bar of 1 cm diameter, when tested under an axial pull of 10 kN, was found to reduce its diameter by 0.0003 cm. The modulus of rigidity of the rod is 53 GPa. Find the Poisson's ratio, modulus of elasticity, and bulk modulus.

SOLUTION

Given data: Axial pull = 10 kN

Diameter = 10 mm

Change in diameter = 0.003 mm

$$\text{Axial stress, } \sigma = \frac{10 \times 10^3}{\frac{\pi}{4}10^2} = 127.32 \text{ MPa.}$$

Let ε_l and ε_d be the strain in length direction and we know that lateral direction, respectively.

$$E = 2N(1 + \mu) \tag{2.12}$$

$$E = \frac{\sigma}{\varepsilon_l}$$

$$\mu = \frac{\varepsilon_d}{\varepsilon_l}$$

$$\varepsilon_d = \frac{0.00003}{1} = 0.00003$$

$$G = 53 \times 10^3 \text{ MPa.}$$

Substituting the above value in equation (2.12)

$$\frac{\sigma}{\varepsilon_l} = 2G\left(1 + \frac{\varepsilon_d}{\varepsilon_l}\right)$$

$$\Rightarrow \quad \frac{127.32}{\varepsilon_l} = 2 \times 53 \times 10^3\left(1 + \frac{0.0003}{\varepsilon_l}\right)$$

$$\Rightarrow \quad \varepsilon_l + 0.0003 = 0.0012$$

$$\Rightarrow \quad \varepsilon_l = 0.0009$$

$\therefore$ Poisson's ratio, $\mu = \dfrac{0.0003}{0.0009} = 0.33$

Modulus of elasticity $(E) = \dfrac{127.32}{0.0009} = 141.47$ GPa.

Bulk modulus $(K) = \dfrac{E}{3(1-2\mu)} = 141.47$ GPa.

PROBLEM 2.3 **Objectives 1,2**

A bar of mild steel of 20 mm diameter is subjected to an axial pull of 50 kN. The increase in the length over a gauge length of 200 mm is measured to be 0.16 mm. The decrease in the diameter is 0.0048 mm. From the above data, determine the modulus of elasticity (E) and Poisson's ratio.

SOLUTION

Given data: Diameter $(d) = 20$ mm

Axial pull = 50 kN

Increase in length $(\delta l) = 0.16$ mm

Gauge length $(l) = 200$ mm

Decrease in diameter $\delta d = 0.0048$ mm

Axial stress developed $\sigma = \dfrac{P}{A} = \dfrac{50{,}000}{\dfrac{\pi}{4}(20^2)} = 159.15$ MPa

Longitudinal strain $\varepsilon_l = \dfrac{\delta L}{L} = \dfrac{0.16}{200} = 0.0008$

$\therefore$ Modulus of elasticity $E = \dfrac{\sigma}{\varepsilon_l} = \dfrac{159.15}{0.0008} = 198.9$ GPa

Poisson's ratio $\mu = \dfrac{\varepsilon_d}{\varepsilon_l}$

Lateral strain $\varepsilon_d = \dfrac{\delta d}{d} = \dfrac{0.0048}{20} = 0.00024$

$\therefore$ Poisson's ratio $\mu = 0.3$.

PROBLEM 2.4 **Objectives 1,2**

A block of dimensions 25 mm × 30 mm × 400 mm is subjected to an axial load of 6 t in the length direction. If the deformation in the lateral direction is restricted, estimate the effective modulus and Poisson's ratio of the material. $E = 75$ GPa and $\mu = 0.33$.

SOLUTION

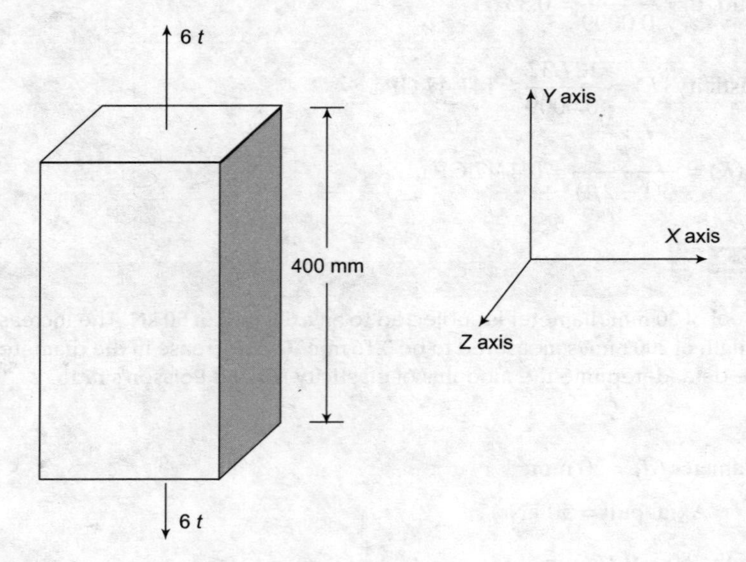

FIGURE 2.8

Axial pull = $P = 6 \times 10^4$ N

Cross-sectional area = 25 × 30 = 750 mm^2

Axial stress = $\sigma_y = \dfrac{6 \times 10^4}{750} = 80$ MPa.

From the given data, strain in x and z direction is restricted.

$$\Rightarrow \qquad \varepsilon_x = 0, \varepsilon_x = 0$$

$$\sigma_y = 80 \text{ MPa.}$$

As ε_x and ε_z are zero, σ_x and σ_z exist.

$$\therefore \qquad \varepsilon_x = 0$$

$$\Rightarrow \qquad \frac{\sigma_x}{E} - \mu\frac{\sigma_y}{E} - \mu\frac{\sigma_z}{E} = 0$$

$$\Rightarrow \qquad \sigma_x - \mu\sigma_z = \mu\sigma_y \tag{2.13}$$

$$\varepsilon_z = 0$$

$$\Rightarrow \qquad \sigma_z - \mu\sigma_x = \mu\sigma_y \tag{2.14}$$

Solving equations (2.13) and (2.14)

$$-\mu^2\sigma_z + \mu\sigma_x = \mu^2\sigma_y \tag{2.15}$$

$$\Rightarrow \qquad \sigma_z = \sigma_y \times \frac{\mu + \mu^2}{1 - \mu^2}$$

$$\sigma_z = \sigma_y \times \frac{\mu}{1-\mu}$$

$$\sigma_x = \sigma_y \times \frac{\mu}{1-\mu}$$

$$\varepsilon_y = \frac{\sigma_y}{E} - \mu\frac{\sigma_x}{E} - \mu\frac{\sigma_z}{E}$$

$$\Rightarrow \qquad \varepsilon_y = \frac{\sigma_y}{E}\left\{1 - \frac{\mu^2}{1-\mu} - \frac{\mu^2}{1-\mu}\right\} = \frac{\sigma_y}{E}\left\{\frac{1-\mu-2\mu^2}{1-\mu}\right\}$$

$$\therefore \qquad \varepsilon_y = \frac{\sigma_y}{E}\left\{\frac{(1+\mu)(1-2\mu)}{(1-\mu)}\right\}.$$

$$\text{Effective modulus} = \frac{\sigma_y}{\varepsilon_y} = \frac{E(1-\mu)}{(1+\mu)(1-2\mu)}$$

$$E_{\text{eff}} = \frac{75(1-0.33)}{(1+0.33)(1-2\times 0.33)} = 115 \text{ GPa}.$$

$$\text{Effective Poisson's ratio} = \frac{\varepsilon_x \text{ or } \varepsilon_z}{\varepsilon_y} = 0.$$

PROBLEM 2.5 **Objective 4**

A bar of uniform cross-section suffered a longitudinal strain of 1/700. Determine the volumetric strain of the bar if $\mu = \frac{1}{3}$.

SOLUTION

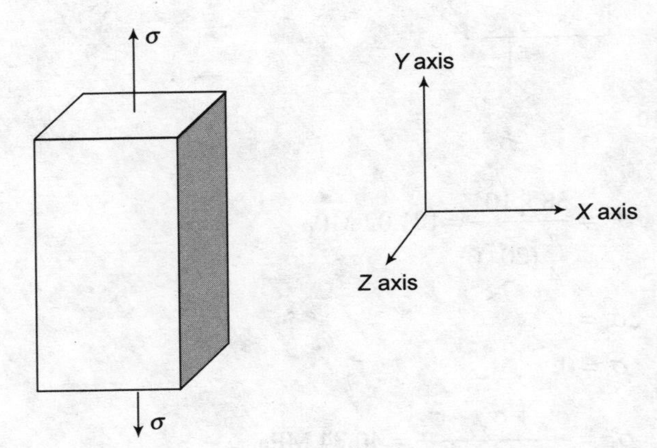

FIGURE 2.9

Given that $$\varepsilon_x = \frac{1}{700}$$

$$\sigma_y = \sigma_z = 0.$$

$\therefore$ $$\varepsilon_y = \varepsilon_z = -\frac{1}{700}$$

$$\varepsilon_y = \varepsilon_z = -\frac{1}{2100}.$$

Volumetric strain $$e_v = \varepsilon_x + \varepsilon_y + \varepsilon_z$$

$$= \frac{1}{700} - \frac{1}{2100} - \frac{1}{2100} = \frac{1}{2100}$$

Volumetric strain $= \dfrac{1}{2100}$.

PROBLEM 2.6

Objective 4

Estimate the change in volume of a 2-m-long 20 mm diameter bar subjected to an axial load of 38 kN. Take E = 200 GPa and $\mu = 0.25$.

SOLUTION

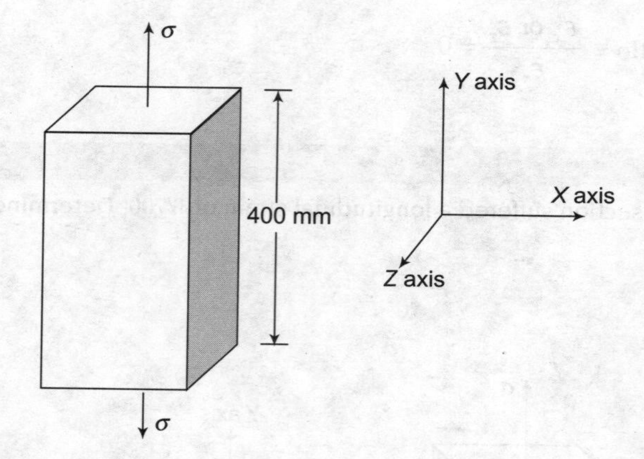

FIGURE 2.10

$$\sigma_y = \frac{38 \times 10^3}{\frac{\pi}{4}(20^2)} = 121.02 \text{ MPa}$$

$$\sigma_x = 0$$

$$\sigma_z = 0$$

$$\sigma_v = \frac{\sigma_x + \sigma_y + \sigma_z}{3} = 40.34 \text{ MPa}.$$

$$\text{Bulk modulus } (K) = \frac{E}{3(1-2\mu)} = 133.33 \text{ GPa.}$$

$$\text{As bulk modulus } K = \frac{\sigma_v}{\frac{\delta v}{V}}$$

$$\text{Volumetric strain } e_v = \frac{\delta v}{V} = \frac{\sigma_v}{K} = \frac{40.34}{133.33\times 10^3} = 0.0003$$

$$\text{Change in volume} = 0.0003 \times \frac{\pi}{4}(20^2)\times 2000 = 190.1 \text{ mm}^3 = 0.19 \text{ cc.}$$

PROBLEM 2.7 **Objective 4**

A steel cube of size 300 mm × 200 mm × 250 mm is subjected to forces as shown in Figure 2.11. If modulus of elasticity of the steel cube is 200 GPa and Poisson's ratio is 0.33, determine the change in the dimensions and volume of the block.

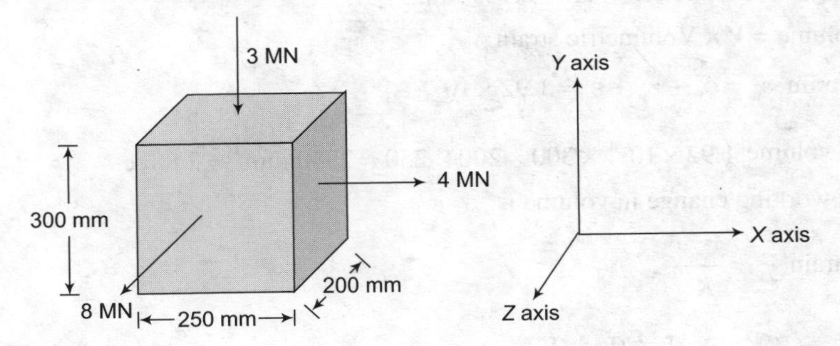

Figure 2.11

SOLUTION

Given data:

$$\text{Normal stress in } X \text{ direction, } \sigma_x = \frac{4\times 10^6}{300\times 200} = 66.67 \text{ MPa (T)}$$

$$\text{Normal stress in } Y \text{ direction, } \sigma_y = \frac{-3\times 10^6}{250\times 200} = -60 \text{ MPa (C)}$$

$$\text{Normal stress in } Z \text{ direction, } \sigma_z = \frac{8\times 10^6}{300\times 250} = 106.67 \text{ MPa (T)}$$

Change of dimension in X direction $\Delta x = \varepsilon_x \times 250$

$$\text{Strain in } X \text{ direction } \varepsilon_x = \frac{\sigma_x}{E} - \mu\frac{\sigma_y}{E} - \mu\frac{\sigma_z}{E}$$

$$\varepsilon_x = \frac{1}{2\times10^5}\{66.67 - 0.33\,(-60) - 0.33\times106.67\} = 2.57\times10^{-4}$$

$\therefore$ $$\Delta x = 0.064 \text{ mm.}$$

Similarly, $\varepsilon_y = \frac{\sigma_y}{E} - \mu\frac{\sigma_x}{E} - \mu\frac{\sigma_z}{E}.$

$\therefore$ Strain in Y direction $= \frac{1}{2\times10^5}\{-60 - 0.33\times66.67 - 0.33\times106.67\} = -0.000586.$.

Decrease in Y direction $= 0.000586\times300 = 0.176$ mm.

Strain in Z direction, $\varepsilon_z = \frac{\sigma_z}{E} - \mu\frac{\sigma_x}{E} - \mu\frac{\sigma_y}{E}$

$$\varepsilon_x = \frac{1}{2\times10^5}\{106.67 - 0.33\,(-60) - 0.33\times(-60)\} = 5.22\times10^{-4}.$$

Change of length in Z direction $= \delta Z = 0.104$ mm.

Change in volume $= V\times$ Volumetric strain

Volumetric strain $e_v = \varepsilon_x + \varepsilon_y + \varepsilon_z = 1.92\times10^{-4}.$

$\therefore$ Change in volume $1.92\times10^{-4}\times300\times200\times250 = 2880\text{ mm}^3 = 2.88$ cc.

Other way of working change in volume is,

Volumetric strain $e_v = \frac{\sigma_v}{K}$

$$\sigma_v = \frac{\sigma_x + \sigma_y + \sigma_z}{3} = 37.78.$$

Bulk modulus $K = \frac{E}{3(1-2\mu)} = 196.08$ GPa

$$e_v = \frac{37.78}{196.08\times1000} = 1.92\times10^{-4}.$$

$\therefore$ Change in volume $= 1.92\times10^{-4}\times300\times200\times250 = 2880\text{mm}^3 = 2.88$ cc.

PROBLEM 2.8

Objective 3

Prove that, Poisson's ratio for the material of a body is 0.5, if its volume does not change when stressed. Prove also that Poisson's ratio is zero when there is no lateral deformation when a member is axially stressed.

SOLUTION

Consider a body subjected to direct stress of σ

Given that change of volume = 0.

$$\Rightarrow \qquad \delta V = 0$$

$$\Rightarrow \qquad \frac{\delta V}{V} = \text{Volumetric strain} = 0.$$

Bulk modulus $K = \dfrac{\sigma}{\dfrac{\delta V}{V}}$

$$\Rightarrow \qquad K = \infty \text{ as } \frac{\delta V}{V} = 0.$$

We know that $E = 3K(1 - 2\mu)$

E = Modulus of elasticity

μ = Poisson's ratio

$$(1-2\mu) = \frac{E}{3K} = \frac{E}{3(\infty)} = 0$$

$$(1 - 2\mu) = 0$$

$$\mu = 0.5; \text{ hence proved.}$$

PROBLEM 2.9 **Objective 1,2**

A metallic rod of 1 cm diameter, when tested under an axial pull of 10 kN was found to reduce its diameter by 0.0003 cm. The modulus of rigidity for the rod is 51 kN/mm^2. Find the Poisson's ratio, modulus of elasticity, and bulk modulus.

SOLUTION

Diameter of the metallic rod = 1 cm = 10 mm

Axial pull P = 10 kN = 10,000 N

Change in diameter = 0.0003 cm = 0.003 mm

Rigidity modulus (N) = 0.51×10^5 N/mm^2

Let μ be Poisson's ratio of the material.

$$\mu = \frac{(\delta d / d)}{(\delta L / L)} = \frac{(0.003/10)}{(\delta L / L)} = \text{ where } (\delta L / L) \text{ is linear strain.}$$

$$\mu\left(\frac{\delta L}{L}\right) = 0.0003. \qquad (2.16)$$

$$E = \text{Modulus of elasticity} = \frac{\sigma}{\dfrac{\delta L}{L}}$$

$$\sigma = \text{Normal stress} = \frac{10 \times 10^3}{\dfrac{\pi}{4}(10^2)} = 127.32 \text{ MPa}$$

$$\frac{\delta L}{L} = \frac{127.32}{E} \tag{2.17}$$

Using equation (2.17) in (2.16) $\mu \dfrac{127.32}{E} = 0.0003$

$$\Rightarrow \qquad \mu = E\frac{0.0003}{127.32}.$$

We know that $E = 2N(1 + \mu)$

$$E = 2 \times 0.51 \times 10^5 \left[1 + E\frac{0.0003}{127.32}\right].$$

Solving the above equation, $E = \dfrac{1.02 \times 10^5}{0.7597} = 1.34 \times 10^5$ MPa

$$E = 2N(1 + \mu)$$

$$\Rightarrow \qquad \mu = 0.316$$

and

$$E = 2K(1 - 2\mu)$$

$$\Rightarrow \qquad K = \frac{E}{3(1 - 2\mu)} = 1.216 \times 10^5 \text{ MPa}$$

$$= 121.6 \text{ GPa.}$$

SUMMARY

- **Poisson's ratio is the ratio of lateral strain to the longitudinal strain.**
- **The ratio of change in volume to the original volume is known as volumetric strain.**
- **Bulk modulus is ratio of normal stress to the volumetric strain.**
- **The relation between Young's modulus and bulk modulus is given by**
  $$E = 3K(1 - 2\mu)$$
- **The relation between modulus of elasticity and modulus of rigidity is given by**
  $$E = 2N(1 + \mu)$$

OBJECTIVE TYPE QUESTIONS

1. The number of elastic constants in an isotropic and homogeneous material is
(a) 1 (b) 2 (c) 3 (d) 21

2. Poisson's ratio is the ratio of
(a) Longitudinal stress to longitudinal strain
(b) Lateral stress to lateral strain
(c) Longitudinal strain to lateral strain
(d) Lateral strain to longitudinal strain

3. The ratio of rigidity modulus to elastic modulus of a material is 0.4. Poisson's ratio of the same material is
(a) 1/2 (b) 1/3 (c) 1/4 (d) 1

4. The ratio of rigidity modulus to elastic modulus of a material is 0.4. The ratio of elastic modulus to bulk modulus of the material is
(a) 1/2 (b) 1/3 (c) 1/4 (d) 3/2

5. A cube of size 100 mm × 100 mm × 100 mm subjected to compressive stress of 50 MPa on one face and 200 MPa tensile stress on the other face. What is the force on the third face such that the volumetric strain in the body is zero?
(a) 150 MPa compressive (b) 150 MPa compressive
(c) 200 MPa compressive (d) Zero

6. The range of Poisson's ratio is
(a) (0 to ∞) (b) (0 to 1) (c) (–1 to 1) (d) (–1 to 0.5)

7. A bar is subjected to axial longitudinal stress of 160 MPa. Volumetric strain of the bar if modulus of elasticity and Poisson's ratio of the material are 200 GPa and 0.25, respectively
(a) 160×10^{-3} (b) 200×10^{-3} (c) 16×10^{-3} (d) 20×10^{-3}

8. The diagonal of a steel plate of size 200 mm × 200 mm, thickness 10 mm is increased by $2\sqrt{2}$mm. What is the shear strain induced in the plate?
(a) 0.001 (b) 0.01 (c) 0.005 (d) 0.0005

9. A plate is subjected to a shear stress of 20 MPa. The complimentary shear stress is (MPa)
(a) 10 (b) 5 (c) 100 (d) 20

10. A square plate is subjected to pure shear stress of 20 MPa. What is the intensity of normal stress (MPa) developed along the diagonal direction of the plate? The Poisson's ratio of the plate material is 0.3.
(a) 10 (b) 20 (c) 6 (d) 66.67

11. A cube of size 100 mm × 100 mm × 100 mm subjected to hydrostatic stress (same compressive stress on all faces) of 4 MPa. If the linear strain on each side noticed as 1 microstrains, the bulk modulus of the material is
(a) 133 GPa (b) 4 GPa (c) 133 MPa (d) 4 MPa

Solutions for Objective Questions

Sl. No.	1.	2.	3.	4.	5.	6.	7.	8.	9.	10.
Answer	(b)	(d)	(c)	(d)	(a)	(d)	(c)	(c)	(d)	(b)

Sl.No.	11.
Answer	(a)

EXERCISE PROBLEMS

1. Determine the changes in length and thickness of a steel bar which is 3 m long, 20 mm wide, and 10 mm thick subjected to an axial pull of 15 kN in the direction of 3 m length. Take $E = 210$ GPa and Poisson's ratio = 0.3.

2. A cube of size 100 mm is confined on four faces and receives a compressive force of 150 kN. Estimate the stresses developed on all the three faces, if $E = 200$ GPa and Poisson's ratio = 0.3. Also find the change in the volume of the cube.

3. Determine the value of Poisson's ratio of a rectangular bar of length 1 m, breadth 30 mm, and depth 20 mm, under an axial compressive load of 300 kN. The decrease in length is given as 2.55 cm and increase in breadth is 0.02 mm. Also determine the rigidity modulus.

   Ans: 0.267; 78.95 GPa.

4. A bar of 18 mm in diameter was subjected to tensile load of 30 kN and the extension measured on 200 mm gauge length was 0.075 mm and change in diameter was 0.00224 mm. Calculate Poisson's ratio and values of the three moduli.

   Ans: 0.332; $E = 314.38$ GPa; $N = 118$ GPa; $K = 311.88$ GPa.

5. A steel bar of 300 mm long, 50 mm wide, and 20 mm thick is subjected to a pull of 250 kN in the direction of its length. Determine the change in volume. Take $E = 200$ GPa and $\mu = 0.25$.

   Ans: $e_v = 6.25 \times 10^{-4}$; $\Delta V = 187.5$ mm^3

6. Calculate the modulus of rigidity and bulk modulus of a cylindrical bar of diameter of 20 mm and length 1 m, if the longitudinal strain in a bar during a tensile test is three times the lateral strain. Find the change in volume, if the bar is subjected to a hydrostatic pressure of 90 N/mm^2. Take $E = 200$ GPa.

7. A wooden member 300 mm × 180 mm × 50 mm is subjected to a force of 4 kN (tensile), 3 kN (tensile), and 5 kN (compressive) along x, y, and z directions, respectively. Determine the change in the sides and volume of the block. Take $e = 20$ GPa and Poisson's ratio = 0.2.

8. A bar of cross-section 16 mm × 16 mm is subjected to an axial pull of N. The lateral dimension of the bar is found to be changed to 15.994 mm × 15.994 mm. If the modulus of rigidity of the material is 80 GPa, determine the Poisson's ratio and modulus of elasticity.

9. A piece of material is subjected to three perpendicular stresses and the strains in the three directions are in the ratio 1:2:3. If Poisson's ratio is 0.25, find the ratio of the stresses, if $E = 200$ GPa.

10. Consider a 100 mm × 100 mm steel plate subjected to transverse biaxial tensile stress of 20 MPa in the x direction and 40 MPa in the y direction.

    (a Assuming the bar to be in a state of plane stress, determine the strain in z direction and the elongations of the plate in the x and y directions.

    (b) Assuming the bar to be in a state of plane strain, determine the strain in z direction and the elongations of the bar in the x and y directions. Take $E = 80$ GPa and $v = 0.25$.

11. A material is subjected to a longitudinal tensile. While it is free to contract laterally in one direction, half the lateral contraction in the other direction is prevented. Find the ratio between the Young's modulus and modified modulus of elasticity.

12. A rectangular aluminum plate similar has the following dimensions. $a = 50$ mm, $b = 75$ mm, and thickness = 5 mm. Uniformly distributed tensile force of 25 kN acts on 50 mm × 5 mm face and a compressive force of 37.5 kN on 75 mm × 5 mm face. Determine the strains in the three principal directions. Let $E =$ 150 GPa and $v = 0.35$.

13. A cube of iron, the length of whose edge is 100 cm, is subjected to a uniform pressure of 100 MPa on two opposite faces; the other faces are prevented by lateral pressure from extending more than 0.4 cm. Determine the pressure on these faces. $E = 200$ GPa and $v = 0.25$.

14. A bar of rectangular cross-section is in tension under an axial stress of 250 MPa. If Poisson's ratio is 0.25 for the material, what stresses must be applied to the side faces to prevent any change in cross-sectional dimension? Determine the axial strain, by the introduction of these lateral stresses.

15. A bar of 16 mm diameter is subjected to a pull of kN. The extension measured on gauge length of 200 mm is 0.5 mm and change in diameter is 0.003 mm. Calculate E, Poisson's ratio, and bulk modulus of the bar.

16. For a body having Young's modulus 195 GPa and Poisson's ratio 0.28, calculate the bulk modulus. Hence estimate the change in volume of 100 mm × 100 mm × 100 mm block of same material when immersed in water to a depth of 1 km. The density of water is 9.81 kN/cu.m.

17. A load of 100 kN is applied to the bar of 18 mm. The bar which is 400 mm long is elongated by 0.7 mm. Determine the modulus of elasticity of the material. If the Poisson's ratio of the material is 0.3, find the change in diameter.

CHAPTER 3

SHEAR FORCE AND BENDING MOMENT

UNIT OBJECTIVE

This chapter explains how to analyze and draw shear force diagram (SFD) and bending moment diagram (BMD) for all structural elements/members. It also explains in detail various of supports, reactions, etc. The presentation attempts to help the student achieve the following:

Objective 1: Identify different types of supports and reactions.

Objective 2: Determine shear force (SF) and bending moment (BM) at a location.

Objective 3: Draw the SFD and BMD of elements.

3.1 INTRODUCTION

In the earlier chapters, we have discussed about the internal stresses and strains developed in the members subjected to axial forces. The members subjected to transverse forces are referred as beams. In this chapter, we discuss the internal reactions in a member, as presented in Figure 3.1. Direction parallel to X axis is referred as longitudinal direction. Direction parallel to Y axis is referred as transverse direction. In addition, direction parallel to Z axis is referred as lateral direction. If the support in Figure 3.1 is along Z axis, the transverse direction will be parallel to Z axis and lateral direction will be parallel to Y axis.

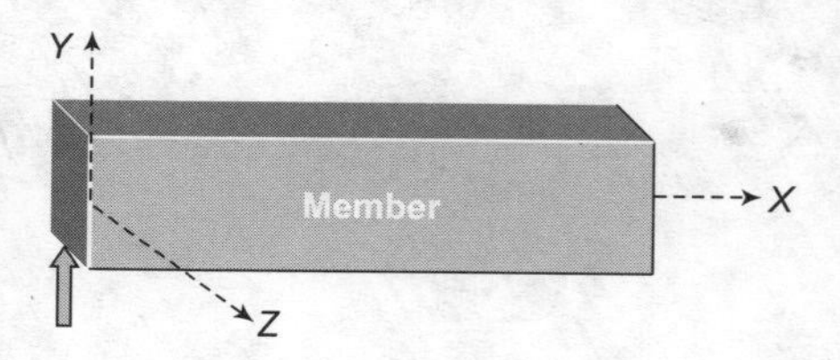

FIGURE 3.1 Different axes of member.

The internal reactions at a section in beam, due to external forces or tractions or body forces, can be presented as shown in Figure 3.2. For clarity, internal moments are represented by double arrows, which follows right-hand thumb rule. As per right-hand thumb rule, thumb represents the axis of the couple and direction of four fingers represents the direction of the couple.

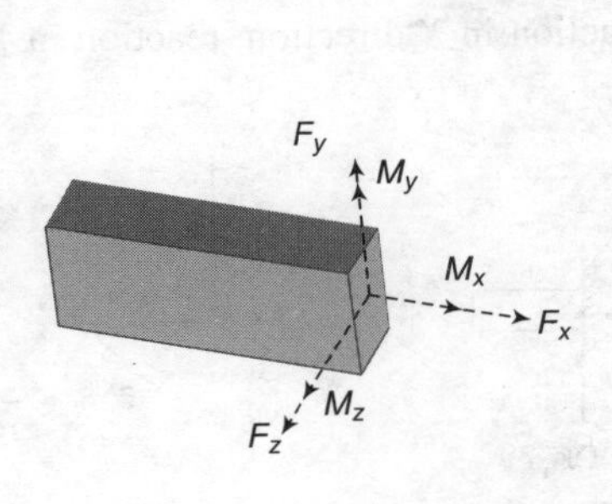

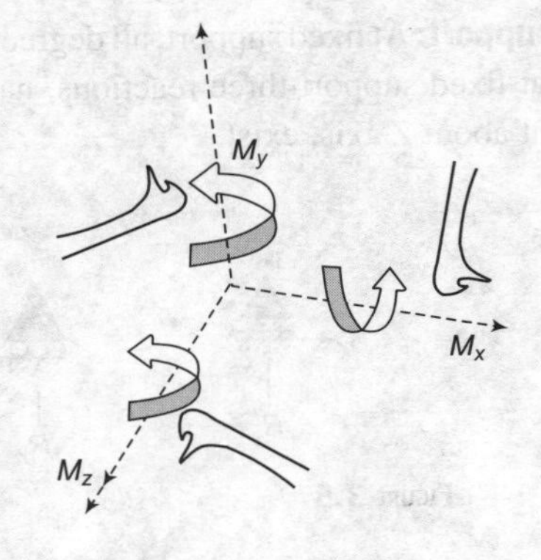

FIGURE 3.2 Right-hand thumb rule.

F_x = Axial force or normal thrust

F_y = Shear force

F_z = Shear force

M_x = Twisting moment

M_y = Lateral bending moment (BM)

M_z = Bending moment

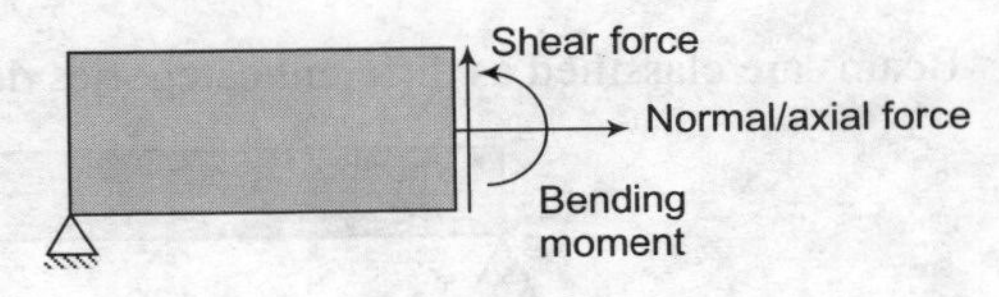

FIGURE 3.3

In case of planar members and coplanar loading, at any section in a member there exist three forces namely, moment, SF and normal force. Using three equilibrium equations, the three internal actions can be determined. Members are to be designed to resist these forces. The main objective of this chapter is to determine these forces.

3.2 DIFFERENT TYPES OF SUPPORTS AND BEAMS

The different types of supports are classified based on the possible deformations (degrees of freedom allowed). The diagrammatic representation of these supports is shown in Figure 3.4.

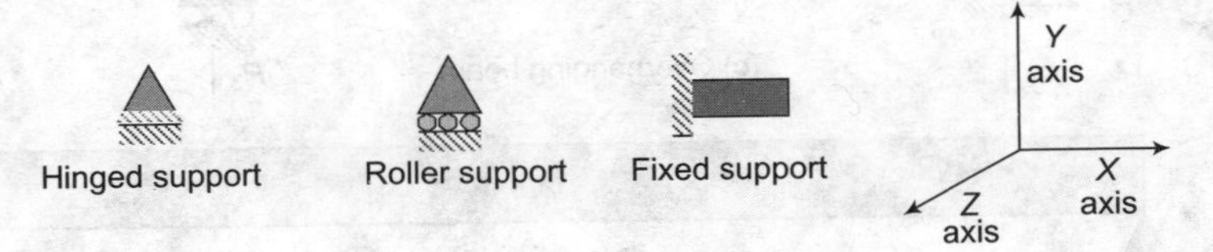

FIGURE 3.4

Hinged support: At hinged support, the rotation in the *XY* plane, that is, about *Z* axis is allowed, whereas the linear displacement in *X* and *Y* directions is restrained. Hence, these two restraints generate two reactions in *X* and *Y* directions. The example of hinged support is the hinge used for connecting the doors to the door frames.

Roller support: At hinged support, the rotation in the *XY* plane, that is, about *Z* axis, as well as linear displacement along the support (*X* direction in the present case) is allowed and *Y* direction is restrained. Hence the single restraint generates single reaction in *Y* direction. The example of roller support is a plank resisting on smooth spherical rollers.

Fixed support: At fixed support, all degrees of freedom (i.e., displacement and rotation) are restrained. Thus, at fixed support three reactions, namely reaction in *X* direction, reaction in *Y* direction, and moment about *Z* axis, exist.

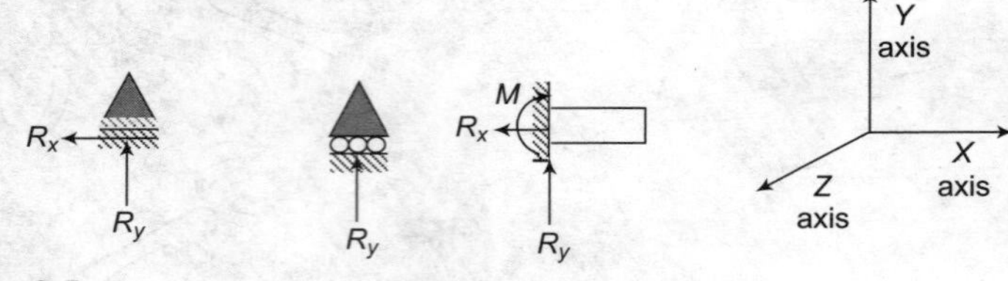

FIGURE 3.5

3.3 DIFFERENT TYPES OF BEAMS

Beams are classified as different categories depending on the end supports.

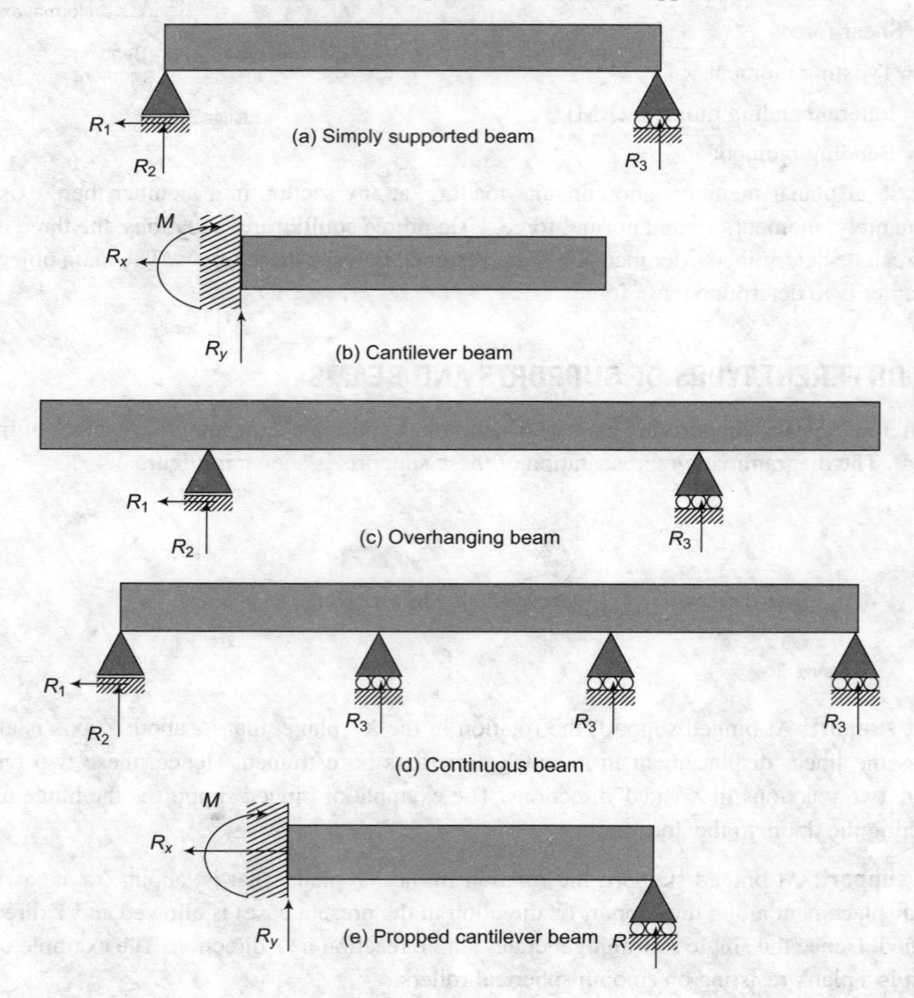

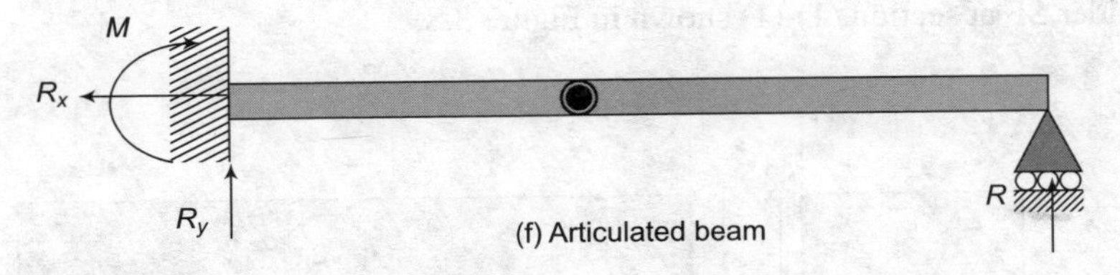

FIGURE 3.6

3.4 DIFFERENT TYPES OF LOADS

Loads are classified as

(a) Concentrated loads
(b) Uniformly distributed loads (UDLs)
(c) Distributed loads
(d) Couples

The above-mentioned loads were shown in Figure 3.7 on typical beam.

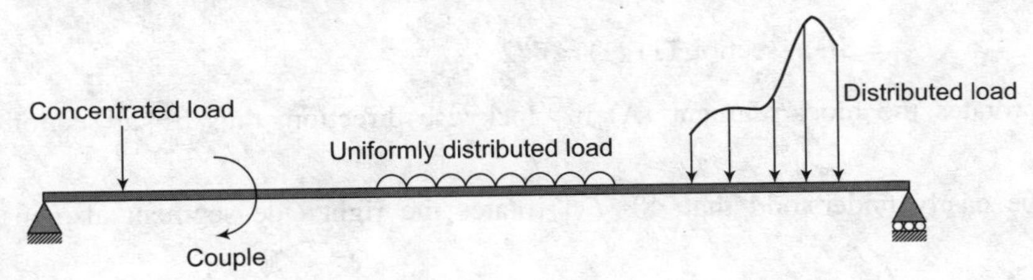

FIGURE 3.7

3.5 SHEAR FORCE AND BENDING MOMENT

SF at a section in a beam/member is the algebraic sum of the transverse loads either to the right or left of the section considered. The shear at a section can be easily found by using equilibrium equations. A shear force diagram (SFD) is the diagrammatic representation of variation of SF along the length of the beam.

Sign convention: The SF that tries to rotate the body/member/beam in clockwise direction is positive. The SF that rotates the member in anticlockwise direction is negative. For example, consider a simply supported beam, subjected to concentrated load at mid span.

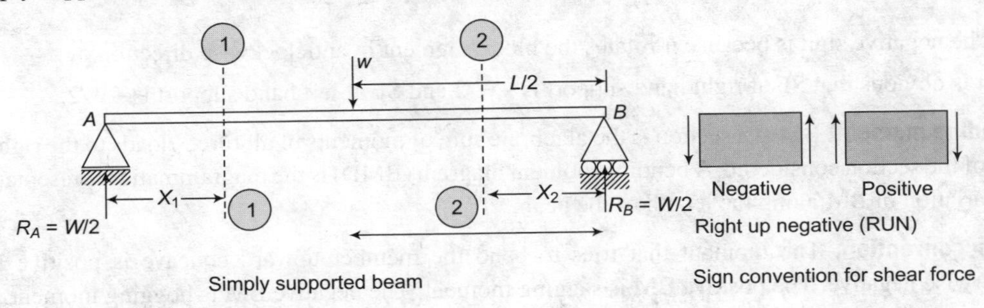

FIGURE 3.8

Let us consider SF at section (1)-(1) shown in Figure 3.8.

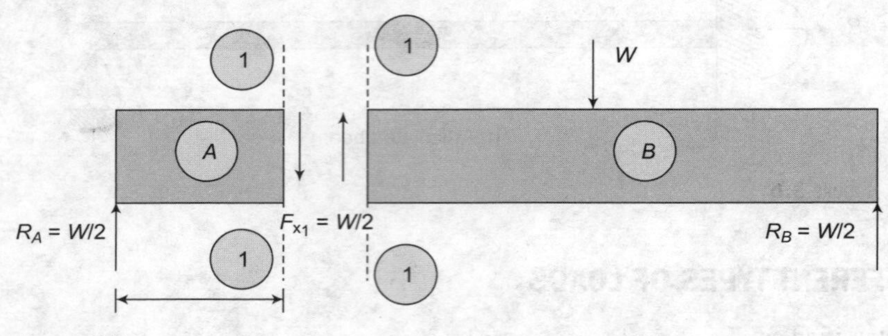

FIGURE 3.9

SF at section (1)-(1) is $W/2$, because by definition, it is the algebraic sum of transverse forces either to the left or to the right of the section.

SF at section (1)-(1) can also be obtained considering the equilibrium of segment (A) or segment (B).

$$F_{X1} = \text{SF at section (1)-(1)} = W/2$$

SF F_{X1} rotates the block/segment (A) in clockwise direction, thus SF at section (1)-(1) is positive.

It can be easily understood that, SF F_{X1} rotates the right side segment also in clockwise direction.

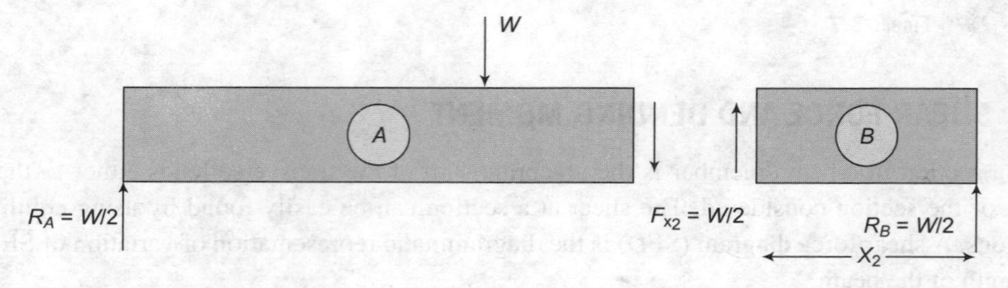

FIGURE 3.10

By definition SF at section (2)-(2)

$$F_{x2} = -W/2.$$

The negative sign is because it rotates the block/segment in anticlockwise direction.

It is obvious that SF at right hand support is –W/2 and SF at left hand support is $+W/2$.

Bending moment: BM at a section is the algebraic sum of moments of all forces/loads to the right or left of the section considered. A bending moment diagram (BMD) is the diagrammatic representation of variation of BM along the length of the beam.

Sign convention: The moment that tries to bend the member upward concave is positive BM otherwise negative. The positive BM is sagging moment. The negative BM is hogging moment.

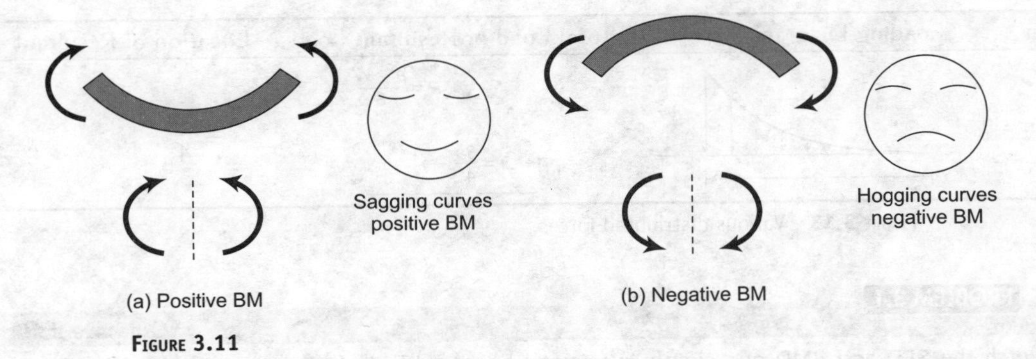

FIGURE 3.11

Distributed forces: For calculation purpose, the distributed forces may be converted as concentrated loads and applied at the centroid of the distributed load diagram. For better understanding, consider parabolic loading shown in Figure 3.12, resultant of a distributed force system is the area of distributed force diagram.

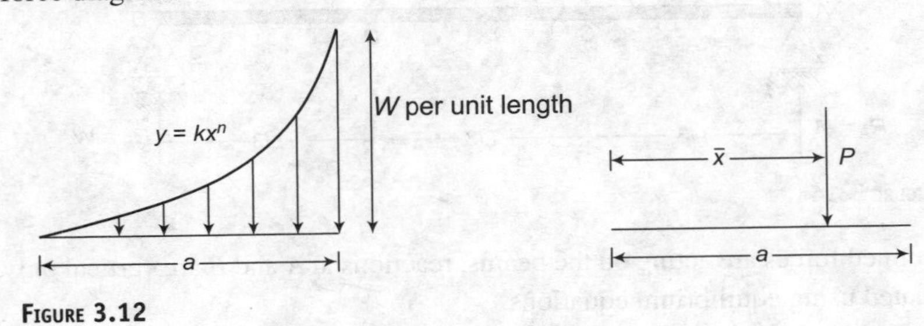

FIGURE 3.12

Resultant of a distributed force system is the area of the distributed load diagram and the resultant acts at the centroid of the loading diagram.

$$P = \frac{1}{n+1} Wa$$

$$\bar{x} = \frac{n+1}{n+2} a.$$

For example:

n =	Loading Diagram	Total Load or Resultant Load	Location of Resultant
n = 0	W, a	$P = a \times W$, $\bar{x} = \frac{a}{2}$	$\bar{x} = \frac{a}{2}$
n = 1	W, a	$P = \frac{a \times W}{2}$, $\bar{x} = \frac{2a}{3}$	$\bar{x} = \frac{2a}{3}$

FIGURE 3.13 (*Contd.*)

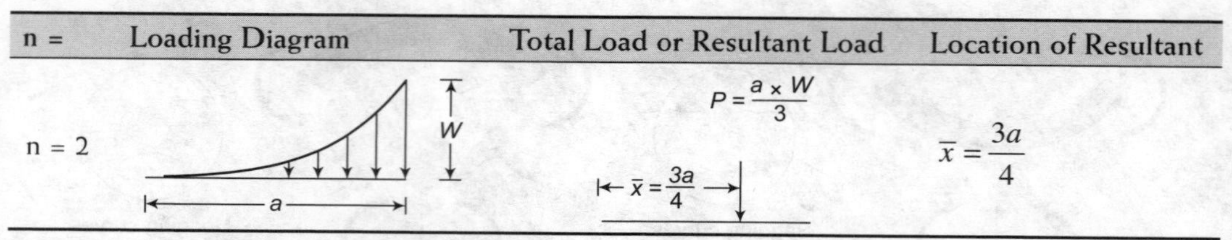

n =	Loading Diagram	Total Load or Resultant Load	Location of Resultant
n = 2		$P = \frac{a \times W}{3}$	$\bar{x} = \frac{3a}{4}$

FIGURE 3.13 Various distributed forces.

PROBLEM 3.1

Objective 2, 3

Sketch the SFD and BMD of a simply supported beam subjected to concentrated loads of same intensity acting at middle third points of the span.

SOLUTION

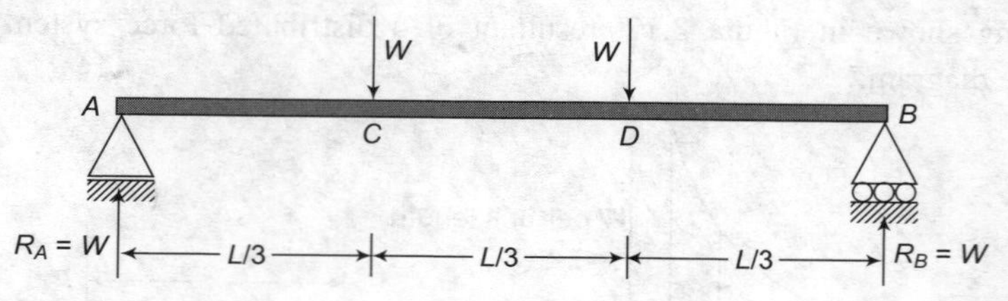

FIGURE 3.14

Because no inclined forces are acting on the beams, reactions at A and B are vertical only. Reactions are to be evaluated using equilibrium equations.

Sum of vertical forces is zero

$$\Rightarrow \quad R_A + R_B = 2\text{ W}$$

Sum of the moments about $A = 0$

$$\Rightarrow \quad R_B(L) = W \times \frac{L}{3} + W \times \frac{2L}{3}$$

$$\Rightarrow \quad R_B = W$$

$$\Rightarrow \quad R_A = W$$

Variations of SF and BM in the region DB: Consider a section located at a distance 'x' from B. From equilibrium condition,

$$\Rightarrow \quad F_V = 0 \Rightarrow F_x = R_B = W$$

$$\Rightarrow \quad \sum M = 0 \Rightarrow M_X = R_B X = WX.$$

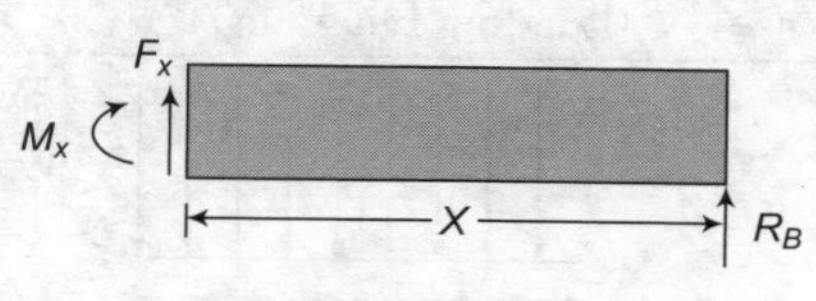

FIGURE 3.15

SF(V_x) at section considered

$$V_X = -W$$

$$\text{SF at } X = 0 \Rightarrow V_B = -W$$

$$\text{SF at } X = \frac{L}{3} = V_D = -W$$

BM

$$M_X = WX$$

$$\text{BM at } X = 0 \Rightarrow M_B = 0$$

$$\text{BM at } X = \frac{L}{3} = M_D = \frac{WL}{3}.$$

In the region *DC*, consider a section located at a distance *x* from *B*. Limits of *x* are *L*/3, 2*L*/3

$$V_X = -R_B + W$$

$$\text{SF} = 0 \text{ at } X = \frac{L}{3} \text{ and } X = \frac{2L}{3}$$

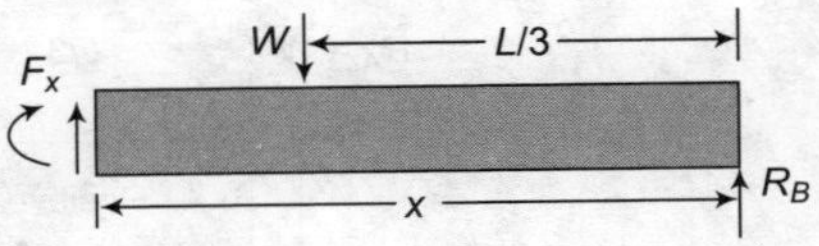

$$M_X = R_B X - W\left(\frac{X-L}{3}\right)$$

$$\text{BM at } X = \frac{L}{3} \Rightarrow M_D = \frac{WL}{3}$$

$$\text{BM at } X = \frac{2L}{3} = M_C = \frac{WL}{3}$$

FIGURE 3.16

In the region *CA*, consider a section located at a distance '*x*' from *B*. Limits are from 2*L*/3 and *L*.

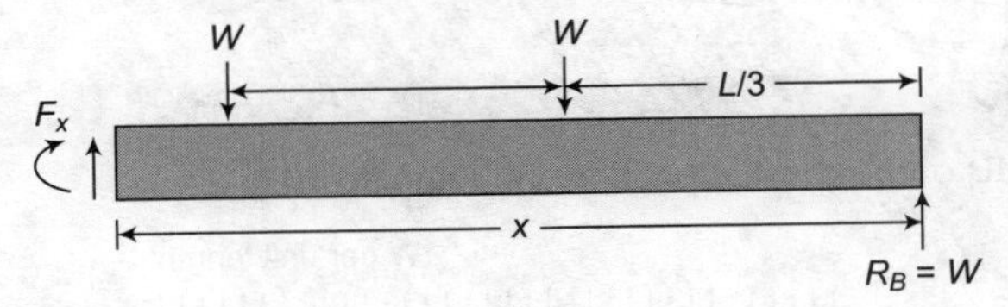

FIGURE 3.17

SF

$$V_X = -R_B + W + W$$

$$\Rightarrow \quad VX = W$$

$$\text{SF at } X = \frac{2L}{3} \Rightarrow V_C = W$$

$$\text{SF at } X = L = V_A = W$$

BM

$$M_X = W(L - X).$$

The above expression can be obtained considering the from left support *A*,

$$\Rightarrow \quad V_X = R_A = W$$

$$M_X = R_A\,(L - X)$$

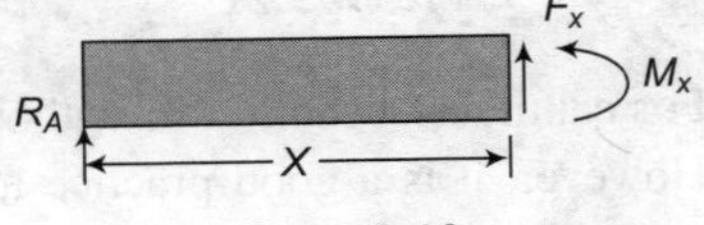

FIGURE 3.18

$$\text{at } X = \frac{2L}{3} \Rightarrow M_C = \frac{WL}{3}$$

$$\text{at } X = L = M_A = 0$$

SFD and BMD

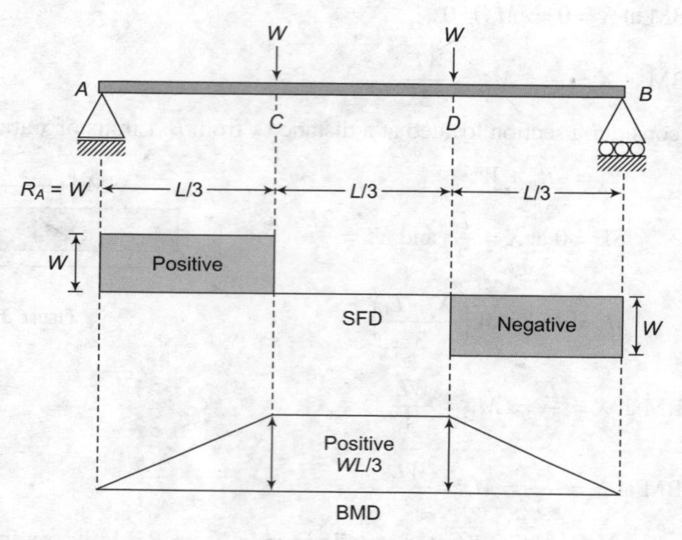

FIGURE 3.19

PROBLEM 3.2

Objective 2, 3

Sketch the SFD and BMD of the cantilever shown in Figure 3.20

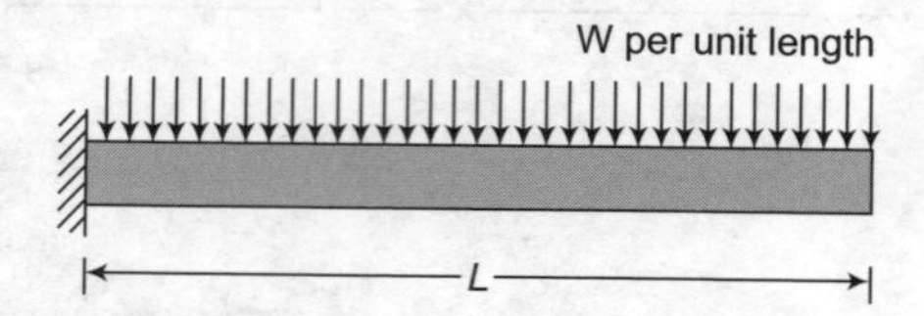

FIGURE 3.20

SOLUTION

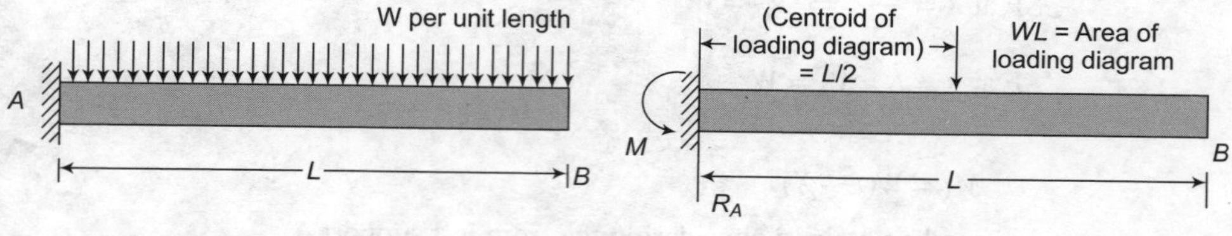

FIGURE 3.21

For cantilever beam, even without evaluating reaction at the fixed end, solution can be obtained. However, it is a good practice to determine the end reactions also. The distributed force can be converted into a concentrated load for finding the reactions only.

$$\sum F_V = 0 \Rightarrow R_A = WL$$

$$M = \frac{WL^2}{2}.$$

For SF and BM: consider a section located at a distance 'x' from B.

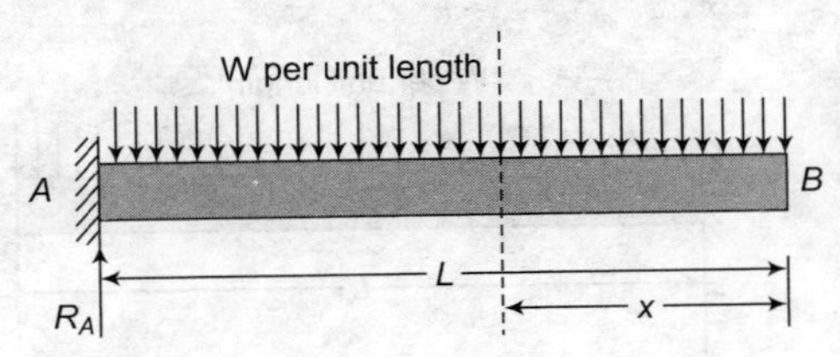

FIGURE 3.22

Converting the distributed loads into concentrated loads within the region considered

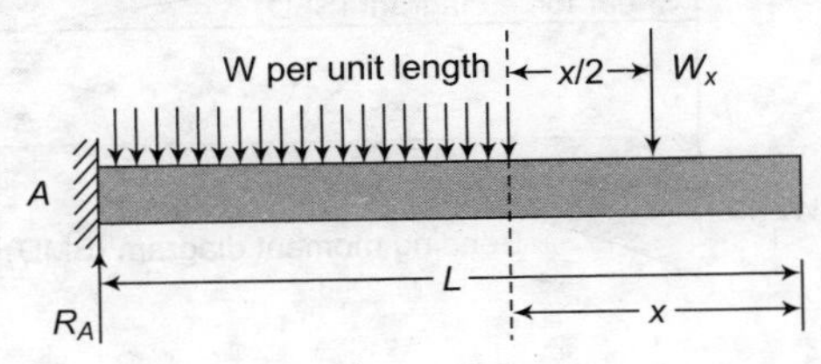

FIGURE 3.23

To find the vertical reaction at A, use the equilibrium equation that sum of vertical forces is equal to zero.

$$\sum V = 0$$

$$R_A - WL = 0$$

$$R_A = WL$$

To find the fixing moment at A, use the equilibrium equation that sum of moments of all forces is equal to zero.

$$M_A - \frac{WL \times L}{2} = 0;$$

$$M_A = \frac{WL^2}{2}$$

SFD and BMD Calculations

Consider a section located at a distance 'x' from B.

$$V_X = WX$$

at $X = 0$; $\quad V_B = 0$

at $X = L$; $\quad V_A = WL$

$$\text{BM} = M_X = -\frac{WX(X)}{2} = -\frac{WX^2}{2}$$

at $X = 0$; $\quad M_B = 0$

at $X = L$; $\quad M_A = -\frac{WL^2}{2}$

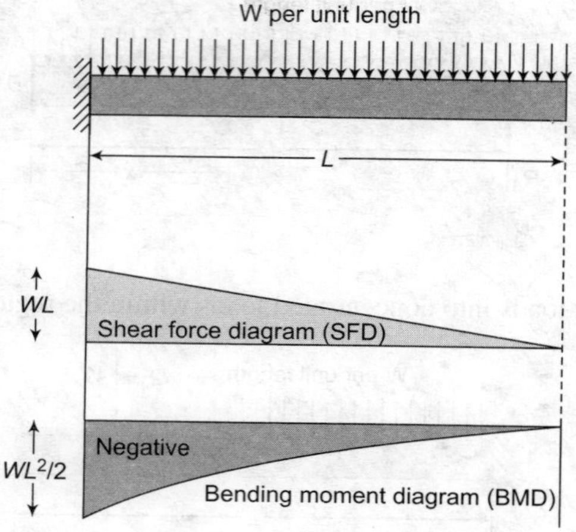

FIGURE 3.24

Many times, a student may face difficulty in deciding the shape of the BMD, if the variation of BM is nonlinear. In the present case, the BMD can be drawn in two ways as shown in Figure 3.25. The correct shape of BMD can be arrived by taking the help of $\frac{dM}{dx}$, that is, SF.

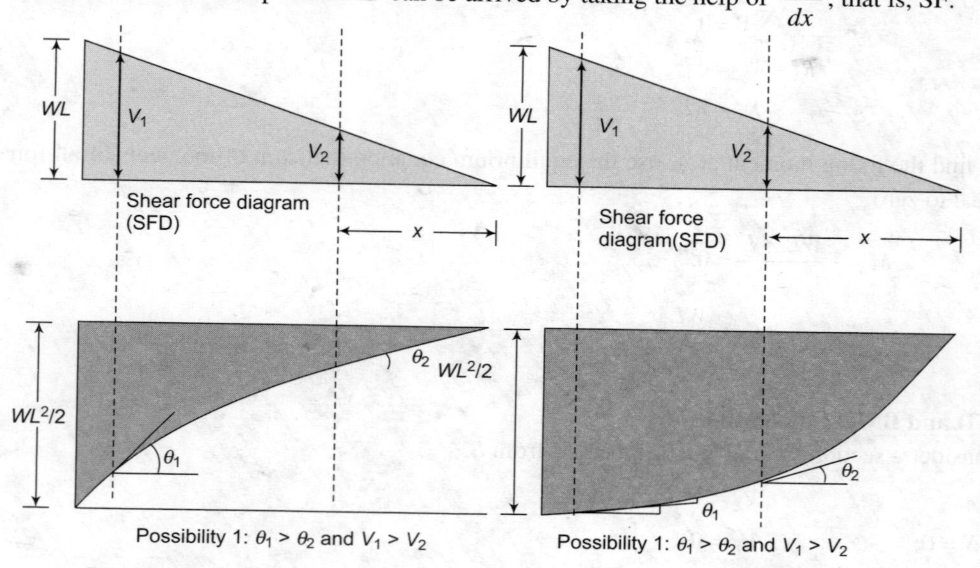

FIGURE 3.25

From Figure 3.25, it is clear that SF is increasing with 'x'. Hence in the BMD, slope of BM curve should also increase with 'x'. In the first possibility, (a) $\theta_1 > \theta_2$ in the BMD and $V_1 > V_2$, whereas in the second possibility (b) $\theta_1 < \theta_2$ in the BMD and $V_1 > V_2$. For a correct BMD, curve slope of BMD and value of SF should have same variation. Therefore, the option (a) is correct and (b) is not correct. The similar argument can be put up ever while drawing the SFD too.

PROBLEM 3.3 **Objective 2, 3**

Sketch SFD and BMD for the cantilever beam shown in Figure 3.26.

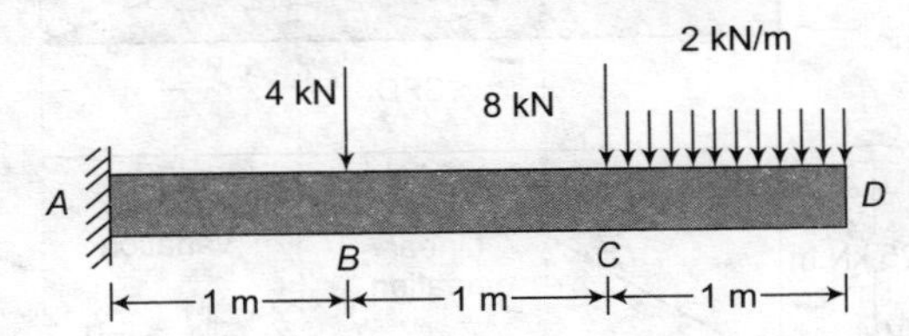

FIGURE 3.26

SOLUTION

Sum of the vertical forces is zero.

$$\Rightarrow \quad R_A = 4 + 8 + 2 \times 1 = 14 \text{ kN}$$

$$\Rightarrow \quad M_A = 4 \times 1 + 8 \times 2 + 2 \times 1(2.5) = 25 \text{ kN·m}$$

SF and BM calculation can be put in a tabular form for convenience.

Region	'x' Measured from	Limits for 'x'	Expression for SF (kN)	Expression for BM (kN·m)	Remarks
DC	D	0–1	Vx = 2 × x = 2x At x = 0; VD = 0 and at x = 1; Vc = −2 kN	$M_x = -2 \times x \times \left(\frac{x}{2}\right) = x^2$ At x = 0; MD = 0 and at x = 1; Mc = −1 kN·m	Variation of SF is linear Variation of BM is parabolic
CB	D	1–2	Vx = 2 × 1 + 8 = 10 At x = 1; Vc = 10 and at x = 2; VB = −10 kN	Mx = 2 × 1 (x − 0.5) − 8(x − 1) = −10x + 9 At x = 1; MC = −1 and at x = 2; MB = −11 kN·m	Variation of SF is constant Variation of BM is linear
BA	D	2–3	Vx = 2 × 1 + 8 + 4 = 14 At x = 2; VB = 14 and at x = 3; VA = −14 kN	Mx = −2 × 1 × (x − 0.5) − 8(x − 1) = −14x + 17 At x = 2; MC = −9 and at x = 3; MA = −25 kN·m	Variation of SF is constant Variation of BM is linear

SFD and BMD

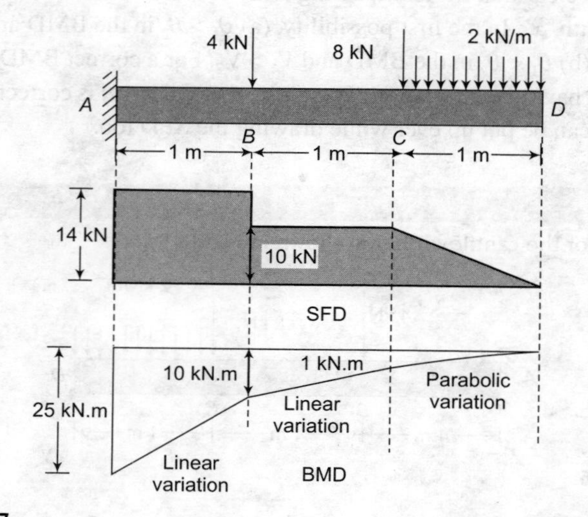

FIGURE 3.27

PROBLEM 3.4

Objective 2, 3

Sketch BMD and SFD for the overhanging beam shown in Figure 3.28.

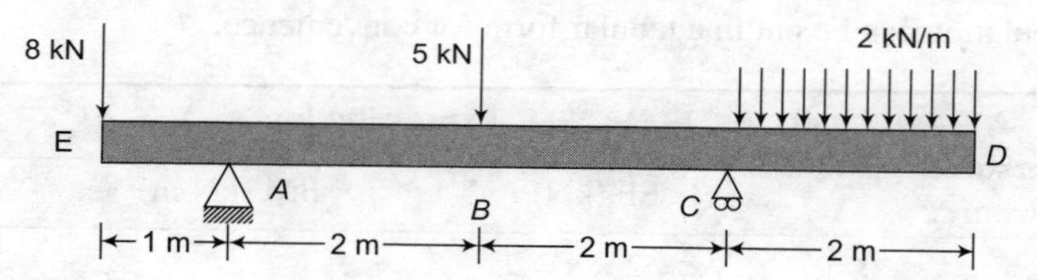

FIGURE 3.28

SOLUTION

First the reaction at the supports A and C are to be found from equilibrium equations.

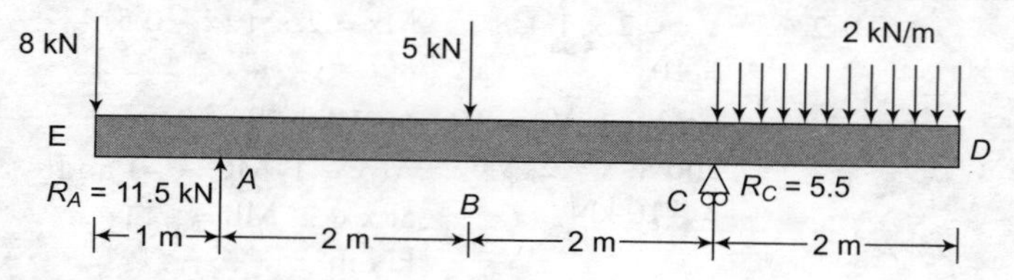

FIGURE 3.29

Sum of moments about A is equal to zero.

$$\Rightarrow \quad 2 \times 2 \times 5 + 5 \times 2 - 8 \times 1 = R_C \times 4$$

$$\Rightarrow \quad R_C = 5.5 \text{ kN}$$

Sum of the vertical forces is equal to zero.

$\Rightarrow$ $$R_A + R_C = 17$$

$$R_C = 5.5 \text{ kN}$$

$\Rightarrow$ $$R_A = 11.5 \text{ kN}$$

SF and BM Calculations

Region	'x' Measured from	Limits for 'x'	Expression for SF (kN)	Expression for BM (kN·m)	Remarks
DC	*D*	0–2	$Vx = 2 \times x = 2x$ At $x = 0$; VD = 0 and at $x = 2$; $Vc = -4$ kN	$M_x = -2 \times x \times \left(\frac{x}{2}\right) = x^2$ At $x = 0$; MD = 0 and at $x = 2$; Mc = −4 kN·m	Variation of SF is linear Variation of BM is parabolic
CB	*D*	2–4	$Vx = 2 \times 2 - 5.5 = -1.5$ At $x = 2$; Vc = −1.5 kN and at $x = 4$; VB = −1.5 kN	$Mx = -2 \times 2 \times (x - 1) + 5.5\,(x - 2)$ $= 1.5x - 7$ At $x = 2$; MC = −4 and at $x = 4$; MB = −1 kN·m	Variation of SF is constant Variation of BM is linear
BA	*D*	4–6	$Vx = 2 \times 2 - 5.5 + 5 = 3.5$ At $x = 4$; VB = 3.5 kN and at $x = 6$; VA = 3.5 kN	$Mx = -2 \times 2 \times (x - 1) + 5.5\,(x - 2)$ _ $5(x - 4)$ $= -3.5x + 13$ At $x = 4$; MC = −1 and at $x = 6$; MA = −8 kN·m	Variation of SF is constant Variation of BM is linear
EA	*E*	0–1	$Vx = -8$ At $x = 0$; $VE = 0$ and at $x = 1$; VA = −8 kN	$Mx = -8 \times x$ At $x = 0$; ME = 0 and at $x = 1$; MA = −8 kN·m	Variation of SF is constant Variation of BM is linear

SFD and BMD

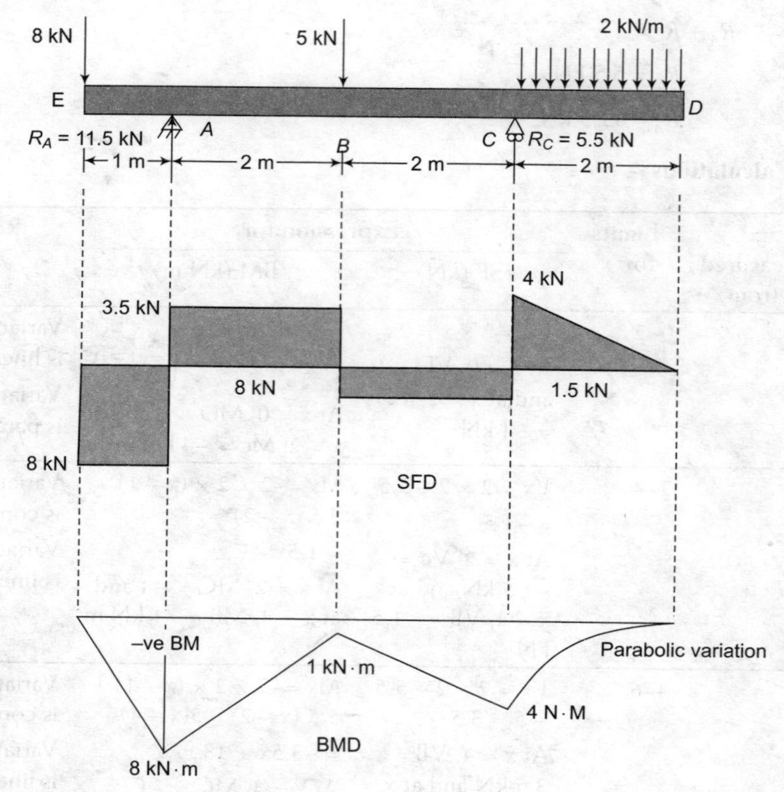

FIGURE 3.30

PROBLEM 3.5

Objective 2, 3

Sketch SFD and BMD for a simply supported beam subjected to UDL of intensity W per unit length.

SOLUTION

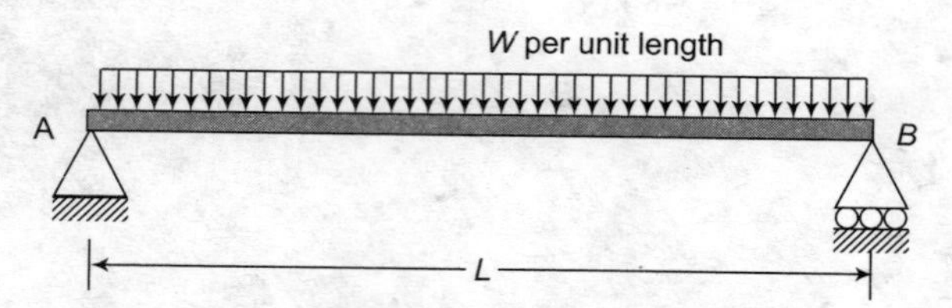

FIGURE 3.31

Sum of the moments of all forces about A is equal to zero.

$$\Rightarrow \qquad R_B \times L = W(L)\,\frac{L}{2}$$

$$\Rightarrow \qquad R_B = \frac{WL}{2}$$

Sum of the vertical forces is equal to zero.

$$\Rightarrow \quad R_A + R_B = WL$$

$$\Rightarrow \quad R_B = \frac{WL}{2}$$

$$\therefore \quad R_A = \frac{WL}{2}$$

Consider a section located 'x' distance from B

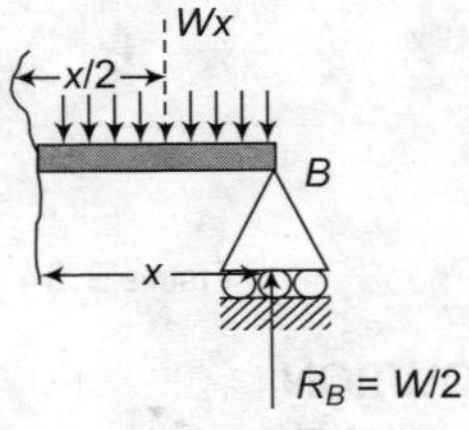

SF at the section under consideration is $V_x = -\frac{W}{2} + Wx$

FIGURE 3.32

At $x = 0$; $V_B = -\frac{W}{2}$

At $x = L$; $V_A = \frac{W}{2}$ and at $x = \frac{L}{2}$; $V_C = 0$.

BM at the section under consideration is $M_x = -\frac{Wx}{2} + Wx\frac{x}{2}$

At $x = 0$; $M_B = 0$

At $x = L$; $M_A = 0$ and between A and B the variation of BM is parbolic.

BM in a member is either maximum or minimum, where the SF (which is the first derivative of BM) is zero.

$$\therefore \quad M_{max} = M_C = \frac{WL^2}{8} \text{ where } \left(V_C = 0 \text{ and } x = \frac{L}{2}\right).$$

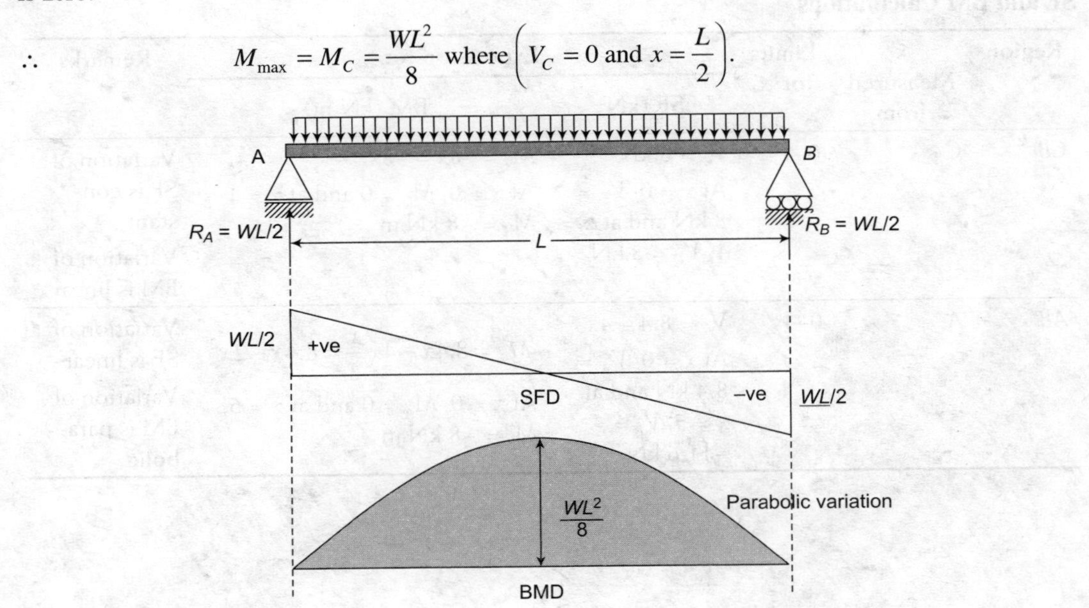

FIGURE 3.33

PROBLEM 3.6

Objective 2, 3

Sketch the SFD and BMD for the beam shown in Figure 3.34. Determine the maximum BM. Also locate the points of contra flexure.

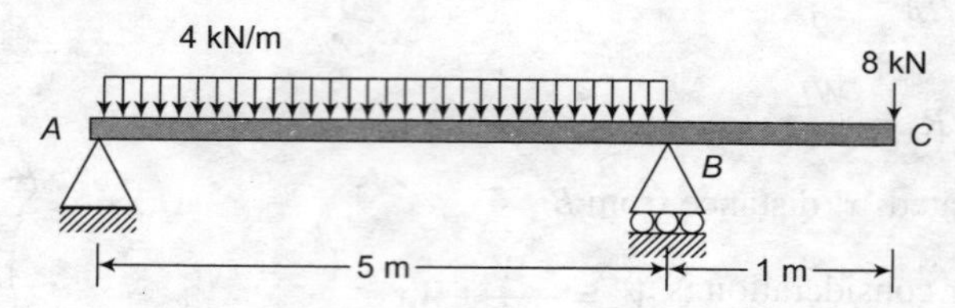

FIGURE 3.34

SOLUTION

Sum of the moments of all forces about A is equal to zero.

$$\Rightarrow \quad R_B \times 5 - 4 \times 5 \times \frac{5}{2} - 8 \times 6 = 0$$

$$\Rightarrow \quad R_B = 19.6 \text{ kN}$$

Sum of the vertical forces is equal to zero.

$$\Rightarrow \quad R_A + R_B = 28$$

$$R_B = 28 - 19.6$$

$$\therefore \quad R_A = 8.4 \text{ kN}$$

SF and BM Calculations

Region	'x' Measured from	Limits for 'x'	Expression for SF (kN)	Expression for BM (kN·m)	Remarks
CB	*C*	0–1	$V_x = 8$ kN At x = 0; V_C = 8 kN and at x = 1; V_C = 8 kN	$M_x = -8x = -8x$ At x = 0; M_C = 0 and at x = 1; M_B = −8 kN·m	Variation of SF is constant. Variation of BM is linear
AB	*A*	0–4	$V_x = 8.4 - 4x$ At x = 0; V_A = 8.4 kN and at x = 5; V_B = −11.6 kN	$M_x = 8.4x - 4x\frac{x}{2} = 8.4x - 2x^2$ At x = 0; M_A = 0 and at x = 5; M_B = −8 kN·m	Variation of SF is linear Variation of BM is parabolic

SFD and BMD

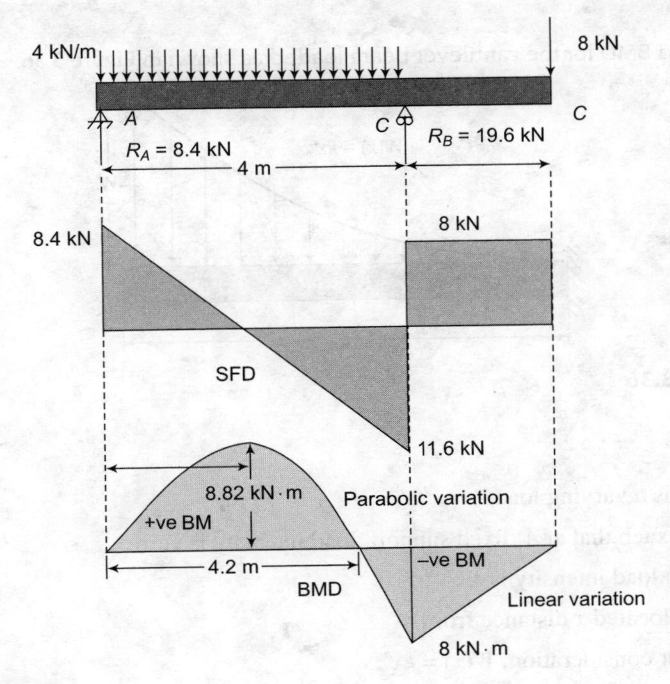

FIGURE 3.35

For maximum BM, equate the expression for SF to zero. SF is zero in the region *AB*, as it is evident from the expression for SF.

For maximum BM, $V_x = 0$,

$$V_x = R_A - 4x$$

$$8.4 - 4x = 0$$

$$x = 2.1 \text{ m}$$

BM is maximum at point located 2.1 m from the support *A*

$$M_{max} = 8.4x - 2x^2$$

$$x = 2.1,\ M_{max} = 8.82 \text{ kN-m}$$

At point of contraflexure, BM changes its sign. This location is important especially in the case of members with different strengths in compression and tension like 'reinforced concrete elements'.

For point of contraflexure, equate BM to zero in the region AB

$$8.4x - 2x^2 = 0$$

$$x = 0 \text{ or } x = 4.2 \text{ m}$$

$x = 4.2$ m is the point in Figure 3.35, where BM is changing its sign, but at $x = 0$, BM is not changing its sign. Hence, point of contraflexure exists at a distance 4.2 m from the support *A*.

PROBLEM 3.7

Objective 2, 3

Sketch the SFD and BMD for the cantilever beam loaded as shown in Figure 3.36.

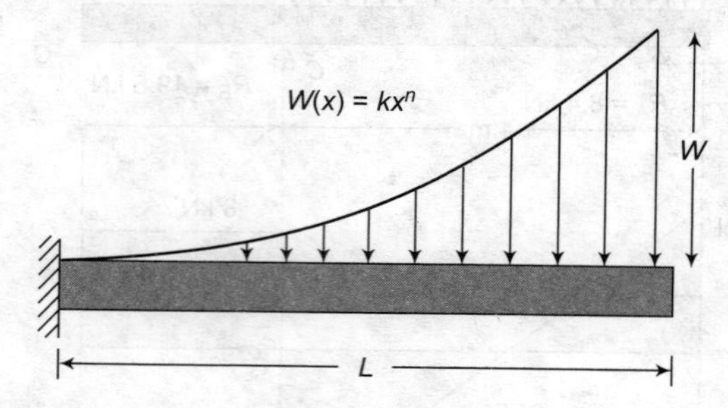

FIGURE 3.36

SOLUTION

The given loading is a varying load.

The conditions are such that at A, fixed support, load intensity is zero.

At B (free end), the load intensity is W.

Consider a section located x distance from A.

At the section under consideration, $W(x) = kx^n$.

Substituting the mentioned conditions, at $x = L$, $W(x) = W$;

$$\Rightarrow \qquad k = \frac{W}{L^n}$$

Consider a strip of width δx as shown in Figure 3.37.

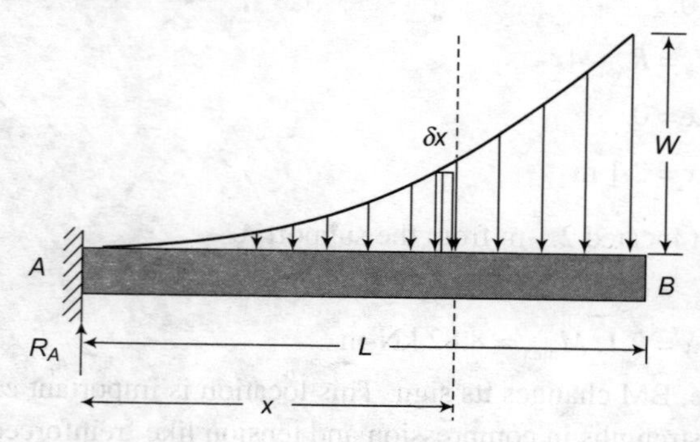

FIGURE 3.37

To find reaction at A, use equilibrium equation such that sum of vertical forces is equal to zero.

$$\xrightarrow{\text{yields}} R_A - \int_0^L W(x)dx = 0$$

$$\xrightarrow{\text{yields}} R_A = \int_0^L \frac{W}{L^n} x^n dx = \frac{1}{n+1} WL.$$

To find the reaction moment at A, use equilibrium equation such that sum of moments of all forces about A is equal to zero.

$$\xrightarrow{\text{yields}} M_A - \int_0^L W(x) x dx = 0$$

$$\xrightarrow{\text{yields}} M_A = \int_0^L \frac{W}{L^n} x^{n+1} dx = \frac{1}{n+2} WL^2$$

To find SF and BM at a section located 'x' distant from A, consider the left side of the section to make the calculations simple.

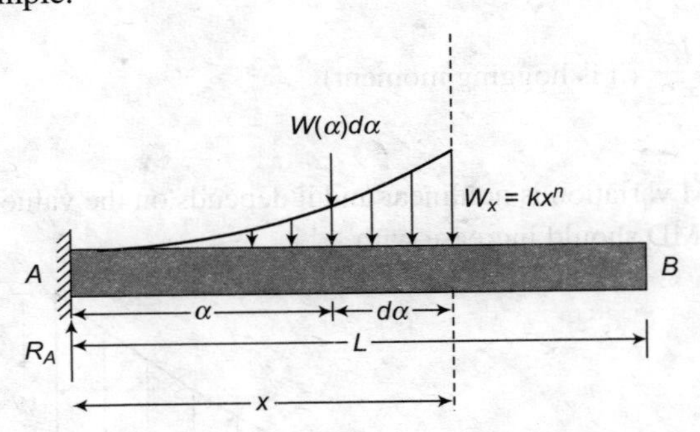

FIGURE 3.38

To find SF and BM at section located 'x' distant from A, consider an elemental strip of thickness '$d\alpha$', located 'α' distant from A.

Elemental force on the elemental strip = $W(\alpha)d\alpha = \frac{W}{L^n}\alpha^n d\alpha$

Moment of this elemental force at a section 'x' distant from A = $(x - \alpha)W(\alpha)d\alpha$

Then, SF at section x distant from A is $V_x = R_A - \int_0^x W(\alpha)d\alpha = \frac{1}{n+1}WL - \int_0^x \frac{W}{L^n}\alpha^n d\alpha$

$$\xrightarrow{\text{yields}} V_x = \frac{1}{n+1}WL - \frac{Wx^{n+1}}{(n+1)L^n} = \frac{WL}{n+1}\left\{1 - \left(\frac{x}{L}\right)^{n+1}\right\}$$

At $x = 0$, SF at $A = \frac{WL}{n+1}$

At $x = L$, SF at $B = 0$.

Between A and B, the SF variation depends on the value 'n'. Slope of SF is loading, thus the tangent to the SFD should increase with 'x'.

Then, BM at section x distant from A is

$$M_x = -M_A + R_A x - \int_0^x (x-\alpha)W(\alpha)d\alpha$$

$$= -\frac{1}{n+2}WL^2 + \frac{1}{n+1}WLx - \int_0^x \frac{W}{L^n}\alpha^n (x-\alpha)dx$$

$$\xrightarrow{\text{yields}} M_x = -\frac{1}{n+2}WL^2 + \frac{1}{n+1}WLx - \frac{Wx^{n+2}}{(n+1)(n+2)L^n}$$

$$\xrightarrow{\text{yields}} M_x = \frac{WL^2}{(n+2)(n+1)}\left\{-(n+1)+(n+2)\left(\frac{x}{L}\right)-\left(\frac{x}{L}\right)^{n+2}\right\}$$

At $x = 0$, BM at $A = -\dfrac{WL^2}{n+2}$ (it is hogging moment)

At $x = L$, BM at $B = 0$.

Between A and B, the BM variation is nonlinear and it depends on the value 'n'. Slope of BM is SF, thus the tangent to the BMD should increase with 'x'.

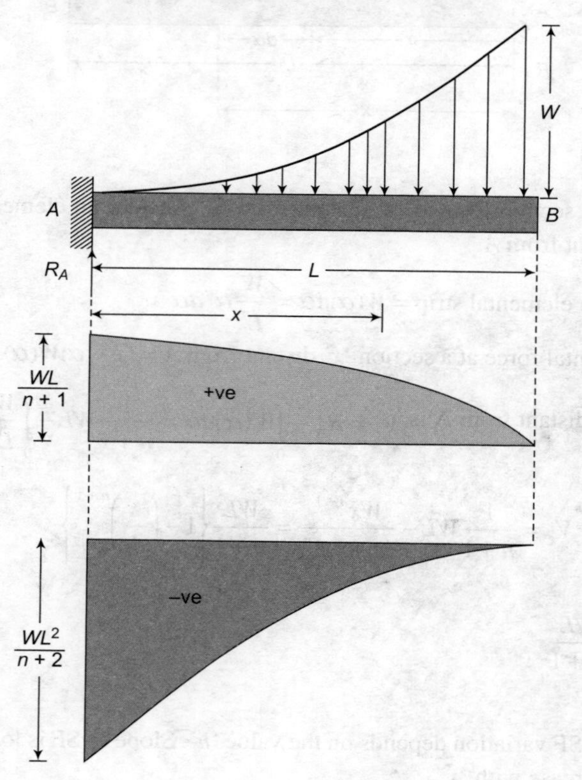

FIGURE 3.39

PROBLEM 3.8

Objective 2, 3

A double cantilever beam of length L carries overhanging portions of 'a'; each carries UDL over its entire span. Determine the length of the overhanging portion 'a' in terms of beam length L for the following conditions:

(a) Maximum –ve BM is equal to maximum +ve BM
(b) Maximum +ve BM is zero or no +ve BM
(c) Maximum –ve BM is zero or no –ve BM
(d) In the above three cases, for which case the design BM is less.

SOLUTION

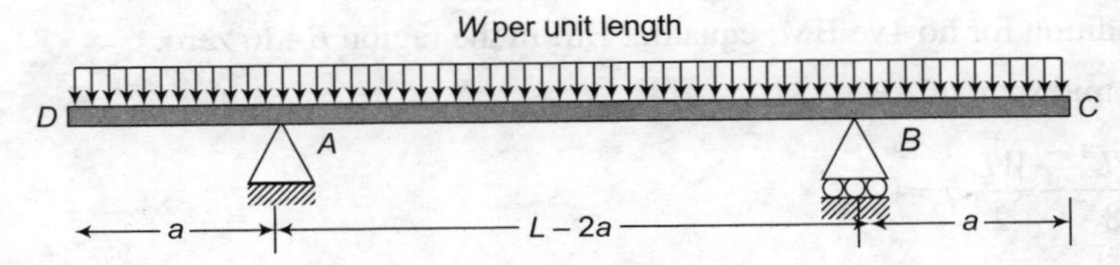

FIGURE 3.40

Vertical reaction at A and B = $R_A = WL/2$

$$R_B = WL/2$$

BM in the region BC, consider a section located at a distant 'x' from C

$$M_X = -\frac{WX^2}{2}$$

BM in the region BA, consider a section located at a distant 'x' from C

$$M_X = -\frac{WX^2}{2} + \frac{WL}{2}(X - a).$$

Maximum negative BM occurs at support B, that is, $x = a$.

$$M_{\max} = -\frac{Wa^2}{2}.$$

For maximum positive BM, which occurs in the region BA

$$\frac{dM}{dx} = -Wx + \frac{WL}{2} = 0$$

$$\Rightarrow \qquad x = \frac{L}{2}$$

$$M_{\max} = -\frac{WL^2}{8} + \frac{WL^2}{4} - \frac{WL}{2}a$$

$$= \frac{WL^2}{8} - \frac{WL}{2}a$$

Case (a): Max +ve BM = Max –ve BM

As we are equating the maximum +ve BM and maximum –ve BM, the sign associated with negative BM can be dropped down.

$$\frac{WL^2}{8} - \frac{WL}{2}a = \frac{Wa^2}{2}$$

$$\Rightarrow \quad 4a^2 + 4la - L^2 = 0$$

$$\Rightarrow \quad a = 0.207 \text{ L}$$

$$\text{BM} = 0.021\, WL^2.$$

Case (b): Condition for no +ve BM; equating BM in the region *BA* to zero.

This occurs at mid span of *AB*

$$\frac{WL^2}{8} - \frac{WL}{2}a = 0$$

$$\Rightarrow \quad a = \frac{L}{4}$$

$$\text{BM} = \frac{Wa^2}{2} = \frac{WL^2}{32} = 0.03125\, WL^2.$$

Case (c): Maximum –ve BM = 0

$$\frac{Wa^2}{2} = 0$$

$$\Rightarrow \quad a = 0$$

$$\text{BM} = \frac{WL^2}{8} = 0.125\, WL^2.$$

Case (d): From considering the above three cases, the design BM is minimum for case (a), for economical design maximum positive BM.

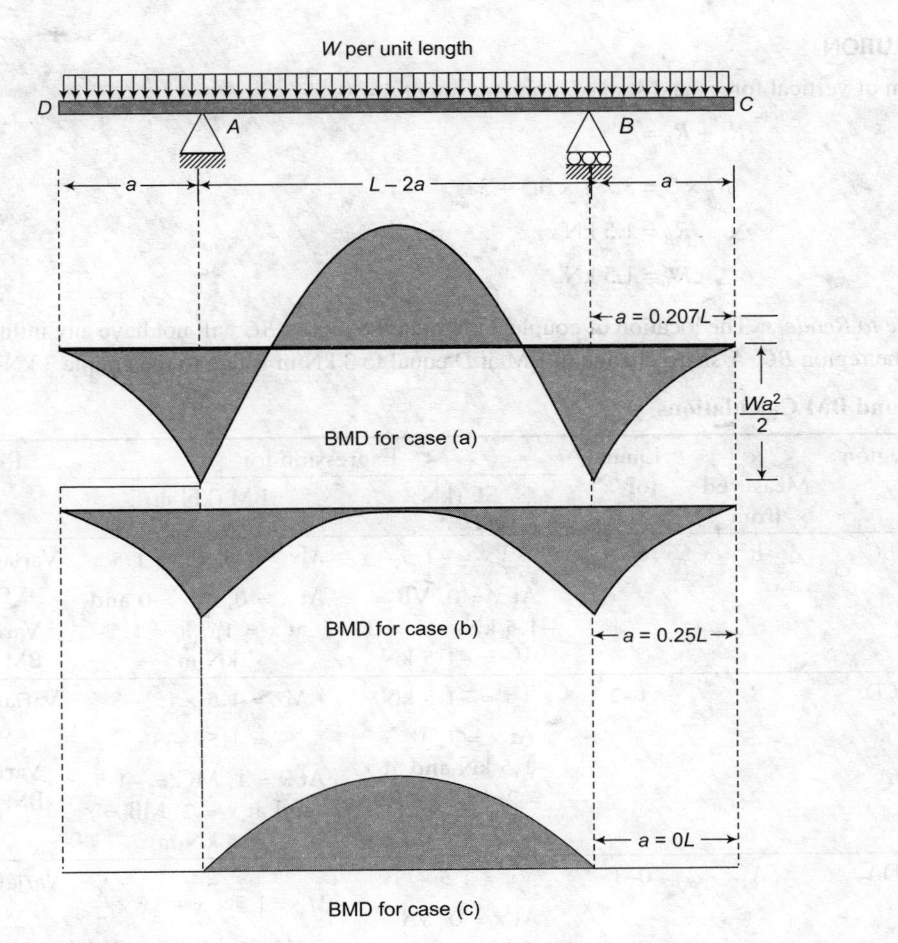

FIGURE 3.41

PROBLEM 3.9 Objective 2, 3

Sketch BMD and SFD for simply support beam shown in Figure 3.42.

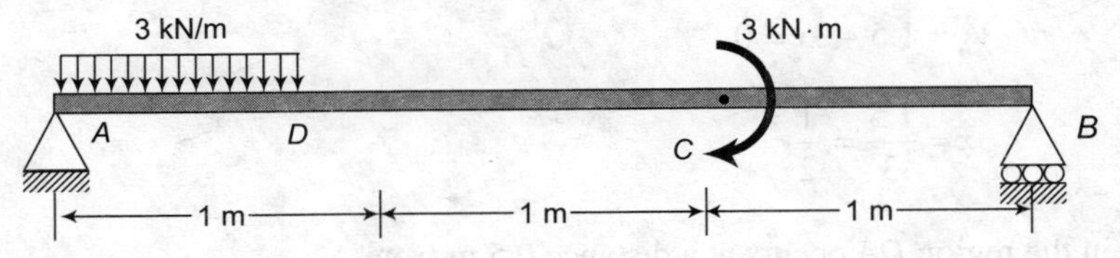

FIGURE 3.42

SOLUTION

Sum of vertical forces and sum of moments about $A = 0$.

$$R_A + R_B = 3$$

$$\Rightarrow \quad R_B \times 3 = 3 \times 1 \times 0.5 + 3$$

$$\Rightarrow \quad R_B = 1.5 \text{ kN}$$

$$R_A = 1.5 \text{ kN.}$$

Note to Readers: The location of couple 3 kN·m in the region *BC* will not have any influence on SF in the region *BC*. A sharp change in BM at *D* equal to 3 kN·m is due to the couple 3 kN·m.

SF and BM Calculations

Region	'x' Measured from	Limits for 'x'	Expression for SF (kN)	Expression for BM (kN·m)	Remarks
BC	*B*	0–1	$Vx = -1.5$ At $x = 0$, $VB = -1.5$ kN; at $x = 1$, $V_C = -1.5$ kN	$Mx = 1.5 \times x = 1.5\,x$ At $x = 0$, $MB = 0$ and at $x = 1$, $Mc = 1.5$ kN·m	Variation of SF is constant Variation of BM is linear
CD	*B*	1–2	$Vx = -1.5$ kN At $x = 1$, $V_C = -1.5$ kN and at $x = 2$; $VD = -1.5$ kN	$Mx = 1.5 \times (x) - 3$ $= 1.5x - 3$ At $x = 1$, $MC = -1.5$ and at $x = 2$, $MB = 1.5$ kN·m	Variation of SF is constant Variation of BM is linear
DA	*A*	0–1	$Vx = 1.5 - 3x$ At $x = 0$, $VA = 1.5$ kN and at $x = 1$, $VD = -1.5$ kN	$M_x = 1.5 \times x - 3x \times \frac{x}{2}$ $= 1.5x - 0.75 \times 2$ At $x = 0$, $MA = 0$ and at $x = 1$, $MA = 0$ kN·m	Variation of SF is linear Variation of BM is parabolic

For maximum BM in the region *DA*, equate SF in the region *DA* to zero.

$$V_x = 1.5 - 3x = 0$$

$$x = \frac{1.5}{3} = \frac{1}{2}$$

Maximum BM in the region *DA* occurs at a distance 0.5 m from A.

Maximum BM in the region $DA = M = 1.5\left(\frac{1}{2}\right) - 3\left(\frac{1}{2}\right)\left(\frac{1}{2}\right)\frac{1}{2} = \frac{3}{8}$ kN·m

SFD and BMD

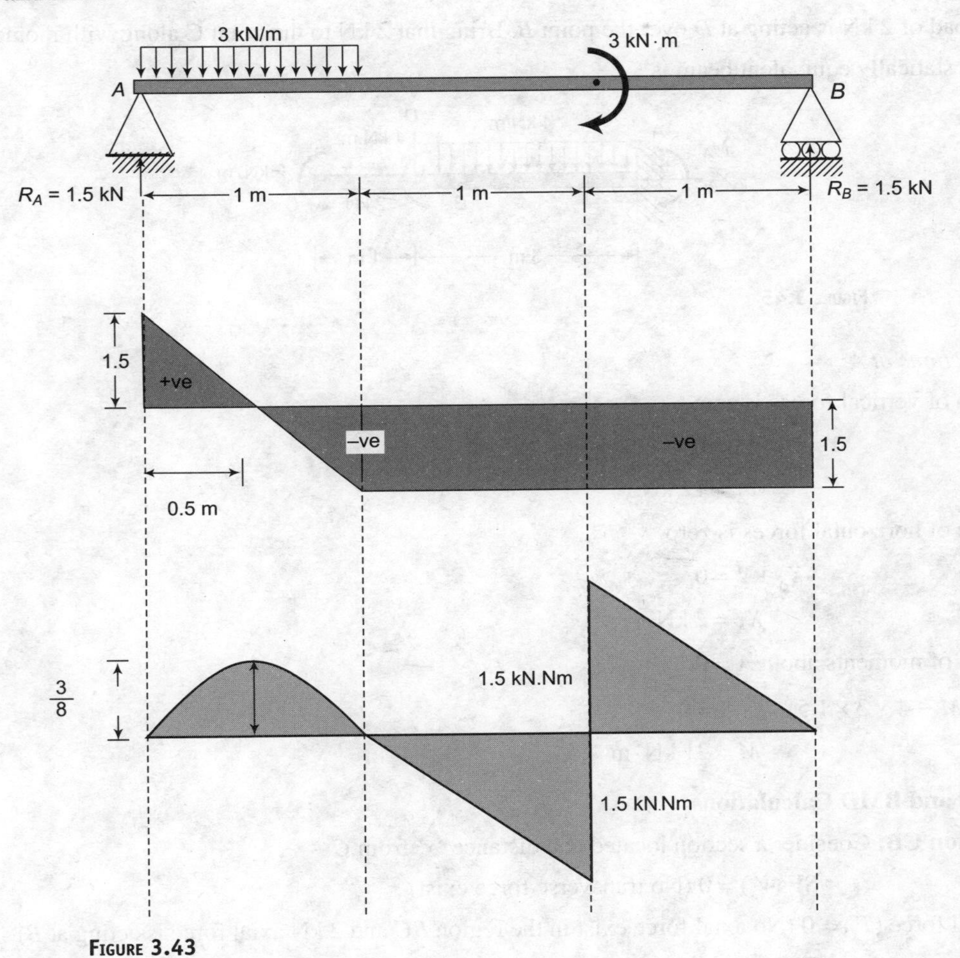

FIGURE 3.43

PROBLEM 3.10 **Objective 2, 3**

Sketch the SFD, BMD, and axial force diagram (AFD) for the beam shown in Figure 3.44.

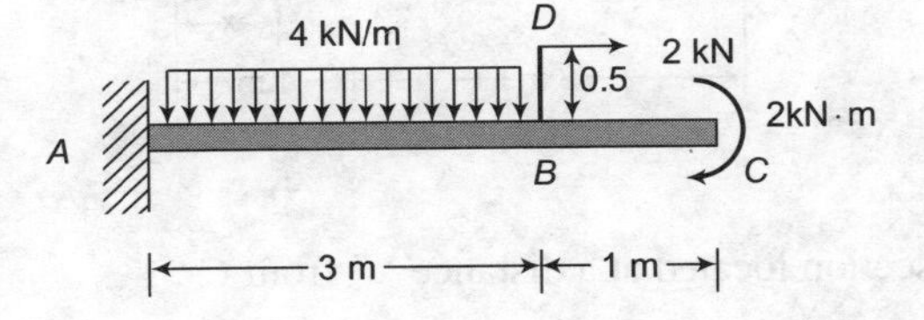

FIGURE 3.44

SOLUTION

A load of 2 kN is acting at D over the point B. Bring that 2 kN to the point C along with moment.

The statically equivalent beam is

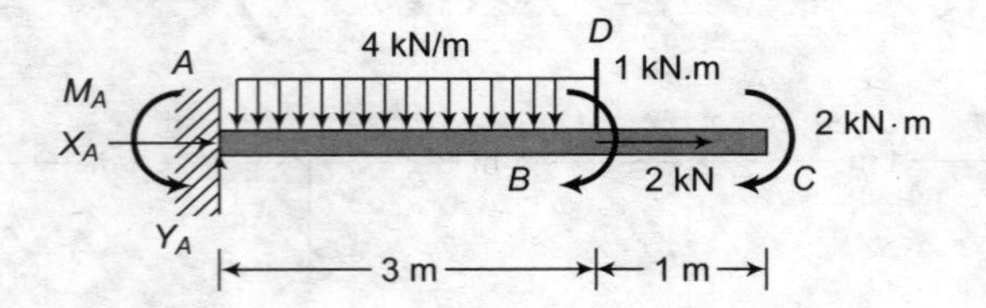

FIGURE 3.45

Reactions at A:

Sum of vertical forces is zero

$$Y_A - 4 \times 3 = 0$$

$$Y_A = 12 \text{ kN.}$$

Sum of horizontal forces is zero

$$X_A + 2 = 0$$

$$X_A = 2 \text{ kN.}$$

Sum of moments about $A = 0$

$$\Rightarrow \quad M - 4 \times 3 \times 1.5 - 2 - 1 = 0$$

$$\Rightarrow \qquad M = 21 \text{ kN}\cdot\text{m}$$

SFD and BMD Calculations

Region CB: Consider a section located at a distance 'x' from C

SF (V_x) = 0 (No transverse force exist)

Axial force (T_x) = 0 (No axial force exist in the region BC and 2 kN axial force is acting at B)

BM (M_x) = −2 kN·m (creates hogging moment)

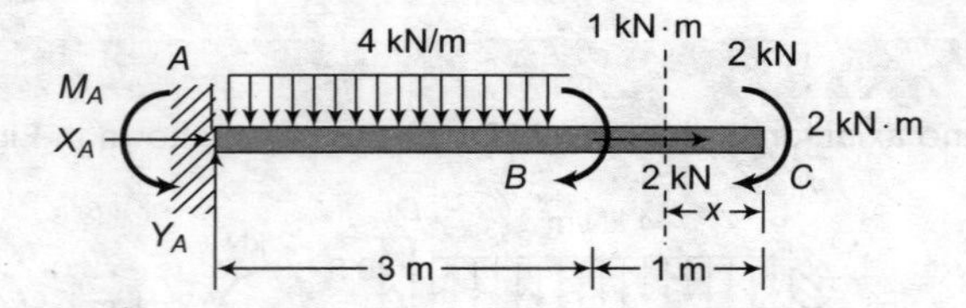

FIGURE 3.46

Region AB: Consider a section located at a distance 'x' from C

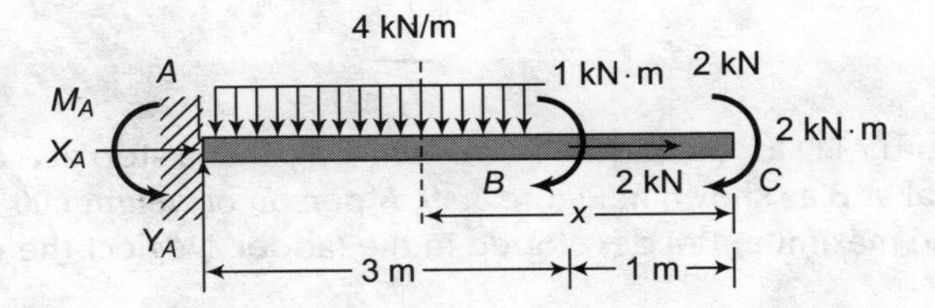

FIGURE 3.47

$$V_X = 4x$$

$$Tx = 2 \text{ kN (Tension)}$$

$$M_x = -2 - 1 - 4 \times (x-1)\frac{(x-1)}{2}$$

$$x = 1, \quad M_B = -3 \text{ kN-m}$$

$$x = 4, \quad M_A = -21 \text{ kN-m.}$$

In Figure 3.48, *AFD* indicates that it is the axial force diagram.

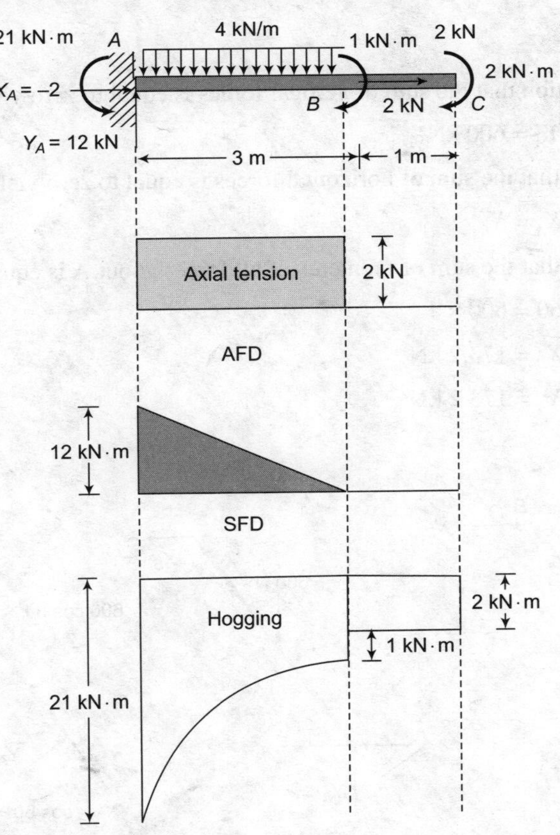

FIGURE 3.48

PROBLEM 3.11

Objective 2, 3

Sketch the SFD, BMD, and AFD for the ladder AB resisting against a step at A on the rough horizontal floor and smooth vertical at B as shown in Figure 3.49. A person of weight 600 N stands at the middle of ladder. Determine the maximum BM developed in the ladder. Neglect the weight of the ladder.

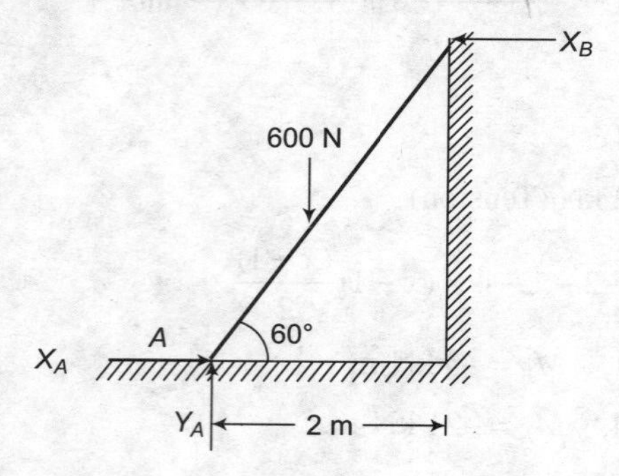

FIGURE 3.49

SOLUTION

The equilibrium condition that the sum of vertical forces is equal to zero yields

$$\Rightarrow \qquad Y_A = 600 \text{ kN}$$

Equilibrium condition that the sum of horizontal forces is equal to zero yields

$$\Rightarrow \qquad X_A = X_B$$

Equilibrium condition that the sum of moments of all forces about A is equal to zero yields

$$\Rightarrow \qquad X_B \times 2 \tan 60 = 600 \times 1$$

$$\Rightarrow \qquad X_B = 173.2 \text{ kN}$$

$$\Rightarrow \qquad X_A = 173.2 \text{ kN}$$

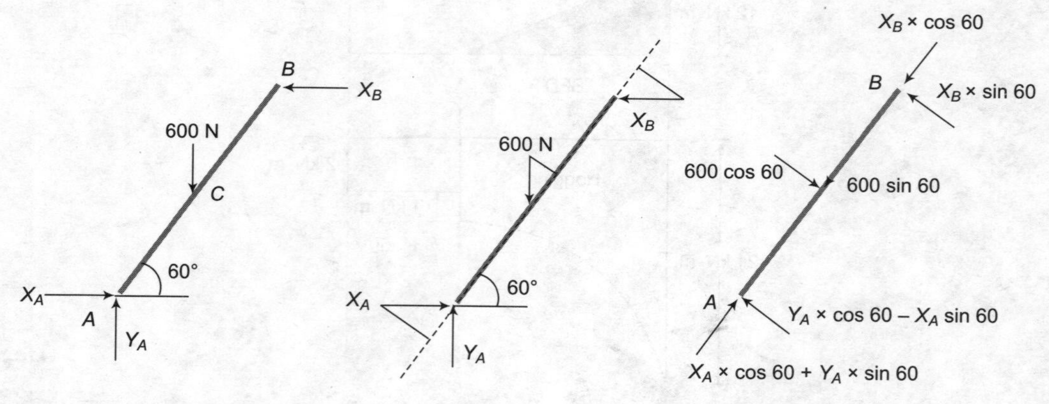

FIGURE 3.50

Resolving the forces at points A, B, and C along and across the ladder yields

At A: Force along the ladder = $X_A \times \cos 60 + Y_A \times \sin 60 = 173.21\cos 60 + 600 \sin 60 = 606.22$ N

Force across the ladder = $Y_A \times \cos 60 - X_A \times \sin 60 = 600 \cos 60 - 173.21 \sin 60 = 150$ N

At B: Force along the ladder = $X_B \times \cos 60 = 173.21\cos 60 = 86.61$ N

Force across the ladder = $X_B \times \sin 60 = 1732.21\sin 60 = 150$ N

At C: Force across the ladder = 600 cos 60 = 300 N

Force along the ladder = 600 sin 60 = 519.2 N.

Region BC: Consider a section located at a distance 'x' from B along the ladder.

SF V

$$V_x = -150 \text{ N}$$

At $x = 0$, $\quad V_B = -150$ N

At $x = 2$ m, $\quad V_C = -150$ N

Axial force T

$$T_x = -86.2 \text{ N}$$

At $x = 0$, $\quad T_B = -86.6$ N

At $x = 2$, $\quad T_C = -86.6$ N

BM M

$$M_B = 0$$

At $x = 0$, $\quad M_B = 0$

At $x = 2$, $\quad M_C = 300 \text{ N} \cdot \text{m}$

Region CA:

SF

$$V_x = 150 \text{ N}$$

$$V_A = V_C = 150 \text{ N}$$

Axial force

$$T_x = -606.2 \text{ N}$$

$$T_A = T_C = -606.2 \text{ N}$$

BM

$$M_x = 0$$

At $x = 0$, $\quad M_A = 0$

At $x = 2$, $\quad M_C = 300 \text{ N} \cdot \text{m}$

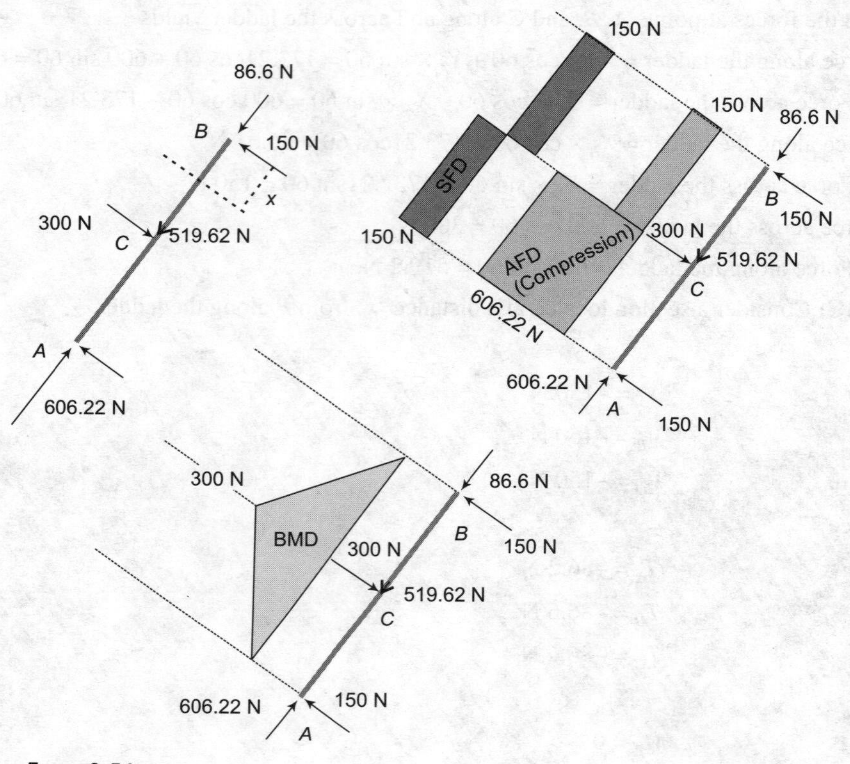

FIGURE 3.51

PROBLEM 3.12

Objective 2, 3

Sketch SFD and BMD for the beam ACB shown in Figure 3.52. End 'A' is hinged and CD is the supporting strut. A strut is an axially loaded member.

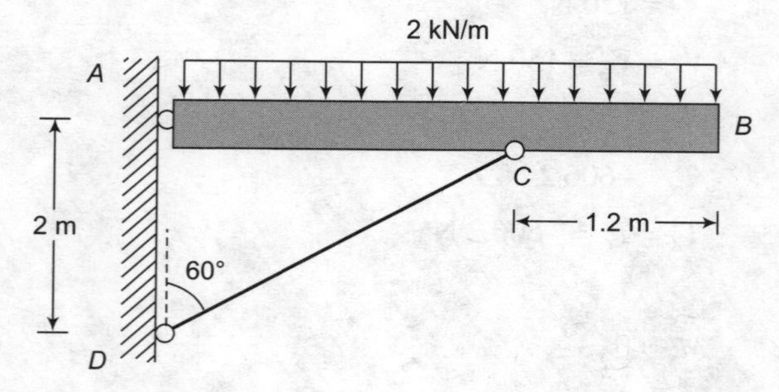

FIGURE 3.52

SOLUTION

Free body diagram (FBD) of the beam *ACB* is shown in Figure 3.53.

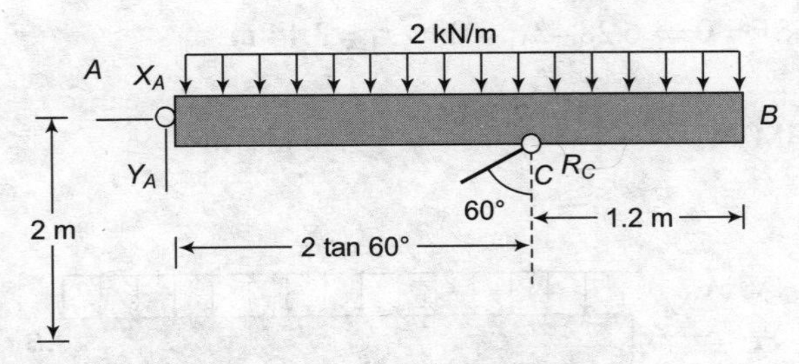

FIGURE 3.53

Sum of the vertical forces is zero.

$$\Rightarrow \qquad Y_A + R_C \cos 60 = 2 \times 2 \tan 60 + 2 \times 1.2$$

$$\Rightarrow \qquad Y_A + 0.5\, R_C = 9.328 \text{ kN}$$

Sum of the horizontal forces is zero.

$$X_A + R_C \cos 30 = 0$$

Sum of the moments about A is zero.

$$\Rightarrow \qquad R_C \cos 60 \times 2 \tan 60 = \frac{2}{2}(2 \tan 60 + 1.2)^2$$

$$\Rightarrow \qquad 1.732 R_C = 21.754$$

$$\Rightarrow \qquad R_C = 12.560 \text{ kN}$$

$$X_A = -10.877 \text{ kN}$$

$$Y_A = 6.280 \text{ kN}$$

The AFD, SFD, and BMD can be drawn even without writing the expression

At B, $T_B = 0$; $V_B = 0$; $M_B = 0$

At C, left side (just right of C)

$$T_C = 0; \quad V_C = 2 \times 1.2 = 2.4; \quad M_C = -2 \times 1.2 \times \frac{1.2}{2} - 1.44 \text{ kN-m.}$$

Just to the left side of C

$$T_C = R_C \cos 30 = 12.56 \cos 30 = 10.877 \text{ kN}$$

$$V_C = 2 \times 1.2 - R_C \cos 60 = 2.4 - 12.56 \cos 60 = -3.88 \text{ kN}$$

$$M_C = -\frac{2 \times (1.2)^2}{2} = -1.44 \text{ kN}\cdot\text{m}$$

At A:

$$T_A = -X_A = 10.877$$

$$V_A = 6.28 \text{ kN}$$

$$M_A = 0$$

$$SF = 0 \Rightarrow 6.28 - 2x_1 = 0 \Rightarrow x_1 = 3.14 \text{ m}$$

$$\text{Max BM} = \frac{6.28(3.14) - 2(3.14)^2}{2} = 9.86 \text{ kN}\cdot\text{m}$$

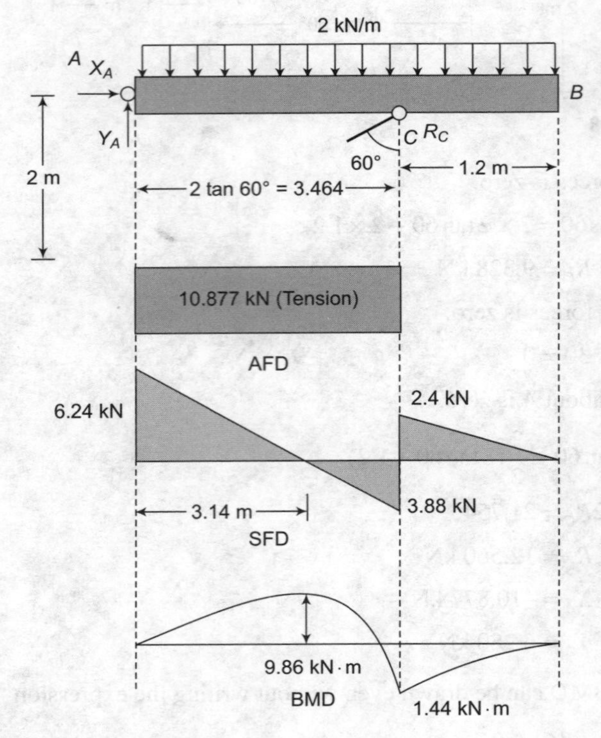

FIGURE 3.54

PROBLEM 3.13

Objective 2, 3

Sketch the AFD, SFD, and BMD for the frame shown in Figure 3.55.

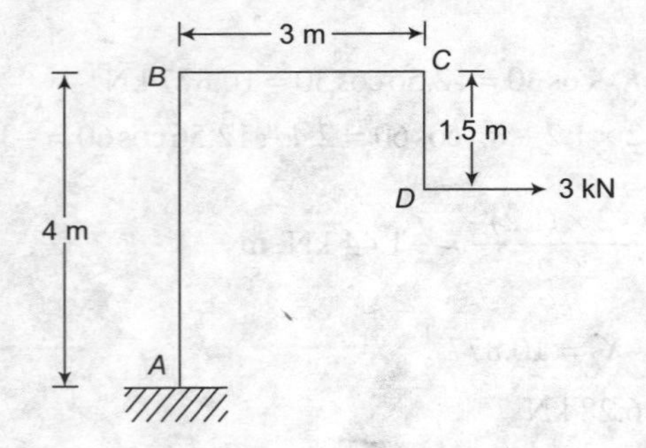

FIGURE 3.55

To draw AFD, SFD, and BMD of the total structure, draw FBD of individual elements, namely *DC*, *CB*, and *BA*.

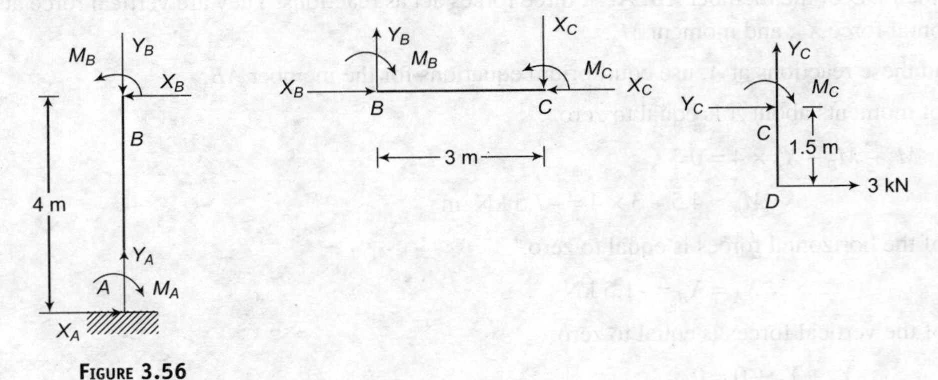

FIGURE 3.56

Take the FBD of the member *CD*. At *C*, three forces act as reactions. They are vertical force at *C* Y_C, horizontal force X_C, and moment M_C.

To find these reactions at *C*, use equilibrium equations for the member *CD*.

Sum of moments about *C* is equal to zero.

$$\Rightarrow \quad M_C - 3 \times 1.5 = 0$$

$$\Rightarrow \quad M_C = 4.5 \text{ kN} \cdot \text{m}$$

Sum of the horizontal forces is equal to zero.

$$X_C + 1.5 = 0$$

$$\Rightarrow \quad X_C = -1.5 \text{ kN}.$$

Sum of the vertical forces is equal to zero.

$$Y_C + 0 = 0$$

$$\Rightarrow \quad Y_C = 0.$$

Take the FBD of the member *BC*. At *B*, three forces act as reactions. They are vertical force at *B* Y_B, horizontal force X_B, and moment M_B.

To find these reactions at *B*, use equilibrium equations for the member *BC*.

Sum of moments about *B* is equal to zero.

$$M_B - M_C + Y_C \times 3 = 0$$

$$\Rightarrow \quad M_B = M_C = 4.5 \text{ kN} \cdot \text{m}$$

Sum of the horizontal forces is equal to zero.

$$X_B = X_C = -1.5 \text{ kN}$$

Sum of the vertical forces is equal to zero.

$$Y_C + Y_B + 0 = 0$$

$\Rightarrow \quad Y_B = 0.$

Take the FBD of the member *AB*. At *A*, three forces act as reactions. They are vertical force at *A* Y_A, horizontal force X_A, and moment M_A.

To find these reactions at *A*, use equilibrium equations for the member *AB*.

Sum of moments about *A* is equal to zero.

$$M_A - M_B - X_B \times 4 = 0$$

$$\Rightarrow \quad M_A = 4.5 - 3 \times 4 = -7.5 \text{ kN} \cdot \text{m}$$

Sum of the horizontal forces is equal to zero.

$$X_A = X_B = -1.5 \text{ kN}$$

Sum of the vertical forces is equal to zero.

$$Y_A + Y_B + 0 = 0$$

$$\Rightarrow \quad Y_A = 0.$$

To draw AFD, SFD, and BMD of the total structure ABCD, draw AFD, SFD, and BMD for individual members *CD*, *BC*, and *AB*.

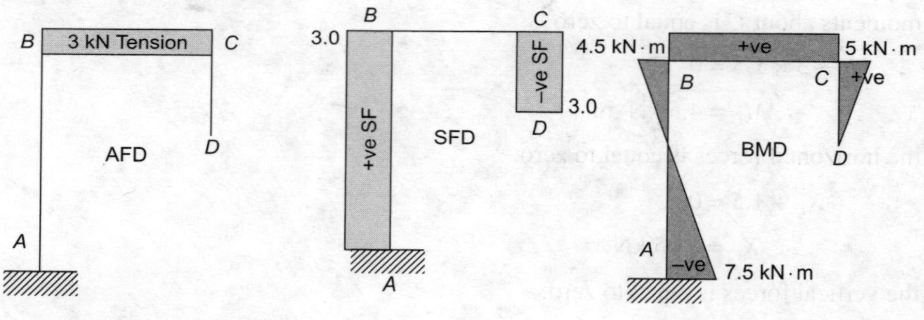

FIGURE 3.57

3.6 RELATIONSHIP BETWEEN LOAD, SF, AND BM

Better understanding of the relationship between these quantities help us a lot, while drawing the SFD and BMDs.

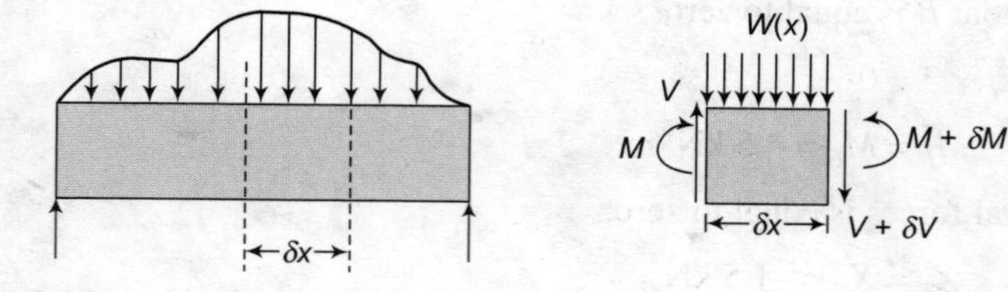

FIGURE 3.58

Consider an elemental portion of beam subjected to typical loading. FBD of the elemental portion is taken as shown in Figure 3.58. As the elemental portion is very small $W(x)$, over this portion can be taken as uniform. Applying equilibrium condition:

Sum of the vertical forces is equal to zero.

$$\Rightarrow V - W(x)\ \delta x - V - \delta V = 0$$

$$W(x) = \frac{\delta v}{\delta x}$$

$$V = \int W(x)dx.$$

This indicates that area of loading diagram yields SF.

Sum of the moments about left end of the FBD portion is equated to zero.

This yields

$$M + W(x)\delta x\frac{\delta x}{2} - (M + \delta M) + (V - \delta V)\delta x = 0.$$

Neglecting higher order terms such as $(x)\delta x\frac{\delta x}{2}$ and $(\delta V)\delta x$, the above expression reduces to

$$(\delta M) + (V)\ \delta x = 0$$

and

$$V = \frac{dM}{dx}.$$

Otherwise the expressions relating load, BM, and SF can be written as

$$M = \int V dx.$$

This indicates that area of SFD yields BM.

$$W(x) = \frac{d^2M}{dx}.$$

First derivate of moment variation gives the SF. Second derivate of moment variation gives the loading. First derivate of SF variation gives the loading.

Important things to remember

Area of SFD gives BM.

Slope of BMD gives the SF.

PROBLEM 3.15 **Objective 2, 3**

Sketch the SFD and BMD for the beam ABCD. A is hinged to the vertical wall while the cable EF holds the member through a rigid prop. There acts a UDL of intensity 3 kN/m over BC and a concentrated load of 4 kN at D.

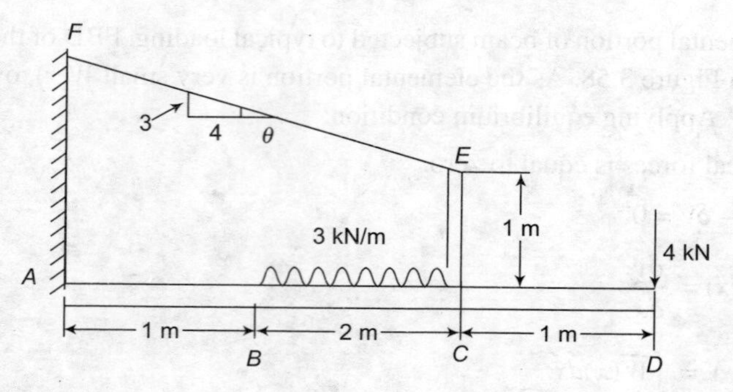

FIGURE 3.59

SOLUTION

Draw the FBD of the beam *ABCD*.

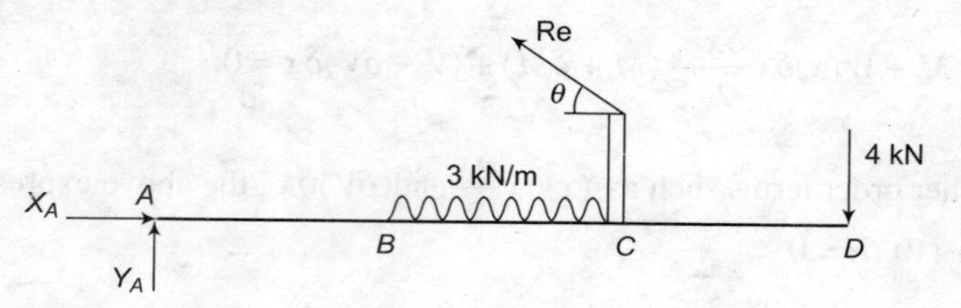

FIGURE 3.60

Apply equilibrium equations.

Sum of vertical forces, horizontal forces, and moment about *A* is zero.

$\Rightarrow \quad Y_A + R_e \sin\theta = 3\times 2 + 4 = Y_A + 0.6R_e = 10$

$\Rightarrow \quad X_A - R_e \cos\theta = 0 \Rightarrow X_A = 0.8R_e$

$\Rightarrow \quad 3\times 2\times 2 + 4\times 4 = R_e \cos\theta \times 1 + R_e \sin\theta \times 3$

$\Rightarrow \quad R_e = 10.769 \text{ kN}$

$\Rightarrow \quad X_A = 8.615 \text{ kN}$

$\Rightarrow \quad Y_A = 3.538 \text{ kN}$

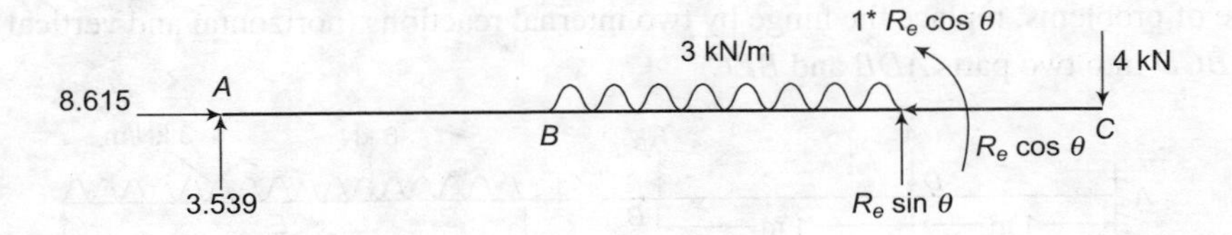

Figure 3.61

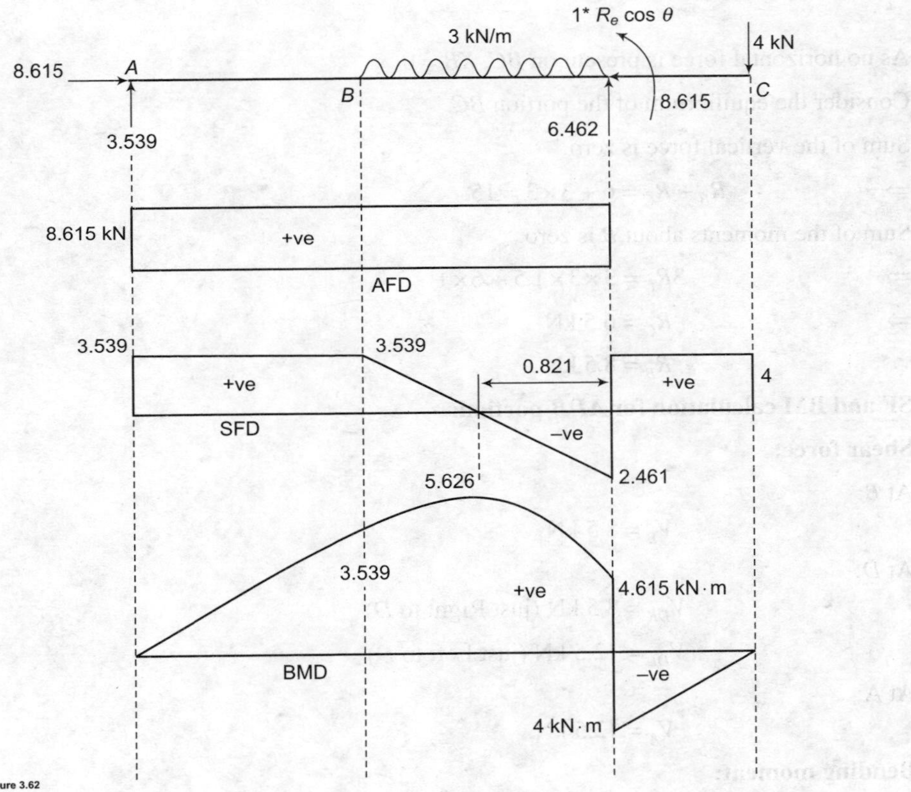

Figure 3.62

PROBLEM 3.16

Objective 2, 3

Sketch the SFD and BMD of the articulated beam shown in Figure 3.63.

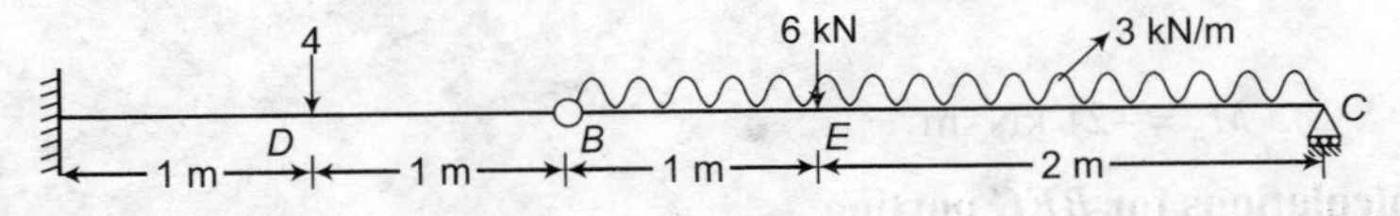

FIGURE 3.63

SOLUTION

An articulation is an internal hinge in a beam. An articulation is a joint capable of transferring shear and axial force but not moment. Thus, at an articulation or internal hinge BM is zero. To analyze this type of problems, replace the hinge by two internal reactions (horizontal and vertical). Split the beam *ABCD* into two parts *ADB* and *BEC*.

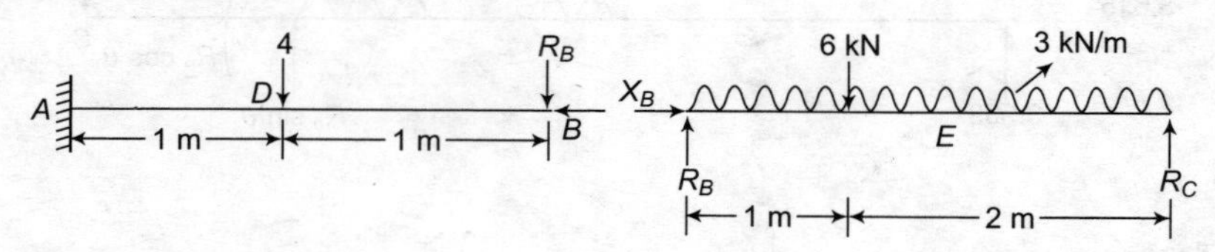

FIGURE 3.64

As no horizontal force is present on *BC*, *XB* = 0

Consider the equilibrium of the portion *BC*.

Sum of the vertical force is zero.

$$\Rightarrow \quad R_B + R_C = 6 + 3 \times 3 = 15.$$

Sum of the moments about *B* is zero.

$$\Rightarrow \quad 3R_C = 3 \times 3 \times 1.5 + 6 \times 1$$

$$\Rightarrow \quad R_C = 6.5 \text{ kN}$$

$$\therefore \quad R_B = 8.5 \text{ kN}$$

SF and BM calculation for *ADB* portion:

Shear force:

At *B*:

$$V_B = 8.5 \text{ kN}$$

At *D*:

$$V_{DR} = 8.5 \text{ kN (just Right to } D)$$

$$V_{DL} = 12.5 \text{ kN (just Left to } D)$$

At A:

$$V_A = 12.5 \text{ kN.}$$

Bending moment:

At B:

$$M_B = 0$$

At D:

$$M_D = -8.5 \text{ kN} \cdot \text{m}$$

At A:

$$M_A = -21 \text{ kN} \cdot \text{m}$$

SF and BM calculations for *BEC* portion:

Shear force:

At *C*:

$$V_C = -6.5 \text{ kN}$$

At *E*:

$$V_{EL} = -6.5 + 3\times2$$
$$V_{ER} = -6.5 + 3\times2 + 6$$
$$= 5.5 \text{ kN}$$

At *B*:

$$V_B = -6.5 + 3\times + 6$$
$$= 8.5 \text{ kN}.$$

Bending moment:

At *C*:

$$M_C = 0$$

At *E*:

$$M_E = 6.5\times2 - 3\times\frac{2^2}{2}$$
$$= 7 \text{ kN}\cdot\text{m}$$

At *B*:

$$M_B = 6.5\times3 - 3\times3\times1.5 - 6\times1$$
$$= 0$$

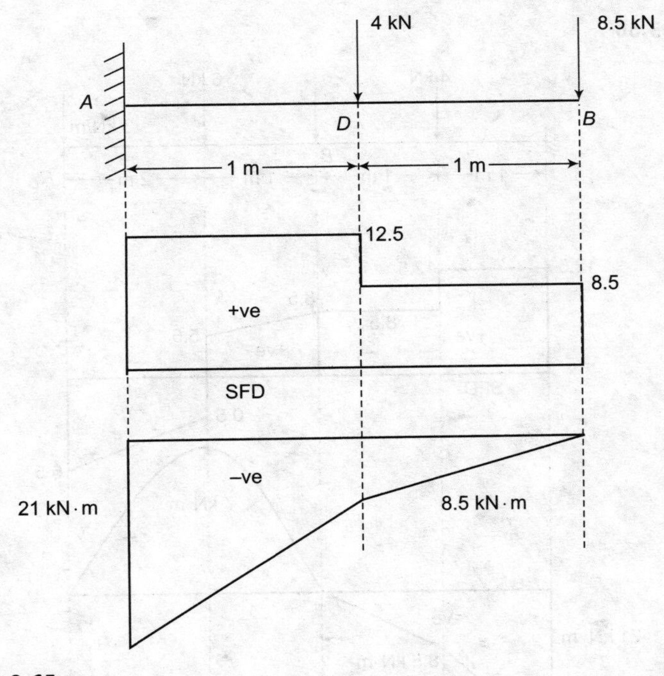

FIGURE 3.65

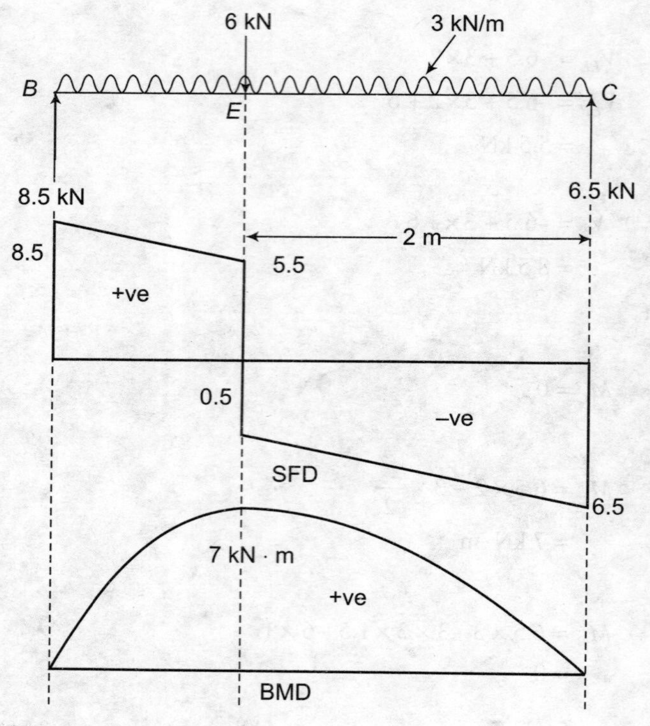

FIGURE 3.66

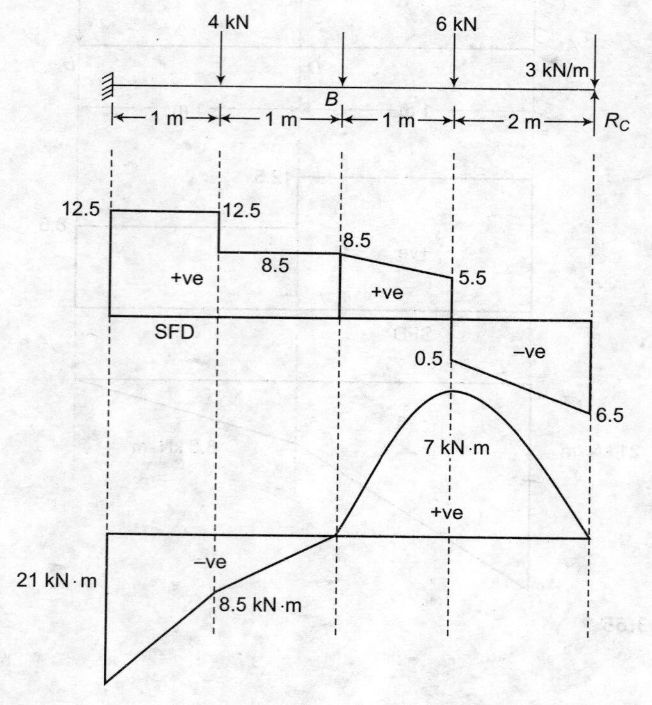

FIGURE 3.67

The BMD and SFD of the total structure can be obtained by joining the SFD and BMD of ADB and BEC portions as shown in Figure 3.67.

PROBLEM 3.17 **Objective 2, 3**

A simply supported beam of length 5 m carries a uniformly increasing load of 800 N/m run at one end to 1600 N/m run at the other end. Draw the SFD and BMD for the beam.

SOLUTION

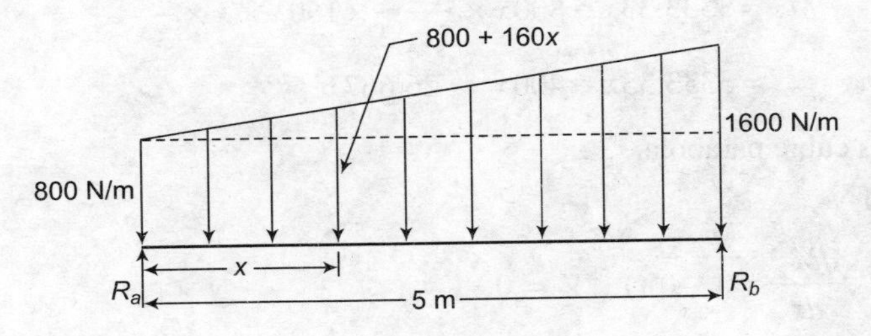

FIGURE 3.68

$$\sum V = 0 \Rightarrow R_a + R_b = \text{Area of loading diagram}$$

$$= \left(\frac{800+1600}{2}\right) \times 5 = 6000$$

Taking moments about A, $\Sigma M_A = 0$

$\Rightarrow$ $R_b \times 5 =$ Moment of area of loading diagram about A

Divide the loading diagram into a rectangular portion and triangular portion.

$$R_b \times 5 = 800 \times 5 \times \frac{5}{2} + \frac{1}{2} \times 800 \times 5 \times \left(\frac{2}{3} \times 5\right)$$

$$R_b = 3333.33\,\text{N}; \quad R_a = 6000 - R_b$$

$\Rightarrow$ $$R_a = 2666.67\ \text{N}$$

Consider a section located X distance from A load at this section.

$$W(x) = 800 + \frac{160 - 800}{5} x = 800 + 160x$$

SF at this section

$$V_x = R_a - \text{Area of loading diagram upto } X$$

$$= 3333.33 - 800 \times x \times -\frac{1}{2}(160 \times x \times x)$$

$$= 3333.33 - 800x - 80x^2$$

SF at $x = 0 \Rightarrow V_A = 3333.33$ N

SF at $x = 5 \Rightarrow V_B = -2666.67$ N

Between A and B, SF variation is parabolic.

BM at the section located X distance from A.

$$M_x = R_a x - \text{Moment of area of loading diagram upto } X$$

$$M_x = 3333.33x - 800x \times \frac{x}{2} - \frac{1}{2} \times 160x \times x \times \frac{x}{3}$$

$$= 3333.33x - 400x^2 - 26.667x^3$$

Variation of BM is cubic parabola.

For maximum BM,

$$\frac{dM_x}{dx} = 0 \quad \text{(or)} \quad V_x = 0$$

$$\Rightarrow 3333.33 - 800x - 80x^2 = 0$$

$$\Rightarrow \quad x = 3.165 \text{ m}$$

$$M_{\max} = 3333.33(3.165) - 400(3.165)^2 - 26.667(3.165)^3 = 5697.65 \text{ N}\cdot\text{m}$$

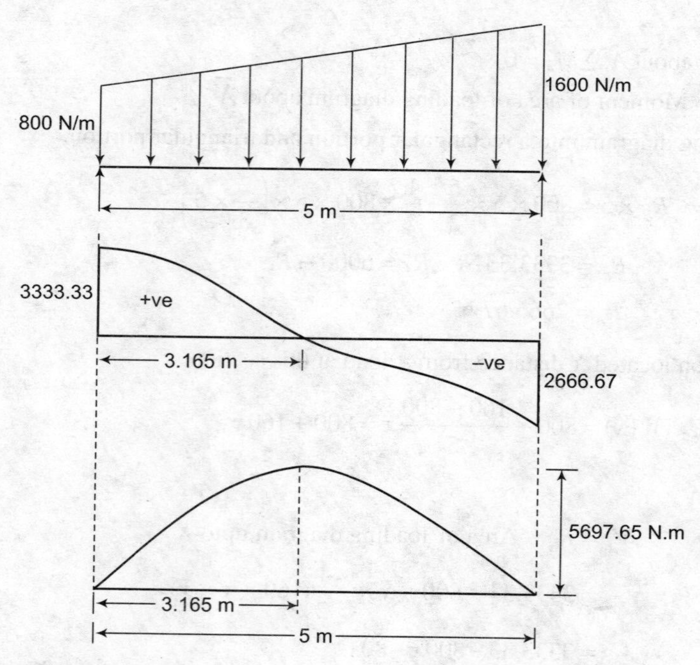

FIGURE 3.69

SUMMARY

- **SF at a section is the resultant vertical force to the right or left of the section.**
- **The diagram that shows the variation of SF along the length of a beam is known as SFD.**
- **BM at a section is algebraic sum of moments of all the forces acting to the left or right of the section.**
- **The diagram that shows the variation of BM along the length of a beam is known as BMD.**
- **If the end portion of beam is extended beyond the support, then it is known as overhanging beam.**
- **The SF changes suddenly at a section where there is a vertical point load.**
- **The SF between any two vertical loads remains constant.**
- **BM is maximum at a section where SF is zero after changing its sign.**

OBJECTIVE TYPE QUESTIONS

1. A concentrated force F is applied (perpendicular to the plane of the figure) on the tip of the bent bar shown in the figure. The equivalent load at a section close to the fixed end is: [GATE 1999]

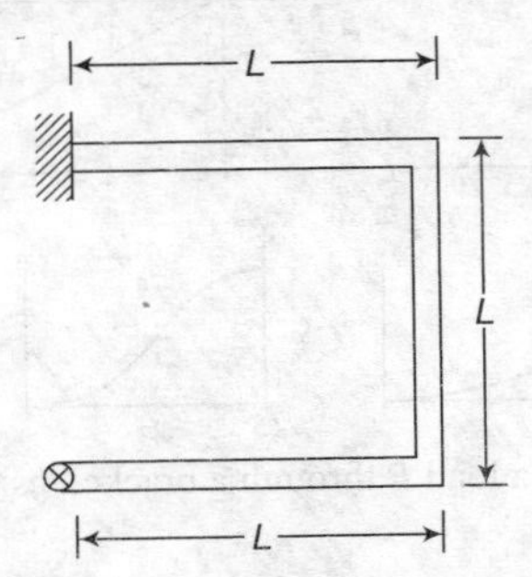

(a) Force F (b) Force F and BM FL
(c) Force F and twisting moment FL (d) Force F BM FL, and twisting moment FL

2. The SF in a beam subjected to pure positive bending is [GATE-1995]
(a) Positive (b) Zero (c) Negative

3. Two identical cantilever beams are supported as shown, with their free ends in contact through a rigid roller. After the load P is applied, the free ends will have [GATE-2005]

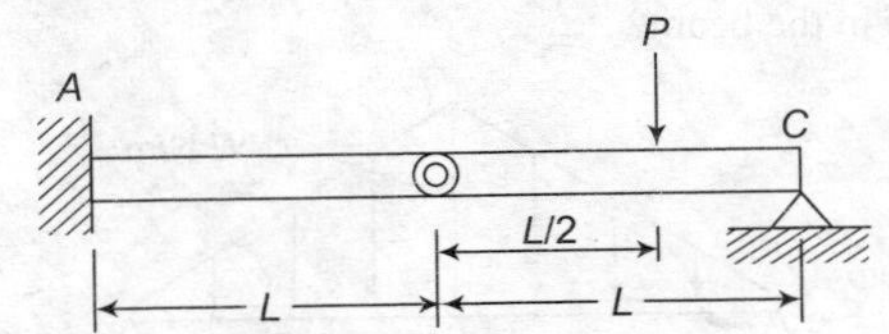

(a) Equal deflections but not equal slopes
(b) Equal slopes but not equal deflections
(c) Equal slopes as well as equal deflections
(d) Neither equal slopes nor equal deflections

4. A beam is made up of two identical bars *AB* and *BC*, by hinging them together at *B*. The end *A* is built-in (cantilevered) and the end *C* is simply-supported. With the load *P* acting as shown, the BM at *A* is: [GATE-2005]

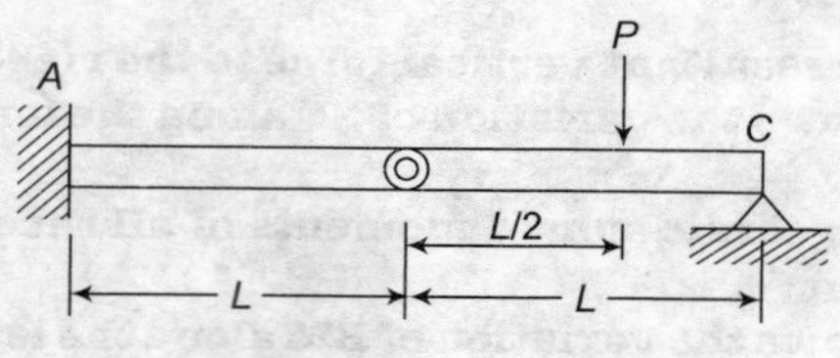

(a) Zero (b) $PL/2$
(c) $3PL/2$ (d) Intermediate indeterminate

5. The shapes of the BMD for a uniform cantilever beam carrying a UDL over its length is: [GATE-2001]

(a) A straight line (b) A hyperbola (c) An ellipse (d) A parabola

6. A cantilever beam carries the antisymmetric load shown, where ω_0 is the peak intensity of the distributed load. Qualitatively, the correct BMD for this beam is: [GATE-2005]

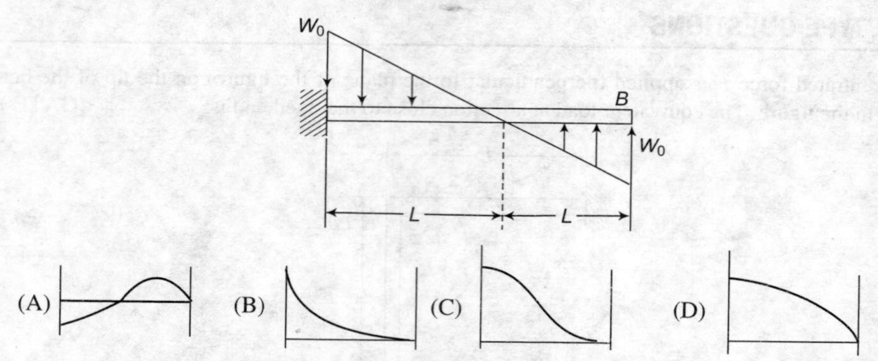

7. A simply supported beam carries a load *P* through a bracket, as shown in the figure. The maximum BM in the beam is [GATE-2000]

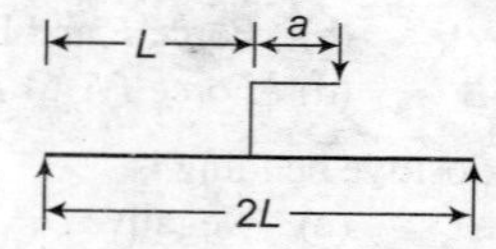

(a) $PL/2$ (b) $PL/2 + aP/2$ (c) $PL/2 + aP$ (d) $PL/2 - aP$

8. A simply supported beam is subjected to a distributed loading as shown in the diagram given below: What is the maximum SF in the beam? [IES-2004]

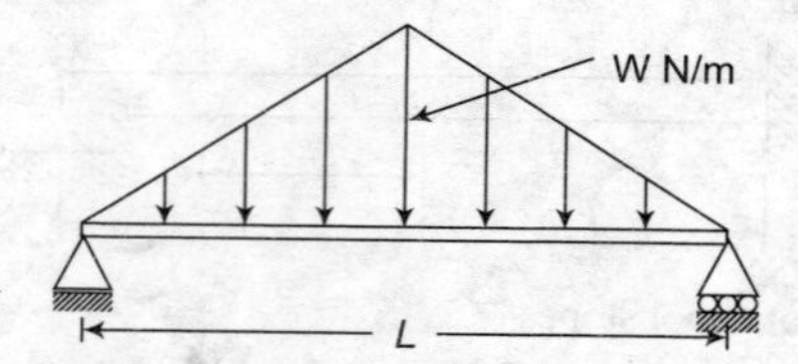

(a) $WL/3$ (b) $WL/2$ (c) $WL/3$ (d) $WL/6$

9. A lever is supported on two hinges at *A* and *C*. It carries a force of 3 kN as shown in the above figure. The BM at *B* will be [IES-1998]

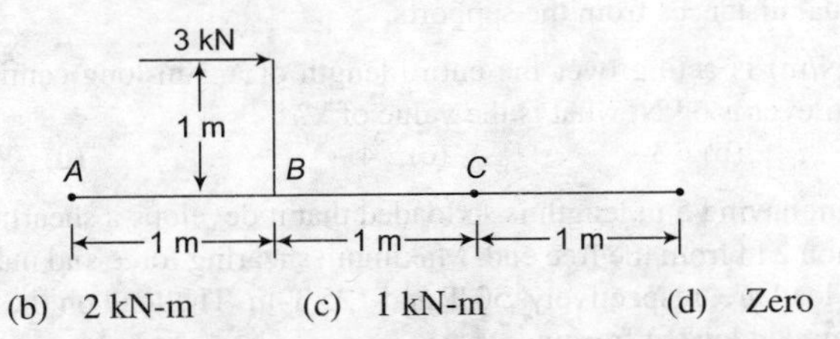

(a) 3 kN-m (b) 2 kN-m (c) 1 kN-m (d) Zero

10. A beam subjected to a load *P* is shown in the given figure. The BM at the support *AA* of the beam will be [IES-1997]

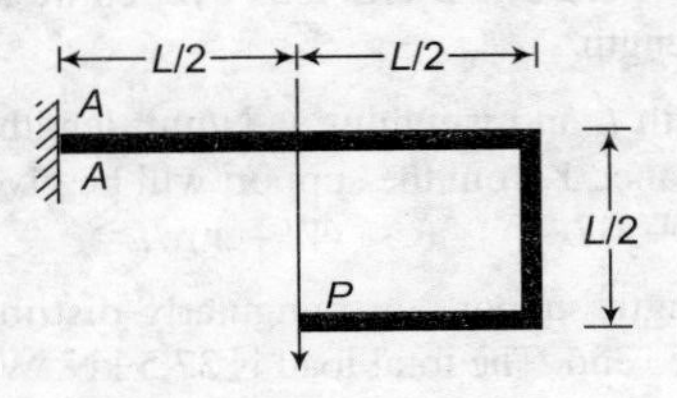

(a) *PL* (b) *PL*/2 (c) 2*PL* (d) Zero

11. The BM(M) is constant over a length segment (I) of a beam. The shearing force will also be constant over this length and is given by [IES-1996]

(a) M/l (b) $M/2l$ (c) $M/4l$ (d) None of the above

12. A rectangular section beam subjected to a BM M varying along its length is required to develop same maximum bending stress at any cross-section. If the depth of the section is constant, then its width will vary as [IES-1995]

(a) M (b) $M^{1/2}$ (c) M^2 (d) $1/M$

13. Consider the following statements:

If at a section distant from one of the ends of the beam, *M* represents the BM. *V* is the SF and *w* is the intensity of loading, then

1. $dM/dx = V$
2. $dV/dx = s$
3. $dw/dx = y$ (the deflection of the beam at the section)

Select the correct answer using the codes given below: [IES-1995]

(a) 1 and 3 (b) 1 and 2 (c) 2 and 3 (d) 1, 2, and 3

14. The given figure shows a beam *BC* simply supported at *C* and hinged at *B* (free end) of a cantilever *AB*. The beam and the cantilever carry forces of 100 kg and 200 kg, respectively. The BM at *B* is: [IES-1995]

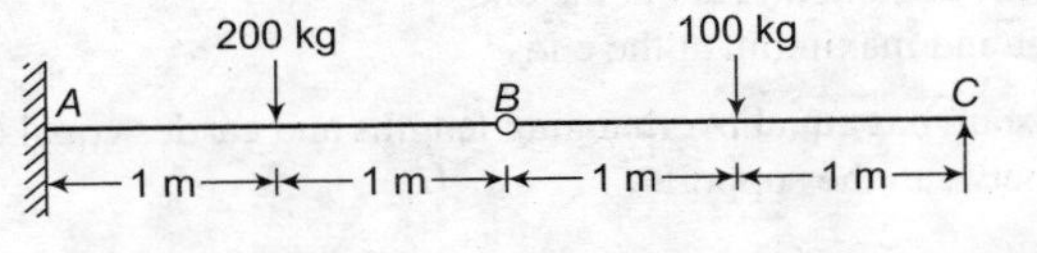

(a) Zero (b) 100 kg-m (c) 150 kg-m (d) 200 kg-m

15. If the SF acting at every section of a beam is of the same magnitude and of the same direction then it represents a [IES-1996]

(a) Simply supported beam with a concentrated load at the center.

(b) Overhung beam having equal overhang at both supports and carrying equal concentrated loads acting in the same direction at the free ends.

(c) Cantilever subjected to concentrated load at the free end.
(d) Simply supported beam having concentrated loads of equal magnitude and in the same direction acting at equal distances from the supports.

16. A UDL x (in kN/m) is acting over the entire length of a 3-m-long cantilever beam. If the SF at the midpoint of cantilever is 6 kN, what is the value of x? [IES-2009]
(a) 2 (b) 3 (c) 4 (d) 5

17. A cantilever beam having 5 m length is so loaded that it develops a shearing force of 20 T and a BM of 20 T-m at a section 2 m from the free end. Maximum shearing force and maximum BM developed in the beam under this load are, respectively, 50 T and 125 T-m. The load on the beam is: [IES-1995]
(a) 25 T concentrated load at free end
(b) 20 T concentrated load at free end
(c) 5 T concentrated load at free end and 2 T/m load over entire length
(d) 10 T/m UDL over entire length

18. A vertical hanging bar of length L and weighing w N/unit length carries a load W at the bottom. The tensile force in the bar at a distance Y from the support will be given by [IES-1992]
(a) $W + wL$ (b) $W + w(L - y)$ (c) $(W + w)y/L$ (d) $W + W/w\,(L - y)$

19. A cantilever beam of 2 m length supports a triangularly distributed load over its entire length, the maximum of which is at the free end. The total load is 37.5 kN. What is the BM at the fixed end? [IES-2007]
(a) 50×10^6 N·mm (b) 12.5×10^6 N·mm
(c) 100×10^6 N·mm (d) 25×10^6 N·mm

20. Assertion (A): If the BM along the length of a beam is constant, then the beam cross-section will not experience any shear stress. Reason (R): The SF acting on the beam will be zero everywhere along the length. [IES-1998]
(a) Both *A* and *R* are individually true and *R* is the correct explanation of *A*
(b) Both *A* and *R* are individually true but *R* is NOT the correct explanation of *A*
(c) *A* is true but *R* is false
(d) *A* is false but *R* is true

21. Assertion (A): If the BMD is a rectangle, it indicates that the beam is loaded by a uniformly distributed moment all along the length. Reason (R): The BMD is a representation of internal forces in the beam and not the moment applied on the beam. [IES-2002]
(a) Both *A* and *R* are individually true and *R* is the correct explanation of *A*
(b) Both *A* and *R* are individually true but *R* is NOT the correct explanation of *A*
(c) A is true but *R* is false
(d) A is false but *R* is true

22. If a beam is subjected to a constant BM along its length, then the SF will [IES-1997]
(a) Also have a constant value everywhere along its length
(b) Be zero at all sections along the beam
(c) Be maximum at the center and zero at the ends
(d) Zero at the center and maximum at the ends

23. A simply supported beam has equal overhanging lengths and carries equal concentrated loads *P* at ends. BM over the length between the supports [IES-2003]
(a) Is zero
(b) Is a nonzero constant
(c) Varies uniformly from one support to the other
(d) Is maximum at mid-span

24. The BMD for the case shown below will be q as shown in [IES-1992]

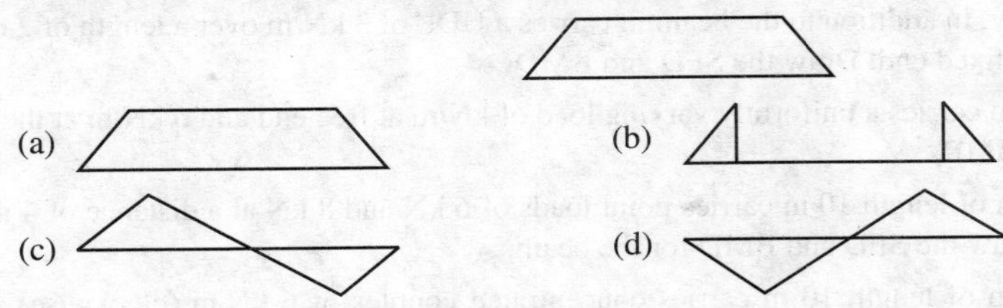

25. Which one of the following portions of the loaded beam shown in the given figure is subjected to pure bending? [IES-1999]

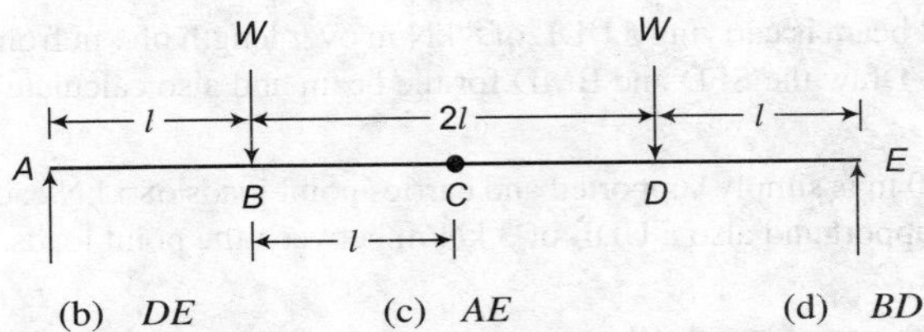

(a) AB (b) DE (c) AE (d) BD

26. Constant BM over span l will occur in [IES-1995]

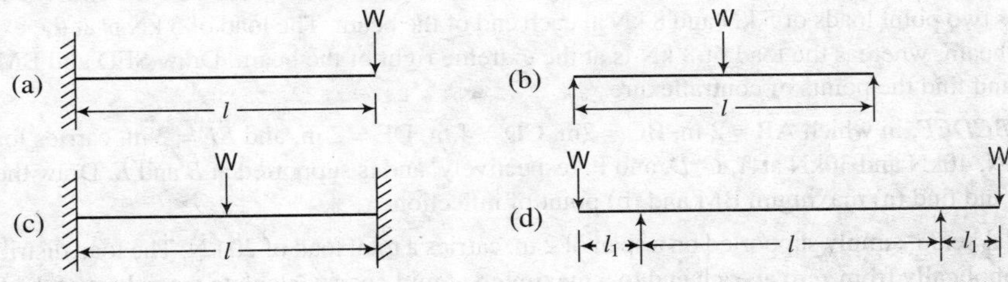

27. A beam having uniform cross-section carries a UDL of intensity q per unit length over its entire span, and its mid-span deflection is δ. The value of mid-span deflection of the same beam when the same load is distributed with intensity varying from $2q$ unit length at one end to zero at the other end is: [IES-1995]
(a) 1/3 δ (b) 1/2 δ (c) 2/3 δ (d) δ

Solutions for Objective Questions

Sl. No.	1.	2.	3.	4.	5.	6.	7.	8.	9.	10.
Answer	(c)	(b)	(a)	(b)	(d)	(d)	(c)	(d)	(a)	(b)

Sl.No.	11.	12.	13.	14.	15.	16.	17.	18.	19.	20.
Answer	(d)	(a)	(b)	(a)	(c)	(c)	(d)	(b)	(a)	(a)

Sl.No.	21.	22.	23.	24.	25.	26.	27.
Answer	(d)	(b)	(b)	(a)	(d)	(d)	(d)

EXERCISE PROBLEMS

1. A 4-m-long cantilever is loaded with a UDL of 2 kN/m, which runs over a length of 2 m from the free end. It also carries a point load of 4 kN at a distance of 1 m from the free end. Draw the SFD and BMD.

2. A cantilever of length 6 m carries two point loads of 12 kN and 18 kN at a distance of 1 m and 4 m from the fixed end, respectively. In addition to the beam, it carries a UDL of 3 kN/m over a length of 2 m at a distance of 3 m from the fixed end. Draw the SFD and BMD.
3. A cantilever of length 6 m carries a uniformly varying load of kN/m at free end and 6 kN·m at the fixed end. Draw the SFD and BMD.
4. A simply supported beam of length 10 m carries point loads of 6 kN and 8 kN at a distance of 4 m and 8 m from the left end. Draw the SFD and BMD for the beam.
5. A simply supported beam of length 10 m carries concentrated couples of 6 kN·m (clockwise) and 8 kN·m (anticlockwise) at a distance of 4 m and 8 m from the left end. Draw the SFD and BMD for the beam.
6. A simply supported beam is carrying a UDL of 8 kN·m over length of 4 m from the right end. The length of the beam is 6 m. Draw the SFD and BMD for the beam and also calculate the maximum BM on the section.
7. A beam of length 10 m is simply supported and carries point loads of 5 kN each at a distance of 4 m and 6 m from the left support and also a UDL of 3 kN/m between the point loads. Draw SFD and BMD for the beam.
8. A simply supported beam of length 10 m rests on supports 8 m apart, the right end is overhanging by 1.5 m and the left end is overhanging by 0.5 m. The beam carries a UDL of 5 kN/m over the entire length. It also carries two point loads of 5 kN and 8 kN at each end of the beam. The load of 5 kN is at the extreme left of the beam, whereas the load of 8 kN is at the extreme right of the beam. Draw SFD and BMD for the beam and find the points of contraflexure.
9. A beam *ABCDEF*, in which AB = 2 m, BC = 2m, CD = 4 m, DE = 2 m, and *EF* = 2 m, carries loads of 50kN, 50kN, 40kN and 30kN at *A*, *C*, *D*, and F, respectively, and is supported at *B* and *E*. Draw the SFD and BMD and find (a) maximum BM and (b) point of inflection.
10. A horizontal beam, simply supported on a span of 2 m, carries a total load of 20 kN. The load distribution varies parabolically from zero at each end to a maximum at mid-span. Calculate the values of the BM at intervals of 0.25 m and plot the BMD and SFD.
11. A beam *ABCD* is 20 m long and is simply supported at *B* and *D*, 16 m apart. *A* concentrated load of 18 kN at *A* and a total distributed load of 100 kN, which varies linearly from p kN/m at the center C to q kN/m at *D*, spread from *C* to *D*. Find the values of p and q for the reactions at *B* and *D* to be equal. Also plot SFD and BMD.
12. An inclined ladder of length 3.5 m rests on a rough horizontal floor and smooth vertical wall. If the ladder is subjected to a UDL load of 5 kN/m, plot the AFD, SFD, and BMD. The ladder is inclined 60° to the horizontal.
13. Sketch the AFD, SFD, and BMD for the frame shown in Figure 1.

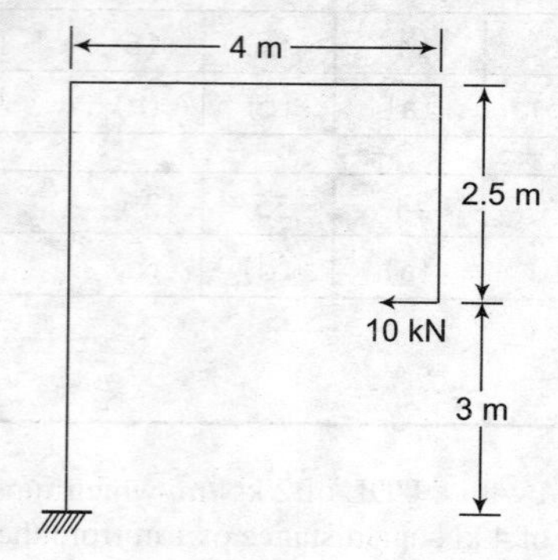

FIGURE 1

14. Sketch the AFD, SFD, and BMD for the frame shown in Figure 2.

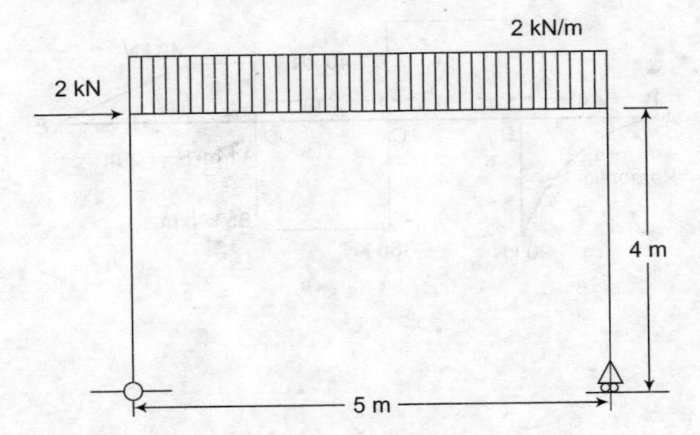

FIGURE 2

15. A horizontal girder 10 m long is hinged at one end 'A' and freely on a roller support at distance of 7 m from the hinged end. The beam carries a UDL of intensity 1000 N/m from the end A for a length of 5 m, a point load 2000 N inclined at 30° to the vertical at free end, 6 m from the end a point load of 3000 N inclined at 45° to the vertical at the point E, 1 m from the hinged end. Determine the support reactions. Draw the AFD, SFD, and BMD.

16. Sketch the AFD, SFD, and BMD for the frame shown in Figure 3. 18 kN is 2 m from the bottom of the left column and the internal hinge is as shown below the horizontal beam.

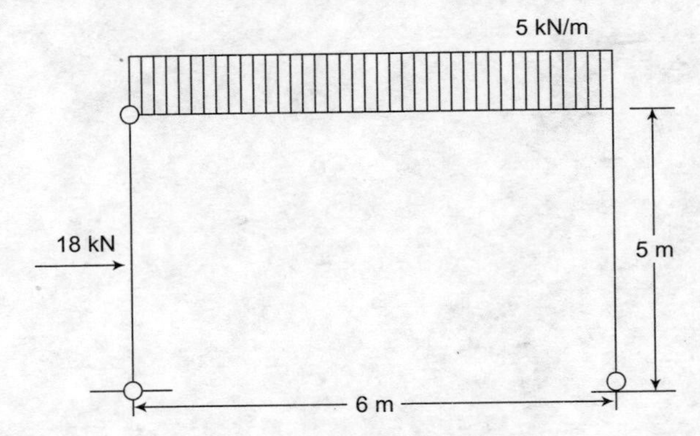

FIGURE 3

17. Sketch the AFD, SFD, and BMD for the frame shown in Figure 4.

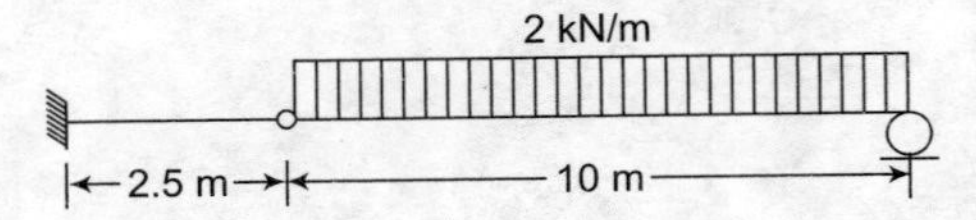

FIGURE 4

18. Sketch the loading diagram and BMD from the SFD given in Figure 5.

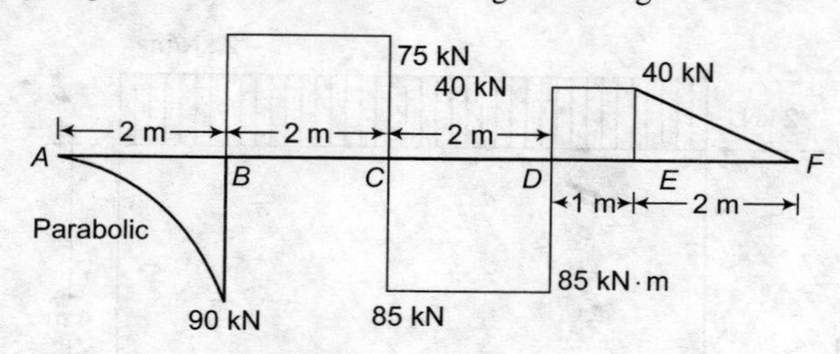

FIGURE 5

CHAPTER 4

BENDING STRESS

UNIT OBJECTIVE

This chapter provides information regarding bending, bending stress, and derivations for the formulae, etc. The presentation attempts to help the student achieve the following:

Objective 1: Determine the bending stress of the material.

Objective 2: Determine the deflection and radius of curvature.

Objective 3: Determine the stress in the extreme fibers and dimensions.

Objective 4: Determine the moment of resistance (MR) of the sections.

4.1 INTRODUCTION

In the previous chapter, we have seen how to draw shear force diagrams (SFDs) and bending moment diagrams (BMDs). The question that arises is that, how these bending moment (BM) and shear force (SF) should be handled in design and what type of stresses develops due to these internal forces? In this chapter, we will discuss about the quantification of stress developed due to BM. The stress that develops due to bending is normal stress generally referred as flexural stress or bending stress.

4.2 ASSUMPTIONS IN THEORY OF BENDING

1. Material of the beam is homogeneous, elastic, and isotropic.
   A material is said to be elastic, if the force–displacement relationship follows the same path during the loading as well as during the unloading. In homogeneous materials, the elastic properties such as modulus of elasticity and Poisson's ratio is same everywhere. A material is said to be isotropic, if the elastic properties are same in all directions.

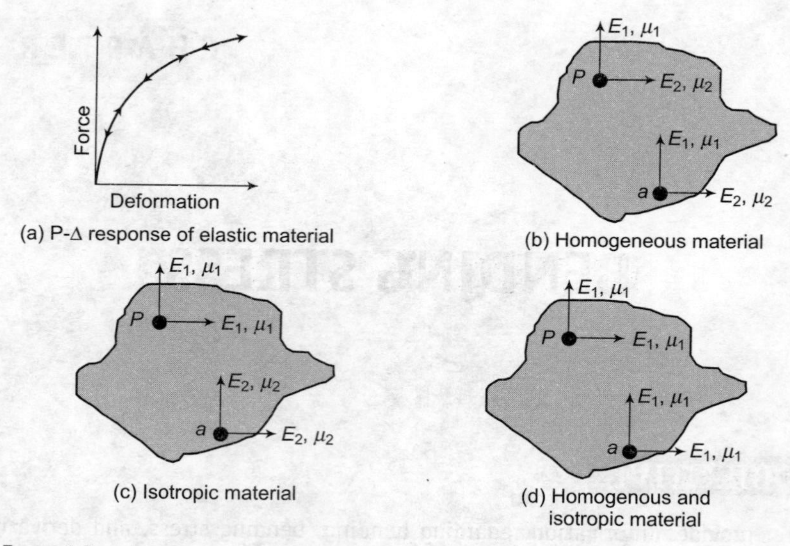

(a) P-Δ response of elastic material (b) Homogeneous material

(c) Isotropic material (d) Homogenous and isotropic material

FIGURE 4.1

2. Material of the beam obeys Hooke's law, that is, force–displacement relationship is linear.

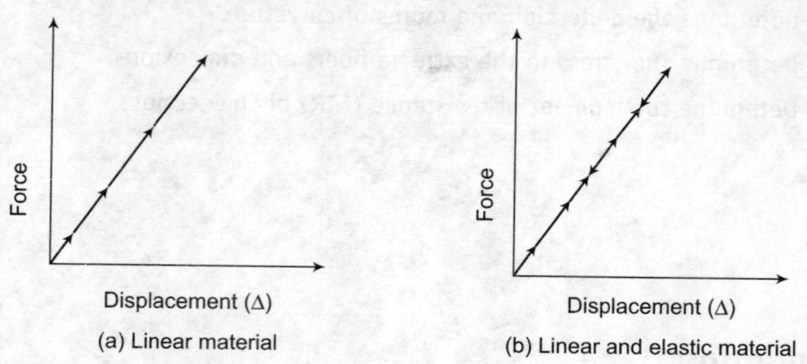

(a) Linear material (b) Linear and elastic material

FIGURE 4.2

3. Plane sections normal to the longitudinal remains plane even after bending. This is the most important assumption; it means that the strain variation is linear across the depth of the beam cross-section.

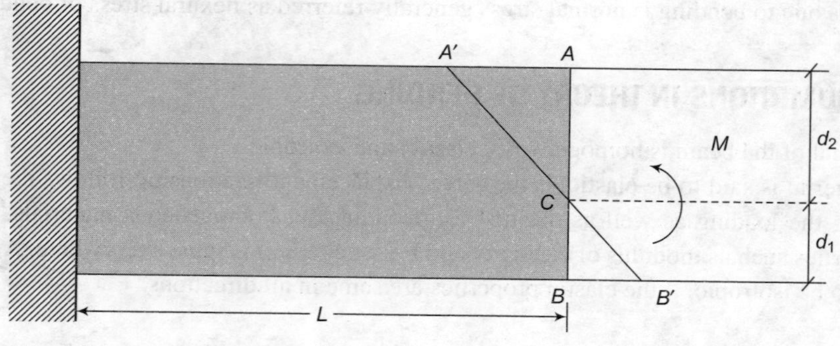

FIGURE 4.3

In Figure 4.3, cantilever beam is subjected to moment M. AB is a plane section, before the application of the couple. After the application of the couple, the plane section AB has rotated about C, taken position $A'B$. As the $A'B'$ is also a straight line, the displacement of any point on AB is proportional to its distance from C. Hence, the strain at any point on the cross-section AB is proportional to its distance from C.

$$\Rightarrow \quad \frac{\left(\dfrac{B'B}{L}\right)}{d_1} = \frac{\left(\dfrac{A'A}{L}\right)}{d_2}$$

4. The beam is subjected to pure bending.
   Pure bending means the beam should be subjected to constant BM and no SF and axial force act on it, for example,

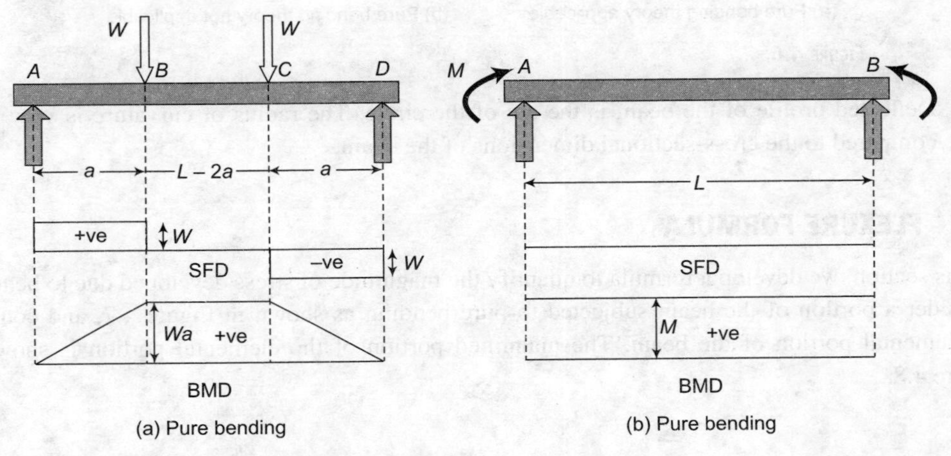

(a) Pure bending (b) Pure bending

FIGURE 4.4 Pure Bending.

In the beam shown in Figure 4.4 (a), the portion BC of the beam $ABCD$ is subjected to pure bending. In Figure 4.4 (b), the beam AB is subjected to pure bending.

5. Modulus of elasticity of the beam is same both in compression and tension.
6. The cross-section of the beam is symmetric about the plane of loading and beams were shown in Figure 4.5. If the loading is out of plane, then the loading creates twisting of the member cross-section. Thus, the cross-section should be symmetrical about the plane of loading.

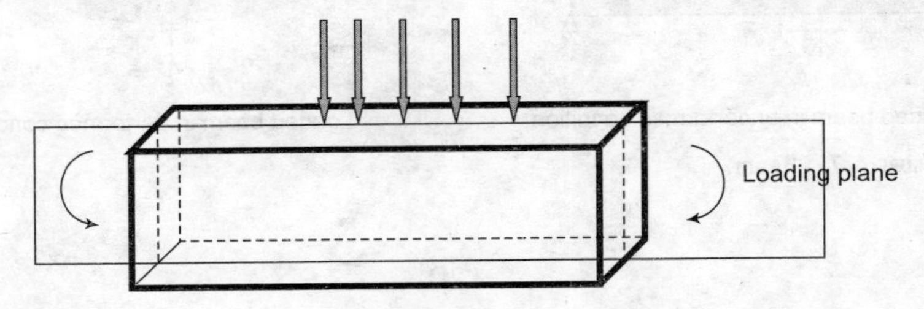

FIGURE 4.5

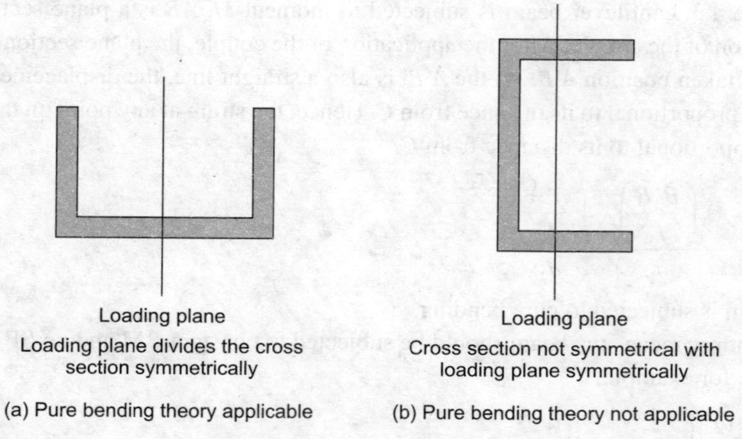

FIGURE 4.6

7. Deflected profile of the beam is the arc of the circle. The radius of curvature is very large compared to the cross-sectional dimensions of the beam.

4.3 FLEXURE FORMULA

In this section, we develop a formula to quantify the magnitude of stress developed due to bending. Consider a portion of the beam subjected to pure bending as shown in Figure 4.7, and consider an elemental portion of the beam. The magnified portion of the elemental portion is shown in Figure 4.8.

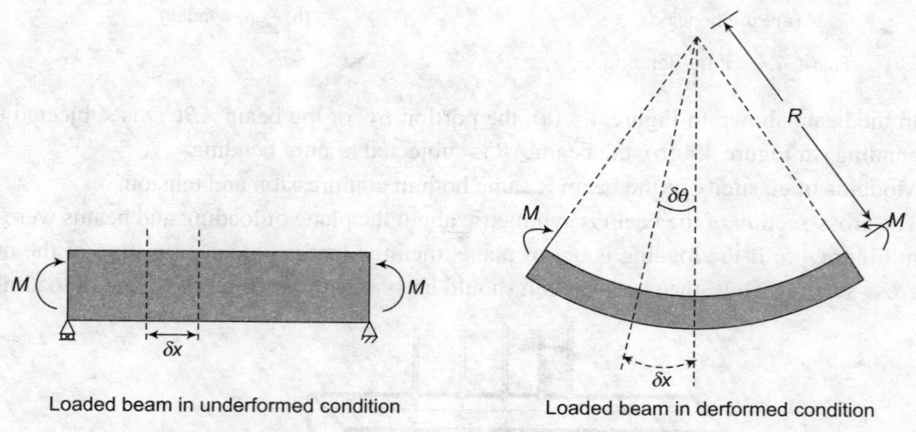

FIGURE 4.7 Beam.

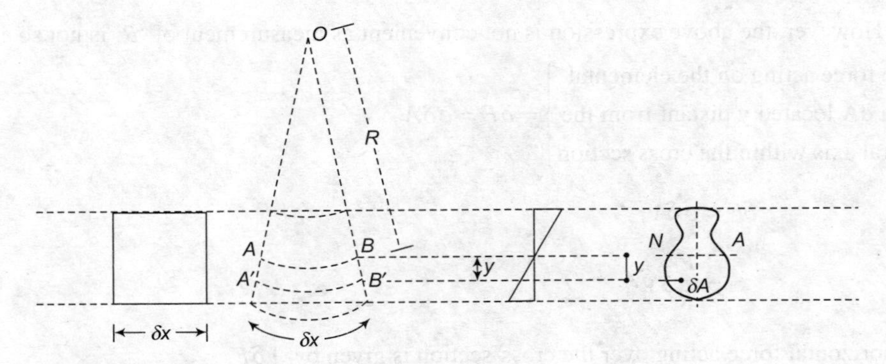

FIGURE 4.8

From the deflected profile of the beam, it is clear that the top fibers of the beam are compressed, whereas the bottom side fibers are subjected to elongation. Thus within the beam, there exists one layer, which is subjected to neither extension nor compression due to bending. This layer is called *neutral surface*, within the cross-section of the beam it is referred as *neutral axis*. At neutral axis level, strain is zero. Consider an elemental area 'δA' located at a distance 'y' from the neutral axis within the cross-section.

Let AB be the length of the fiber 'δx' before bending. After applying moment, let the fiber length AB be extended to $A'B'$.

Strain at the level y distant from the neutral axis $= \dfrac{A'B' - AB}{AB}$.

Let 'R' be the radius of curvature of elastic profile of the beam

$$A'B' = (R + y)\delta\phi$$

$$AB = \delta x = Rd\phi \quad (\because \text{ at } y = 0\text{, i.e., at neutral axis length of the layer does not change})$$

$\therefore$ Strain at a distance y from neutral axis $= \in = \dfrac{(R + y)\delta\phi - R\,\delta\phi}{(R)\cdot\delta\phi}$

$$\Rightarrow \qquad \in = \frac{y}{R}. \tag{4.1}$$

From Eq. (4.1), it is clear that, the strain variation is linear, if radius of curvature is constant. As per the seventh assumption, radius of curvature is constant. Strain variation is linear across the depth of the cross-section; this confirms the second assumption that plane section remains plane.

Let 'E' be the modulus of elasticity of the material of the beam, then

Normal stress developed at a fiber located y distance from neutral axis $= \sigma = \in E = \dfrac{y}{R}E$

$$\therefore \qquad \frac{\sigma}{y} = \frac{E}{R}. \tag{4.2}$$

From Eq. (4.2), it can be concluded that stress variation is linear with the depth of the beam cross-section. However, the above expression is not convenient as measurement of 'R' is not so easy.

$$\left.\begin{array}{r}\text{The force acting on the elemental}\\ \text{area dA located y distant from the}\\ \text{neutral axis within the cross section}\end{array}\right\} = \delta F = \sigma \delta A$$

i.e.,
$$\delta F = \sigma \delta A$$
$$= \frac{E}{R} y \delta A.$$

Net horizontal force acting over the cross-section is given by $\int \delta F$.

$$\int \delta F = \int_A \frac{E}{R} y \delta A$$

As per the fourth assumption, the net horizontal force/force normal to the cross-section of the beam is zero.

$$\Rightarrow \qquad \int \delta F = 0$$

$$\Rightarrow \qquad \int_A \frac{E}{R} y \delta A = 0.$$

E cannot be equal to zero as the material of the body is a solid and possesses elastic modulus.

The radius of curvature 'R' of the elastic curve cannot be infinity, as the material of the body is not rigid.

$$E \neq 0 \text{ and } R \neq \infty, \text{ hence } \int y \delta A = 0. \qquad (4.3)$$

We know that $\int y \delta A = Y_c A$, in which Y_c is the position of the centroid of the cross-section from neutral axis.

$$\int y \delta A = 0$$
$$CY_c = 0.$$

This means that neutral and centroidal axes of the beam cross-section coincide. To quantify stress 'σ' in terms of external moment, equate the internal couple developed due to elemental forces within the cross-section to the external BM.

$$\text{Internal moment} = \int_A \delta F\, y$$

$$= \int_A \left(\frac{E}{R} y \delta A\right) y = \frac{E}{R} \int_A y^2 dA \left(\because \frac{E}{R} \text{ is constant}\right).$$

M is the external moment.

$$\therefore \qquad M = \frac{E}{R}\int y^2 dA$$

$\int y^2 \cdot dA$ is the second moment of area (I).

$$\therefore \qquad M = \frac{E}{R} I$$

$$\text{Or} \qquad \frac{M}{I} = \frac{E}{R}. \tag{4.3}$$

The limiting moment or MR of the cross-section is given by

$$\text{MR} = [\sigma_{\text{allowable}}]\left[\frac{I}{y_{\max}}\right]. \tag{4.4}$$

Allowable stress depends on the material; $\frac{I}{y_{\max}}$ depends on the cross-section. This term $\frac{I}{y_{\max}}$ is referred as 'section modulus'. More the section modulus, more is the MR. Section modulus is denoted by the letter 'Z', units are mm^3. An engineer can design better cross-sections, having higher section modulus keeping the cross-sectional area same.

Curvature of a beam = $\frac{1}{R}$

$$\therefore \qquad \frac{1}{R} = \frac{M}{EI} = \frac{1}{Ey}\frac{My}{I} = \frac{\sigma}{Ey} = \frac{\in}{y}$$

$$\text{or} \qquad \frac{1}{R} = \frac{\in}{y}$$

⇒ Curvature of a beam is nothing but the slope of strain variation diagram across the depth of the beam.

From Eqs. (4.2) and (4.3)

$$\frac{M}{I} = \frac{E}{R} = \frac{\sigma}{y}$$

M = BM acting at the section under consideration

I = Second moment of area about neutral axis or centroidal axis

E = Modulus of elasticity of the material of the beam

R = Radius of curvature of the deflected profile of the beam

σ = Normal stress developed due to bending over a fiber located at a distance y from the neutral axis.

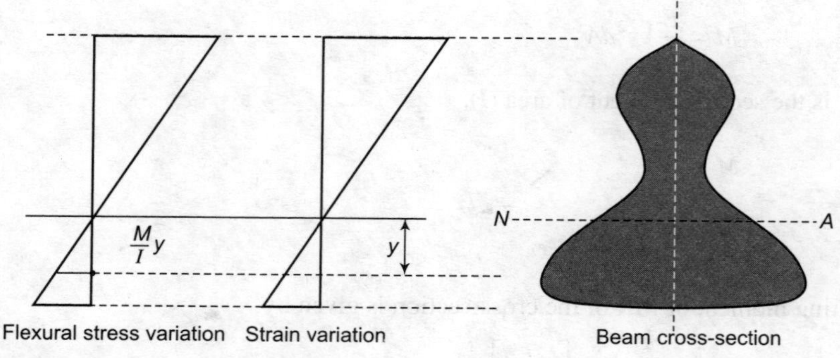

FIGURE 4.9

In the case of bending, the fiber far away from neutral axis is critical. If the maximum stress within the cross-section reaches the limiting value, the failure of the cross-section takes place.

PROBLEM 4.1

Objective 1

Determine the maximum stress induced in a metallic strip of thickness 10 mm wounds around a drum of radius 1 m. The modulus of elasticity of the material of the metallic strip is 80 GPa.

SOLUTION

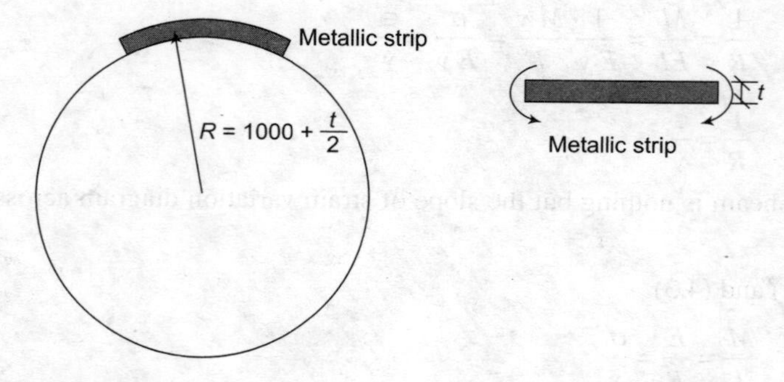

FIGURE 4.10

The bent profile of the metallic strip is arc of circle, thus subjected to pure bending.

Radius of curvature of the strip, $R = 1000 + \dfrac{10}{2} = 1005$ mm

We know that $\dfrac{E}{R} = \dfrac{\sigma}{y}$

For maximum normal stress, $\sigma_{\max} = \dfrac{E}{R} y_{\max}$

$$\sigma_{\max} = \frac{E}{R}\left(\frac{t}{2}\right)$$

$$= \frac{80 \times 1000}{1005} \times 5 = 398 \text{ MPa.}$$

PROBLEM 4.2 **Objective 2**

A simply supported beam of length 3 m is subjected to end couples of intensity 8 kN-m. The cross-section of the beam is 200 mm wide and 50 mm thick. If $E = 200$ GPa, determine (a) maximum stress in the beam, (b) deflection at midspan, (c) radius of curvature, and (d) deflection at (1/4)th span.

SOLUTION

The beam is subjected to pure bending. Thus, the deflection profile of the beam is an area of circle of radius R.

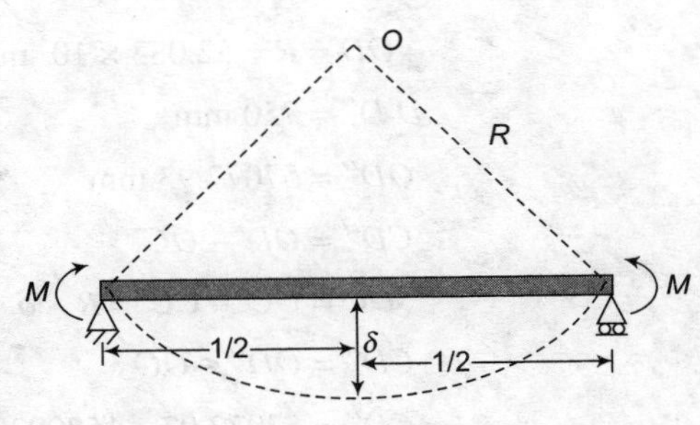

FIGURE 4.11

We know, $$\frac{M}{I} = \frac{E}{R}$$

or $$\frac{1}{R} = \frac{M}{EI}$$

I = Moment of inertia (MI) or second moment of area about Neutral Axis.

$$= \frac{50^3 \times 200}{12} = 2.0833 \times 10^6 \text{ mm}^4.$$

$$\text{Curvature of the beam} = \frac{1}{R} = \frac{8 \times 10^6}{2 \times 10^5 \times 2.0833 \times 10^6}$$

$$\Rightarrow \quad \frac{1}{R} = 1.92 \times 10^{-5} \text{ / mm}$$

$$\Rightarrow \quad R = 0.52083 \times 10^5 \text{ mm.}$$

Let δ be the central deflection. As the deflected profile of the beam is a part of the circle,

$$(2R - \delta)\delta = \left(\frac{l}{2}\right)^2 \Rightarrow 2R\delta - \delta^2 = \frac{l^2}{4}.$$

Neglecting δ^2 as δ is very small.

$$\delta = \frac{l^2}{8R} = \frac{3000 \times 3000}{8} \times 1.92 \times 10^{-5} = 21.6 \text{ mm.}$$

$$\text{Maximum stress} = \sigma_{max} = \frac{M}{I} y_{max}$$

$$y_{max} = \frac{50}{2} = 25 \text{ mm}$$

$$\sigma_{max} = \frac{8 \times 10^6}{2.0833 \times 10^6} \times 25 = 96 \text{ MPa.}$$

Deflection at (1/4)th span:

From Figure 4.12,

Deflection at (1/4)th span = CD

From $\Delta ODD'$

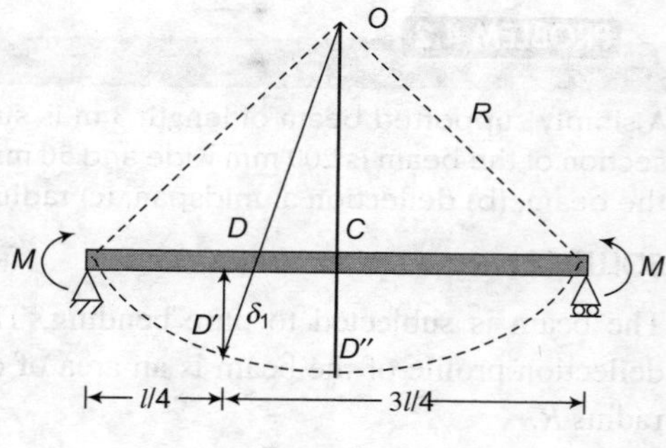

FIGURE 4.12

$$OD'' = \sqrt{(OD')^2 - (D'D'')^2}$$

$$OD = R = 52.083 \times 10^3 \text{ mm}$$

$$D'D'' = 750 \text{ mm}$$

$$OD'' = 57077.93 \text{ mm}$$

$$CD'' = OD' - OC$$

$$OC = OC' - CC' = R - \delta$$

$$CD'' = OD'' - OC$$

$$\therefore \quad CD'' = 52077.93 - \{520833 - 21.6\} = 16.20 \text{ mm.}$$

Maximum stress in the beam = 96 MPa

Deflection at midspan = 21.6 mm

Deflection at (1/4)th span = 16.2 mm

Radius of curvature = 52.083 mm

PROBLEM 4.3

Objective 3

A beam of semicircular cross-section is subjected to a sagging BM of 6 kN-m. If the radius of the semicircle is 125 mm, determine extreme fiber stresses. Also sketch the variation of flexural strength across the depth.

SOLUTION

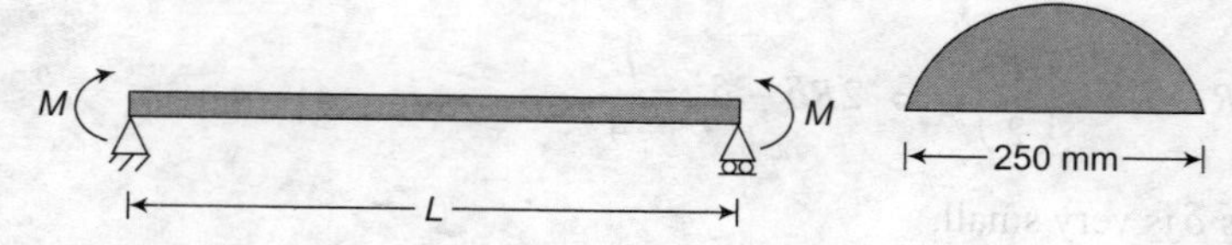

FIGURE 4.13

Maximum normal stress = $\sigma_{max} = \frac{M}{I} y_{max}$

Normal stress = $\sigma = \frac{M}{I} y$

$$M = 6 \times 10^6 \text{ kN-m}$$

I = second moment of area about Center of gravity

Neutral axis is located at a distance of $\frac{4r}{3\pi} = 53.05$ mm from the base.

$$\left.\begin{array}{r}\text{Second moment area of}\\ \text{semicircle about base}\end{array}\right\} I_b = \frac{\pi d^4}{128} = 95.874 \times 10^6 \text{ mm}^4$$

Applying parallel-axis theorem,

$$I_b = I_{CG} + Ah^2$$

$$95.874 \times 10^6 = I + \frac{\pi(250)^2}{8} \times (53.052)^2$$

$$I = 29.375 \times 10^6 \text{ mm}^4$$

$$\sigma = \frac{6 \times 10^6}{29.37 \times 10^6} \cdot y = 0.2043y$$

$$\sigma = 0.2043y$$

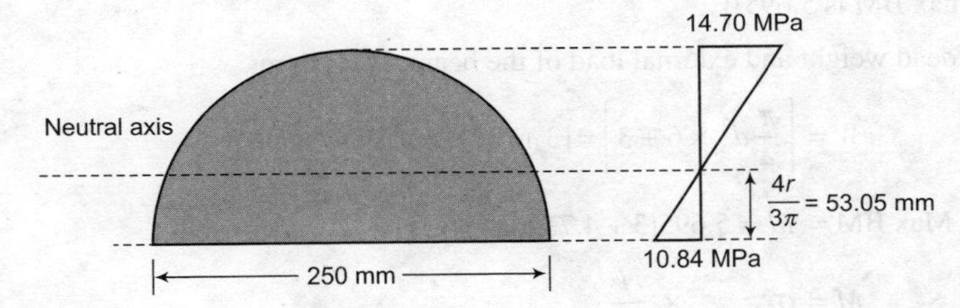

FIGURE 4.14

$$\text{Stress at the top fiber} = \sigma_1 = 0.2043 \times 71.95$$
$$= 14.70 \text{ MPa (Compressive)}$$

$$\text{Stress at the bottom fiber} = \sigma_b = 0.2043 \times 53.05$$
$$= 10.84 \text{ MPa (Tensile).}$$

PROBLEM 4.4 **Objective 3**

An overhanging beam *ABC* of span between support 3 m and overhanging of 1.5 m carries a uniformly distributed load (*UDL*) of 3 kN/m (excluding self-weight of the beam). The cross-section of the beam is circular. Determine the diameter of the beam if the allowable stress is limited to 18 MPa. Density of the material of the beam is 6 kN/m^3.

SOLUTION

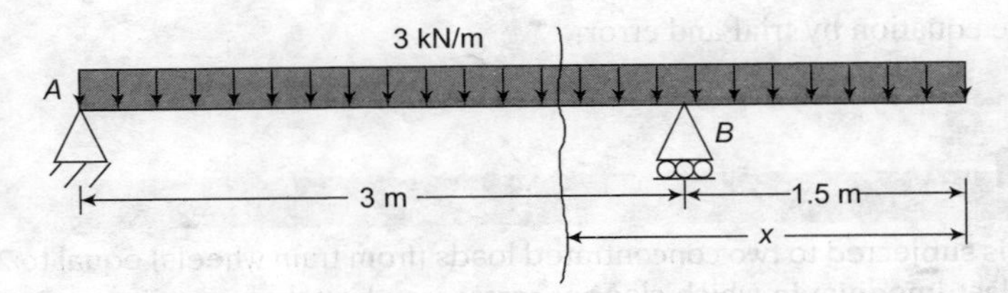

FIGURE 4.15

Sum of the moments about $A = 0$

$$\Rightarrow \quad R_b \times 3 = W \times 4.5 \times \frac{4.5}{2}$$

$$\Rightarrow \quad R_b = 3.375\ \text{W}$$

$$R_a = 1.125\ \text{W}$$

For max BM, SF = 0

$$\Rightarrow \quad R_b - Wx = 0$$

$$\Rightarrow \quad 3.375W - Wx = 0$$

$$x = 3.375 \text{ m from } C.$$

$$\text{Max +ve BM} = 3.375 \times 3.375\ \text{W} - \frac{W}{2}(3.375)^2 = 5.69\ \text{W}$$

$$\text{Max –ve BM} = -W\frac{(1.5)^2}{2} = -1.125\ \text{W}$$

∴ Absolute max BM is $5.695W$.

'W' includes dead weight and external load of the beam. 'd' is in mt.

$$W = \left[\frac{\pi}{4}d^2 \times 6 + 3\right] = [3 + 4.7124d^2]\,\text{kN/m}$$

$$\therefore \quad \text{Max BM} = M = 5.695\{3 + 4.7124d^2\}$$

$$M = \sigma_{\text{allowable}} \times \frac{I}{y_{\max}}$$

For circular cross-section,

$$Z = \frac{I}{y_{\max}} = \frac{\frac{\pi d^4}{64}}{\left(\frac{d}{2}\right)}$$

$$\therefore \quad Z = \frac{\pi d^3}{32}; \ \sigma_{\text{allowable}} = 18 \times \frac{10^6 N}{m^2} = 18000\ \text{kN/m}^2$$

$$5.695\{3 + 4.7124d^2\} = 18000 \times \frac{\pi d^3}{32}$$

$$26.837d^2 + 17.085 = 1767.14d^3.$$

Solving the above equation by trial and error,

$$d = 0.218 \text{ m}.$$

PROBLEM 4.5

Objective 3

A railway sleeper is subjected to two concentrated loads (from train wheels) equal to 250 kN each. The reaction from ballast (medium in which sleeper rests) may be taken as uniform. Determine breadth and depth of sleeper, if the allowable stress in it is limited to 10 MPa. Take $\frac{\text{breadth}}{\text{depth}} = 1.2$, $L = 1.7$ m, and $a = 0.6$ m.

SOLUTION

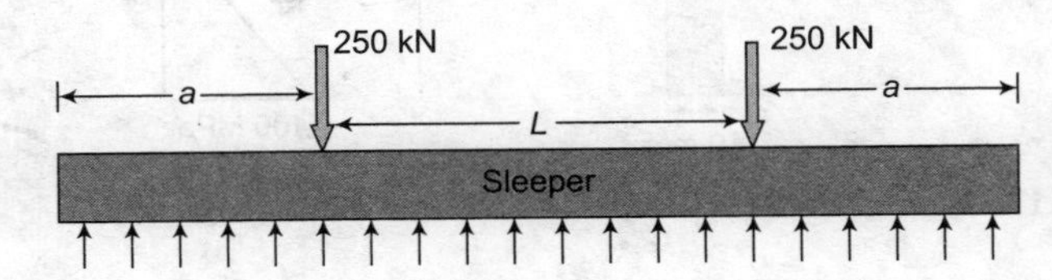

FIGURE 4.16

Let q be the upward reaction from the ballast.

Then $$q = \frac{250 \times 2}{2.9} = 172.41 \text{ kN/m}$$

$$\sigma_{max} = 10 \text{ MPa.}$$

For designing the cross-sectional dimensions, equate the maximum BM to MR of the cross-section.

$$\text{Max –ve BM at } C = 172.41 \times \left(\frac{2.9}{2}\right)\left(\frac{2.9}{4}\right) - 250 \times \frac{1.7}{2} = -31.25 \text{ kN m}$$

$$\text{Max +ve BM at } D = 172.41 \times \left(\frac{0.6^2}{2}\right) = 31.03 \text{ kN m}$$

$\therefore$ Design moment = 31.25 kN m

$$M = \sigma_{max} \times Z; \; Z = \frac{\frac{bd^3}{12}}{\left(\frac{d}{2}\right)} = \frac{bd^2}{6}$$

$$\Rightarrow \quad 31.25 \times 10^6 = 10 \times \frac{bd^2}{6}$$

And it is given that, $\frac{b}{d} = 1.2$

$$\therefore \quad 31.25 \times 10^6 = 10 \times \frac{1.2\, d^3}{6}$$

$$\Rightarrow \quad d = 250 \text{ mm}; \; b = 300 \text{ mm.}$$

PROBLEM 4.6 **Objective 3**

Determine the flange width of a T section shown in Figure 4.17 if the cross-section is designed for balanced failure. The allowable stress in compression is 65 MPa, while the same in tension is 100 MPa. The member is subjected to sagging moment. Also determine the MR of the cross-section.

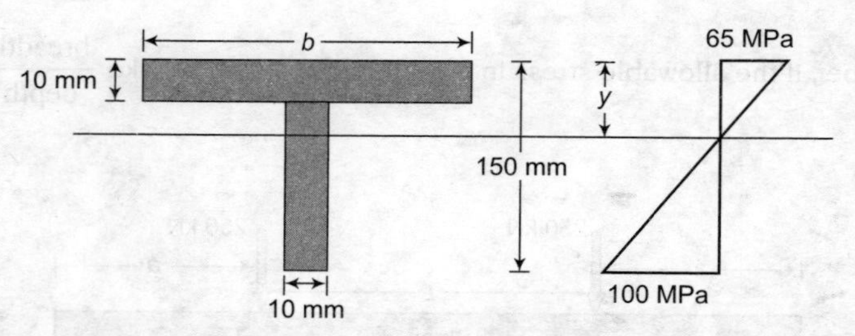

FIGURE 4.17

SOLUTION

For balance failure, the top fibers as well as the bottom should reach their allowable stresses simultaneously. Thus at failure, the stress block is as shown in Figure 4.17. Let 'y' be the location of neutral axis or centroidal axis of the cross-section, then

$$\frac{65}{y} = \frac{100}{150 - y}$$

$\Rightarrow$ $y = 59.1$ mm.

$\therefore$ Centroid of cross-section should be 59.1 mm from top.

We know, $$y_c = \frac{A_1 y_1 + A_2 y_2}{A_1 + A_2}$$

$$y_c = 59.1 = \frac{b \times 10 \times 5 + 140 \times 10 \times \left(\frac{140}{2} + 10\right)}{140 \times 10 + b \times 10}$$

$\therefore$ $8274 + 59.1b = 5b + 11{,}200$

$\Rightarrow$ $b = 54.09$ mm

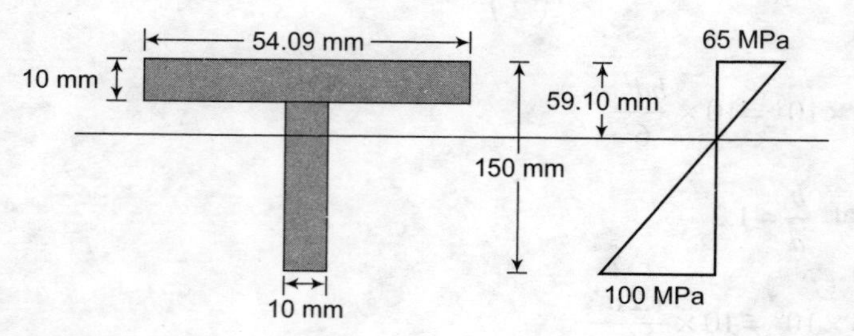

FIGURE 4.18

MR:

$$I = \frac{54.09 \times 10^3}{12} + 54.0 \times 10 \times (59.1 - 5)^2 + \frac{10 \times 140^3}{12} + 140 \times 10 \times (80 - 59.1)^2$$

$$= 4.4858 \times 10^6 \text{ mm}^4$$

Considering the fiber stress at the bottom most fiber

$$M = \sigma_{\text{all.}} \times \frac{I}{y_{\max}}$$

$$= 100 \times \frac{4.4858 \times 10^6}{90.9}$$

$$= 4.935 \times 10^6 \text{ N} \cdot \text{mm} = 4.935 \text{ kN m}$$

If the same cross-section is subjected to hogging moment, then maximum stress at the bottom most fiber and top most fiber is 100 MPa.

Then MR

$$M_{\text{hogging}} = 65 \times \frac{4.4858 \times 10^6}{90.9} = 3.208 \times 10^6 \text{ Nm}$$

$$= 3.208 \text{ kN m}$$

PROBLEM 4.7 **Objective 3**

A rectangular cross-section is to be cut from a circular cross-section of diameter '*d*'. Determine the breadth and height of the strongest beam that can be obtained from the circular cross-section.

SOLUTION

For a cross-section to be strong under flexure, the MR shall be maximum. If the material is constant, then section modulus must be maximum for a strong section.

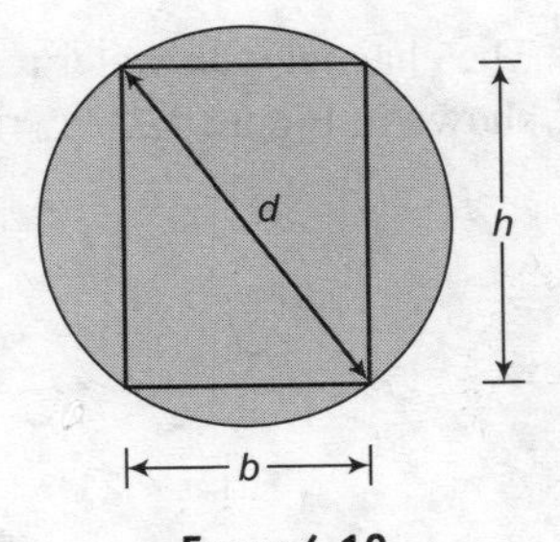

FIGURE 4.19

$$\left.\begin{array}{r}\text{Section modulus}\\ \text{of rectangular section}\end{array}\right\} = Z = \frac{I}{y_{\max}} = \frac{\frac{bh^3}{12}}{\left(\frac{h}{2}\right)} = \frac{bh^2}{6}. \quad (4.4)$$

From Figure 4.19, it is clear that, the diagonal *AC* of the rectangle is the diameter of the circle.

$$\therefore \quad d^2 = b^2 + h^2$$

$$\Rightarrow \quad h^2 = b^2 - d^2. \quad (4.5)$$

Using Eqs. (4.5) and (4.4)

$$Z = \frac{1}{6} b[d^2 - b^2].$$

For max section modulus, $\frac{dZ}{db} = 0$

$$\Rightarrow \quad \frac{1}{6}[d^2 - 3b^2] = 0 \quad (\because \text{ diameter of the circle is constant})$$

$$\Rightarrow \qquad b = \frac{d}{\sqrt{3}}$$

$$\Rightarrow \qquad h = \sqrt{\frac{2}{3}}\, d.$$

PROBLEM 4.8 **Objective 1**

A channel section shown in Figure 4.20 is used for storing water. The channel with closed ends rests on two supports separated by a distance of 6 m. To what level water can be stored in the channel. If the allowable tensile stress in material of the channel is limited to 1 MPa, specific gravity of channel material is 21 (density 21 kN/m^3). Density of water is 10 kN/m^3.

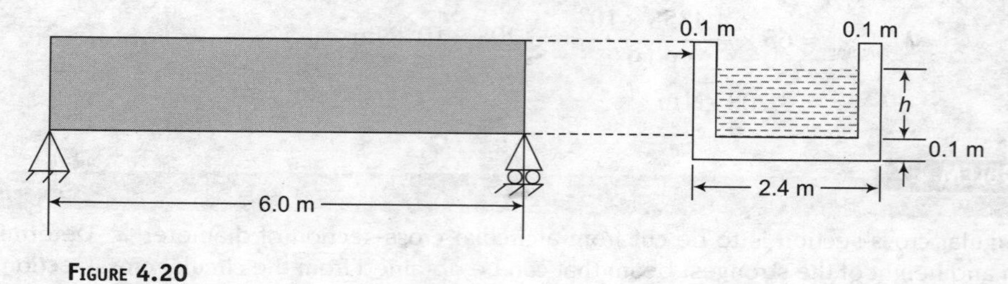

FIGURE 4.20

SOLUTION

The channel with water resisting on two rigid supports with end wall may be represented as a beam shown in Figure 4.21.

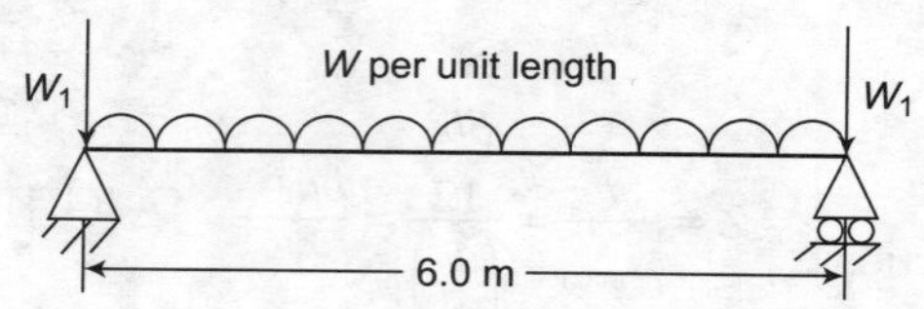

FIGURE 4.21

$w_1 \rightarrow$ Weight of the closing walls (end walls of the channel). This does not contribute to BM.

$w \rightarrow$ (Self-weight of the channel + weight of water present in the channel) per unit length

Let 'h' be the height/depth of water in the channel (units mt)

Then, $w = \{[2 \times 0.9 \times 0.1 + 2.2 \times 0.1]21 + 2.2 \times h \times 10\}$

$$= (8.4 + 22h).$$

MR of the cross-section:

$$M = \sigma_{\text{all.}} \frac{I}{y}$$

Here, the allowable stress in bottom fiber is given as 1 MPa (tension).

$$\therefore \qquad M = \sigma \frac{I}{y_b}.$$

Let $\bar{y}$ be the centroid of the channel from bottom.

$$\bar{y} = \frac{900 \times 100 \times 2 \times 450 + 2200 \times 100 \times 50}{2 \times 900 \times 100 + 2200 \times 100} = 230 \text{ mm}$$

$$I = 2\left\{\frac{900^3 \times 100}{12} + 900 \times 100(450 - 230)^2\right\} + \frac{100^3 \times 2200}{12} + 200 \times 100(230 - 50)^2$$

$$= 2.8137 \times 10^4 \text{ mm}^4$$

$$y_b = 230 \text{ mm}$$

$$\therefore \quad \text{MR} = M = 1.0 \times \frac{2.8137 \times 10^{10}}{230} = 122.49 \text{ kN m}$$

Equating the MR to maximum BM in the beam; Max BM $= \dfrac{wl^2}{8}$

$$\therefore \quad [8.4 + 22h] \cdot \frac{6^2}{8} = 12.49$$

$$\Rightarrow \quad h = 0.855 \text{ m}$$

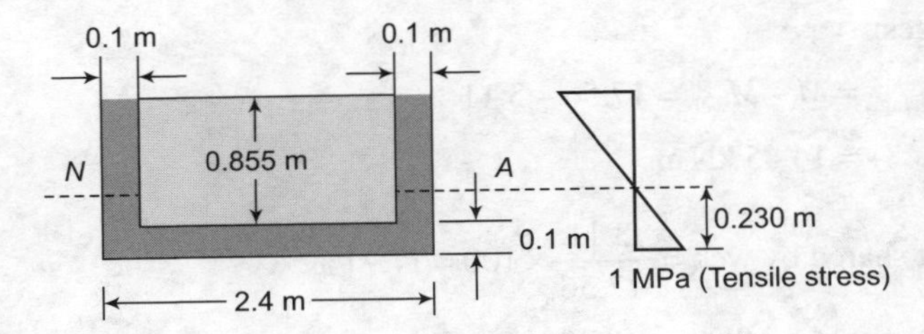

FIGURE 4.22

If the depth of the water is more than 855 mm, the structure cannot withstand the BM due to water and self-weight of the channel and fails.

PROBLEM 4.9 **Objective 4**

Determine the MR of an I section, with flange dimension 100 mm × 10 mm, web dimension 180 × 8 mm. The allowable stress in the material is limited to 80 MPa. Also determine the percentage moment resisted by the flange and web.

SOLUTION

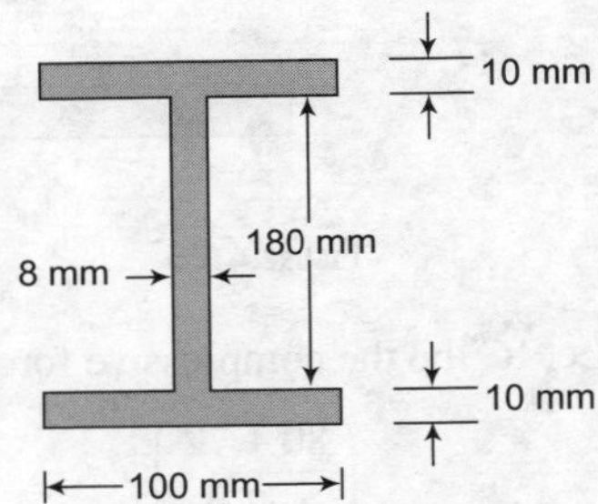

FIGURE 4.23

$$MI = I = \frac{1}{12}\{100 \times 200^3 - 92 \times 180^3\}$$

$$= 2.19547 \times 10^7 \text{ mm}^4$$

$$\text{Section modulus, } Z = \frac{2.19547 \times 10^7}{100} = 2.19547 \times 10^5 \text{ mm}^3$$

$$\text{MR} = \sigma_{\text{all.}} \times Z$$

$$= 80 \times 2.19547 \times 10^5$$

$$= 17.56 \times 10^6 \text{ N} \cdot \text{mm} = 17.56 \text{ kN} \cdot \text{m}$$

MR shared by web portion:

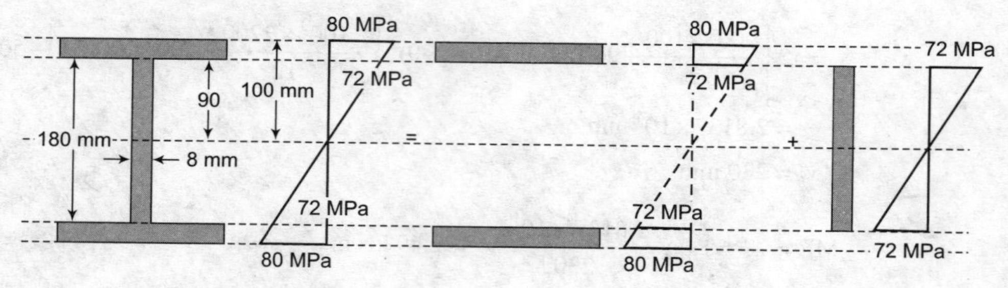

FIGURE 4.24

$$M_{\text{web}} = 72 \times \frac{1}{6} \times 8 \times 180^2$$

$$= 3.11 \times 10^6 \text{ N mm}$$

$$= 3.11 \text{ kN m}$$

MR shared by flanges:

$$M_{\text{flange}} = M - M_{\text{web}} = 17.56 - 3.11$$

$$= 14.45 \text{ kN m}$$

Percentage moment shared by web $= \dfrac{3.11}{17.56} \times 100 = 17.71\%$

Percentage moment shared by flange = 82.29%.

The other way of determining the moment shared by the flanges:

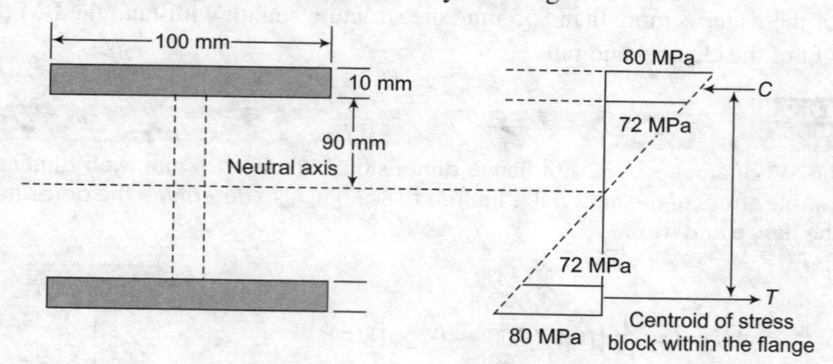

FIGURE 4.25

Let 'C' be the compressive force acting in the top flange.

$$\text{Then, } C = \underbrace{\left(\frac{80 + 72}{2}\right)}_{\text{average stress}} \times 100 \times 10 = 76{,}000 \text{ N}$$

This 'C' acts at the centroid of the stress block within the top flange. Tensile force in the bottom flange = 76, 000 N, acts at the centroid of the stress block within the bottom flange.

Centroid of the stress block (trapezium with opposite sides as 72 MPa, 80 MPa, and height 10 mm)

$$\text{in the bottom flange} = \frac{2 \times 72 + 80}{72 + 80} \times \left(\frac{10}{3}\right) = 4.91 \text{ mm.}$$

$$\therefore \quad \left.\begin{array}{c}\text{Moment shared}\\ \text{by the flange}\end{array}\right\} = C \times (200 - 4.91 - 4.91)$$

$$= 14.45 \times 10^6 \text{ N mm}$$

$$= 14.45 \text{ kN m.}$$

PROBLEM 4.10

Objective 4

A rectangular cross-section of 200 × 100 mm is used to resist BM. If a circular hole of diameter 75 mm is introduced at the centroid of the cross-section, estimate the percentage reduction in the moment carrying capacity of the section.

SOLUTION

Let 'σ' be the allowable bending stress of the material of the beam.

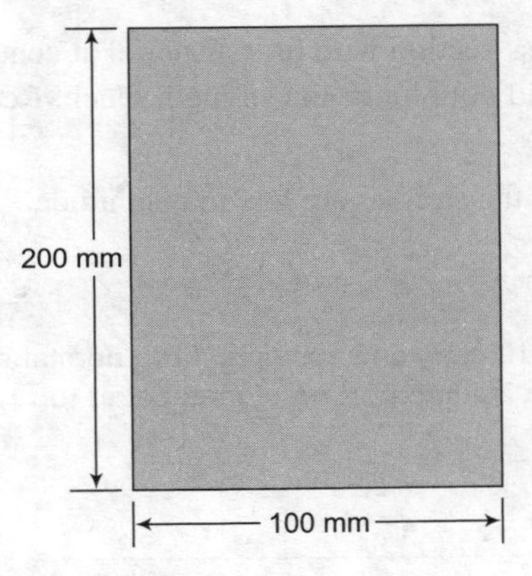

200 mm
75 mm
100 mm

FIGURE 4.26

$$\left.\begin{array}{c}\text{MR}\\ \text{of rectangular section}\end{array}\right\} = M = \sigma \times \frac{1}{6} bh^2$$

$$M_a = 0.667 \times 10^6 \sigma. \tag{4.6}$$

$$\left.\begin{array}{c}\text{MR}\\ \text{of rectangular section with}\\ \text{hole at a center}\end{array}\right\} = M_b = \sigma \times Z_b$$

$$Z_b = \frac{I}{y_{\max}}$$

$$I_b = \frac{1}{12} \times 100 \times 200^3 - \frac{\pi}{64}(75)^4$$

$$= 65.11 \times 10^6 \text{ mm}^4$$

$$Z_b = \frac{6.5114 \times 10^7}{100} = 0.651 \times 10^6 \text{ mm}^3$$

$$\therefore \quad M_b = 0.651 \times 10^6 \, \sigma. \tag{4.7}$$

$$\left.\begin{array}{c}\text{\% reduction in the strength}\\ \text{with the hole}\end{array}\right\} = \frac{M_a - M_b}{M_a} \times 100 = 2.35\%.$$

$$\text{\% reduction in the cross-sectional area} = \frac{A_a - A_b}{A_a} \times 100$$

$$\text{Reduction in the cross-sectional area} = \frac{\frac{\pi}{4}(75)^2}{200 \times 100} \times 100 = 22.09\%.$$

From the above figures, it can be concluded that cross-section with little material at centroids is more economical. Reduction of about 22% materials did not reduce the strength much. Reduction in strength is only 2.35%.

Hint: Always remove material in a cross-section where the stresses are less in magnitude.

PROBLEM 4.11 **Objective 4**

Determine the MR if triangular cross-section of breadth 100 mm and 100 mm, if the normal stress is limited to 27 MPa. Also estimate MR of the same section if some portion of material at top (apex) is removed as defined by α. Take $\alpha = 3$ mm.

SOLUTION

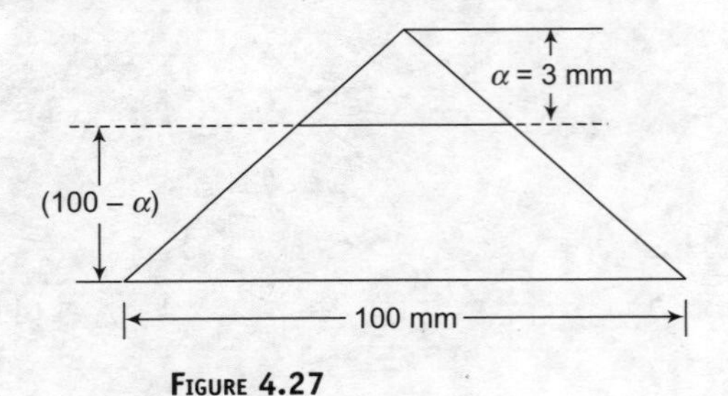

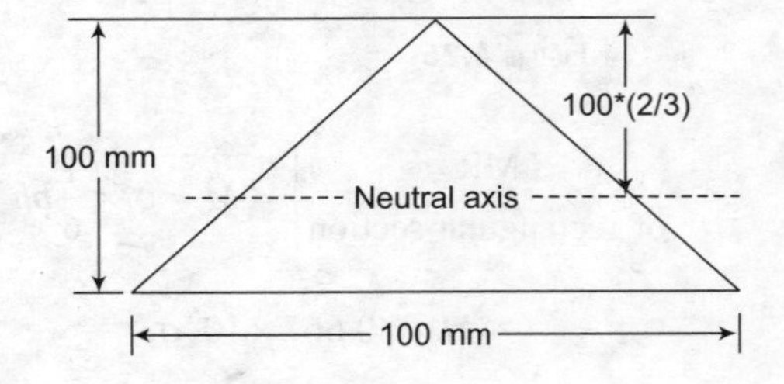

FIGURE 4.27

$$\left.\begin{array}{c}\text{MR of}\\ \text{triangular section}\end{array}\right\} M = \sigma \times Z$$

$$I = \frac{100 \times 100^3}{36} = 2.78 \times 10^6 \text{ mm}^4$$

$$y_{max} = 66.67 \text{ mm}$$

$$z = 41{,}700 \text{ mm}^3$$

$$\therefore \quad M = 27 \times \frac{2.78 \times 10^6}{66.67} = 1.125 \times 10^6 \text{ N mm} = 1.125 \text{ kN m}.$$

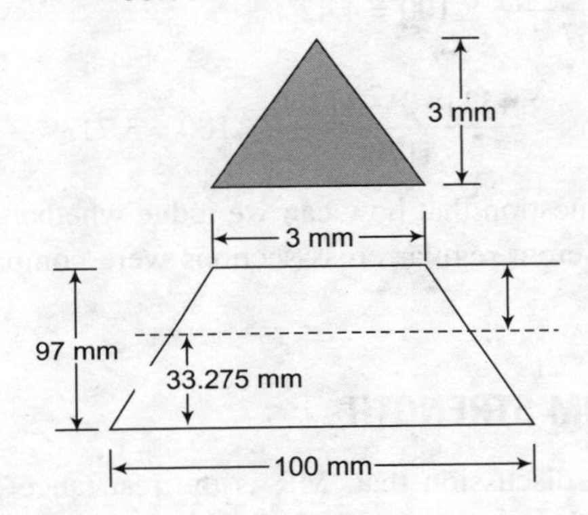

FIGURE 4.28

If the material at the top of the cross-section is removed as shown in Figure 4.28

$$\bar{y} = \frac{2 \times 3 + 100}{3 + 100} \times \frac{97}{3} = 33.275 \text{ mm}$$

$$\therefore \quad y_t = 97 - 33.275 = 63.725 \text{ mm}$$

$$I_{\text{base}} = \frac{3 \times 97^3}{3} + \frac{(100 - 3) \times 97^3}{12} = 8.29 \times 10^6 \text{ mm}^4$$

$$I_{\text{NA}} = I_{\text{base}} - A\bar{y}^2; \; A = \frac{100 + 3}{2} \times 97 = 4995.5 \text{ mm}^2$$

$$I_{\text{NA}} = 8.29 \times 10^6 - 4995.5 \times 33.275^2$$

$$= 2.756 \times 10^6 \text{ mm}^4.$$

$$\therefore \quad \text{Section modulus} = Z = \frac{I}{y_{max}} = \frac{2.756 \times 10^6}{63.725}$$

$$\Rightarrow \quad Z = 43248.33 \text{ mm}^3.$$

$$\text{MR} = \sigma_{\text{all.}} \times Z = 27 \times 43248.33$$

$$= 1.1677 \times 10^6 \text{ N mm}$$

$$\text{MR} = 1.1677 \text{ kN m}.$$

If we look at the values of MR in the first case and second case, it can be found that removal of material has improved the MR or bending strength. The reason for this is removal of material at the apex has reduced the *MI* to a little extent but reduced 'y_{max} or y_t' more, thus increasing the section modulus.

$$\% \text{ reduction in } I\ (MI) = \frac{2.78 - 2.756}{2.78} \times 100 = 0.863\%$$

$$\% \text{ reduction } y_{max} = \frac{66.67 - 63.725}{66.67} \times 100 = 4.42\%$$

$$\% \text{ increase in section modulus} = \frac{43248.3 - 41700}{41700} \times 100 = 3.713\%.$$

Now for a designer, it is a question that how can we judge whether a cross-section is efficient in resisting flexure or not? Different regular cross-sections were compared and are presented in the next section of the chapter.

4.4 BEAM OF UNIFORM STRENGTH

We have seen in our earlier discussion that, MR is the resistance offered by the cross-section, whereas BM is the moment due to external loads. In a prismatic beam (cross-section is constant along the length), the MR is constant. This MR should be more than or equal to the BM for safe design. Consider the case of a prismatic simply beam subjected to concentrated load at midspan.

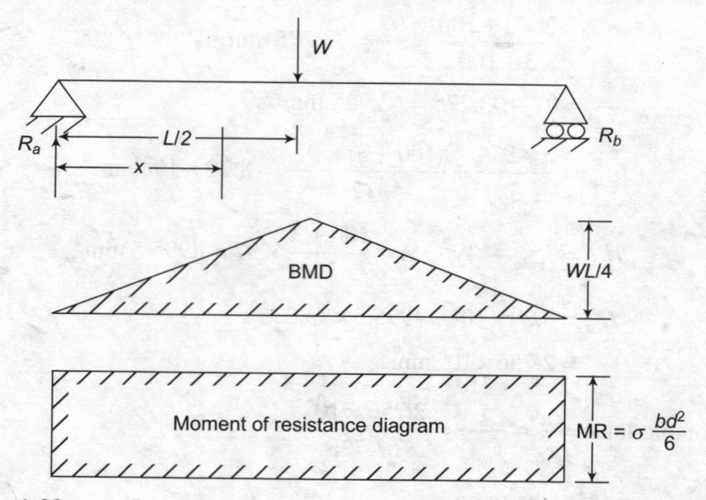

FIGURE 4.29

For design of the beam in Figure 4.29, $MR = \frac{WL}{4}$ (i.e., MR = max Bm). At point '*C*', the cross-section of the beam is just sufficient to resist BM.

∴ The cross-section at other section can be reduced, such that at any section BM is just equal to the MR at that section. For the beam considered in Figure 4.29, the cross-sectional width can be reduced as shown in Figure 4.30.

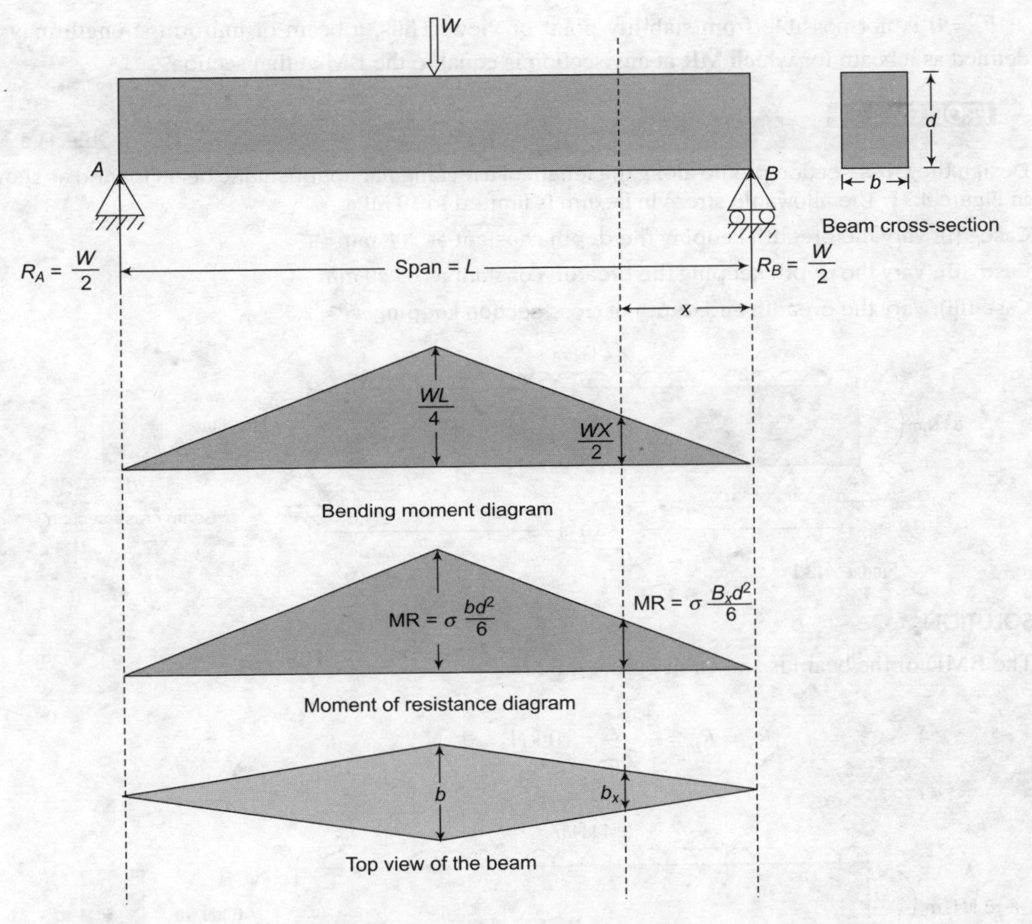

FIGURE 4.30

$$\frac{1}{6}\sigma bd^2 = \frac{WL}{4}$$

$$\text{Breadth at midspan, } b = \frac{6WL}{4\sigma bd^2} = \frac{3WL}{2\sigma bd^2}$$

$$\left.\begin{array}{r}\text{BM at section located} \\ x \text{ distance from } B\end{array}\right\} = M_x = \frac{W}{2}x.$$

For economical design, MR = BM.

If the breadth of cross-section is adjusted, then

$$\frac{1}{6}\sigma b_x d^2 = \frac{Wx}{2}$$

$$\therefore \text{ at } x = 0;\ b_x = 0 \text{ and at } x = \frac{L}{2};\ b_x = \frac{3WL}{2\sigma d^2}$$

$b_x = 0$ is not possible from stability point of view. Thus, a beam of uniform strength may be defined as a beam for which MR at any section is equal to the BM at that section.

PROBLEM 4.12

Objective 1

Design the cross-section profile along the length of a rectangular nonprismatic beam loaded as shown in Figure 4.31. The allowable stress in flexure is limited to 10 MPa.

Case (i): Vary the breadth keeping the depth constant as 208 mm

Case (ii): Vary the depth keeping the breadth constant as 83.20 mm

Case (iii): Vary the breadth and depth of cross-section keeping $d/b = 2.5$.

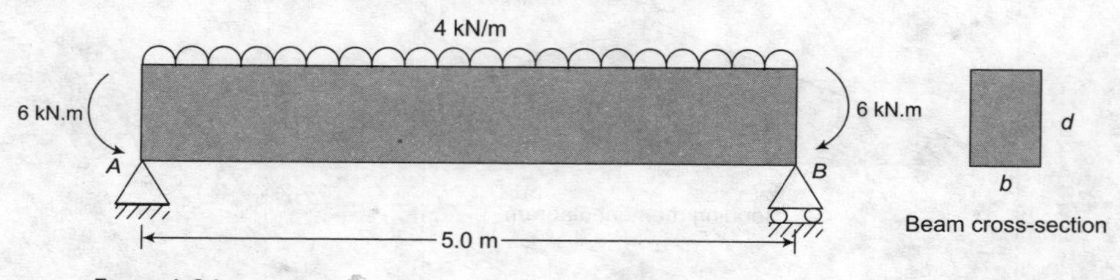

FIGURE 4.31

SOLUTION

The BMD of the beam is shown in Figure 4.32.

$$R_a = R_b = \frac{4 \times 5}{2} = 10 \text{ kN}$$

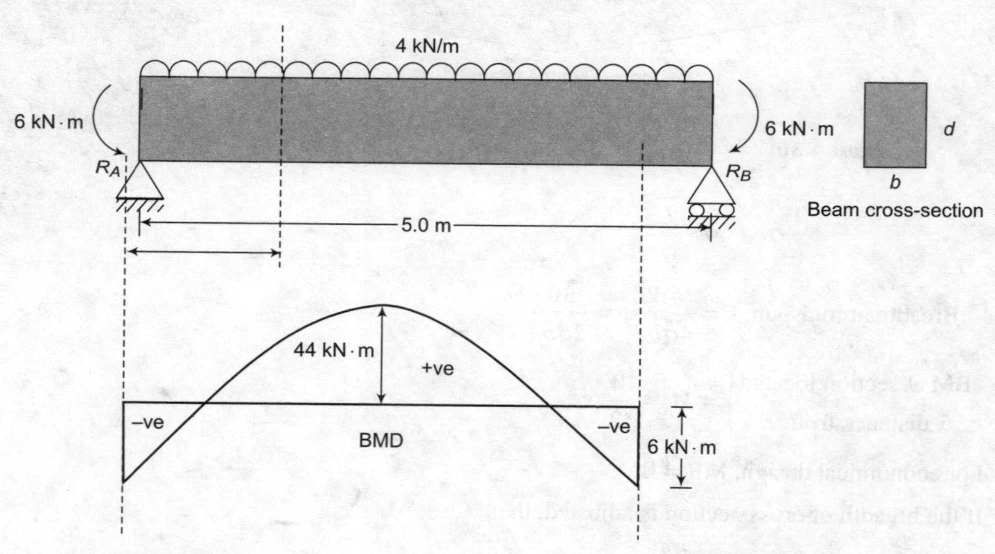

FIGURE 4.32

Let BM at a section 'x' distant from A be M_x.

$$M_x = R_a \times x - W\frac{x^2}{2} - 6 = 10x - 4\times\frac{x^2}{2} - 6$$

For point of contraflexure $M_x = 10x - 2x^2 - 6 = 0$

$$= x^2 - 5x + 3 = 0$$

$$X = 4.303 \text{ m or } 0.697 \text{ m.}$$

Two point of inflection, hence two values of x.

Maximum sagging BM occurs at midspan as the loading is symmetric.

$$\text{Maximum sagging BM} = \frac{4\times5\times5}{8} - 6 = 6.5 \text{ kN}\cdot\text{m}$$

Case (i): Depth of the section be kept constant and is given as 208 mm. To get beam of uniform strength, breadth 'b' is varied. Let b_x be the breadth at any section located 'x' distant from A.

$$\text{BM } M_x = (10x - 2x^2 - 6)\text{kN}\cdot\text{m}$$

For uniform strength condition, BM (M_x) is equal to the MR

$$\text{MR} = \sigma\times\frac{1}{6}\times b_x\times d^2$$

$$\text{MR} = 10\times\frac{1}{6}\times b_x\times(208)^2$$

Equating BM to MR

$$\text{MR} = 0.721\times10^5\, b_x$$

$$\left|(10x - 2x^2 - 6)\right|\times10^6 = 0.721\times10^5\, b_x$$

Here, absolute value of BM shall be taken as the member dimensions cannot have negative values.

$$b_x = 13.868\left|(10x - 2x^2 - 6)\right|$$

At $x = 2.5$, $b_x = 86.86$ mm

At $x = 0$, $b_x = 90.14$ mm

Breadth variation along the length must resemble the BM variation for uniform strength. The plan of beam (top view) is shown in Figure 4.33. Beam depth is constant at all places, equal to 208 mm.

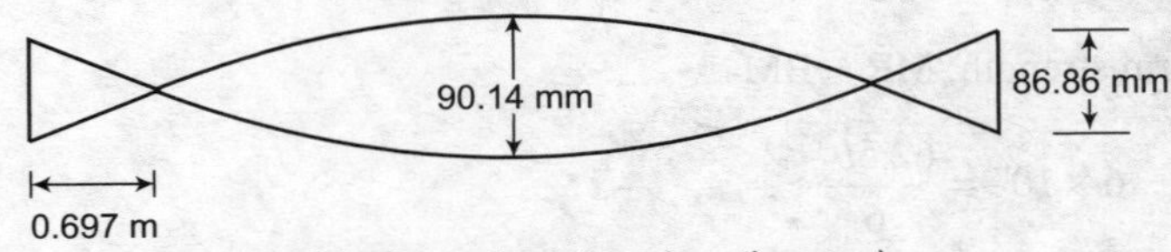

FIGURE 4.33

Case (ii): Keep the breadth constant as 83.20 mm, the beam depth is varied. Let the depth of the beam at a section located 'x' distant from support 'A'

$$MR = \sigma \times \frac{1}{6} \times b \times d_x^{\,2}$$

$$= 10 \times \frac{1}{6} \times 83.20 \times d_x^{\,2}$$

For beam of uniform strength, MR = BM

$$\left|(10x - 2x^2 - 6)\right| \times 10^6 = \frac{10 \times 83.20}{6} \times d_x^2$$

$$d_x = \sqrt{\frac{\left|(10x - 2x^2 - 6)\right| \times 10^6}{10 \times 83.20} \times 6}$$

At $x = 0$, $d_x = 208$ mm; $b = 83.20$ mm

At $x = 2.5$ m, $d_x = 216.51$ mm; $b = 83.20$ mm.

A 3D view of the beam is shown below in Figure 4.34.

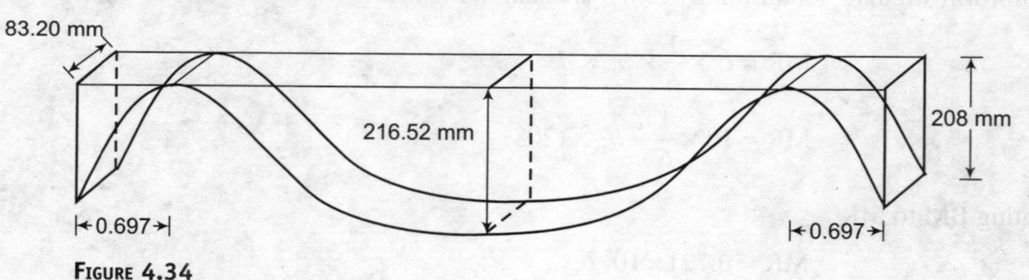

FIGURE 4.34

Case (iii): The cross-section of beam is to be designed at any section by its breadth b and depth d. Take $d/b = 2.5$.

The cross-section at A, $MR = \sigma_y \dfrac{bd^2}{6}$

$$MR = 10 \times \frac{b(2.5b)^2}{6}$$

$$MR = \frac{62.5b^3}{6}$$

For beam of uniform strength, MR = BM

$$6 \times 10^6 = \frac{62.5b^3}{6}$$

$$b = 83.20 \text{ mm}$$

$$d = 2.5 \times 83.20 = 208 \text{ mm}$$

At midspan BM = 6.5 kN m

$$6.5 \times 10^6 = \frac{62.5b^3}{6}$$

Breadth of the beam required at midspan = b = 85.45 mm and hence the depth = 2.5 × 85.45 = 213.63 mm.

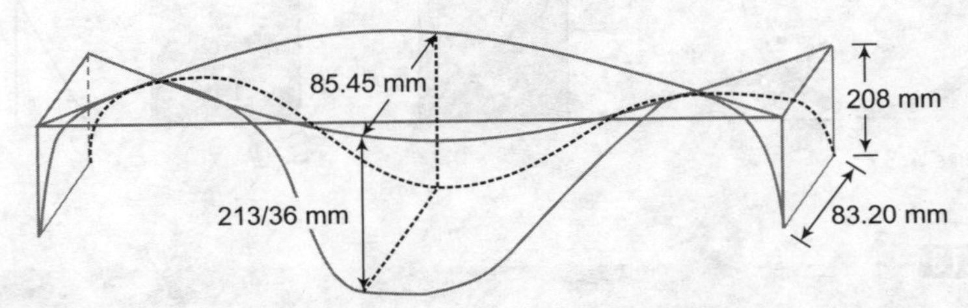

FIGURE 4.35

4.5 ECONOMIC SECTION

Economical sections are those, in which all fibers reach the ultimate stress simultaneously. This means that stress at all point within the cross or most of the point within the cross-section should reach the corresponding failure stress simultaneously. In case of bending, the material near the neutral axis is little stressed. Let us consider three cross-sections circular, rectangular, and *I* section.

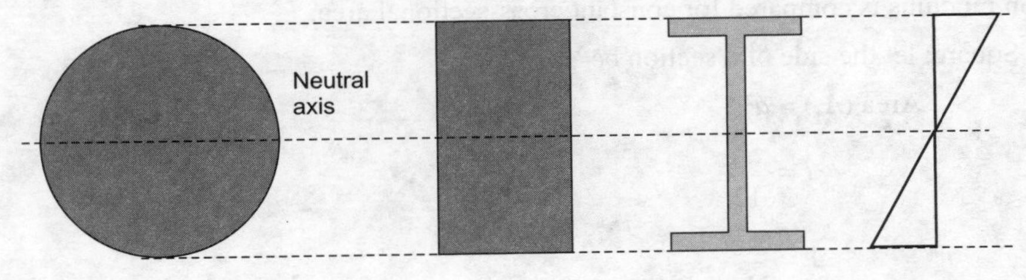

FIGURE 4.36

In Figure 4.36, it can be observed that, in circular section, very little material at top and bottom is highly stressed, whereas at centroid (at neutral axis) more material is located, which is less stressed. Thus, a circular cross-section is less efficient in resisting bending. Let us consider the case of rectangular cross-section. In rectangular section, the material available at locations of high stress (top and bottom) is more and the material available near the neutral axis (where the stress is zero) is also more. Thus, rectangular cross-section is more efficient in resisting flexure than circular cross-section. However, the rectangular section is not completely efficient in resisting flexure.

In the case of *I* section, more material is present at top and bottom, where the flexural stress is high. In the web, material is less, where the flexural stresses are relatively low. Thus, *I* section is the most efficient section in resisting bending. To make a rectangular section most efficient, remove the material near the neutral axis, that is, convert a rectangular section into a hollow/voided rectangular section. This can be observed in the example problem, where reduction in c/s area by 22% reduces the flexural resistance by only 2.35%. Thus, to make a circular section more efficient in flexure, introduce circular hole at centroid. This indicates that a hollow circular cross-section is more efficient in resisting flexure than a solid circular cross-section.

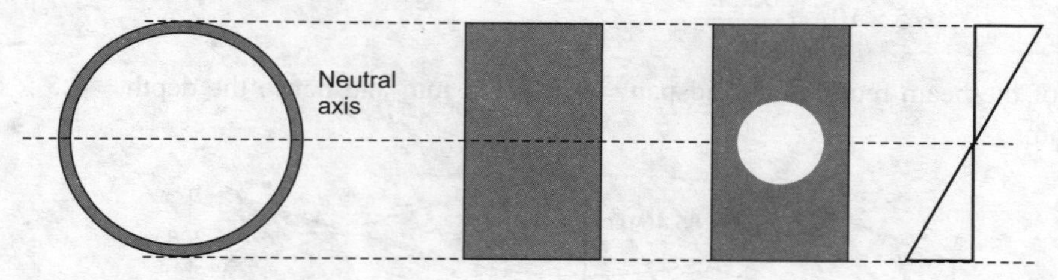

FIGURE 4.37

PROBLEM 4.13

Objective 4

Compare the flexural efficiency of the following sections

(i) Square section
(ii) Rectangular cross-section
(iii) Circular
(iv) Hollow circular
(v) Square section oriented such that diagonal is the plane of bending

SOLUTION

If the material is the same, then higher the section modulus higher will be bending strength. Thus, the section modulus is compared for constant cross-sectional area.

Case (i): Square: let the side of a section be 'a'

$$\text{Area } (A_1) = a^2$$

$$I = \frac{a^4}{12}$$

$$Y = \frac{a}{2}$$

$$Z = \frac{a^4}{12} \times \frac{2}{a} = \frac{a^3}{6}$$

$$Z_1 = 0.1667a^3$$

Case (ii): Consider rectangular section with depth to breadth ratio 'x'

$$Z_2 = \frac{I}{Y}$$

$$A_2 = bd$$

We know that $d/b = x$ and $A_1 = A_2$

$$A_1 = a^2;\ A_2 = d^2/x$$

Equating $A_1 = A_2 \Rightarrow a^2 = d^2/x$ or $d = \sqrt{x}a$

Section modulus $Z_2 = \frac{I}{Y} = 1/6\ bd^2$

$\Rightarrow$ $1/6 \cdot d^3/x$ but $d = \sqrt{x}a$

Hence $Z_2 = 1/6 \times a^3 \times \sqrt{x} = 0.1667a^3 \times \sqrt{x}$

If $x > 1$, then $Z_2 > Z_1$

Therefore, if depth of section is more than the breadth, then the corresponding cross-section will be more efficient than square cross-section. If x is less than 1, that is, breadth is more compared to depth, the corresponding cross-section is less efficient in resisting flexure compared to a square cross-section.

Case (iii): Circular section: let 'D' be diameter of the section

$$A_3 = \frac{\pi}{4}D^2 \,;\, I = \frac{\pi}{64}D^4 \,;\, y = \frac{D}{2}$$

$$Z_3 = \frac{\frac{\pi D^4}{64}}{\frac{D}{2}} = \frac{\pi D^3}{32}$$

$$A_1 = A_3 = a^2 D = \frac{2}{\sqrt{\pi}}a$$

$$Z_3 = \frac{\pi}{32} \times \frac{8}{\pi\sqrt{\pi}}a^3 = 0.141a^3$$

Z_3 is less than Z_1. Thus, a circular section is less efficient in resisting bending than a square section.

Case (iv): Let the inner diameter to outer diameter ratio of the hollow circular section be 'x'

Then, $$A_4 = \frac{\pi}{4}(d_o^2 - d_i^2) = \frac{\pi}{4}d_o^2(1 - x^2)$$

$$A_4 = A_1 = a^2$$

$\Rightarrow$ $$d_0 = \frac{2}{\sqrt{\pi}}a\sqrt{(1 - x^2)}$$

$$I_4 = \frac{\pi}{64}(d_0^4 - d_i^4) = \frac{\pi}{64}d_o^4(1 - x^4)$$

$$y_4 = \frac{d_o}{2}$$

$$Z_4 = \frac{\pi}{32}d_o^{\,3}(1 - x^4)$$

$$d_o = \frac{2a}{\sqrt{\pi}}\sqrt{(1 - x^2)}$$

$$\therefore \qquad Z_4 = \frac{\pi}{32} \times \frac{8}{\pi\sqrt{\pi}} \left| \frac{(1-x^4)}{(1-x^2)\sqrt{(1-x^2)}} \right| a^3 = \frac{1}{4\sqrt{\pi}} \left[\frac{1-x^2}{\sqrt{(1-x^2)}} \right] a^3$$

For a hollow section, $x < 1$, $Z_4 > Z_3$

Thus, a hollow circular section is more efficient in resisting flexure than a circular section.

Case (v): diamond section

$$A_5 = A_1 = a^2$$

$$I_5 = \frac{a^4}{12}; \quad y_5 = \frac{a}{\sqrt{2}}$$

$$Z_5 = \frac{\frac{a^4}{12}}{\frac{a}{\sqrt{2}}} = 0.1179a^3$$

A square section is more efficient in resisting bending than the same section when diagonal is oriented in the plane of bending.

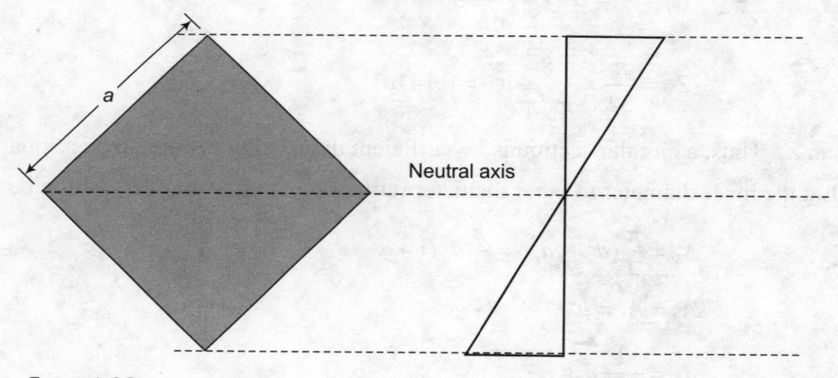

FIGURE 4.38

PROBLEM 4.14

Objective 3

Compare the bending resistance of an *I* section shown in Figure 4.39 with a square section of same cross-sectional area.

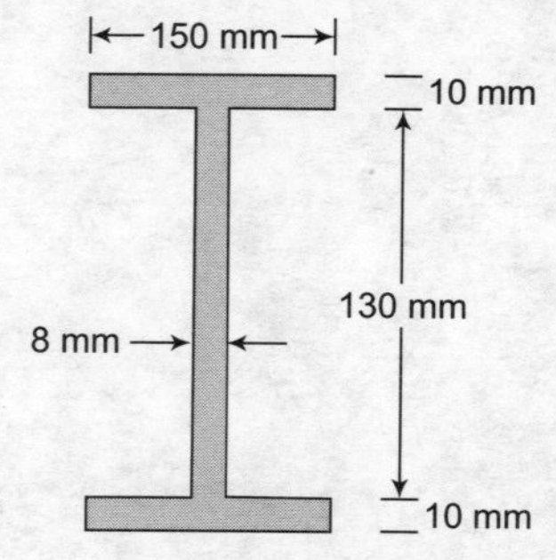

FIGURE 4.39

SOLUTION

Cross-sectional area of *I* section is 4040 mm²

Cross-sectional area of square section is 4040 mm²

Side of square section is 63.561 mm

Section modulus of *I* section

$$Z_1 = \frac{I}{y}$$

$$I = \frac{150 \times 150^3}{12} - \frac{142 \times 130^3}{12} = 16.190 \times 10^6 \text{ mm}^4$$

$$Z_1 = \frac{16.19 \times 10^6}{75} = 2.1586 \times 10^5 \text{ mm}^3$$

Section modulus of square section

$$Z_2 = \frac{1}{6} \times a^3$$

$$Z_2 = 0.428 \times 10^5 \text{ mm}^3$$

$$\frac{Z_1}{Z_2} = 5.044$$

In the present case, I section is 5.044 times stronger than square section of same area (same weight).

PROBLEM 4.15 **Objective 3**

A square cross-section of side 'a' is oriented with its diagonal in the plane of bending. The top and bottom portions of the cross-section are removed to improve the MR, as shown in Figure 4.40.

Show that for maximum strength, $h = \frac{1}{9}\left(\frac{a}{\sqrt{2}}\right)$.

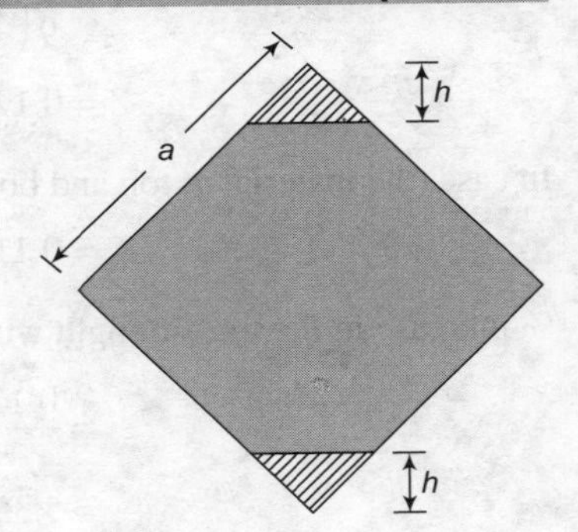

FIGURE 4.40

SOLUTION

In the ΔAEF, $EF = 2h$ (Angle $AFE = 45°$)

$$MI = \frac{[\sqrt{2}a - 2h]^3 \times 2h \times 2h}{12} + 4 \times \frac{(\frac{a}{\sqrt{2}} - h)^4}{12}$$

$$MI = \frac{1}{48}(\sqrt{2}a - 2h)^4 + \frac{1}{6}h(\sqrt{2}a - 2h)^3$$

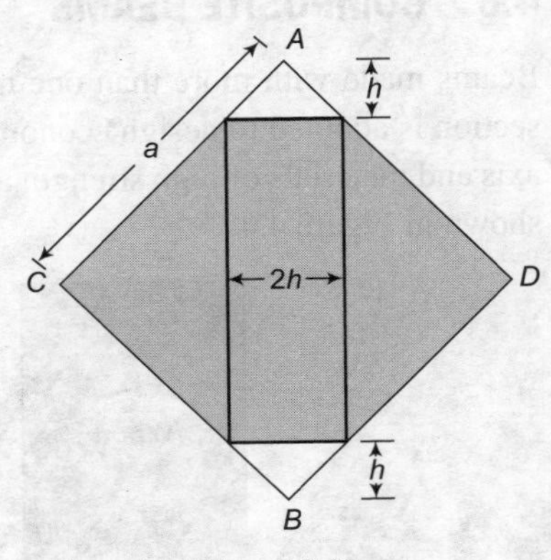

FIGURE 4.41

Section modulus

$$Z = \frac{I}{y_{\max}}$$

$$y_{\max} = \frac{a}{\sqrt{2}} - h$$

$$Z = \frac{1}{24}(\sqrt{2}a - 2h)^2 + \frac{h}{\sqrt{3}}(\sqrt{2}a - 2h)^2.$$

For maximum section modulus $dZ/dh = 0$

$$\frac{dZ}{dh} = -\frac{2\times 3}{24}(\sqrt{2}a - 2h)^2 + \frac{1}{3}(\sqrt{2}a - 2h)^2 - \frac{2\times 2h}{3}(\sqrt{2}a - 2h) = 0$$

$(\sqrt{2}a - 2h)$ cannot be zero

$$\therefore \quad -\frac{1}{4}(\sqrt{2}a - 2h) + \frac{1}{3}(\sqrt{2}a - 2h) - \frac{4h}{3} = 0$$

$$\Rightarrow \quad \frac{\sqrt{2}a - 2h}{12} - \frac{4h}{3} = 0$$

$$\Rightarrow \quad \sqrt{2}a - 2h = 16h$$

$$\Rightarrow \quad \sqrt{2}a = 18h$$

$$\Rightarrow \quad h = \frac{a}{9\sqrt{2}}$$

$$h = \frac{1}{9}\left(\frac{a}{\sqrt{2}}\right), \quad Z = \frac{1}{24}\left(\frac{8}{9}\sqrt{2}a\right)^3 + \frac{a}{27\sqrt{2}a}\left(\frac{8}{9}\sqrt{2}a\right)^2$$

$$Z = 0.12416a^3$$

In case, the material at top and bottom is not removed.

$$Z_0 = 0.11785a^3$$

% increase in flexural strength with the removal of material

$$= \frac{0.12416 - 0.11785}{0.11785} \times 100$$

$$= 5.35\%$$

4.6 COMPOSITE BEAMS

Beams made with more than one material are referred as composite beams. Generally, composite section is adopted to design economical section by providing materials of low strength near neutral axis and materials of high strength at the extreme positions. The examples of composite sections are shown in Figure 4.42

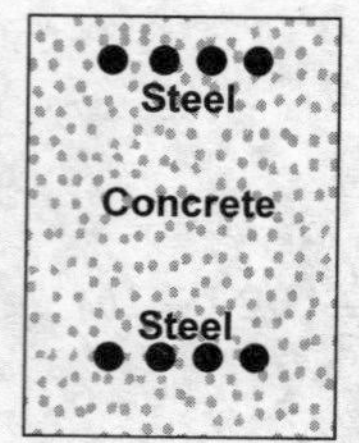

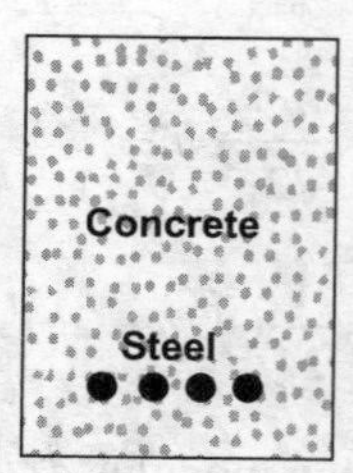

FIGURE 4.42

For the stress analysis of composite sections subjected to flexure, it is assumed that

i. perfect bond exists between different materials.
ii. plane section normal to the longitudinal axis remains plane even after bending, that is, strain variation is linear.

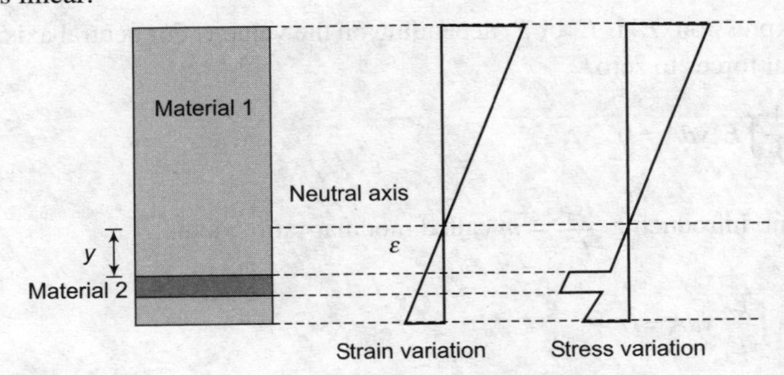

FIGURE 4.43

Strain over a fiber located 'y' distance from neutral axis is given by

$$\varepsilon = \frac{y}{R} \text{ (from simple bending derivation)}$$

Let E_1 and E_2 be the modulus of the materials (1) and (2), respectively. Radius of curvature is constant.

$$\therefore \qquad \sigma = \frac{E}{R} \times y.$$

Consider an elemental area within the cross-section of the composite section. Elemental force on the elemental area = $\sigma \times \delta A$

Total horizontal force = $\int \sigma \times dA$

$$\int_{A_1} \frac{E_1}{R} y dA + \int_{A_2} \frac{E_2}{R} y dA$$

Total moment of all elemental forces about NA = $M = \int \sigma y \times dA$

Total moment of all elemental forces about NA

$$M = \int_{A_1} \frac{E_1}{R} y^2 dA + \int_{A_2} \frac{E_2}{R} y^2 dA$$

$$= \frac{E_1 I_1}{R} + \frac{E_2 I_2}{R}$$

Radius of curvature is constant along the depth of the cross-section.

$$\therefore \qquad \frac{1}{R} = \frac{M}{E_1 I_1 + E_2 I_2}$$

$I_1 = MI$ of portion (1) about NA

$I_2 = MI$ of portion (2) about NA

Stress at any location 'y' distance NA

$$\sigma = Ey\frac{M}{E_1 y_1 + E_2 y_2}$$

In the above expression 'E' is E_1 or E_2 depending on the value y. For neutral axis, portion equates the total horizontal forces to zero.

$$\Rightarrow \frac{1}{R}\int E_1 ydA + \frac{1}{R}\int E_2 ydA = 0$$

As R is constant, introducing $\frac{E_2}{E_1} = m$ called modular ratio, yields

$$\int ydA + \int \frac{E_2}{E_1} ydA = 0$$

$$\Rightarrow \quad \int_{A_1} ydA + \int_{A_2} y(mdA) = 0.$$

The above equation represents that neutral axis coincides with the centroid of the cross-section, in which portion 'A_2' is modified as mA_2. With this modification, y should not be affected. Thus, it is convenient to modify the breadth of the portion (2) by multiplying with 'm'.

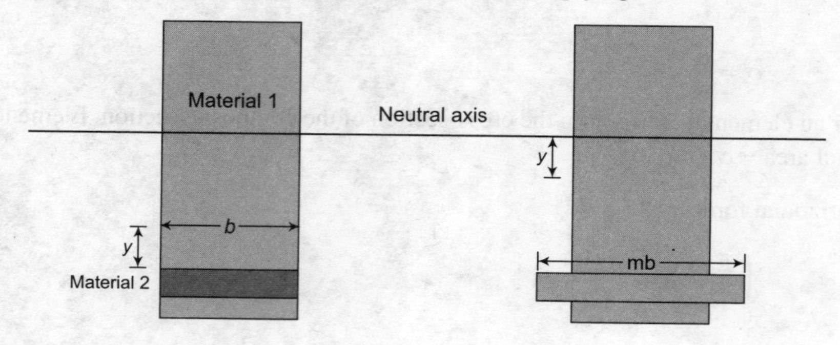

FIGURE 4.44

If we use the modifying factor the modular as $\frac{E_2}{E_1}$, the cross-section will be converted into a material of single modulus E_1.

Stress over a fiber located in the region (1) located y distant from NA

$$\sigma = \frac{E_1 My}{E_1 I_1 + E_2 I_2}$$

$$\sigma = \frac{My}{I_1 + mI_2}$$

$I_1 + mI_2$ is second moment area of transformed section

$$\sigma = \frac{M}{I_t} y$$

Stress over a fiber located in the region (2) located

$$\frac{E_2 M y}{E_1 I_1 + E_2 I_2} = \frac{E_2 M y}{E_1 \left(I_1 + \frac{E_2}{E_1} I_2 \right)}$$

y distant from NA $\sigma = m \dfrac{My}{I_1 + MI_2}$

$$= m \frac{My}{I_t}$$

From the expression, it is clear that flexure formulae $\sigma = \dfrac{M}{I} y$ can be used for finding the stress. In place of 'I', the second moment area of transformed section is to be used. Stress in the modified regions is to be multiplied by the modular ratio to quantify the stress in the actual material.

PROBLEM 4.16 **Objective 1**

Sketch the variation of bending stress across the depth of a composite section of two materials wood and steel as shown in Figure 4.45, subjected to 20 kN/m; $E_s = 200$ GPa and $E_w = 20$ GPa.

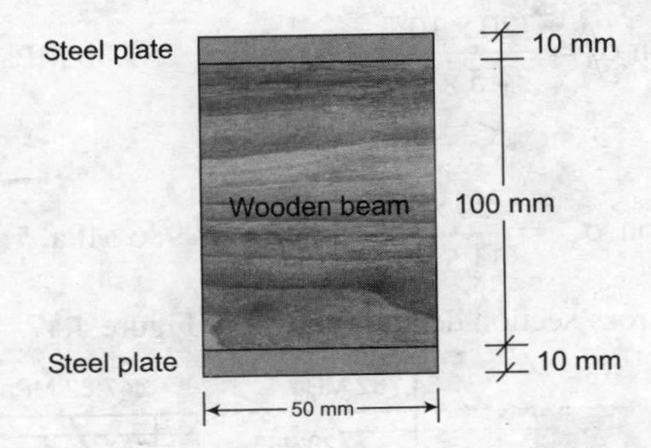

FIGURE 4.45

SOLUTION

Let us convert the steel into wood, so that the total cross-section is of wood.

$$\text{Modular ratio } m = \frac{E_S}{E_W} = \frac{200}{20} = 10$$

Breadth in the region where steel is present should be multiplied by 'm'. The transformed section is shown in Figure 4.46.

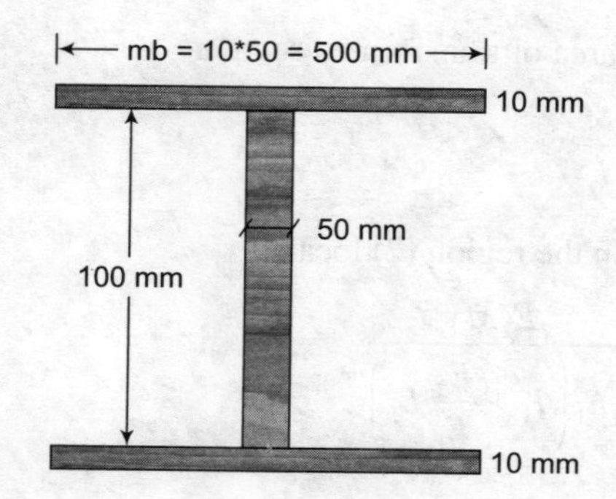

FIGURE 4.46

NA is located 60 mm from top of the section. MI of the transformed section is

$$I = \frac{1}{12}(500 \times 120^3 - 450 \times 100^3)$$

$$I = 34.5 \times 10^6 \text{ mm}^4$$

Stress at the top most fiber

$$\sigma_t = \frac{20 \times 10^6}{34.5 \times 10^6} \times 60$$

$$\sigma_t = 34.782 \text{ MPa}$$

Stress in steel at top fiber = 34.782 × 10 = 347.82 MPa

Stress in steel at the junction $\sigma_{js} = \dfrac{20 \times 10^6}{34.5 \times 10^6} \times 50 \times 10 = 289.86$ MPa

Junction of steel and wood

Stress in wood at the junction $\sigma_{jw} = \dfrac{20 \times 10^6}{34.5 \times 10^6} \times 50 = 28.986$ MPa

Stress variation across the cross-section depth is shown in Figure 4.47.

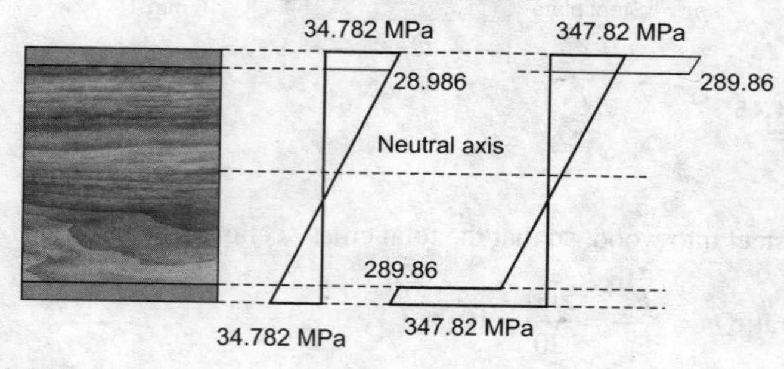

FIGURE 4.47

PROBLEM 4.17

Objective 1

Sketch the variation of bending stress across the depth of two composite sections shown in Figure 4.48. $E_S = 200$ GPa; $E_b = 100$ GPa; $E_a = 50$ GPa. BM acting on the section is 30 kN·m.

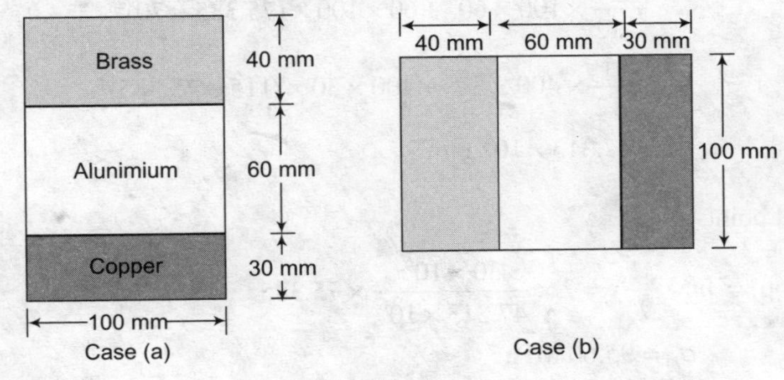

FIGURE 4.48

SOLUTION

Let the materials steel and brass be converted into aluminum.

$$\text{Modular ratio for steel } m_s = \frac{E_s}{E_a} = \frac{200}{50} = 4$$

Breadth of steel portion is $= m_s \times b = 4 \times 100 = 400$ mm

$$\text{Modular ratio for brass } m_b = \frac{E_b}{E_a} = \frac{100}{50} = 2$$

Breadth of steel portion is $= m_s \times b = 2 \times 100 = 200$ mm

Case (a): Transformed section of case (a) is shown in Figure 4.49.

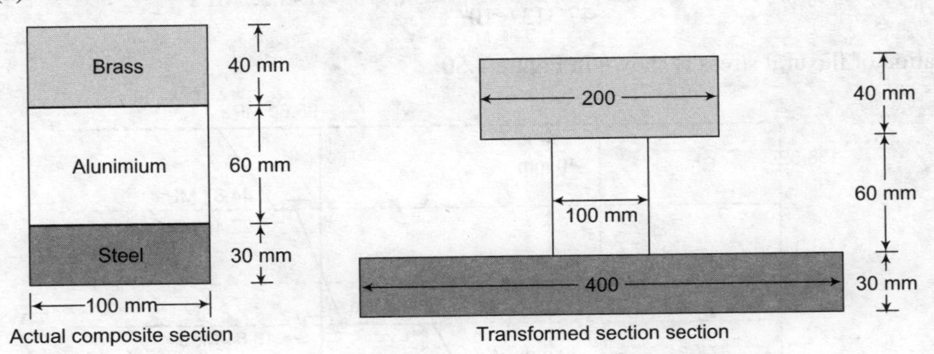

FIGURE 4.49

Let $\bar{y}$ be the centroid of the transformed from top.

$$\bar{y} = \frac{40 \times 200 \times 20 + 60 \times 100 \times (40 + 30) + 400 \times 30 \times (40 + 60 + 15)}{200 \times 40 + 60 \times 100 + 400 \times 30}$$

$$= \frac{1960000}{26000} = 75.385 \text{ mm}$$

Second moment area

$$I = \frac{1}{12} \times 200 \times 40^3 + 200 \times 40 \times (75.385 - 20)^2$$
$$+ \frac{1}{12} \times 100 \times 60^3 + 60 \times 100 \times (75.385 - 70)^2$$
$$+ \frac{1}{12} \times 400 \times 30^3 + 400 \times 30 \times (115 - 75.385)^2$$
$$I = 47.313 \times 10^6 \text{ mm}^4$$

Stress at different points

Stress in brass (top) = $m_b \cdot \frac{M}{I} y = 2 \times \frac{30 \times 10^6}{47.313 \times 10^6} \times 75.385$

$$\sigma_{bt} = 95.60 \text{ MPa}$$

Stress in brass (bottom) at junction = $\sigma_{bb} = 2 \times \frac{30 \times 10^6}{47.313 \times 10^6} \times 35.385 = 44.87$ MPa

Stress in aluminum (top) = $\sigma_{at} = \frac{30 \times 10^6}{47.313 \times 10^6} \times 35.385 = 22.435$ MPa

Stress in aluminum (bottom) = $\sigma_{ab} = \frac{30 \times 10^6}{47.313 \times 10^6} \times 24.615 = 15.61$ MPa

Stress in steel (top) = $\sigma_{st} = 4 \times \frac{30 \times 10^6}{47.313 \times 10^6} \times 24.615 = 62.43$ MPa

Stress in steel (bottom) = $\sigma_{sb} = 4 \times \frac{30 \times 10^6}{47.313 \times 10^6} \times 54.615 = 138.52$ MPa

Variation of flexural stress is shown in Figure 4.50.

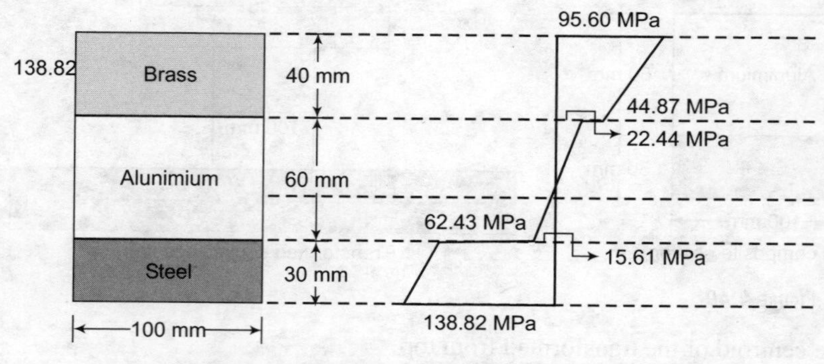

Flexural stress variation across the depth of the cross-section

FIGURE 4.50

Case (b): The transformed section is shown in Figure 4.51.

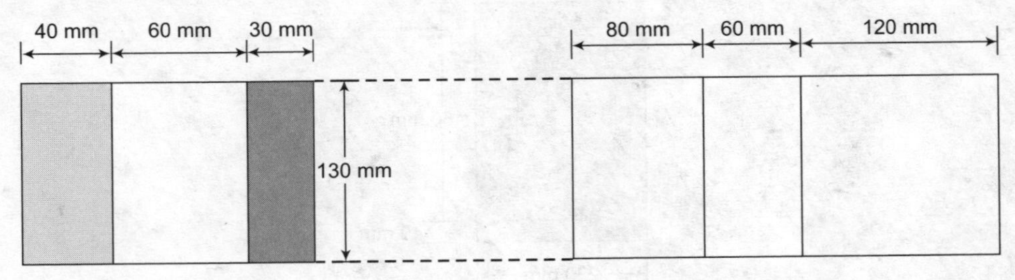

FIGURE 4.51

$$I = \frac{130^3 \times 260}{12} = 47.062 \times 10^6 \text{ mm}^4$$

$$\text{Stress at top fiber and bottom} = \frac{30 \times 10^6}{47.602 \times 10^6} \times 65 = 40.96 \text{ MPa}$$

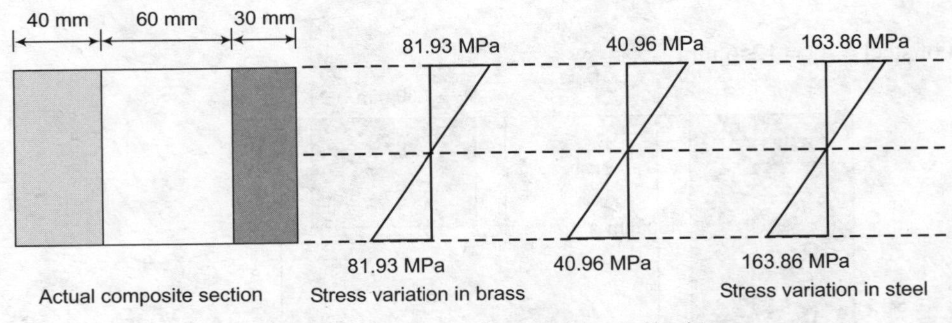

FIGURE 4.52

Stress at bottom as well as at top in brass portion

$$m_b \cdot \frac{M}{I} y = 2 \times \frac{30 \times 10^6}{47.602 \times 10^6} \times 65 = 81.93 \text{ MPa}$$

Stress at bottom as well as at top in steel portion

$$= 4 \times \frac{30 \times 10^6}{47.602 \times 10^6} \times 65 = 163.86 \text{ MPa}$$

PROBLEM 4.18 **Objective 1**

Determine the stresses developed in a reinforced concrete beam subjected to a BM of 36 kN·m. Neglect the concrete present in the tension. To resist tension steel of cross-sectional area 1256 mm² is provided 40 mm from the bottom of the beam. Take $E_s/E_c = 7$, breadth of the section is 230 mm. Depth of the section is 500 mm. Depth of the beam to the center of reinforcement is 460 mm.

SOLUTION

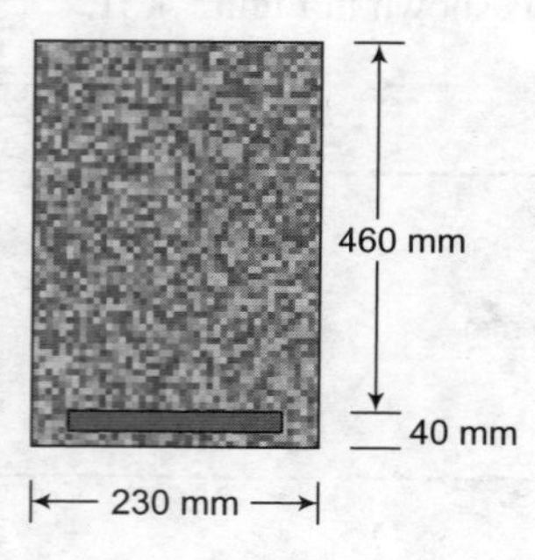

FIGURE 4.53

The transformed section of the beam cross-section is shown in Figure 4.54. Resistance of concrete in tension region is very poor and hence contribution of concrete in tension region is ignored. Thus, it is assumed that no concrete is present in the tension region. Only reinforcement takes the tensile stress. The concrete present in the tension is only to provide bond between concrete in compression region and reinforcement in the tension region.

Modular ratio $m = \dfrac{E_s}{E_c} = 7$

Area of steel = A_{st} = 1256 mm^2

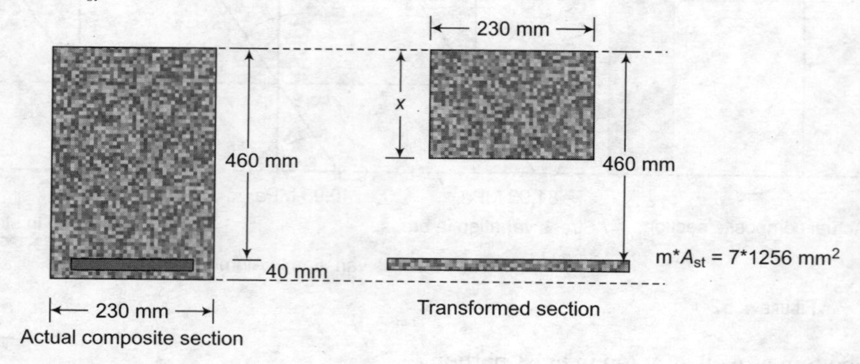

FIGURE 4.54

Let 'x' be the position of neutral axis, that is, centroidal axis of the transformed section. It is to be noted that tensile strength of concrete is ignored or concrete in tension zone is ignored.

Take the reference axis as neutral axis itself then

$$y_c = 0 = \frac{A_1 y_1 + A_2 y_2}{A_1 + A_2} \Rightarrow A_1 y_1 + A_2 y_2 = 0$$

$$\Rightarrow \quad 230 \times x \times \frac{x}{2} - m \times A_{st}(460 - x) = 0$$

$$\Rightarrow \quad 115x^2 - 7 \times 1256(460 - x) = 0$$

$$\Rightarrow \quad 115x^2 + 8792x - 4044320 = 0$$

$$\Rightarrow \quad x = 153.16 \text{ mm}$$

MI of the transformed section (here MI of steel about its own centroidal axis is neglected)

$$I = \frac{230 \times 153.16^3}{3} + 7 \times 1256 \times (460 - 153.16)^2 = 11.032 \times 10^8 \text{ mm}^4$$

Stress in the top most fiber in concrete

$$\sigma_{ct} = \frac{36 \times 10^6}{11.032 \times 10^8} \times 153.16 = 5 \text{ MPa (comp)}$$

Stress in the steel fiber

$$\sigma_s = 7 \times \frac{36 \times 10^6}{11.032 \times 10^8}(460 - 153.16) = 70.09 \text{ MPa}$$

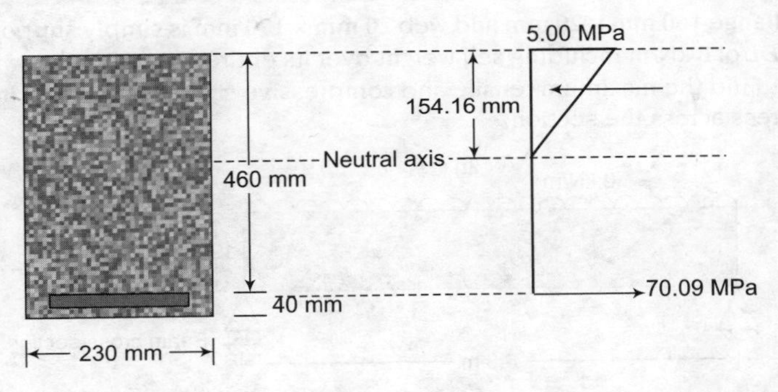

FIGURE 4.55

PROBLEM 4.19

Objective 4

Find the dimensions of the strongest rectangular beam that can be cut out of a log of wood 2.6 m diameter.

SOLUTION

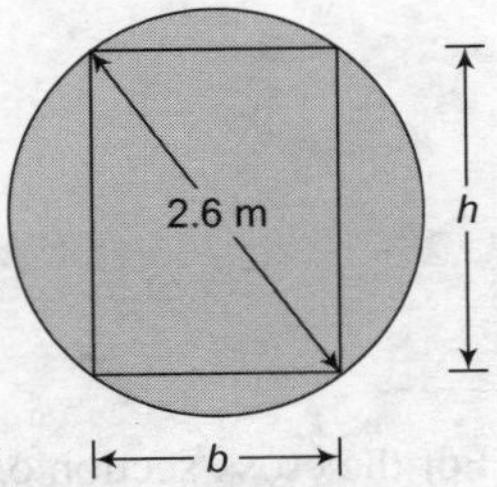

Diameter of the circular log = 2.6 m

Let b and h be the dimensions of the rectangular cross-section that can be cut from a circular log. For a cross-section to be strong in flexure, its section modulus should be maximum.

FIGURE 4.56

Section modulus of rectangular section

$$Z = \frac{I}{Y_{\max}} = \frac{\frac{bh^3}{12}}{\frac{h}{2}} = \frac{bh^2}{6}$$

We know that, $b^2 + h^2 = 2.6^2$

$$\Rightarrow \qquad h^2 = 2.6^2 - b^2$$

$$\therefore \qquad Z = \frac{b(2.6^2 - b^2)}{6}$$

For Z to be maximum,

$$\frac{dZ}{db} = 0$$

$$\Rightarrow \quad 2.6^2 - 3b^2 = 0$$

$$\Rightarrow \quad b = \sqrt{\frac{2.6^2}{3}} = 1.50 \text{ m}$$

$$h = \sqrt{2.6^2 - 1.5^2} = 2.12 \text{ m}$$

PROBLEM 4.20

Objective 1

A *T* beam having flange 160 mm × 20 mm and web 20 mm × 170 mm is simply supported over a span of 6.5 m. It carries *UDL* of 6 kN/m including self weight over its entire span, together with appoint load of 40 kN at midspan. Find the maximum tensile and compressive stresses occurring in the beam section and sketch the stress across the section.

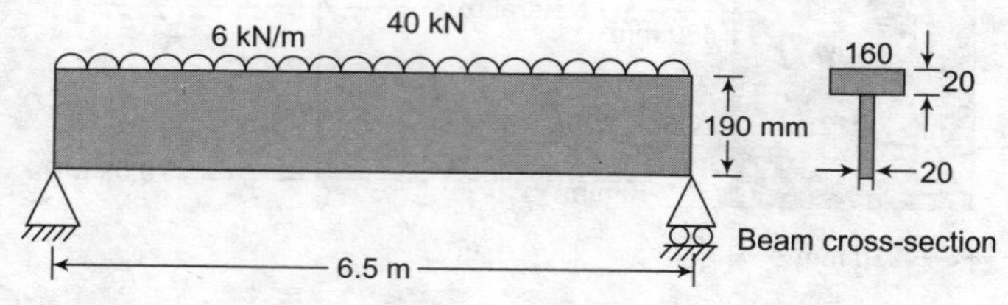

FIGURE 4.57

SOLUTION

Maximum BM in the beam $= \dfrac{wl^2}{8} + \dfrac{WL}{4}$

w = *UDL*

W = concentrated load at midspan

$$M_{\max} = \frac{6 \times 6.5^2}{8} + \frac{40 \times 6.5}{4} = 96.6875 \text{ kN m}$$

$$= 96.6875 \times 10^6 \text{ N} \cdot \text{mm}$$

For the cross-section of the beam,

$$\bar{x} = \frac{160 \times 20 \times 10 + 170 \times 20 \times (20 + \frac{170}{2})}{160 \times 20 + 170 \times 20}$$

$$\bar{x} = 58.94 \text{ mm}$$

$I = MI$ about NA

$$= \frac{160 \times 20^3}{12} + 160 \times 20(58.94 - 10)^2$$

$$+ \frac{170^3 \times 20}{12} + 170 \times 20\left(20 + \frac{170}{2} - 58.94\right)^2$$

$$= 23.172 \times 10^6 \text{ mm}^4$$

Bending stress $\sigma = \frac{M}{I} y$

$$\sigma = \frac{96.6875 \times 10^6}{23.1728 \times 10^6} y$$

y is location of fiber from neutral axis, where σ is to be found.

$\Rightarrow \qquad \sigma = 4.173y$

At top fiber $\sigma_{top} = 4.173 \times 58.94 = 245.96$ MPa (comp)

At bottom fiber $\sigma_b = 4.173(190 - 58.94) = 546.91$ MPa (tensile)

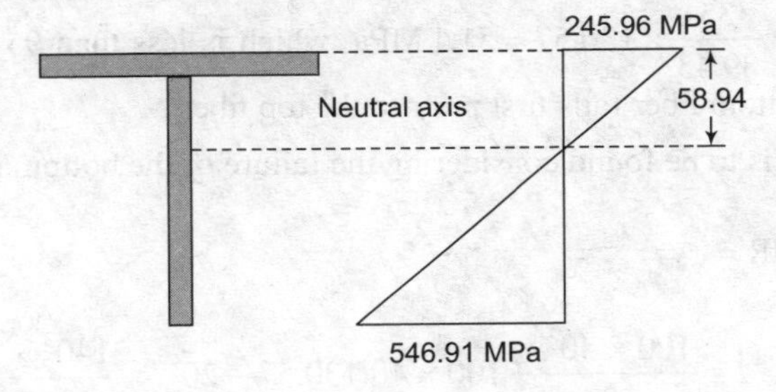

Figure 4.58

PROBLEM 4.21 **Objective 3**

A cast iron beam has I section with top flange 100 mm × 40 mm, web 140 mm × 20 mm, and bottom flange 180 mm × 40 mm. If tensile stress does not exceed 35 MPa and compressive stress 95 MPa, what is the maximum *UDL* the beam can carry over a simply supported span of 6.5 m, if the larger flange is in tension.

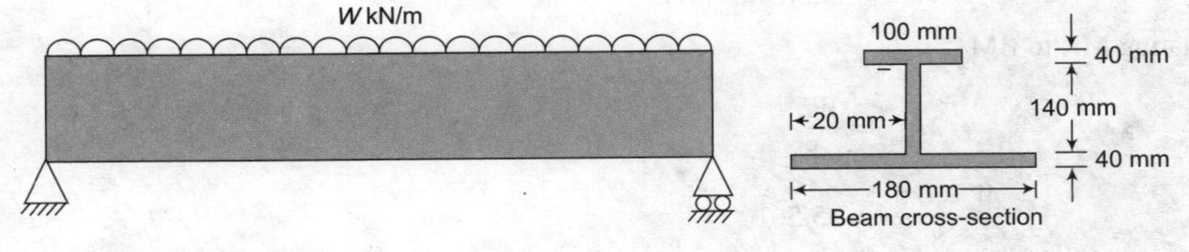

Figure 4.59

SOLUTION

As the beam is simply supported one, it is subjected to sagging BM. Bottom fibers are subjected to tension and top fibers are subjected to compression.

Let $\bar{y}$ be the location centroidal neutral axis of the cross-section from top

$$\bar{y} = \frac{100 \times 40 \times 20 + 140 \times 20 \times 110 + 180 \times 40 \times 200}{100 \times 40 + 140 \times 20 + 180 \times 40}$$

$$= 130.57 \text{ mm}$$

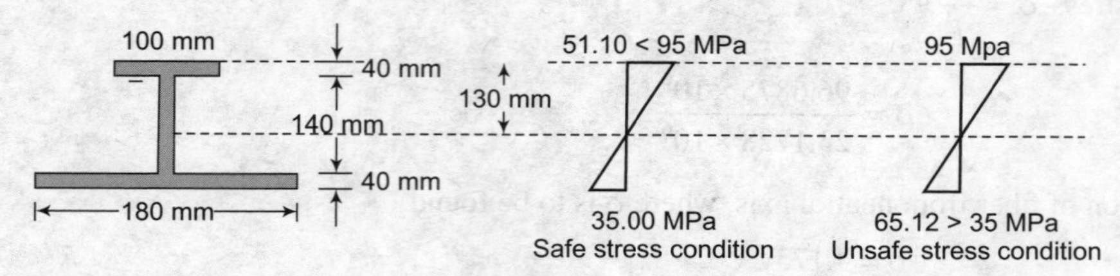

FIGURE 4.60

For the condition of failure, if the bottom fiber is 35 MPa (failure stress in tension) compressive stress at the top fiber = $\frac{35}{89.43} \times 130.57 = 51.1$ MPa which is less than 95 MPa (failure stress in compression). Hence bottom fiber fails first prior to the top fiber.

MR of the cross-section is to be found considering the failure of the bottom fiber.

$$\text{MR} = \sigma \frac{I}{y_b}$$

$$I = \frac{100 \times 40^3}{12} + 100 \times 40(130.57 - 20)^2 + \frac{140^3 \times 20}{12} + 140 \times 20$$

$$(130.57 - 110)^2 + \frac{180 \times 40^3}{12} + 180 \times 40(130.57 - 200)^2$$

$$= 90.862 \times 10^6 \text{ mm}^4$$

$$\text{MR} = 35 \times \frac{90.862 \times 10^6}{89.43}$$

$$= 35.561 \text{ kN m}$$

Equating MR to BM

$$M_{\max} = \frac{WL^2}{8}$$

$$\frac{W \times 6.5^2}{8} = 35.561$$

$$W = 6.733 \text{ kN/m.}$$

PROBLEM 4.22 **Objective 4**

A water main 110 mm internal diameter is made of mild steel plate 12 mm thick and is running full. If it is freely supported at the ends, find the maximum permissible span, if the bending stress is not to exceed 5MPa. Unit weight of steel is 81 kN/m³ and unit weight of water is 10 kN/m³.

SOLUTION

d = inner diameter of the water main = 110 mm

t = thickness of the water main = 12 mm

Weight of water and weight of water main are the loads on the water main resisting between two supports.

Density of water = 10 kN/m³

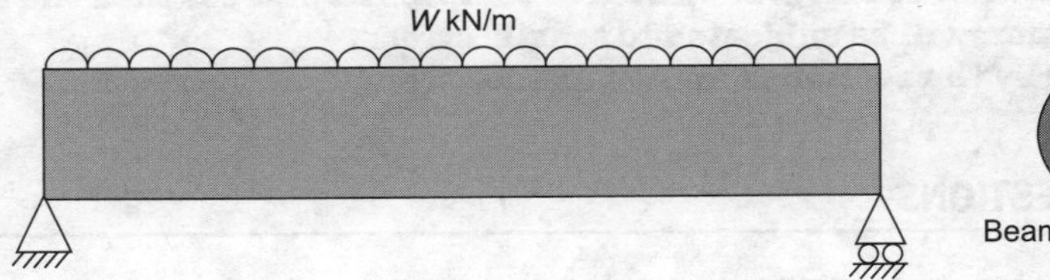

FIGURE 4.61

W_1 = Weight of water in the water main per m = $\frac{\pi}{4}(0.11)^2 \times 1 \times 10 = 0.095$ kN/m

W_2 = Weight of water main = $\pi \times (0.11)\frac{12}{1000} \times 81 = 0.336$ kN/m

$$W = W_1 + W_2 = 0.431 \text{ kN/m}$$

Maximum BM; $M = 0.431 \times l^2/8$

Bending stress is limited to 5 MPa.

Outer diameter of the water main = 110 + 2 × 12 = 134 mm

Inner diameter of the water main = 110 mm

MI of the cross-section $I = \frac{\pi}{64}(134^4 - 110^4)$

$$= 8.64 \times 10^6 \text{ mm}^4$$

$$Y = 134/2 = 67 \text{ mm}$$

Max normal stress $\sigma = \frac{M}{I}y$

Max normal stress is limited to 5 MPa.

$$5 = \frac{\left(\frac{0.431 \times l^2}{8}\right) \times 10^6}{8.64 \times 10^6} \times 67$$

$$\Rightarrow \quad 0.431 \times \frac{l^2}{8} = 0.645$$

$$\Rightarrow \quad l = 3.46 \text{ m}$$

Span between the supports shall be less than 3.46 m.

SUMMARY

- **The stresses produced due to constant BM are known as bending stresses.**
- **The bending stress in any layer is directly proportional to the distance of the layer from the neutral axis of the member.**
- **The bending stress on the NA is zero.**
- **If the top layer of the section is subjected to compressive stress then the bottom layer of the section will be subjected to tensile stress.**
- **The MR offered by the section is known as the strength of the section.**

OBJECTIVE TYPE QUESTIONS

1. Two beams, one having square cross-section and another circular cross-section, are subjected to the same amount of BM. If the cross-sectional area as well as the material of both the beams are the same then [GATE-2003]

(a) Maximum bending stress developed in both the beams is the same
(b) The circular beam experiences more bending stress than the square one
(c) The square beam experiences more bending stress than the circular one
(d) As the material is same both the beams will experience same deformation

2. Beam *A* is simply supported at its ends and carries *UDL* of intensity w over its entire length. It is made of steel having Young's modulus *E*. Beam *B* is cantilever and carries a *UDL* of intensity *w*/4 over its entire length. It is made of brass having Young's modulus *E*/2. The two beams are of same length and have same cross-sectional area. If *sA* and *sB* denote the maximum bending stresses developed in beams A and B, respectively, then which one of the following is correct? [IES-2005]

(a) sA/sB
(b) $sA/sB < 1.0$
(c) $sA/sB > 1.0$
(d) sA/sB depends on the shape of cross-section

3. Consider the following statements in case of beams: [IES-2002]

1. Rate of change of SF is equal to the rate of loading at a particular section
2. Rate of change of BM is equal to the SF at a particular section.
3. Maximum SF in a beam occurs at a point where BM is either zero or BM changes sign

Which of the above statements are correct?

(a) 1 alone (b) 2 alone (c) 1 and 2 (d) 1, 2, and 3

4. A T-section beam is simply supported and subjected to a *UDL* over its whole span. Maximum longitudinal stress at [IES-2011]

(a) Top fiber of the flange
(b) The junction of web and flange
(c) The mid-section of the web
(d) The bottom fiber of the web

5. Two beams of equal cross-sectional area are subjected to equal BM. If one beam has square cross-section and the other has circular section, then [IES-1999]

(a) Both beams will be equally strong
(b) Circular section beam will be stronger
(c) Square section beam will be stronger
(d) The strength of the beam will depend on the nature of loading

6. Assertion (A): For structures steel, I beams preferred to other shapes. Reason (R): In I beams, a large portion of their cross-section is located far from the neutral axis. [IES-1992]
   (a) Both A and R are individually true and R is the correct explanation of A
   (b) Both A and R are individually true but R is NOT the correct explanation of A
   (c) A is true but R is false
   (d) A is false but R is true

Solutions for Objective Questions

Sl.No.	1.	2.	3.	4.	5.	6.
Answer	(b)	(d)	(c)	(d)	(b)	(a)

EXERCISE PROBLEMS

1. Obtain an expression for the section modulus of an equilateral triangular section of side 'a'.
2. Compare the flexural strength of beams with following cross-sections of equal weight, same material, and same length. Show the results with reference to the I section.
   (a) I section with 150 mm × 20 mm flanges and 12 mm × 260 mm web;
   (b) rectangular section with depth equal to twice of its width;
   (c) square section; and (d) circular section
3. A beam of I section of MI 95 cm^4 and depth 14 cm is freely supported at its ends. Over what span can a uniform load of 500 kg/m can be carried if the maximum stress is 60 N/mm². What additional central load can be carried when the maximum stress is 90 N/mm².
4. A steel plate of 100 mm wide and 15 mm thick is bent into circular arc of 8 m radius. Determine the maximum bending stress induced and the BM, which can produce this stress. $E = 200$ GPa.
5. Determine the allowable superimposed UDL on a span 4 m simply supported beam with symmetrical I section consisting of 150 mm × 20 mm flanges and 20 mm × 150 mm web, if the allowable bending stress is 150 MPa and the unit weight of beam material is 78.5 kN/m³. Find the percentage of BM resisted by web and flanges.
6. The cross-section of a cast iron beam is an unsymmetrical I section with 150 mm × 20 mm top flange, 20 mm × 200 mm web, and 250 mm × 20 mm bottom flange. The allowable bending stresses are 30 MPa in tension and 50 MPa in compression.

   Determine the MR of the section,
   (i) When it is used as simply supported beam
   (ii) When it is used as cantilever beam
7. A symmetrical double over hanging beam of symmetrical I section of MI 954 cm^4 and depth 14 cm. The overhanging portion is 1.2 m and the middle span is 6.6 m. Determine the safe UDL that can be placed in beam, if the maximum stress is limited to 60 N/mm².
8. A rectangular beam of size 20 mm × 400 mm is used for resisting flexural loads (sagging in nature). For economy sake, a circular (symmetrical about loading plane) of size 80 mm has been introduced in the beam cross-section, through the beam length. The allowable stresses in the material of the beam are 60 MPa (compression) and 90 MPa (tension). Determine the location fail simultaneously. Also estimate the flexural strength of the cross-section for this configuration.
9. A timber beam 72 mm wide by 144 mm deep is to be reinforced by bonding strips of aluminum alloy 72 mm wide on top and bottom faces, over the whole length of the beam. If the MR of the composite beam is to be four times that of the timber alone with the same value of maximum stress in the timber,

determine the thickness of alloy strip and the ratio of maximum stresses in alloy and timber.

10. Determine the MR due to sagging moment of an unsymmetrical I section of top flange 100 mm × 10 mm, web 250 mm × 8 mm, and bottom flange 200 mm × 10 mm. The allowable tensile stress is 20 MPa and the compressive is 40 MPa.

CHAPTER 5

FLEXURAL SHEAR STRESS

UNIT OBJECTIVE

This chapter deals with the flexural shear stresses of composite sections and various types of sections. The presentation attempts to help the student achieve the following:

Objective 1: Determine shear stress variations of a triangular, circular, square, H, I, and composite sections subjected to varying bending moment (BM).

Objective 2: Determine shear stress variations of composite sections subjected to varying BM.

Objective 3: Evaluate average and maximum shear stress of different sections subjected to varying BM.

5.1 INTRODUCTION

In the previous chapter, we have seen how to quantify stresses developed due to bending moment (BM). In developing the expression for bending stresses, it was assumed that BM is constant and shear force (SF) is zero. But this is a rare condition; generally, in all practical problems SF and BM occur together. Thus, the bending is not uniform bending but it is varying flexure. The varying BM introduces SF there by shearing stresses in addition to the normal stress due to flexure. In this chapter, we quantify the shearing stresses produced due to varying BM.

5.2 EXPRESSION FOR FLEXURAL SHEAR STRESS

Consider a beam subjected to varying moment. Take an elemental strip of length 'δx' in the longitudinal direction of the beam as shown in Figure 5.1.

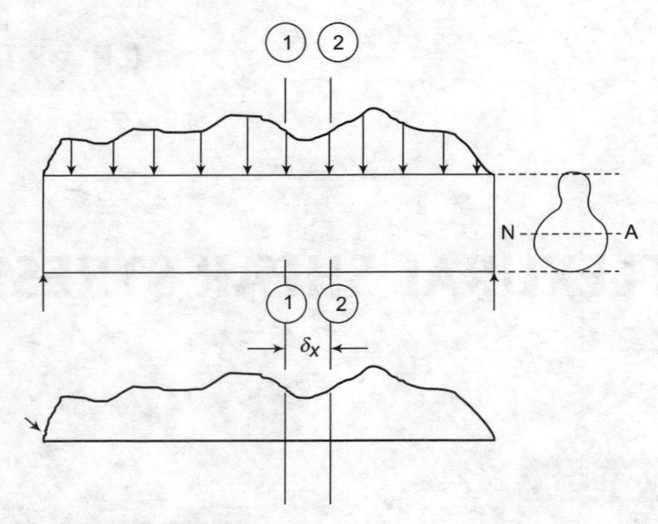

FIGURE 5.1

At section (1)-(1), the BM is M, while the BM at section (2)-(2) is $M + \delta M$. At section (1)-(1), normal stress develops due to BM.

The normal stress at section (1)-(1) is $\frac{M}{I}\xi$, in which ξ is the location of elemental area δA with in the cross-section from neutral axis (NA).

The normal stress at the section (2)-(2) is $\frac{M + \delta M}{I}\xi$.

The normal stresses (bending) at sections (1)-(1) and (2)-(2) were shown in Figure 5.2.

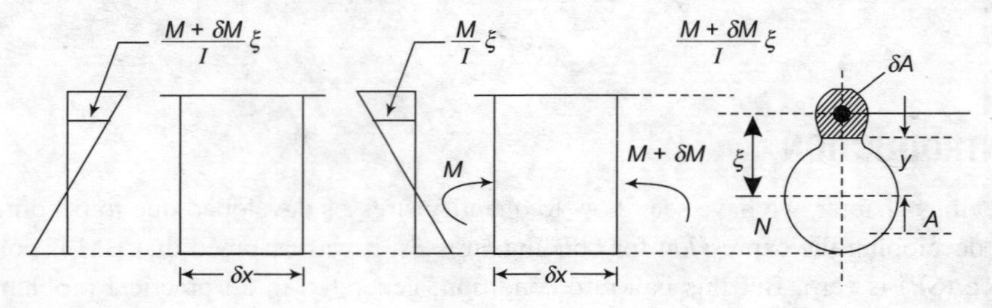

FIGURE 5.2

Let us consider the equilibrium of force acting at sections (1)-(1) and (2)-(2) just above the fiber considered, where shearing stress are to be found, that is, y distance from NA.

Stress at section (1)-(1) at ξ distance from NA $= \frac{M}{I}\xi$

Force on the elemental area within the cross-section located ξ distance from NA $= \frac{M}{I}\xi\,\delta A$

Total force acting on shaded area of cross-section shown in Figure 5.2

$$F_1 = \int_y^{y_t} \frac{M}{I}\xi\, dA.$$

Similarly, the total force acting on the shaded area of the cross-section (2)-(2)

$$F_2 = \int_y^{y_t} \frac{M+\delta M}{I}\xi\, dA.$$

The free body diagram (*FBD*) of the shaded portion over length δx is shown in Figure 5.3.

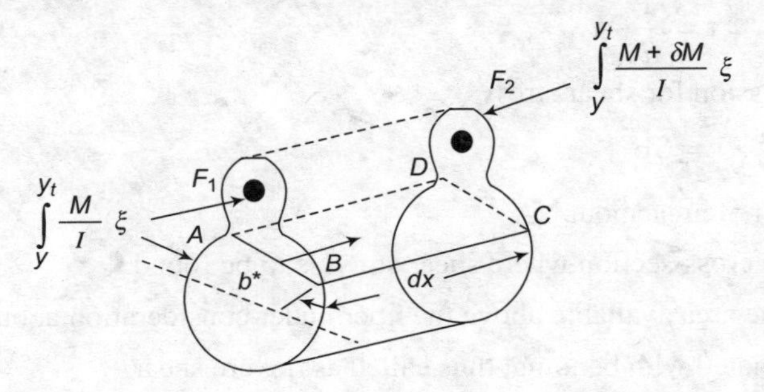

FIGURE 5.3

For equilibrium, sum of forces in the horizontal direction/longitudinal direction of the element should be zero.

From the *FBD*, it is clear that there should be force equal to $F_2 - F_1$ to maintain equilibrium. Thus, force cannot act on the surface, and hence it has to act along the plane *ABCD*.

The force acting along the plane

$$= F_2 - F_1$$

$$= \int_y^{y_t} \frac{\delta M}{I}\xi\, dA$$

$$\Rightarrow \text{SF on plane } ABCD = \int_y^{y_t} \frac{\delta M}{I}\xi\, dA$$

$$\text{Shear stress acting on plane } ABCD = \frac{F_2 - F_1}{(\delta X)b_*}$$

in which b^* is the breadth available over the fiber considered, located at y distance from NA, where shear stress is to be determined.

$$\tau = \frac{\int_y^{y_t} \frac{\delta M}{I}\xi}{\delta X \times b_*} dA = \frac{\delta M}{\delta X} \times \frac{1}{Ib_*} \times \int_y^{y_t} \xi\, dA$$

δM is constant over the length δx.

b_* does not depend on dA. We know that $\frac{dM}{dx} = V$, the SF acting over the cross-section

$$\tau = \frac{V}{Ib_*}\int_{y}^{y_t} \xi dA.$$

This shear stress τ is called horizontal shear stress. The horizontal shear stress induces vertical shear over the cross-section called vertical shear. The vertical shear stress and the horizontal shear stress are same in magnitude as they are complimentary to each other.

$$\tau = \frac{V}{Ib_*}Ay.$$

In the above expression for shear stress

$$V = \text{SF}$$

I = second moment of area about NA

b_* = breadth of the cross-section where shear stress is to be found.

Ay = moment of the area available above the fiber under consideration about the NA.

This shear is associated with bending, thus called as flexure shear.

PROBLEM 5.1

Objective 3

Sketch the variation of shear stress across the depth of a rectangular section. Determine the maximum shear stress in terms of average shear stress.

SOLUTION

Let 'V' be the SF acting on the section.

$$\text{SF} = V$$

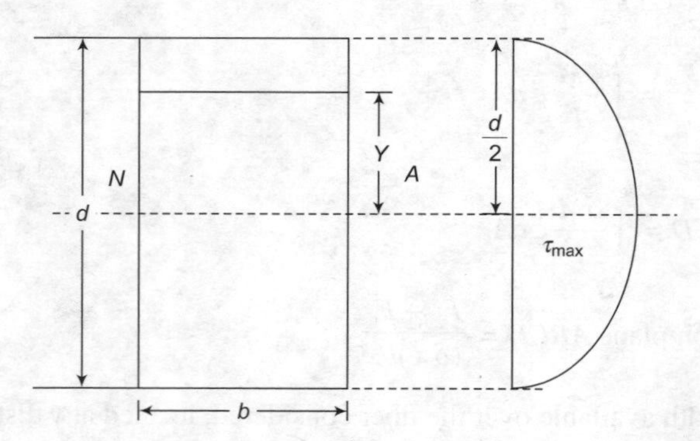

FIGURE 5.4

$$I = \frac{bd^3}{12}$$

$$b_* = b$$

$$Ay = b\left[\frac{d}{2} - y\right]\frac{1}{2}\left[\frac{d}{2} + y\right]$$

$$\tau = \frac{6V}{bd^3}\left\{\frac{d^2}{4} - y^2\right\}.$$

From the above expression, it is clear that the variation is parabolic at $y = \pm\frac{d}{2}$, that is, at extreme fibers, $\tau = 0$

At $y = 0$, that is, at centroid $= \tau_{max}$

$$\tau_{max} = \frac{6V}{bd^3} \times \frac{d^2}{4}$$

$$\Rightarrow \qquad \tau_{max} = 1.5\frac{V}{bd}$$

$\frac{V}{bd}$ is called average shear stress.

$$\therefore \qquad \tau_{max} = 1.5 \times \tau_{ave}.$$

PROBLEM 5.2

Objective 1

Sketch the variation of shear stress across the depth of a circular section subjected to varying BM.

SOLUTION

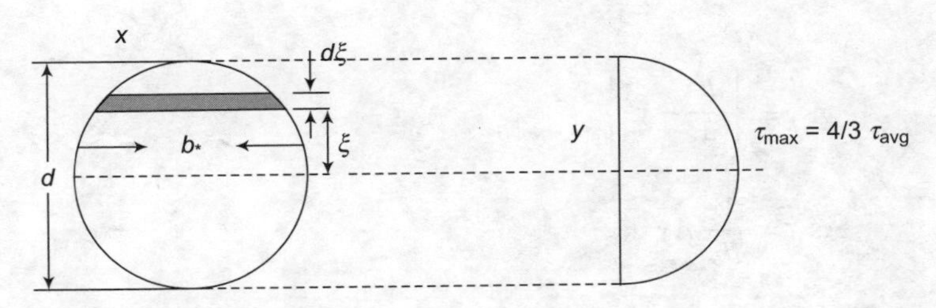

FIGURE 5.5

$$SF = V$$

$$\left.\begin{array}{l}\text{Second moment of area} \\ \text{Or moment of inertia (MI)}\end{array}\right\} = I = \frac{\pi d^4}{64}.$$

Consider a fiber located at 'y' distance from NA, where the shear stress is to be found.

$$b_* = 2 \times \sqrt{\frac{d^2}{4} - y^2}$$

To determine $A\ y$, consider an elemental strip located at '&' distance from NA as shown in Figure 5.6. Area of this element $dA = b \times \xi d\xi$

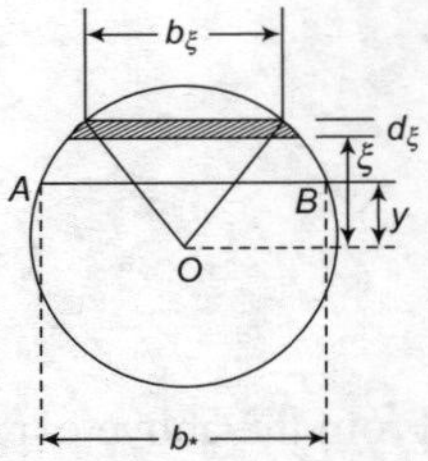

$$b_\xi = \sqrt{\frac{d^2}{4} - \xi^2}$$

FIGURE 5.6

$\therefore$ $$A\bar{y} = \int_y^{d/2} \xi dA$$

$$= \int_y^{d/2} 2\sqrt{\frac{d^2}{4} - \xi^2 d\xi}$$

Put $\frac{d^2}{4} - \xi^2 = t$ at $\xi = \frac{d}{2}; t = 0$

at $\xi = y$; $\quad t = \left(\frac{d^2}{4}\right) - y^2$

then $\quad -2\xi\, d\xi = dt$

or $\quad 2\xi\, d\xi = -dt$

$\therefore$ $$A\bar{y} = \int_{\frac{d^2}{4}-y^2}^{0} -\sqrt{t}\,dt$$

$$A\,\bar{y} = \left\{\frac{-t^{3/2}}{3/2}\right\}_{\frac{d^2}{4}-y^2}^{0}$$

$$A\,\bar{y} = \left\{\frac{d^2}{4} - y^2\right\}^{3/2}$$

$$\text{As } \tau = \frac{V}{Ib_*} A\bar{y}$$

$$\tau = \frac{V}{\frac{\pi}{64} d^4\, 2\sqrt{\frac{d^2}{4} - y^2}} \frac{2}{3}\left\{\frac{d^2}{4} - y^2\right\}^{3/2}$$

$$= \frac{64V}{3\pi d^4}\left\{\frac{d^2}{4} - y^2\right\}$$

From the above equation, it can be inferred that variation shear stress across the depth parabolic.

At $y = \pm\frac{d}{2}$; $\quad \tau = 0$

At $y = 0$; $\quad \tau = \tau_{max} = \frac{16V}{3\pi d^2}$

$$\tau_{max} = \frac{16V}{3\pi d^2} = \frac{4}{3}\frac{V}{\left[\frac{\pi}{4}d^2\right]}$$

$\frac{V}{\frac{\pi}{4}d^2}$ is average shear stress.

$$\therefore \qquad \tau_{max} = \frac{4}{3}\tau_{ave}.$$

PROBLEM 5.3

Objective 1

Sketch the shear stress variation across the depth of a square section. The diagonal of the square is in the plane of bending.

SOLUTION

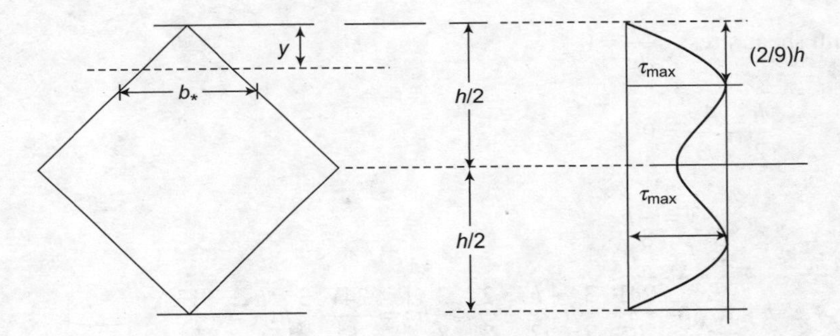

FIGURE 5.7

Let 'h' be the diagonal height of the cross-section, consider a fiber located at 'y' distance from the apex of the cross-section as shown in Figure 5.7. Where shear stress is to be determined

$$SF = V$$

$$MI = 2\left\{h\frac{\left(\frac{h}{12}\right)^3}{12}\right\} = \frac{h^4}{48}$$

$$b_* = 12\frac{y}{\left(\frac{h}{2}\right)} = 2y$$

$$A\bar{y} = \frac{1}{2}b_*\, y\left[\frac{h}{2} - \frac{2}{3}y\right]$$

$$\therefore \qquad \tau = \frac{V}{Ib_*} A\bar{y}$$

$$\tau = \frac{V}{\frac{h^4}{48} b_*} y\left\{\frac{h}{2} - \frac{2}{3}y\right\} = \frac{24V}{h^4} y\left\{\frac{h}{2} - \frac{2}{3}y\right\}$$

The variation of shear across the depth is parabolic. The expression mentioned above is applicable only in the top region. However, same response can be extended for the bottom portion also because of cross-sectional symmetry.

$$\tau = \frac{24V}{h^4} y\left(\frac{h}{2} - \frac{2}{3}y\right)$$

At $y = 0$; $\tau = 0$ and at $y = \frac{h}{2}$; $\tau = \frac{2V}{h^2}$

For maximum shear stress, $\frac{d\tau}{dy} = 0$

$$\Rightarrow \qquad \frac{h}{2} - \frac{4y}{3} = 0$$

$$\Rightarrow \qquad y = \frac{3}{8}h$$

$$\therefore \qquad \tau_{max} = \frac{24V}{h^4}\frac{3}{8}h\left\{\frac{h}{2} - \frac{2}{3} \times \frac{3}{8}h\right\} = \frac{24V}{h^4}\frac{3}{8}h\frac{h}{4} = \frac{9V}{4h^2}$$

$$\tau_{ave} = \frac{V}{\left(\frac{h^2}{2}\right)} = \frac{2V}{h^2}$$

$$\therefore \qquad \tau_{max} = \frac{9}{8}\tau_{ave}.$$

Shear stress at mid height $= \tau_{ave} = \frac{2V}{h^2}$.

PROBLEM 5.4

Objective 1

Sketch the variation of shear stress across the depth of a triangular section subjected to varying BM in Figure 5.8.

SOLUTION

Let the breadth and height of the triangular section be b and h, respectively.

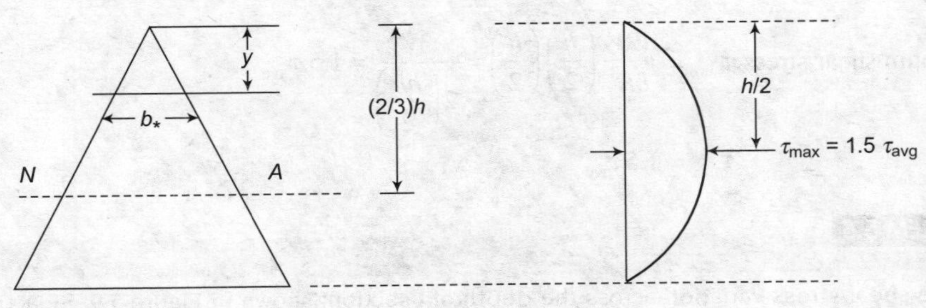

FIGURE 5.8

Consider a fiber located at '*y*' distance from the apex as shown in Figure 5.8.

Let b^* be the breadth of fiber, where shear stress is to be found.

$$\text{SF} = V$$

$$\text{MI about NA} = \frac{bh^3}{36}$$

$$b_* = \frac{b}{h} y$$

$A\bar{y}$ = moment of the shaded area about NA

$$= \left\{\frac{2}{3}h - \frac{2}{3}y\right\} \times \frac{1}{2} y b_*$$

$$= \frac{1}{3} b \times y\{h - y\}$$

$$\therefore \qquad \tau = \frac{V}{\frac{bh^3}{36} \times b_*} \cdot \frac{1}{3} b_* y(h - y)$$

$$\tau = \frac{12V}{bh^3} y(h - y)$$

$\therefore$ Shear stress variation is parabolic.

At $y = 0$ and h; $\tau = 0$

$$\text{At } y = \frac{2h}{3}; \qquad \tau_{\text{NA}} = \frac{12V}{bh^3} \frac{2h}{3} \frac{h}{3} = \frac{4}{3} \frac{V}{\left(\frac{bh}{2}\right)} = \frac{4}{3} \tau_{\text{ave}}$$

$$\text{For maximum shear, } \frac{d\tau}{dy} = 0$$

$$\Rightarrow \qquad (h - 2y) = 0$$

$$\Rightarrow \qquad y = \frac{h}{2}$$

$$\therefore \text{ Maximum shear stress } \tau_{max} = \frac{12V}{bh^3}\left(\frac{h}{2}\right)\left(\frac{h}{2}\right) = \frac{3V}{2\left(\frac{bh}{2}\right)} = 1.5\,\tau_{ave}$$

$$\tau_{max} = 1.5\tau_{ave}$$

PROBLEM 5.5

Objective 1

Sketch the shear stress variation across the depth of I section, shown in Figure 5.9. SF acting on the section is 100 kN. Estimate shear resisted by the web and flange portions.

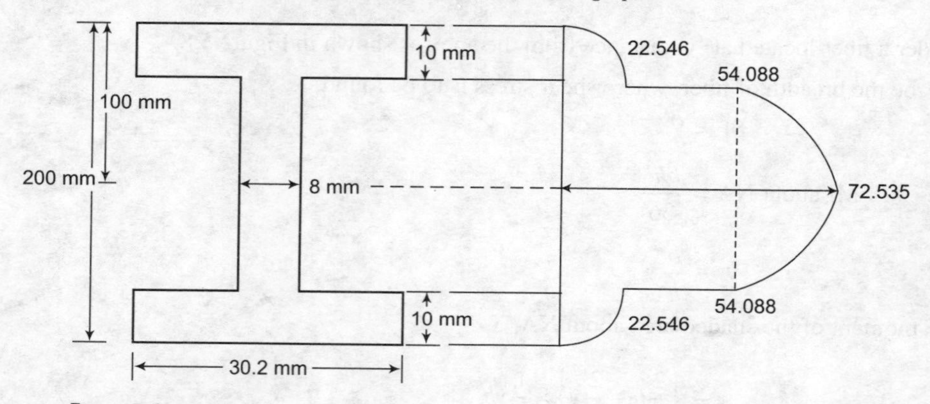

FIGURE 5.9

SOLUTION

$$\text{SF} = V = 100 \text{ kN}$$

$$I = \frac{1}{12}\{100 \times 200^3 - 92 \times 180^3\}$$
$$= 21.955 \times 10^6 \text{ mm}^4.$$

Consider a fiber located at 'y' distance from NA in the flange portion of the cross-section.

$$b_* = 100 \text{ mm}$$

$$A\overline{Y} = 100 \times (100 - y)\frac{1}{2}(100 + y)$$

$$= \frac{100}{2}(100^2 - y^2)$$

$$\therefore \quad \tau = \frac{V}{Ib_*}A\overline{Y}$$

$$= \frac{1008 \times 10^3}{21.955 \times 10^6 \times 100} \times \frac{100}{2}(100^2 - y^2)$$

$$= 22.774 \times 10^{-4}(100^2 - y^2)$$

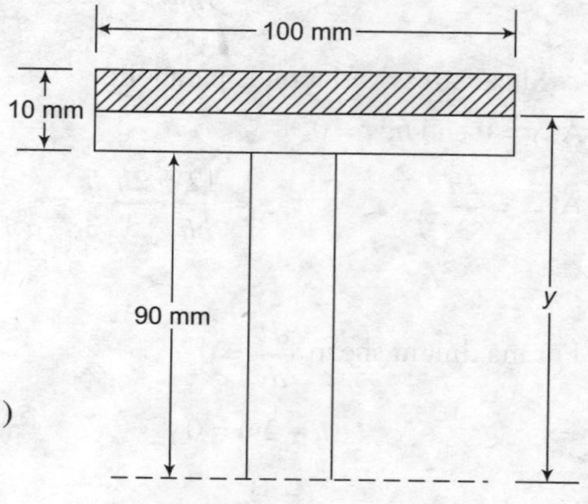

FIGURE 5.10

At $y = 100$; $\tau = 0$

At $y = 10$; $\tau = 0$

At $y = 10$; $\tau = 22.546$ MPa (variation is parabolic).

Consider a fiber located at 'y' distance from NA within the web of the cross-section.

$$A\overline{Y} = A_1Y_1 + A_2Y_2$$

$$= 100 \times 10 \times 95 + 8 \times (90 - y)\frac{1}{2}(90 + y)$$

$$= 95 \times 10^3 + 4(90^2 - y^2)$$

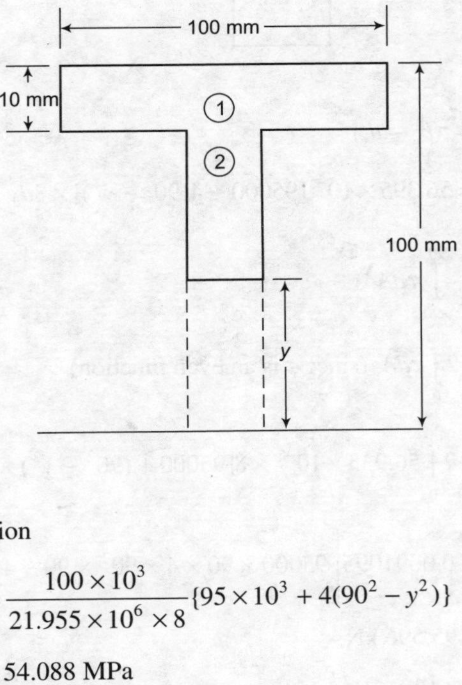

100 mm

10 mm

1

2

100 mm

y

FIGURE 5.11

Shear stress in the web portion

$$\tau = \frac{100 \times 10^3}{21.955 \times 10^6 \times 8}\{95 \times 10^3 + 4(90^2 - y^2)\}$$

At $y = 90$ mm; $\tau_{jw} = 54.088$ MPa

At $y = 0$; $\tau_{max} = 72.535$ MPa

Variation of shear stress is parabolic in the web portion too.

The variation of shear stress is shown in Figure 5.9.

SF Resisted by Web:

Shear stress at a fiber located at 'y' distance from NA is given by

$$\tau = 56.935 \times 10^{-5}\{95000 + 4(90^2 - y^2)\}.$$

Consider elemental area shown in Figure 5.12

SF over the elemental area = $\tau \cdot dA$

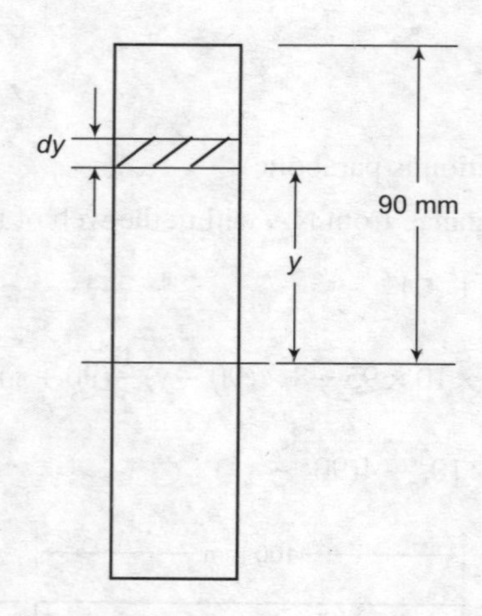

FIGURE 5.12

$$= \tau(b_w - d_y)$$

$$= 56.395 \times 10^{-5}[95000 + 4(90^2 - y^2)] \times 8dy$$

$$\text{Total shear in the web} = \int_{-90}^{90} \tau dA$$

$$V_W = 2\int_0^{90} \tau dA \text{ (since } \tau \text{ is an even function)}$$

$$\therefore \quad V_W = 2\int_0^{90} 56.935 \times 10^{-5} \times 8[95000 + (90^2 - y^2) \times 4]dy$$

$$= 0.0091095\left[95000 \times 90 \times 4 \times 90^2 \times 90 - 4 \cdot \frac{90^3}{3}\right]$$

$$= 95.596 \text{ kN.}$$

$\therefore$ SF resisted by the flange = $100 - V_w$

$V_f = 4.404$ kN

$\therefore$ % Shear resisted by web = 95.596%

$\therefore$ % Shear resisted by flange = 4.404%

From the above figures, it is clear that, SF is resisted by mainly the web of an *I* section. Thus in *I* section, flanges resist bending while the web resists the SF.

PROBLEM 5.6

Objective 1

Sketch the variation of shear stress across the depth of the *H* section in Figure 5.13. SF acting on the cross-section is 100 kN.

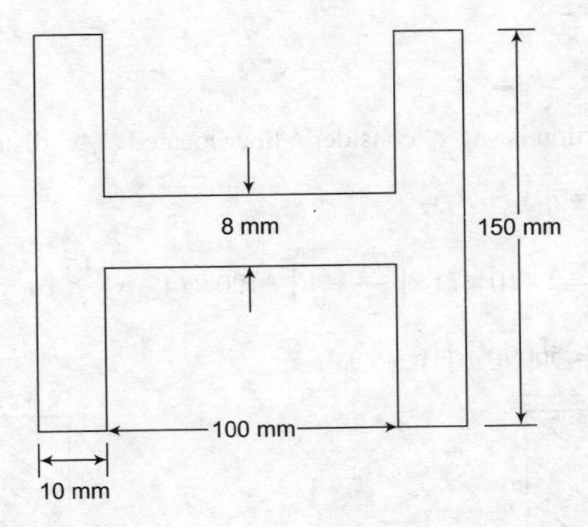

FIGURE 5.13

SOLUTION

$$SF = V = 100 \text{ kN}$$

$$b^* = 2 \times 10 = 20 \text{ mm}$$

$$A\overline{Y} = 2 \times 10 \times (75 - y)\frac{1}{2}(75 + y) = 10(75^2 - y^2)$$

$$I = \frac{1}{12}\{2 \times 10 \times 150^3 + 100 \times 8^3\} = 5.6293 \times 10^6 \text{ mm}^4$$

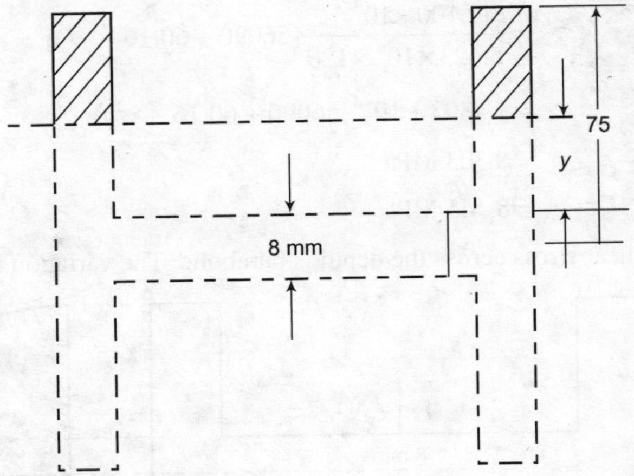

FIGURE 5.14

Shear stress in the flange portion $= \dfrac{V}{Ib^*}A\overline{Y}$

$$\tau = \frac{100 \times 10^3}{5.6293 \times 10^6 \times 20} \times 10[75^2 - y^2]$$

$$= 8.88 \times 10^{-3}\,[75^2 - y^2]$$

At $y = 75$; $\tau = 0$

At $y = 4$; $\tau_j = 49.82$ MPa.

Shear stress within the portion is say τ_w consider a fiber located at 'y' distance from NA

$$A\bar{y} = A_1 y_1 + A_2 y_2$$

$$= 2 \times 10 \times 71 \times \left(\frac{71}{2} + 4\right) + 120 \times (4 - y)\frac{1}{2}(4 + y)$$

$$= 56090 + (16 - y^2)$$

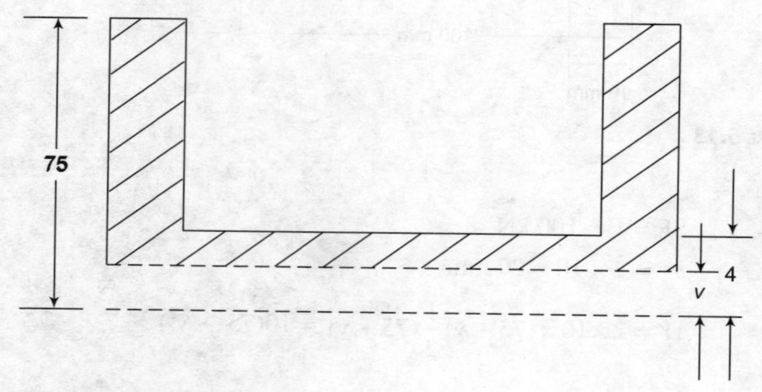

FIGURE 5.15

$$\therefore \qquad \tau_w = \frac{100 \times 10^3}{5.623 \times 10^6 \times 120}[56090 + 60(16 - y^2)]$$

$$\tau_w = 1.4803 + 10^{-4}\{56090 + 60(16 - y^2)\}$$

At $y = 4$; $\qquad \tau_{jw} = 8.303$ MPa

At $y = 0$; $\qquad \tau_{w\max} = 8.445$ MPa

The variation of shear stress across the depth is parabolic. The variation is shown in Figure 5.16.

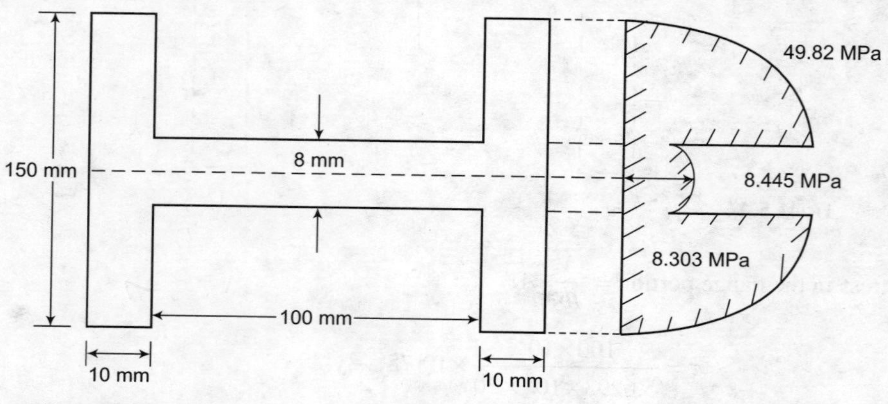

FIGURE 5.16

PROBLEM 5.7

Objective 2

Three wooden planks of size 10 mm × 25 mm are nailed together to form a beam of size 100 mm wide and 75 mm deep. The cross-section is subjected to a vertical SF of 750 N. If the allowable shear stress in nail is limited to 400 N, estimate the spacing of the nails.

SOLUTION

Unless the three planks are nailed together, three planks cannot act together. Otherwise each plank bends individually.

The vertical shear produces horizontal shear stress equal to $\frac{V}{Ib_*}A\bar{y}$. The shear is to be resisted by the nails. As long as the nails are capable of resisting the shear due $\frac{V}{Ib_*}A\bar{y}$ three planks act together. Let 's' be the spacing of the nails in the longitudinal direction of the beam.

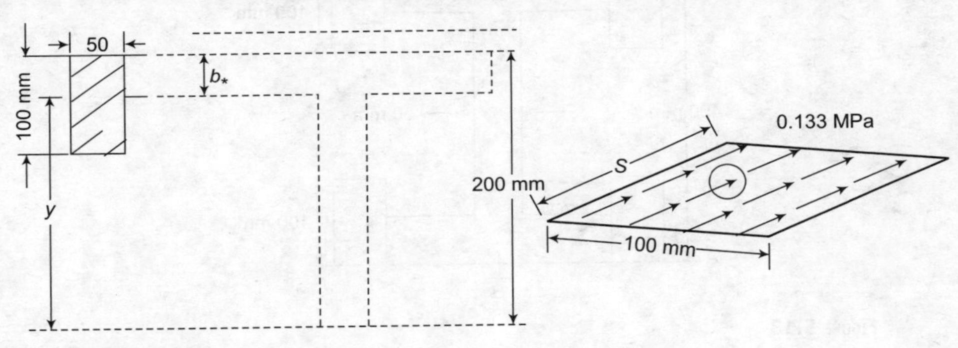

FIGURE 5.17

Let 'τ' be the shear stress at the junction of top and middle plank due to vertical shear 'V'

$$\tau = \frac{V}{Ib_*}A\bar{y}$$

Given that $V = 750$ N

$$I = \frac{100\times(75)^3}{12}\times 3.5156\times 10^6 \text{ mm}^4$$

$$b_* = 100$$

$$A\bar{y} = 100\times 25\times\left(\frac{25}{2}+\frac{25}{2}\right) = 62500$$

$$\therefore \qquad \tau = \frac{750}{3.5156\times 10^6\times 100} = 0.133 \text{ MPa}$$

The nail has to resist the shear stress in the region $s \times 100$

$\therefore$ Shear to be resisted by each nail = $\tau \times s \times 100$

The shear on each nail should be less than the shear resistance of the nail, that is, 400 N. For complete action of three planks

$$\tau \times s \times 100 < 400$$

$$\Rightarrow \quad 0.133 \times s \times 100 < 400$$

$$\Rightarrow \quad s < 30 \text{ mm.}$$

30 mm spacing is too small. Hence, provide nail in two rows. So the spacing of nails can be increased to 60 mm.

PROBLEM 5.8

Objective 1

A built-up wooden cross-section is subjected to a vertical shear of 4 kN. If the SF in the nail, located at places *A* and *B* as shown in Figure 5.18, is limited to 250 N, determine the spacing of the nails.

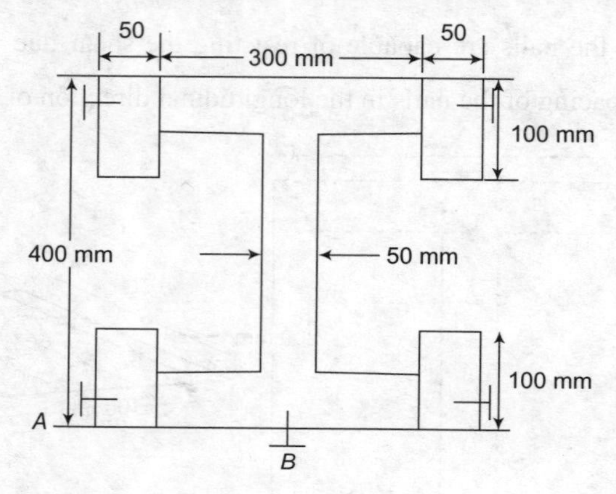

FIGURE 5.18

SOLUTION

The nails placed at '*A*' are subjected to longitudinal shear. In case of open section $A\bar{y}$ is to be calculated for the area available from the free outstanding element also.

$$\text{SF} = V = 4 \text{ kN.}$$

$$MI = \text{second moment of area about NA}$$

$$= \frac{1}{12}\{400 \times 400^3 - 250 \times 300^3 - 100 \times 200^3\}$$

$$= 1.5042 \times 10^9 \text{ mm}^4$$

Shear stress at the location '*A*' for thin walled open section b^* shall be taken as thickness of the thin element.

$$\therefore \quad b_* = 50 \text{ mm}$$

$$A\bar{y} = 100 \times 50 \times (200 - 50)$$

$$= 75 \times 10^4$$

$\therefore$ shear stress at the junction where the nails were provided (location *A*) $= \tau_A = \dfrac{V}{Ib_*}A\bar{y}$

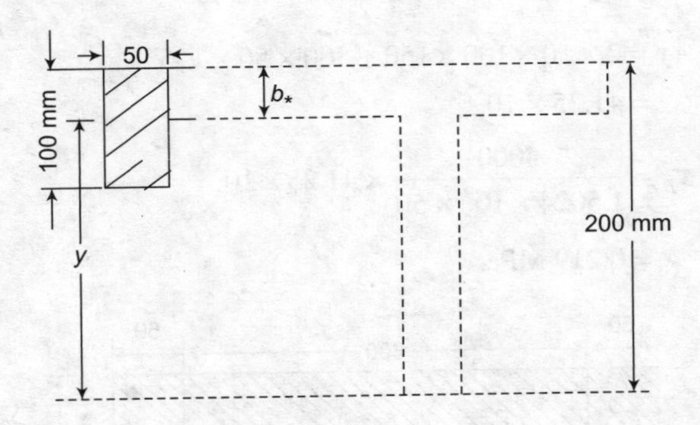

FIGURE 5.19

$$\tau_A = \frac{4 \times 10^3}{1.55042 \times 10^9 \times 50} \times 75 \times 10^4 = 0.04 \text{ MPa.}$$

Let S_A be the spacing of the nails, then

Force coming on to each nail at $A = F_A = S_A \times \tau \times 50$.

This force should be less than the shear resistance of the nail 250 N (given)

$\therefore \qquad S_A \times 0.04 \times 50 < 250$

$\Rightarrow \qquad S_A = 125$ mm.

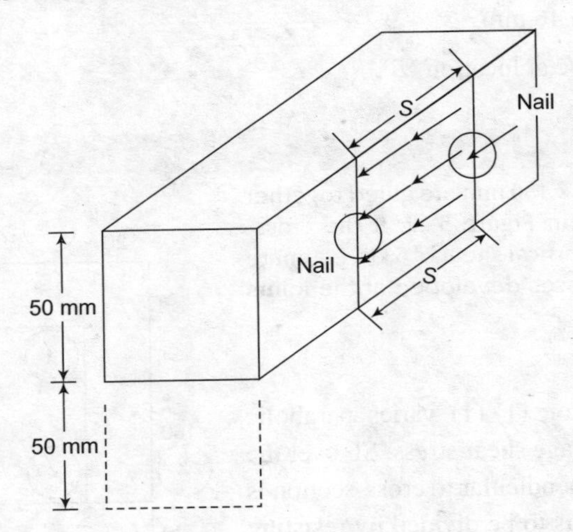

FIGURE 5.20

Let S_B be the spacing of the nails provided at location 'B'. The longitudinal shear stress at this location be τ_B.

$$\tau_B = \frac{V}{Ib_*} A\bar{y}$$

$$b_* = 50 \text{ mm}$$

$$A\bar{y} = 2 \times 50 \times 100 \times 150 + 300 \times 50 \times 175$$
$$= 41.25 \times 10^5$$

$$\tau_B = \frac{4000}{1.5024 \times 10^9 \times 50} \times 41.25 \times 10^5$$
$$= 0.219 \text{ MPa.}$$

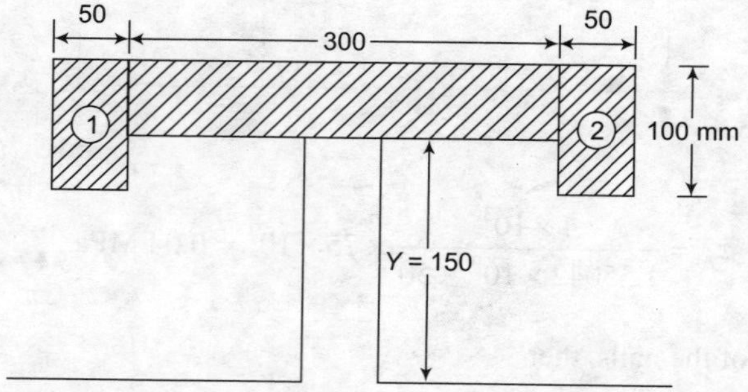

FIGURE 5.21

The SF on the nail should be less than 400 N

$\therefore$ $$S_B \times 50 \times 0.219 < 400$$
$$S_B < 36.46 \text{ mm.}$$

Provide nails at 30 mm c/c at location 'B'.

PROBLEM 5.9

Objective 2

Four planks of size 20 mm × 180 mm are glued together to form a section shown in Figure 5.22. If the cross-section is subjected to a vertical shear of 5 kN, estimate the average shearing stresses developed at the joints shown.

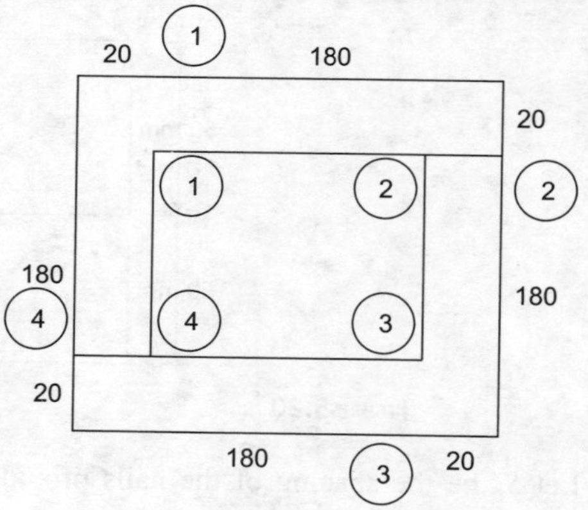

SOLUTION

The shear stress at section (1)-(1) varies parabolic ally. Thus, to find the average shear stress, SF over the section for unit width perpendicular to cross-section is to be calculated. This SF is to be divided by resisting area to get average shear stress.

FIGURE 5.22

Given that,

$$\text{SF} = V = 5000 \text{ N}$$

$$I = \frac{1}{12}\{200^4 - 160^4\} = 78.72 \times 10^6 \text{ mm}^4$$

Consider a fiber located at 'y' distance from NA in the top flange portion.

$$b_* = 200 \text{ mm}$$

$$A\bar{y} = 200 \times (100 - y)\frac{1}{2}(100 + y)$$

$$A\bar{y} = 100\{100^2 - y^2\}$$

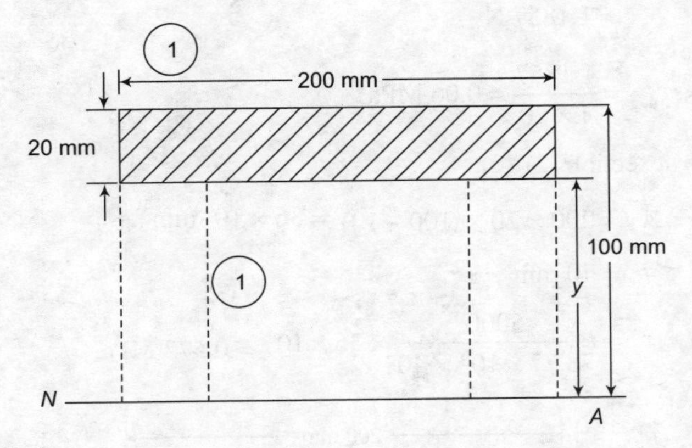

FIGURE 5.23

$$\therefore \qquad \tau = \frac{5000}{78.72 \times 10^6 \times 200} \times 100 \times [100^2 - y^2]$$

$$\tau = 3.176 \times 10^{-5}\{100^2 - y^2\}$$

At $y = 20$, $\tau = 0.305$ MPa.

SF over a depth of 20 mm

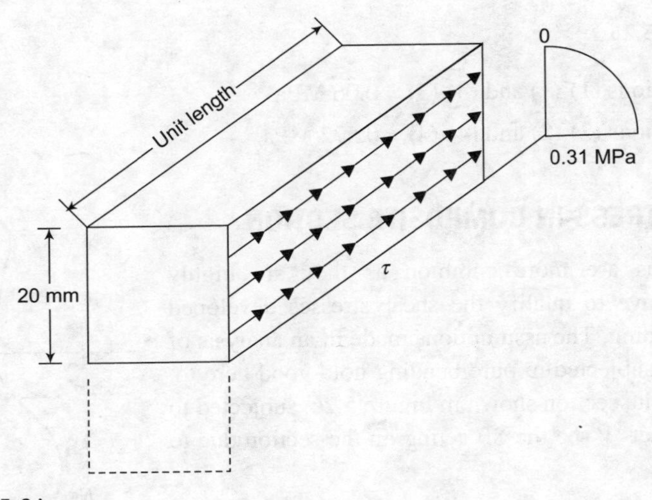

FIGURE 5.24

$$F_1 = \int_{80}^{100} 3.176 \times 10^{-5}\{100^2 - y^2\} \times 1 \times dy$$

$$= 3.176 \times 10^{-5}\{100^2(100-80) - \frac{1}{3}(100^3 - 80^3)\}$$

$$= 1.1857 \text{ N}$$

Average shear stress $\tau_{avg} = \dfrac{1.1857}{1 \times 20} = 0.06$ MPa

Average shear stress at section (2)-(2):

$$A\bar{y} = 200 \times 20 \times (100 - 10) = 36 \times 10^4 \text{ mm}^3$$

$$b_* = 40 \text{ mm}$$

$$\tau_2 = \frac{5000}{78.72 \times 10^6 \times 40} \times 36 \times 10^4 = 0.572 \text{ MPa.}$$

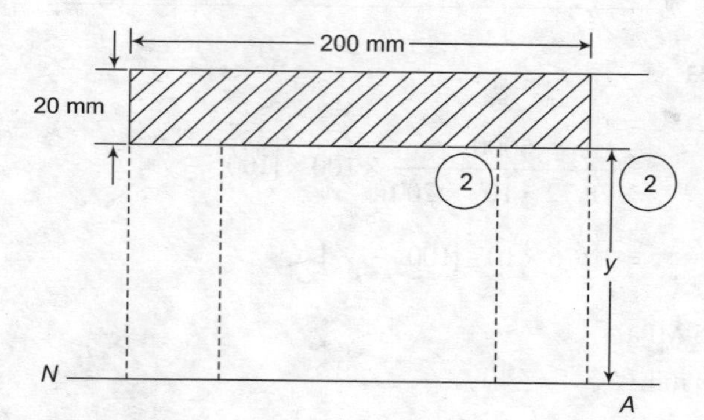

FIGURE 5.25

Shear stress at sections (1)-(1) and (3)-(3) = 0.06 MPa.

Shear stress at sections (2)-(2) and (4)-(4) = 0.572 MPa.

5.3 SHEAR STRESS IN COMPOSITE SECTION

Composite sections are more common as they are highly economical. We have to qualify the shear stresses developed due to varying bending. The assumptions made in an analysis of composite section subjected to pure bending hold good here to. Consider a composite section shown in Figure 5.26, subjected to varying bending. Let 'V' be the SF acting on the section due to varying moment.

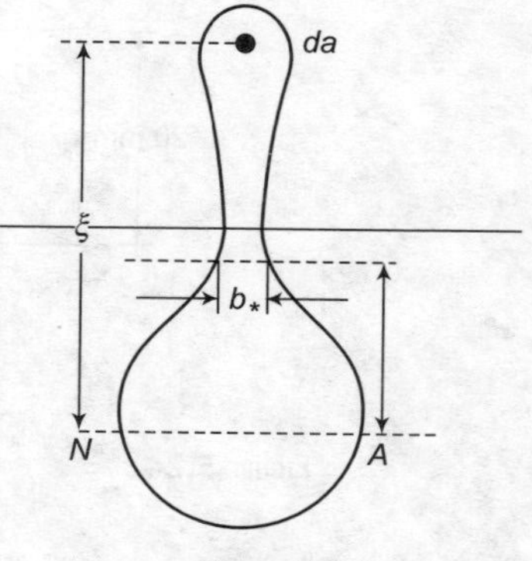

FIGURE 5.26

In a composite section bending stress at any fiber located at 'ξ' distance from NA is given by

$$\sigma = \frac{E_i M \cdot \xi}{E_1 I_1 + E_2 I_2 + \cdots},$$

in which 'E_i' is the modulus of elasticity of the material, where the bending stress 'σ' is to be found.

If the numerator and denominator are divided by 'E_r' modulus of elasticity of reference material into which the total cross-section has been converted (transformed section)

Then,
$$\sigma = \frac{\frac{E_i}{E_r} \cdot M \cdot \xi}{\frac{E_1}{E_r} I_1 + \frac{E_2}{E_r} I_2 + \cdots}$$

$$\sigma = \frac{m_i M \xi}{m_1 I_1 + m_2 I_2 + \cdots}.$$

The denominator is the *MI* or second moment area of the transformed section.

m_i = modular ratio of the material, in which the 'σ' is to be determined.

$$\therefore \quad \sigma = m_i \frac{M\xi}{I_t}$$

I_t = *MI* of the transformed section.

To obtain expression for shear stress, consider an element of length of the beam 'δx' in longitudinal direction.

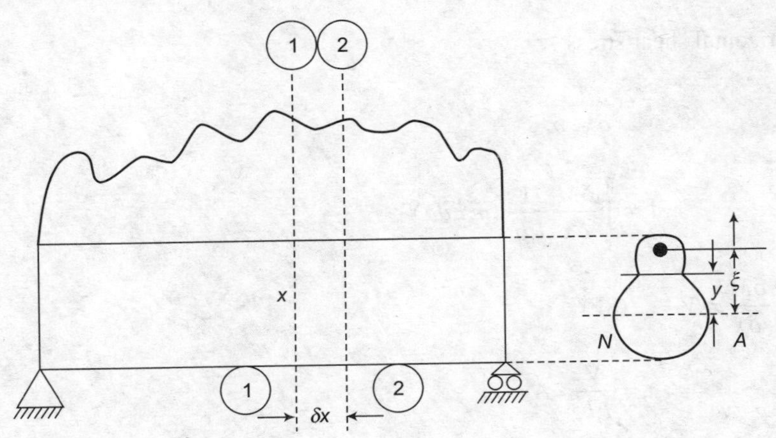

FIGURE 5.27

Consider the *FBD* portion available above the fiber under consideration between the sections (1)-(1) and (2)-(2) shown in Figure 5.28. BM at sections (2)-(2) be $M + \Delta M$.

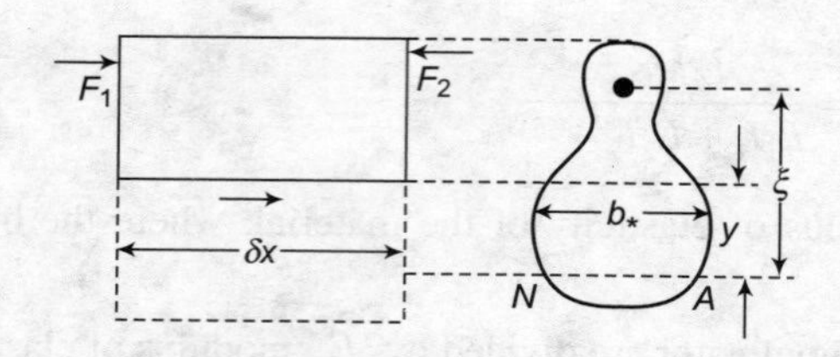

FIGURE 5.28

Consider an elemental area 'dA' located at 'ξ' distance from NA.

Elemental force at a section (1)-(1) $= \dfrac{m_i M \xi}{I_t} dA$

Total force in shaded portion of Figure 5.28 at section (1)-(1) $F_1 = \displaystyle\int_y^{y_t} \frac{M\xi}{I_t}(m_i dA)$.

Similarly, total force in the shaded portion of Figure 5.28 at section (2)-(2) $F_2 = \displaystyle\int_y^{y_t} \frac{M + \delta M}{I_t} \cdot \xi(m_i dA)$.

Therefore, net horizontal force that keeps the shaded portion between sections (1)-(1) and (2)-(2) is $F + F_2 - F_1$.

$$\therefore \qquad F = F_2 - F_2$$

$$= \int_y^{y_t} \frac{\delta M}{I_t} \xi(m_i dA)$$

$\therefore$ The horizontal shear stress 'τ'

$$\tau = \frac{F}{\delta x \cdot b_*}$$

$$\therefore \qquad \tau = \int_y^{y_t} \frac{\delta M}{\delta x} \frac{1}{I_t b_*}(m_i \xi dA)$$

We know $\dfrac{\delta M}{\delta x} = V$ (SF)

$$\tau = \frac{V}{I_t b_*} \cdot A\bar{y}$$

In the above expression, it is the MI of transformed section

$$A\bar{y} = \int_y^{y_t} \xi[m_i dA]$$

$\therefore A\overline{Y}$ is the moment of the area available above the fiber under consideration, about NA in the transformed section as the term $m_i dA$ is present in the integral. b_* is the breadth of the fiber, where shear stress is to be found in the cross-section, but not the transformed section. Therefore, use transformed section for second moment of area and $A\overline{Y}$ terms only in the expression.

PROBLEM 5.10 **Objective 2**

Determine the shear stress at the interface of a composite section shown in Figure 5.29, subjected to a vertical SF of 100 kN. Take E_1 = 200 GPa, E_2 = 100 GPa, and E_3 = 40 GPa.
SF = V = 100 kN

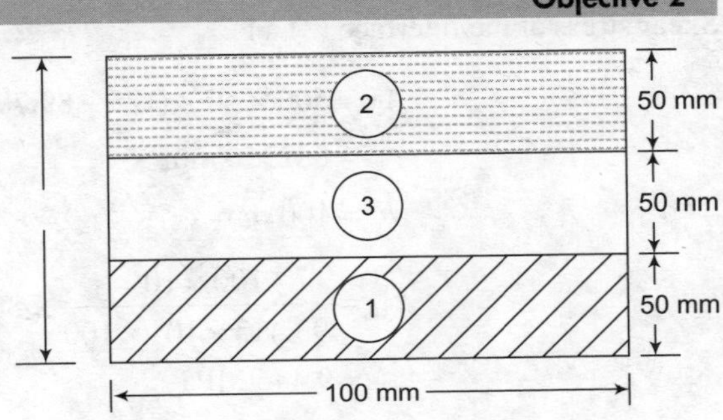

FIGURE 5.29

SOLUTION

Let us convert the cross-section into a uniform cross of material (3).

$$\therefore \quad m_1 = \frac{E_1}{E_2} = \frac{200}{40} = 5$$

$$m_2 = \frac{E_2}{E_3} = \frac{100}{40} = 2.5$$

$$m_3 = \frac{E_3}{E_1} = 1.0.$$

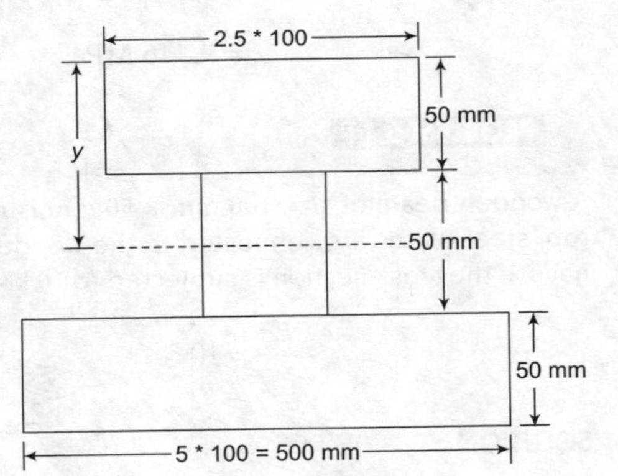

FIGURE 5.30

$\therefore$ The transformed section is shown in Figure 5.30.

$$\bar{y} = \frac{250 \times 50 \times 25 + 100 \times 50 \times 75 + 500 \times 50 \times 125}{250 \times 50 + 100 \times 50 + 500 \times 50}$$

= 89.706 mm.

Moment of inertia

$$I_t = \frac{1}{12} \times 250 \times 50^3 + 250 \times 50 \times [89.706 - 25]^2$$

$$+ \frac{1}{12} \times 100 \times 50^3 + 100 \times 50 \times \quad [89.706 - 25]^2$$

$$+ \frac{1}{12} \times 500 \times 50^3 + 500 \times 50\{125 - 89.706\}^2$$

$$= 93.413 \times 10^6 \text{ mm}^4$$

Shear stress at the interface between the materials (2) and (3).

$A\bar{y}$ = moment of the area above the interface in the transformed section above NA.

$$= 250 \times 50 \times \{89.706 - 25\}$$

$$= 808825 \text{ mm}^3$$

$b_* =$ breadth at the interface in the actual cross-section

$= 100$ mm (not 250 mm)

$$\therefore \quad \tau_{2-3} = \frac{100 \times 10^3}{93.413 \times 10^6 \times 100} \times 808825$$

$= 8.659$ MPa.

Shear stress at the interface (3)–(1):

$$A\bar{y} = 500 \times 50 \times (125 - 89.706)$$

$$= 882350 \text{ mm}^3$$

$$b_* = 100 \text{ mm}$$

$$\tau_{31} = \frac{100 \times 10^3}{93.413 \times 10^6 \times 100} \times 882350$$

$= 9.446$ MPa

$$\tau_{31} = \frac{100 \times 10^3}{93.413 \times 10^6 \times 100} \times 882350$$

$= 9.446$ MPa.

PROBLEM 5.11

Objective 2

A wooden beam of size 100 mm × 50 mm is reinforced with a steel plate of size 50 mm × 10 mm. At top, steel plates are connected to the wooden member through nails. Determine the spacing of the nails, if the cross-section is subjected to 10 kN and shear resistance of the nail is limited to 4 kN, take.

$$\frac{E_s}{E_w} = 10$$

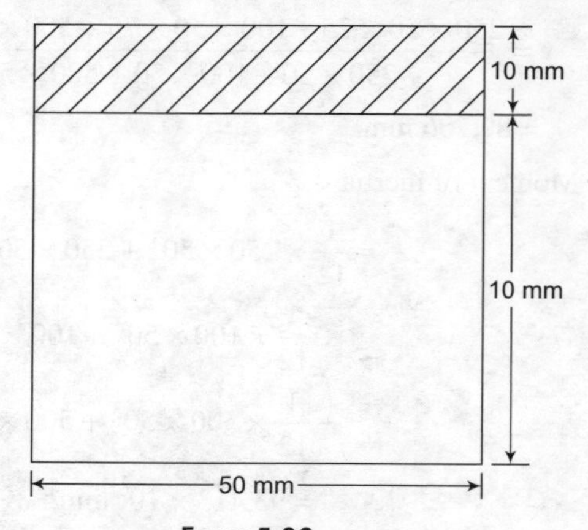

FIGURE 5.32

SOLUTION

$$\text{SF} = V = 10 \text{ kN.}$$

The transformed section is shown in Figure 5.32.

$\frac{E_S}{E_W} = 10$, convert the cross-section into wood.

Let $\bar{y}$ be the location of the centroid of the transformed section.

$$\bar{y} = \frac{500 \times 10 \times 5 + 50 \times 100 \times 60}{500 \times 10 + 50 \times 100} = 32.5 \text{ mm}$$

$I_t = MI$ of the transformed section

$$= \frac{1}{3}\{50 \times 110^3 + 450 \times 10^3\} - 10 \times 32.5^2 = 11.771 \times 10^6.$$

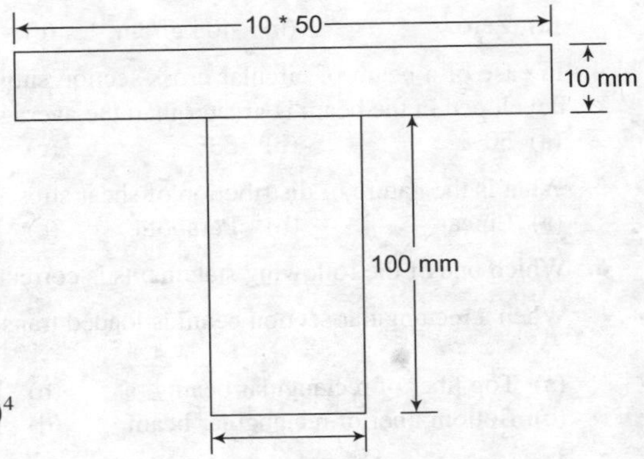

Let τ be the shear stress at the interface.

$$\tau = \frac{V}{I_t b_*} A\bar{y}$$

$$A\bar{Y} = 500 \times 10 \times (32.5 - 5)$$

$$= 13.75 \times 10^4 \text{ mm}^4$$

$$b_* = 50 \text{ mm}$$

$$\tau = \frac{10 \times 10^3}{11.771 \times 10^6 \times 100} \times 13.75 \times 10^4$$

$$= 1.168 \text{ MPa.}$$

FIGURE 5.33

Let '*S*' be the spacing of the nails.

SF on the nail due to shear stress $= \tau \times s \times b_* = 58.410$ s.

SF on the nail should be less than 4 kN.

$\therefore$ $58.41 \text{ s} < 4000$

$s < 68.48$ mm.

Provide nails at 60 mm c/c along the length of member.

SUMMARY

- **The stresses produced due to constant BM are known as bending stresses.**
- **The bending stress in any layer is directly proportional to the distance of layer from the NA.**
- **The bending stress on the neutral is zero.**
- **If the top layer of the section is subjected to compressive stress then bottom layer of the section will be subjected to tensile stress.**
- **Total moment of resistance (MR) of composite beam is the sum of MR of individual section.**

OBJECTIVE TYPE QUESTIONS

1. The ratio of average shear stress to the maximum shear stress in a beam with a square cross-section is: [GATE-1994, 1998]

(a) 1 (b) 2 (c) 3 (d) 2

2. At a section of a beam, SF is *F* with zero BM. The cross-section is square with side *a*. Point A lies on NA and point *B* is midway between NA and top edge, that is, at distance *a*/4 above the NA. If t_A and t_B denote shear stresses at points A and B, then what is the value of t_A/t_B? [IES-2005]

(a) 0 (b) 3/4 (c) 4/3 (d) None of above

3. A wooden beam of rectangular cross-section 10 cm deep by 5 cm wide carries maximum SF of 2000 kg. Shear stress at NA of the beam section is [IES-1997]

(a) Zero (b) 40 kgf/cm^2 (c) 60 kgf/cm^2 (d) 80 kgf/cm^2

4. In case of a beam of circular cross-section subjected to transverse loading, the maximum shear stress developed in the beam is greater than the average shear stress by [IES-2006, 2008]

(a) 50% (b) 33% (c) 25% (d) 10%

5. What is the nature of distribution of shear stress in a rectangular beam? [IES-1993, 2004, 2008]

(a) Linear (b) Parabolic (c) Hyperbolic (d) Elliptic

6. Which one of the following statements is correct?

When a rectangular section beam is loaded transversely along the length, shear stress develops on [IES-2007]

(a) Top fiber of rectangular beam (b) Middle fiber of rectangular beam
(c) Bottom fiber of rectangular beam (d) Every horizontal plane

7. The transverse shear stress acting in a beam of rectangular cross-section, subjected to a transverse shear load, is: [IES-1995, GATE-2008]

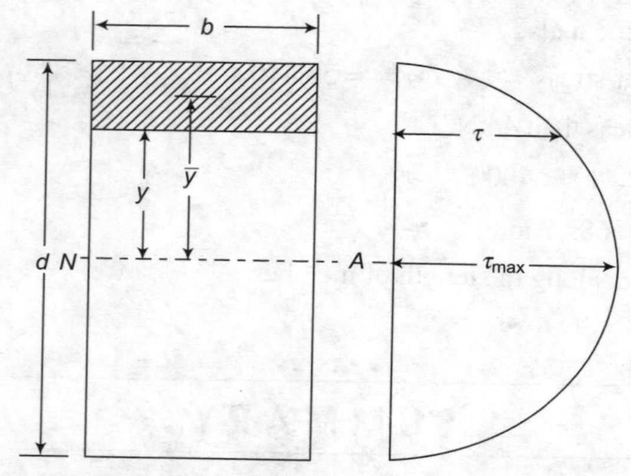

(a) Variable with maximum at the bottom of the beam
(b) Variable with maximum at the top of the beam
(c) Uniform
(d) Variable with maximum on the

8.

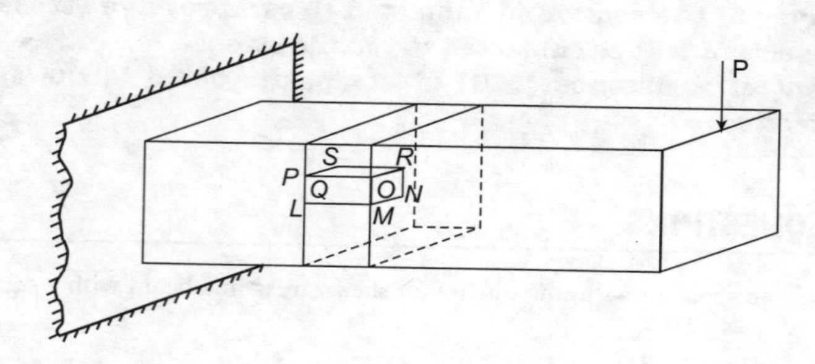

A cantilever is loaded by a concentrated load P at the free end as shown. The shear stress in the element *LMNOPQRS* is under consideration. Which of the following figures represents the shear stress directions in the cantilever? [IES-2002]

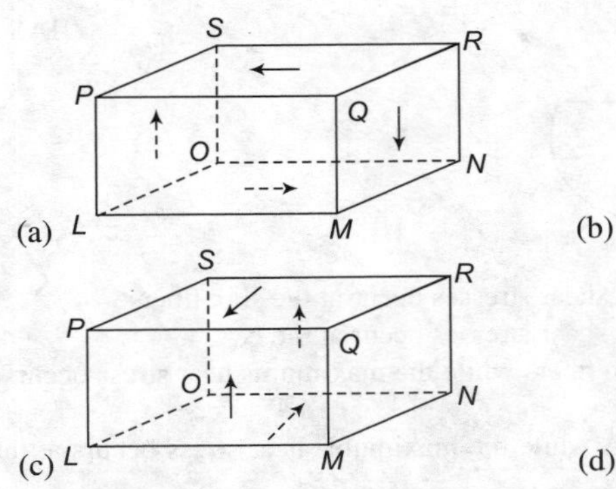

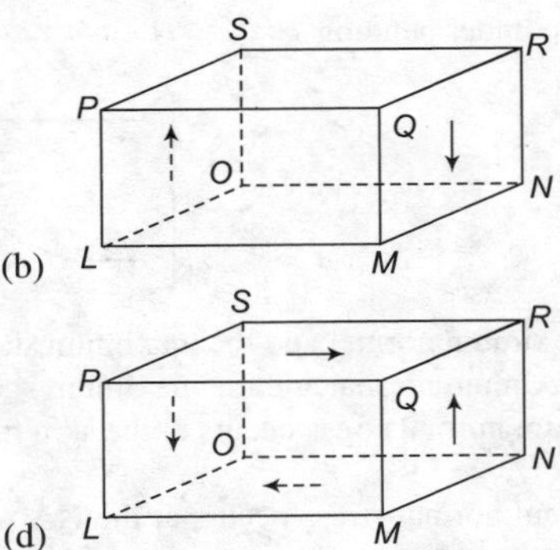

9. In *I* section of a beam subjected to transverse SF, the maximum shear stress is developed. [IES-2008]

(a) At the center of the web
(b) At the top edge of the top flange
(c) At the bottom edge of the top flange
(d) None of the above

10. The given figure (all dimensions are in mm) shows an *I* section of the beam. The shear stress at point *P* (very close to the bottom of the flange) is 12 MPa. The stress at point *Q* in the web (very close to the flange) is: [IES-2001]

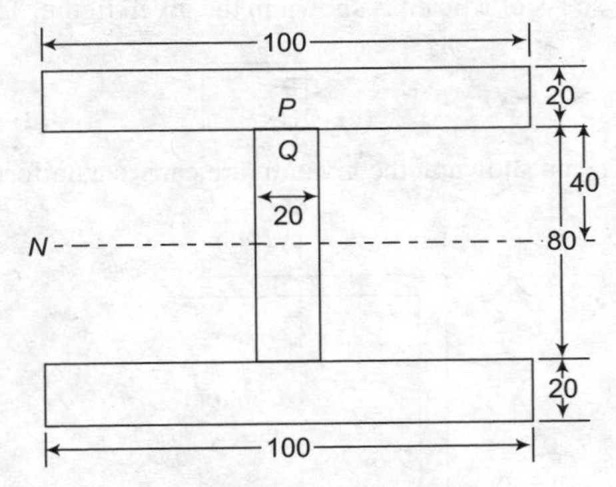

(a) Indeterminable due to incomplete data (b) 60 MPa
(c) 18 MPa (d) 12 MPa

11. Consider the following statements: [IAS-2007]
Two beams of identical cross-section but of different materials carry same BM at a particular section, then
1. The maximum bending stress at that section in the two beams will be same.
2. The maximum shearing stress at that section in the two beams will be same.
3. Maximum bending stress at that section will depend upon the elastic modulus of the beam material.
4. Curvature of the beam having greater value of *E* will be larger.

Which of the statements given above are correct?
(a) 1 and 2 only (b) 1, 3, and 4 (c) 1, 2, and 3 (d) 2, 3, and 4

12. In a loaded beam under bending [IAS-2003]

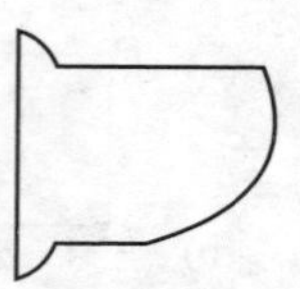

(a) Both the maximum normal and the maximum shear stresses occur at the skin fibers
(b) Both the maximum normal and the maximum shear stresses occur at the NA
(c) The maximum normal stress occurs at the skin fibers while the maximum shear stress occurs at the NA
(d) The maximum normal stress occurs at the NA while the maximum shear stress occurs at the skin fibers

13. Select the correct shear stress distribution diagram for a square beam with a diagonal in a vertical position: [IAS-2002]

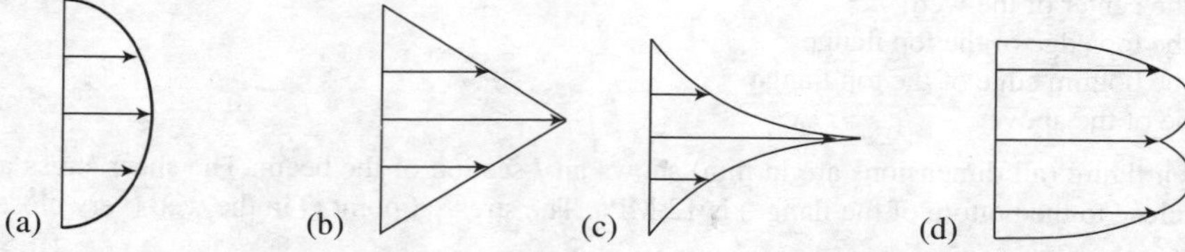

14. The distribution of shear stress of a beam is shown in the given figure. The cross-section of the beam is: [IAS-2000]

(a) 1 (b) T (c) (d)

15. A channel section of the beam shown in the given figure carries a uniformly distributed load. [IAS-2000]

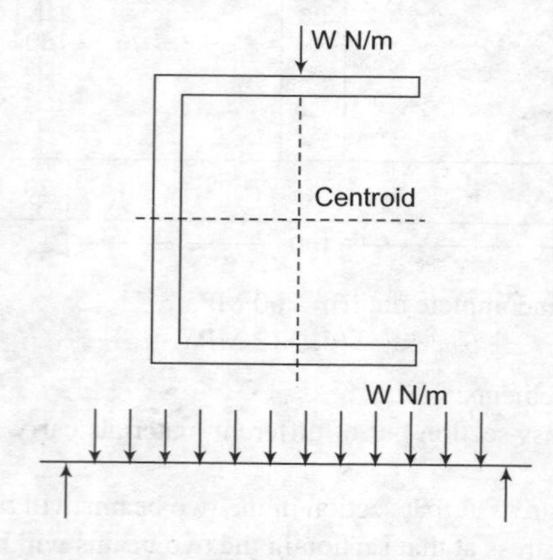

Assertion (A): The line of action of the load passes through the centroid of the cross-section. The beam twists besides bending.

Reason (R): Twisting occurs since the line of action of the load does not pass through the web of the beam.

(a) Both A and R are individually true and R is the correct explanation of A
(b) Both A and R are individually true but R is NOT the correct explanation of A
(c) A is true but R is false
(d) A is false but R is true

Solutions for Objective Questions

Sl. No.	1.	2.	3.	4.	5.	6.	7.	8.	9.	10.
Answer	(b)	(c)	(c)	(b)	(b)	(b)	(d)	(d)	(a)	(b)

Sl.No.	11.	12.	13.	14.	15.
Answer	(a)	(b)	(d)	(b)	(c)

EXERCISE PROBLEMS

1. A circular section of diameter 100 mm is subjected to an SF of 100 kN due to varying BM. Determine the maximum intensity of shear stress included the cross-section of the beam.
2. The cross-section of a beam is '*I*' section, for which each flange is 200 mm × 20 mm and web is 200 mm × 20 mm. The maximum SF acting on the beam is 40 kN. Draw the shear stress distribution across the section, by making all circular values. Find the percentage of SF carried by web and flange.
3. The cross-section of a beam is '*T*' shape with 300-mm-wide × 50-mm-thick flange and 50-mm-thick × 350–mm-deep web. Sketch the variation of shear stress across the cross-section, if it is subjected to an SF of 400 kN. Indicate the values at all circular points in terms of average shear stress.
4. Find the percentage of BM and SF resisted by flanges and web of the symmetrical I section with 200 mm × 20 mm flanges and 10 mm × 230 mm web.
5. A prismatic member with square cross-section is used as a simply supported beam with one of its diagonal in the plane of bending. Draw the shear stress distribution by marking all silent values. The beam is subjected to maximum SF of 200 kN. The size of cross-section is 120 mm × 120 mm.
6. Sketch the variation of shear stress for the *H* section if an SF of 100 kN acts on it. Verticals are 10 mm × 200 mm while horizontal connector is 20 mm × 150 mm.
7. A laminated wooden beam is built up by gluing together three 100-mm-wide and 50-mm-deep boards to form a solid beam of 100 mm wide and 150 mm deep. The allowable shear stress in glued joint is 20 MPa. If the beam is simply a supported beam used for a span of 6 m, determine the allowable concentrated load *P* at midspan of the beam.

CHAPTER 6

ANALYSIS OF TRUSSES

UNIT OBJECTIVE

This chapter deals with the analysis of different types of trusses using different methods. The presentation attempts to help the student achieve the following:

Objective 1: Identify the forces in statically determinate truss.

Objective 2: Determine the forces in various types of truss members using methods such as methods of joints, method of sections, tension coefficient method, and graphical analysis.

6.1 INTRODUCTION

Frameworks of axially loaded members arranged to resist external forces are called trusses. The members are generally welded or riveted at the joints, but for the sake of simplification of calculations, the joints are assumed to be hinges. The general configurations of trusses are shown in Figure 6.1.

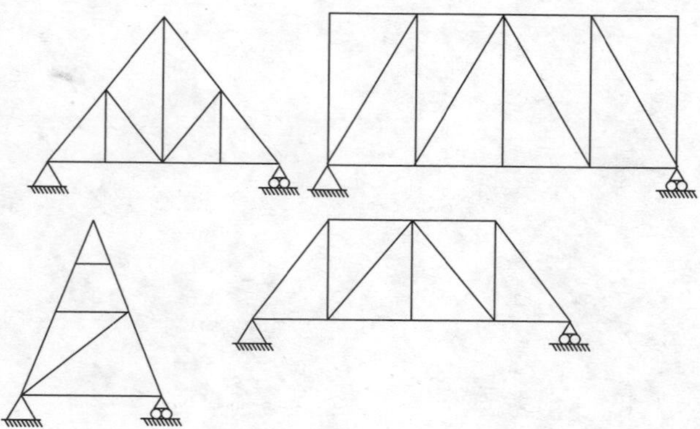

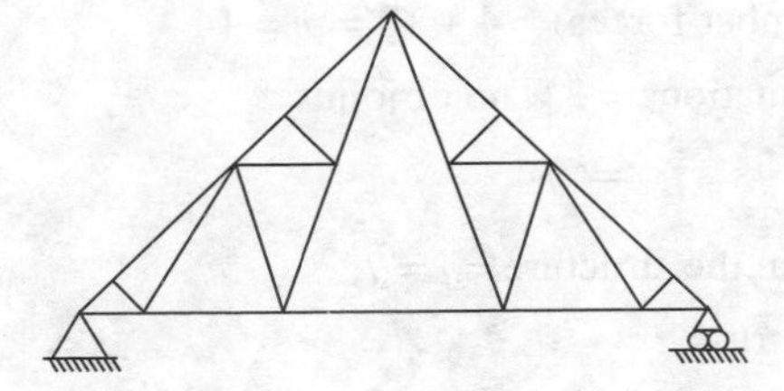

FIGURE 6.1

6.2 TYPES OF TRUSSES

The different types of trusses are classified into three categories. They are

1. Statically determinate trusses
2. Statically indeterminate trusses
3. Statically unstable trusses or imperfect trusses.

This classification is based on the members present in the truss and the number of equilibrium equations that can be used to analyze the truss. Analysis of statically determinate trusses only will be dealt in this chapter. Let us consider a simple triangulation shown in Figure 6.2.

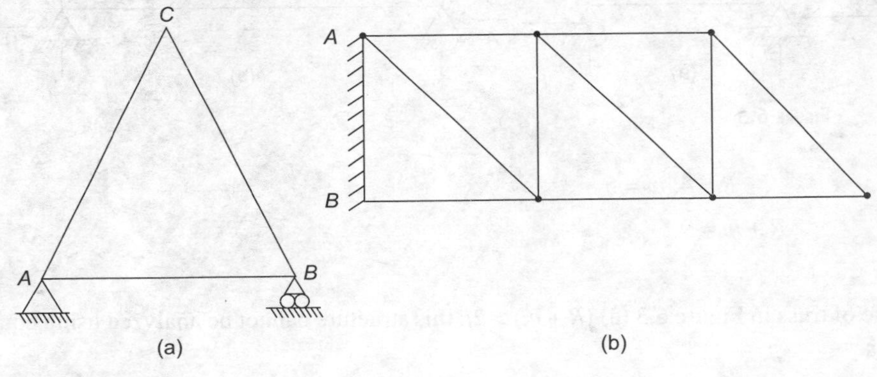

FIGURE 6.2

In Figure 6.2(a), the truss has three members. Each member carries axial force (compression to tension), thus there are three internal forces. The three external reactions are namely two reactions (vertical and horizontal) at A and one reaction at B.

The total number of internal and external reactions is 6 = (3 + 3). There are three joints in the triangulation. The number of equilibrium equations at each and every joint is 2 (i.e., sum of vertical forces = 0). Hence the total equilibrium equations are 6 (i.e., 3 × 2). As the total number of reactive forces is equal to the total number of equilibrium equation, the structure can be analyzed using static equilibrium equations alone. Thus, these types of structures are called statically determinate structures. Consider the truss shown in Figure 6.3 (b).

Number of external reactions (R) = 4 (i.e., two at A and two at B)

Number of members (member forces) = 4 + 10 = m = 14

Number of equilibrium equations = 2 × no of joints

$$= 2j$$

Number of joints present in the structure = j = 7.

For statically determinate truss,

$$2j = m + R$$

Number of equilibrium equations = Number of unknown forces.

If the number of unknown forces is more than the number of equilibrium equations, the corresponding truss is called as statically indeterminate truss. If $m + R > 2j$, then the truss falls under the category of statically indeterminate truss. Consider the truss shown in Figure 6.3.

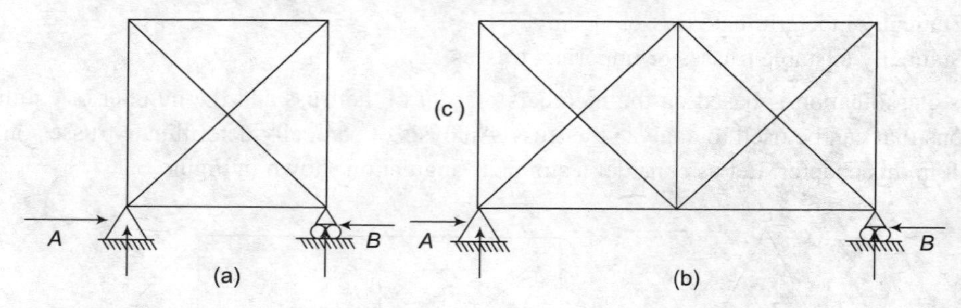

FIGURE 6.3

$$R = 4;\ m = 5$$

$$\therefore \quad R + m = 9$$

$$j = 4$$

In case of truss in Figure 6.3 (a) $(R + m) > 2j$, this structure cannot be analyzed using equilibrium equations.

In Figure 6.3 (b),

Number of members = m = 11

Number of reactions (external) = R = 3

Number of joints = j = 6

$$2j = 12$$

$$(m + R) > 2j$$

The trusses in which the total number of unknown reactive forces (m + R) is less than the number of equilibrium equations ($2j$), the corresponding truss is referred as statically imperfect truss. They are also called statically unstable trusses. Consider the case of truss present in Figure 6.4 (a) and 6.4 (b), for these structures $(m + R) < 2j$.

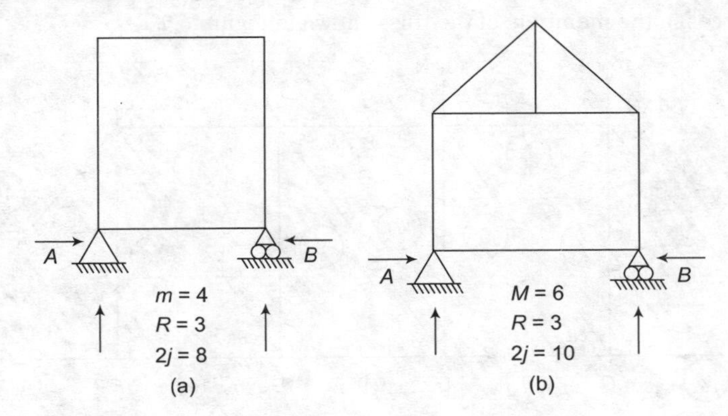

FIGURE 6.4

The trusses falling under this category are not stable, thus cannot be used for load resisting.

6.3 ANALYSIS OF STATICALLY DETERMINATE TRUSSES

The assumptions made in the analysis of trusses are:

1. Members of the truss are subjected to axial forces only.
2. All joints of the trusses are frictionless hinges.
3. Load act at the joints of the truss.

Various methods of analyzing the trusses are:

(a) Method of joints
(b) Method of sections
(c) Tension coefficient method
(d) Graphical analysis.

6.4 METHOD OF JOINTS

In this method, the forces in the members of the truss are determined from the selected joints. At each and every joint, equilibrium equations are used to find the member forces. While selecting the joint, it is to be kept in mind that the number of unknown member forces should not be more than two. In case of simply supported trusses, the support reactions are to be evaluated first, then the two member forces. In case of cantilever trusses, the support reactions need not be found.

PROBLEM 6.1

Objective 1

Determine the forces in the members of the truss shown in Figure 6.5.

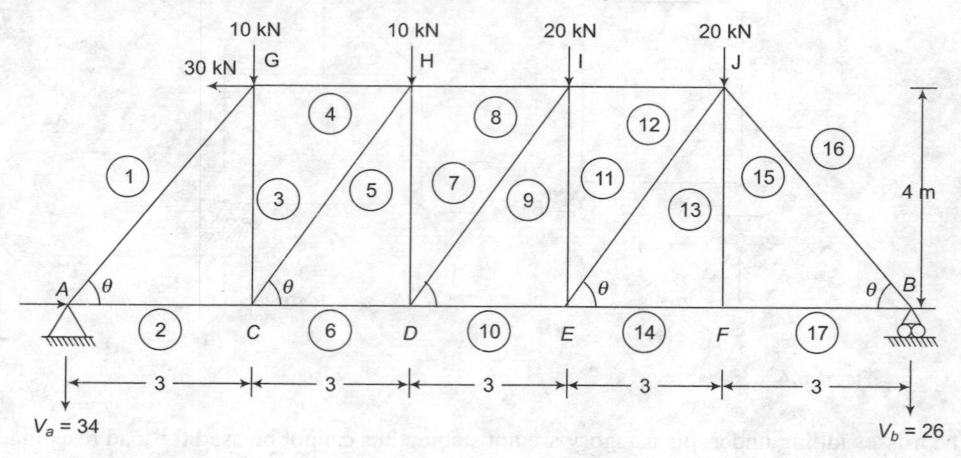

FIGURE 6.5

SOLUTION

Let θ be the inclination of the bar AG. Then $\tan\theta = \frac{4}{3}$

$\Rightarrow$ $\cos\theta = 0.6$ and $\sin\theta = 0.8$

Determination of reactions at A and B

Sum of the horizontal forces = 0

$$\Rightarrow \quad H_a = 30 \text{ kN} \tag{6.1}$$

Sum of vertical forces = 0

$$\Rightarrow \quad V_a + V_b = 60 \text{ kN} \tag{6.2}$$

Sum of moments about $A = 0$

$$\Rightarrow \quad V_b \times 15 = 10 \times 3 + 10 \times 6 + 20 \times 9 + 20 \times 12 - 30 \times 4 = 390$$

$$\Rightarrow \quad V_b = 26 \text{ kN} \tag{6.3}$$

$$V_a = 34 \text{ kN (from Eq. (2))}.$$

To find the forces in the members of the truss, assume that, all members carry tensile force (force away from joint), and a negative value indicates compressive force in the member.

Consider joint A: Unknown forces are F_1 and F_2 (let F_1, F_2 be the forces in the members (1) and (2), respectively).

Sum of the forces in y direction = 0

$$(\sum F_y = 0)$$

$$\Rightarrow \quad V_a + F_1 \sin\theta = 0$$

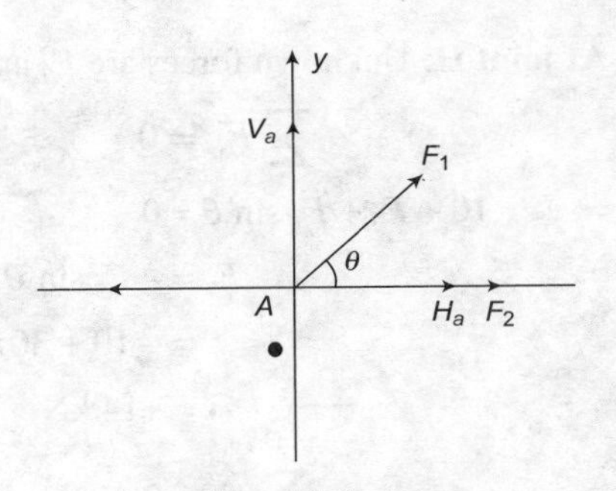

FIGURE 6.6

$$\Rightarrow \qquad F_1 = -\frac{34}{0.8} = -42.5 \text{ kN}$$

Sum of the forces in x direction = 0

$$(\sum F_x = 0)$$

$$\Rightarrow \quad H_a + F_2 + F_1 \cos\theta = 0$$

$$F_2 = -H_a - F_1 \cos\theta$$

$$= -30 + 42.5 \times 0.6 = -4.5 \text{ kN}.$$

At joint G: Unknown forces are F_3 and F_4.

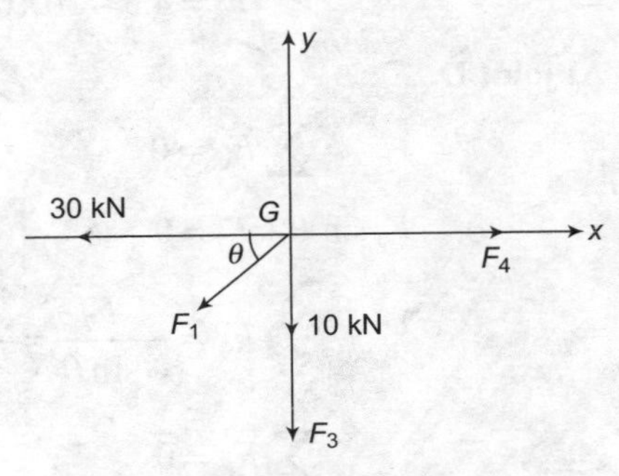

FIGURE 6.7

$$\sum F_y = 0$$

$$\Rightarrow \quad -F_1 \sin\theta - 10 - F_3 = 0$$

$$\Rightarrow \qquad F_3 = -10 - F_1 \sin\theta$$

$$= -10 + 42.5 \times 0.8$$

$$= +24 \text{ kN}$$

$$\sum F_x = 0$$

$$\Rightarrow \qquad F_4 = F_1 \cos\theta + 30$$

$$= -42.5 \times 0.6 + 30$$

$$= 4.5 \text{ kN}.$$

At joint C: Unknown forces are F_5 and F_6.

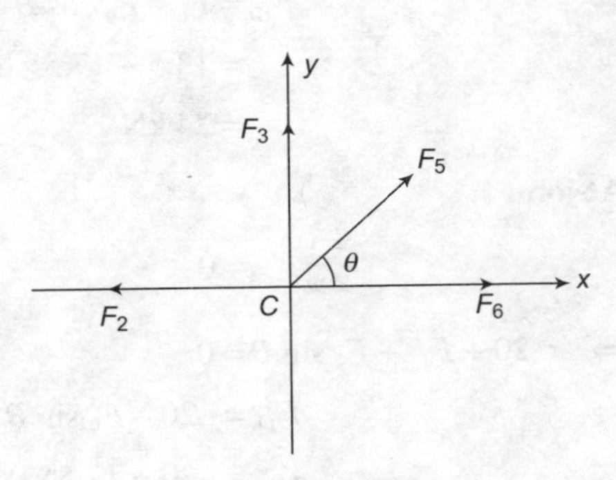

FIGURE 6.8

$$\sum F_y = 0$$

$$\Rightarrow \quad F_3 + F_5 \sin\theta = 0$$

$$\sum F_x = 0$$

$$\Rightarrow \quad F_5 \cos\theta + F_6 - F_2 = 0$$

$$\Rightarrow \qquad F_6 = F_2 - F_5 \cos\theta$$

$$= -4.5 + 30 \times 0.6$$

$$= 13.5 \text{ kN}$$

$$\Rightarrow \qquad F_5 = -\frac{F_3}{\sin\theta} = -\frac{24}{08} = -30 \text{ kN}$$

$$\Rightarrow \qquad \sum F_x = 0$$

$$\Rightarrow \quad F_4 + F_5 \cos\theta = F_8$$

$$\Rightarrow \qquad F_8 = 4.5 - 30 \times 0.6 = 13.5 \text{ kN}$$

At joint H: Unknown forces are F_7 and F_8.

$$\sum F_y = 0$$

$$\Rightarrow \quad 10 + F_7 + F_5 \sin\theta = 0$$

$$\Rightarrow \quad F_7 = -F_5 \sin\theta - 10$$

$$= -10 + 30 \times 0.8$$

$$= +14 \text{ kN}$$

$$\sum F_x = 0$$

$$\Rightarrow \quad F_4 + F_5 \cos\theta = F_8$$

$$\Rightarrow \quad F_8 = 4.5 - 30 \times 0.6 = 13.5 \text{ kN}$$

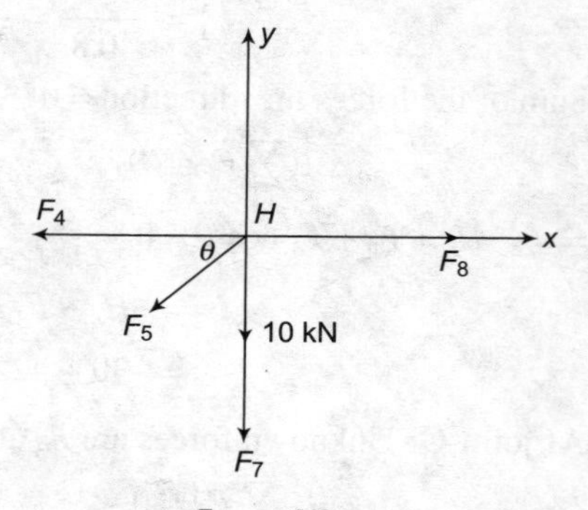

FIGURE 6.9

At joint D:

$$\sum F_y = 0$$

$$\Rightarrow \quad F_9 \sin\theta + F_7 = 0$$

$$\Rightarrow \quad F_9 = -\frac{F_7}{\sin\theta} = -\frac{14}{0.8} = 17.5 \text{ kN}$$

$$\sum F_x = 0$$

$$\Rightarrow \quad F_9 \cos\theta + F_{10} = F_6$$

$$\Rightarrow \quad F_{10} = F_6 - F_9 \cos\theta$$

$$= 13.5 + 17.5 \times 0.6$$

$$= 24 \text{ kN}$$

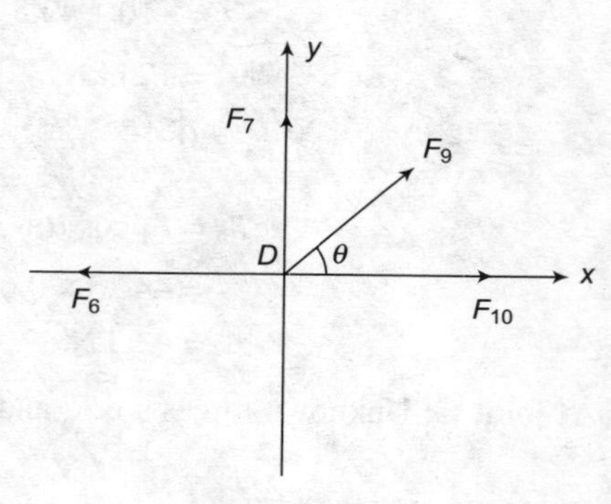

FIGURE 6.10

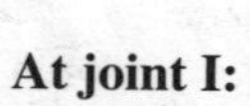

At joint I:

$$\sum F_y = 0$$

$$\Rightarrow \quad 20 + F_{11} + F_9 \sin\theta = 0$$

$$\Rightarrow \quad F_{11} = -20 - F_9 \sin\theta$$

$$= -20 + 17.5 \times 0.8$$

$$= -6 \text{ kN}$$

$$\sum F_x = 0$$

$$\Rightarrow \quad F_{12} = F_8 + F_9 \cos\theta$$

$$= -13.5 - 17.5 \times 0.6$$

$$= -24 \text{ kN}$$

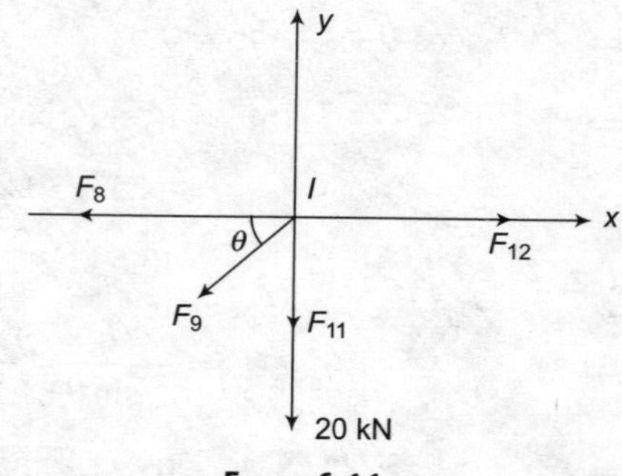

FIGURE 6.11

At joint E:

$$\sum F_y = 0$$

$\Rightarrow \qquad F_{11} + F_{13} \sin \theta = 0$

$$F_{13} = -\frac{F_{11}}{\sin\theta} = -\left(-\frac{6}{0.8}\right) = 7.5 \text{ kN}$$

$$\sum F_x = 0$$

or

$\Rightarrow \qquad F_{13} \cos \theta + F_{14} = F_{10}$

$$F_{14} = F_{10} - F_{13} \cos \theta$$
$$=24 - 7.5(0.6)$$
$$= 19.5 \text{ kN}$$

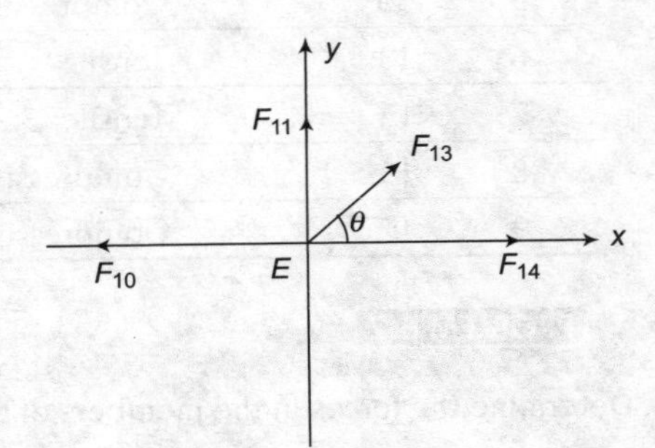

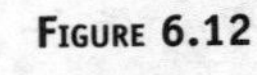

FIGURE 6.12

At joint J:

$$\sum F_x = 0$$

$\Rightarrow \qquad F_{16} \cos \theta = F_{13} \cos \theta + F_{12}$

$$= 7.5 \cos \theta + (-24)$$
$$= -19.5 \text{ kN}$$

$\Rightarrow \qquad F_{16} = -32.5 \text{ kN}$

$$\sum F_y = 0$$

$\Rightarrow \qquad 20 + F_{15} + F_{13} \sin \theta + F_{16} \sin \theta = 0$

$\Rightarrow \qquad F_{15} = -20 - 7.5 \sin \theta + 32.5 \sin \theta$

$$= 0$$

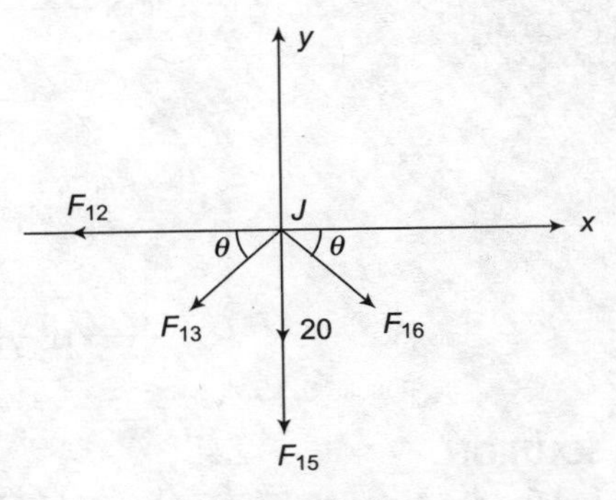

FIGURE 6.13

At joint F:

$$\sum F_x = 0$$

$\Rightarrow \qquad F_{14} = F_{17}$

$\therefore \qquad F_{17} = 19.5 \text{ kN}$

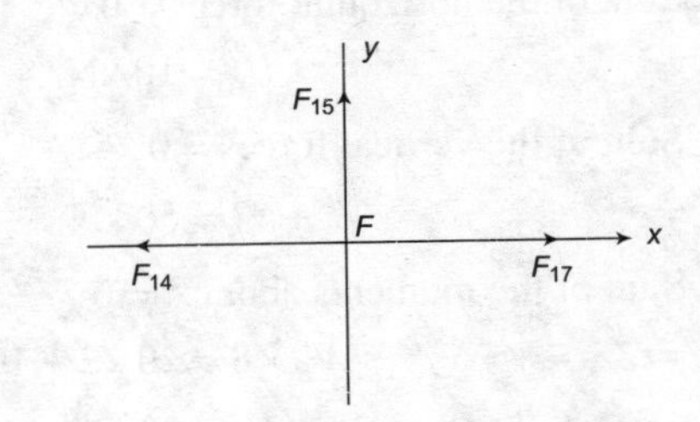

FIGURE 6.14

Table 6.1 The forces in various members with magnitude and nature are presented below

Member	Magnitude	Nature	Member	Magnitude	Nature
1	42.5	Compressive	10	24	Tensile
2	4.5	Compressive	11	6	Compressive
3	24	Tensile	12	24	Compressive
4	4.5	Tensile	13	7.5	Tensile
5	30	Compressive	14	19.5	Tensile
6	13.5	Tensile	15	0	-
7	14	Tensile	16	32.5	Compressive
8	13.5	Compressive	17	19.5	Tensile
9	17.5	Compressive			

PROBLEM 6.2

Objective 2

Determine the forces in the members of the truss shown in Figure 6.15. Use the method of joints.

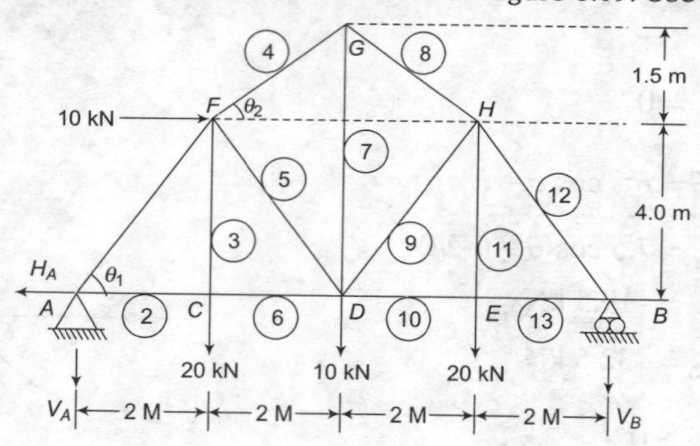

FIGURE 6.15

SOLUTION

Let the reactions at the supports A and B be V_A, H_A, and V_B.

Sum of the horizontal forces = 0 (6.4)

$$\Rightarrow \qquad H_A = 10 \text{ kN}$$

Sum of the vertical forces = 0 (6.5)

$$\Rightarrow \qquad V_A + V_B = 50 \text{ kN}$$

Sum of the moments about $A = 0$ (6.6)

$$\Rightarrow \qquad V_B \times 8 = 20\times2 + 10\times4 + 20\times6 + 10\times4$$

$$\Rightarrow \qquad V_B = 30 \text{ kN}$$

$$V_A = 20 \text{ kN}$$

$$\tan\theta_1 = \frac{4}{2} \Rightarrow \theta_1 = 63.435°$$

$$\sin \theta_1 = 0.894$$

$$\cos \theta_1 = 0.447$$

$$\tan \theta_2 = \frac{1.5}{2} = \frac{3}{4}$$

$$\sin \theta_2 = 0.6$$

$$\cos \theta_2 = 0.8$$

At joint A:

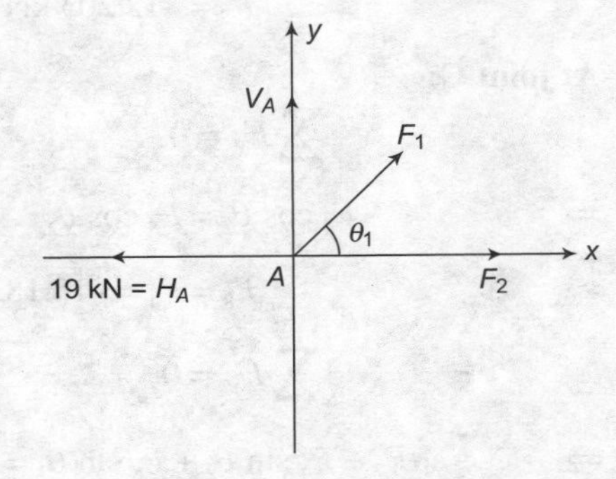

FIGURE 6.17

$$\sum F_y = 0$$

$$\Rightarrow \quad V_A + F_1 \sin \theta_1 = 0$$

$$\Rightarrow \quad F_1 = -\frac{20}{0.894} = -22.371 \text{ kN}$$

$$\sum F_x = 0$$

$$\Rightarrow \quad F_2 + F_1 \cos \theta_1 = 10$$

$$\Rightarrow \quad F_2 = 10 + 22.371 \times (0.447)$$

$$= 20 \text{ kN}$$

At joint C:

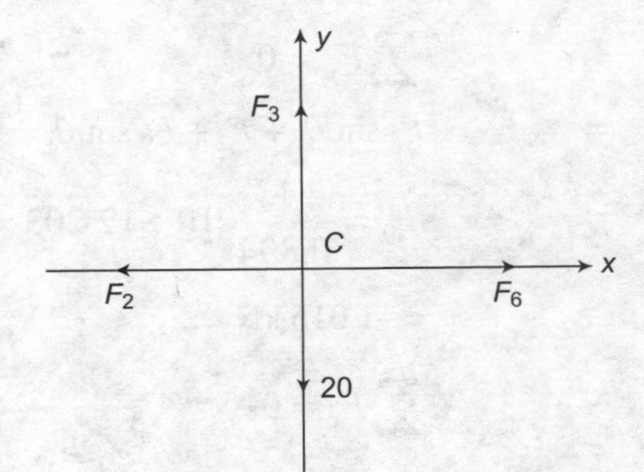

FIGURE 6.17

$$\sum F_y = 0$$

$$\Rightarrow \quad F_3 = 20 \text{ kN}$$

$$\sum F_x = 0$$

$$\Rightarrow \quad F_2 = F_6 = 20 \text{ kN}$$

At joint F:

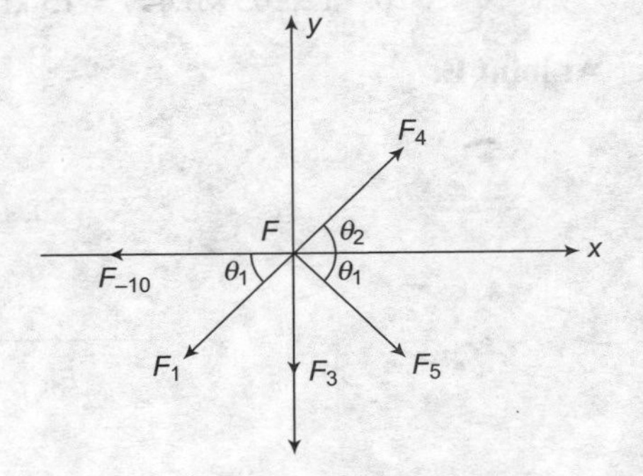

FIGURE 6.18

$$\Rightarrow \quad F_5 \cos \theta_1 + F_4 \cos \theta_2 = F_1 \cos \theta_1 - 10$$

$$0.447\, F_5 + 0.8\, F_4 = -22.371 \cos \theta_1 - 10$$

$$= -20 \qquad (6.7)$$

$$\sum F_y = 0$$

$$\Rightarrow \quad F_5 \sin \theta_1 - F_4 \sin \theta_2 = -F_3 - F_1 \sin \theta_1$$

$$0.894\, F_5 - 0.6\, F_4 = -20 + 22.371 \sin \theta$$

$$= 0 \qquad (6.8)$$

Eq. (6.7) × 2 – Eq. (6.8)

$$\Rightarrow \quad 0.894 F_5 + 1.6 F_4 = -40$$

$$0.894 F_5 - 0.6 F_4 = 0$$

$$2.2 F_4 = -40$$

$$\Rightarrow \quad F_4 = -18.182 \text{ kN}$$

$$F_5 = -12.203 \text{ kN}$$

At joint G:

$$\sum F_x = 0$$

$$\Rightarrow \quad F_4 \cos \theta_2 = F_8 \cos \theta_2$$

$$\Rightarrow \quad F_8 = F_4 = -18.182 \text{ kN}$$

$$\sum F_y = 0$$

$$\Rightarrow \quad F_7 + F_4 \sin \theta_2 + F_8 \sin \theta_2 = 0$$

$$F_7 = 2 \times 18.182 \times 0.6 = 21.818 \text{ kN}$$

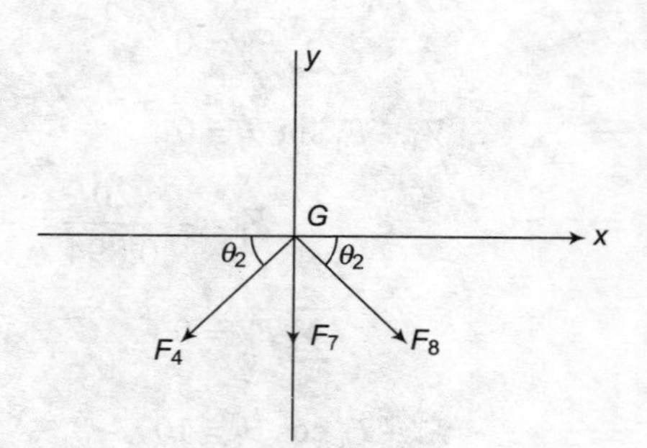

FIGURE 6.19

At joint D:

$$\sum F_y = 0$$

$$\Rightarrow \quad F_5 \sin \theta_1 + F_7 + F_9 \sin \theta_1 = 10$$

$$\Rightarrow \quad F_9 = \frac{1}{0.894}\{10 + 12.203 \times 0.894 - 21.818\}$$

$$= -1.016 \text{ kN}$$

$$\sum F_x = 0$$

$$\Rightarrow \quad F_{10} = F_6 + F_5 \cos \theta_1 - F_9 \cos \theta_1$$

$$20 - 12.203 \times 0.447 = 15 \text{ kN}$$

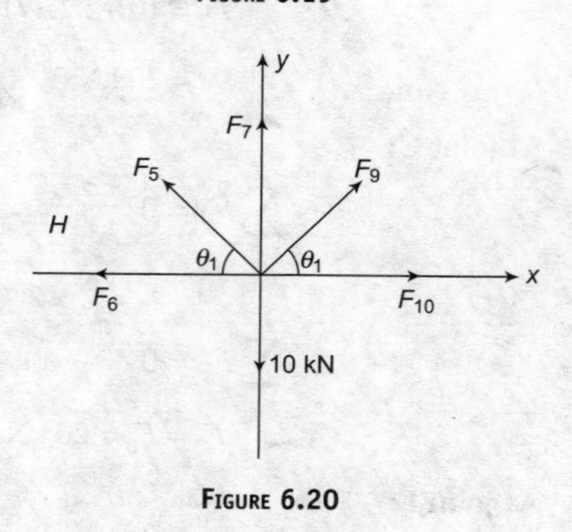

FIGURE 6.20

At joint E:

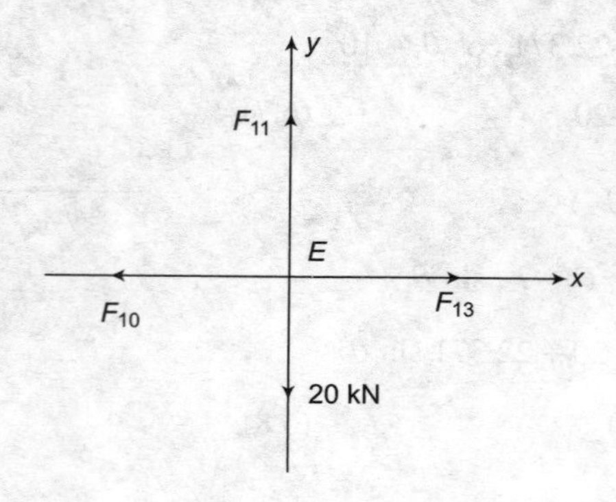

FIGURE 6.21

$$\sum F_y = 0 \Rightarrow F_{11} = 20 \text{ kN}$$
$$\sum F_x = 0 \Rightarrow F_{13} = F_{10} = 15 \text{ kN}$$

At joint H:

$$\sum F_x = 0$$

$$\Rightarrow \quad F_{12} \cos \theta_1 = F_9 \cos \theta_1 + F_8 \cos \theta_2$$
$$= -1.016(0.447) - 18.182 \times 0.8$$
$$= -15$$
$$\Rightarrow \quad F_{12} = -33.557 \text{ kN}$$

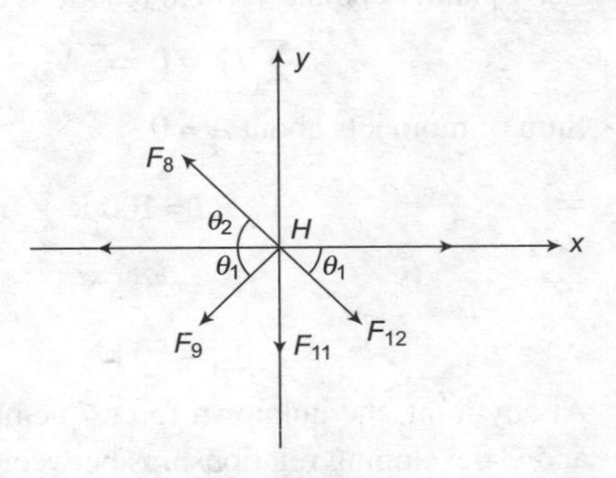

FIGURE 6.22

Table 6.2 The forces in the members are tabulated below

Member	Magnitude	Nature	Member	Magnitude	Nature
1	22.371	Compressive	8	18.182	Compressive
2	20.0	Tensile	9	1.016	Compressive
3	20	Tensile	10	15.0	Tensile
4	18.182	Compressive	11	20	Tensile
5	12.203	Compressive	12	33.557	Compressive
6	20	Tensile	13	15	Tensile
7	21.818	Tensile			

PROBLEM 6.3 **Objective 2**

Determine forces in the members of the truss shown in Figure 6.23. Use the method of joints. *ABC* is an equilateral triangle EDF is also an equilateral triangle.

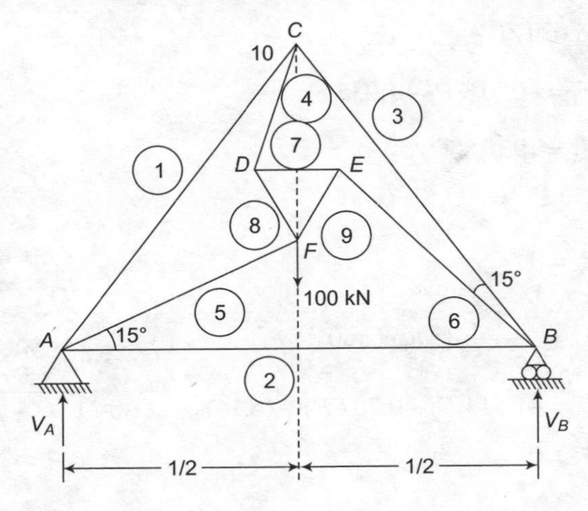

FIGURE 6.23

SOLUTION

Let V_A and V_B be the vertical reactions at A and B, respectively.

$$\sum F_V = 0 \Rightarrow V_A + V_B = 100 \text{ kN} \tag{6.9}$$

Sum of moments about $A = 0$

$$\Rightarrow \quad V_B \times l = 100 \times \frac{l}{2}$$

$$\Rightarrow \quad V_B = 50 \text{ kN}$$

$$\therefore \quad V_A = 50 \text{ kN} \tag{6.10}$$

At any joint, the unknown forces/member actions are more than two. Hence consider joints A, B, and C developing relationships between F_1, F_2, and F_3.

At joint A:

$$\sum F_y = 0$$

$$\Rightarrow \quad V_A + F_1 \sin 60° + F_5 \sin 15° = 0$$

$$\Rightarrow \quad F_5 = \left[-50 - \frac{V_3}{2} F_1\right] \frac{1}{\sin 15°}$$

$$F_5 = 193.185 - 3.345\ F_1 \quad (6.11)$$

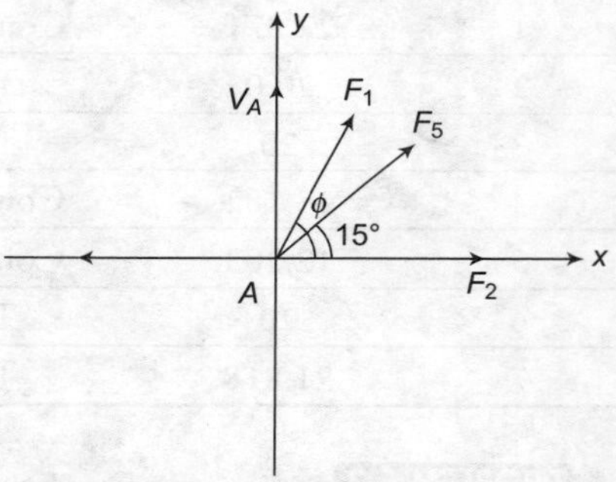

$$\sum F_x = 0$$

$$\Rightarrow \quad F_2 + F_5 + \cos 15° + F_1 \cos 60° = 0$$

$$F_5 = \frac{1}{\cos 15°}\{-F_2 - F_1 \cos 60°\}$$

$$= -0.518\ F_1 - 1.035\ F_2 \tag{6.12}$$

FIGURE 6.24

Equating Eqs. (6.11) and (6.12)

$$-193.185 - 3.346\ F_1 = -0.518\ F_1 - 1.035\ F_2$$

$$\Rightarrow \quad 1.035\ F_2 - 2.828\ F_1 = 193.185 \tag{6.13}$$

At Joint B:

$$\sum F_x = 0$$

$$\Rightarrow \quad F_6 = \frac{1}{\cos 45°}\{-F_3 \cos 60° - F_2\}$$

$$= -1.414\ F_2 - 0.707\ F_3 \quad (6.14)$$

$$\sum F_y = 0$$

$$\Rightarrow \quad F_6 = \frac{1}{\sin 45°}\{-50 - F_3 \sin 60°\}$$

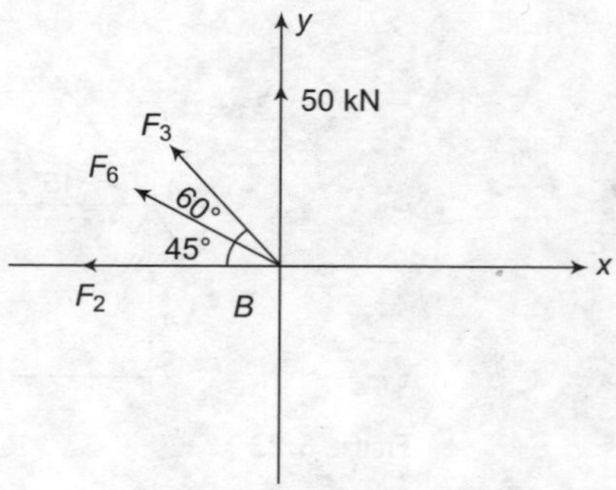

FIGURE 6.25

$$= -70.711 - 1.225\,F_3 \qquad (6.15)$$

From Eqs. (6.14) and (6.15)

$$\Rightarrow \quad -1.414\,F_2 - 0.707\,F_3 = -70.711 - 1.225\,F_3$$

$$\Rightarrow \quad 1.414\,F_2 - 0.518\,F_3 = 70.711 \qquad (6.16)$$

At joint C:

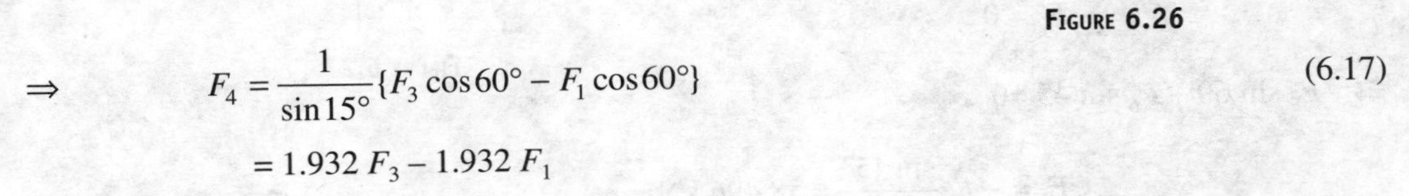

FIGURE 6.26

$$\sum F_x = 0$$

$$\Rightarrow \quad F_4 = \frac{1}{\sin 15^\circ}\{F_3 \cos 60^\circ - F_1 \cos 60^\circ\} \qquad (6.17)$$

$$= 1.932\,F_3 - 1.932\,F_1$$

$$\sum F_y = 0$$

$$\Rightarrow \quad F_4 = +\frac{1}{\cos 15^\circ}\{-F_1 \sin 60^\circ - F_3 \sin 60^\circ\}$$

$$= -0.897\,F_1 - 0.897\,F_3 \qquad (6.18)$$

From Eqs. (6.17) and (6.18)

$$1.932\,F_3 - 1.932\,F_1 = -0.897\,F_1 - 0.897\,F_3$$

$$\Rightarrow \quad 1.035\,F_1 = 2.829\,F_3$$

$$F_1 = 2.733\,F_3 \qquad (6.19)$$

Using Eq. (6.19) in Eq. (6.13)

$$1.035\,F_2 - 7.73\,F_3 = 193.185 \qquad (6.20)$$

Eq. (8) × 0.732

$$\Rightarrow \quad 1.035\,F_2 - 0.379\,F_3 = 51.758 \qquad (6.21)$$

Solving Eqs. (6.20) and (6.21)

$$-7.351\,F_3 = 141.427 \Rightarrow F_3 = -19.239 \text{ kN}$$

$$\therefore \quad F_2 = 42.963 \text{ kN} \quad \Rightarrow \quad F_1 = -2.733 \times 19.239 = -52.580 \text{ kN}$$

Using the value of F_1 and F_2 in Eq. (6.12)

$$F_5 = -17.23 \text{ kN}$$

Using the value of F_2 and F_3 in Eq. (6.14)

$$F_6 = -47.148 \text{ kN}$$

Using F_3 and F_1 values in Eq. (6.19)

$$F_4 = 64.422 \text{ kN}$$

At joint D:

$$\sum F_y = 0$$

$$\Rightarrow \quad F_4 \cos 15° = F_8 \sin 60°$$

$$\Rightarrow \quad F_8 = 64.422 \frac{\cos 15°}{\sin 60°}$$

$$= -52.6 \text{ kN}$$

FIGURE 6.27

At joint E:

$$\sum F_y = 0$$

$$\Rightarrow \quad F_9 \sin 60 + F_6 \sin 45 = 0$$

$$\Rightarrow \quad F_9 = -F_6 \frac{\sin 45°}{\sin 60°} = 38.496 \text{ kN}$$

$$\sum F_x = 0$$

Can be used to cross-check the force in the member (7)

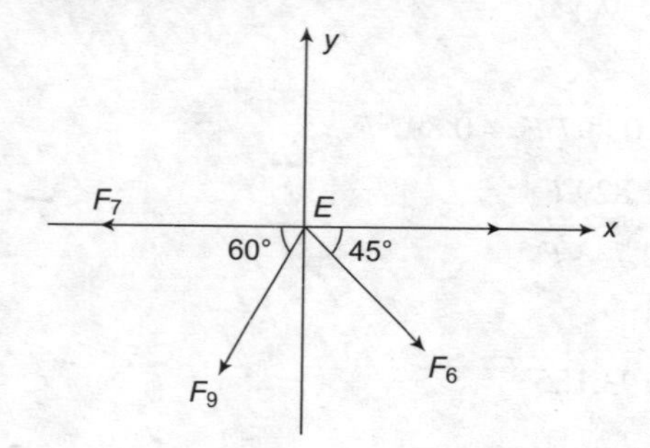

FIGURE 6.28

Table 6.3 The forces in various members are tabulated below

Member Designation	Magnitude of the Force (kN)	Nature of the force
1	52.580	Compressive
2	42.963	Tensile
3	19.239	Compressive
4	64.422	Tensile
5	17.230	Compressive
6	47.148	Compressive
7	52.60	Compressive
8	71.853	Tensile
9	38.496	Tensile

PROBLEM 6.4 **Objective 1**

Determine the forces in the members of the truss shown in Figure 6.29, using the method of joints.

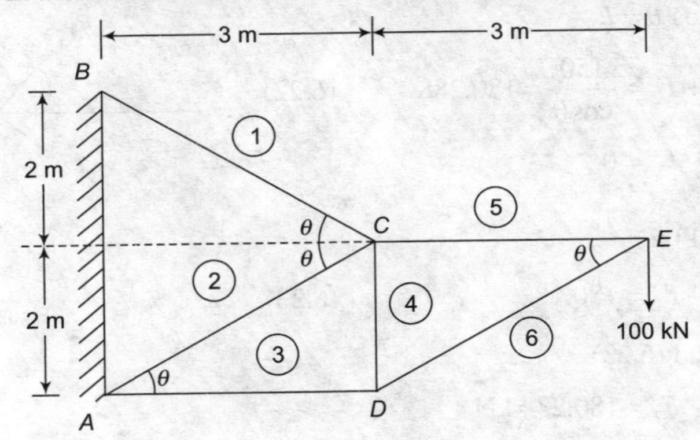

FIGURE 6.29

SOLUTION

For cantilever type truss, the reactions need not be calculated.

$$\tan\theta = \frac{2}{3}$$

$\Rightarrow$ $\sin\theta = 0.555$

$\cos\theta = 0.832$

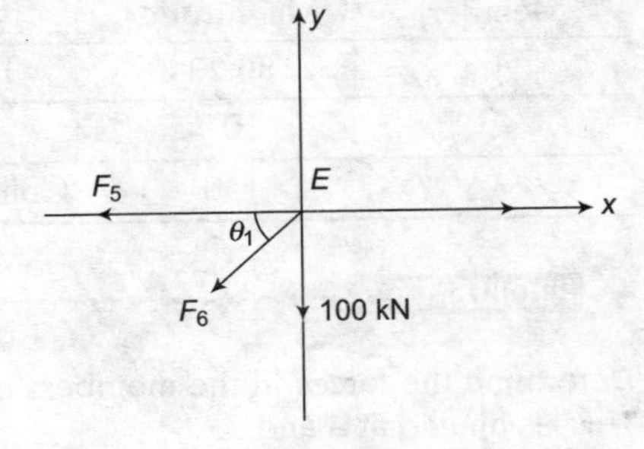

FIGURE 6.30

At joint E:

$$\sum F_y = 0 \Rightarrow F_6 \sin\theta + 100 = 0$$

$\Rightarrow$ $F_6 = -180.18$ kN

$$\sum F_x = 0 \Rightarrow F_6 \cos\theta + F_5 = 0$$

$\Rightarrow$ $F_5 = 150$ kN

At joint D:

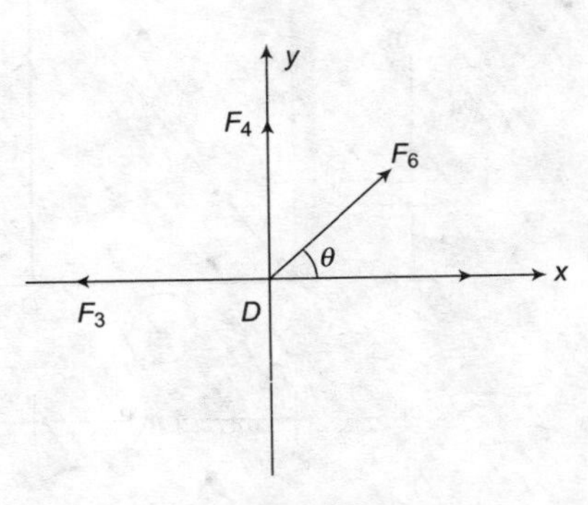

FIGURE 6.31

$$\sum F_y = 0$$

$\Rightarrow$ $F_4 + F_6 \sin\theta = 0$

$\Rightarrow$ $F_4 = 180.18 \sin\theta$

$= 100$ kN

$$\sum F_x = 0$$

$\Rightarrow$ $F_3 = F_6 \cos 60 = -180.18\ \{0.832\}$

$= -150$ kN

At joint C:

$$\sum F_x = 0$$

$$\Rightarrow \quad F_1 \cos\theta + F_2 \cos\theta = F_5$$

$$F_1 + F_2 = \frac{150}{\cos\theta} = 180.288 \qquad (6.22)$$

$$\sum F_y = 0$$

$$\Rightarrow \quad (F_1 - F_2)\sin\theta = F_4$$

$$F_1 - F_2 = 180.18 \qquad (6.23)$$

FIGURE 6.32

Solving Eqs. (6.22) and (6.23)

$$F_1 = 180.234 \text{ kN}$$

$$F_2 = 0.054 \text{ kN} \propto 0$$

(Actual value is 0. Error occurred due to rounding of digits)

Table 6.4 Forces in the different members of the truss are shown below

Member	Magnitude	Nature	Member	Magnitude	Nature
1	180.234	Tensile	4	100	Tensile
2	0	-	5	150	Tensile
3	150	Compressive	6	180.18	Compressive

PROBLEM 6.5

Objective 2

Determine the forces in the members of the truss shown in Figure 6.33, using the method of joints. Truss is hinged at A and B.

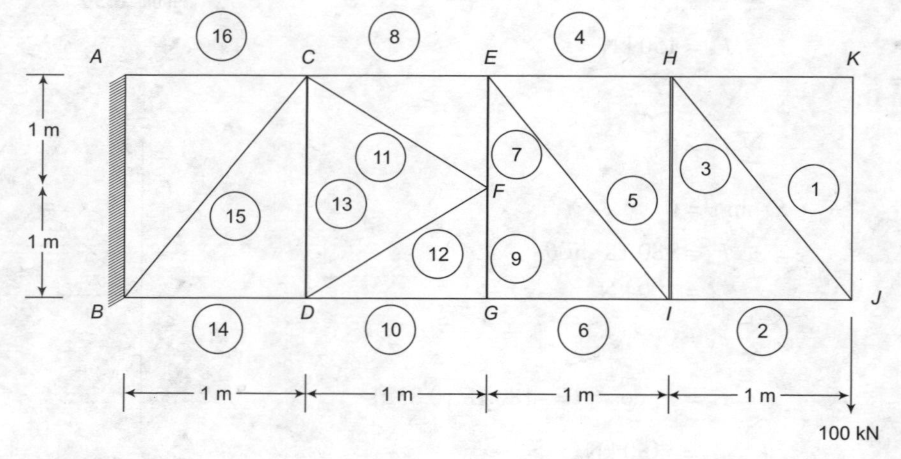

FIGURE 6.33

SOLUTION:

At joint *K*, only two members are present without any external load. Hence force in these two members is zero.

$$\tan\alpha_1 = \frac{2}{1}$$

$$\sin\,\alpha_1 = 0.894$$

$$\cos\,\alpha_1 = 0.447$$

$$\tan\alpha_2 = \frac{1}{1} \Rightarrow \alpha_2 = 45°$$

$$\tan\alpha_3 = \frac{2}{1.5}$$

$$\sin\,\alpha_2 = 0.8$$

$$\cos\,\alpha_2 = 0.6$$

At joint J:

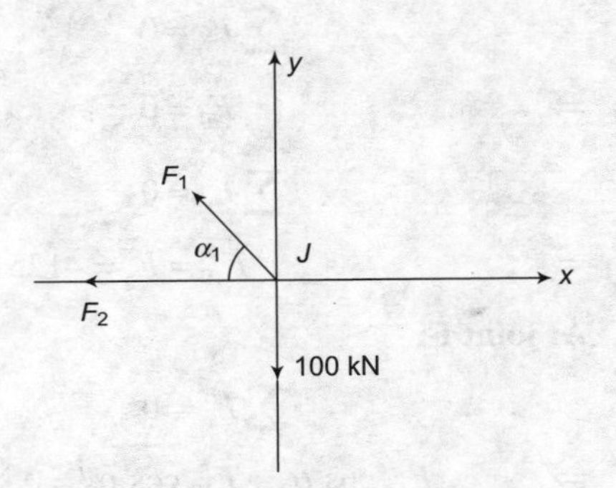

FIGURE 6.34

$$\sum F_y = 0$$

$$\Rightarrow \quad F_1 \sin\,\alpha_1 = 100$$

$$\Rightarrow \quad F_1 = 111.857\text{ kN}$$

$$\sum F_x = 0$$

$$\Rightarrow \quad F_2 + F_1 \cos\,\alpha_1 = 0$$

$$\Rightarrow \quad F_2 = -50\text{ kN}$$

At joint H:

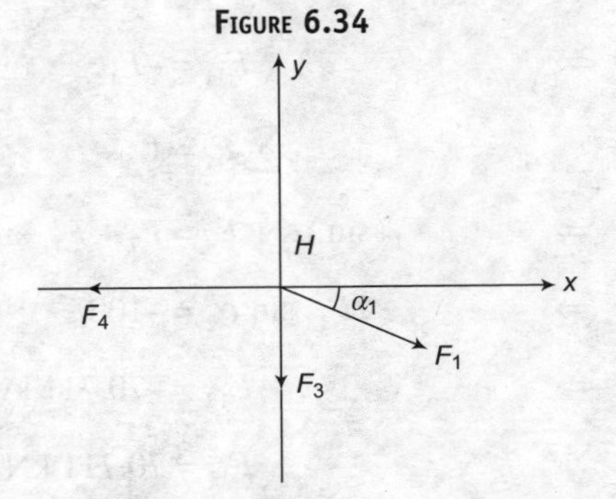

FIGURE 6.35

$$\sum F_x = 0$$

$$\Rightarrow \quad F_4 = F_1 \cos\,\alpha_1 = 50\text{ kN}$$

$$\sum F_y = 0$$

$$\Rightarrow \quad F_3 + F_1 \sin\,\alpha_1 = 0$$

$$\Rightarrow \quad F_3 = -111.857(0.894)$$

$$= -100\text{ kN}$$

At joint I:

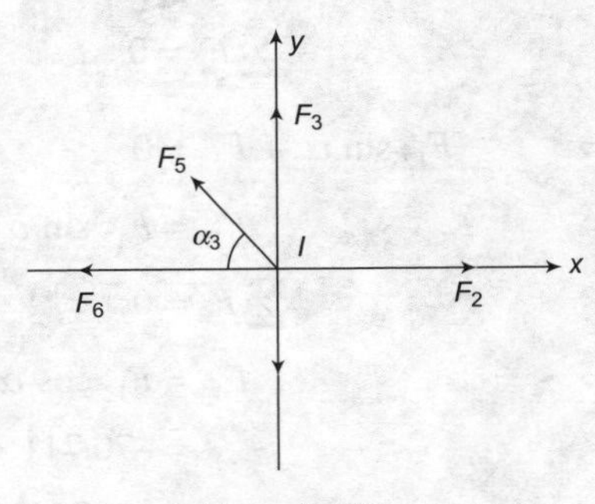

FIGURE 6.36

$$\sum F_y = 0$$

$$\Rightarrow \quad F_3 + F_5 \sin\,\alpha_3 = 0$$

$$\Rightarrow \quad F_5 = -\frac{F_3}{\sin\alpha_3} = \frac{100}{0.8}$$

$$= 125\text{ kN}$$

At joint E:

$$\sum F_x = 0$$

$\Rightarrow$ $$F_8 = F_4 + F_4 \cos \alpha_3 = 50 + 125 \times 0.6$$

$$= 125 \text{ kN}$$

$$\sum F_y = 0$$

$\Rightarrow$ $$F_7 = F_5 \sin \alpha_3 = -125 \times 0.5 = -100 \text{ kN}$$

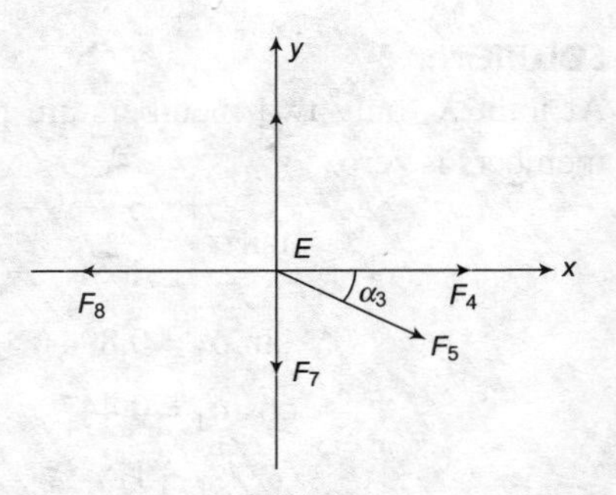

FIGURE 6.37

At joint G:

$$\sum F_y = 0$$

$\Rightarrow$ $$F_9 = 0$$

$$\sum F_x = 0$$

$\Rightarrow$ $$F_{10} = F_6 = -125 \text{ kN}$$

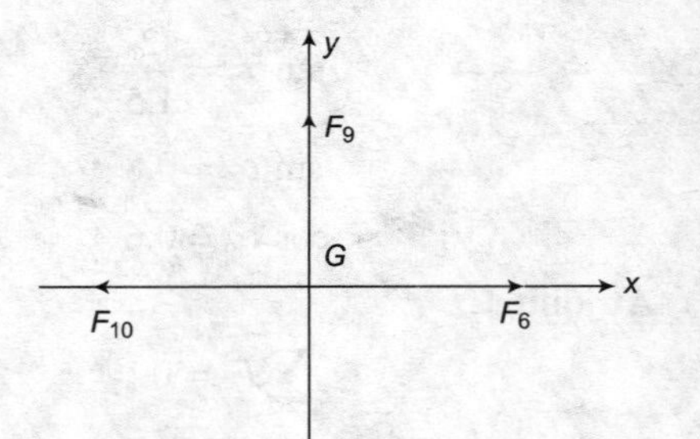

FIGURE 6.38

At joint F:

$$\sum F_x = 0$$

$\Rightarrow$ $$F_{11} \cos \alpha_2 + F_{12} \cos \alpha_2 = 0$$

$\Rightarrow$ $$F_{11} = -F_{12}$$

$$\sum F_y = 0$$

$\Rightarrow$ $$F_{12} \sin \alpha_2 + F_9 = F_7 + F_{11} \sin \alpha_2$$

$\Rightarrow$ $$2F_{12} \sin \alpha_2 = -100$$

$\Rightarrow$ $$F_{12} = -70.711 \text{ kN}$$

$\therefore$ $$F_{11} = 70.711 \text{ kN}$$

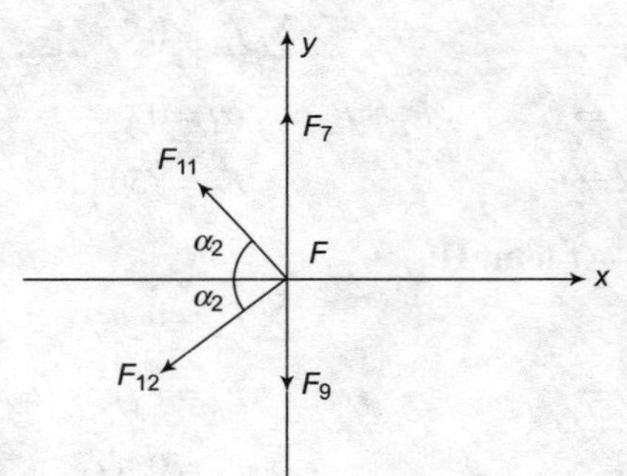

FIGURE 6.39

At joint D:

$$\sum F_y = 0$$

$\Rightarrow$ $$F_{12} \sin \alpha_2 + F_{13} = 0$$

$\Rightarrow$ $$F_{13} = F_{12} \sin \alpha_2 = 50 \text{ kN}$$

$$\sum F_x = 0$$

$\Rightarrow$ $$F_{14} = F_{12} \cos \alpha_2 + F_{10}$$

$$= -70.711 \times \cos \alpha_2 - 125$$

$$= -175 \text{ kN}$$

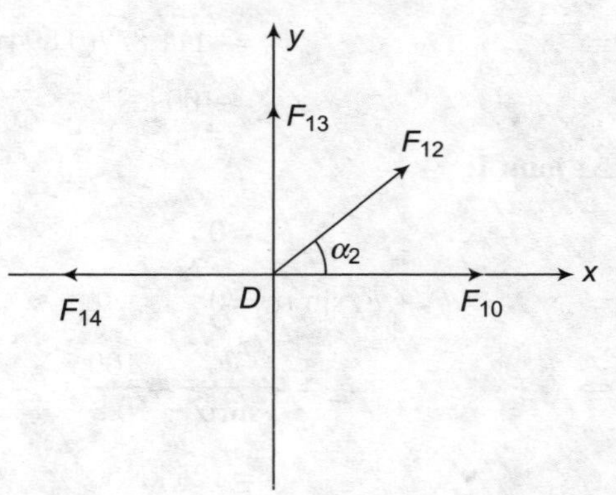

FIGURE 6.40

At joint C:

$$\sum F_y = 0$$

$$\Rightarrow \quad F_{11} \sin \alpha_2 + F_{13} + F_{15} \sin \alpha_1 = 0$$

$$\Rightarrow \quad F_{15} = \frac{1}{0.894}\{-70.711 \sin 45° - 50\}$$

$$= -111.856 \text{ kN}$$

$$\sum F_y = 0$$

$$\Rightarrow \quad F_{16} + F_{15} \cos \alpha_1 = F_8 + F_{11} \cos \alpha_2$$

$$\Rightarrow \quad F_{16} = 125 + 70.711 \cos 45° + 111.856 \cos \alpha_1$$

$$= 225 \text{ kN}$$

FIGURE 6.41

Table 6.5 The forces in various members are tabulated below

Member	Magnitude	Nature	Member	Magnitude	Nature
1	111.857	Tensile	9	0	-
2	50	Compressive	10	125	Compressive
3	100	Compressive	11	70.711	Tensile
4	50	Tensile	12	70.711	Compressive
5	125	Tensile	13	50	Tensile
6	125	Compressive	14	175	Compressive
7	100	Compressive	15	111.856	Compressive
8	125	Tensile	16	225	Tensile

PROBLEM 6.6

Objective 2

Determine the axial force in the members of the truss shown in Figure 6.42, using the method of joints.

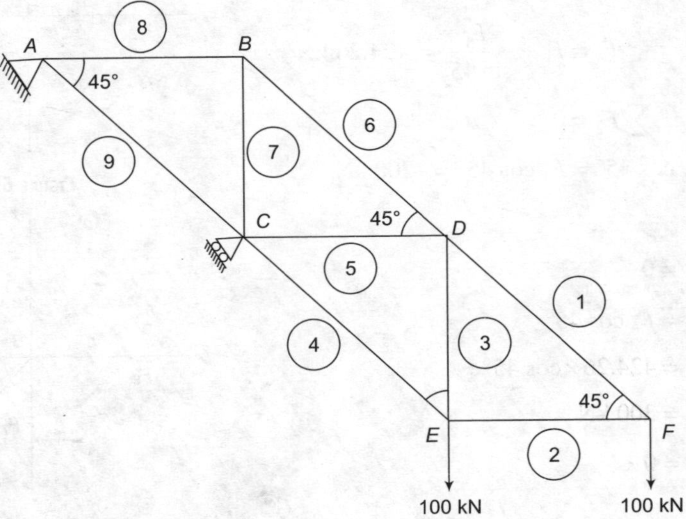

FIGURE 6.42

SOLUTION

This problem can be solved without finding the support reactions. Consider the joint F.

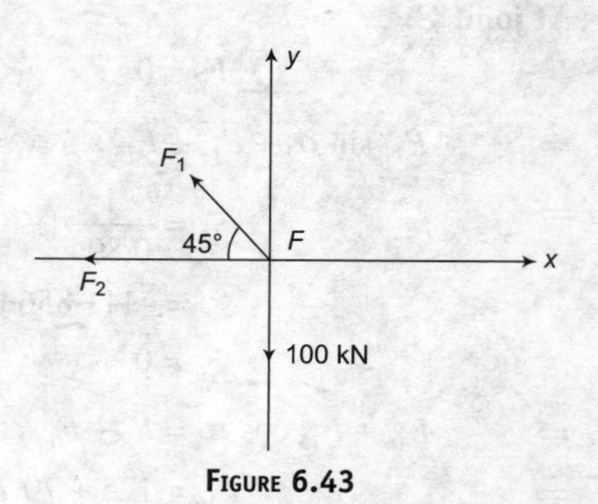

FIGURE 6.43

At joint F:

$$\sum F_y = 0$$

$\Rightarrow$ $F_1 \sin 45° = 100$

$\Rightarrow$ $F_1 = 141.42$ kN

$$\sum F_x = 0$$

$\Rightarrow$ $F_1 \cos 45° + F_2 = 0$

$\Rightarrow$ $F_2 = -100$ kN

At joint E:

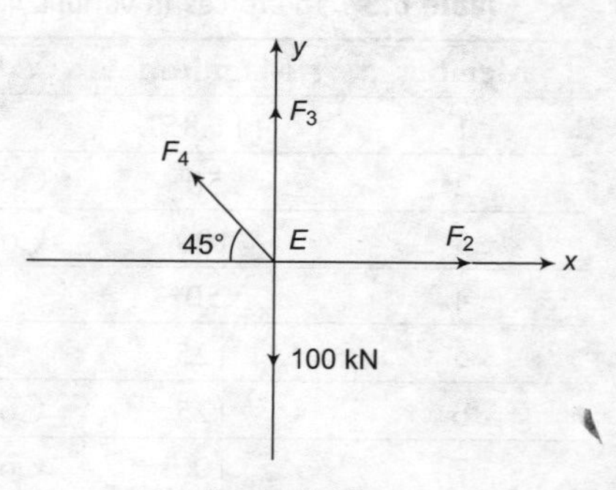

FIGURE 6.44

$$\sum F_x = 0$$

$\Rightarrow$ $F_2 = F_4 \cos 45° = 100$

$\Rightarrow$ $F_4 = -141.42$ kN

$$\sum F_y = 0$$

$\Rightarrow$ $F_4 \sin 45° + F_3 = 100$ kN

$\Rightarrow$ $F_3 = 100 + 141.42 \sin 45° = 200$ kN

At joint D:

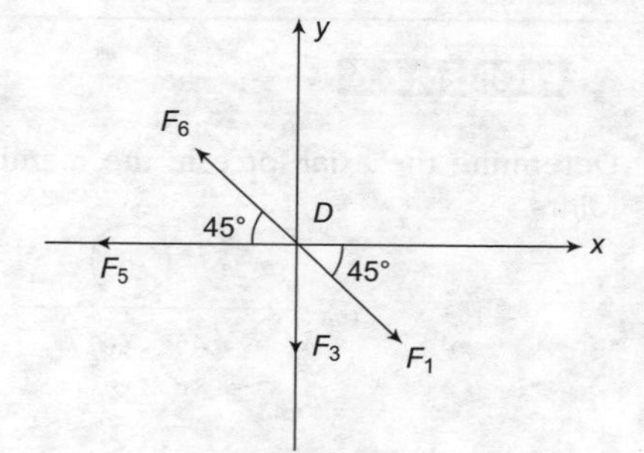

FIGURE 6.45

$$\sum F_y = 0$$

$\Rightarrow$ $F_3 + F_1 \sin 45° = F_6 \sin 45°$

$\Rightarrow$ $F_6 = F_1 + \dfrac{F_3}{\sin 45°} = 424.26$ kN

$$\sum F_x = 0$$

$\Rightarrow$ $F_5 + F_6 \cos 45° = F_1 \cos 45° = -200$ kN

At joint B:

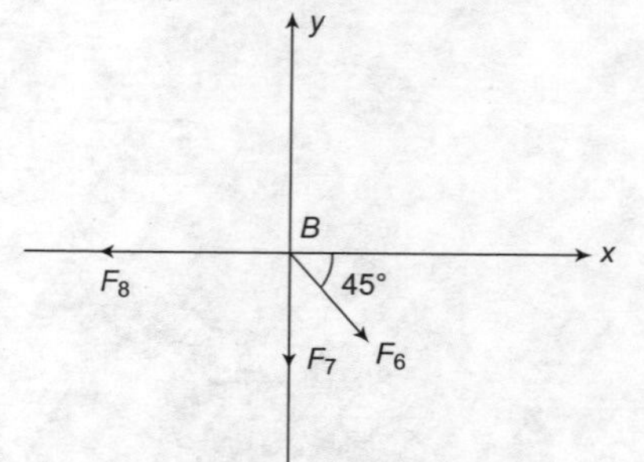

FIGURE 6.46

$$\sum F_x = 0$$

$\Rightarrow$ $F_8 = F_6 \cos 45°$

$= 424.26 \times \cos 45°$

$= 300$ kN

$$\sum F_y = 0$$

$$F_7 + F_6 \sin 45° = 0$$

$$\Rightarrow \quad F_7 = -F_6 \sin 45°$$

$$\Rightarrow \quad = -300 \text{ kN}$$

Resolve the forces perpendicular to the R_C direction at joint C

$$\Rightarrow F_9 + F_7 \cos 45° = F_4 + F_5 \cos 45°$$

$$F_9 = -141.42 + 300 \cos 45° + (-200) \cos 45°$$

$$= -70.709 \text{ kN}$$

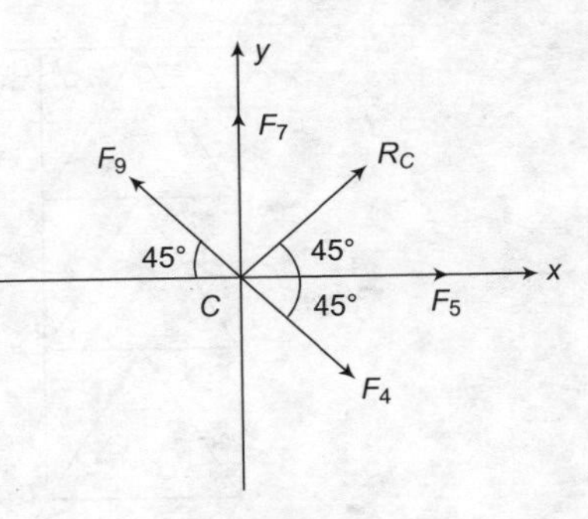

FIGURE 6.47

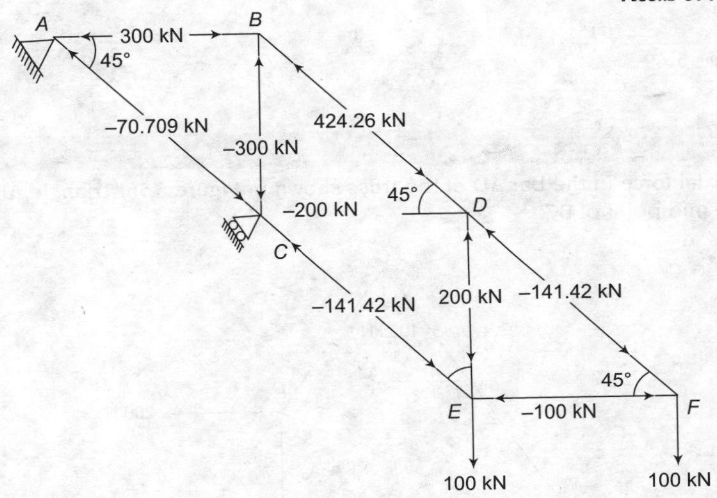

FIGURE 6.48

6.5 METHOD OF SECTIONS

Method of sections is a procedure for determining the forces in the truss members by isolating a portion of the structure by a section and implementing the equilibrium condition on the isolated portion. In method of sections, the section should be chosen in such a way that the section should pass through the members, in which the axial forces are to be found. Generally, a section which is cut not more than three members, so that three equilibrium equations can be used to find the forces. Consider a truss shown in Figure 6.49. A section (1)-(1) is used to separate the structure. Equilibrium of any part of the truss 6.49 (a) or 6.49 (b) can be used. In this method, all unknown forces are assumed to be tensile forces. A negative sign indicates that the force is compressive.

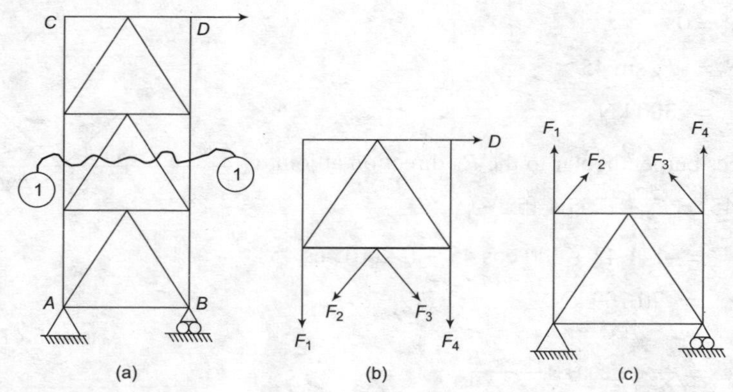

FIGURE 6.49

PROBLEM 6.7

Objective 2

Determine the axial force in the bar AD of the truss shown in Figure 6.50. Triangle ABC is equilateral triangle. D is the mid point of BC.

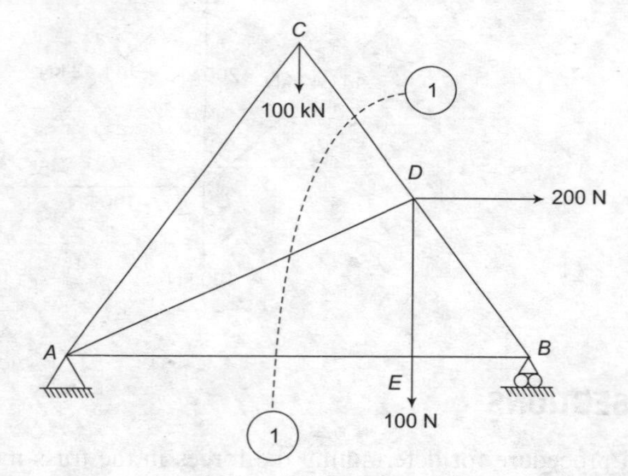

FIGURE 6.50

SOLUTION

Let $AB = L$, then $DB = L/2$

$$EB = DB \cos 60^\circ = L/4$$

Take section (1)-(1), consider the right portion, shown in Figure 6.51.

Taking moments about B (this point is preferred because the unknown forces F_1, F_2, and V_B are passing through this point).

$$F_{AD} \cos 30^\circ \times \frac{l}{2} \sin 60^\circ - 200 \times \frac{l}{2} \sin 60^\circ + F_{AD} \sin 30^\circ \times \frac{l}{4} + 100 \times \frac{l}{4} = 0$$

$$\Rightarrow \qquad F_{\text{AD}}\left(\frac{l}{2}\right) = 86.603l - 25l$$

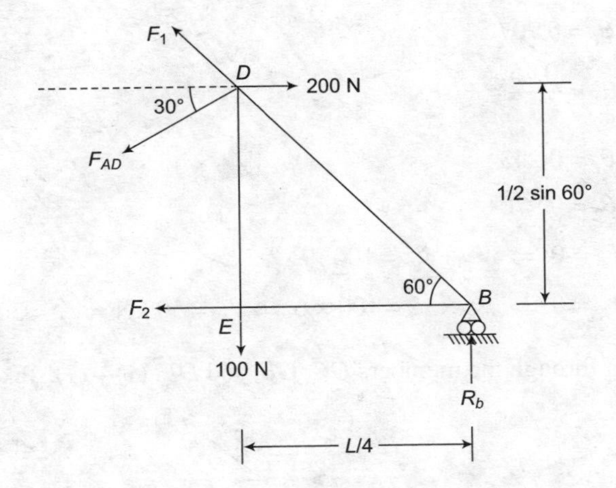

FIGURE 6.51

$$\Rightarrow \qquad F_{AD} = 123.21 \text{ N}$$

It is important to note that, force in the member '*AD*' is influenced by the load 100 N at *C*.

PROBLEM 6.8 **Objective 2**

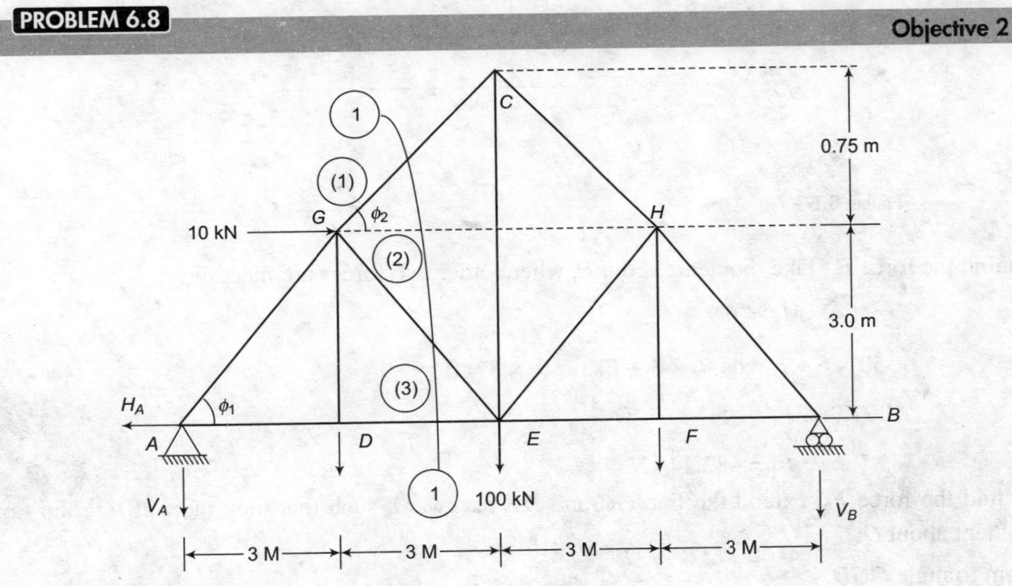

FIGURE 6.52

Using method of sections, determine the force in the members (1), (2), and (3) of truss shown in Figure 6.52.

SOLUTION

Let θ_1 and θ_2 be the inclinations of the members *AG* and *GC*, respectively.

$$\tan \theta_1 = 1$$

$$\sin \theta_1 = 0.707$$

$$\cos\,\theta_1 = 0.707$$

$$\tan\theta_2 = \frac{0.75}{3} = 1/4$$

$$\sin\,\theta_2 = 0.243$$

$$\cos\,\theta_2 = 0.970$$

$$\sum F_y = 0 \Rightarrow V_A + V_B = 100$$

$$\sum M_A = 0 \Rightarrow V_B \times 12 = 100 \times 6 \Rightarrow V_B = 50 \text{ kN} = V_A = 50 \text{ kN}.$$

Take any section passing through the members *GC*, *GE*, and *DE*. (1) – (1). Take the left side portion of section.

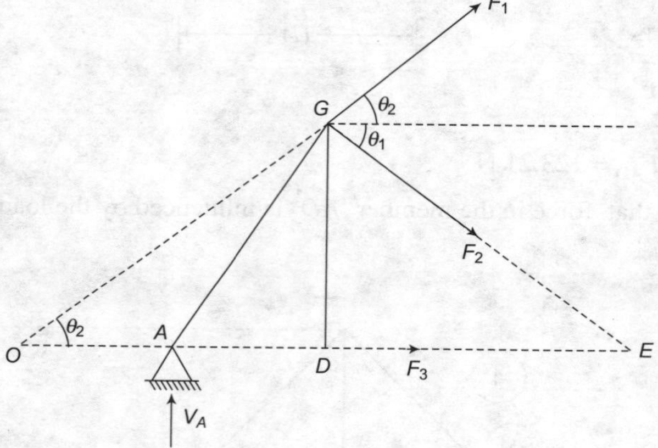

FIGURE 6.53

To find the force F_1, take moments about *E*, where other two forces are meeting.

$$\sum M_E = 0$$

$$\Rightarrow \quad 50 \times 6 + F_1 \cos\,\theta_2 \times 3 + F_1 \sin\,\theta_2 \times 3 = 0$$

$$\Rightarrow \quad 1.213 \times 3F_1 = -50 \times 6$$

$$\Rightarrow \quad F_1 = -82.44 \text{ kN}$$

To find the force F_2, extend the line *GC* and *DE* backward, such that they meet at '*O*' and take moment about *O*.

From triangle *OGD*,

$$\text{OD} = \frac{\text{GD}}{\tan\theta_2} = \frac{3}{(1/4)} = 12 \text{ m}$$

$$\text{OA} = 12 - 3 = 9$$

Taking moments about *O*,

$$\Rightarrow \quad -50 \times 9 - F_2 \sin\,\theta_1 \times OD + F_2 \cos\,\theta_1 \times GD = 0$$

$$\Rightarrow \quad -50 \times 9 + F_2 \times (0.707 \times 12) + F_2 \times (0.707 \times 3) = 0$$

$\Rightarrow \quad F_2 = 42.43$ kN

To find the force F_3, take moments about G

$\Rightarrow \quad V_A \times 3 - F_3 \times 3 = 0$

$\Rightarrow \quad F_3 = V_A$

$= 50$ kN

$\therefore \quad F_1 = 82.44$ kN (Compressive)

$F_2 = 42.43$ kN (Tensile)

$F_3 = 50$ kN

PROBLEM 6.9 **Objective 1**

Determine the forces in the members (1), (2), and (3) of the truss shown in Figure 6.54.

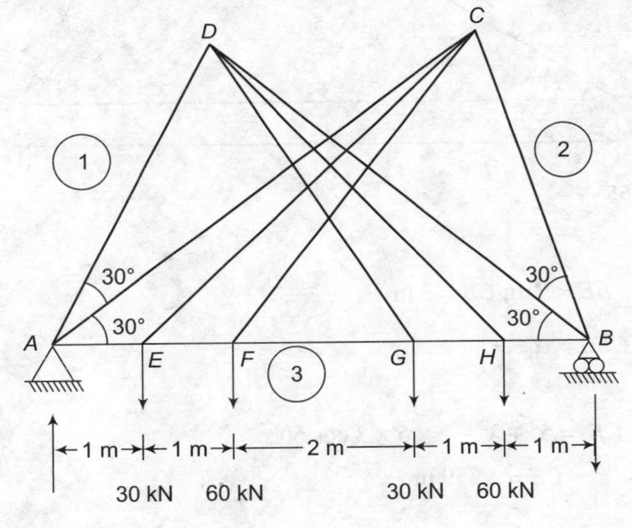

FIGURE 6.54

SOLUTION

Determine the reactions at A and B, respectively.

Let V_A and V_B be the vertical reactions at A and B, respectively.

Sum of vertical forces is equal to zero

$\Rightarrow \quad V_A + V_B = 180$

$$\sum M_A = 0$$

$\Rightarrow \quad 30 \times 1 + 60 \times 2 + 30 \times 4 + 60 \times 5 = V_B \times 6$

$\Rightarrow \quad V_B = 95$ kN

$V_A = 85$ kN

Consider the joint sections at E and F, determine the force in the members CE and CF. Then, go to the joint C and resolve the forces normal to CA, so that the force in the member (2) can be found.

Let α_1 and α_2 be the inclination of the members CE and CF with horizontal, respectively.

From ΔDCB, $\angle C = 90°$

$$\therefore \quad BC = AB \cos 60° = 6 \times \cos 60°$$

$$= 3 \text{ m}$$

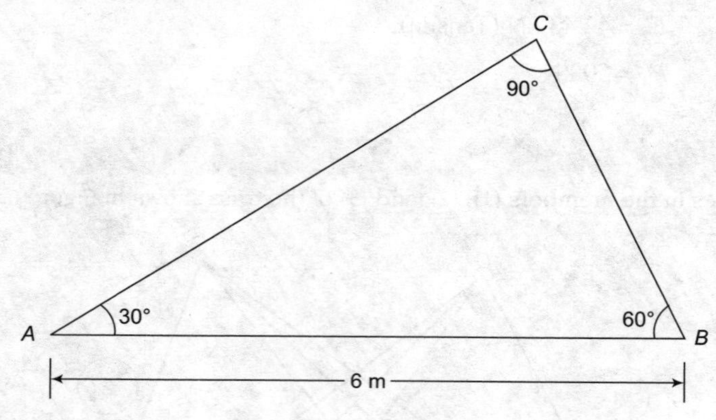

FIGURE 6.55

From ΔBCE,

$$BE = 5 \text{ m}; \; CB = 3 \text{ m}$$

$$\angle CEB = \alpha_1$$

Using cosine rule

$$CE^2 = 5^2 + 3^2 - 2 \times 5 \times 3 \cos 60°$$

$$\Rightarrow \quad CE = 4.359 \text{ m}$$

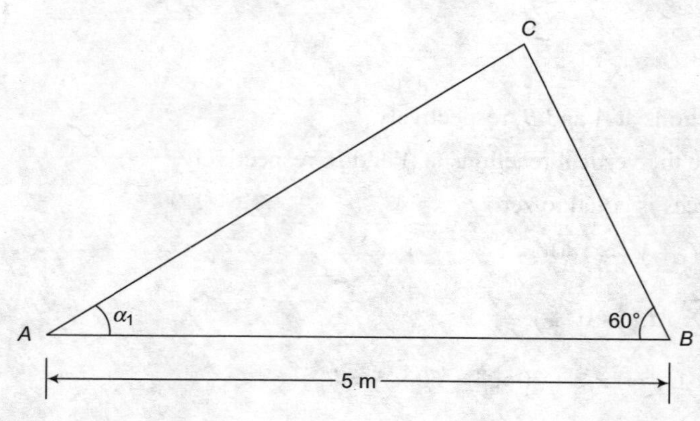

FIGURE 6.56

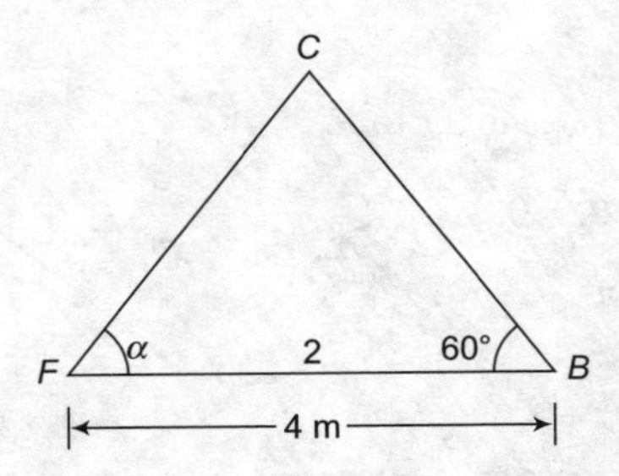

FIGURE 6.57

Applying sine rule

$$\frac{CE}{\sin 60°} = \frac{BC}{\sin \alpha_1}$$

$$\Rightarrow \quad \sin \alpha_1 = \frac{3}{4.359} \sin 60°$$

$$\Rightarrow \quad \sin \alpha_1 = 0.596$$

$$\alpha_1 = 36.587°$$

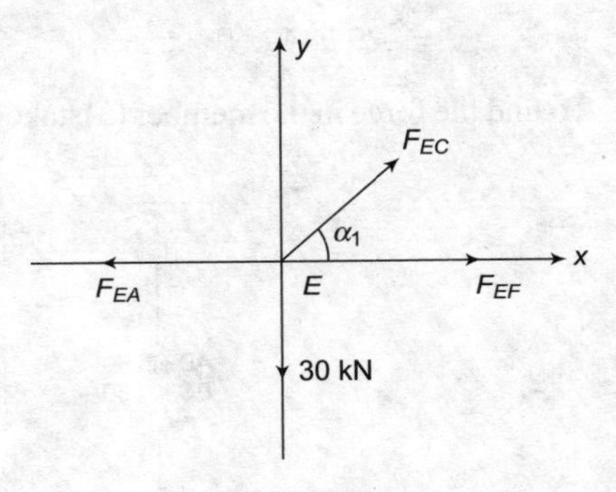

FIGURE 6.58

In ΔCFB

$$CF^2 = 4^2 + 3^2 - 2 \times 4 \times 3 \cos 60°$$

$$CF = 3.606 \text{ m}$$

Applying sine rule,

$$\frac{CF}{\sin 60°} = \frac{3}{\sin \alpha_2}$$

$$\sin \theta_2 = 0.721; \; \theta_2 = 46.102°$$

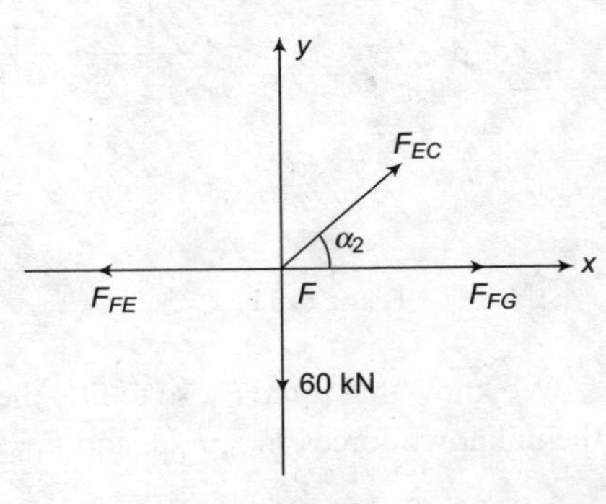

FIGURE 6.59

At joint E:

$$\sum F_y = 0$$

($\therefore$ we do not know the force in the members *EA* and *EF*)

$$\Rightarrow \quad F_{EC} = \frac{30}{\sin \alpha_1} = 50.332 \text{ kN}$$

At joint F:

$$\sum F_y = 0$$

$$\Rightarrow \quad F_{FC} \sin \alpha_2 = 60$$

$$\Rightarrow \quad F_{FC} = \frac{60}{\sin \alpha_2} = 83.267 \text{ kN}$$

At joint C:

Resolving the forces along CB, F_{CA} is to $\perp r$ to F_2

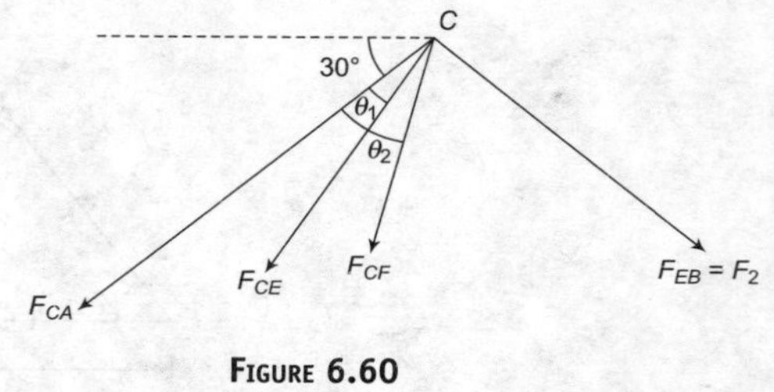

FIGURE 6.60

$$\therefore \quad F_2 + F_{CE}\sin\theta_1 + F_{CF}\sin\theta_2 = 0$$

$$\theta_1 = 36.5870 - 30 = 6.587°$$

$$\theta_2 = 46.102 - 30 = 16.587°$$

$$\Rightarrow \quad F_2 = -50.332\sin(6.587) - 83.267\sin(16.587°)$$

$$= -29.544 \text{ kN}$$

To find the force in the member (3), take a section passing through the members DG, DH, DB, and (2).

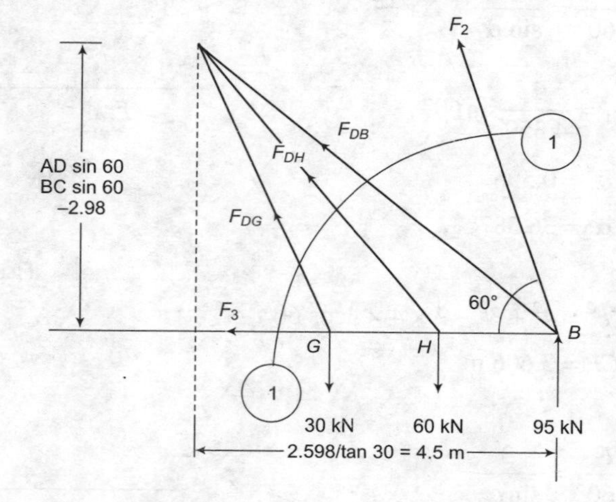

FIGURE 6.61

We know that force F_2, so to find the force in the member (1), and take moments about D, so that the unknown forces F_{DG}, F_{DH}, and F_{DB} can be avoided in the calculations.

$$\Rightarrow \quad F_2\sin 60 \times 4.5 - F_2\cos 60 \times 2.598$$

$$- F_3 \times 2.598 + 95 \times 4.5 - 60 \times 3.5 - 30 \times 2.5 = 0$$

$$\Rightarrow \quad 65.741 - 2.598\, F_3 = 0$$

$$\Rightarrow \quad F_3 = 25.305 \text{ kN}$$

Similarly, to find force in the member (1), take a section passing through the members (1), (3), CA, CE, and CF. Take moments about C

Sum of the moments about $C = 0$

$$\Rightarrow \quad F_1\sin 60 \times 4.5 + 85 \times 4.5 - F_1\cos 60 \times 2.598$$

$$- 30 \times 3.5 - 60 \times 2.5 - F_3 \times 2.598 = 0$$

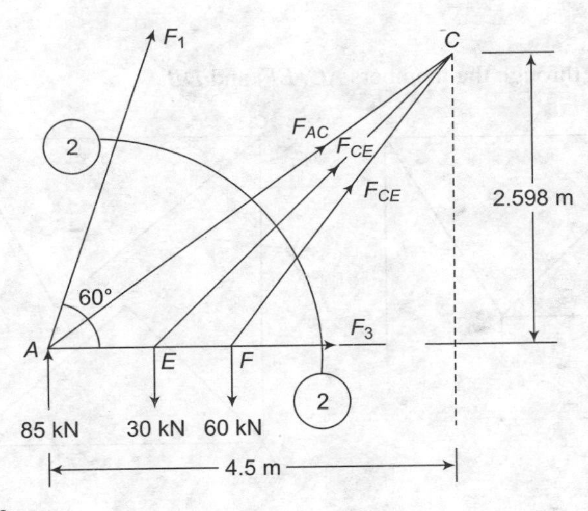

FIGURE 6.62

$$\Rightarrow \quad 2.598\, F_1 + 127.5 - 25.305 \times 2.598 = 0$$

$$\Rightarrow \quad F_1 = -23.771 \text{ kN}$$

$$\therefore \quad F_1 = 23.771 \text{ kN (Compressive)}$$

$$F_2 = 29.544 \text{ kN (Compressive)}$$

$$F_3 = 25.305 \text{ kN (Tensile)}.$$

PROBLEM 6.10 Objective 1

Determine the forces in the members (1), (2), and (3) of the truss shown in Figure 6.63.

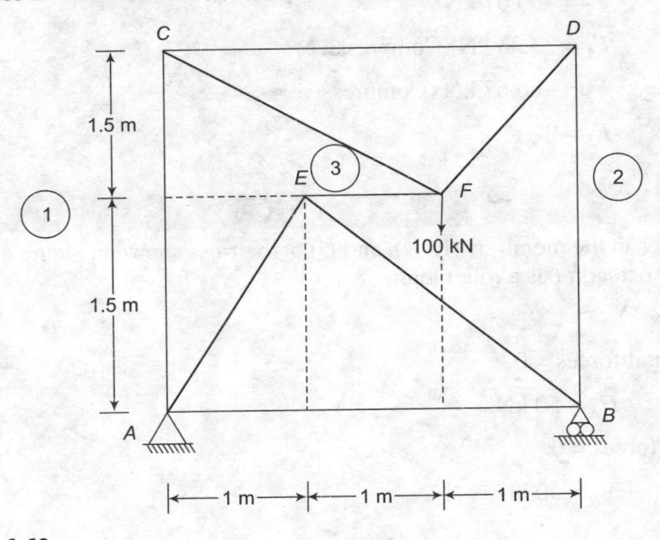

FIGURE 6.63

SOLUTION

Take a section passing through the members AC, EF, and DB.

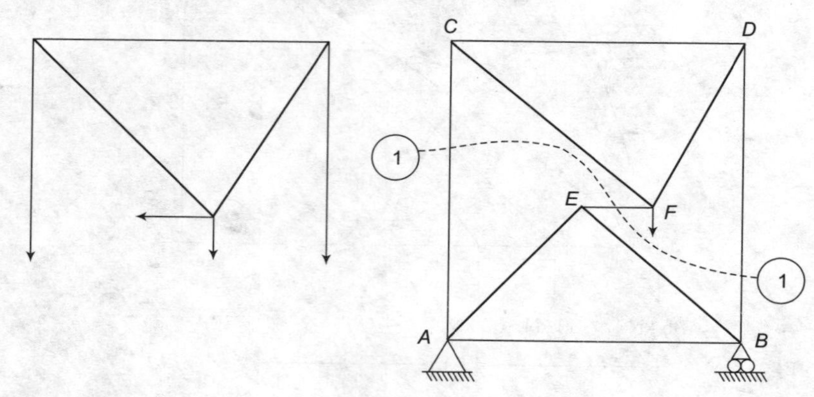

FIGURE 6.64

Sum of all horizontal forces is equal to zero.

$$\Rightarrow \qquad F_3 = 0$$

Sum of the vertical forces = 0

$$\Rightarrow \qquad F_1 + F_2 = -100 \text{ kN} \qquad (6.24)$$

Taking moments about D

$$F_1 \times 3 + 100 \times 1 = 1 \qquad (6.25)$$

$$\Rightarrow \qquad F_1 = -33.33 \text{ kN}$$

$$\therefore \qquad F_2 = -66.67 \text{ kN}$$

$$F_1 = 33.33 \text{ kN (Compressive)}$$

$$F_2 = -66.67 \text{ kN (Compressive)}$$

$$F_3 = 0$$

PROBLEM 6.11

Objective 1

Determine the forces in the members (1), (2), and (3) of the truss shown in Figure 6.65. *ABCDEF* is half portion of a regular octagon *B* is a roller joint.

SOLUTION

Sum of the horizontal forces = 0

$$\Rightarrow \qquad H_A = 50 \text{ kN}$$

Use of the vertical forces = 0

$$V_A + V_B = 50 \text{ kN}$$

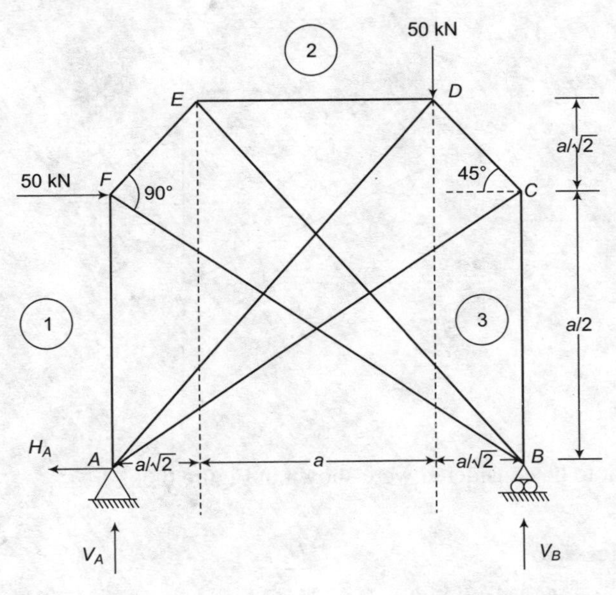

FIGURE 6.65

Taking moments about A,

$$50\,a + 50\left(\frac{a}{\sqrt{2}} + a\right) - V_B(a + \sqrt{2a}) = 0$$

$\Rightarrow$ $$V_B = 45.71 \text{ kN:} \quad V_A = 4.289 \text{ kN}$$

Different sections that to be considered were shown in Figure 6.66.

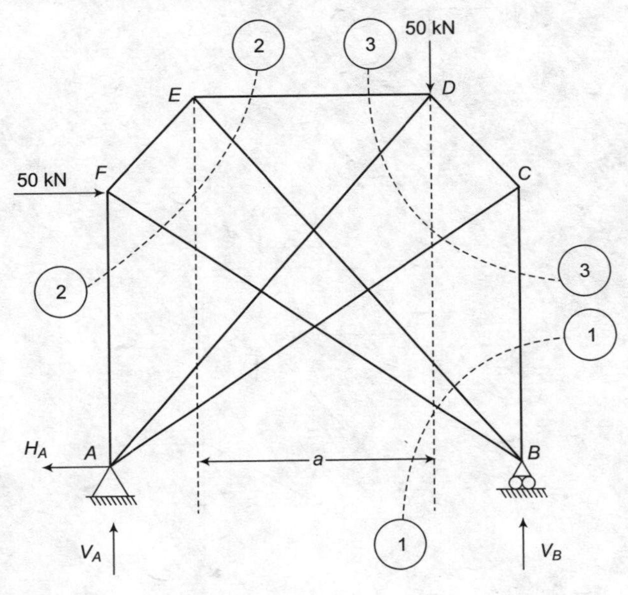

FIGURE 6.66

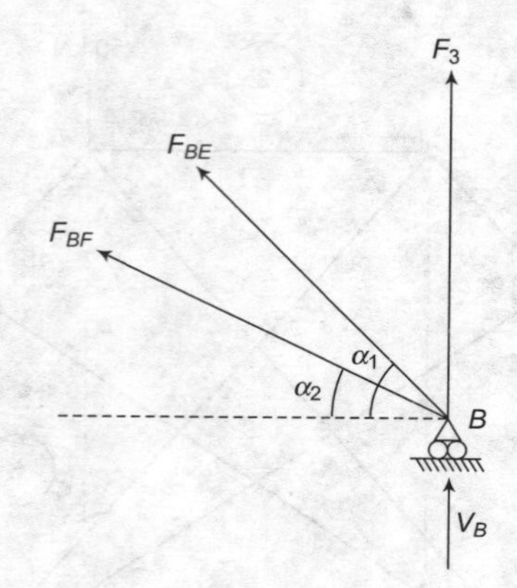

FIGURE 6.67

Different sections that to be considered were shown in Figure 6.66.

Section (1)-(1)

Sum of horizontal forces = 0

$$\Rightarrow \qquad F_{\mathrm{BE}} \cos \alpha_1 + F_{\mathrm{BF}} \cos \alpha_2 = 0 \tag{6.26}$$

Section (2)-(2)

Sum of the horizontal forces = 0

$$\Rightarrow \qquad 50 + F_{\mathrm{BE}} \cos \alpha_1 + F_{\mathrm{BF}} \cos \alpha_2 + F_2 = 0 \tag{6.27}$$

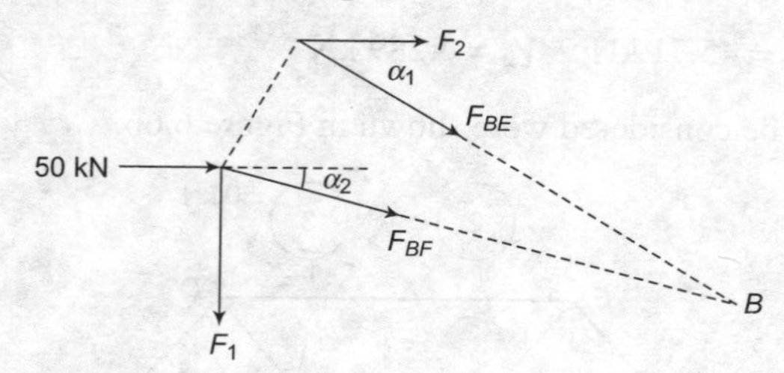

FIGURE 6.68

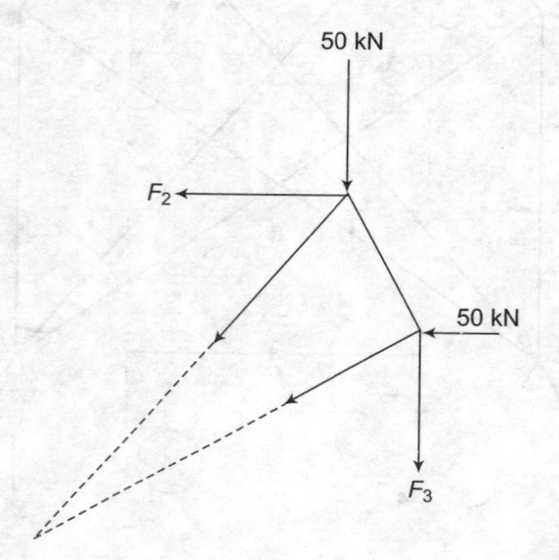

FIGURE 6.69

Substituting Eqs. (6.26) and (6.27)

$$\Rightarrow \quad 50 + F_2 = 0$$

$$\Rightarrow \quad F_2 = -50 \text{ kN}$$

Taking moments about B:

$$\Rightarrow \quad F_2\left(\frac{a}{2}+\frac{a}{V_2}\right)+50\left(\frac{a}{2}\right)-F_1(a+\sqrt{2})=0$$

$$\Rightarrow \quad -60.355 + 25 - 2.4142F_1 = 0$$

$$\Rightarrow \quad F_1 = -14.645 \text{ kN}$$

Section (3)-(3)

Taking moments about A,

$$F_3(a+a\sqrt{2})+50\left(a+\frac{a}{\sqrt{2}}\right)-F_2\left(\frac{a}{2}+\frac{a}{\sqrt{2}}\right)=0 \tag{6.28}$$

$$\Rightarrow \quad 145.711 + 2.414F_3 = 0$$

$$\Rightarrow \quad F_3 = -60.355 \text{ kN}$$

$$\therefore \quad F_1 = 14.645 \text{ kN (Compressive)}$$

$$F_2 = 50 \text{ kN (Compressive)}$$

$$F_3 = -60.355 \text{ kN (Compressive)}$$

6.6 TENSION COEFFICIENT METHOD

This method is like method of joints, developed mainly to determine the forces in all members of trusses using computer programming. Tension coefficient of a member is the tensile force, and it carries per unit length of the member. The units are kN/m or N/mm.

$$\text{Tension coefficient } T = \frac{\text{Tensile force } (F) \text{ in the member}}{\text{Length of the member } (L)}$$

$$T_{AB} = \frac{F_{AB}}{L_{AB}}$$

Negative value of tension coefficient represents compressive force in the member. Let us consider a member 'AB' carrying for 'F_{AB}' in the x–y plane shown in Figure 6.70.

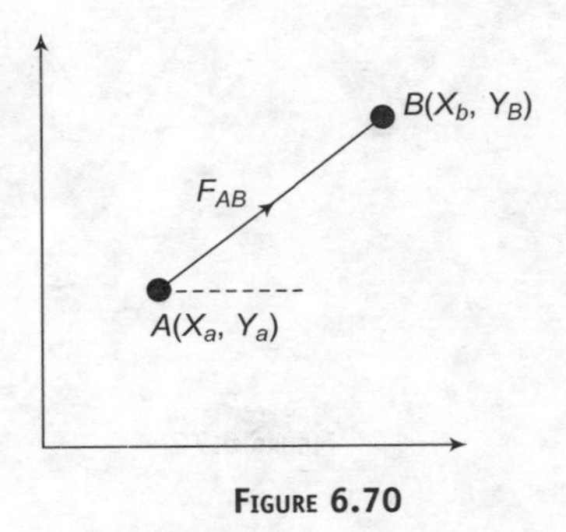

FIGURE 6.70

Component of force F_{AB} along x axis is given by

$$F_{xAB} = F_{AB} \cos\theta$$

$$F_{xAB} = \frac{F_{AB}}{L_{AB}}(x_b - x_a)$$

$$= T_{AB}\,(x_b - x_a)$$

The component of force F_{AB} in y direction is given by

$$F_{yAB} = F_{AB} \sin \theta$$

$$F_{xAB} = \frac{F_{AB}}{L_{AB}}(y_b - y_a)$$

$$= T_{AB}\,(y_b - y_a)$$

Expressing the components of a force along x–y direction in terms of the coordinates and tension coefficient will be of more use in the truss analysis. This tension coefficient method is more used in space truss analysis.

PROBLEM 6.12

Objective 2

Determine the force in the members of the truss shown in Figure 6.71, using tension coefficient method.

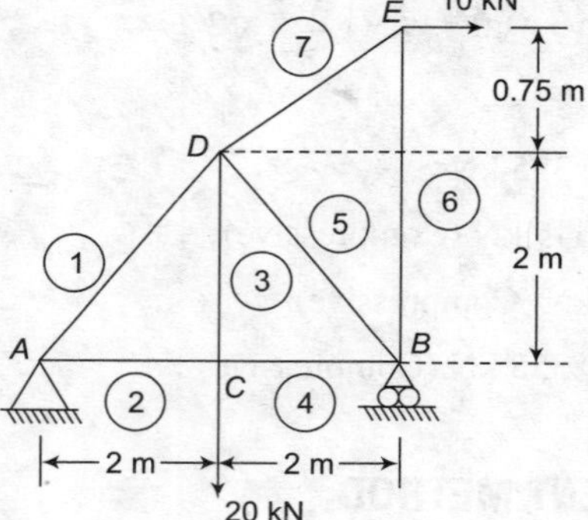

FIGURE 6.71

SOLUTION

Taking 'A' as origin, determine the coordinate of the various points.

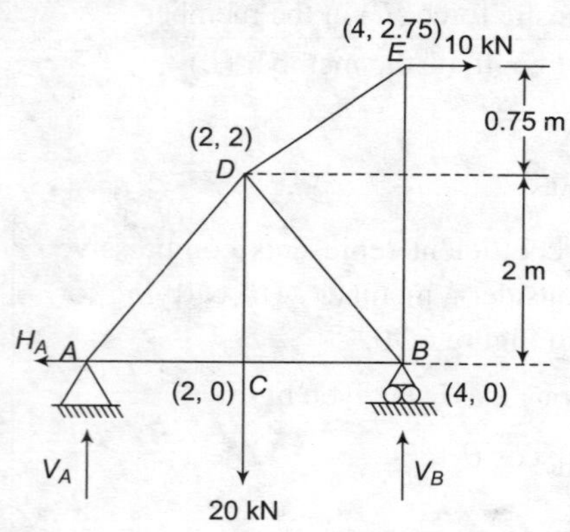

FIGURE 6.72

Let V_A, H_A, and V_B be the reactions at A and B

$$\sum F_x = 0 \tag{6.29}$$

$$\Rightarrow \quad H_A = 10 \text{ kN}$$

$$\sum F_y = 0$$

$$\Rightarrow \quad V_A + V_B = 20 \text{ kN} \tag{6.30}$$

Taking moments about A,

$$20\,(2) + 10\,(2.75) - V_B\,(4) = 0 \tag{6.31}$$

$$\Rightarrow \quad V_B = 16.875 \text{ kN}$$

$$V_A = 3.125 \text{ kN}$$

Consider the joint equilibrium of joint E:

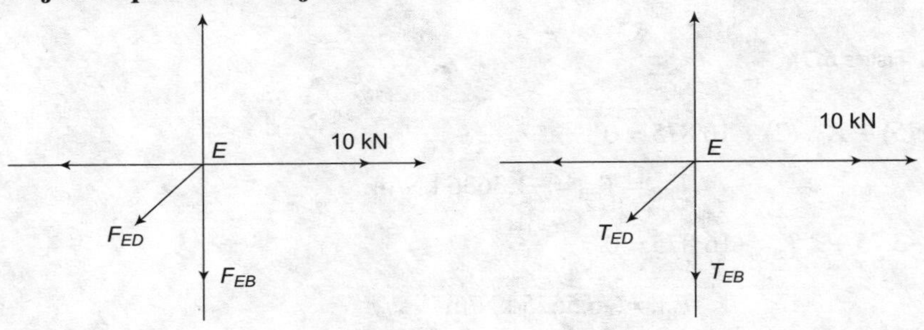

FIGURE 6.73

$$\sum F_y = 0$$

$$\Rightarrow T_{\text{ED}}(x_d - x_e) + T_{\text{EB}}(x_b - x_e) + 10 = 0$$

$$\Rightarrow \quad T_{\text{ED}}\,(2 - 4) + 10 = 0 \tag{6.32}$$

$$\Rightarrow \quad T_{\text{ED}} = \frac{10}{2} = 5 \text{ kN/m}$$

$$\sum F_y = 0$$

$$T_{\text{ED}}\,(y_d - y_e) + T_{\text{EB}}\,(y_b - y_e) = 0 \tag{6.33}$$

$$\Rightarrow \quad T_{\text{ED}}\,(2 - 2.75) + T_{\text{EB}}\,(0 - 2.75) = 0$$

$$\Rightarrow \quad -0.75T_{\text{ED}} - 2.75T_{\text{EB}} = 0$$

$$\Rightarrow \quad T_{\text{EB}} = -1.3636 \text{ kN/m}$$

Consider the joint B:

$$\sum F_y = 0$$

$$\Rightarrow \quad T_{BE}\,(y_e - y_b) + T_{BD}\,(y_d - y_b) + 16.875 = 0$$

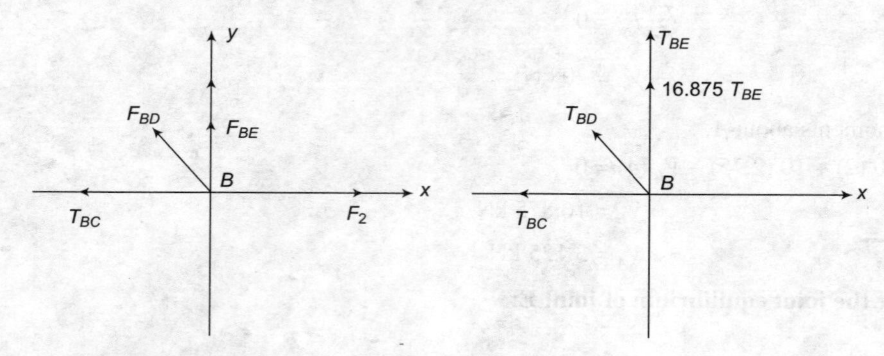

FIGURE 6.74

$$\Rightarrow \quad T_{BE}\,(2.75) + T_{BD}\,(2) + 16.875 = 0 \tag{6.34}$$

$$T_{BE} = T_{EB} = -1.3636 \text{ kN/m}$$

$$-3.75 + 2\,T_{BD} + 16.875 = 0$$

$$\Rightarrow \quad T_{BD} = -6.5625 \text{ kN/m}$$

$$\sum F_x = 0$$

$$\Rightarrow \quad T_{BC}\,(x_c - x_b) + T_{BD}\,(x_d - x_b) = 0$$

$$\Rightarrow \quad -2T_{BC} - 6.5625(-2) = 0$$

$$\Rightarrow \quad T_{BC} = 6.5625 \text{ kN/m}$$

At joint D:

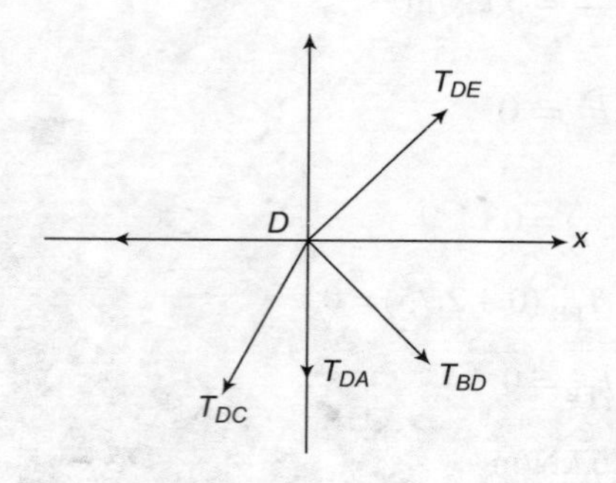

FIGURE 6.75

$$\sum F_x = 0$$

$\Rightarrow$ $T_{DE}(x_e - x_d) + T_{BD}(x_b - x_d) + T_{DA}(x_a - x_d) = 0$

$\Rightarrow$ $T_{DE} = T_{ED} = 5$ kN/m

$\Rightarrow$ $T_{BD} = T_{DB} = -6.5625$ kN/m

$\Rightarrow$ $5(2) - 6.5625(2) + T_{DA}(-2) = 0$

$\Rightarrow$ $T_{DA} = -1.5625$ kN/m

$$\sum F_y = 0$$

$\Rightarrow$ $T_{DE}(y_e - y_d) + T_{BD}(y_b - y_d) + T_{DA}(y_a - y_d) + T_{DC}(y_c - y_d) = 0$

$\Rightarrow$ $5(0.75) - 6.5625(-2) + (-1.5625)(-2) + T_{DC}(-2)$

$\Rightarrow$ $T_{DC} = 10$ kN/m

At joint C:

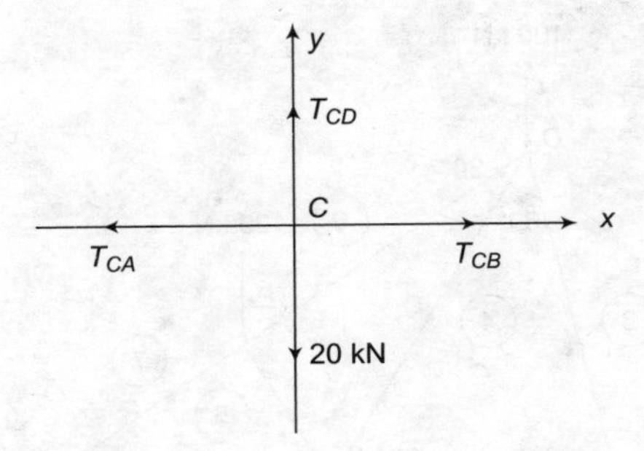

FIGURE 6.76

$$\sum F_y = 0$$

$\Rightarrow$ $T_{CD}(y_d - y_c) - 20 = 0$

$\Rightarrow$ $T_{CD}(2) = 20$

$\Rightarrow$ $T_{CD} = 10$

$$\sum F_x = 0$$

$\Rightarrow$ $T_{CA}(x_a + x_c) + T_{CB}(x_b - x_c) = 0$

$\Rightarrow$ $T_{CB} = T_{BC} = 6.5625$ kN/m

$\Rightarrow$ $T_{CA}(-2) + 6.5625(2) = 0$

$\Rightarrow$ $T_{CA} = 6.5625$ kN/m

Table 6.6 The forces in all members of the trusses are shown below

S.No.	Member	Length (m)	Tension Coefficient	Force (kN)	Nature
1	AD	$2\sqrt{2}$	−1.5625	4.419	Compressive
2	AC	2	6.5625	13.125	Tensile
3	DC	2	10.0	20.0	Tensile
4	CB	2	6.5625	13.125	Tensile
5	BD	$2\sqrt{2}$	−6.5625	18.562	Compressive
6	BE	2.75	−1.3636	3.75	Compressive
7	ED	2.136	5.0	10.68	Tensile

PROBLEM 6.13

Objective 2

Determine the forces in the members of the truss, using tension coefficient method, shown in Figure 6.77.

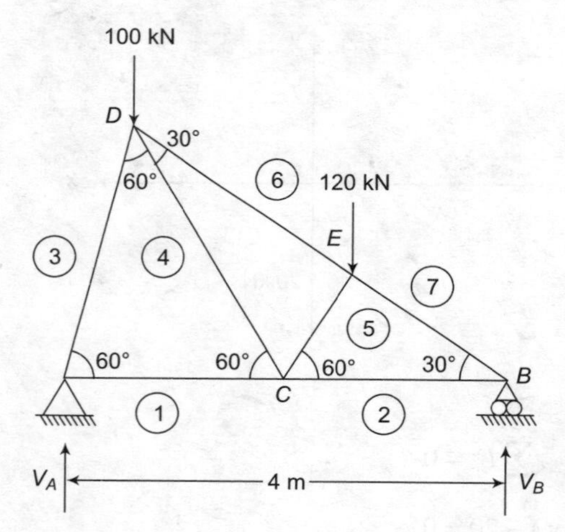

FIGURE 6.77

SOLUTION

Let V_A and V_B be the reactions at A and B, respectively.

$$\sum F_y = 0$$

$$V_A + V_B = 220 \text{ kN}$$

Taking moments about A,

$$4\ V_B = 100 \times 4 \cos 60° \times \cos 60° + 120\ (4 - 1.732 \cos 30°) \tag{6.35}$$

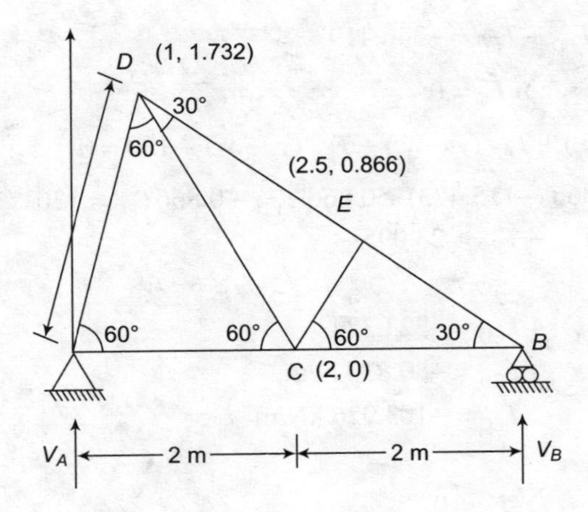

FIGURE 6.78

$$4\, V_B = 400$$
$$V_B = 100 \text{ kN}$$
$$V_A = 120 \text{ kN}$$

Coordinates of different points.

$$A = (0, 0)$$
$$B = (2, 0)$$
$$C = (4, 0)$$
$$D = (1, 1.732)$$
$$E = (2.5, 0.866)$$

At joint B:

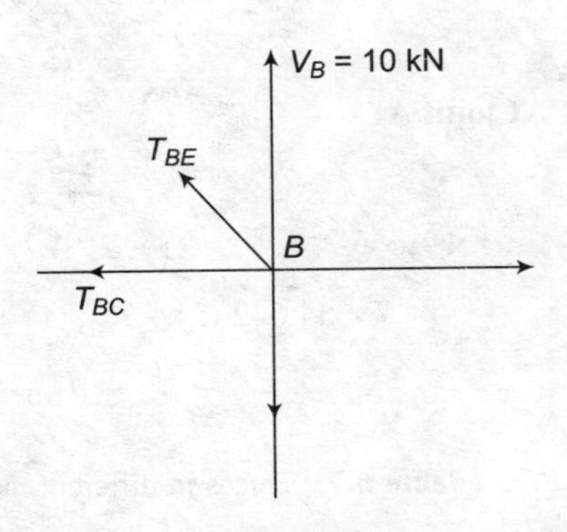

FIGURE 6.79

$$\sum F_y = 0$$
$$T_{\text{BE}}\,(y_e - y_b) + 100 = 0$$
$$T_{\text{BE}}\,(0.866) + 100 = 0$$
$$T_{\text{BE}} = -115.473 \text{ kN/m}$$
$$\sum F_x = 0$$
$$T_{\text{BE}}\,(x_e - x_b) + T_{\text{BC}}\,(x_c - x_b) = 0$$
$$-1.5\, T_{\text{BE}} - 2\, T_{\text{BC}} = 0$$
$$T_{\text{BC}} = 86.605 \text{ kN/m}$$

At joint E:

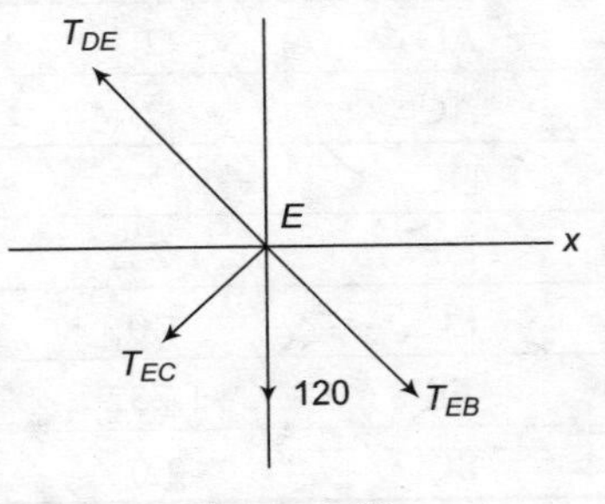

FIGURE 6.80

$$\sum F_x = 0$$
$$T_{\text{EB}}\,(x_b - x_e) + T_{\text{ED}}\,(x_d - x_e) + T_{\text{EC}}\,(x_c - x_e) = 0$$
$$1.5\, T_{\text{EB}} - 1.6\, T_{\text{ED}} + T_{\text{EC}}\,(-0.5) = 0$$
$$1.5\,(-115.473) - 1.5\, T_{\text{ED}} - 0.5\, T_{\text{EC}} = 0$$

$$3\,T_{ED} + T_{EC} = -346.419 \tag{6.36}$$

$$\sum F_y = 0$$

$$T_{ED}\,(y_d - y_e) + T_{EC}\,(y_c - y_e) + T_{EB}\,(y_b - y_e) - 120 = 0$$

$$-0.866\,(-115.473) + 0.866\,T_{ED} - 0.866T_{EC} = 120$$

$$T_{ED} - T_{EC} = 23.095 \tag{6.37}$$

Eqs. (2.36) + (2.37)

$$4\,T_{ED} = -323.324$$

$$T_{ED} = -80.83 \text{ kN/m}$$

$$T_{EC} = -103.926 \text{ kN/m}$$

At joint C:

$$\sum F_y = 0$$

$$T_{CE}\,(y_e - y_c) + T_{CD}\,(y_d - y_c) = 0$$

$$0.866\,T_{CE} + 1.732\,T_{CD} = 0$$

$$T_{CD} = -T_{CE}/2 = 51.963 \text{ kN/m}$$

FIGURE 6.81

$$\sum F_x = 0$$

$$T_{CB}\,(x_b - x_c) + T_{CE}\,(x_e - x_c) + T_{CD}\,(x_d - x_c) + T_{CA}\,(x_a - x_c) = 0$$

$$173.21 - 103.926\,(0.5) + 51.963\,(-1) + T_{CA}\,(-2) = 0$$

$$T_{CA} = 34.642 \text{ kN/m}$$

At joint A:

$$\sum F_x = 0$$

$$T_{AC}\,(x_c - x_a) + T_{AD}\,(x_d - x_a) = 0$$

$$34.642\,(2) + T_{AD}\,(1) = 0$$

$$T_{AD} = -69.284 \text{ kN/m}$$

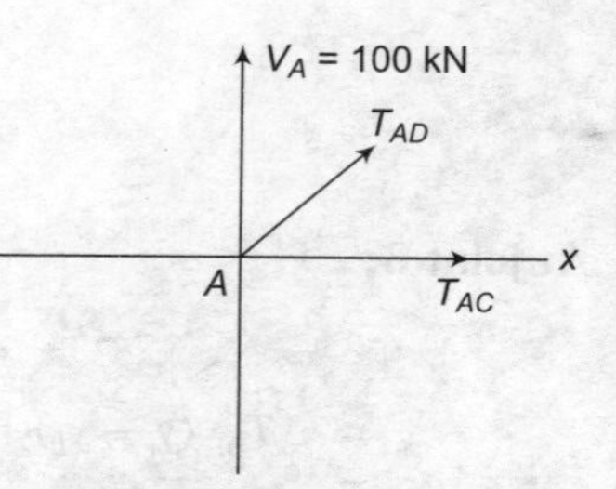

FIGURE 6.82

Table 6.7 Forces in different members of the truss

Member	Length	Tension Coefficient	Force	Nature
AD	2	−69.284	138.568	Compressive
AC	2	34.642	69.284	Tensile
DC	2	51.963	103.926	Tensile
DE	1.732	−80.831	140.0	Compressive
EC	1.0	−103.926	103.926	Compressive
EB	1.732	−115.473	200	Compressive
BC	2.0	86.605	173.21	Tensile

SUMMARY

- **The relation between number of joints (j) and number of members (n) in a perfect frame is given by $n = 2j - 3$.**
- **The reaction on a roller support is at right angles to the roller base.**
- **The forces in the members of a frame are determined by:**
  (i) Method of joints (ii) Method of sections (iii) Graphical method
- **The force in a member will be compressive if the member pushes the joint to which it is connected, whereas the force in the member will be tensile if the member pulls the joint to which it is connected.**
- **If three forces act at a joint and two of them are along the same straight line, then the third force would be zero.**
- **Method of section is mostly used, when the forces in a few members of a truss are to be determined.**
- **If a truss carries horizontal loads, then the support reaction at the hinged end will consist of (i) horizontal reaction and (ii) vertical reaction.**

OBJECTIVE TYPE QUESTIONS

1. The pin-jointed 2D truss is loaded with a horizontal force of 15 kN at joint S and another 15 kN vertical force at joint U, as shown. Find the force in member RS (in kN) and report your answer taking tension as positive and compression as negative. [GATE 2013]

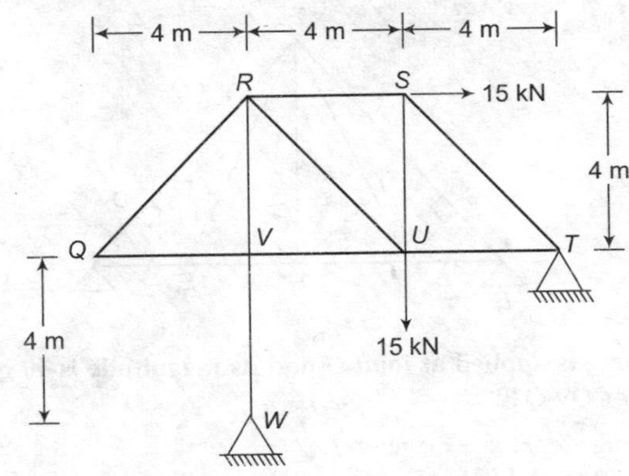

(a) 1 to 2.5 (b) 3.5

(c) 5.5 (d) zero

2. For the truss shown in the below figure, the force in the member QR is

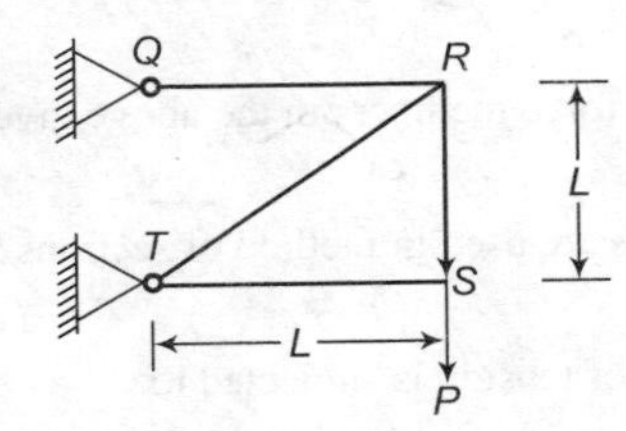

(a) Zero (b) $P_2\sqrt{2P}$ (c) P (d) $\sqrt{2}P$

3. How many equilibrium equations do we need to solve generally on each joint of a truss?
(a) 1 (b) 2 (c) 3 (d) 4

4. If a member of a truss is in compression, then what will be the direction of force that it will apply to the joints?
(a) Outward (b) Inward (c) Depends on case (d) No force will be there

5. If a member of a truss is in tension, then what will be the direction of force that it will apply to the joints?
(a) Outward (b) Inward (c) Depends on case (d) No force will be there

6. What should be ideally the first step to approach to a problem using method of joints?
(a) Draw free body diagram (FBD) of each joint
(b) Draw FBD of overall truss
(c) Identify zero force members
(d) Determine external reaction forces

7. What should be the angle (in degrees) in the given system (part of a bigger system) if both of the members have to be a zero force member?

(a) 22.5 (b) 45 (c) 67.5 (d) Any angle

In the above figure, force is applied at joint C and its magnitude is 10 N with downward direction. This question is used for Q9-Q12.

8. Which of the following are 0 force members?
(a) FG, HI, HJ (b) HI, HJ, AE (c) HI, HJ, HE (d) HI, HJ, FH

9. What will be the magnitude of force (in N) transmitted by FI?
(a) 0 (b) 1 (c) 2 (d) 3

10. What will be the magnitude of force (in N) transmitted by IC?
(a) 0 (b) 1 (c) 2 (d) 3

11. What is the total number of zero force members in the above given system?
(a) 7 (b) 8 (c) 9 (d) 10

12. How many equilibrium equations are used in method of sections?
(a) 2 (b) 4 (c) 3 (d) 5

13. In trusses, a member in the state of tension is subjected to:
(a) Push (b) Pull (c) Lateral force (d) Either pull or push

14. In method of sections, what is the maximum number of unknown members through which the imaginary section can pass?
(a) 1 (b) 2 (c) 3 (d) 4

15. Method of substitute members is used for which type of trusses?
(a) Complex (b) Compound (c) Simple (d) Simple and compound

16. First step to solve complex truss using method of substitute members is to convert it into unstable simple truss.

State whether the above statement is true or false.
(a) True (b) False

17. On differentiating V with respect to X, we will get:
(a) W (b) –W (c) M (d) None of these

18. If a member of a truss is in compression, then what will be the direction of force that it will apply to the joints?
(a) Outward (b) Inward (c) Depends on case (d) No force will be there

Solutions for Objective Questions

Sl. No.	1.	2.	3.	4.	5.	6.	7.	8.	9.	10.
Answer	(0)	(c)	(b)	(a)	(b)	(c)	(d)	(a)	(a)	(a)

Sl.No.	11.	12.	13.	14.	15.	16.	17.	18.
Answer	(c)	(c)	(b)	(c)	(a)	(b)	(b)	(a)

EXERCISE PROBLEMS

1. Determine the forces in the members of the trusses shown below using the tension coefficient method, the method of joints, and the method of sections. All members are of same length.

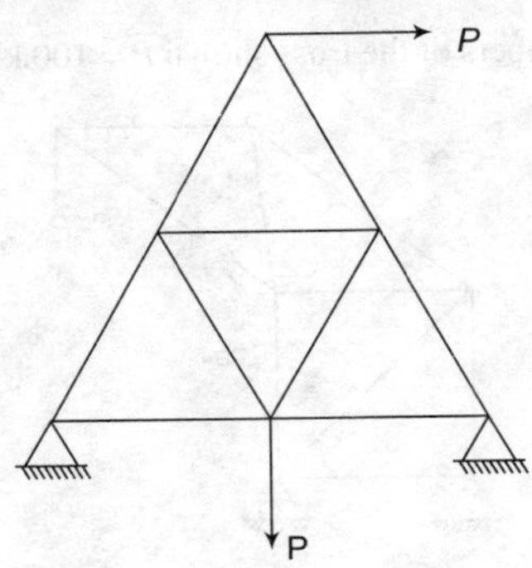

FIGURE 1

2. Determine the forces in the members of the truss shown in Figure 2 using the method of joints. Height of the truss is 3.5 m.

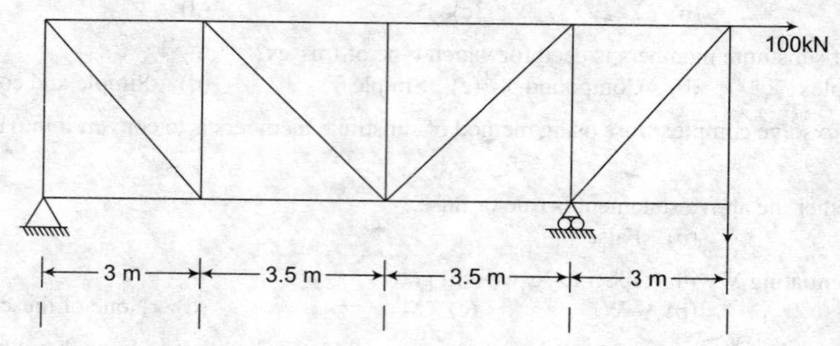

FIGURE 2

3. Determine the forces in the members circled of the truss shown using the method of sections.

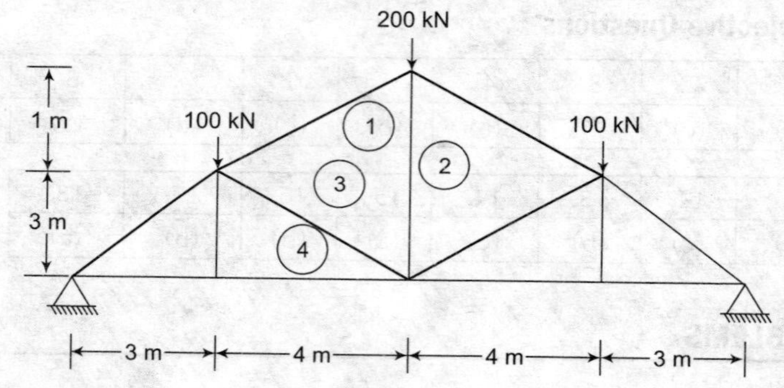

FIGURE 3

4. Determine the forces in the members of the truss shown $P = 100$ kN. Horizontal members are 3 m length and vertical members are 4 m.

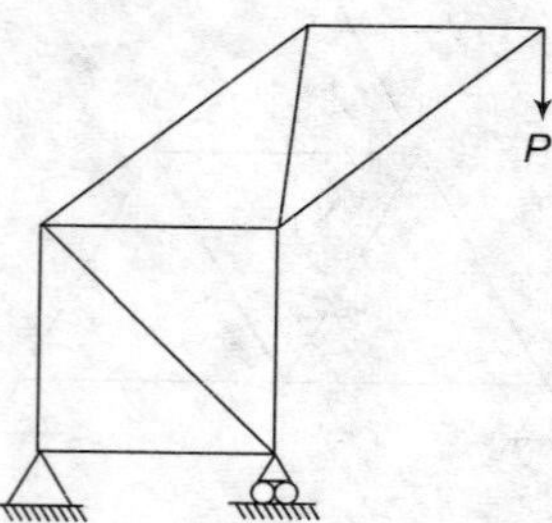

FIGURE 4

5. Determine the forces in the members of the truss shown. Each member length is 4.5 m.

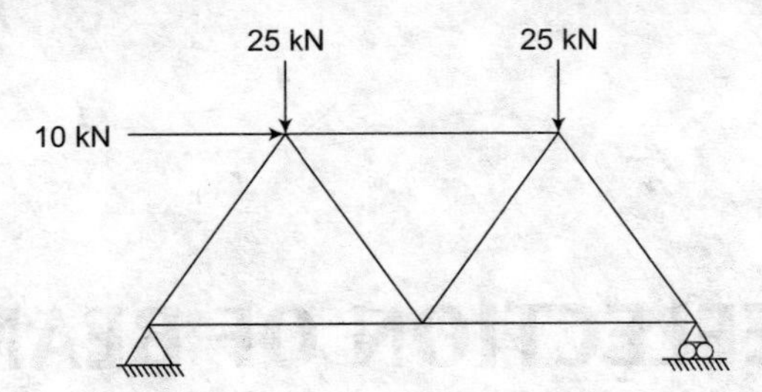

FIGURE 5

6. Determine the forces in the members of the truss shown. Vertical load is 50 kN and the horizontal load is 25 kN.

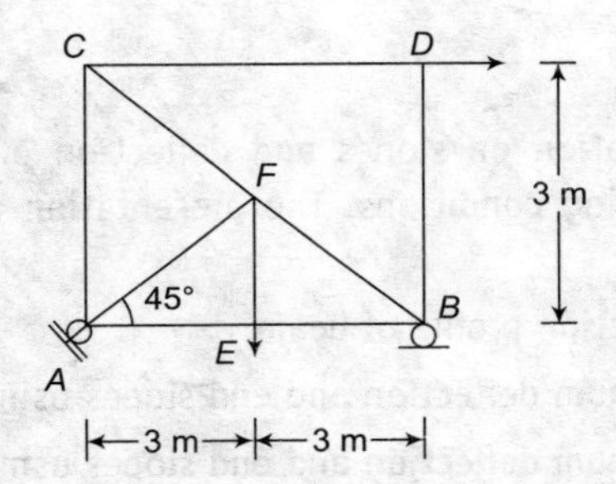

FIGURE 6

CHAPTER 7

DEFLECTION OF BEAMS

UNIT OBJECTIVE

This chapter provides information on slopes and deflection of beams with various support conditions and different loading conditions. The presentation attempts to help the student achieve the following:

Objective 1: Obtain the deflection profile of beam.

Objective 2: Determine maximum deflection and end slopes using Macaulay's method.

Objective 3: Determine maximum deflection and end slopes using moment area method.

Objective 4: Determine maximum deflection and end slopes using conjugate beam method.

Objective 5: Determine maximum deflection and end slopes using superposition technique.

Objective 6: Determine the load carried by beam, based on allowable deflection.

7.1 INTRODUCTION

The beams are subjected to transverse loading deform. The deformed profile of a beam is referred to as the elastic curve of the beam. The deflection calculations are necessary as high deflections cause:

1. Undesirable psychological effects (in case of floor beams of buildings).
2. Misalignment of machine parts.
3. Distress in brittle material structures.

Deflection calculations are also necessary in the analysis of statistically indeterminate structures. The information about the deflection characteristics of members is needed in the dynamic analysis of structures. The different methods of calculating deflections are

1. Double integration method.
2. Macaulay's method.
3. Moment area method.

4. Conjugate beam method.
5. Energy methods.

7.2 MOMENT-CURVATURE RELATIONSHIP

The elastic curve of a beam subjected to constant moment (pure bending) assumes an arc of a circle. If the bending moment (BM) is varying, associated with shear force (SF) also, the elastic curve varies. Thus, the radius of curvature also varies. In chapter 4, we have seen that

$$\frac{M}{I} = \frac{E}{R}$$

$$\Rightarrow \qquad M = EI\left(\frac{1}{R}\right),$$

in which EI is flexural rigidity, M is moment, and $1/R$ is curvature.

EI is always a positive quantity, thus to the tune of moment, curvature also have sense like positive/negative.

Consider a curve AB shown in Figure 7.1. Let us develop an expression for the radius of curvature (R) or curvature ($1/R$) of the curve AB from the fundamentals of mathematics. Consider an elemental position 'ds' of the curve AB. For this elemental position, radius of curvature can be taken as constant, say 'R'. '$\delta\theta$' is the deviation between the tangents drawn at P and Q. PC and QC are the normals drawn to the tangents at P and Q. Thus, $\angle PCQ$ is also '$\delta\theta$'.

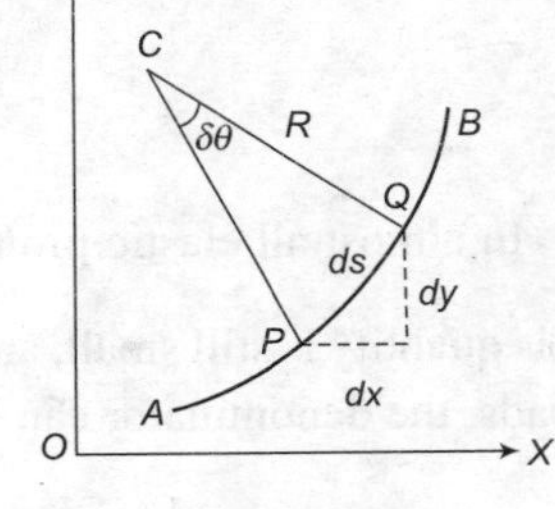

FIGURE 7.1

'Tangent at Q is required'

'Tangent at P is required'

$$PQ = ds = R,\ \delta\theta = R{\cdot}d\theta$$

$$\therefore \qquad \frac{1}{R} = \frac{d\theta}{ds}$$

$$\therefore \qquad \text{Curvature} = \frac{d\theta}{ds} \qquad (7.1)$$

We know that, if 'θ' is the inclination of the tangent with the positive x axis measured in the counterclockwise direction.

$$\frac{dy}{dx} = \tan\theta \qquad (7.2)$$

Differentiate with respect to 'x' on both sides

$$\frac{d^2y}{dx^2} = \sec^2\theta \times \frac{d\theta}{dx}$$

$$\frac{d^2y}{dx^2} = \sec^2\theta \times \frac{d\theta}{ds} \times \frac{ds}{dx}$$

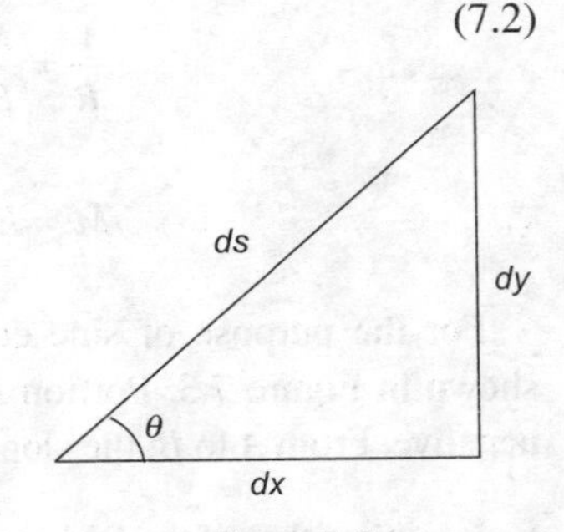

FIGURE 7.2

From the elemental part of curve, it is clear that $\frac{ds}{dx} = \sec\theta$

$$\therefore \qquad \frac{d^2y}{dx^2} = \sec^3\theta \times \frac{d\theta}{ds} \qquad (7.3)$$

$$\tan\theta = \frac{dy}{dx}\sec^3\theta = [\sec^2\theta]^{3/2} = \{1+\tan^2\theta\}^{3/2} = \left\{1+\left(\frac{dy}{dx}\right)^2\right\}^{3/2}.$$

∴ Eq. (7. 3) can be written as

$$\frac{d^2y}{dx^2} = \left\{1+\left(\frac{dy}{dx}\right)^2\right\}^{3/2} \cdot \frac{d\theta}{ds}.$$

From Eq. (7. 1), $1/R$ = curvature = $\dfrac{d\theta}{ds}$

$$\frac{1}{R} = \frac{\dfrac{d^2y}{dx^2}}{\left\{1+\left(\dfrac{dy}{dx}\right)^2\right\}^{3/2}}.$$

Mathematically,

$$\frac{1}{R} = \frac{\dfrac{d^2y}{dx^2}}{\left\{1+\left(\dfrac{dy}{dx}\right)^2\right\}^{3/2}}.$$

In almost all elastic profiles of beams, $\dfrac{dy}{dx}$ is very small; (being measured in radians) square of this quantity is still small, and hence can be neglected. Thus, in case of beams subjected to transverse loads, the denominator can be equated to one. Thus

$$\frac{1}{R} = \frac{d^2y}{dx^2}$$

From the strength of the material principles

$$\frac{1}{R} = \frac{M}{EI}$$

$$\therefore \qquad M = EI\frac{d^2y}{dx^2}. \tag{7. 4}$$

For the purpose of sine convention, consider the curve ABC as shown in Figure 7.3. Portion AB is hogging curve, for which BM is negative. From A to B, the slope (dy/dx) is decreased with positive 'x'.

It implies that, if the BM is negative, $\left[\left(\dfrac{dy}{dx}\right)_B - \left(\dfrac{dy}{dx}\right)_A\right]$ is negative

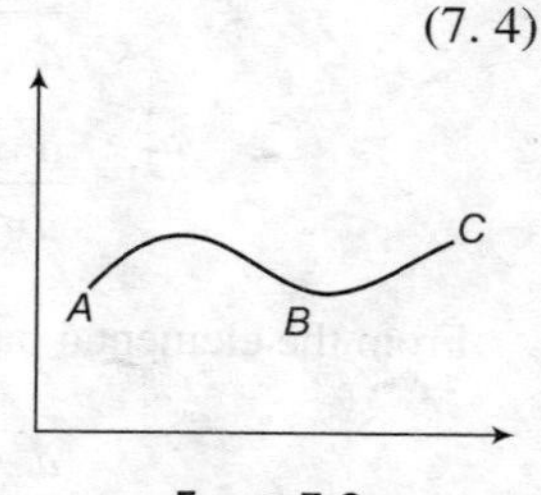

FIGURE 7.3

or $\dfrac{d^2y}{dx^2}$ is negative.

Consider the portion '*BC*' of the curve. It is sagging curve, subjected to sagging BM (sign is +ve). From point *D* to point *C*, slope is increasing with positive *x*. Hence Eq. (7. 4) can be comfortably used to determine the deflections. Unfortunately for general loading of any member, the deflections are downward. Hence, if *y* axis is chosen downward, then Eq. (7. 4) is to be readjusted as

$$EI\frac{d^2y}{dx^2} = -M. \tag{7.5}$$

In solving the above differential equation, two constants of integration appear. These integration constants are to be found from the boundary conditions consistent with the support condition such as that

1. Hinged support, vertical deflection $y = 0$
2. Roller support, vertical deflection $y = 0$
3. Fixed support, vertical deflection $y = 0$ slope $\left(\dfrac{dy}{dx}\right) = 0$

We can present the number of relations between loadings. SF, BM, slope, and deflection as

$$EI\frac{d^2y}{dx^2} = M$$

$$EI\frac{d^3y}{dx^3} = \frac{dM}{dx} = V \qquad \text{(SF)}$$

$$EI\frac{d^4y}{dx^4} = \frac{dV}{dx} = W \qquad \text{(Loading)}$$

$$\left(\frac{dy}{dx}\right) = \text{slope}$$

PROBLEM 7.1 **Objective 1**

Determine the maximum deflection in simply supported beam subjected to uniformly distributed load (UDL) over entire span.

SOLUTION

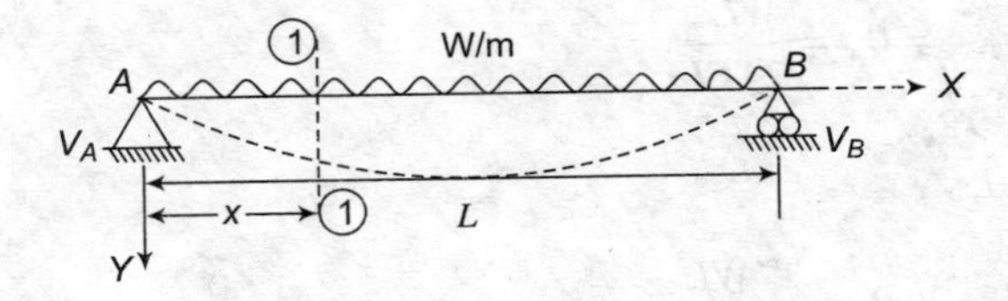

FIGURE 7.4

Using equilibrium equation

$$V_A = V_B = \frac{WL}{2}.$$

Consider a section located at a distance 'x' from A

$$M_X = \frac{WL}{2} \times x - \frac{Wx^2}{2}$$

We know $EI\frac{d^2y}{dx^2} = -M$

$$\Rightarrow \quad EI\frac{d^2y}{dx^2} = -\frac{WL}{2} \times x + \frac{Wx^2}{2}$$

$$\Rightarrow \quad EI\frac{dy}{dx} = -\frac{WL}{2\times 2} \times x^2 + \frac{Wx^3}{6} + C_1$$

$$\Rightarrow \quad EI \times y = -\frac{WL}{12} \times x^3 + \frac{Wx^4}{24} + C_1x + C_2$$

$$\Rightarrow \quad y = \frac{1}{EI}\left\{-\frac{WL}{12} \times x^3 + \frac{Wx^4}{24} + C_1x + C_2\right\}.$$

As per boundary conditions,
At $x = 0$, $y = 0$ (hinged support)

$$\Rightarrow \quad C_2 = 0$$

At $x = L$, $y = 0$ (roller support)

$$\Rightarrow \quad C_1 = \frac{WL^3}{24}$$

$$\Rightarrow \quad y = \frac{1}{EI}\left\{-\frac{WL}{12} \times x^3 + \frac{Wx^4}{24} + \frac{WL^3}{24}x\right\}$$

$$\Rightarrow \quad \frac{dy}{dx} = \frac{1}{EI}\left\{-\frac{WL}{2} \times x^2 + \frac{Wx^3}{6} + \frac{WL^3}{24}\right\}$$

Slope at support A, $x = 0$

$$\left[\frac{dy}{dx}\right]_{x=0} = \theta_A = \frac{WL^3}{24EI}$$

Slope at support B, $x = L$

$$\left[\frac{dy}{dx}\right]_{x=L} = \theta_B = -\frac{WL^3}{24EI}$$

For maximum deflection $\dfrac{dy}{dx} = 0$

$$\Rightarrow \qquad -3Lx^2 + 2x^3 + \frac{L^3}{2} = 0$$

$$\Rightarrow \qquad -6Lx^2 + 4x^3 + L^3 = 0$$

$$\Rightarrow \qquad x = \frac{L}{2}$$

Maximum deflection $(y_{max})_{X=L/2} = \dfrac{1}{EI}\left(-\dfrac{WL^4}{96} + \dfrac{WL^4}{384} + \dfrac{WL^4}{48}\right)$

$$(y_{max})_{X=L/2} = \frac{1}{EI}\left(5\frac{WL^4}{384}\right)$$

$$\theta_A = \frac{WL^3}{24EI} \text{ and } \theta_B = -\frac{WL^3}{24EI}$$

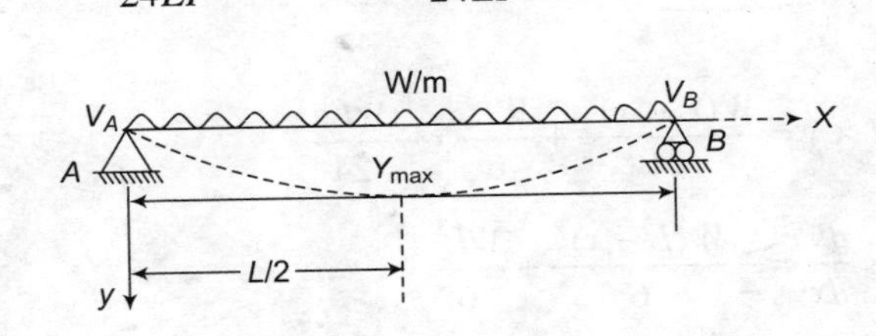

FIGURE 7.5

From the above expression, it is clear that higher the flexural rigidity (EI) lower will be the deflection. To reduce the deflection either modulus of elasticity of the material is to be increased or second moment of inertia (MI) (I) about neutral axis (NA) is to be increased.

PROBLEM 7.2 **Objective 1**

Drive the deflection profile of cantilever beam shown in Figure 7.6.

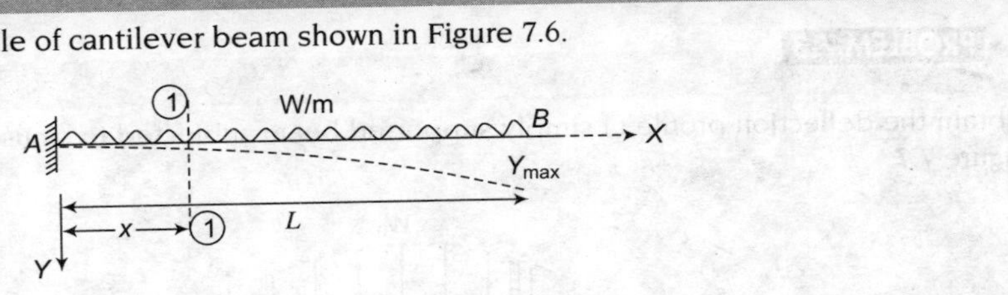

FIGURE 7.6

SOLUTION

Consider the section located at a distance 'x' from A

$$M_x = -\frac{W(L-x)^2}{2}.$$

We know $$EI\frac{d^2y}{dx^2} = -M_x$$

$$\therefore \quad EI\frac{d^2y}{dx^2} = \frac{W(L-x)^2}{2}$$

$$\Rightarrow \quad EI\frac{dy}{dx} = -\frac{W(L-x)^3}{6} + C_1$$

$$\Rightarrow \quad EI \cdot y = \frac{W(L-x)^4}{24} + C_1x + C_2$$

C_1 and C_2 are to be found; the boundary condition that at fix support, slope, and deflection is zero.
At $x = 0$, $y = 0$

$$\Rightarrow \quad C_2 = -\frac{W(L)^4}{24}$$

At $x = 0$, $$\frac{dy}{dx} = 0$$

$$\Rightarrow \quad C_1 = \frac{W(L)^3}{6}$$

$$\Rightarrow \quad EI \times y = \frac{W(L-x)^4}{24} + \frac{WL^3}{6}x - \frac{WL^4}{24}$$

$$EI \times \frac{dy}{dx} = -\frac{W(L-x)^3}{6} + \frac{WL^3}{6}$$

At $x = L$, $$(y_{\max})_{X=l} = \frac{1}{EL}\left(\frac{WL^4}{8}\right)$$

$$(\theta_{\max})_{X=l} = \frac{dy}{dx} = \frac{1}{EL}\left(\frac{WL^3}{6}\right)$$

PROBLEM 7.3

Objective 1

Obtain the deflection profile of simply supported beam subjected to sinusoidal loading as shown in Figure 7.7.

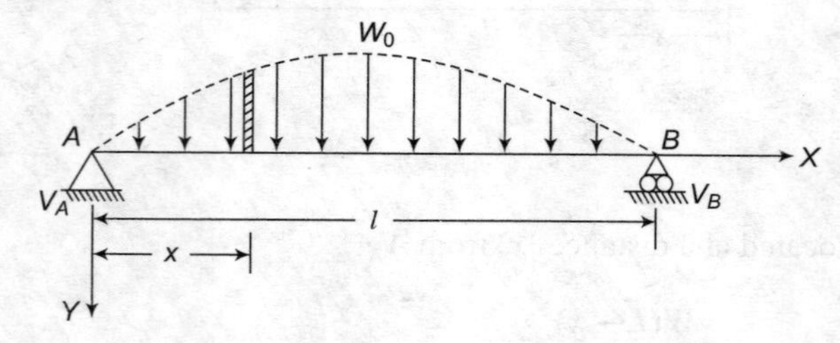

FIGURE 7.7

SOLUTION

As loading is sinusoidal, at section 'x' distant from A

$$w(x) = w_0 \sin\frac{\pi x}{l}.$$

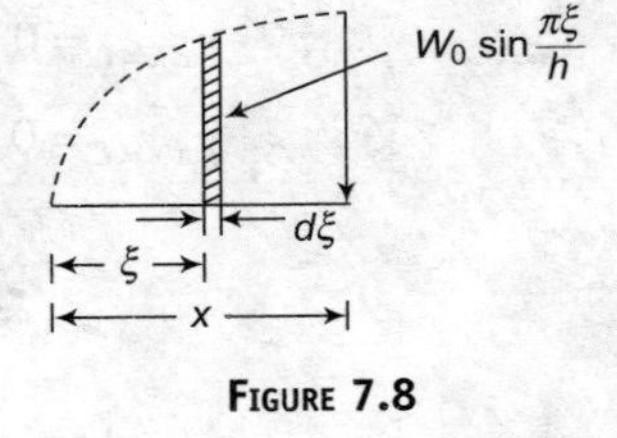

FIGURE 7.8

Let V_A and V_B be the reaction at A and B as loading is symmetric.

$$V_A = V_B = \frac{1}{2}\int_0^l w(x)\cdot dx = \frac{1}{2}\int_0^l w_0 \sin\frac{\pi x}{l}dx$$

$$= \frac{w_0}{2\frac{\pi}{l}}\left[-\cos\frac{\pi x}{l}\right]_0^l = w_0\frac{l}{2\pi}(2) = w_0\frac{l}{\pi}$$

From BM at section 'x' distance from support A

$$M_X = w_0\frac{lx}{\pi} - \int_0^x (x-\xi)\times w_0 \sin\frac{\pi\xi}{l}d\xi$$

$$= w_0\frac{lx}{\pi} - \int_0^x (x-\xi)\times w_0 d\left[\frac{-\cos\frac{\pi\xi}{l}}{\frac{\pi}{l}}\right]$$

$$\leq w_0\frac{l}{\pi} + w_0\frac{l}{\pi}\left\{\left[(x-\xi)\cos\frac{\pi\xi}{l}\right]_0^x - \int_0^x \cos\frac{\pi\xi}{l}(-1)d\xi\right\}$$

$$= w_0\frac{l}{\pi}x + w_0\frac{l}{\pi}\left\{\left(\frac{\sin\frac{\pi\xi}{l}}{\frac{\pi}{l}}\right)_0^X\right\}$$

$$= w_0\frac{lx}{\pi} - w_0\frac{l}{\pi}x + w_0\frac{l^2}{\pi^2}\left[\sin\frac{\pi x}{l} - 0\right]$$

$$= w_0\frac{l^2}{\pi^2}\sin\frac{\pi x}{l}$$

We know that $\quad EI\times\frac{d^2y}{d^2x} = -M_x$

$$\Rightarrow \quad EI\times\frac{d^2y}{dx^2} = -w_0\frac{l^2}{\pi^2}\sin\frac{\pi x}{l}$$

$$\Rightarrow \quad EI\times\frac{dy}{dx} = w_0\frac{l^2}{\pi^2}\frac{\cos\frac{\pi x}{l}}{\left(\frac{\pi}{l}\right)} + c_1 = w_0\frac{l^3}{\pi^3}\cos\frac{\pi x}{l} + c_1$$

$$EI\times y = w_0\frac{l^4}{\pi^4}\sin\frac{\pi x}{l} + c_1 x + c_2.$$

Applying the boundary conditions, that at $x = 0$, $y = 0$ and $x = L$, $y = 0$

$$y_{\text{at } x=0} = 0 \Rightarrow c_2 = 0$$

$$y_{\text{at } x=L} = 0 \Rightarrow c_1 = 0$$

$$y = \frac{1}{EI} w_0 \frac{l^4}{\pi^4} \sin \frac{\pi x}{l}$$

$$\frac{dy}{dx} = \frac{1}{EI} w_0 \frac{l^3}{\pi^3} \cos \frac{\pi x}{l}$$

$$M_x = w_0 \frac{l^2}{\pi^2} \sin \frac{\pi x}{l}$$

$$V_x = \frac{dM}{dx} = w_0 \frac{l}{\pi} \cos \frac{\pi x}{l}$$

$$w(x) = -w_0 \sin \frac{\pi x}{l}$$

At $x = L/2$, $y = y_{\max}$

$$\therefore \quad y_{\max} = w_0 \frac{L^4}{\pi^4 EI}$$

At $x = 0$, $\quad \theta_A = w_0 \dfrac{L^3}{\pi^3 EI}$

At $x = L$, $\quad \theta_B = -w_0 \dfrac{L^3}{\pi^3 EI}$

At $x = L/2$, $\quad M_x = W_0 \dfrac{L^2}{\pi^2}$

At $x = 0$, $\quad V_A = W_0 \dfrac{L}{\pi}$

At $x = L$, $\quad V_B = -W_0 \dfrac{L}{\pi}$

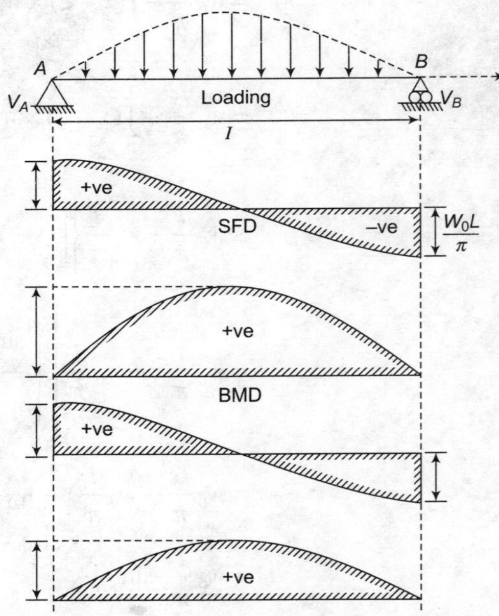

FIGURE 7.9

PROBLEM 7.4

Objective 1

Determine the maximum deflection and slope of a cantilever beam subjected to triangular load *c* shown in Figures 7.10 and 7.11.

(a)

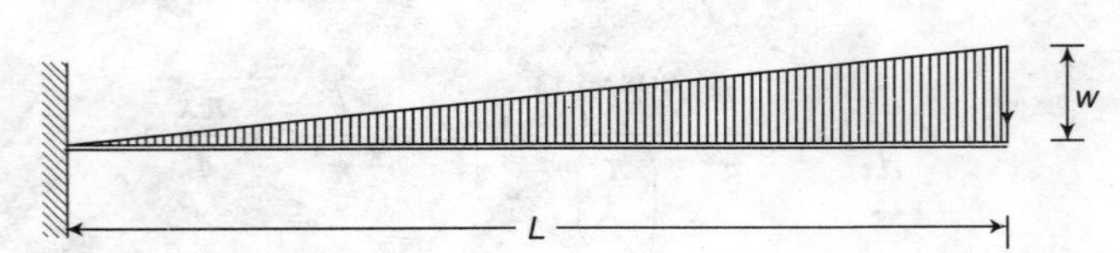

FIGURE 7.10

b)

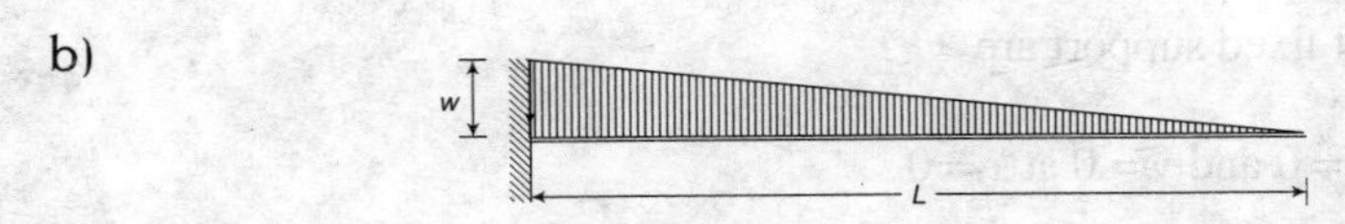

FIGURE 7.11

SOLUTION

Case (a): consider a section located at a distance 'x' from A

$$w(x) = \frac{w}{l}x$$

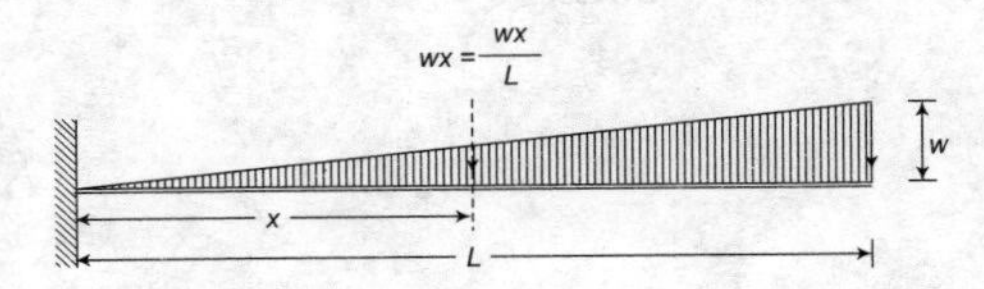

FIGURE 7.12

BM at the section considered

M_x = Moment of the loading diagram about x

$$M_x = -\left\{\frac{w_x + 2w}{w + w_x} \times \frac{(l-x)}{3} \times \frac{w + w_x}{2}(l-x)\right\}$$

$$= -(2w + w_x)\frac{(l-x)^2}{6} = \frac{-1}{6}\left[2w + \frac{w}{l}x\right](l-x)^2 = \frac{-w}{6}\left\{2 + \frac{x}{l}\right\}(l-x)^2$$

We know that

$$EI \times \frac{d^2y}{dx^2} = -M_x$$

$$EI \times \frac{d^2y}{dx^2} = \frac{w}{6}\left\{2 + \frac{x}{l}\right\}(l-x)^2 = \frac{w}{6}\left(2l^2 + 2x^2 - 4lx + xl - 2x^2 + \frac{x^3}{l}\right)$$

$$= \frac{w}{6}\left\{2L^2 - 3Lx + \frac{x^3}{L}\right\}$$

$$\Rightarrow \quad EI \times \frac{dy}{dx} = \frac{w}{6}\left\{2L^2x - 3l\frac{x^2}{2} + \frac{x^4}{4l}\right\} + c_1$$

$$\Rightarrow \quad EI = \frac{w}{6}\left\{2l^2\frac{x^2}{2} - 3l\frac{x^3}{6} + \frac{x^5}{20l}\right\} + c_1x + c_2$$

The boundary conditions at fixed support are

$$\frac{dy}{dx} = 0 \text{ at } x = 0 \text{ and } y = 0 \text{ at } x = 0$$

$$C_1 = 0 \text{ and } C_2 = 0$$

$$\Rightarrow \quad EI\ y = \frac{w}{6}\left\{L^2x^2 - L\frac{x^3}{2} + \frac{x^5}{20L}\right\}$$

$$\Rightarrow \quad y = \frac{w}{6EI}\left\{L^2x^2 - L\frac{x^3}{2} + \frac{x^5}{20L}\right\}$$

For maximum deflection

at $x = L$, $\quad y = y_{\max}$

$$y_{\max} = \frac{11wl^4}{120EI}$$

For maximum slope $x = L$, $\dfrac{dy}{dx} = \gamma_{\max}$

$$\therefore \quad \theta_{\max} = \frac{wl^3}{8EI}$$

Case (b):

Consider a section located 'x' distance from 'A'

$$\omega_x = \frac{w}{l}(l - x)$$

BM at the section considered is

$$M_x = -[\text{Moment of loading diagram about the section considered}]$$

$$M_X = \frac{-1}{2}w_x(L - x)\frac{1}{3}(L - x)$$

$$= \frac{-1}{6}w_x(L - x)^2$$

$$= \frac{-1}{6l}w\ (L - x)^3$$

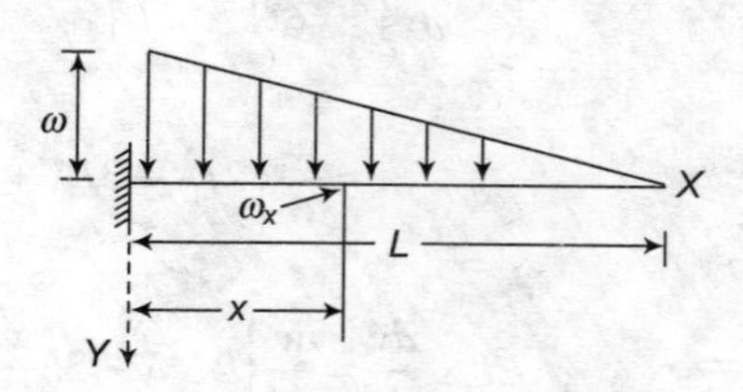

FIGURE 7.13

We know that

$$EI\frac{d^2y}{dx^2} = -M_x$$

$$EI\frac{d^2y}{dx^2} = \frac{1}{6L}w(L-x)^3$$

$$EI\frac{dy}{dx} = \frac{-1}{24L}w(L-x)^4 + c_1$$

$$EI\ y = \frac{+1}{120L}w(L-x)^5 + c_1x + c_2$$

As per the boundary condition, that is

At $x = 0$, $y = 0$, and $\frac{dy}{dx} = 0$ being fixed support

$$c_1 = \frac{wL^3}{24}$$

$$c_2 = \frac{-wL^4}{120}$$

$$y = \frac{1}{EI}\left\{\frac{+1}{120L}w(L-x)^5 + \frac{wL^3}{24}x - \frac{wL^4}{120}\right\}$$

$$\frac{dy}{dx} = \frac{1}{EI}\left\{\frac{-1}{24L}w(L-x)^4 + \frac{wL^3}{24}\right\}$$

At $x = L$, $\quad y = y_{max}$

$$y_{max} = \frac{wL^4}{30EI}$$

At $x = L$, $\quad \gamma = \gamma_{max}$

$$\therefore \quad \theta_{max} = \frac{wL^3}{24EI}$$

Law of superposition can be understood easily from the above two cases. By algebraically adding cases (a) and (b), the solution for the cantilever beam subjected to UDL can be obtained.

$$Y_B = \frac{11WL}{120EI};\ \theta_B = \frac{WL^3}{8EI}$$

$$Y_B = \frac{WL^3}{30EI};\ \theta_B = \frac{WL^3}{24EI}$$

$$Y_B = \frac{WL^4}{8EI};\ \theta_B = \frac{WL^3}{6EI}$$

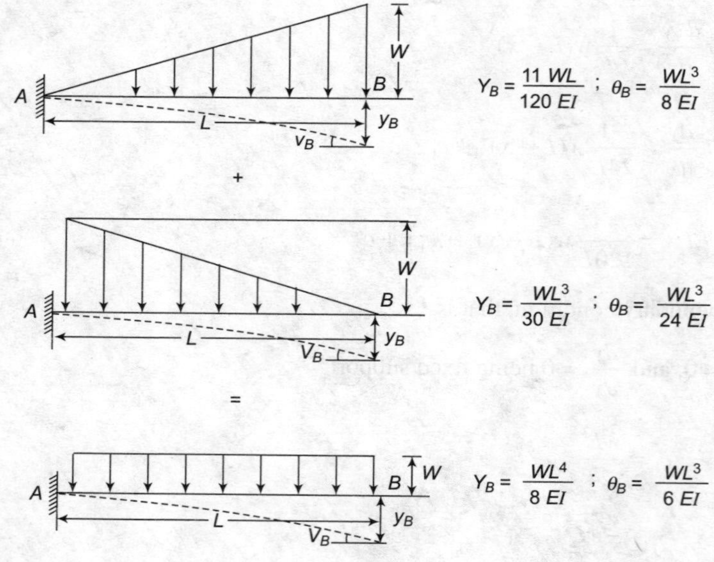

FIGURE 7.14

7.3 MACAULAY'S METHOD

In all problems done in the earlier section on this chapter, the loading is uniform/continuous over the entire span of the beam. If discreet load acts on the beam, then double integration technique adopted becomes more tedious. Writing the single expression for the moment for the entire beam will not be possible, if discrete load is present. To handle this problem, Macaulay's method is adopted. In this method, discrete load is handled using *singularity function*.

A singularity function of 'x' is written as $|(x - x_0)^n|$, in which 'n' is the integer, such that if $\boldsymbol{x > x_0}$ the value $|(x - x_0)^n|$ is equal to $(x - x_0)^n$, otherwise $|(x - x_0)^n|$ is zero.

That is, $$|(x - x_0)^n| = (x - x_0)^n \text{ if } x > x_0$$

$$|(x - x_0)^n| = 0 \text{ if } x \leq x_0$$

In other words, if the term in the special braces is negative, its value is taken zero or it is neglected. Using the singularity by function, let us write the expression for the BM of the following case:

Case (1): Point load section should be chosen in such a way that it should cover the beam.

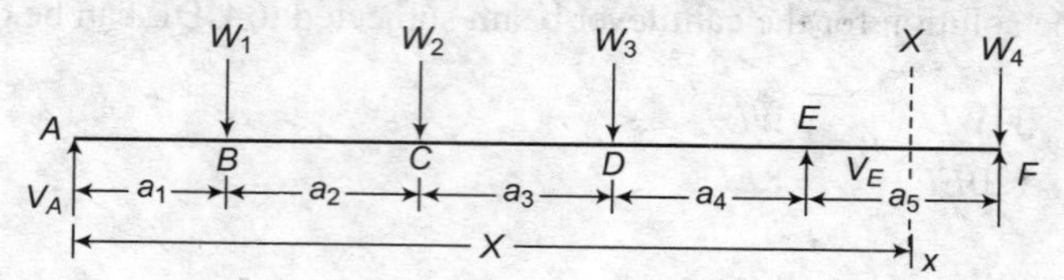

FIGURE 7.15

$$M_x = V_A x - W_1 \ |(x - a_1)| - W_2 |(x - a_1 - a_2)| - W_3 |(x - a_1 - a_2 - a_3)| + V_E$$
$$|(x - a_1 - a_2 - a_3 - a_4)|.$$

The above expression can be comfortably used for BM at any place within the beam. For example, for BM expression of region CD, terms associate with the V_E and W_3 vanish.

$$M_x = V_A x - W_1 - (x - a_1) - W_2(x - a_1 - a_2).$$

Case (2): UDLs

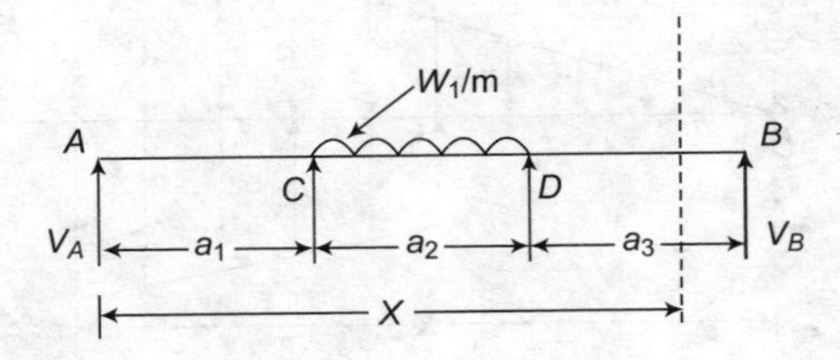

FIGURE 7.16

In the beam shown in the figure, the UDL is within the region CD only. Thus, if we write a general expression for the BM at the section shown, we cannot include W_1, associated with x. Hence extend the UDL up to the section taken and apply upward (–ve) load from D to the section. Then, the adjusted loading is as presented in Figure 7.17.

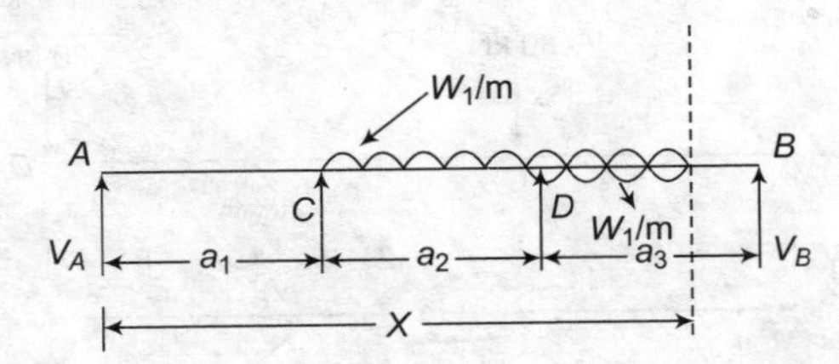

FIGURE 7.17

Upward and downward loads of the same intensity make no difference in the final BM values.

$$M_x = V_A x - W_1 \left| \frac{\langle x - a_1 \rangle^2}{2} \right| + \left| \frac{\langle x - a_1 - a_2 \rangle^2}{2} \right|.$$

In the above expression, if the region is AC, where the BM is to be obtained, then the terms associated with $W_1(\uparrow)$ and $W_1(\downarrow)$ vanish. Thus

$$M_x = V_A x$$

Similarly, BM in the region CD

$$M_x = V_A x - W_1 \frac{(x - a_1)^2}{2}.$$

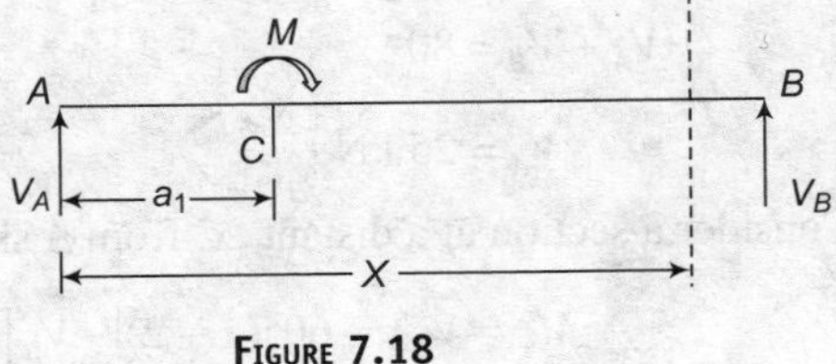

FIGURE 7.18

Both case (3) moments/couples

Same procedure can be adopted as in the case of point load, but the power of the multiples in the special braces should be handled by substituting zero.

$$M_x = V_A x - M_1 \left| \langle x - a_1 \rangle \right|^0$$

Case (4): Varying load

It also should be handled as a UDL. The varying load should be extended up to the section considered and upward loading to compensate the excess loading.

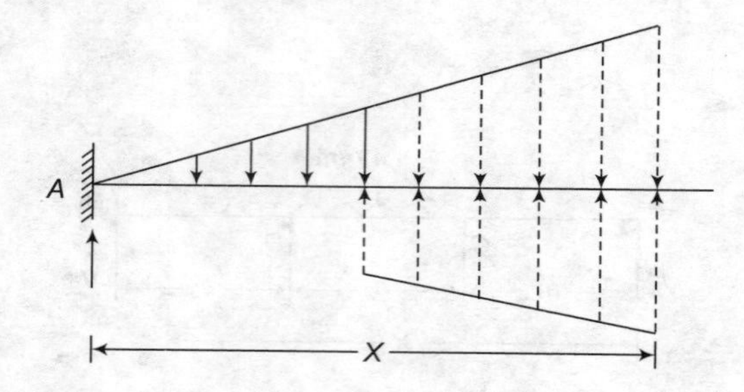

FIGURE 7.19

PROBLEM 7.5

Objective 2

Determine the maximum deflection of the overhanging beam shown in Figure 7.20. Take $EI = 1 * 10^4$ kN·m^2.

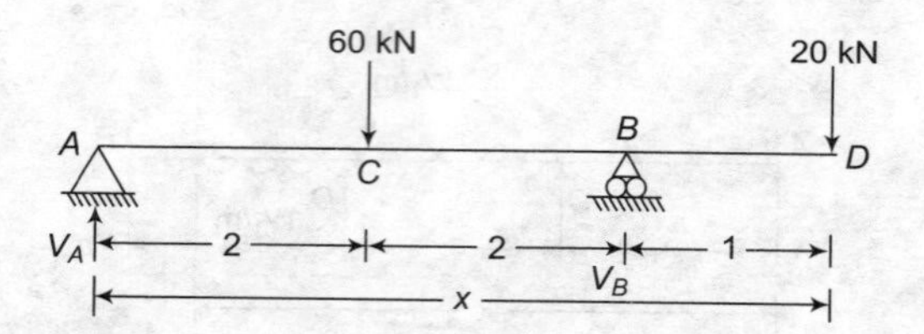

FIGURE 7.20

SOLUTION

Let V_A and V_B be the reactions at the supports A and B, respectively.

Taking moments about A

$$V_B \times 4 = 60 \times 2 + 20 \times 5$$

$$\Rightarrow \quad V_B = 55 \text{ kN}$$

$$V_A + V_B = 80$$

$$\Rightarrow \quad V_A = 25 \text{ kN}$$

Consider a section at a distant 'x' from A shown in Figure 7.19.

$$M_x = V_A x - 60 \left|\langle x-2 \rangle\right| - V_B \left|\langle x-4 \rangle\right|$$

$$= 25x - 60 \left|\langle x-2 \rangle\right| - 55 \left|\langle x-4 \rangle\right|$$

We know, $EI \times \dfrac{d^2 y}{dx^2} = -M$

$$\therefore \qquad EI\frac{d^2y}{dx^2} = -25x + 60\left|\langle x-2\rangle\right| - 55\left|\langle x-4\rangle\right|$$

$$EI\frac{dy}{dx} = -25\frac{x^2}{2} + 60\left|\frac{\langle x-2\rangle^2}{2}\right| - 55\left|\frac{\langle x-4\rangle^2}{2}\right| + C_1$$

$$EI\,(y) = -25\frac{x^3}{6} + 60\left|\frac{\langle x-2\rangle^3}{6}\right| - 55\left|\frac{\langle x-4\rangle^3}{6}\right| + C_1x + C_2$$

Applying boundary condition as

At $x = 0$; $\quad y = 0$ (Hinged support)

$\Rightarrow \quad C_2 = 0$

At $x = 4$; $\quad y = 0 \quad$ (Roller support)

$\Rightarrow \quad 0 = -186.67 + 4C_1$

$\Rightarrow \quad C_1 = 46.67$

$$\therefore \qquad y = \frac{1}{EI}\left\{-25\frac{x^3}{6} + 60\left|\frac{\langle x-2\rangle^3}{6}\right| - 55\left|\frac{\langle x-4\rangle^3}{6}\right| + 46.67x\right\}$$

At $x = 2$;
$$y_C = \frac{1}{EI}\left\{-25\frac{2^3}{6} + 46.67\times 2\right\} = \frac{60}{EI} = \frac{60}{1\times 10^4}$$

$$y_C = 0.006 \text{ m or } 6 \text{ mm}$$

At $x = 5$; $\quad y = y_0$

$$y_D = \frac{1}{EI}\left\{-25\frac{5^3}{6} + 60\left|\frac{3^3}{6}\right| - 55\left|\frac{1^3}{6}\right| + 46.67\times 5\right\} = \frac{1}{EI}\left|-26.67\right|$$

$$y_D = -0.00267 \text{ m (upward)}$$

$$= \text{upward } 2.67 \text{ mm}$$

The deflection profile of the beam is as shown in Figure 7.21.

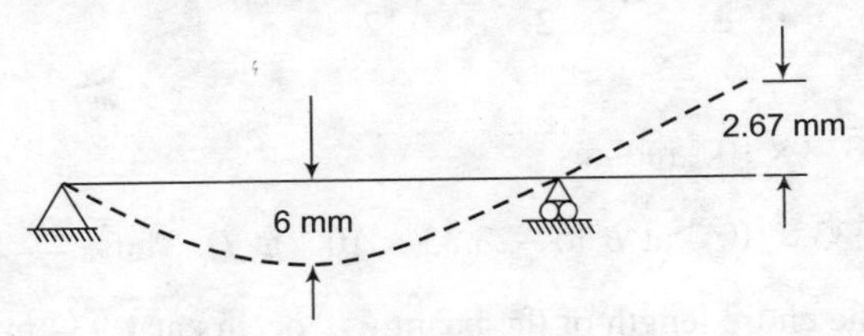

FIGURE 7.21

For the maximum deflection, $\dfrac{dy}{dx} = 0$

$$\frac{dy}{dx} = \frac{1}{EI}\left\{-25\frac{x^2}{2} + 60\left|\frac{\langle x-2\rangle^2}{2}\right| - 55\left|\frac{\langle x-4\rangle^2}{2}\right| + 46.67\right\}$$

To equate $\frac{dy}{dx}$ to zero, the above expression cannot be used directly; $\frac{dy}{dx}$ should be equated to zero indifferent regions, AC, CB, and BD.

In the region AC, $x < 2$

$$\frac{dy}{dx} = \frac{1}{EI}\left\{-25\frac{x^2}{2} + 46.67\right\} = 0$$

$$\Rightarrow \qquad x = 1.932$$

$$y_{max} = \frac{1}{EI}\left\{-25\frac{1.932^3}{6} + 46.67 \times 1.932\right\} = 0.006011 \text{ m}$$

$$= 6.011 \text{ mm}$$

To establish maximum deflection as 6.011 mm, 1.932 m from A, we have to examine the regions CB and CD also. To check (or verify) this, $\frac{dy}{dx}$ in these region need not be equated to zero.

$$\frac{dy}{dx} \text{ at } x = 2 \text{ m is } \theta_C = \frac{1}{EI}\left\{-25\frac{2^2}{2} + 46.67\right\} = -3.33 \times 10^{-4} \text{ radians}$$

$$\frac{dy}{dx} \text{ at } x = 4 \text{ m is } \theta_B = \frac{1}{EI}\left\{-25\frac{4^2}{2} + 60 \times \frac{2^2}{2} + 46.67\right\} = -33.33 \times 10^{-4} \text{ radians}$$

From C to B, there is no change in shape of sign of slope. Hence in the region CB, $\frac{dy}{dx}$ will not be equal to zero. Thus, y_{max} do not exist in this region.

In the region BD,

$$\theta_B = -33.33 \times 10^{-4} \text{ radians}$$

$$\theta_D = \frac{1}{EI}\left\{-25 \times \frac{5^2}{2} + 60 \times \frac{3^2}{2} - 55 \times \frac{1^2}{2} + 46.67\right\}$$

At $x = 5$ m

$$= -23.338 \times 10^{-4} \text{ radians}$$

'θ' is varying from -33.33×10^{-4} at B to -23.33×10^{-4} at D. Thus, $\frac{dy}{dx}$ cannot be zero in this region too. Hence over the entire length of the beam y_{max} occurs at 1.932 m from A and is equal to 6.011 mm.

PROBLEM 7.6

Objective 2

Using the Macaulay's method, determine the deflection at different salient points of the beam shown in Figure 7.22. Take, $EI = 1 \times 10^5$ kN·m^2.

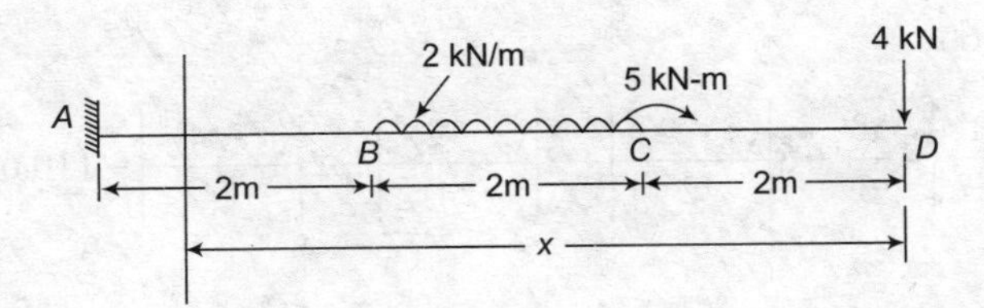

FIGURE 7.22

SOLUTION

Take 'D' as origin, consider a section located 'x' distant from D. Apply upward and downward UDL of intensity 2 kN/m from B up to the section considered.

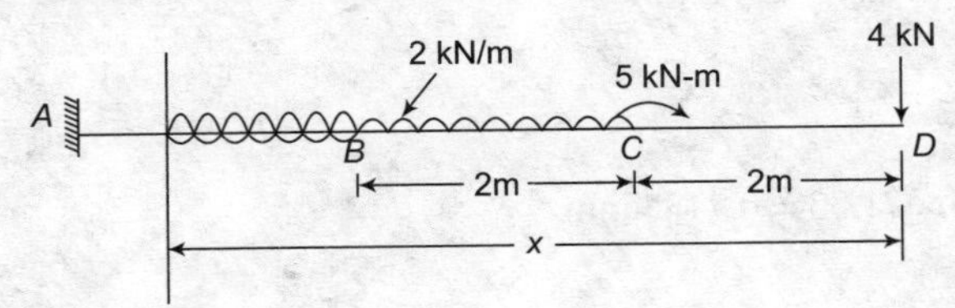

FIGURE 7.23

$$M_x = -4x - 5\left|\langle x-2\rangle^0\right| - 2\left|\frac{\langle x-2\rangle^2}{2}\right| + 2\left|\frac{\langle x-4\rangle^2}{2}\right|$$

We know that, $EI\dfrac{d^2y}{dx^2} = -M_x$

$$\therefore \qquad EI\frac{d^2y}{dx^2} = 4x + 5\left|\langle x-2\rangle^0\right| + 2\left|\frac{\langle x-2\rangle^2}{2}\right| - 2\left|\frac{\langle x-4\rangle^2}{2}\right|$$

$$EI\frac{dy}{dx} = 4\frac{x^2}{2} + 5\left|\frac{\langle x-2\rangle^1}{1}\right| + 2\left|\frac{\langle x-2\rangle^3}{6}\right| - 2\left|\frac{\langle x-4\rangle^3}{6}\right| + C_1$$

$$EI(y) = 4\frac{x^3}{6} + 5\left|\frac{\langle x-2\rangle^2}{2}\right| + 2\left|\frac{\langle x-2\rangle^4}{24}\right| - 2\left|\frac{\langle x-4\rangle^4}{24}\right| + C_1x + C_2$$

At fixed support, that is, at A, slope is zero and deflection is zero.

$\therefore$ At $x = 6$, $\dfrac{dy}{dx} = 0$

$$\Rightarrow \qquad 0 = 110.667 + C_1$$

or $$C_1 = -110.667$$

At $x = 6$, $y = 0$;

$$\Rightarrow \qquad 0 = -460 + C_2$$

or $\qquad C_2 = 460$

$$\therefore \qquad y = \frac{1}{EI}\left\{4\frac{x^3}{6} + 5\left|\frac{\langle x-2\rangle^2}{2}\right| + 2\left|\frac{\langle x-2\rangle^4}{24}\right| - 2\left|\frac{\langle x-4\rangle^4}{24}\right| - 110.667x + 460\right\}$$

At $x = 0$, $y = y_D$

$$y_D = y_{\max} = \frac{460}{1\times 10^5} = 0.0046 \text{ m} = 4.6 \text{ mm}$$

At $x = 2$, $y = y_c$

$$y_c = \frac{1}{EI}\times 244 = 2.44 \text{ mm}$$

At $x = 4$, $y = y_B$

$$y_B = \frac{1}{EI}\times 71.33 = 0.7133 \text{ mm}$$

Similarly, $\qquad \frac{dy}{dx} = \frac{1}{EI}\left(4\frac{x^2}{2} + 5\left|\langle x-2\rangle\right| + 2\left|\frac{\langle x-2\rangle^3}{6}\right| - 2\left|\frac{\langle x-4\rangle^3}{6}\right| - 110.66\right)$

At $x = 0$, $\qquad \frac{dy}{dx} = \theta_D$

$$\theta_D = -110.67/EI = -0.0011 \text{ radians}$$

At $x = 2$, $\qquad \frac{dy}{dx} = \theta_C$

$$\theta_C = -0.00103 \text{ radians}$$

At $x = 4$, $\qquad \frac{dy}{dx} = \theta_B$

$$\theta_B = -0.00066 \text{ radians}$$

PROBLEM 7.7

Objective 2

Determine the maximum deflection and end slopes of the beam loaded as shown in Figure 7.24. Also sketch the elastic curve of the beam $EI = 2\times 10^3$ kN·m^2

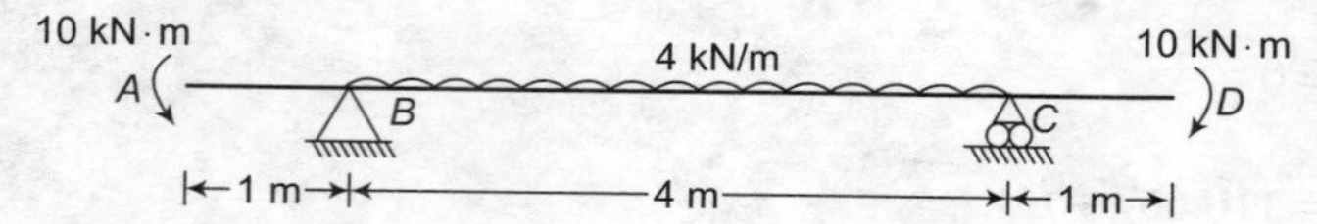

FIGURE 7.24

SOLUTION

Let V_B and V_C be the vertical reactions at B and C, respectively.

Sum of the vertical forces = 0

$$\Rightarrow \qquad V_B + V_C = 4 \times 4 = 16 \tag{7.6}$$

Taking moments about B,

$$\Rightarrow \qquad 10 - 4 \times 4 \times 2 + V_C * 4 - 10 = 0$$

$$\Rightarrow \qquad V_C = 8 \text{ kN}$$

$$V_B = 8 \text{ kN}$$

Consider a section located 'x' distant from A as shown in Figure 7.25.

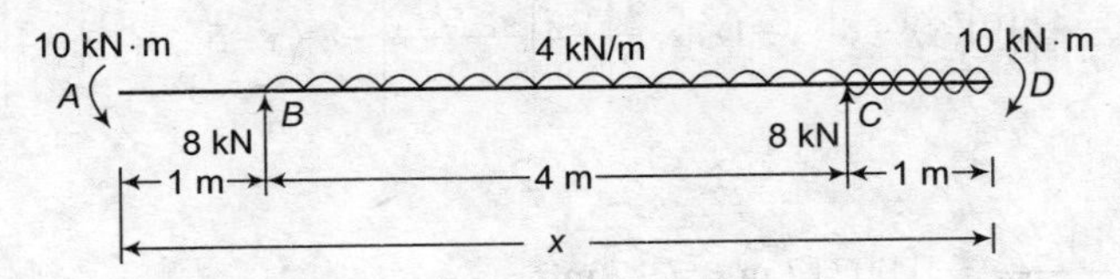

FIGURE 7.25

$$M_x = -10\left|\langle x\rangle^0\right| + 8\left|\langle x-1\rangle\right| - 4\left|\frac{\langle x-1\rangle^2}{2}\right| + 8\left|\langle x-5\rangle\right| + 4\left|\frac{\langle x-5\rangle^2}{2}\right|$$

We know that, $EI\dfrac{d^2y}{dx^2} = -M_x$

$$\therefore \qquad EI\frac{d^2y}{dx^2} = 10\left|\langle x\rangle^0\right| - 8\left|\langle x-1\rangle\right| + 4\left|\frac{\langle x-1\rangle^2}{2}\right| - 8\left|\langle x-5\rangle\right| - 4\left|\frac{\langle x-5\rangle^2}{2}\right|$$

$$EI\frac{dy}{dx} = 10\left|\langle x\rangle\right| - 8\left|\langle x-1\rangle^2 \times \frac{1}{2}\right| + 4\left|\frac{\langle x-1\rangle^3}{6}\right| - 8\left|\langle x-5\rangle^2 \times \frac{1}{2}\right| - 4\left|\frac{\langle x-5\rangle^3}{6}\right| + C_1$$

$$EI\,(y) = 10\left|\langle x\rangle^2 \times \frac{1}{2}\right| - 8\left|\langle x-1\rangle^3 \times \frac{1}{6}\right| + 4\left|\frac{\langle x-1\rangle^4}{24}\right| - 8\left|\langle x-5\rangle^3 \times \frac{1}{6}\right| - 4\left|\frac{\langle x-5\rangle^4}{24}\right| + C_1x + C_2$$

From the boundary condition that,
At $x = 1$, $y = 0$ (Hinged support)

$$\Rightarrow \qquad C_1 + C_2 = -5 \tag{7.7}$$

At $x = 5$, $y = 0$ (Roller support)

$$\Rightarrow \qquad 5C_1 + C_2 = -82.33 \tag{7.8}$$

From Eqs. (1) and (3)

$$4\,C_1 = -77.33$$

$$\Rightarrow \qquad C_1 = -19.333$$

$$\Rightarrow \qquad C_2 = 14.333$$

$$y = \frac{1}{EI}\left\{\begin{aligned} &10\left|\langle x\rangle^2 \times \frac{1}{2}\right| - 8\left|\langle x-1\rangle^3 \times \frac{1}{6}\right| + 4\left|\frac{\langle x-1\rangle^4}{24}\right| \\ &-8\left|\langle x-5\rangle^3 \times \frac{1}{6}\right| - 4\left|\frac{\langle x-5\rangle^4}{24}\right| - 19.333x + 14.333 \end{aligned}\right\}$$

and

$$\frac{dy}{dx} = \frac{1}{EI}\left\{10\left|\langle x\rangle\right| - 8\left|\langle x-1\rangle^2 \times \frac{1}{2}\right| + 4\left|\frac{\langle x-1\rangle^3}{6}\right| - 8\left|\langle x-5\rangle^2 \times \frac{1}{2}\right| - 4\left|\frac{\langle x-5\rangle^3}{6}\right| - 19.333\right\}$$

At $x = 0$, $y = y_A$

$$\therefore \qquad y_A = \frac{14.333}{EI} = 0.00717 \text{ m} = 7.17 \text{ mm}$$

At $x = 3$ (at midspan), $y = y_{\max}$ (from symmetry)

$$\therefore \qquad y_{\max} = \frac{-6.67}{EI} = -0.00333 \text{ m}$$

Deflection at midspan = 3.33 mm upward. From symmetry of loading and beam, it can be concluded that, $Y_A = Y_D$.

Slope at A, i.e., at $x = 0$, $\frac{dy}{dx} = \theta_A$

$$\therefore \qquad \theta_A = -\frac{19.333}{EI} = -0.00967 \text{ radians}$$

At $x = 1$, $\qquad \frac{dy}{dx} = \theta_B$

$$\therefore \qquad \theta_B = -\frac{9.333}{EI} = -0.000467 \text{ radians}$$

A cross check at $x = 3$; $\frac{dy}{dx} = 0$; owing the symmetry,

At $x = 3$, $\qquad \frac{dy}{dx} = 0$

Deflection profile of the beam is shown in Figure 7.26.

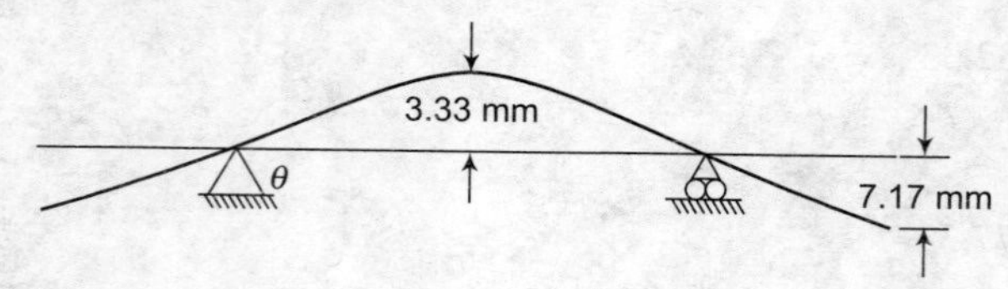

FIGURE 7.26

PROBLEM 7.8 **Objective 2**

Determine the maximum deflection of the beam subjected to loading as shown in Figure 7.27. What are called rigid body deformation? Explain with the help of the solution of this problem.

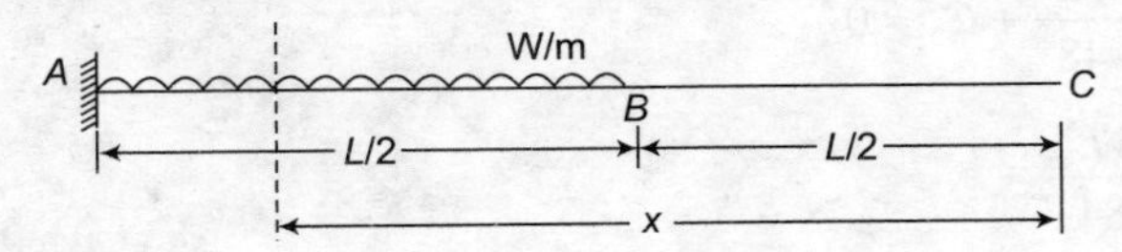

FIGURE 7.27

SOLUTION

Take 'C' as origin. Consider a section located 'x' distant from C.

$$M_x = -W\left|\frac{\left\langle x-\frac{L}{2}\right\rangle^2}{2}\right|$$

We know that, $EI\dfrac{d^2y}{dx^2} = -M_x$

$$EI\frac{d^2y}{dx^2} = W\left|\frac{\left\langle x-\frac{L}{2}\right\rangle^2}{2}\right|$$

$$EI\frac{dy}{dx} = W\left|\frac{\left\langle x-\frac{L}{2}\right\rangle^3}{6}\right| + C_1$$

$$EI(y) = W\left|\frac{\left\langle x-\frac{L}{2}\right\rangle^4}{24}\right| + C_1x + C_2$$

From the boundary condition,

At $x = L$, $y = 0$ and $x = L$, $\dfrac{dy}{dx} = 0$

At $x = L$, $\quad \dfrac{dy}{dx} = 0$

$$\Rightarrow \quad 0 = \frac{WL^3}{48} + C_1$$

$$\Rightarrow \qquad C_1 = -\frac{WL^3}{48}$$

At $x = L$, $y = 0$

$$\Rightarrow \qquad \frac{WL^4}{384} - \frac{WL^4}{48} + C_2 = 0$$

$$\Rightarrow \qquad C_2 = \frac{7WL^4}{384}$$

$$\therefore \qquad y = \frac{1}{EI}\left\{W\left|\frac{\left\langle x - \frac{L}{2}\right\rangle^4}{24}\right| - \frac{WL^3}{48}x + \frac{7WL^4}{384}\right\}$$

and

$$\frac{dy}{dx} = \frac{1}{EI}\left\{W\left|\frac{\left\langle x - \frac{L}{2}\right\rangle^3}{12}\right| - \frac{WL^3}{48}\right\}$$

At $x = L/2$, $y = y_B$, and $\frac{dy}{dx} = \theta_B$

$$y_B = \frac{WL^4}{128EI} \quad \text{and} \quad \theta_B = -\frac{WL^3}{48EI}$$

At $x = 0$, $y = y_C$, and $\frac{dy}{dx} = \theta_C$

$$y_C = \frac{7WL^4}{384EI} \quad \text{and} \quad \theta_C = -\frac{WL^3}{48EI}$$

From the above value, it is clear that slope at B and slope at C are same. It implies that slope between B and C is constant. The deflection profile of the beam was shown in Figure 7.28.

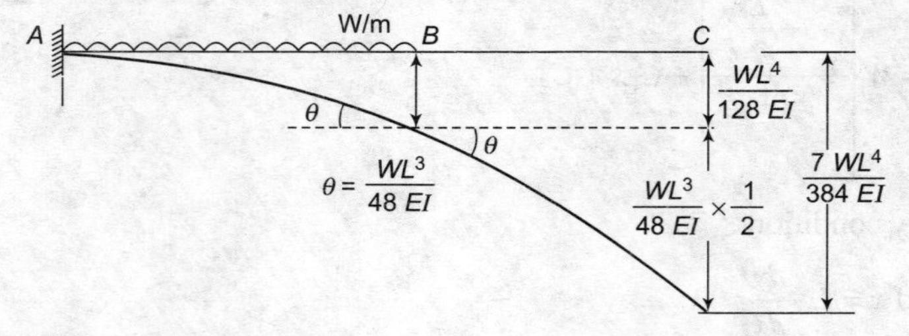

FIGURE 7.28

In the region BC, deflection increases (changes) with constant slope. Thus, the deformations in this region are called rigid body deformations. The BM and SF in the region BC is zero. The stress developed due to BM and SF is zero. Hence, the region BC is not subjected to any stress but undergoes deformation.

7.4 MOMENT AREA METHOD

Generally, this method is useful in finding deflections or slopes at particular locations and nonprismatic beams. This method takes the advantage of BM diagrams (BMDs). Consider a beam shown in Figure 7.29 subjected to arbitrary loading. Let EI be the flexural rigidity of the beam. The BMD and elastic curve of the beam were also presented.

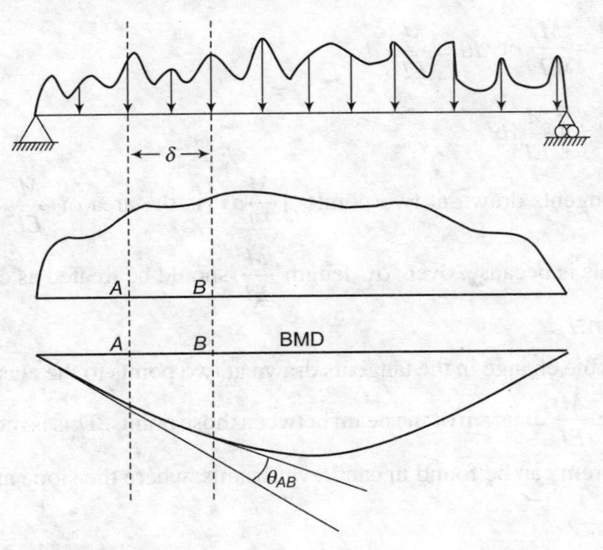

FIGURE 7.29 Elastic curve.

Consider two sections (1)-(1) and (2)-(2) separated by a distance 'δx' as shown in Figure 7.30. The exaggerated profile of 'δx' portion is shown in Figure 7.30.

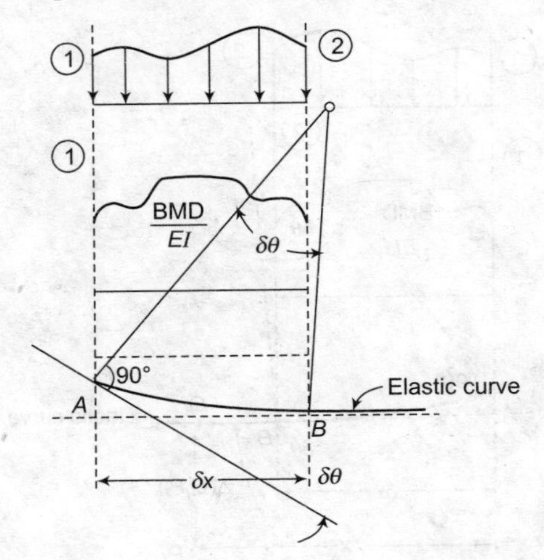

FIGURE 7.30

OA and *OB* are the normals drawn to the tangents drawn at *A* and *B* to the elastic curve at sections (1)-(1) and (2)-(2), respectively. Let '*R*' be the radius of the curvature.

We know that $\dfrac{1}{R} = \dfrac{M}{EI}$ and also, $\dfrac{d\theta}{ds} = \dfrac{1}{R}$

as δx is very small, $ds \cong dx$

$$\therefore \qquad \frac{d\theta}{dx} = \frac{M}{EI} \text{ or } d\theta = \frac{M}{EI} \cdot dx$$

$$\Rightarrow \qquad \theta = \int \frac{M}{EI} dx$$

θ is the change of tangents drawn at two points $\int \frac{M}{EI} dx$ is the area of $\frac{M}{EI}$ diagram between the points (1) and (2). This is because, over 'δx' length $\frac{M}{EI}$ should be treated as constant.

Moment Area Theorem 1

Thus, we can say that the change in the tangents drawn at two points to the elastic curve of a beam is equal to the area of the $\frac{M}{EI}$ diagram of the beam between those points. This is moment area theorem 1.

Utility of this theorem can be found in cantilever beams, where the slope at fixed end is zero.

Moment Area Theorem 2

The vertical intercept between two tangents, drawn at points P and Q to the elastic curve, on a given line is the moment area of $\frac{M}{EI}$ diagram between the points P and Q about the same line.

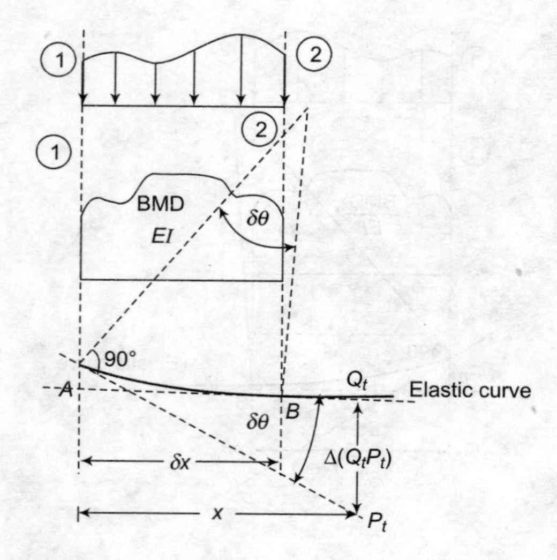

FIGURE 7.31

Vertical intercept $\Delta(Q_tP_t) = x\delta\theta$

$$\Delta(Q_tP_t) = x\frac{M}{EI}dx$$

$$\Delta(Q_tP_t) = \text{Element vertical intercept}$$

$$\text{Vertical intercept} = \int\frac{M}{EI}\cdot x \times dx$$

$$= \text{moment of } \frac{M}{EI} \text{ diagram about the reference line}$$

If the deflection profile of the beam is not known, proper sign convention is to be adopted to use these theorems.

Sign convention Sign convention is very important, if moment area method is adopted for analysis.

A positive value for change in slope θ_{AB} is the tangent at right side; point B is measured in anticlockwise direction.

A negative value for change in slope θ_{AB} is the tangent at right side, and point B is measured in clockwise direction.

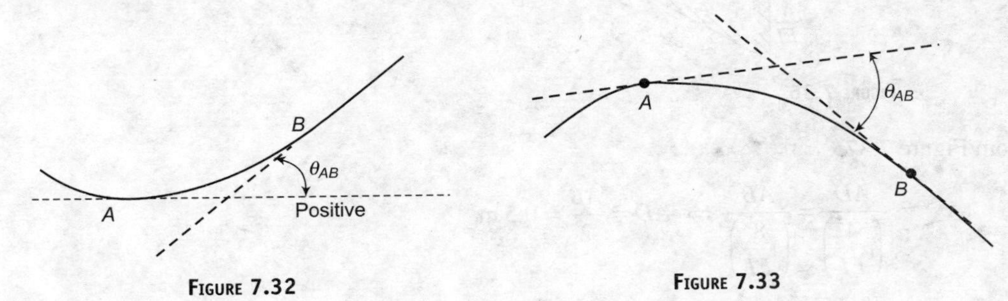

FIGURE 7.32

FIGURE 7.33

The vertical intercept at any point is positive if the point lies above the reference tangent from which intercept measured.

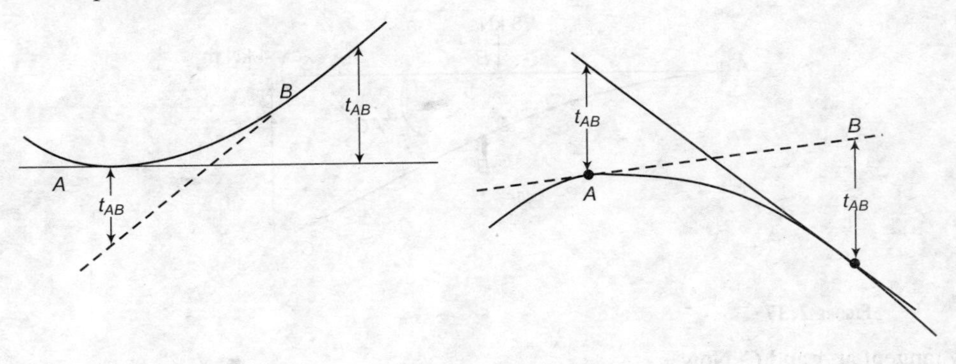

FIGURE 7.34

PROBLEM 7.9

Objective 3

Determine the slope and deflection of the cantilever beam loaded as shown in Figure 7.35. Take $EI = 1 \times 10^4$ kN-m^2.

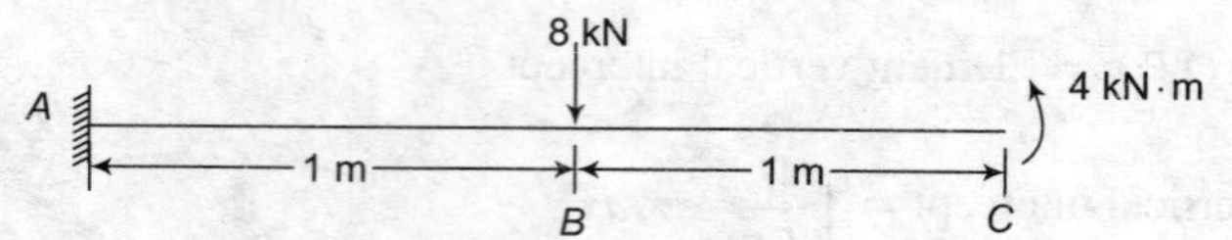

FIGURE 7.35

SOLUTION

To solve this problem, using moment area method, draw BMD and then $\frac{M}{EI}$ diagram.

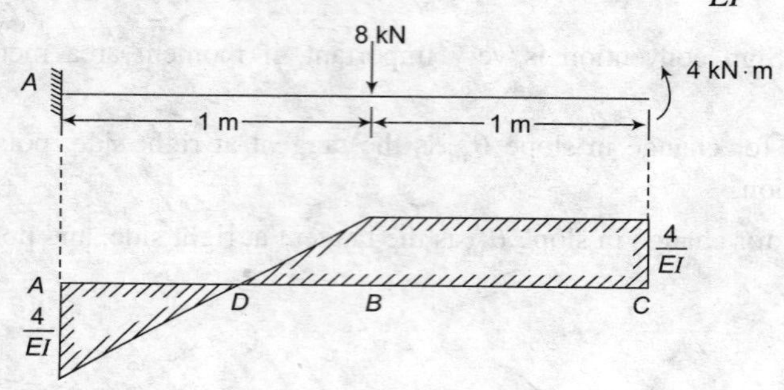

FIGURE 7.36

From Figure 7.37.

$$\frac{AD}{\left(\frac{4}{EI}\right)} = \frac{AB}{\left(\frac{8}{EI}\right)} \Rightarrow AD = \frac{AB}{2} = 0.5 \text{ m}$$

It is not easy to assume the correct deflection profile of the beam; thus, assume the deflection profile as shown in Figure 7.37. Negative value of the deflections indicated that the assume profile is wrong.

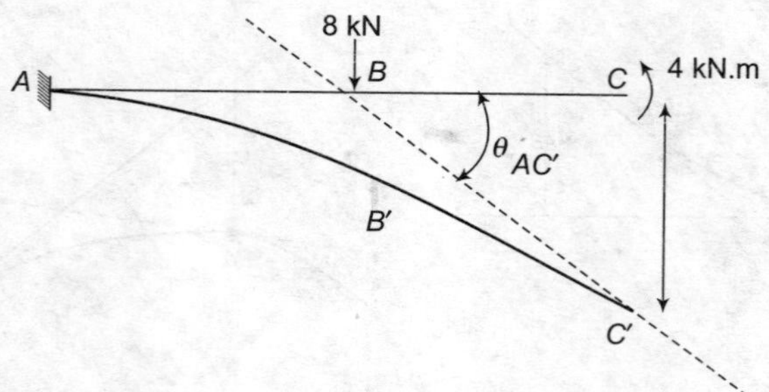

FIGURE 7.37

Draw tangent at A and C. Now

$\theta_{AC'}$ = change of slope measured from tangent at A to tangent at C (right side), thus negative.

$$= -\left\{\frac{M}{EI} \text{ diagram between A and C}\right\}$$

$$= -\frac{1}{EI}\left\{4\times 1 + 4\times\frac{1}{2}\times 0.5 - 4\times 0.5\times\frac{1}{2}\right\} = \frac{4}{EI}$$

As $\theta_{AC'}$ is –ve , the slope at C should be in the other direction.

CC' = vertical intercept between the tangents drawn at A and C, at C.

As the right side, point C is below the tangent drawn at A, the sign should be negative.

$$CC' = -\left\{\text{moment of } \frac{M}{EI} \text{ diagram between A and C, about C}\right\}$$

$$= \frac{1}{EI}\left\{(4\times 1\times 0.5) + \left(4\times 0.5\times 0.5\times\left[1+\frac{0.5}{3}\right]\right)\left[4\times 0.5\times 0.5\left[2-\frac{0.5}{3}\right]\right]\right\}$$

$$= \frac{1}{EI}\{2 + 1.1667 - 1.833\}$$

$$= -\frac{1.333}{EI}$$

As CC' is negative, which is the vertical deflection at C, the assumed deflection profile is wrong; hence, deflection at 'C' should be upward. Let us consider the deflection at D.

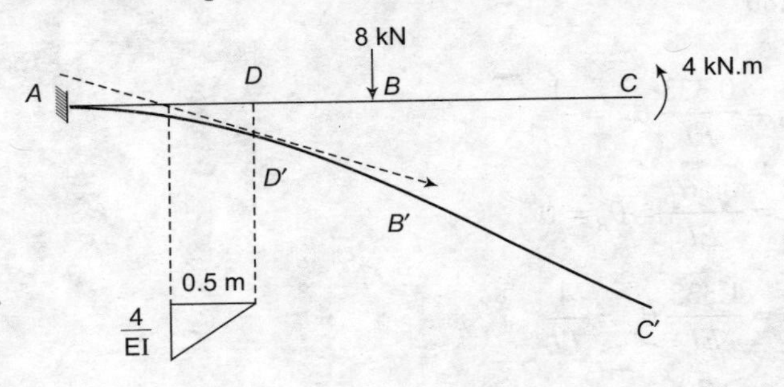

FIGURE 7.38

y_D = deflection at $D = DD'$

DD' = vertical intercept between the tangent drawn at A and D, at D

$$= \text{moment of } \frac{M}{EI} \text{ diagram between } A \text{ and } D\text{, about } D$$

$$DD' = -\left\{\frac{-4}{EI}\times\frac{1}{2}\times 0.5\times\frac{2}{3}\times 0.5\right\} = \frac{0.333}{EI}.$$

As DD' is +ve, the assumed deflection is correct. Here, the point to be understood is that the deflection assumed at 'D' (downward) is correct and the deflection assumed at 'C' (downward) is wrong.

For deflection at B, take moment of $\frac{M}{EI}$ diagram about E between A and B.

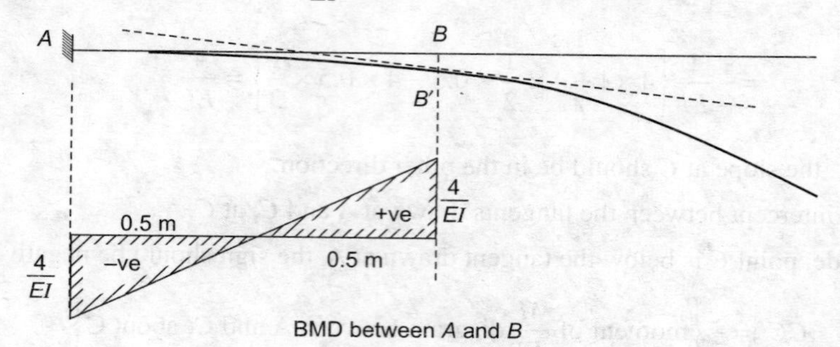

FIGURE 7.39

$$BB' = Y_B = -\left\{\frac{4}{EI}\times\frac{1}{2}\times 0.5\times\frac{0.5}{3} - \frac{4}{EI}\times\frac{1}{2}\times 0.5\left(1-\frac{0.5}{3}\right)\right\} = \frac{0.667}{EI}$$

⇒ Assumed direction of deflection is correct.

∴ The correct deflection profile of the beam is shown in Figure 7.40.

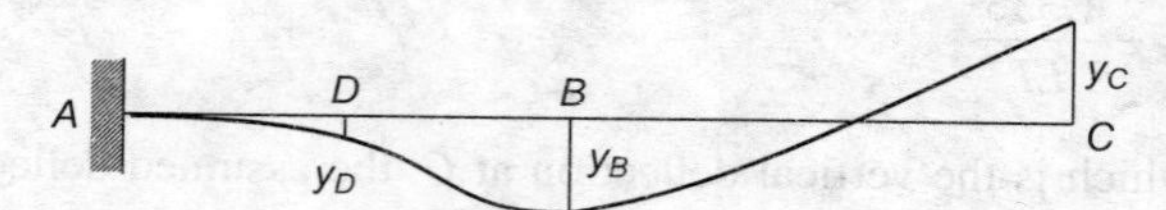

FIGURE 7.40

$$y_D = \frac{0.333}{EI}\quad \theta_D = \frac{1}{EI}$$

$$y_B = \frac{0.667}{EI} = \theta_B = 0$$

$$y_C = \frac{1.333}{EI}\quad \theta_C = \frac{4}{EI}$$

PROBLEM 7.10

Objective 3

Determine the deflection at the tip of a cantilever beam shown in Figure 7.41. Use moment area theorem.

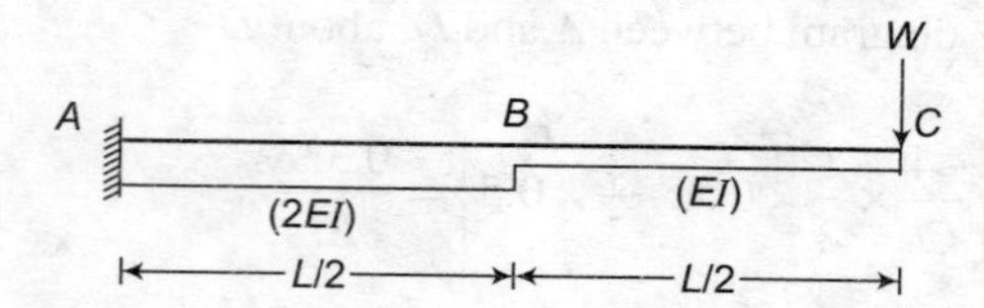

FIGURE 7.41

SOLUTION

The BMD and $\frac{M}{EI}$ diagram of the beam shown in Figure 7.42.

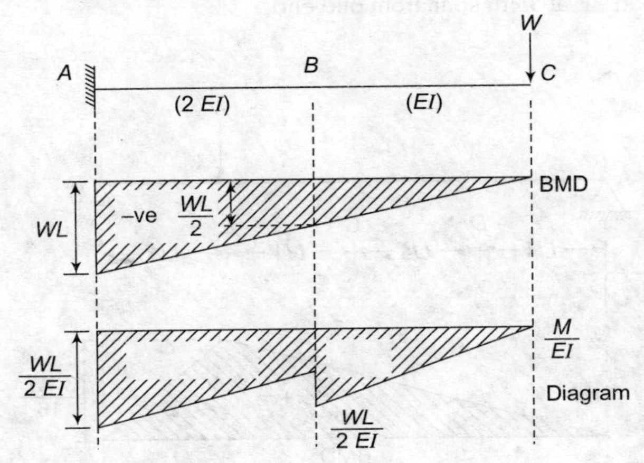

FIGURE 7.42

For the deflection at the tip (y_C), draw tangent at A and C. Tangent at A is horizontal because the slope at A is zero being fixed. Measure the vertical intercept between the tangent drawn, at A and C, at C.

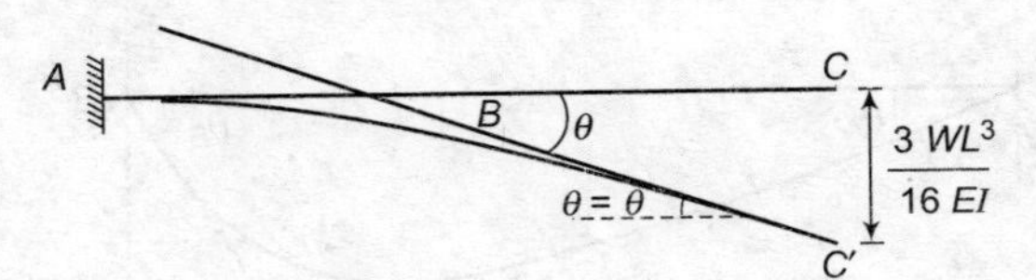

FIGURE 7.43

$CC' = -$(moment of M/EI diagram between the points A and C, about C)

$$= -\left\{-\left[\frac{1}{2}\times\frac{WL}{2EI}\times\frac{L}{2}\times\frac{2}{3}\times\frac{L}{2}+\frac{WL}{4EI}\times\frac{L}{2}\times\left(L-\frac{L}{4}\right)+\frac{WL}{4EI}\times\frac{L}{2}\times\frac{1}{2}\times\left(L-\frac{1}{3}\times\frac{L}{2}\right)\right]\right\}$$

$$= \frac{3WL^3}{16EI}$$

To find at C, measure the deviation the tangents drawn at A and C.

$$\theta_{AC} = \theta_C = -\left\{-\left(\frac{1}{2}\times\frac{L}{2}\times\frac{WL}{2EI}+\frac{WL}{4EI}\times\frac{L}{2}+\frac{WL}{4EI}\times\frac{L}{2}\times\frac{1}{2}\right)\right\}$$

$$= \frac{5WL^2}{16EI}$$

PROBLEM 7.11

Objective 3

Determine the deflection at 1/4th span, midspan and end slopes of a simply supported beam, subjected to point load '*W*' at 3/4th span from one end.

SOLUTION

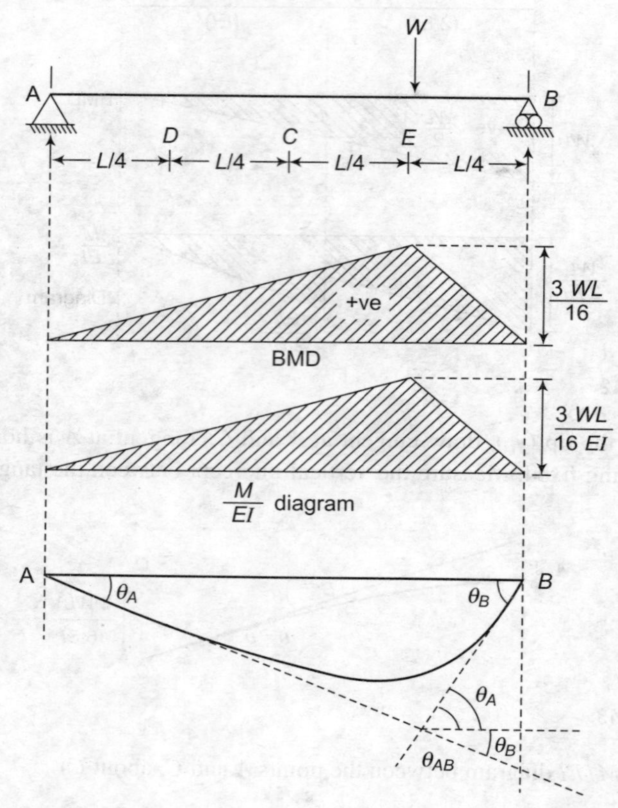

FIGURE 7.44

θ_{AB} = change of slope measured format tangent at 'A' to the tangent at *B*. Point '*B*' is on right side direction of slope is anticlockwise (Thus, the sign is +ve).

= Area of $\dfrac{M}{EI}$ diagram between points A and B

$$\theta_{AB} = \left\{\frac{3WL}{16EI} \times \frac{1}{2} \times L\right\} = \frac{3WL^2}{32EI}$$

$$\Rightarrow \quad \theta_A + \theta_{AB} = \frac{3WL^2}{32EI}$$

In case of simply supported, where point of zero slope location is difficult, second moment area theorem shall be used to determine the slope at *A* and *B*.

BB' = vertical intercept between the tangents drawn at A and B at B. The intercept BB' is above the tangent AB'. Thus, it is +ve.

$$BB' = \text{moment of } \frac{M}{EI} \text{ diagram about } B.$$

$$= \left\{\frac{3WL}{16EI} \times \frac{1}{2} L \left(\frac{L}{4} + L\right)\frac{1}{3}\right\} = \frac{5WL^3}{128EI}$$

From $AB'B$,

$$\tan\theta_A = \frac{BB'}{AB} \text{ as } \theta_A \text{ is very small.}$$

$$\tan\theta_A = \theta_A = \frac{\frac{5WL^3}{128EI}}{L} = \frac{5WL^2}{128EI}$$

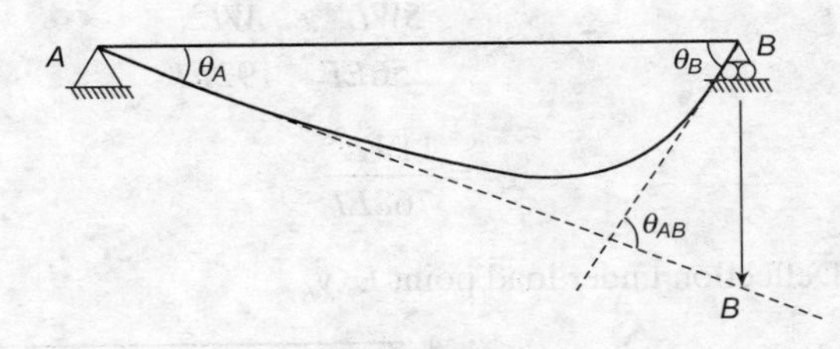

FIGURE 7.45

$$\theta_A = \frac{5WL^2}{128EI}$$

$$\Rightarrow \qquad \theta_B = \theta_{AB} - \theta_A$$

$$= \frac{3WL^2}{32EI} - \frac{5WL^2}{128EI} = \frac{7WL^2}{128EI}$$

Deflection at 1/4th span from left support A: Consider the deflection profile between A and D. Draw tangents at A and D.

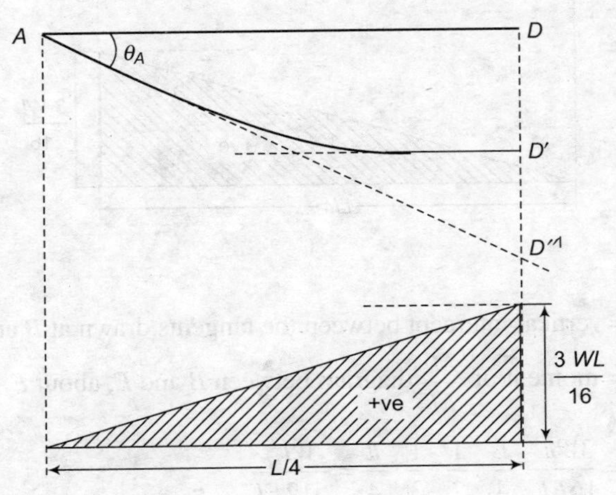

FIGURE 7.46

DD' = vertical intercept between drawn at A and D, at D

$$= \text{Moment of } \frac{M}{EI} \text{ diagram between points } A \text{ and } D \text{ about } D.$$

$$= \frac{WL}{16EI} \times \frac{L}{4} \times \frac{1}{2}\left\{\frac{1}{3} \times \frac{1}{4}\right\} = \frac{WL^3}{1536EI}$$

From $\Delta ADD'$, $DD' = AD \times \tan \theta_A = \frac{L}{4} \times \frac{5WL^2}{128EI} = \frac{5WL^3}{512EI}$

The deflection y_D at $D = DD'' = DD' - D'D''$

$$= \frac{5WL^3}{512EI} - \frac{WL^3}{1536EI} = \frac{5WL^3}{256EI}$$

Deflection at C, $y_C = CC' - C'C''$

$$y_c = \frac{5WL^3}{256EI} - \frac{WL^3}{192EI}$$

$$y_C = \frac{11WL^3}{768EI}$$

Deflection under load point E, y_E

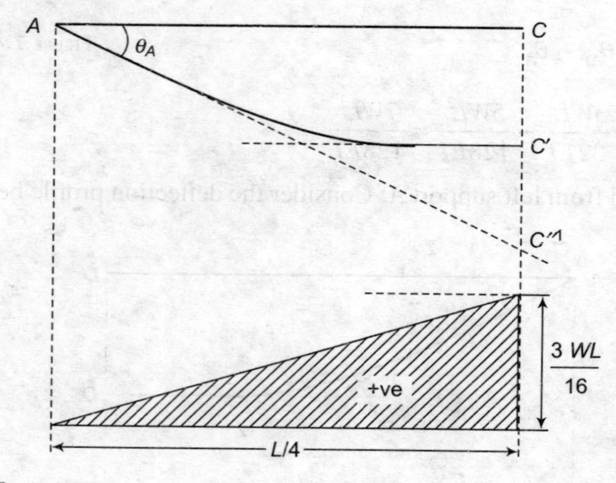

FIGURE 7.47

$E'E''$ = vertical intercept between the tangents drawn at B and E, at E

= moment of $\frac{M}{EI}$ diagram between B and E, about E

$$= \frac{3WL}{16EI} * \frac{L}{4} * \frac{1}{2} * \frac{1}{3} * \frac{L}{4} = \frac{WL^3}{512EI}$$

From $\Delta ACC'$, $CC' = \frac{L}{2} \tan \theta_A$

$$\frac{L}{2} \times \frac{5WL^2}{128EI}$$

Deflection at C, $y_C = CC' - C'C''$

$$y_C = \frac{5WL^3}{256EI} - \frac{WL^3}{192EI}$$

$$y_C = \frac{11WL^3}{768EI}$$

Deflection under load point E, Y_E:

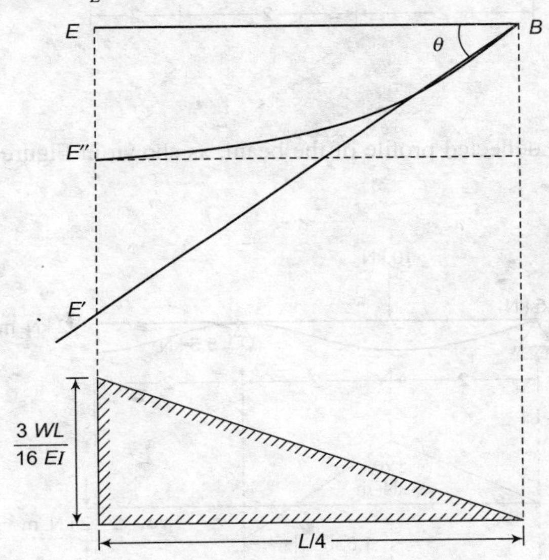

FIGURE 7.48

$E'E''$ = vertical intercept between the tangents drawn at B and E, at E.

$= \text{moment of } \dfrac{M}{EI}$ diagram between B and E about E.

$$= \frac{3WL}{16EI} \times \frac{L}{4} \times \frac{1}{2} \times \frac{1}{3} \times \frac{L}{4}$$

$$= \frac{WL3}{512EI}$$

$$EE'' = \frac{L}{4} \times \tan\theta_B$$

$$= \frac{L}{4} \times \frac{7WL^2}{128EI} = \frac{7WL^3}{512EI}$$

$$Y_E = EE'' = EE' - E'E''$$

$$= \frac{7WL^3}{512EI} - \frac{WL^3}{512EI} = \frac{3WL^3}{256EI}$$

PROBLEM 7.12

Objective 3

Determine the deflection at mid span and tip of the overhang of the beam loaded as shown in Figure 7.49, using moment area method. Take $EI = 10^4$ kN·m^2

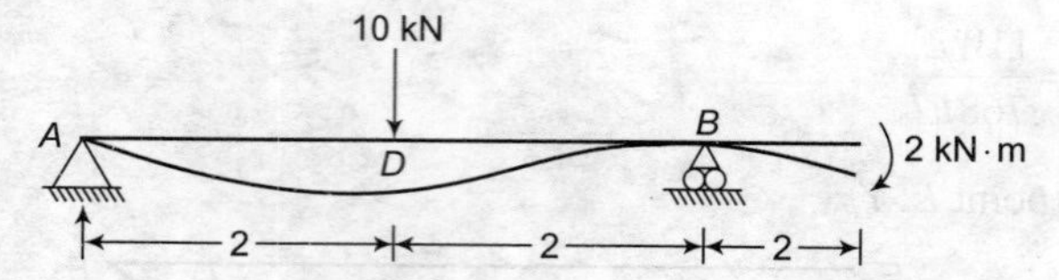

FIGURE 7.49

SOLUTION

Sketch the approximate, deflected profile of the beam, as shown in Figure 7.49. Sketch the BMD and $\frac{M}{EI}$ diagram.

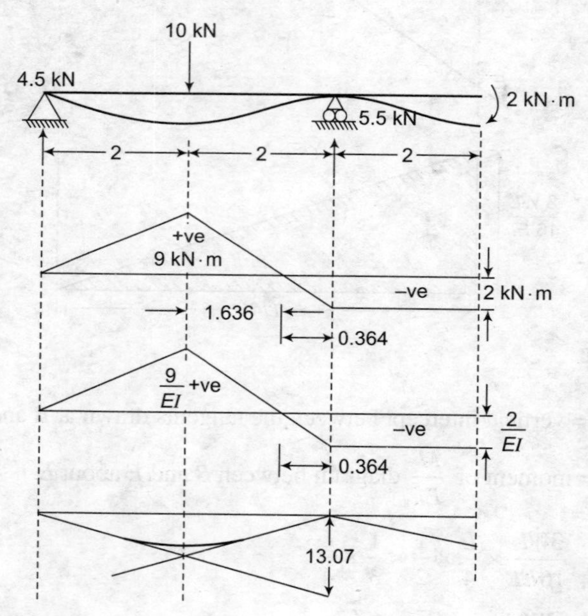

FIGURE 7.50

From Figure 7.49,

BB' = vertical intercept between the tangents drawn at A and B, at B

= moment of $\frac{M}{EI}$ diagram between A and B, about B

$$= \left\{\frac{1}{2}\times\frac{9}{EI}\times 3.636\times\left(0.364+\frac{1.636+3.636}{3}\right)-\frac{1}{2}\times\frac{2}{EI}\times 0.364\times\frac{0.364}{3}\right\}=\frac{34.665}{EI}$$

From $\Delta AB'B$,

$$\tan\theta_A = \frac{BB'}{AB} = \frac{\frac{34.665}{EI}}{4} = \frac{8.666}{EI} \text{ radians}$$

θ_{AB} = change of slope between the tangents drawn at A and B

$$= \text{area of } \frac{M}{EI} \text{ diagram between the points } A \text{ and } B$$

$$= \left\{\frac{9}{EI}\times\frac{1}{2}\times 3.636 - \frac{1}{2}\times 0.364\times\frac{2}{EI}\right\} = \frac{15.998}{EI}$$

$$\theta_{AB} = \theta_A + \theta_B \Rightarrow \theta_B = \frac{7.332}{EI}$$

Deflection at D:

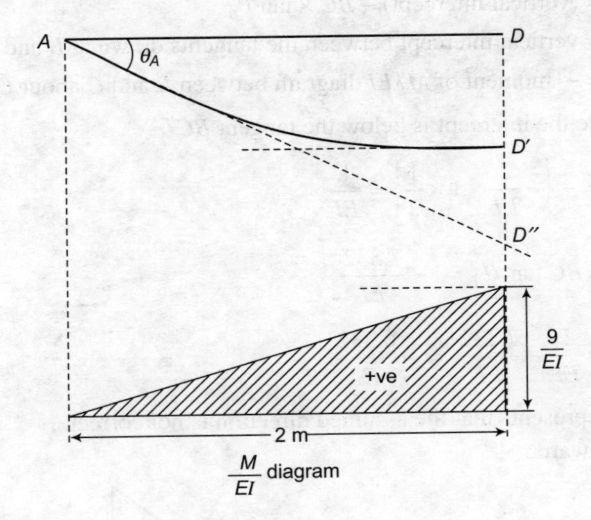

FIGURE 7.51

$D'D''$ = vertical intercept between the tangents drawn at A and D, at D

$$= \text{moment of } \frac{M}{EI} \text{ diagram between } A \text{ and } D \text{ about } D$$

$$= \frac{9}{EI}\times 2\times\frac{1}{2}\times\frac{2}{3} = \frac{6}{EI}$$

$$DD'' = AD \tan\theta_A$$

$$= 2\times\frac{8.666}{EI} = \frac{17.332}{EI}$$

Deflection at D, $Y_D = DD'' - D'D''$

$$= \frac{11.332}{EI}$$

Deflection at C: Consider the portion BC of the beam

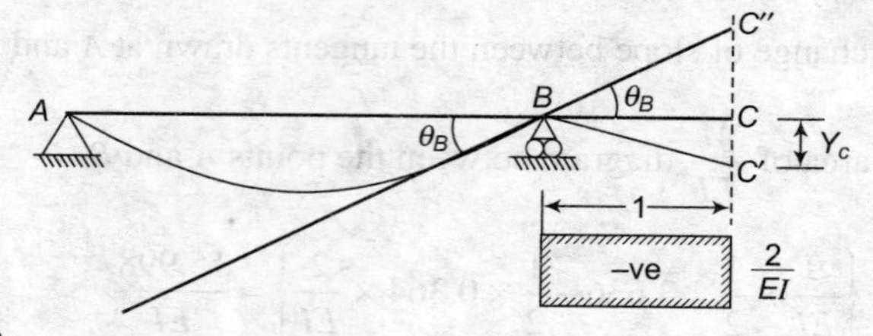

FIGURE 7.52

From $\Delta BCC''$ and $\Delta BCC'$

$$y_C = C''C' - CC''$$

$$= \text{(vertical intercept)} - BC \times \tan\theta_B$$

$\therefore$ $C''C'$ = vertical intercept between the tangents drawn at B and C', at C

$$= -\{\text{moment of } M/EI \text{ diagram between } B \text{ and } C \text{ about } C\}$$

The –ve sign is because the intercept is below the tangent BC''

$$= -\left\{-\frac{2}{EI} \times 1 \times \frac{1}{2}\right\} = \frac{1}{EI}$$

$$CC'' = BC \tan\theta_B = 1 \times \frac{7.332}{EI}$$

$$\therefore \quad y_C = \frac{1}{EI} - \frac{7.332}{EI} = -\frac{6.332}{EI}$$

Negative value of y_C represents that the assumed direction is not correct.

$\therefore$ deflection at C is upward.

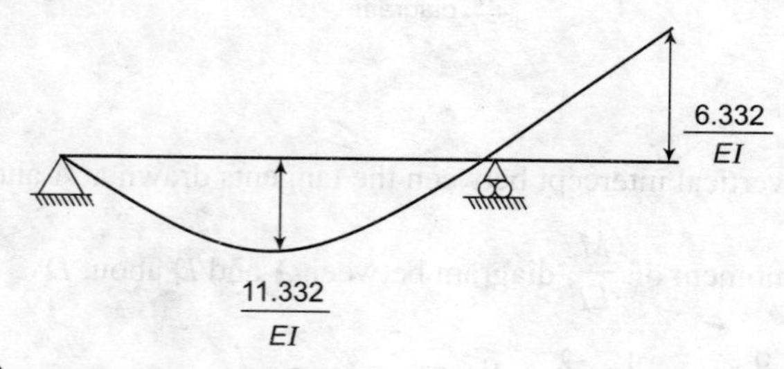

FIGURE 7.53

$$\therefore \quad y_C = \frac{6.332}{10^4} \times 10^3 = 0.6332 \text{ mm}$$

$$Y_D = \frac{11.332}{10^4} \times 10^3 = 1.133 \text{ mm}$$

PROBLEM 7.13 **Objective 3**

Determine the midspan deflection of simply supported beam showing Figure 7.54. BMD and M/EI diagrams were also given.

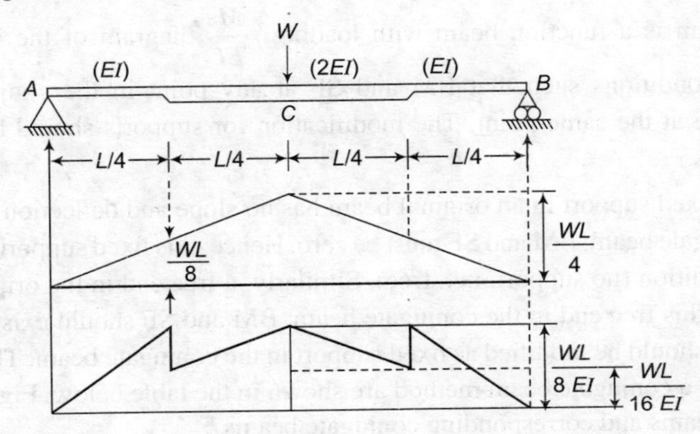

FIGURE 7.54

SOLUTION

Consider the elastic curve of the beam as the loading is symmetric and beam is symmetric; the slope at mid-point is zero, $\theta_A = \theta_B$.

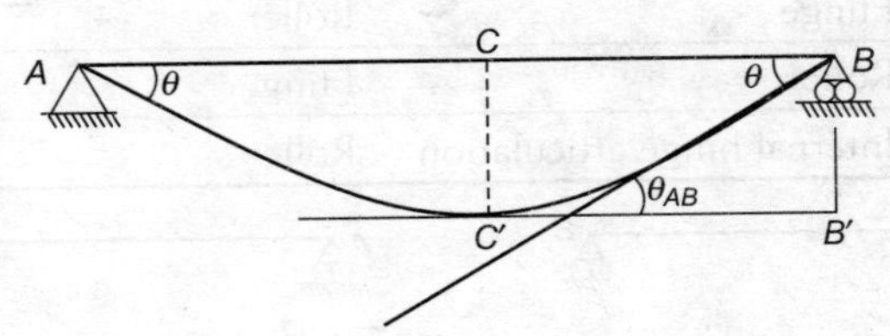

FIGURE 7.55

As slope at C' is zero, $C'B'$ (tangent at C') is a horizontal line, and BB', the vertical intercept between C and B at B, is equal to CC' (Y_C)

$$Y_C = BB'$$

$$= \text{moment of } \frac{M}{EI} \text{ diagram between } B \text{ and } C \text{ about } B$$

$$= \left\{\frac{WL}{8EI}\times\frac{L}{4}\times\frac{1}{2}\times\frac{2}{3}\times\frac{L}{4}+\frac{WL}{16EI}\times\frac{L}{4}\left(\frac{L}{4}+\frac{L}{8}\right)+\frac{WL}{16EI}\times\frac{L}{4}\times\frac{1}{2}\left(\frac{L}{2}-\frac{1}{3}\times\frac{L}{4}\right)\right\}=\frac{3WL^3}{256EI}$$

Slope at B, $\quad \theta_B = \theta_{CB} = \text{area of } \dfrac{M}{EI} \text{ diagram between } B \text{ and } C$

$$= \left\{\frac{WL}{8EI}\times\frac{1}{2}\times\left(\frac{L}{4}+\frac{L}{2}\right)-\frac{WL}{16EI}\times\frac{L}{4}\times\frac{1}{2}\right\}=\frac{5WL^2}{128EI}$$

7.5 CONJUGATE BEAM METHOD

This is other method of evaluating deflections of beams, where simple BM and estimate force calculations are enough to estimate deflections and slopes.

A conjugate beam is a function beam with loading, $\frac{M}{EI}$ diagram of the original beam, and modified support conditions such that BM and SF at any point in the conjugate beam yields deflection and slope at the same point. The modification for supports should be in tune with the boundary conditions.

For example, a fixed support in an original beam has no slope and deflection. Thus, at this fixed support in the conjugate beam, BM and SF must be zero. Hence, this fixed support should be replaced by support free condition (no support, i.e., free). Similarly, a free end in the original has deflection and slope. Thus, at this free end in the conjugate beam, BM and SF should exist. Hence a free end in an original beam should be modified as fixed support in the conjugate beam. The modifications of different supports in a conjugate beam method are shown in the table below. Figure 7.55 shows the sketch of original beams and corresponding conjugate beams.

Original Beam	Conjugate Beam
Fixed	Free
Free	Fixed
Hinge	Roller
Roller	Hinge
Internal hinge/articulation	Roller

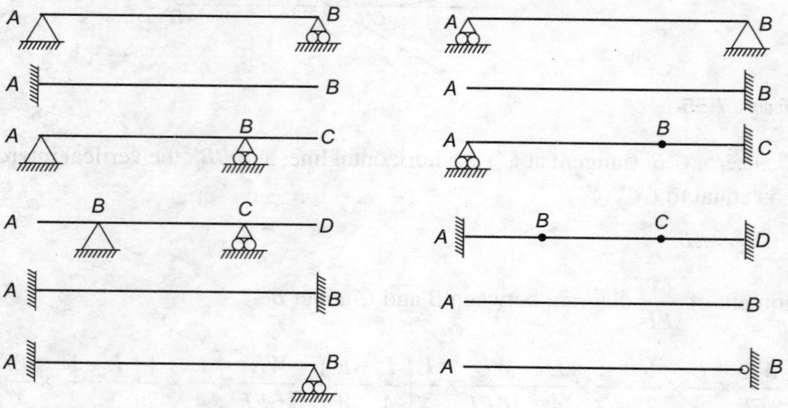

FIGURE 7.55

Sign Convention Positive BMD of the original beam will be taken as downward loading in the conjugate beam. Similarly, negative BMD is taken as upward loading in the conjugate beam.

This method is mostly used to estimate deflections at required locations and when the flexural rigidity of the beam varies, that is, for nonprismatic members.

PROBLEM 7.14 **Objective 4**

Determine the deflection and slope at the free end of a cantilever subjected to point load at the free end. Also find the deflection at ½ span.

SOLUTION

Sketch the BMD and $\frac{M}{EI}$ diagram for the cantilever as shown in Figure 7.56.

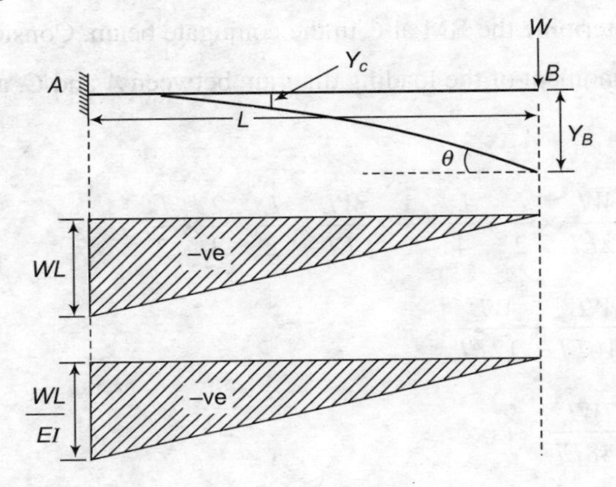

FIGURE 7.56

The conjugate beam with loading is shown in Figure 7.57. Negative BMD of the original beam yielded upward loading in the conjugate beam.

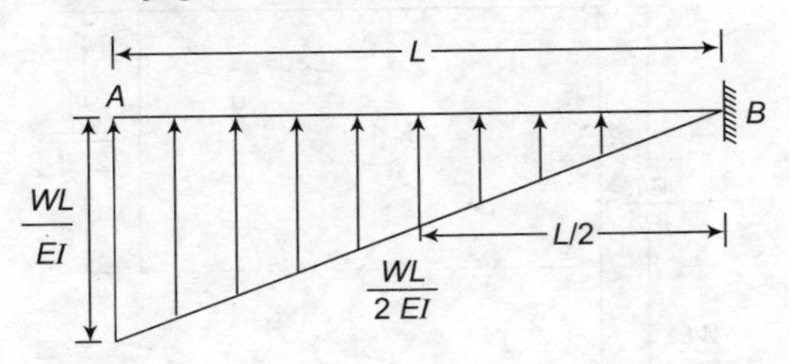

FIGURE 7.57

For the cantilever shown in Figure 7.57, SF at B is

$$V_B = \text{area of loading diagram}$$

$$= \frac{WL}{EI} \times \frac{1}{2} \times L = \frac{WL^2}{2EI}$$

$$\therefore \qquad \theta_B = \frac{WL^2}{2EI}$$

Beam moment at B, in the conjugate beam

$$M_B = \text{moment of area of loading diagram about } B$$

$$= \frac{WL}{EI} \times \frac{1}{2} \times L \times \frac{2L}{3}$$

$$= \frac{WL^3}{3EI}$$

$\therefore$ Deflection at B in the original beam $Y_B = \frac{WL^3}{3EI}$

For deflection at 'C', determine the BM at C in the conjugate beam. Consider the left side portion.

$$M_C = \text{moment of the loading diagram between } A \text{ and } C\text{, about } C.$$

$$= A_1 x_1 + A_2 x_2$$

$$= \frac{WL}{2EI} \times \frac{L}{2} \times \frac{L}{4} + \frac{1}{2} \times \frac{WL}{EI} \times \frac{L}{2} \times \frac{2}{3} \times \frac{L}{2}$$

$$= \frac{WL^3}{16EI} + \frac{WL^3}{12EI}$$

$$= \frac{7WL^3}{48EI}$$

Deflection at C, $Y_C = \frac{7WL^3}{48EI}$

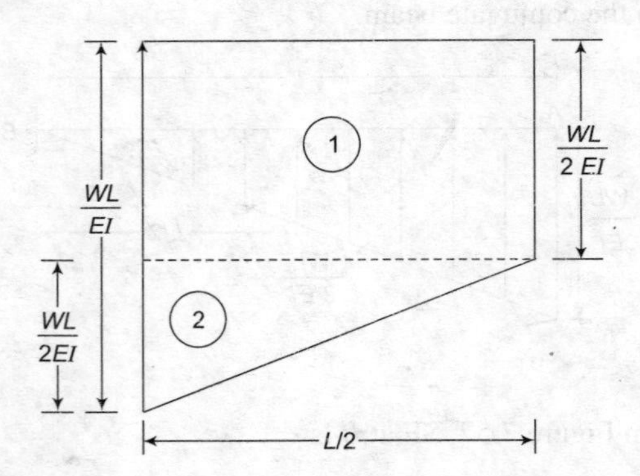

FIGURE 7.58

PROBLEM 7.15 **Objective 4**

Determine the deflection at midspan, and at the overhanging tips of the beam shown in Figure 7.59. Use the conjugate beam method.

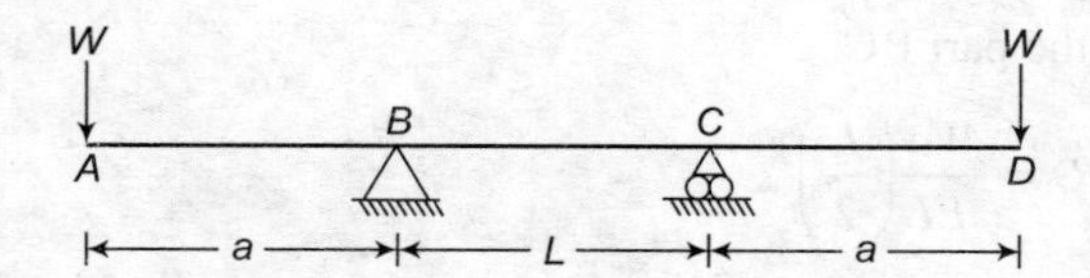

FIGURE 7.59

SOLUTION

Sketch the BMD and $\dfrac{M}{EI}$ diagram of the beam.

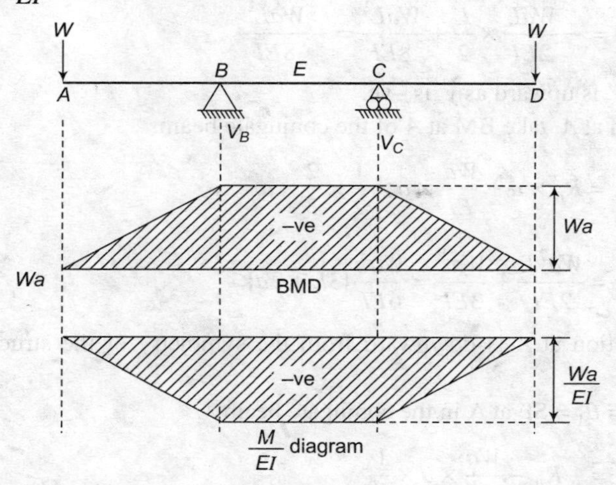

FIGURE 7.60

Figure 7.61 shows the conjugate beam with loading as $\dfrac{M}{EI}$ diagram. End A and D free so they should be fixed. B and C are intermediate supports thus they should be made internal hinges.

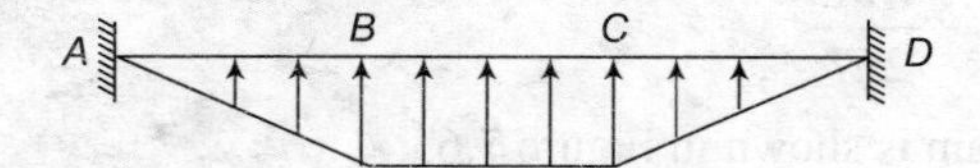

FIGURE 7.61

Now the conjugate beam is an articulated beam, about which, we have ahead discussed in Chapter 3. Device the structure into three parts, transferring the reaction at the articulation to the adjacent parts.

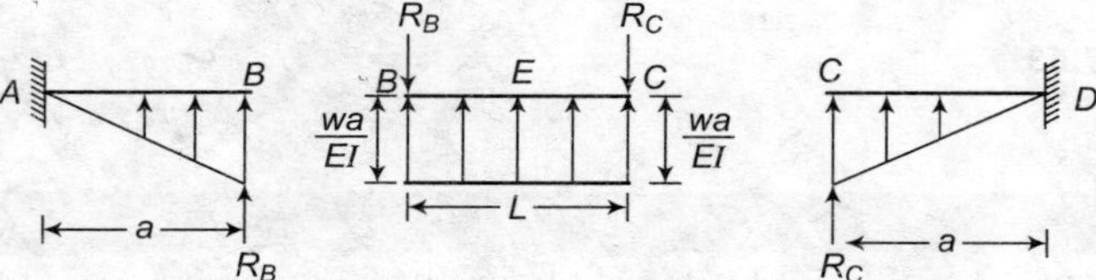

FIGURE 7.62

Take the equilibrium of the part BC.

$$R_B = R_C = \frac{Wa}{EI}\left(\frac{L}{2}\right)$$

∴ To obtain the deflection at the midspan of BC, that is, *E*.

$$y_e = \text{BM at } `E\text{' in the conjugate beam}$$

$$= -R_C\frac{L}{2} + \frac{Wa}{EI} \times \frac{L}{2} \times \frac{L}{4}$$

$$= -\frac{WaL}{2EI} \times \frac{L}{2} + \frac{WaL^2}{8EI} = -\frac{WaL^2}{8EI}$$

⇒ Deflection at '*E*' is upward as y_e is –ve.

To find the deflection at *A*, take BM at *A* of the conjugate beam.

$$y_A = R_B \times a + \frac{Wa}{EI} \times a \times \frac{1}{2} \times \frac{2}{3} \times a$$

$$= \frac{Wa^2L}{2EI} + \frac{Wa^3}{3EI} = \frac{Wa^2}{6EI}\{3L + 2a\}$$

As y_A is +ve, deflection at *A* is downward from the symmetry of the structure and we can say deflection at *A*.

$$\text{Slope at } A = \theta_A = \text{SF at A in the conjugate beam}$$

$$= -R_B - \frac{Wa}{EI} \times a \times \frac{1}{2}$$

$$= -\frac{WaL}{2EI} - \frac{Wa^2}{2EI} = -\frac{Wa}{2EI}[L + a]$$

$$\text{Slope at } B = \theta_B = \text{SF at } B \text{ in the conjugate beam}$$

$$= -\frac{WaL}{2EI}$$

Deflected profile of the beam is shown in Figure 7.63.

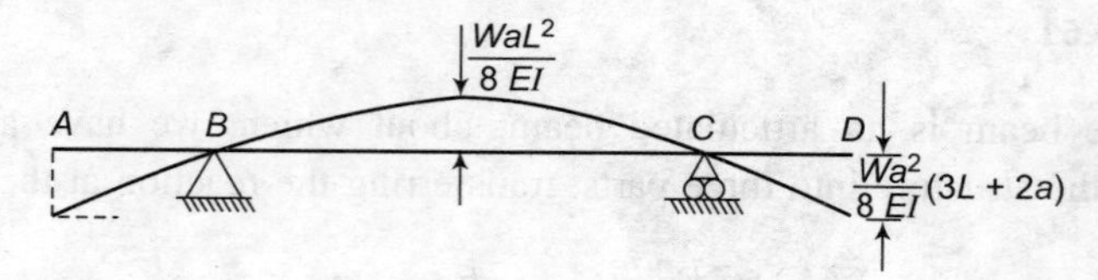

FIGURE 7.63

PROBLEM 7.16

Objective 4

For a fixed beam with concentrated load at midspan, the end moments are determined as *WL*/8 as shown in Figure 7.64. Using conjugate beam technique, determine deflection at midspan of the beam.

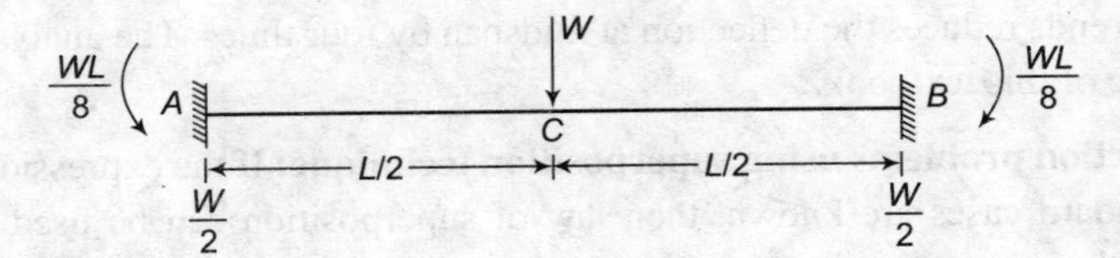

FIGURE 7.64

SOLUTION

For the given beam and loading, sketch the BMD and $\frac{M}{EI}$ diagram. The corresponding conjugate beam with $\frac{M}{EI}$ diagram as loading is shown in Figure 7.65.

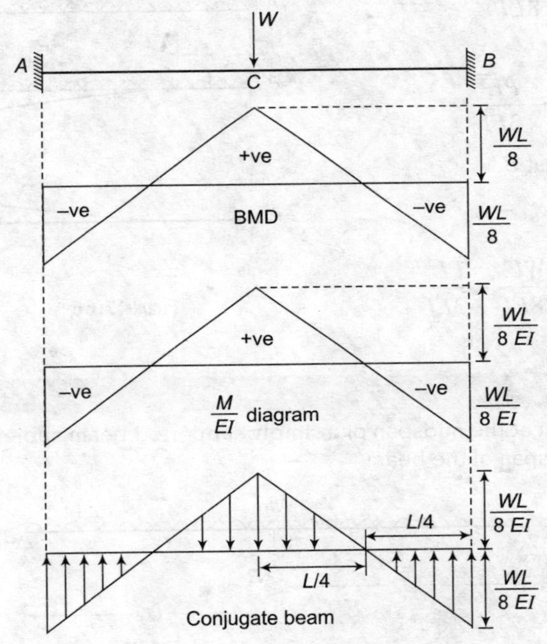

FIGURE 7.65

For determining deflection at midspan C, BM at C in the conjugate beam is to be found.

y_C = BM at C is the conjugate beam

$$= \frac{WL}{8EI} \times \frac{1}{2} \times \frac{L}{4}\left\{\frac{L}{2} - \frac{1}{3} \times \frac{L}{4}\right\} - \frac{WL}{8EI} \times \frac{1}{2} \times \frac{L}{4}\left\{\frac{L}{4} \times \frac{1}{3}\right\}$$

$$= \frac{WL^3}{64EI}\left\{\frac{1}{2} - \frac{1}{12} - \frac{1}{12}\right\} = \frac{WL^3}{192EI}$$

In the absence of fixidity at the ends, moment disappears; thereby, it becomes a simply supported beam. For a simply supported beam, with similar loading midspan deflection is $\frac{WL}{48EI}$. This implies that fixidity at the ends reduces the deflection at midspan by four times. The analysis of fixed beams is out of the scope of this textbook.

Solution to deflection problems using superposition technique: If the expressions for deflections and slope of standard cases are known, then law of superposition can be used to determine the deflections. Take the case of cantilever subjected to loading as shown in Figure 7.66. Cantilever is subjected to UDL and at an upward point load 'P' at free end.

The problem of cantilever with two types of loads can be converted into two cantilever problems. One is with UDL acting downward, and the other is with end point load acting in the upward direction. End up deflection for the both cases is known as

$$y_{B1} = \frac{WL^4}{8EI}$$

and

$$y_{B2} = -\frac{PL^3}{3EI}$$

Thus, the net deflection at B.

$$y_B = y_{B1} - y_{B2}$$

$$y_B = \frac{WL^4}{8EI} - \frac{PL^3}{3EI}$$

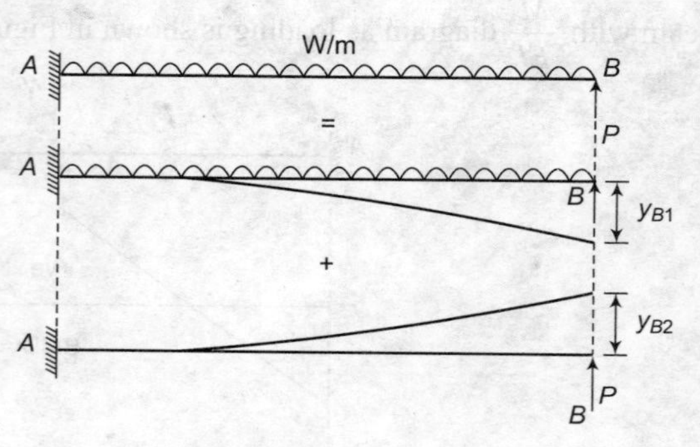

FIGURE 7.66

PROBLEM 7.17

Objective 5

Determine the deflection at the midspan of a simply supported beam subjected to UDL. A spring of stiffness 'K' is at the midspan of the beam.

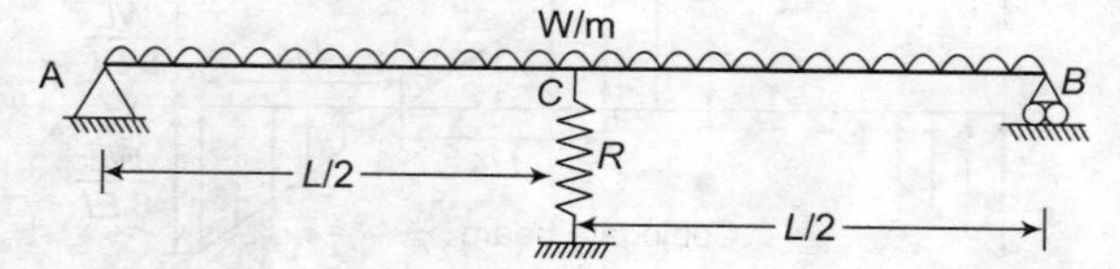

FIGURE 7.67

SOLUTION

Let 'R' be the reaction in the spring. Then, divide the loading on the structure into two categories and algebraically sum up the two solutions to get the final solution.

We know that $y_{C1} = \dfrac{5WL^4}{384EI}(\downarrow)$ (a standard case)

$$y_{C2} = \frac{RL^3}{48EI}(\uparrow)$$

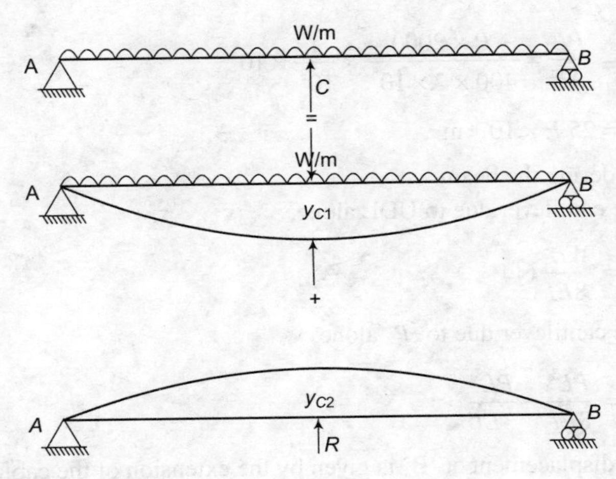

FIGURE 7.68

∴ Net downward deflection at C

$$y_C = y_{C1} - y_{C2} = \frac{5WL^4}{384EI} - \frac{RL^3}{48EI}$$

If 'y_C' is the net deflection at C, the force in the spring should be $y_C \cdot$K

$$\Rightarrow \qquad R = y_C \times \text{K}$$

Substituting the value of 'R' in the expression for 'y_C'

$$y_C = \frac{5WL^4}{384EI} - \frac{y_C KL^3}{48EI}$$

$$\Rightarrow \qquad y_C = \frac{\dfrac{5WL^4}{384EI}}{\left(1 + \dfrac{KL^3}{48EI}\right)}$$

PROBLEM 7.18 **Objective 5**

A cantilever beam carrying 4 kN/m UDL over its 4 m entire span carries a steel cable of 2 m length and 400 mm^2 cross-sectional area. To what extent the deflection at tip of the cantilever has decreases. Also determine the percentage decrease in the fixed end moment due to the cable. Take EI of the beam as 3×10^6 kN·m^2 and $E_{cable} = 2 \times 10^5$ MPa.

SOLUTION

Let 'P' be the tension in the cable (kN). Then, the vertical displacement of the cable BC at B.

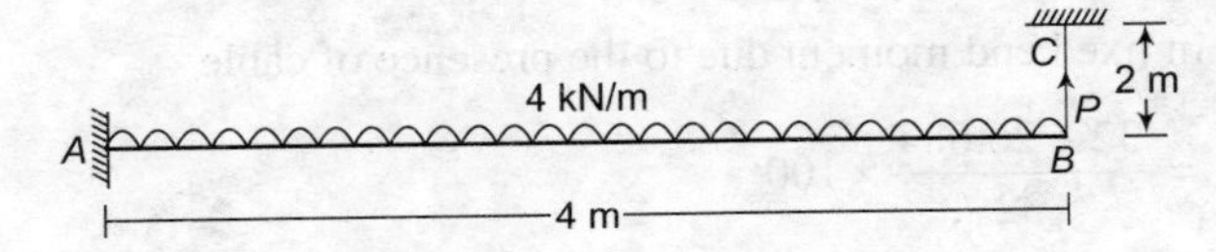

FIGURE 7.69

$$\Delta_B = \frac{PL}{AE} = \frac{P \times 2000}{400 \times 2 \times 10^5} \times \frac{1}{10^3} \times 10^3$$

$$= 25\,P \times 10^{-6}\text{ m}$$

Deflection at B, considering the beam

Deflection at B, in the cantilever due to UDL alone,

$$y_{B1} = \frac{WL^4}{8EI}(\uparrow)$$

Deflection at B, in the cantilever due to 'P' alone,

$$y_{B2} = \frac{PL^3}{8EI} - \frac{PL^3}{3EI}$$

However, the vertical displacement at 'B' is given by the extension of the cable as

$$\Delta_B = 25P \times 10^{-6}$$

Equating Δ_B and y_B for displacement compatibility,

$$\Delta_B = y_B$$

$$\therefore \quad 25P \times 10^{-6} = \frac{WL^4}{8EI} - \frac{PL^3}{3EI} = \frac{4 \times 4^4}{8 \times 3 \times 10^6} - P \times \frac{4^3}{3 \times 3 \times 10^6}$$

$$25\,P \times 10^{-6} = 42.67 \times 10^{-6} - P\{7.11 \times 10^{-6}\}$$

$$\Rightarrow \quad P = 1.329\text{ kN}$$

$$\Delta_B = 0.033\text{ mm}$$

$$\therefore \text{ BM at A} = \frac{WL^2}{2} - PL = 26.684\text{ kN·m}$$

In the absence of cable,

$$\Delta_B = \frac{WL^4}{8EI} = \frac{4 \times 4^4}{8 \times 3 \times 10^6} \times 10^3 = 0.0427\text{ mm}$$

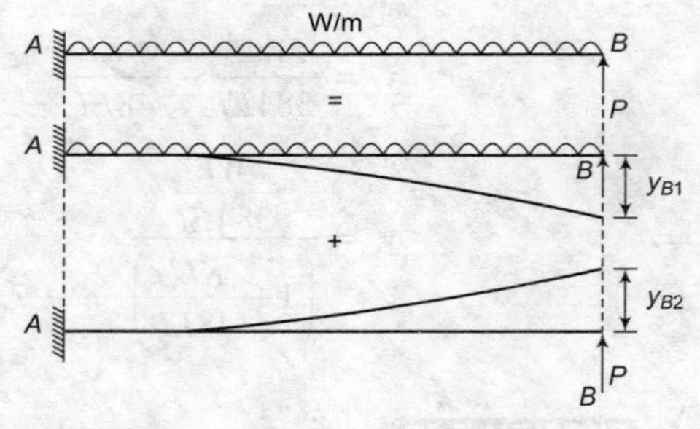

FIGURE 7.70

$$\text{BM at A, } M_A = \frac{4 \times 4^2}{2} = 32\text{ kN/m}$$

Percentage decrease in deflection at 'B' due to the presence of cable

$$= \frac{0.0427 - 0.033}{0.0427} \times 100$$

$$= 29.39\%$$

Percentage decrease in fixed end moment due to the presence of cable

$$= \frac{32 - 26.684}{32} \times 100$$

$$= 16.613\%$$

PROBLEM 7.19 **Objective 5**

Determine the deflection at the tip of the cantilever loaded as shown in Figure 7.71.

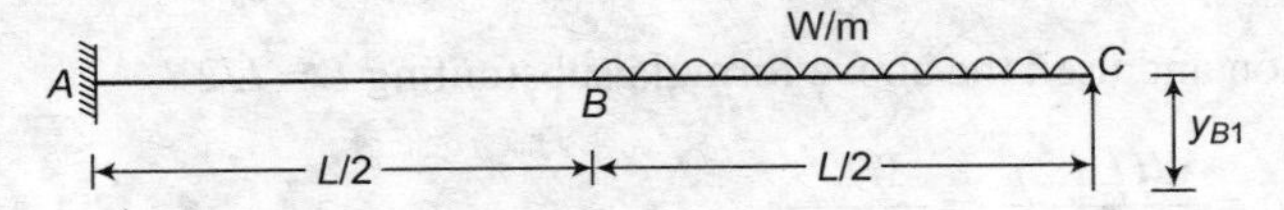

FIGURE 7.7.1

SOLUTION

The deflection at the tip of the cantilever is shown in Figure 7.72. By converting the loading into standard loading as shown in Figure 7.72:

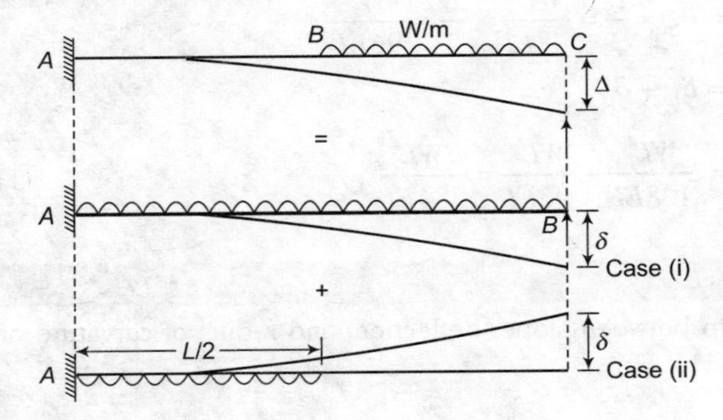

FIGURE 7.72

Downward deflection at C in case (i)

$$\delta_1 = \frac{WL^4}{8EI}(\downarrow) \text{ (standard case)}$$

Upward deflection at C, in case (ii)

$$\delta_2 = \frac{WL^4}{128EI} + \frac{WL^3}{48EI} \times \frac{L}{2}$$

$$\delta_2 = \frac{WL^4}{128EI} + \frac{WL^4}{96EI} = \frac{7WL^4}{384EI}(\uparrow)$$

$$\text{Deflection at } C = \delta_1 - \delta_2 = \frac{41WL^4}{384EI}$$

δ_2 Calculation:

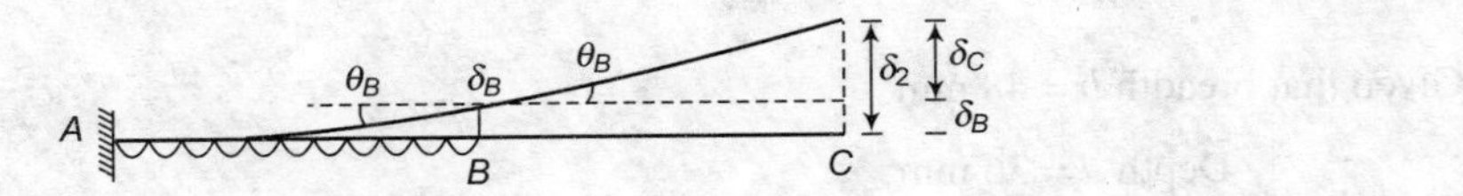

FIGURE 7.73

For cantilever beam subjected to UDL, deflection at the tip

$$= \frac{WL^4}{8EI}, \text{ and slope at the tip} = \frac{WL^3}{6EI}$$

Consider '*AB*' portion and used the above formula substituting $L = L/2$

$$\therefore \qquad \delta = \frac{W(L/2)^4}{8EI} = \frac{WL^4}{128EI}$$

$$\theta = \frac{W(L/2)^3}{6EI} = \frac{WL^3}{48EI}$$

'*BC*' portion in the above case deform like a rigid body (slope is constant).

$$\therefore \qquad \delta_C' = \theta_B \times \frac{L}{2} = \frac{WL^3}{48EI} \times \frac{L}{2} = \frac{WL^3}{96EI}$$

$$\because \qquad \delta_C = \delta_B + \delta_C'$$

$$= \frac{WL^4}{128EI} + \frac{WL^4}{96EI} = \frac{7WL^4}{384EI}$$

PROBLEM 7.20 **Objective 1**

Derive the relationship between slope, deflection, and radius of curvature of a simply supported beam.

SOLUTION

1. Derive relationship between deflection, slope, and curvature.
2. Derive $EI \frac{d^2y}{dx^2} = -M$
3. Derive the governing differential equation for deflection and BM in a beam.

PROBLEM 7.21 **Objective 6**

A 300-mm-long curvature of rectangular section 48 mm wide and 36 mm deep carries a UDL. Calculate the value of load *w*, if the maximum deflection in the cantilever is not to exceed 1.5 mm. Take $EI = 70 \times 10^9$ GN/m^2.

SOLUTION

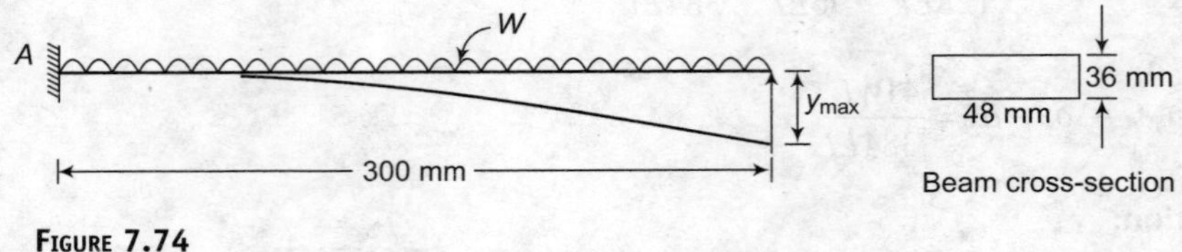

FIGURE 7.74

Given that breadth $b = 48$ mm

Depth $d = 36$ mm

Span $L = 300$ mm

$$\text{MI about NA} = \frac{bd^3}{12} = \frac{48 \times 36^3}{12} = 186624 \text{ mm}^4$$

Modulus of elasticity $E = 70$ GPa $= 70 \times 10^3$ MPa

$$y_{\max} = \frac{WL^4}{8EI}$$

$y_{\max}$ is limited to 1.5 mm.

$$\text{The units of } W \text{ are } \frac{\text{N}}{\text{m}} = \frac{W}{1000} \text{ N/mm}$$

$$1.5 = \frac{\frac{W}{1000} \times (300)^4}{8 \times 70 \times 10^3 \times 186624}$$

$$W = 19353.6 \text{ N/m}$$

$\therefore$ The allowable UDL on the cantilever is 19.35 kN/m.

PROBLEM 7.22

Objective 1

A beam AB of span L carries a distributed load of varying intensity from zero at A to *w* per unit length at B. Measuring *x* from the end A, establish the equation for the deflection curve of the beam.

SOLUTION

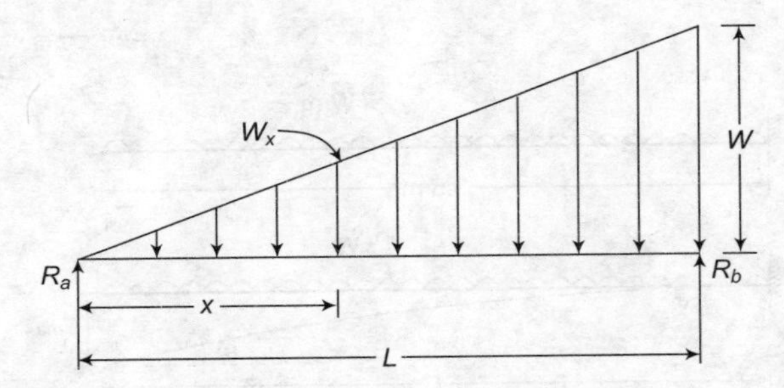

FIGURE 7.75

$$\sum V = 0 \Rightarrow R_a + R_b = \frac{1}{2} \times W \times L$$

Taking moments about A, $\sum M_A = 0$

$$\Rightarrow \qquad R_b \times L = \frac{1}{2} \times W \times L \times \frac{2L}{3}$$

$$R_b = \frac{WL}{3}, \; R_a = \frac{WL}{6}$$

$$M_x = +R_a x - \frac{1}{2} \times W \times \frac{x}{L} \times x \times \frac{x}{3} = \frac{WL}{6}x - \frac{Wx^3}{6L}$$

$$EI\frac{d^2y}{dx^2} = -M_x = -\frac{WL}{6}x + \frac{Wx^3}{6L}$$

$$EI\frac{dy}{dx} = -\frac{WL}{6}\frac{x^2}{2} + \frac{Wx^4}{24L} + C_1$$

Integrating on both sides, $EIy = -\frac{WLx^3}{36} + \frac{Wx^5}{120L} + C_1x + C_2$

Boundary condition at $x = 0$, $y = 0$; at $x = L$, $y = 0$

At $x = 0, y = 0 \Rightarrow C_2 = 0$

At $x = L, y = 0 \Rightarrow \frac{WL^4}{36} - \frac{WL^4}{120} + C_1L = 0 \Rightarrow C_1 = \frac{7WL^3}{360}$

Expression for deflection is given by, $y = \frac{1}{EI}\left\{-\frac{WLx^3}{36} + \frac{Wx^5}{120L} + \frac{7WL^3x}{360}\right\}$

PROBLEM 7.23

Objective 1

A 3.5-m-long cantilever carries a UDL over the entire length. If the slope at the free end is one degree, what is the deflection at the free end.

SOLUTION

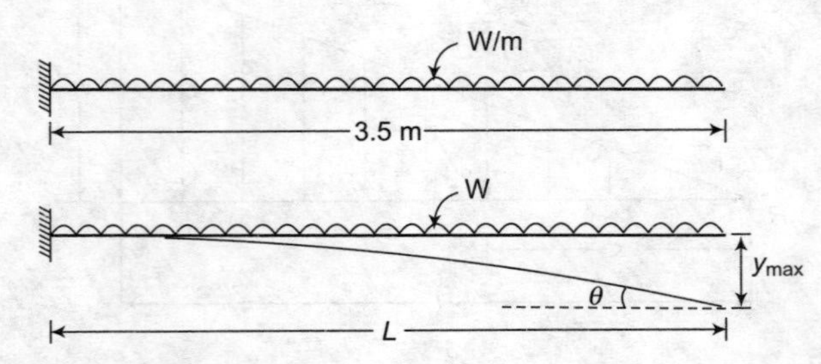

FIGURE 7.76

$$\theta = \frac{WL^3}{6EI} \text{ and } y_{max} = \frac{WL^4}{8EI}$$

Given that rotation at free end $\theta = 1° = \frac{1}{180}\pi$ radians

$$\Rightarrow \quad \frac{WL^3}{6EI} = \frac{\pi}{180} = 0.01745 \text{ radians}$$

$$y_{\max} = \frac{WL^3}{6EI}\frac{L}{\left(\frac{8}{6}\right)} = \frac{WL^3}{6EI} \times \left(\frac{6L}{8}\right)$$

$$\therefore \qquad y_{\max} = 0.01745 * \frac{6L}{8} = \frac{0.01745 \times 6 \times 3.5}{8} = 0.0458 \text{ mm}$$

PROBLEM 7.24 **Objective 1**

A 3-m-long cantilever is loaded with a point load of 450 N at the free end. If the section is rectangular 80 mm (wide) × 160 mm (deep), and $E = 10$ GN/m^2, calculate slope and deflection.

(i) At the free end of the cantilever
(ii) At a distance of 0.55 m from the free end.

SOLUTION

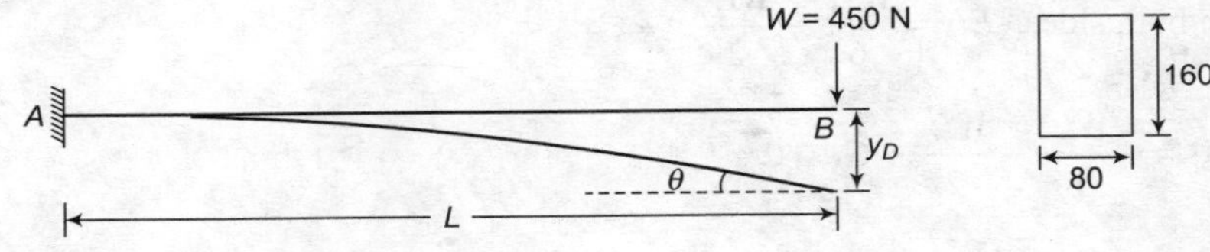

FIGURE 7.77

$$Y_B = \frac{WL^3}{3EI}$$

$$\theta_B = \frac{WL^2}{2EI}$$

$$l = \frac{1}{12} \times 80 \times 160^3 = 27.307 \times 10^6 \text{ mm}^4$$

$$E = 10 \text{ GPa} = 10 \times 10^3 \text{ N/mm}^2$$

$$L = 3 \text{ m} = 3000 \text{ mm}$$

$$y_B = \frac{450 \times 3000^3}{3 \times 1 \times 10^4 \times 27.307 \times 10^6} = 14.83 \text{ mm}$$

$$\theta_B = \frac{450 \times 3000^2}{2 \times 1 \times 10^4 \times 27.307 \times 10^6} = 0.0074 \text{ radians}$$

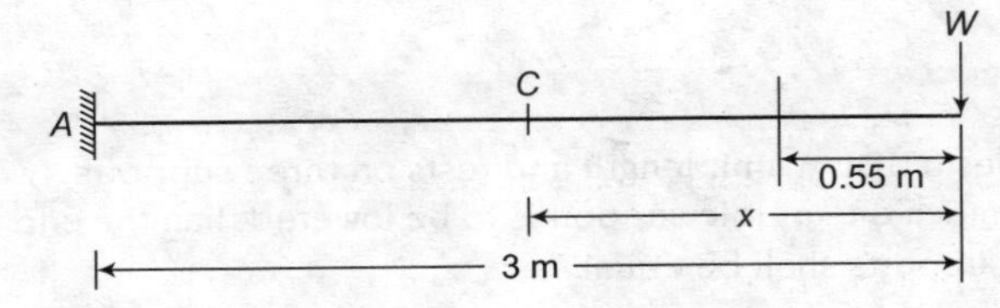

FIGURE 7.78

$$EI\frac{d^2y}{dx^2} = -M_x = -Wx$$

$$\therefore \quad EI\frac{d^2y}{dx^2} = Wx$$

Integrating on both sides, $EI\dfrac{dy}{dx} = \dfrac{Wx^2}{2} + C_1$

At $x = L$, $\dfrac{dy}{dx} = 0$

$$\Rightarrow \quad C_1 = -\frac{WL^2}{2}$$

Integrating on both sides, $EIy = \dfrac{Wx^3}{6} - \dfrac{WL^2}{2}x + C_2$

At $x = L, y = 0 \Rightarrow \dfrac{WL^3}{6} - \dfrac{WL^3}{2} + C_2 = 0$

$$\Rightarrow \quad C_2 = \frac{WL^3}{3}$$

$$y = \frac{1}{EI}\left[\frac{Wx^3}{6} - \frac{WL^2}{2}x + \frac{WL^3}{3}\right]$$

For $x = 0.55$ m = 550; $L = 3$ m = 3000 mm; $W = 450$ N

$$y_C = \frac{450}{1\times10^4\times27.307\times10^6}\left\{\frac{550^3}{6} - \frac{3000^2\times550}{2} + \frac{3000^3}{3}\right\} = 10.8 \text{ mm}$$

$$\theta_C = \left(\frac{dy}{dx}\right)_{at\ x = 550\text{ mm}} = \frac{1}{EI}\left\{\frac{Wx^2}{2} - \frac{WL^2}{2}\right\}$$

$$= \frac{450}{1\times10^4\times27.307\times10^6}\left\{\frac{550^2}{2} - \frac{3000^2}{2}\right\}$$

$$= -0.00717 \text{ radians}$$

PROBLEM 7.25 **Objective 6**

A beam of length L carries a UDL W/unit length and rests on three supports, two at the ends and one in the middle, find how much the middle support is to be lowered than the end ones in order that the pressures on the three supports shall be equal.

SOLUTION

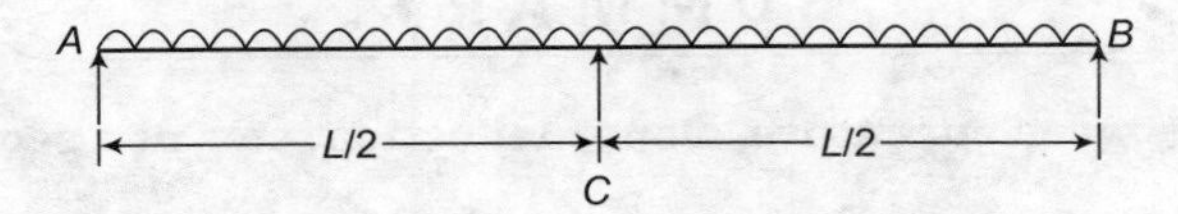

FIGURE 7.79

By how much amount the central support C shall be lowered so that the reaction of all three supports is same.

It means that, $R_a = R_b = R_c = R = \dfrac{WL}{3}$

Pictorially the given problem can be represented as

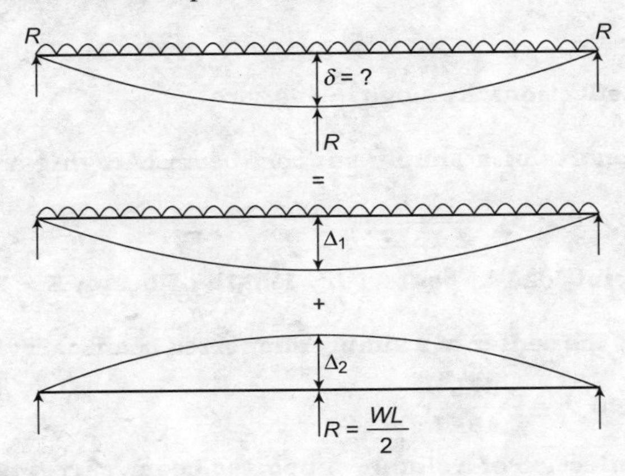

FIGURE 7.80

Δ_1 = downward deflection of the beam without central support

Δ_2 = upward deflection at central support reaction = $\dfrac{WL}{3}$

$$\Delta_1 = \frac{5WL^4}{384EI}$$

$$\Delta_2 = \frac{RL^3}{48EI} = \frac{WL^4}{144EI} \quad \left(\because R = \frac{WL}{3}\right)$$

$$\Delta = \Delta_1 - \Delta_2 = \frac{5WL^4}{384EI} - \frac{WL^4}{144EI} = \frac{7WL^4}{1152EI}$$

SUMMARY

- **The relation between curvature, slope, deflection, etc., at a section is given by:**
  Deflection = y

  Slope = $\frac{dy}{dx}$

  BM = $EI\frac{d^2y}{dx^2}$

  SF = $EI\frac{d^3y}{dx^3}$

  W = $EI\frac{d^4y}{dx^4}$

- **For maximum deflection, the slope $\frac{dy}{dx}$ is zero.**
- **Slope at the supports of a simply support beam carrying a point load at the center is given by:** $\theta_A = \theta_B = -\frac{WL^2}{16EI}$

  in which W = point load at center; L = length of beam; E = Young's modulus; $I = MI$
- **The deflection at the center of a simply supported beam carrying a point load at the center is given** $y_C = -\frac{WL^3}{48EI}$
- **The slope and deflection of a simply supported beam, carrying a UDL of w/unit length over the entire span, are given by,**

  $\theta_A = \theta_B = -\frac{WL^2}{24EI}$ **and** $y_C = \frac{5}{384}\frac{WL^2}{EI}$

OBJECTIVE TYPE QUESTIONS

1 A lean elastic beam of given flexural rigidity EI is loaded by a single force F as shown in figure. How many boundary conditions are necessary to determine the deflected center line of the beam? [GATE-1999]

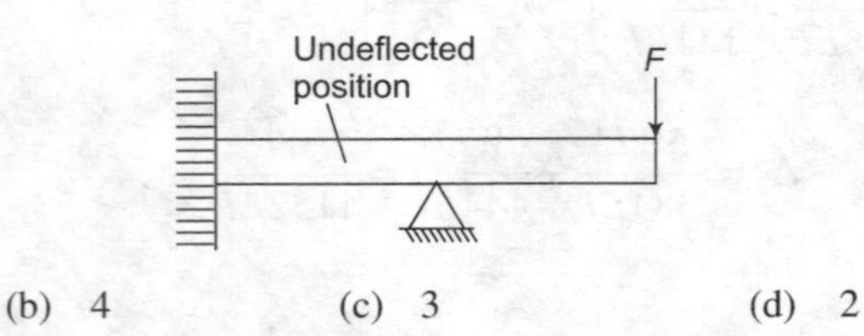

(a) 5 (b) 4 (c) 3 (d) 2

2 A simply supported beam carrying a concentrated load W at midspan deflects by $\delta 1$ under the load. If the same beam carries the load W such that it is distributed uniformly over entire length and undergoes a deflection $\delta 2$ at the midspan. The ratio $\delta 1$: $\delta 2$ is: [IES-1995; GATE-1994]

(a) 2:1 (b) $\sqrt{2}:1$ (c) 1:1 (d) 1:2

3 A simply supported laterally loaded beam was found to deflect more than a specified value. [GATE-2003]

Which of the following measures will reduce the deflection?
(a) Increase the area MI
(b) Increase the span of the beam
(c) Select a different material having lesser modulus of elasticity
(d) Magnitude of the load to be increased

4 A cantilever beam of length L is subjected to a moment M at the free end. The MI of the beam cross-section about the NA is I and the Young's modulus is E. The magnitude of the maximum deflection is [GATE-2012]

(a) $\frac{ML^2}{2EI}$ (b) $\frac{ML^2}{EI}$ (c) $\frac{2ML^2}{EI}$ (d) $\frac{4ML^2}{EI}$

5 Consider the following statements: in a cantilever subjected to a concentrated load at the free end [IES-2003]

1. The bending stress is maximum at the free end
2. The maximum shear stress is constant along the length of the beam
3. The slope of the elastic curve is zero at the fixed end

Which of these statements are correct?
(a) 1, 2, and 3 (b) 2 and 3 (c) 1 and 3 (d) 1 and 2

6 A cantilever of length L, moment of inertia I, and Young's modulus E carries a concentrated load W at the middle of its length. The slope of cantilever at the free end is: [IES-2001]

(a) $\frac{WL^2}{2EI}$ (b) $\frac{WL^2}{4EI}$ (c) $\frac{WL^2}{8EI}$ (d) $\frac{WL^2}{16EI}$

7 The two cantilevers A and B shown in the figure have the same uniform cross-section and the same material. Free end deflection of cantilever A is δ. The value of midspan deflection of the cantilever B is: [IES-2000]

(a) $\frac{1}{2}\delta$ (b) $\frac{2}{3}\delta$ (c) δ (d) 2δ

8 A cantilever beam of rectangular cross-section is subjected to a load *W* at its free end. If the depth of the beam is doubled and the load is halved, the deflection of the free end as compared to original deflection will be: [IES-1999]
(a) Half (b) One-eighth (c) One-sixteenth (d) Double

9 simply supported beam of constant flexural rigidity and length 2L carries a concentrated load P at its midspan and the deflection under the load is δ. If a cantilever beam of the same flexural rigidity and length L is subjected to load P at its free end, then the deflection at the free end will be: [IES-1998]

(a) $\frac{1}{2}\delta$ (b) δ (c) 2δ (d) 4δ

10 Two identical cantilevers are loaded as shown in the respective figures. If slope at the free end of the cantilever in Figure *E* is θ, the slope at free and of the cantilever in Figure *F* will be: [IES-1997]

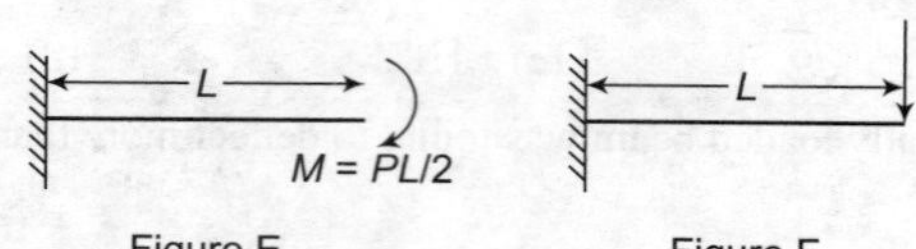

Figure E Figure F

(a) $\frac{1}{3}\theta$ (b) $\frac{1}{2}\theta$ (c) $\frac{2}{3}\theta$ (d) θ

11 A cantilever beam carries a load W uniformly distributed over its entire length. If the same load is placed at the free end of the same cantilever, then the ratio of maximum deflection in the first case to that in the second case will be: [IES-1996]

(a) 3/8 (b) 8/3 (c) 5/8 (d) 8/5

12 The given figure shows a cantilever of span L subjected to a concentrated load P and a moment M at the free end. Deflection at the free end is given by [IES-1996]

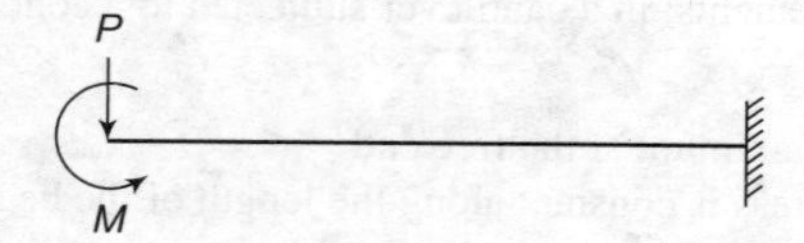

(a) $\frac{PL^2}{2EI}+\frac{ML^2}{3EI}$ (b) $\frac{ML^2}{2EI}+\frac{PL^3}{3EI}$ (c) $\frac{ML^2}{3EI}+\frac{PL^3}{2EI}$ (d) $\frac{ML^2}{2EI}+\frac{PL^3}{48EI}$

13 Maximum deflection of a cantilever beam of length l carrying UDL w per unit length will be: [IES- 2008]

(a) $\frac{wl^4}{(EI)}$ (B) $\frac{wl^4}{(4EI)}$ (c) $\frac{wl^4}{(8EI)}$ (d) $\frac{wl^4}{(384EI)}$

(Where E = modulus of elasticity of beam material and I = MI of beam cross-section)

14 Assertion (A): In a simply supported beam subjected to a concentrated load P at midspan, the elastic curve slope becomes zero under the load.

Reason (R): The deflection of the beam is maximum at midspan. [IES-2003]

(a) Both A and R are individually true and R is the correct explanation of A
(b) Both A and R are individually true but R is NOT the correct explanation of A
(c) A is true but R is false
(d) A is false but R is true

15 A 2-m-long beam BC carries a single concentrated load at its midspan and is simply supported at its ends by two cantilevers AB = 1 m long and CD = 2 m long as shown in the figure. [IES-1997]

The SF at end A of the cantilever AB will be

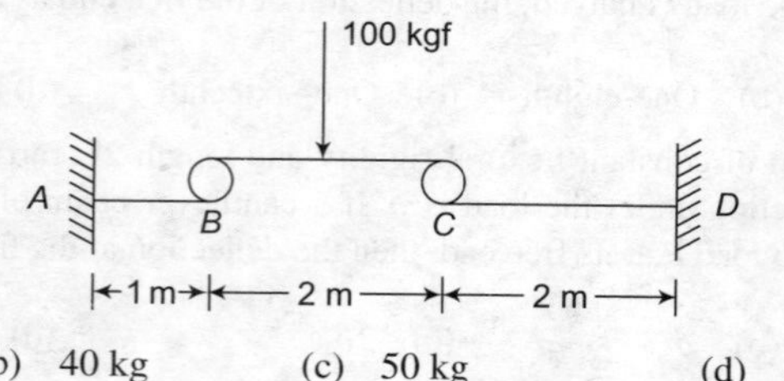

(a) Zero (b) 40 kg (c) 50 kg (d) 60 kg

16 A simply supported beam of rectangular section 4 cm by 6 cm carries a midspan concentrated load such that the 6 cm side lies parallel to line of action of loading; deflection under the load is δ. If the beam is now supported with the 4 cm side parallel to line of action of loading, the deflection under the load will be: [IES-1993]

(a) 0.44 δ (b) 0.67 δ (c) 1.5 δ (d) 2.25 δ

17 Which one of the following is represented by the area of the SF diagram from one end up to a given location on the beam? [IAS-2004]

(a) BM at the location (b) Load at the location
(c) Slope at the location (d) Deflection at the location

18 In a cantilever beam, if the length is doubled while keeping the cross-section and the concentrated load acting at the free end the same, the deflection at the free end will increase by [IAS-1996]

(a) 2.66 times (b) 3 times (c) 6 times (d) 8 times

19 By conjugate beam method, the slope at any section of an actual beam is equal to: [IAS-2002]

(a) EI times the SF of the conjugate beam
(b) EI times the BM of the conjugate beam
(c) SF of conjugate beam
(d) BM of the conjugate beam

20 $I = 375 \times 10^{-6}$ m^4; $l = 0.5$ m; $E = 200$ GPa. Determine the stiffness of the beam shown in the above figure [IES-2002]

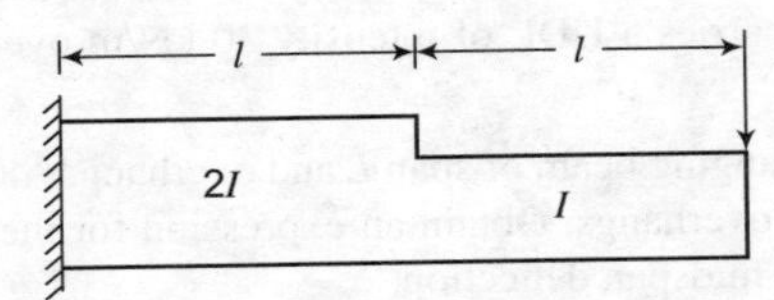

(a) 12×10^{10} N/m (b) 10×10^{10} N/m (c) 4×10^{10} N/m (d) 8×10^{10} N/m

Solutions for Objective Questions

Sl. No.	1.	2.	3.	4.	5.	6.	7.	8.	9.	10.
Answer	(d)	(d)	(a)	(a)	(b)	(c)	(c)	(c)	(c)	(d)

Sl.No.	11.	12.	13.	14.	15.	16.	17.	18.	19.	20.
Answer	(a)	(b)	(c)	(a)	(c)	(d)	(a)	(d)	(c)	(c)

EXERCISE PROBLEMS

1. Obtain an expression for the central deflection of a simply supported beam subjected to concentrated load at 1/4th span.

2. A prismatic beam ABC of 8 m long is simply supported at A and B as shown below. It is subjected to a concentrated load of 10 kN at C, a couple 40 kN·m anticlockwise at D and a UDL of 20 kN/m over EB. Find the deflection at the free end *C*. Determine the slope at supports. The flexural rigidity of the beam is $EI = 48 \times 10^{12}$ N-mm^2.

2 m 2 m 2 m 2 m
A D E B C

FIGURE 1

3. Determine the deflection at the free end of a cantilever beam subjected to end couple of 5 kN/m and UDL of 2 kN/m. The span of the beam is 2.5 m. Take a circular cross-section of 100 mm in diameter and modulus of elasticity $E = 200$ GPa. If a concentric borehole of diameter 50 mm is cut in the cross-section of the beam, what would be the percentage increase in free end deflection?
4. A simply supported prismatic beam of 6 m span is subjected to a central concentrated load of 60 kN. Find the maximum deflection and slope using moment area method. If the flexural rigidity of the end quarter spans were reduced to half their values, what would be the percentage increase in the deflection?
5. Estimate the flexural rigidity required for the shown cantilever beam if the maximum deflection is limited to 2 mm.

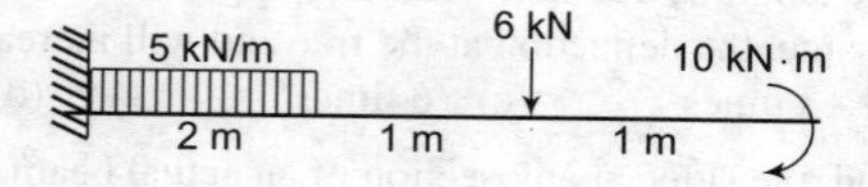

FIGURE 2

6. Using moment area theorem, find an expression for the slope at the supports and midspan deflection of a simply supported beam of span 'l' subjected to midspan couple M.
7. A horizontal cantilever 2 m long has its free end attached to a vertical tie of 3 m long and 300 mm^2 in cross-sectional area. Determine the load taken by the tie rod and the maximum deflection of the cantilever, if the cantilever carries a UDL of intensity 30 kN/m over its span. Take EI of the beam as 1200 kN/m^2.
8. A symmetrical double overhanging beam of span L and overhangs 'a' each, is subjected to concentrated loads at the free ends of the overhangs. Obtain an expression for the deflection of the beam and hence deduce an expression for the midspan deflection.
9. Determine the central span deflection of the beam shown in Figure 3. $EI = 1 \times 10^4$ kNm^2.

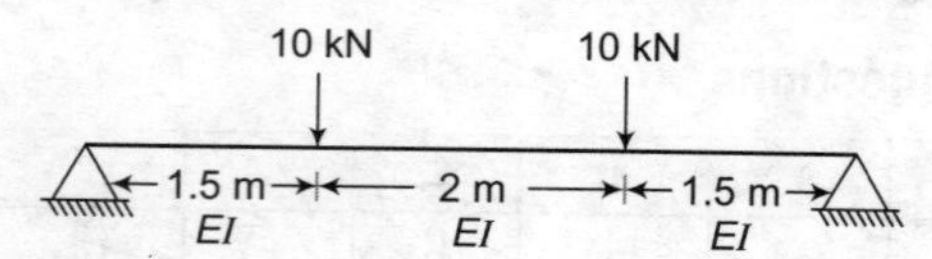

FIGURE 3

CHAPTER 8

ANALYSIS OF CYLINDERS

UNIT OBJECTIVE

This chapter provides information on the stresses induced in thick, thin, and compound cylinders subjected to internal pressure and rise in temperature. The presentation attempts to help the student achieve the following:

Objective 1: Determine stresses in a cylinder due to internal pressure.

Objective 2: Analysis of compound cylinders and shrink fit.

Objective 3: Determine hoop stress and longitudinal stress.

Objective 4: Evaluate shrinkage pressure and shrinkage allowance.

8.1 INTRODUCTION

Cylinders are the most commonly used structural elements used to convey fluids under pressure/gravity, maintain pressure, and store the fluids (e.g., water tanks, gas cylinders). Analysis of cylinders includes the determination of stresses and deformations developed due to internal pressure.

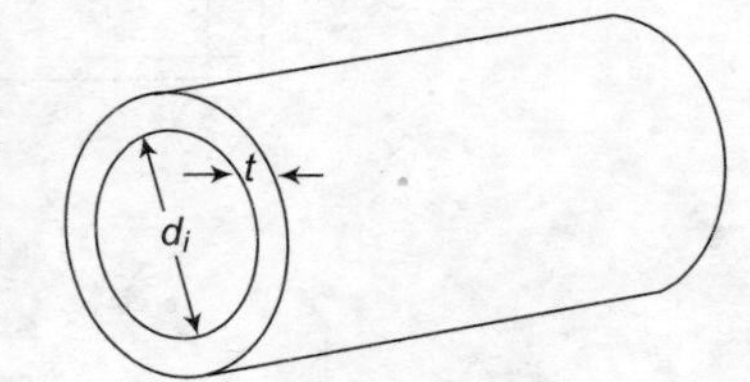

FIGURE 8.1

The cylinders are classified into two categories.

1. Thin cylinders
2. Thick cylinders

Let 'd_i' and 't' be the inner diameter and thickness of the cylinder, respectively. These cylinders are those for which $\frac{d_i}{t} > 20$. In this, inner diameter to thickness ratio is less than 20, and the corresponding cylinders are referred as thick cylinders. In case of thin cylinders, the variation of radial stress across the thickness is neglected. Thus, the thin cylinder analysis is approximate one.

In thick cylinder analysis, the variation of radial pressure across the thickness of the cylinder is not neglected. Thus, the thick cylinder analysis is more accurate than thin cylinder analysis. If the thickness of the cylinder is very less, thick cylinder analysis coincides with the thin cylinder analysis.

8.2 ANALYSIS OF THIN CYLINDERS

Internal pressure in cylinder increases the length and perimeter of the cylinders. Thus, it can be inferred that there exists tensile stress in the length direction as well as in the circumferential direction. The stress in longitudinal direction of cylinder is called longitudinal stress denoted by σ_l, and stress along the circumference or circumferential direction is referred as circumferential stress or hoop stress, denoted by σ_h. These two stresses are shown in Figure 8.2.

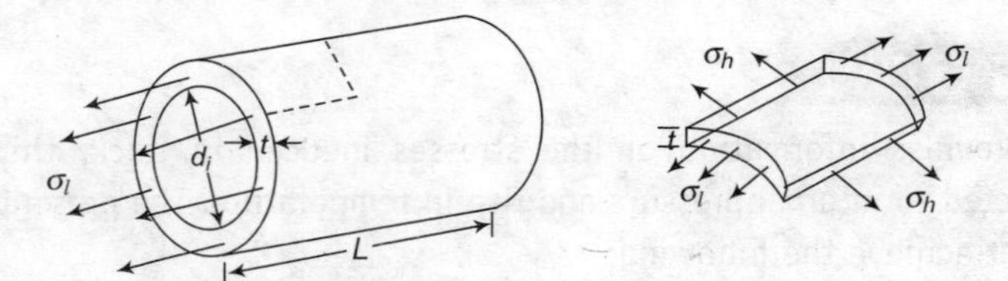

Figure 8.2

The hoop stress (σ_h) is sometimes called as bursting stress. Knowing these stresses, the change in the diameter, length and internal volume of the cylinder can be computed.

8.2.1 Expression for Hoop Stress and Longitudinal Stress

Consider a close cylinder, subjected to internal fluid pressure 'p'. Let 'd' and 't' be the inner diameter and wall thickness of the cylinder, respectively. To evaluate the hoop stress (σ_h), cut the cylinder horizontally by section (1)-(1) as shown in Figure 8.3.

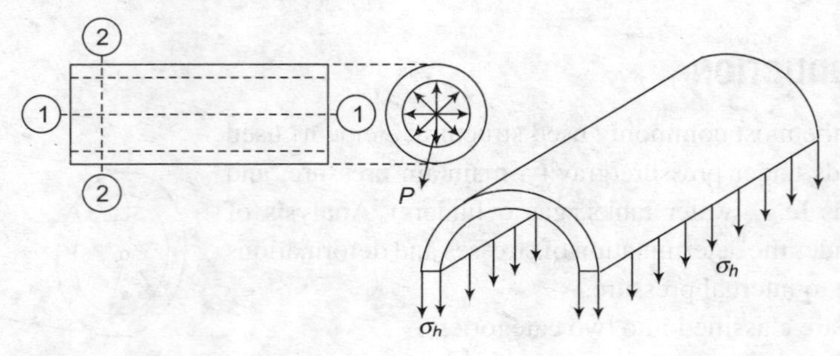

Figure 8.3

The free body diagram (FBD) of the cut section is shown in Figure 8.3. Considering unit length of the cylinders, write the forces acting on it, then FBD is as shown in Figure 8.4.

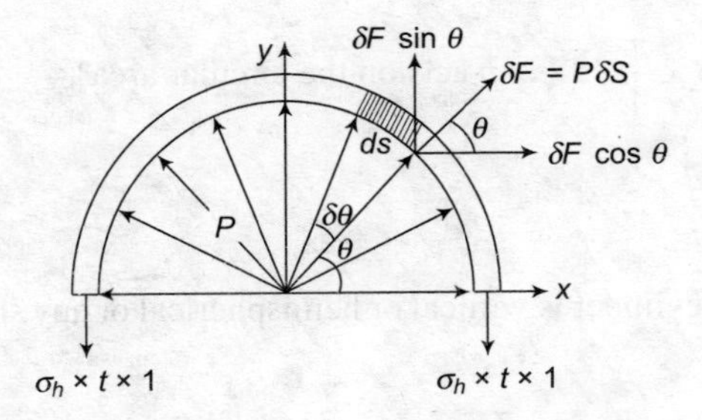

FIGURE 8.4

To implement equilibrium condition, consider an elemental length 'δs' in the circular portion of the cylinder, making an angle of '$\delta\theta$' at the center as shown in Figure 8.4.

Elemental force acting on the elemental strip = $\delta F = (p)\delta s$

$$= p \times \frac{d}{2} \times \delta\theta$$

'δF' is inclined by 'θ' with the x axis.

$\therefore$ Vertical component of $\delta F = \delta F \sin\theta$

$$= p \times \frac{d}{2} \times \sin\theta \times d\theta$$

Total vertical component of pressure = $\int dF$

For equilibrium, sum of the vertical forces = 0

$$\Rightarrow \quad 2 \times \sigma_h \times t \times 1 = \int dF = \int_0^{\pi} \frac{p.d}{2} \sin\theta d\theta$$

$$= \frac{pd}{2}[1 - \cos\theta]_0^{\pi} = pd$$

$$\therefore \quad \sigma_h = \frac{pd}{2t}.$$

For evaluating an expression for longitudinal stress, consider a section (2)-(2) shown in Figure 8.3. FBD of the section (2)-(2) is shown in Figure 8.5.

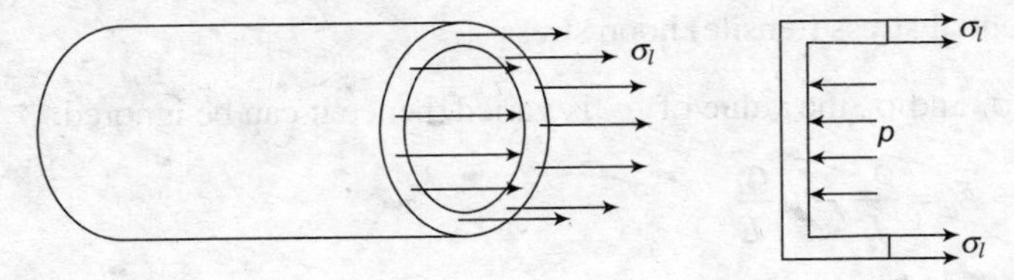

FIGURE 8.5

For equilibrium, sum of horizontal forces must be equal to zero.

$$\sigma_l[\pi d \times t] = p \times \frac{\pi}{4} d^2 \quad \{\because p \text{ acts on the circular area}\}$$

$$\Rightarrow \qquad \sigma_l = \frac{pd}{4t}.$$

Whether the end of the cylinder is vertical or hemispherical or any shape, the resultant of pressure 'p' is $p \times$ exposed area.

Thus, resultant of 'p' in the horizontal direction is $p \times \frac{\pi}{4} d^2$.

Hence, irrespective of the shape of the closed wall, the longitudinal stress 'σ_l' is same $= \sigma_l = \frac{pd}{4t}$.

In a thin cylinder, subjected to internal fluid pressure p, the hoop stress (σ_h) and longitudinal stress σ_l are stresses that need consideration in the design.

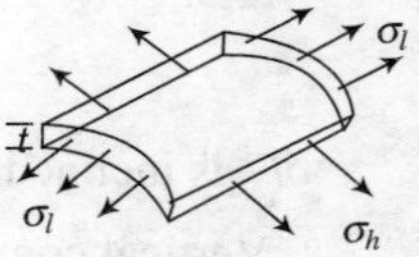

$$\text{Maximum shear stress} = \frac{\sigma_h - \sigma_l}{2}$$

$$\tau_{\max} = \frac{pd}{8t}$$

FIGURE 8.6

Expression for the change in the diameter of the cylinder Let 'δd' be the change in the diameter of the cylinder. Then,

$$\text{Diametrical strain} = \frac{\delta d}{d} = \frac{\delta(\pi d)}{\pi d} = \text{Circumferential strain}$$

$$\text{Thus, circumferential strain} = \varepsilon_h = \frac{\delta d}{d} \tag{8.1}$$

$$\text{We know, } \varepsilon_h = \frac{\sigma_h}{E} - \mu\frac{\sigma_l}{E} - \mu\left[-\frac{p}{E}\right]$$

$p \rightarrow$ Radial stress (compressive)

$\sigma_l \rightarrow$ Longitudinal stress (tensile)

$\sigma_h \rightarrow$ Circumferential stress (tensile) hoop stress

Compared to σ_h and σ_l, the value of 'p' is varied, hence it can be ignored.

$$\therefore \qquad \varepsilon_h = \frac{\sigma_h}{E} - \mu\frac{\sigma_l}{E} \tag{8.2}$$

Equating Eqs. (8.1) and (8.2)

$$\frac{\delta d}{d} = \frac{pd}{2tE} - \mu\frac{pd}{4tE}$$

$$\delta d = d \times \frac{pd}{4tE}\{2 - \mu\}$$

Expression for change in length of cylinder Let 'δL' be the change in the length of the cylinder.

$$\text{Longitudinal strain} = \epsilon_l = \frac{\delta L}{L} \tag{8.3}$$

We know that, $\varepsilon_l = \dfrac{\sigma_l}{E} - \mu\dfrac{\sigma_h}{E} - \left[-\dfrac{p}{E}\right]$

Ignoring p, $\varepsilon_l = \dfrac{\sigma_l}{E} - \mu\dfrac{\sigma_h}{E}$ (8.4)

Equating Eqs. (8.3) and (8.4)

$$\frac{\delta L}{L} = \frac{\sigma_l}{E} - \mu\frac{\sigma_h}{E}$$

$$\varepsilon_l = \frac{\delta L}{L} = \frac{pd}{4tE} - \mu\frac{pd}{2tE}$$

$$\delta L = L\frac{pd}{4tE}\{1 - 2\mu\}$$

Change in the internal volume/capacity of the cylinder Assuming that the end walls are rigid, the internal volume of the cylinder may be expressed as

$$V = \frac{\Pi d^2}{4}L$$

with fluid pressure, d changes and L also changes.

$$\therefore \quad \Delta V = \frac{\pi}{4}[L \times 2d\,\Delta d] + \frac{\pi}{4} \times d^2 \times \Delta L$$

$$\frac{\Delta V}{V} = \frac{\frac{\pi}{4}2d\,L\,\Delta d}{\left(\frac{\pi d^2}{4}\right) \times L} + \frac{\frac{\pi d^2}{4}(\Delta L)}{\frac{\pi}{4}d^2 \times L}$$

$$\Rightarrow \quad \frac{\delta V}{V} = 2 \times \frac{\delta d}{d} + \frac{\delta L}{L}$$

Or

$$\frac{\delta V}{V} = 2\varepsilon_h + \varepsilon_l$$

$$\Rightarrow \quad \frac{\delta V}{V} = 2\left\{\frac{\sigma_h}{E} - \mu\frac{\sigma_l}{E}\right\} + \left\{\frac{\sigma_l}{E} - \mu\frac{\sigma_h}{E}\right\}$$

Or $$\frac{\delta V}{V} = \frac{2}{E}\left[\frac{pd}{2t} - \mu\frac{pd}{4t}\right] + \frac{1}{E}\left[\frac{pd}{4t} - \mu\frac{pd}{2t}\right]$$

$$\Rightarrow \quad \frac{\delta V}{V} = \frac{pd}{4tE}\{5 - 4\mu\}$$

Or $$\delta V = V \times \frac{pd}{4tE}\{5 - 4\mu\}$$

* Change in volume is change in the internal volume of the cylinder, it should not be understood as change in the volume of the material of the cylinder.

PROBLEM 8.1

Objective 1

A closed steel pressure vessel (cylinder) with inner diameter 2 m and length 3.5 m is subjected to an internal pressure of 0.8 MPa. Calculate the following, if wall thickness of the cylinder is 10 mm. $E = 200$ GPa and $\mu = 0.25$.

(a) Maximum normal stress
(b) Change in the diameter
(c) Change in the length
(d) Change in the volume (internal)
(e) Maximum shear stress.

SOLUTION

Given data:

Inner diameter of the cylinder (d) = 2000 mm

Internal pressure (p) = 0.8 N/mm^2

Wall thickness of the cylinder = t = 10 mm

$$\text{Hoop stress} = (\sigma_h) = \frac{pd}{2t} = \frac{0.8 \times 2000}{2 \times 10} = 80 \text{ MPa}$$

$$\text{Longitudinal stress} = \sigma_l = \frac{pd}{4t} = \frac{0.8 \times 2000}{4 \times 10} = 40 \text{ MPa}$$

$$\text{Maximum shear stress} = \frac{\sigma_h - \sigma_l}{2} = \frac{80 - 40}{2} = 20 \text{ MPa}$$

$$\text{Hoop strain } (\in_h) = \frac{\sigma_h}{E} - \mu\frac{\sigma_l}{E} = \frac{1}{2 \times 10^5}\{80 - 0.25 \times 40\} = 3.5 \times 10^{-4}$$

Change in the diameter = $\in_h \times d = 3.5 \times 10^{-4} \times 2000 = 0.7$ mm

$$\therefore \text{ Longitudinal strain } \in_l = \frac{1}{E}\{\sigma_l - \mu\sigma_h\}$$

$$\in_l = \frac{1}{2 \times 10^5}\{40 - 0.25 \times 80\}$$

$$= \frac{20}{2 \times 10^5} = 1 \times 10^{-4}$$

Change in the length of the cylinder

$$\delta L = \in_l \times L$$

$$\delta L = 1 \times 10^{-4} \times 3500$$

$$= 0.35 \text{ mm}$$

Change in the internal volume = $\delta V = (e_v \times V)$

e_v = volumetric strain

$$= 2 \in_h + \in_l$$

$$\therefore \quad (e_v) = 2 \times 2 \times 10^{-4} + 1 \times 10^{-4}$$

$\therefore$ Change in the volume $= 5 \times 10^{-4} \times \frac{\pi}{4}(2000)^2 \times 3500$

$$= 5.498 \times 10^6 \text{ mm}^3$$

(To express increases, in volume generally cm^3 is used as mm^3 unit is very small. It is better to express the cost of the banana as Rs. 2/- rather than 200 paise. Thus, express the increase in internal volume in the cm^3 units).

Increase in the internal volume = δV = 5498 cubic centimetres (cc)

PROBLEM 8.2 **Objective 1**

A thin walled cylindrical pressure vessel has 1 m diameter and 2.5 m long. The cylinder is completely filled with a fluid at atmospheric pressure. Through a small hole an additional fluid of 2000 cc is filled and the hole is closed. Estimate the pressure developed in the vessel, if the wall thickness of the vessel is 8 mm. Take modulus of elasticity and Poisson's ratio of the material of the cylinder are 80 GPa and 0.25, respectively. The bulk modulus of fluid is 2.5 GPa.

SOLUTION

Additional fluid of 2000 cc pumped into the cylinder causes two phenomena.

1. Increase in the internal volume due to increase in the internal pressure.
2. Decrease in the volume of the fluid due to pressure (i.e., compressibility of the fluid)

Because of the above two phenomena, the additional volume of the fluid pumped into the cylinders is equal to the sum of volumetric changes due to increases in the internal volume of the cylinder and decrease in the fluid volume due to pressure.

Let 'p' be the pressure developed in the cylinder.

Increase in internal volume of the cylinder $(\delta V_1) = V \times \frac{pd}{4tE}(5 - 4\mu)$

$$\therefore \quad \delta V_1 = V \times \frac{p \times 1000}{2 \times 8 \times 2 \times 10^5}[5 - 4 \times 0.25]$$

$$= V \times P \times 0.00125$$

Decrease in the fluid volume due to pressure $(\delta V_2) = V \times \frac{P}{k_f}$

in which k_f is the bulk modulus of the fluid.

$$\therefore \quad \Delta V_2 = V \times \frac{p}{13 \times 10^3}$$

$$\Delta V_1 + \Delta V_2 = \Delta V$$

$$\Delta V = v\left\{\frac{pd}{4tE}(5-4\mu) + \frac{p}{k_f}\right\}$$

$$2000 \times 10^3 = \frac{\pi}{4} \times 1000^2 \times 2500\left\{p \times 0.00125 + p \times \frac{1}{2.5 \times 10^3}\right\}$$

$$\Rightarrow \quad p\{0.00165\} \times \frac{\pi}{4} \times 1000^2 \times 2500 = 2000 \times 10^3$$

$$\Rightarrow \quad p = 0.617 \text{ MPa}$$

If the fluid is incompressible, then k_f becomes infinity. In that case $p = 0.814$ MPa.

PROBLEM 8.3

Objective 1

A copper tube with internal diameter 25 mm and thickness 2.5 mm is filled with water at atmospheric pressure. An axial compressive load of 10 kN is applied at ends. Determine the increase in the internal pressure.

Take $E_{cyl.} = 100$ GPa, $\mu_{cyl.} = 0.25$, and $k_f = 25$ GPa

SOLUTION

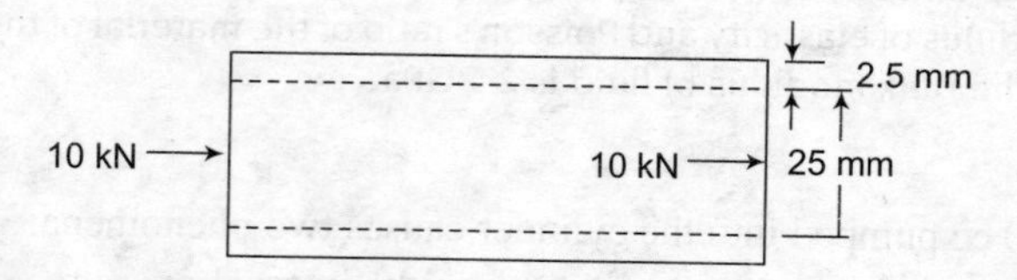

FIGURE 8.7

Let 'p' be the pressure developed in the cylinder due to 10 kN of compressive load. The 10 kN compressive force creates compressive stress in the tube (cylinder) in the longitudinal direction. However, internal fluid pressure develops longitudinal tensile stress.

Longitudinal stress due to 'p' = $\sigma_{l1} = \frac{pd}{4t}$ (tensile)

$$\sigma_{l2} = \frac{10 \text{ kN}}{\pi dt}$$

Longitudinal compressive stress due to 10 kN load $\sigma_{l2} = \frac{10 \times 10^3}{\pi \times 25 \times 2.5}$

Net longitudinal stress $\sigma_l = \dfrac{pd}{4t} - \dfrac{10 \times 10^3}{\pi dt}$ (tensile)

$$= \frac{p \times 25}{4 \times 2.5} - \frac{10 \times 10^3}{\pi \times 25 \times 2.5}$$

$$= 2.5p - 50.93 \text{ (1)}$$

Hoop stress $\sigma_h = \dfrac{pd}{2t} = 5p$

Increase in the internal volume of the cylinder due to only 'p' is $\delta V_1 = v \times \{2 \in_{l1} + \in_l\}$

Hoop strain = $\varepsilon_h = \dfrac{\sigma_h}{E} - \mu \times \dfrac{\sigma_l}{E}$

$$= \frac{1}{E}\{5p - 0.25(2.5p - 50.93)\}$$

$$= \frac{1}{E}\{4.37p + 12.733\}$$

Longitudinal strain, $\varepsilon_l = \dfrac{\sigma_l}{E} - \mu \times \dfrac{\sigma_h}{E}$

$$\varepsilon_1 = \frac{1}{E}\{2.5p - 50.93 - 0.25 \times 5p\}$$

$$= \frac{1}{E}\{1.25p - 50.93\}$$

$$\therefore \quad \delta V_1 = \frac{V}{E}\{2(4.375p + 12.733) + (1.25p - 50.93)\}$$

$$= \frac{V}{E}\{10p - 125.464\}$$

Decrease in the volume of the fluid due to compressibility = $\delta V_2 = V \times \dfrac{p}{k_f}$

As no fluid can move out of the cylinder or into the cylinder,

$$\delta V_1 + \delta V_2 = 0$$

$$\Rightarrow \quad \frac{V}{E}\{10P - 76.396\} + V \times \frac{p}{k_f} = 0$$

$$10p - 76.396 + p\left\{\frac{100}{2.5}\right\} = 0$$

$$\Rightarrow \quad p = 1.528 \text{ MPa}$$

8.3 SHRINK FIT

The design of cylinders mainly depends on hoop stress, as this stress is the maximum stress that is developed in a cylinder. So any mechanism or technique, which reduces the hoop stress, makes the design of cylinders economical. The hoop stress in a cylinder can be reduced by applied radial compressive stress at the outer periphery of the cylinder as shown in Figure 8.8. If p_s is the pressure at the outer periphery of the cylinder, the hoop stress, then hoop stress $\sigma_h = \dfrac{(p - p_s)d}{2t}$.

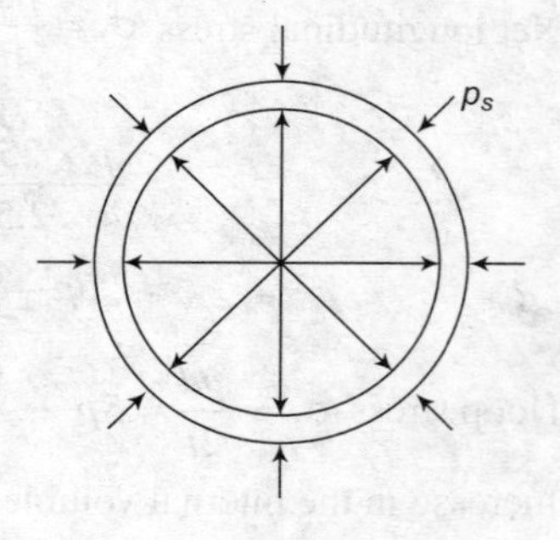

FIGURE 8.8

'p_s' can be applied externally by fitting another cylinder outside, whose inner diameter is slightly less than the outer diameter of the inner cylinder. The outer cylinder is heated so that the outer diameter of the inner cylinder and inner diameter of the outer cylinder match. After fitting outer cylinder over inner one the outer cylinder is called, thereby the outer cylinder applies compression on the inner cylinder. This process is called 'shrink fit'. The pressure that develops at the junction of the cylinders is called shrinkage pressure (p_s). The difference in the inner radius of the outer cylinder and outer radius of the inner cylinder is called shrinkage allowance. It is called that there exists definitely a relationship between the shrinkage pressure and the shrinkage allowance.

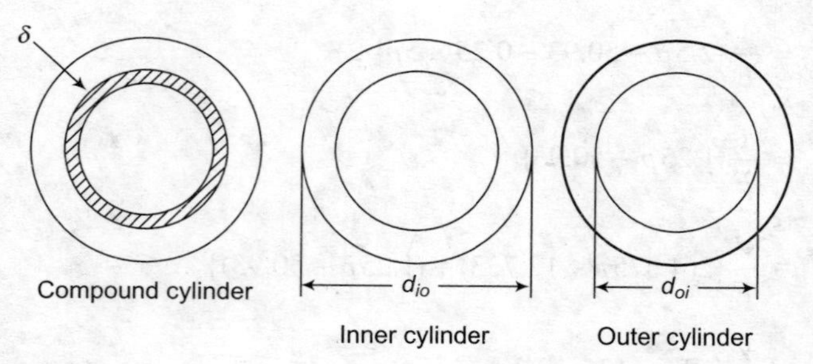

FIGURE 8.9

d_{oi} = inner diameter of the outer cylinder
d_{io} = outer diameter of the inner cylinder

$$\text{Shrinkage allowance } (\delta) = \frac{d_{io} - d_{oi}}{2}$$

Relationship between shrinkage allowance (δ) and shrinkage pressure (p_s): Consider a compound cylinder shown in Figure 8.10.

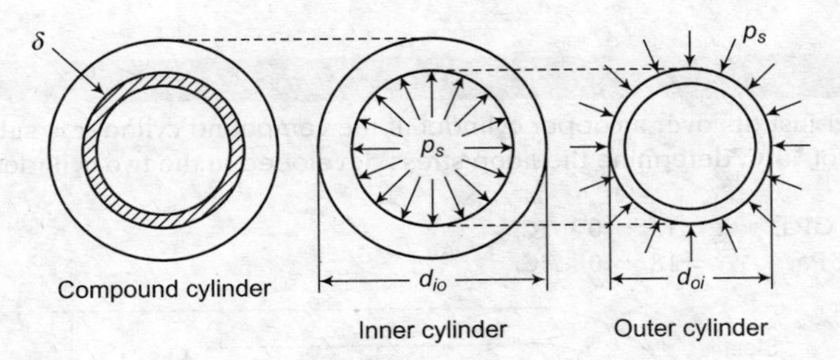

FIGURE 8.10

Let 'p' be the shrinkage pressure at the junction.

Let 't_o' and 't_i' be the thickness of the outer cylinder and inner cylinder, respectively. Let d_o and d_i be the diameter of the outer cylinder and inner cylinder, respectively. In a thin cylinder, inner diameter and outer diameter of a cylinder are considered as mean diameter as the thickness of the cylinder is small compared to the diameter.

Hoop stress in the outer cylinder (tensile) $\sigma_{ho} = \dfrac{p_s d_o}{2t_o}$

Hoop stress in the inner cylinder (compressive) $\sigma_{hi} = \dfrac{p_s d_i}{2t_i}$

The presence of two cylinders reduces the longitudinal stress, thus effect of longitudinal stress on the hoop strain of the cylinders can be ignored.

∴ Hoop strain in the inner cylinder (compressive strain) $\varepsilon_{hi} = \dfrac{p_s d_i}{2t_i E_i}$

in which E_i is the modulus of elasticity of the inner cylinder.

Hoop strain in the outer cylinder (tensile strain) $\varepsilon_{ho} = \dfrac{p_s d_o}{2t_o E_O}$

in which 'E_o' is the modulus of elasticity of outer cylinder.
Increase in the inner diameter of the outer cylinder $(\delta d_o) = \varepsilon_{ho} \times d_o$

$$= \frac{p_s d_o^2}{2t_o E_o}$$

Decrease in the outer diameter of the inner cylinder $(\delta d_i) = \varepsilon_{hi} \times d_i$

$$= \frac{p_s d_i^2}{2t_i E_i}$$

$$\therefore \text{ Shrinkage allowance } = \frac{\delta d_o + \delta d_i}{2}$$

$$\Delta = \frac{p_s}{4}\left\{\frac{d_o^2}{t_o E_o} + \frac{d_i^2}{t_i E_i}\right\}$$

PROBLEM 8.4

Objective 2, 3

A thin steel cylinder just fits over a copper cylinder. If the compound cylinder is subjected to a rise in the temperature of 40°C, determine the hoop stress developed in the two cylinders, due to rise in temperature.

$E_s = 200$ GPa; $\alpha_s = 12 \times 10^{-6}$ /°C,
$E_c = 85$ GPa; $\alpha_c = 18 \times 10^{-6}$ /°C,

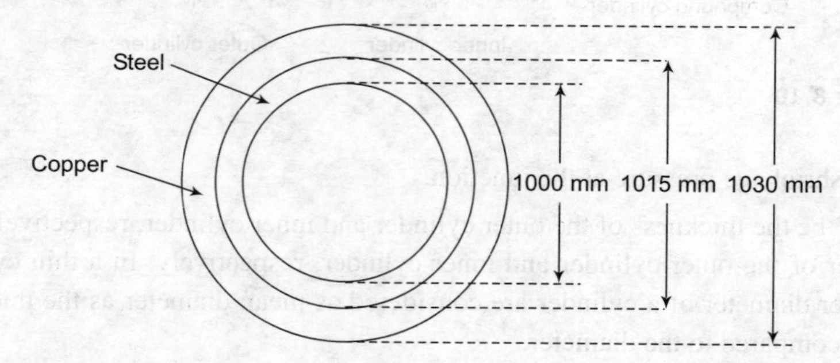

FIGURE 8.11

SOLUTION

Given that

$$\text{Thickness of the outer cylinder } (t_o) = \frac{1030 - 1015}{2} = 7.5 \text{ mm}$$

$$\text{Thickness of the inner cylinder } (t_i) = \frac{1015 - 1000}{2} = 7.5 \text{ mm}$$

Because of rise in temperature, inner diameter of the outer cylinder increases by

$$\delta d_i = \alpha_s \times [d_{\text{oi}}] \times T$$
$$= 12 \times 10^{-6} \times 1015 \times 40$$
$$= 0.487 \text{ mm}$$

Increase in the outer diameter of the inner cylinder

$$= \delta d_i = \alpha_c \times d_{\text{io}} \times T$$
$$= 18 \times 10^{-6} \times 1015 \times 40$$
$$= 0.731 \text{ mm}$$

$\therefore$ Difference the diameters at junction $= 0.731 - 0.487$

$$= 0.244 \text{ mm}$$

$$\therefore \text{ Shrinkage allowance } = \frac{0.244}{2} = 0.122 \text{ mm}$$

Let p_s be the shrinkage pressure at the junction. We know that,

$$\Delta = \frac{p_s}{4}\left\{\frac{d_o^2}{t_o E_s} + \frac{d_i^2}{t_i E_c}\right\}$$

$$\Rightarrow \qquad 0.122 = \frac{p_s}{4}\left\{\frac{1015^2}{7.5 \times 200 \times 10^3} + \frac{1015^2}{7.5 \times 80 \times 10^3}\right\}$$

$$\Rightarrow \qquad p_s = 0.203 \text{ MPa}$$

$\therefore$ Hoop tensile stress in the outer cylinder $= \dfrac{p_s d_o}{2t_o}$

Hoop compressive stress in the inner cylinder $= \dfrac{p_s d_i}{2t_i}$

PROBLEM 8.5

Objective 2, 4

The steel cylinders are used as a compound cylinder with a difference in the common diameter as 0.25 mm. If the common diameter is 100 mm, thickness of the outer cylinder is 2.5 mm and that of the inner is 2 mm, find the shrinkage pressure developed at the junction. Take $E = 200$ GPa. Determine the hoop stress in the two cylinders if a pressure of 21 MPa acts inside the inner cylinder.

SOLUTION

Divide the problem into two cases.

Case (i): Stress calculation due to shrinkage pressure alone.

Case (ii): Stress calculation due to internal pressure alone.

Final stress in the cylinder is the algebraic sum of stresses in case (i) and case (ii).

Case (i): No internal pressure, only shrinkage pressure acts on the cylinders.

Shrinkage allowance $\delta = \dfrac{p_s}{4}\left\{\dfrac{d_o^2}{t_o E_o} + \dfrac{d_i^2}{t_i E_i}\right\}$

Given that,

$$d_i = d_o = 100 \text{ mm}$$

$$\delta = \frac{0.25}{2} = 0.125 \text{ mm}$$

(Difference in diameter is 0.25 mm; thus, difference at junction is $\dfrac{0.25}{2} = 0.125\,\text{mm}$)

$$t_o = 2.5 \text{ mm}$$

$$t_i = 2.0 \text{ mm}$$

$$E_i = E_o = 200 \text{ GPa}$$

$$\therefore \qquad 0.125 = \frac{p_s}{4}\left\{\frac{100^2}{2.5 \times 200 \times 10^3} + \frac{100^2}{2 \times 200 \times 10^3}\right\}$$

$\Rightarrow \qquad p_s = 11.11$ MPa

$\therefore\ \sigma_{\text{hoi}}$ = hoop stress in the outer cylinder for case (i)

$$= \frac{p_s(d)}{2t_o} = \frac{11.11 \times 100}{2 \times 2.5}$$

$$= 222.22 \text{ MPa (tensile)}$$

σ_{h1} = Hoop stress in the inner cylinder for case (i)

$$\frac{p_s d}{2t_i} = 277.78 \text{ MPa (compressive)}$$

Case (ii): Stresses due to internal fluid alone.

$$t = t_o + t_i = 2 + 2.5 = 4.5 \text{ mm}$$

Hoop stress is same in outer as well as inner cylinder $= \sigma_{h2} = \dfrac{pd}{2t}$

$$= \frac{21 \times 100}{2 \times 4.5} = 233.33 \text{ MPa}$$

Hoop stress due to inner pressure is uniform and is tensile.

$\therefore$ Final stress in the outer cylinder $= \sigma_{ho} = \sigma_{ho1} + \sigma_{ho2}$

$$= 222.22 + 233.33$$

$$= 455.56 \text{ MPa}$$

Final stress in the inner cylinder $= \sigma_{hi} = \sigma_{hi1} + \sigma_{hi2}$

$$= -277.77 + 233.33$$

$$= -44.44 \text{ MPa}$$

In the absence of shrink fitting, the hoop stress developed in the cylinder

$$\sigma_h = \frac{pd}{2t}$$

$$= \frac{21 \times 100}{2 \times 2} = 525 \text{ N/mm}^2 \text{ (tensile).}$$

8.4 WIRE WOUND CYLINDERS

The hoop stress in cylinders is reduced by winding the cylinders with high tensile steel wires under tension. Wire wound round the cylinder under tension, induces hoop compression in the walls of the cylinder. This hoop compression walls of the cylinder due to internal pressure. A part from shrinking fitting, wire wounding around the cylinders is another technique to reduce the hoop stress or for economical design. Generally, the wire is wound closely.

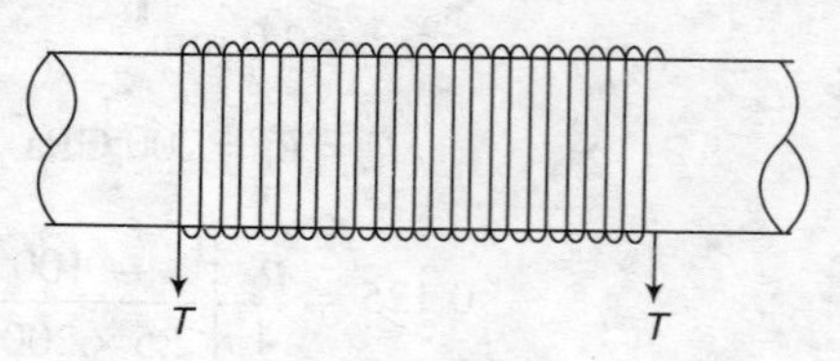

FIGURE 8.12

The analysis of these cylinders is in two stages. In the first stage, the hoop stress in the walls of cylinder is computed only due to tension in wire (no internal fluid pressure). In the second stage, stress in the wire and cylinder is computed due to internal pressure only.

The final stress in the wire and the cylinder are found by the algebraic summation of stresses in two stages.

First stage analysis Let d_ω be the diameter of the wire and $\sigma_{\omega 1}$ be the stress with which the wire wound round the cylinder. Assume that no internal pressure exists. Replace the wire with equivalent thick cylindrical shell. Let this uniform equivalent thickness be 't_ω'. Then, area of the wire over a length of 'd_ω' distance is (because one wire will be available in 'd_ω' distance):

$$\frac{\pi}{4} d_\omega^2 = t_\omega \times d_\omega$$

$$\Rightarrow \qquad t_\omega = \frac{\pi}{4} d_\omega$$

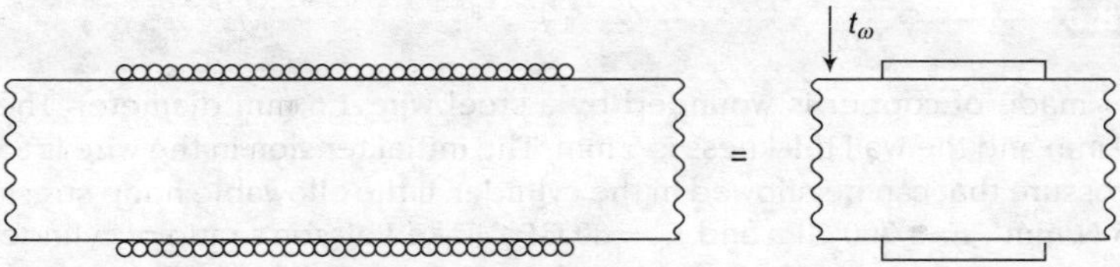

FIGURE 8.13

If more layers of wire present, then 't_ω' should be multiplied by the number of layers of wire. Applying equilibrium equation $\sum V = 0$

$$\Rightarrow \qquad \sigma_{nl} \times t_\omega \times 2 = \sigma_{n1} \times t \times 2$$

$$\sigma_{n1} = \sigma_{\omega 1} \times \left[\frac{t_\omega}{t}\right] \qquad (8.5)$$

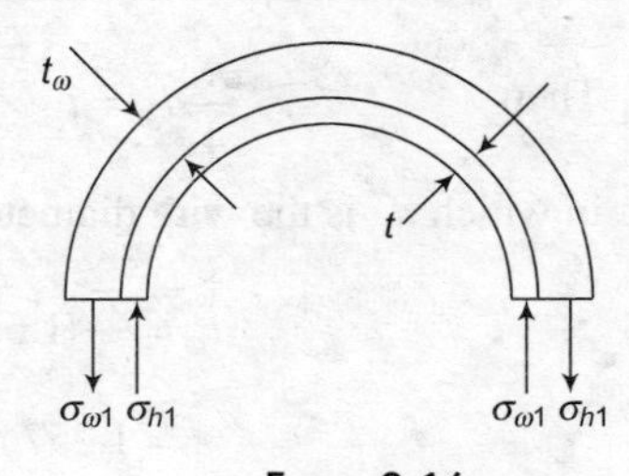

FIGURE 8.14

σ_{n1} is the hoop compressive stress developed in the wall of the cylinder.

Second stage stresses Let σ_{n2} and $\sigma_{\omega 2}$ be stress developed in the cylinder and wire, respectively, due to internal pressure only.

Applying equilibrium condition that sum of vertical force = 0

$$\Rightarrow \qquad \sigma_{h2} \times 2 \times t + \sigma_{\omega 2} \times 2t_\omega = pd \qquad (8.6)$$

in which 'd' is the diameter of the cylinder.

From the above equation, σ_{h2} and $\sigma_{\omega 2}$ cannot be found. The compatibility condition insists that there exists perfect bond/contact between the cylinder and wire.

⇒ hoop strain in the cylinder is equal to the longitudinal strain in the wire.

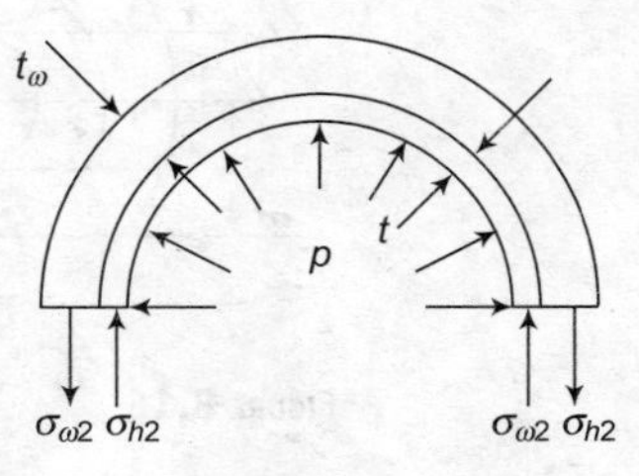

FIGURE 8.15

$$\therefore \quad \frac{\sigma_{h2}}{E_c} - \frac{\sigma_l}{E_c} = \frac{\sigma_{\omega 2}}{E_S} \tag{8.7}$$

in which σ_l = longitudinal stress in the cylinder, which is not influenced by wire winding. In some case, 'σ_l' can be neglected.

$$\sigma_l = \frac{pd}{4t}.$$

Using Eqs. (8.6) and (8.7), σ_{h2} and $\sigma_{\omega 2}$ can be evaluated.

In the third state, final stress in the cylinder and wire can be computed as

$$\sigma_h = -\sigma_{h1} + \sigma_{h2} \text{ (as } \sigma_{h1} \text{ is compressive)}$$

$$\sigma_\omega = \sigma_{\omega 1} + \sigma_{\omega 2}$$

PROBLEM 8.6

Objective 1, 3

A thin cylinder made of copper is wounded by a steel wire 1.6 mm diameter. The diameter of the cylinder is 200 mm and the wall thickness is 5 mm. The initial tension in the wire is 85 MPa. Determine the internal pressure that can be allowed in the cylinder, if the allowable hoop stress in the cylinder is limited to 42.5 N/mm^2. E_S = 200 GPa and E_C = 80 GPa. Take Poisson's ratio of cylinder as 0.25.

SOLUTION

Let 't_ω' be the equivalent thickness of the winding wire.

Then $$\frac{\pi}{4} d_\omega^2 = t_\omega \times d_\omega$$

in which d_ω is the wire diameter.

$$\therefore \quad t_\omega = \frac{\pi}{4}(1.6)$$

$$= 1.257 \text{ mm}$$

Stage I: When the internal pressure is absent.

Given that $\sigma_{\omega 1}$ = 85 MPa

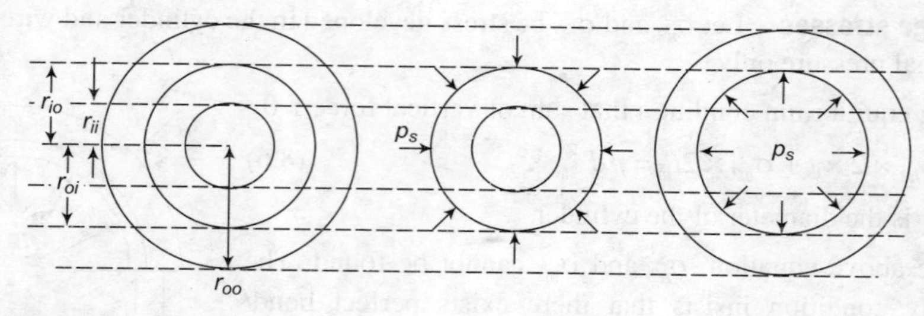

FIGURE 8.16

Applying equilibrium equation,

$$\sum F_V = 0$$

$$\Rightarrow \quad \sigma_{\omega 1} \times t_\omega \times 2 = \sigma_{h1} \times t \times 2$$

$$\Rightarrow \quad 1.257 \times 2 \times 85 = \sigma_{h1} \times 5 \times 2$$

$$\Rightarrow \quad \sigma_{h1} = 21.369 \text{ MPa} \quad \text{(compressive)}.$$

Stage 2: Stress due to internal pressure.

Let 'p' be the internal pressure. Let σ_{n2} and $\sigma_{\omega 2}$ be the stress in the cylinder and steel wire, respectively.

$$\sum F_V = 0$$

$$2 \times t_\omega + \sigma_{\omega 2} + 2 \times t \times \sigma_{h2} = p \times d$$

$$\Rightarrow \quad 2.514\sigma_{\omega 2} + 10\sigma_{h2} = 200p \quad (8.8)$$

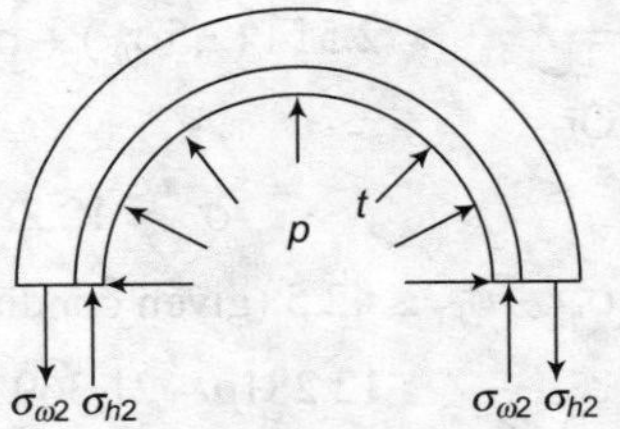

FIGURE 8.17

Applying compatibility condition,

$$\frac{\sigma_{h2}}{E_C} - \mu \times \frac{\sigma_l}{E_C} = \frac{\sigma_{\omega 2}}{E_S}$$

$$\sigma_l = \frac{pd}{4t} = \frac{p \times 200}{4 \times 5} = 10p$$

$$\therefore \quad \frac{E_S}{E_C} = \{\sigma_{h2} - 0.25 \times 10p\} = \sigma_{\omega 2}$$

$$2.5\sigma_{h2} - \sigma_{\omega 2} = 6.25p \quad (8.9)$$

Solving Eqs. (8.8) and (8.9)

$$10\sigma_{h2} - 4\sigma_{\omega 2} = 25p$$

$$10\sigma_{h2} + 2.514\sigma_{\omega 2} = 100p$$

$$-6.514\sigma_{\omega 2} = -175p$$

$$\Rightarrow \quad \sigma_{\omega 2} = 26.865p \quad (8.10)$$

$$\sigma_{h2} = 13.246p \quad (8.11)$$

It is given that final hoop stress in the cylinder is limited to 42.5 MPa.

$$\therefore \quad \sigma_{h2} - \sigma_{h1} = 42.5$$

$$\Rightarrow 13.426p - 12.369 = 42.5$$

$$\therefore \quad p = 4.822 \text{ MPa}$$

Final tress in the wire $= \sigma_{\omega 1} + \sigma_{\omega 2}$

$$= 85 + 26.865p$$

$$= 214.54 \text{ MPa (tensile)}$$

If the longitudinal stress is neglected, then compatibility condition changes to

$$\frac{\sigma_{h2}}{E_C} = \frac{\sigma_{\omega 2}}{E_S}$$

$$\Rightarrow \qquad \sigma_{\omega 2} = 2.5\,\sigma_{h2}$$

$$2.514\sigma_{\omega 2} + 10\sigma_{h2} = 200p \text{ [Eq. (8.8)]}$$

$$\Rightarrow \qquad 2.514\,(2.5\sigma_{h2}) + \sigma_{h2} = 200p$$

Or

$$\sigma_{h2} = 12.281p$$

$$\sigma_{h2} - \sigma_{h1} \le 42.5 \text{ (given condition)}$$

$$\therefore \qquad 12.281p - 21.369 \le 42.5$$

$$\therefore \qquad p \le 5.2 \text{ MPa}$$

If 'σ_l' is neglected, then error percentage in estimating hoop stress is 7.84%.

PROBLEM 8.7

Objective 3

A steel pipe of diameter 500 mm and thickness 10 mm is transmitting water at 0.4 MPa pressure. The pipe is simply supported on two rigid supports, separated by 6.5 m length. Determine the maximum longitudinal stress induced in the cylinder wall.
Take density of water as 1000 kg/m^3 and that of steel is 8000 kg/m^3.

SOLUTION

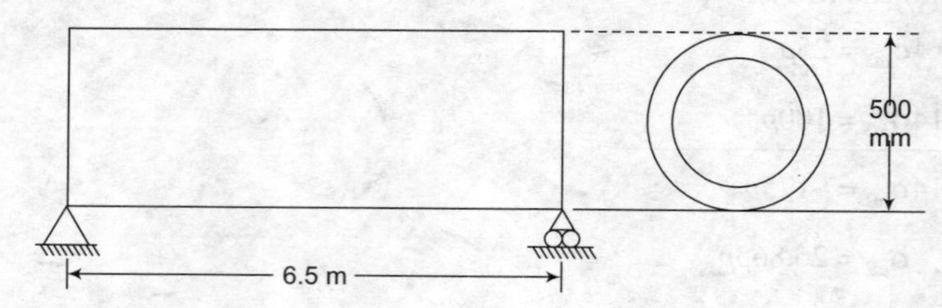

FIGURE 8.18

In the longitudinal direction, the stresses that developed are

(i). Longitudinal stress due to internal pressure

$$\sigma_{l1} = \frac{pd}{4t}$$

(ii). Bending stress due to weight of water and weight of cylinder

$$\sigma_{l2} = \frac{M}{I} \times y$$

Longitudinal stress due to internal fluid pressure $= \dfrac{pd}{4t} = \dfrac{0.4 \times 500}{4 \times 10}$

$$\sigma_{l1} = 5 \text{ MPa (tensile).}$$

Longitudinal stress due to bending

Weight of water per 1m length of the pipe $W_1 = A_p \times \gamma_\omega = \gamma_\omega \times \dfrac{1}{4}\pi D^2$

$$= 1000 \times 9.81 \times \frac{\pi 0.5^2}{4}$$

(9.81 is to convert kg/m^3 into N/m^3) = 1926.19 N/m

$$\therefore \quad W_1 = 1.926 \text{ kN/m}$$

Weight of pipe per 1 m length = $W_2 = \pi D T \times r_p$

$$= 8000 \times 9.81 \times \pi\left(\frac{500}{1000}\right)\left(\frac{10}{1000}\right)$$

$$= 392.4 \text{ N/m}$$

$$W_2 = 0.392 \text{ kN/m}$$

$\therefore$ Total load on the pipe = $W = W_1 + W_2$

$$= 2.318 \text{ kN/m}$$

Maximum bending moment (at midspan) = $M = \dfrac{WL^2}{8}$

$$= \frac{2.318 \times 6.5^2}{8}$$

$$= 12.242 \text{ kN·m}$$

Moment of inertia of the pipe section $= \dfrac{\pi d^3 t}{8} = \dfrac{\pi \times 500^3 \times 10}{8}$

$$= 4.909 \times 10^8 \text{ mm}^4$$

$$y_{\max} = \frac{d}{2} = 250 \text{ mm}$$

As the member is under sagging moment, bottom fiber are subjected to tensile stress due to bending

$$\therefore \quad \sigma_{l2} = \frac{M}{I} y_{\max}$$

$$= \frac{12.242 \times 10^6}{4.909 \times 10^8} \times 250 = 6.234 \text{ MPa}$$

The top fibers of the cylinder are subjected to compressive stress

$\therefore$ Longitudinal stress at bottom fiber at midspan = $\sigma_{l1} + \sigma_{l2}$

$$= 5 + 6.234$$

$$= 11.234 \text{ MPa (tensile)}$$

Longitudinal stress at top fiber at midspan = $\sigma_{l1} - \sigma_{l2}$

$$= 5 - 6.234$$

$$= -1.234 \text{ (compressive).}$$

The variation of resulting longitudinal stress at midspan is shown in Figure 8.19.

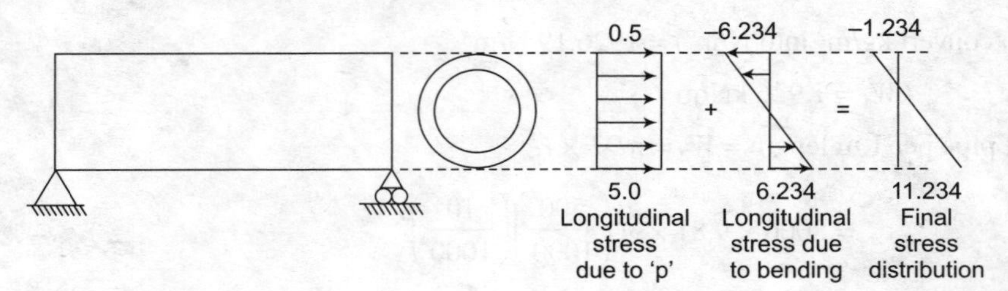

FIGURE 8.19

8.5 THICK CYLINDERS

In thick cylinders, the inner diameter (d_i) to thickness (t) ratio is less than 20. In thin cylinders, the variation of radial stress along the wall thickness is ignored, whereas in the thick cylinders variation in radial stress along the wall thickness is considered. As the thickness of wall is more, ignoring the radial stress variation makes the design more critical and unsafe. Essentially, the analysis of thick cylinders falls under the category of statically indeterminate analysis.

Assumptions

1. Material of the cylinder is homogenous, elastic, and isotropic, and obeys Hook's law.
2. Plane sections remain plane before and after the application of fluid pressure.

The second assumption assumes importance because, cylinder, when subjected to internal fluid pressure expands (deforms) in the longitudinal direction due to 'ε_l'. The plane section remain plane before and after application of internal pressure as shown in Figure 8.20 and is same over the depth of cylinders, thus ε_l is constant. Importance of this condition can be observed in the derivation of expressions for radial stress and hoop stress.

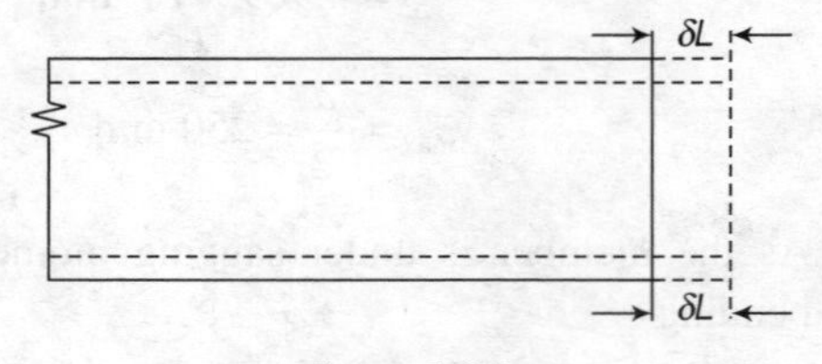

FIGURE 8.20

8.5.1 Expression for Hoop Stress and Radial Stress

Consider a thick pipe subjected to internal pressure as shown in Figure 8.21. Let r_o and r_i be the outer radius and inner radius of the cylinder, respectively. In case of thick cylinders, radial stress variation is considered. The radial stress varies from 'p' to '0' from radius $r = r_i$ to $r = r_o$.

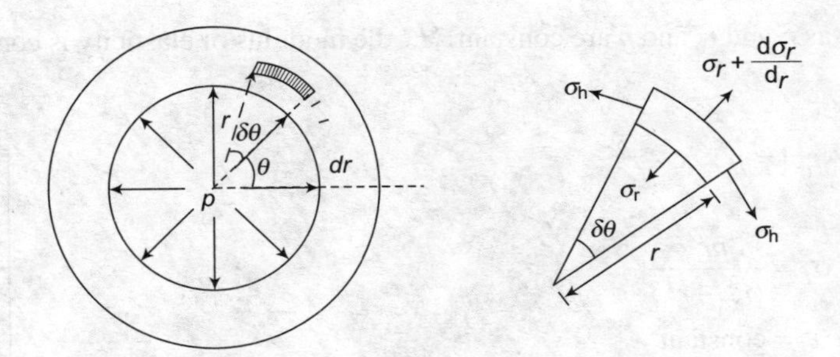

FIGURE 8.21

Consider an element shown in Figure 8.21, located 'r' distant from the center of the cylinder. Let 'dr' be the thickness of the elemental strip. σ_r is the tensile radial stress at radius r and the tensile radial stress at radius $r + dr$ is $\sigma_r + \frac{d\sigma_r}{dr} \times dr$. Let us consider the equilibrium of the elemental strip. The forces on the elemental strip are shown in Figure 8.22.

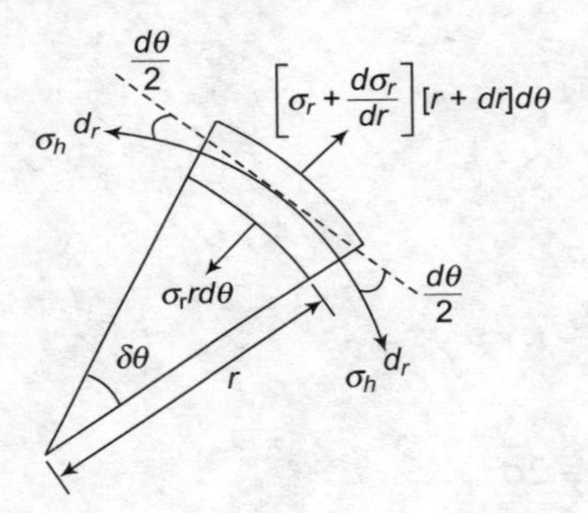

FIGURE 8.22

Resolving the forces along the radial,

$$\left[\sigma_r + \frac{d\sigma_r}{dr} \times dr\right] \times [r + dr]d\theta - \sigma_r \times rd\theta - 2 \times \sigma_h \times dr \times 1 \times \sin\frac{d\theta}{2} = 0$$

$\left(\sin\frac{d\theta}{2} = \frac{d\theta}{2}\right.$ as $d\theta$ is very small) ignoring the higher order terms.

$$\therefore \qquad \frac{d\sigma_r}{dr} \times dr \times r \times d\theta + \sigma_h \times dr \times d\theta - 2 \times \sigma_h \times dr \times \frac{d\theta}{2} = 0$$

$$\Rightarrow \qquad \sigma_h - \sigma_r = \frac{d\sigma_r}{dr} \times r \tag{8.12}$$

Plane sections remain plane even after the application of internal pressure. As per this assumption, longitudinal strain is constant.

'σ_l' is constant as r_i and r_o and p are constant. 'E' the modulus of elasticity is constant, and hence $\frac{\sigma_l}{E}$ is constant.

$$\sigma_l \times \pi[r_o^2 - r_i^2] = p \times \pi \times r^2$$

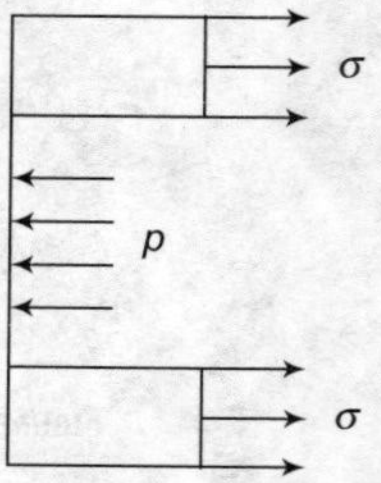

$$\Rightarrow \quad \sigma_l = \frac{pr_i^2}{(r_o^2 - r_i^2)}$$

$$\Rightarrow \quad \varepsilon_l = \text{constant}$$

$$\varepsilon_l = \frac{\sigma_L}{E} - \mu\left(\frac{\sigma_r + \sigma_h}{E}\right) = \text{constant}$$

FIGURE 8.23

$$\sigma_l = \text{longitudinal stress} = \frac{(p)r_i^2}{(r_o^2 - r_i^2)}$$

$$\sigma_r + \sigma_h = \text{constant say 2A}$$

$$\sigma_r + \sigma_h = 2\text{A} \tag{8.13}$$

$$-\sigma_r + \sigma_h = \frac{d\sigma_h}{dr} \times r \tag{8.14}$$

Eq. (8.12) – Eq. (8.13)

$$\Rightarrow \quad 2\sigma_r = 2A - \frac{d\sigma_r}{dr} \times r$$

or
$$\frac{d\sigma_r}{dr} \times r = 2A - 2\sigma_r$$

or
$$\frac{d\sigma_r}{A - \sigma_r} = 2\frac{dr}{r}$$

Integrating on both sides,

$$-\ln[A - \sigma_r] = 2\ln r + \text{constant}$$

$$\Rightarrow \quad \ln\{r^2[A - \sigma_r]\} = \text{constant}$$

or
$$r^2\{A - \sigma_r\} = \text{constant} = B \text{ (say)}$$

$$\therefore \quad \sigma_r = A - \frac{B}{r^2}.$$

A and B are the constants, which can be found from the boundary conditions that at $r = r_i$, $\sigma_r = -p$ (as pressure is compression) and at $r = r_o$, $\sigma_r = 0$ (pressure at outer surface of the cylinder)

$$\sigma_r = A - \frac{B}{r^2}.$$

We know that $\sigma_r + \sigma_h = 2A$

$$\therefore \qquad \sigma_h = A + \frac{B}{r^2}.$$

Hoop stress is maximum when r is minimum possible value, that is, $r = r_i$.

PROBLEM 8.8 **Objective 3**

Sketch the variation of radial stress and hoop stress across the thickness of the walls of a cylinder subjected to internal pressure 'p'. Compare this solution, considering the cylinder as a thin cylinder.

SOLUTION

We know that radial stress

$$\sigma_r = A - \frac{B}{r^2}$$

At $r = r_1$, $\sigma_r = -p$ and at $r = r_o$, $\sigma_r = 0$

$$A - \frac{B}{r_o^2} = 0 \tag{8.14}$$

$$A - \frac{B}{r_i^2} = -p \tag{8.15}$$

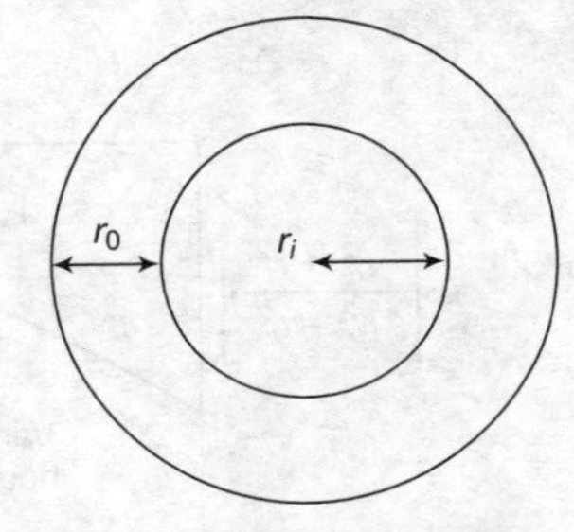

FIGURE 8.24

Eq. (8.14) – Eq. (8.15)

$$B\left\{\frac{1}{r_i^2} - \frac{1}{r_o^2}\right\} = p$$

$$\Rightarrow \qquad B = \frac{p r_o^2 r_i^2}{r_o^2 - r_i^2}.$$

The units of 'B' are 'N'.

$$A = \frac{p r_i^2}{r_o^2 r_i^2}.$$

Units of 'A' are N/mm^2

$$\therefore \qquad \sigma_r = \frac{p r_i^2}{r_o^2 - r_i^2}\left\{1 - \frac{r_o^2}{r^2}\right\}$$

And

$$\sigma_h = A + \frac{B}{r^2}$$

Or

$$\sigma_h = \frac{p r_i^2}{r_o^2 - r_i^2}\left\{1 + \frac{r_o^2}{r^2}\right\}$$

σ_n, maximum at $r = r_i$

$$\therefore \qquad \sigma_{n\max} = p\left\{\frac{r_i^2 + r_o^2}{r_i^2 - r_o^2}\right\}.$$

The above expression for maximum hoop stress reduced to $\dfrac{pd}{2t}$.

When $r_o \to r_i \to r$ when $r_o \to r_i \to r$, $(r_o - r_i) \to t$ (thickness)

Thereby, $\sigma_{n\max} = \dfrac{p \times 2r^2}{2r \times t} = \dfrac{pr}{t}$ or $\dfrac{pd}{2t}$.

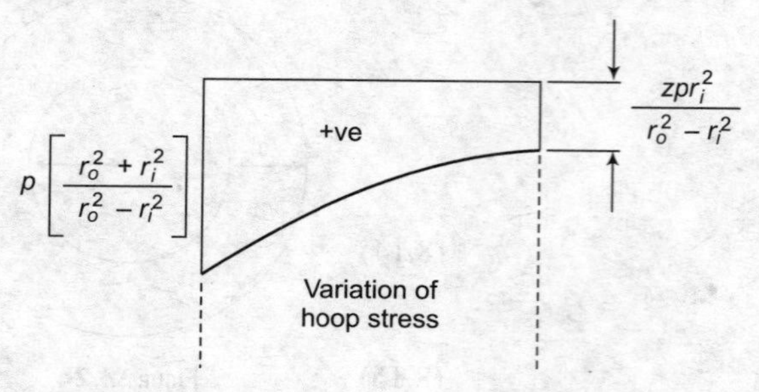

FIGURE 8.25

$$\sigma_{n\max} = p\left\{\frac{r_o^2 + r_i^2}{(r_o^2 - r_i^2)}\right\}$$

We know that $r_o = r_i + t$

$$\therefore \qquad \sigma_{h\max} = p\left\{\frac{r_i^2 + 2r_it + t^2 + r_i^2}{r_i^2 + 2r_it + t^2 - r_i^2}\right\} = \frac{pr_i}{r_it}\left\{\frac{2 + 2\left[\dfrac{t}{r_i}\right] + \left[\dfrac{t}{r_i}\right]^2}{2 + \left(\dfrac{t}{r_i}\right)}\right\}.$$

We know that $2r_i = d$ (inner diameter of the cylinder)

$$\sigma_{h\max} = \frac{pd}{2t}\left\{\frac{2 + 4\left(\dfrac{t}{d}\right) + 4\left(\dfrac{t}{d}\right)^2}{2 + 2\left(\dfrac{t}{d}\right)}\right\}.$$

Taking $\dfrac{t}{d}$ as α

$$\sigma_{h\max} = \frac{pd}{2t}\left\{\frac{2 + \dfrac{4}{\alpha} + \dfrac{4}{\alpha^2}}{2 + \dfrac{2}{\alpha}}\right\} = \frac{pd}{2t}\left\{\frac{1 + \dfrac{2}{\alpha} + \dfrac{2}{\alpha^2}}{1 + \dfrac{1}{\alpha}}\right\}.$$

For $\alpha = 20$, that is, $\dfrac{d}{t} = 20$

$$\sigma_{h\max} = \frac{pd}{2t}\{1.0524\}.$$

If $\dfrac{d}{t} = 20$, the percentage error in estimating the maximum hoop stress in considering the cylinder as thin cylinder is 5.24% (underestimate, thin cylinder analysis).

If

$$\frac{d}{t} = 10, \text{ percentage error} = 10.91\%$$

$$\frac{d}{t} = 40, \text{ percentage error} = 2.56\%.$$

From the above values, it can be understood that, for higher values of $\dfrac{d}{t}$ ratio, thin cylinder analysis can be performed.

Longitudinal stress $\sigma_l = p \times \dfrac{r_i^2}{r_o^2 - r_i^2}$.

Substitute $r_o = r_i + t$;

$$\sigma_l = p \times \frac{r_i^2}{2r_it + t^2} = \frac{pr_i}{2t}\left\{\frac{r_i}{r + \frac{t}{2}}\right\} = \frac{pr_i}{2t}\left\{\frac{1}{1 + \frac{t}{2r_i}}\right\}.$$

Substitute $r_i = \dfrac{d}{2}$

$$\therefore \quad \sigma_l = \frac{pd}{4t}\left\{\frac{1}{1 + \frac{t}{d}}\right\} \text{ or } \sigma_l = \frac{pd}{4t}\left\{\frac{1}{1 + \frac{t}{d}}\right\} \text{ or } \sigma_l = \frac{pd}{4t}\left\{\frac{1}{1 + \frac{1}{\alpha}}\right\} \alpha = \frac{d}{t}.$$

If $\alpha = 10$, $\sigma_l = [0.9091]\dfrac{pd}{4t}$

$$\alpha = 20, \ \sigma_l = [0.9524]\frac{pd}{4t}$$

$$\alpha = 40, \ \sigma_l = [0.9756]\frac{pd}{4t}.$$

With increase in $\dfrac{d}{t}$, σ_l approaches $\dfrac{pd}{4t}$, longitudinal stress considering the cylinder as thin cylinder.

If a thick cylinder is considered as a thin cylinder and analyzed, then hoop stress calculation is not conservative, whereas longitudinal stress calculations are conservative.

PROBLEM 8.9

Objective 1

A thick cylinder of inner diameter 1000 mm is to withstand an internal pressure of 25 MPa, if the allowable stress in tension is not to exceed 100 MPa. Length of the cylinder is 2 m with ends closed with rigid plates. Take $E = 200$ GPa and $\mu = 0.25$. Estimate

(i) increase in the inner diameter
(ii) increase in the outer diameter
(iii) increase in the length
(iv) increase in the internal volume

SOLUTION

Given that $r_i = \dfrac{1000}{2} = 500$ mm

$P = 25$ MPa

The allowable stress in tension is not to exceed 100 MPa.
Maximum tensile stress is the hoop stress at the introdos of the cylinder

$$\sigma_{h\max} = p\left\{\frac{r_i^2 + r_o^2}{r_i^2 - r_o^2}\right\}$$

$$\sigma_{h\max} \leq 100 \text{ MPa.}$$

Let 't' be the wall thickness of the cylinder.

$$p\left\{\frac{r_i^2 + (r_i + t)^2}{(r_i + t)^2 r_i^2}\right\} \leq 100$$

$\therefore$

$$p\left\{\frac{2r_i^2 + 2r_i t + t^2}{2r_i t + t^2}\right\} \leq 100 \Rightarrow p\left\{1 + \left(\frac{2r_i^2}{2r_i t + t^2}\right)\right\} \leq 100$$

$$\Rightarrow \quad 25\left\{1 + \frac{2r_i^2}{(2r_i t + t^2)}\right\} \leq 100 \text{ or } \frac{2r_i^2}{2r_i t + t^2} \leq 3$$

$$\Rightarrow \quad t^2 + 1000t \geq 16.67 \times 10^4$$

$$t^2 + 1000t - 16.67 \times 10^4 \geq 0.$$

Solving the above equation

$$t \geq 145.5 \text{ mm}$$

Provide 150 mm as thickness of the cylinder

$$\therefore \quad r_i = 500 \text{ mm; } r_o = 650 \text{ mm.}$$

Increase in the inner diameter

$$\Delta d_i = \varepsilon_i \times d_i$$

ε_{hi} = hoop strain at $r = r_1$

$$\varepsilon_{hi} = \frac{\sigma_{hi}}{E} - \mu\frac{\sigma_{ri}}{E} - \mu\frac{\sigma_l}{E}$$

σ_{ri} = tensile stress at introdos in the radial direction

$$= -p$$

σ_l = longitudinal stress

$$= \frac{pr_i^2}{r_o^2 - r_i^2} = 25 \times \frac{500^2}{650^2 - 500^2} = 36.23.$$

Hoop stress at $r = r_i$ is

$$\sigma_{hi} = \frac{p[r_o^2 + r_i^2]}{[r_o^2 - r_i^2]}$$

$$\sigma_{hi} = 25\left\{\frac{650^2 + 500^2}{650^2 - 500^2}\right\} = 97.464 \text{ MPa (tensile)}$$

$$\varepsilon_{hi} = \frac{\sigma_{hi}}{E} - \mu\frac{\sigma_{ri}}{E} - \mu\frac{\sigma_l}{E} = \frac{1}{200 \times 10^3}\{97.464 - 0.25(-25) - 0.25(36.23)\}$$

$$= 0.4733 \times 10^{-3}.$$

Increase in the inner diameter Δd_i

$$\Delta d_i = \varepsilon_{hi} \times d_i$$

$$\Delta d_i = 0.4733 \times 10^{-3} \times 1000$$

$$\Delta d_i = 0.4733 \text{ mm}.$$

Let $\in_{ho}$ be the hoop strain at the extrados.
Then,

$$\in_{ho} = \frac{\sigma_{ho}}{E} - \mu - \sigma_{ro} - \mu\frac{\sigma_l}{E}$$

$$\sigma_h = p\left\{\frac{r_o^2 + r^2}{r_o^2 - r_i^2}\right\}\frac{r_i^2}{r^2}$$

At $r = r_o$

$$\sigma_{ho} = \frac{2pr_i^2}{r_o^2 - r_i^2} = \frac{2 \times 25 \times 500^2}{650^2 - 500^2} = 72.464 \text{ MPa}$$

$$\sigma_r = \frac{pr_i^2}{r_o^2 - r_i^2}\left\{1 - \frac{r_o^2}{r_i^2}\right\}$$

At $r = r_o$, $\sigma_{ro} = 0$

$$\sigma_l = 36.23 \text{ MPa (tensile)}$$

$$\varepsilon_{ho} = [72.464 - 0.25 \times 0 - 0.25 \times 36.23]\frac{1}{200 \times 10^3}$$

$$\varepsilon_{ho} = 0.317 \times 10^{-3}.$$

Increase in the external diameter $\Delta d_o = \varepsilon_{ho} \times d_o$

$\Rightarrow$ $$\Delta d_o = 0.317 \times 10^{-3} \times (2 \times 650) = 0.412 \text{ mm}.$$

Longitudinal strain $\varepsilon_l = \frac{\sigma_l}{E} - \mu\frac{\sigma_r}{E} - \mu\frac{\sigma_h}{E}$

ε_l is same everywhere, that is, at $r = r_i$ or r_o

$\therefore$ $$\varepsilon_l = [36.23 - 0.25 \times 0 - 0.25 \times 72464]\frac{1}{200 \times 10^3}$$

$$\varepsilon_l = 0.0906 \times 10^{-3}.$$

Change (increase) in the length $\Delta L = \varepsilon_l \times L$

$\Rightarrow$ $$\Delta L = 0.0906 \times 10^{-3} \times 2000 = 0.181 \text{ mm}.$$

Change in the internal volume

$$\delta V = V \times e_V = V \times (2 \in_{hi} + \in_l)$$

$$\delta V = \frac{\pi}{4} \times 1000^2 \times 2000\{0.4733 \times 10^{-3} \times 2 + 0.0906 \times 10^{-3}\}$$

$$= 1629.23 \times 10^3 \text{ mm}^3$$

$$= 1629.23 \text{ cc}.$$

Increase in the internal volume is 1629.33 cc.

PROBLEM 8.10

Objective 1, 2

The outer diameter of a cylinder is 2 times the inner diameter. Determine the ratio of external pressure to internal pressure applied separately, such that the largest stresses have numerical value if $\mu = 0.25$, determine the same, for the largest strains.

SOLUTION

Let r_o and r_i be the outer and inner radii of the cylinder.

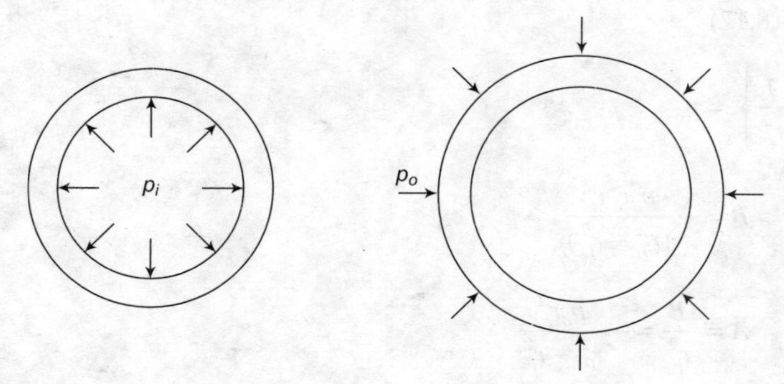

FIGURE 8.26

Given that $\frac{r_o}{r_i} = 2$, we have to determine $\frac{P_o}{P_i}$ for same maximum stress and maximum strain. For internal pressure p_i,

$$\sigma_{h\max} = \frac{p_i[r_o^2 + r_i^2]}{[r_o^2 - r_i^2]}.$$

Given that $\frac{r_o}{r_i} = 2$

$$\therefore \qquad \sigma_{h\max} = p_i\left\{\frac{1+\left(\frac{r_i}{r_o}\right)^2}{1-\left(\frac{r_i}{r_o}\right)^2}\right\} = p_i\left\{\frac{1+\left(\frac{1}{2}\right)^2}{1-\left(\frac{1}{2}\right)^2}\right\}$$

$$\sigma_{h\max} = 1.67\, p_i$$

$$\sigma_{ri} = -p_i$$

$$\sigma_l = \frac{p_i r_i^2}{(r_o^2 - r_i^2)} = \frac{p_i}{\left\{\left(\frac{r_o}{r_i}\right)^2 - 1\right\}} = \frac{p_i}{\{4-1\}} = \frac{p_i}{3}.$$

For external pressure,

$$\sigma_r = A - \frac{B}{r^2}$$

At $r = r_i$, $\sigma_r = 0$ and at $r = r_o$, $\sigma_r = -p_o$

$$A - \frac{B}{r_i^2} = 0 \tag{8.16}$$

$$A - \frac{B}{r_o^2} = -p_o \tag{8.17}$$

Eq. (8.16) – Eq. (8.17)

$$\Rightarrow \quad B\left\{\frac{1}{r_o^2} - \frac{1}{r_i^2}\right\} = p_o$$

$$B = -\frac{p_o r_o^2 r_i^2}{(r_o^2 - r_i^2)}$$

$$\Rightarrow \quad A = \frac{B}{r_i^2} = -\frac{p_o r_o^2}{r_o^2 - r_i^2}$$

$$\therefore \quad \sigma_h = A + \frac{B}{r^2}.$$

For $\sigma_{h\max}$ $r = r_i$

$$\therefore \quad \sigma_{h\max} = -\frac{p_o r_o^2}{r_o^2 - r_i^2} - \frac{p_o r_o^2}{r_o^2 - r_i^2} = \frac{2p_o r_o^2}{r_o^2 - r_i^2} = -\frac{2p_o}{1-\left(\dfrac{r_i}{r_o}\right)^2} = \frac{-2p_o}{1-\left(\dfrac{1}{2}\right)^2}$$

$$= -2.667 p_o.$$

The ratio of $\dfrac{p_o}{p_i}$ for the condition maximum stresses in both cases are same, is

$$[\sigma_{h\max}] \text{ external } P = [\sigma_{h\max}] \text{ internal } P_r$$

$$\Rightarrow \quad 2.667\, p_o = 1.667\, p_i$$

$$\therefore \quad \frac{P_o}{P_i} = 0.625$$

Maximum strain $\in_{\max}$ for only internal pressure (P_i) is given by hoop strain at $r = r_i$

$$\therefore \quad \in_{\max} = \frac{\sigma_{ni}}{E} - \mu\frac{\sigma_r}{E} - \mu\frac{\sigma_l}{E}$$

$$= \frac{1}{E}\left\{1.67 p_i + 0.25 p_i - 0.25 \times \frac{p_i}{3}\right\} = \frac{1.833 p_i}{E}.$$

Maximum strain $\in_{\max}$ for only external pressure P_o is given by hoop strain at $r = r_i$.

$$\sigma_{ni} = 2.667\, p_o$$

$$\sigma_l = 0$$

$$\sigma_{ri} = 0$$

$$\therefore \quad \in_{\max} = -\frac{2.667 p_o}{E}.$$

Equating the maximum strains in both the cases,

$$\frac{1.833 p_i}{E} = 2.667 \frac{p_0}{E}$$

$$\Rightarrow \qquad \frac{p_o}{p_i} = \frac{1.833}{2.667} = 0.6875.$$

8.6 COMPOUND CYLINDER AND SHRINK FIT

Compound cylinders are the cylinders, in which outer cylinder fitted over (after heating) inner cylinder, as in the case of thin cylinders. The diametrical difference at the junction of outer and inner cylinders creates of shrinkage pressure at the junction. This shrinkage pressure causes hoop compressive stresses in the inner cylinder. This compressive stress reduces tensile stress developed due to internal pressure in the inner cylinder, thus allowing higher internal pressure or reducing the wall thickness for a given internal pressure. Shrink fitting helps in economical design of cylinders. **Analysis of compound cylinders:** let 'p_s' be the shrinkage pressure at the junction. The following notation is used for

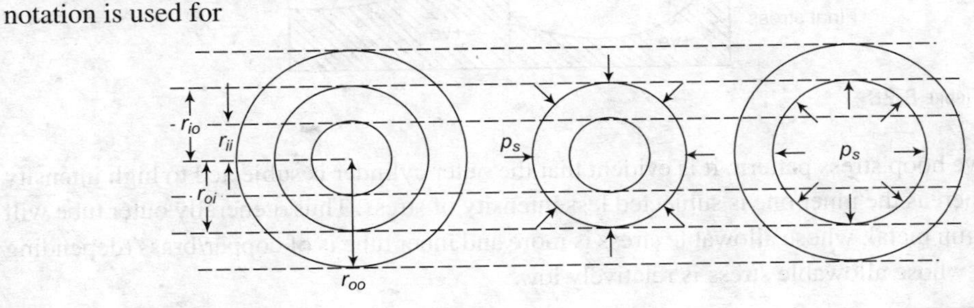

FIGURE 8.27

r_{ii} = inner radius of inner cylinder
r_{io} = outer radius of inner cylinder
r_{oi} = inner radius of outer cylinder
r_{oo} = outer radius of outer cylinder.

The first suffix represents the cylinder, while the second suffix represents whether the fiber is inner or outer. The same notation is even used for hoops stresses, strains, etc. The analysis of compounded cylinders is done in three stages.

Stage 1: Assume no internal fluid pressure; consider only shrinkage pressure p_s and analyze the stresses in the inner cylinder. Refer these stresses as stresses at stage 1.

Stage 2: Analyze the outer cylinder, for shrinkage pressure 'p_s' alone. Refer the stresses obtained in this stage as stresses in stage 2.

Stage 3: Consider the outer tube and inner tube as a single unit. Estimate the stresses for internal fluid pressure (p) alone. Stresses in this stage are stresses in stage 3.

The final stresses are the algebraic sum of stresses obtained in these three stages. The hoop stress variation across the thickness of compound cylinder is shown in Figure 8.28.

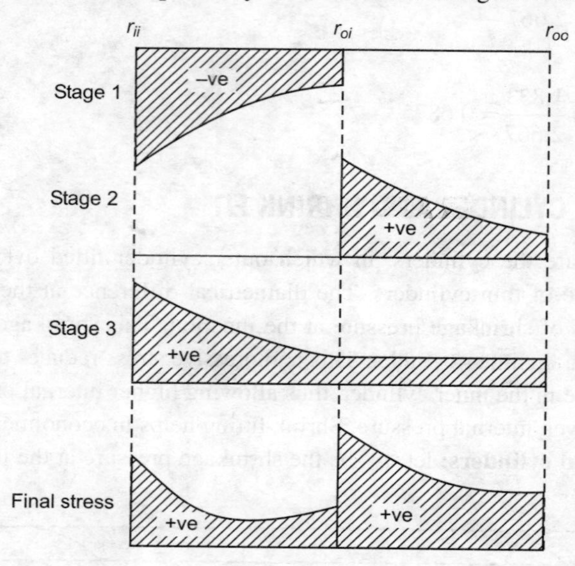

FIGURE 8.28

From the above hoop stress pattern, it is evident that the outer cylinder is subjected to high intensity of stresses, whereas the inner one is subjected less intensity of stress. Thus, generally outer tube will be of steel or gun metal, whose allowable stress is more and inner tube is of copper/brass (depending on necessity), whose allowable stress is relatively low.

PROBLEM 8.11

Objective 3, 4

A compound cylinder is made by shrinking a steel cylinder of outer diameter 200 mm on to another steel cylinder of inner diameter 150 mm. If the shrinkage pressure is 50 N/mm^2 and the internal fluid pressure is 120 MPa, estimate the stresses developed in the outer cylinder as well as inner cylinder. Sketch the variation of hoop stresses and the radial stress along the wall thickness of the cylinder.

SOLUTION

Stresses in the inner cylinder due to only shrinkage pressure:

p_s = shrinkage pressure = 50 N/mm^2

$$r_{ii} = \frac{100}{2} = 50 \text{ mm}$$

$$r_{io} = \frac{150}{2} = 75 \text{ mm}$$

$$\sigma_r = A_1 - \frac{B_1}{r^2}.$$

At $r = 75$ mm, $\sigma_r = -50$; at $r = 50$ mm, $\sigma_r = 0$

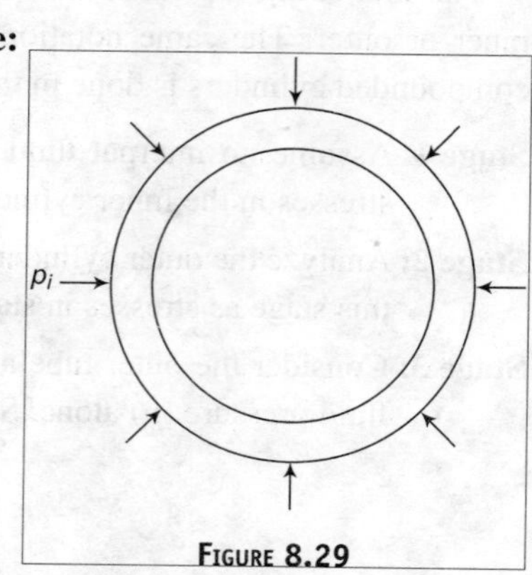

FIGURE 8.29

$$\Rightarrow \qquad A_1 - \frac{B_1}{50^2} = 0 \tag{8.18}$$

$$A_1 - \frac{B_2}{75^2} = -50 \tag{8.19}$$

Solving Eqs. (8.18) and (8.19)

$$B_1 = -225{,}000 \text{ N}; A_1 = -90$$

$$\sigma_h = A_1 + \frac{B_1}{r^2}$$

$$\sigma_{hii1\ r=50]} = -90 - \frac{225000}{50^2} = -180 \text{ MPa.}$$

$$\sigma_{hio1\ r=75]} = -90 - \frac{225000}{75^2} = -130 \text{ MPa.}$$

Stresses due to shrinkage pressure in the outer cylinders:

Shrinkage pressure = 50 MPa

$r_{oi} = 75$ mm

$r_{oo} = 100$ mm

We know that,

$$\sigma_r = A^2 - \frac{B^2}{r_2};$$

At $r = 75$ mm, $\sigma_r = -P_s$

At $r = 100$ mm, $\sigma_r = 0$

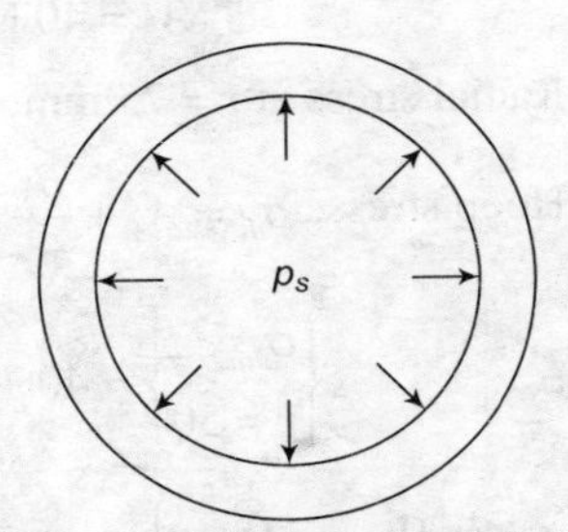

FIGURE 8.30

$$A_2 - \frac{B_2}{100^2} = 0 \tag{8.20}$$

$$A_2 - \frac{B_2}{75^2} = -50 \tag{8.21}$$

Solving Eqs. (8.21) and (8.20)

$$B_2 = 642857 \text{ N}$$

$$A_2 = 64.29 \text{ N/mm}^2$$

$$\sigma_h = A_2 + \frac{B_2}{r^2}$$

$\sigma_{hol\ \text{at}\ r=75]} = 178.57$ N/mm^2

$\sigma_{hool\ \text{at}\ r=100\ \text{mm}]} = 128.57$ N/mm^2.

Stresses in compound cylinder due to internal pressure alone:

Internal pressure = 120 MPa

$$r_{ii} = 50 \text{ mm}$$

$$r_{oo} = 100 \text{ mm}$$

$$\sigma_r = A_3 - \frac{B_3}{r^2}$$

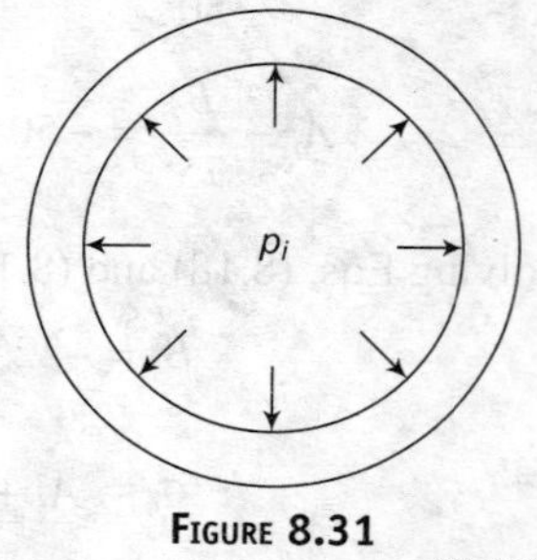

FIGURE 8.31

At $r = 100$ mm, $\sigma_r = 0$ and

At $r = 50$ mm, $\sigma_r = -P_i = -80$

$$\therefore \quad A_3 - \frac{B_3}{100^2} = 0 \tag{8.22}$$

$$A_3 - \frac{B_3}{50^2} = -120 \tag{8.23}$$

Solving Eqs. (8.22) and (8.23)

$$B_3 = 400000$$

$$A_3 = 40 \text{ MPa}$$

Radial stress at $r = 75$ mm; $\sigma_r = -31.11$ N/mm^2

Hoop stress, $\sigma_h = A_3 + \frac{B_3}{r^2}$.

$$\begin{Bmatrix} \sigma_{hii2} \\ r = 50 \end{Bmatrix} = 40 + \frac{400000}{50^2} = 200 \text{ MPa}$$

$$\begin{Bmatrix} \sigma_{hio2} = \sigma_{hoi2} \\ \text{at } r = 75 \end{Bmatrix} = 40 + \frac{400000}{75^2} = 111.11 \text{ MPa}$$

$$\begin{Bmatrix} \sigma_{hoo2} \\ \text{at } r = 50 \end{Bmatrix} = 80 \text{ N/mm}^2.$$

Final stresses: Stresses developed in Inner cylinder and Outer cylinder at different stages are shown in Table 8.1.

Table 8.1 Stresses developed in inner and outer cylinder at different stages

	Inner Cylinder				Outer Cylinder			
	Hoop Stress		Radial Stress		Hoop Stress		Radial Stress	
	r = 50	r = 75	r = 50	r = 75	r = 75	r = 100	r = 75	r = 100
Stage 1	−180	−130	-	−50	-	-	-	-
Stage 2	-	-	-	-	178.57	128.57	−50	0
Stage 3	200	111.11	−120	−31.11	111.11	80	−31.1	0
Final stress	20	−18.89	−120	−81.11	289.68	208.57	−81.11	0

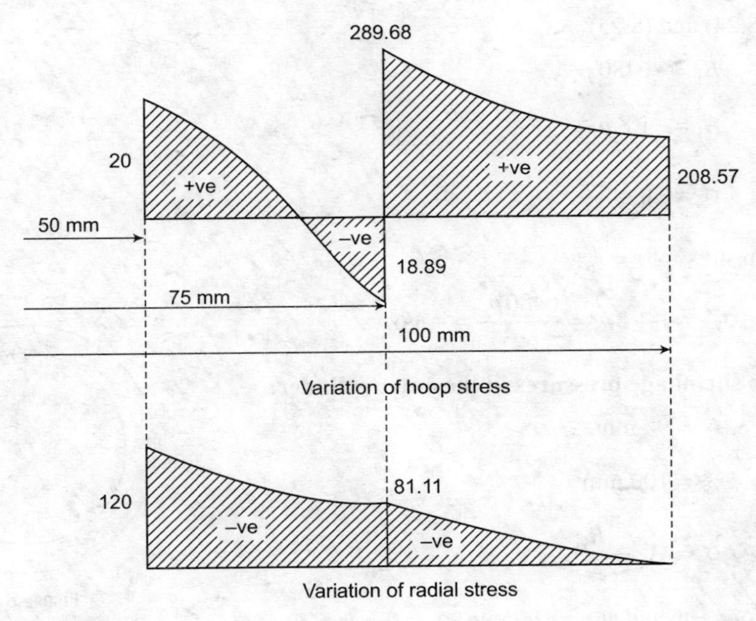

FIGURE 8.32

PROBLEM 8.12

Objective 3, 4

A cylinder of outer diameter 200 mm and inner diameter 180 mm, shrunk on to another cylinder of inner diameter 120 mm and outer diameter 180 mm. Determine the shrinkage pressure required at the junction, for an internal pressure of 80 MPa, such that maximum hoop stresses in the inner and outer cylinders are same.

SOLUTION

Let 'p_s' be the shrinkage pressure.

Stresses in the cylinder due to shrinkage pressure:

$$r_{ii} = 60 \text{ mm}$$

$$r_{io} = 90 \text{ mm}$$

$$\sigma_r = A_1 - \frac{B_1}{r^2}$$

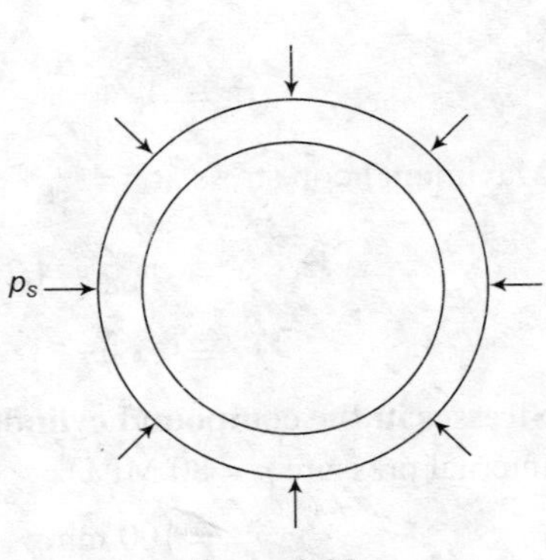

FIGURE 8.33

At $r = 90$ mm, $\sigma_r = -p_s$; at $r = 60$ mm, $\sigma_r = 0$

$$A_1 - \frac{B_1}{90^2} = -p_s \quad (8.24)$$

$$A_1 - \frac{B_1}{60^2} = 0 \quad (8.25)$$

Solving Eqs. (8.24) and (8.25)

$$\Rightarrow \qquad B_1 = -6480\, p_s$$

$$A_1 = -1.8\, p_s$$

$$\sigma_h = A_1 + \frac{B_1}{r^2}.$$

Maximum hoop stress at $r = r_{ii}$

$$\sigma_{hii1} = -1.8p_s = \frac{6480p_s}{60^2} = -3.6p_s.$$

Stresses due to shrinkage pressures in the outer cylinder:

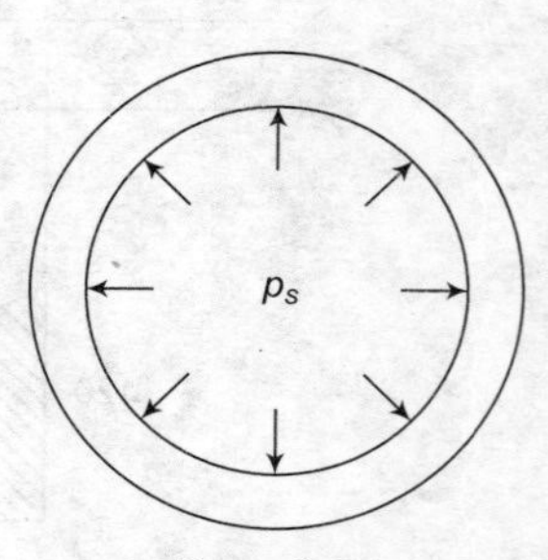

FIGURE 8.34

$$r_{oi} = 90 \text{ mm}$$

$$r_{oo} = 100 \text{ mm}$$

$$\sigma_r = A_2 - \frac{B_2}{r^2}$$

At $r = 90$ mm, $\sigma_r = -p_s$ and at $r = 100$ mm, $\sigma_n = 0$

$$A_2 - \frac{B_2}{100^2} = 0 \tag{8.26}$$

$$A_2 - \frac{B_2}{90^2} = -p_s \tag{8.27}$$

Solving Eqs. (8.26) and (8.27)

$$A_2 = 4231.58p_s$$

$$B_2 = 4.26p_s$$

$$\sigma_h = A_2 + \frac{B_2}{r^2}$$

Maximum hoop stress at $r = r_{oi}$

$$= \sigma_{hoi1} = 4.26p_s + \frac{42631.58p_s}{90^2}$$

$$= 9.52p_s$$

Stresses in the compound cylinder due to internal fluid pressure:

Internal pressure $p = 80$ MPa

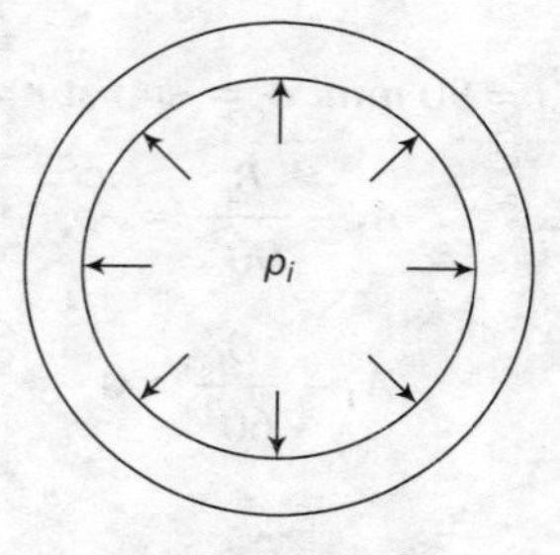

FIGURE 8.35

$$r_{oo} = 100 \text{ mm}$$

$$r_{ii} = 60 \text{ mm}$$

$$\sigma_r = A_3 - \frac{B_3}{r^2}$$

At $r = 60$ mm, $\sigma_r = -80$
At $r = 100$ mm, $\sigma_r = 0$

$$A_3 - \frac{B_3}{100^2} = 0 \tag{8.28}$$

$$A_3 - \frac{B_3}{90^2} = -p_s \tag{8.29}$$

Solving Eqs. (8.28) and (8.29)

$$B_3 = 450000 \text{ N}$$
$$A_3 = 45 \text{ MPa}$$

$$\sigma_h = A_3 + \frac{B_3}{r^2}$$

$$\sigma_{hii2} = 45 + \frac{450000}{60^2} = 170 \text{ MPa}$$

$$\sigma_{hio2} = \sigma_{hoi2} = 45 + \frac{450000}{90^2} = 100.56 \text{ MPa}$$

$$\sigma_{hoo2} = 45 + \frac{450000}{100^2} = 90 \text{ MPa}$$

Maximum hoop stress in the inner cylinder due to shrinkage pressure and internal pressure:

$$\sigma_{hii} = -3.6p_s + 170.$$

Maximum hoop stress in the outer cylinder due to shrinkage pressure and internal pressure:

$$\sigma_{hio} = 9.52p_s + 100.56.$$

Equating the maximum stress in the outer cylinder and inner cylinder

$$-3.6p_s + 170 = 9.52p_s + 100.56$$

$$170 - 100.56 = 13.12p_s$$

$$p_s = 5.293 \text{ MPa}.$$

8.7 RELATIONSHIP BETWEEN SHRINKAGE ALLOWANCE AND SHRINKAGE PRESSURE

It is obvious that shrinkage allowance, which is equal to the difference between the outer radii of inner cylinder (r_{io}) and inner radii of the outer cylinder (r_{oi}), is essential to produce shrinkage pressure.

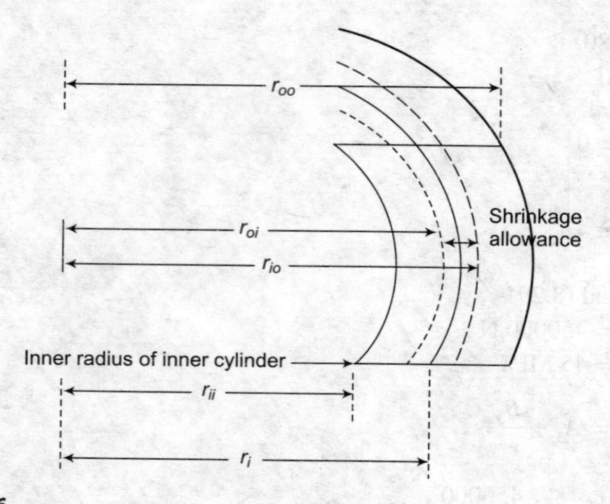

FIGURE 8.36

To obtain a relationship between the shrinkage pressure (p_s) and shrinkage allowance (Δ),

Δ = Decrease in the outer radius of inner cylinder + increase in the inner radius of cylinder

Decrease in the outer radius of the inner cylinder $\Delta r_{io} = \varepsilon_{hio} \times r_{io}$

$\in_{hio}$ is the hoop strain at outer radius of the inner cylinder due to p_s.

r_{ii} = inner radius of inner cylinder

r_{io} = outer radius of inner cylinder

$$\sigma_r = A_1 - \frac{B_1}{r^2}$$

At $r = r_{ii}$, $\sigma_r = 0$ and at $r = r_{io}$, $\sigma_r = -p_s$

$$A_1 - \frac{B_1}{r_{ii}^2} = 0 \tag{8.30}$$

$$A_1 - \frac{B_1}{r_{io}^2} = -p_s \tag{8.31}$$

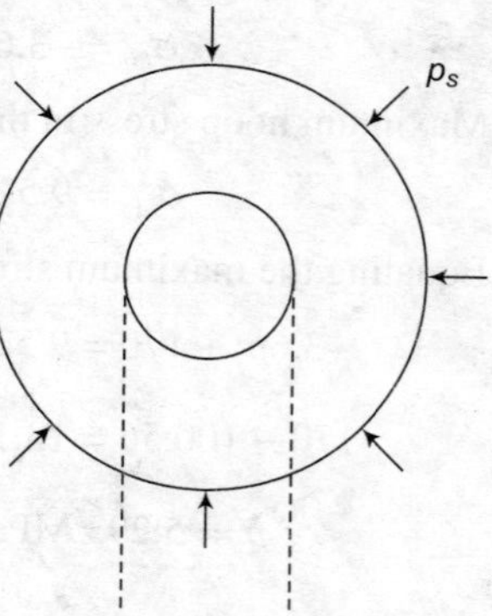

$$B_1 = -p_s \times \frac{r_{io}^2 r_{ii}^2}{(r_{io}^2 - r_{ii}^2)}$$

$$A_1 = -p_s \times \frac{r_{io}^2}{(r_{io}^2 - r_{ii}^2)}$$

p_s

FIGURE 8.37

Hoop stress at the outer radius σ_{nio} of inner cylinder $= A_1 + \dfrac{B_1}{r_{io}^2} = -\, p_s \left[\dfrac{r_{io}^2 + r_{ii}^2}{r_{io}^2 - r_{ii}^2}\right]$

Radial stress at the outer radius of the inner cylinder = $\sigma_{rio} = -\,p_s$.

Let E_i and μ_i be the modulus elasticity and Poisson's ratio of the inner cylinder.

Then, $\in_{hio} = \frac{1}{E_i}\{\sigma_{hio} - \mu_i \sigma_{rio} - \mu \sigma_l\}$

$\sigma_l = 0$ as there is no internal pressure

$$\therefore \qquad \in_{hio} = \frac{1}{E_i}\left\{-p_s\left[\frac{r_{io}^2 + r_{ii}^2}{r_{io}^2 - r_{ii}^2}\right] - \mu_i(-p_s)\right\} = \frac{-p_s}{E_i}\left\{\left[\frac{r_{io}^2 + r_{ii}^2}{r_{io}^2 - r_{ii}^2}\right] - \mu_i\right\}$$

The negative sign in the above expression indicates compressive strain.

$\therefore$ Decrease in the outer radius of the inner cylinder is $\Delta r_{io} = \in_{hio} \times r_{io}$.

$$\therefore \qquad \Delta r_{io} = \frac{p_s r_{io}}{E_i}\left\{\left[\frac{r_{io}^2 + r_{ii}^2}{r_{io}^2 - r_{ii}^2}\right] - \mu_i\right\}. \tag{8.32}$$

Consider the outer cylinder subjected to shrinkage pressure p_s:

$$\sigma_r = A_2 - \frac{B_2}{r^2}$$

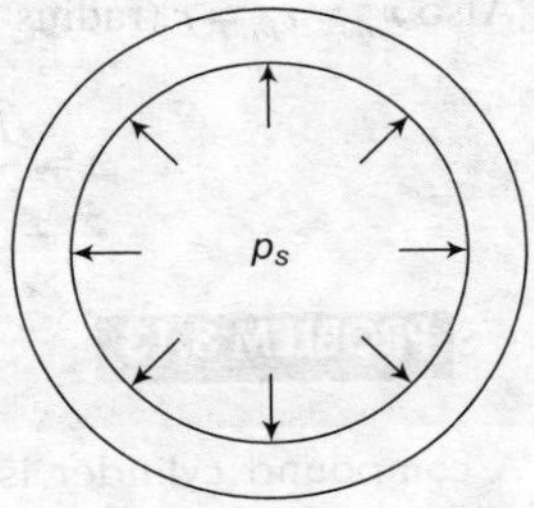

At $r = r_{io}$, $\sigma_r = -p_s$

At $r = r_{oo}$, $\sigma_r = 0$

$$\Rightarrow \qquad A_2 - \frac{B_2}{r_{io}^2} = -p_s \tag{8.33}$$

$$A_2 - \frac{B_2}{r_{oo}^2} = 0 \tag{8.34}$$

p_s

FIGURE 8.38

Solving Eqs. (8.33) and (8.34)

$$B_2 = p_S\left[\frac{r_{oo}^2 r_{oi}^2}{r_{oo}^2 - r_{oi}^2}\right]$$

$$A_2 = p_s\left[\frac{r_{oi}^2}{r_{oo}^2 - r_{oi}^2}\right]$$

Hoop stress $\sigma_h = A_2 + \frac{B_2}{r^2}$,

Hoop stress at the inner radius of the outer cylinder, $r = r_{oi} = \sigma_{hoi} = p_s\left[\frac{r_{oo}^2 + r_{oi}^2}{r_{oo}^2 - r_{oi}^2}\right]$

Radial stress at the inner radius of the outer cylinder $\sigma_{roi} = -p_s$

Hoop strain at the inner radius of the outer cylinder $= \in_{hoi} = \frac{\sigma_{hoi}}{E_o} - \mu_O\frac{\sigma_{rio}}{E_o} - \mu_O - \frac{\sigma_1}{E_o}$

$\sigma_1 = 0$

$$\therefore \qquad \varepsilon_{hoi} = \frac{1}{E_o}\left\{P_s\left[\frac{r_{oo}^2 + r_{oi}^2}{r_o^2 - r_{oi}^2}\right] - \mu_o(-P_s)\right\} = \frac{P_s}{E}\left\{\left[\frac{r_{oo}^2 + r_{oi}^2}{r_{oo}^2 - r_{oi}^2}\right] + \mu_o\right\}$$

Increase in the inner radius of the outer cylinder, $\Delta_{rio} = \in_{noi} \times r_{oi}$

$$\Delta r_{io} = \frac{p_s r_{oi}}{E_o}\left\{\left(\frac{r_{oo}^2 + r_{oi}^2}{r_{oo}^2 - r_{oi}^2}\right) + \mu_o\right\} \tag{8.35}$$

Shrinkage allowance $\Delta = \Delta r_{io} + \Delta r_{oi}$.

$$\Delta = p_s\left\{\frac{r_{io}}{E}\left[\frac{r_{io}^2 + r_{ii}^2}{r_{io}^2 - r_{ii}^2}\right] - \frac{\mu_i r_{io}}{E_i} + \frac{r_{oi}}{E_o}\left[\frac{r_{oo}^2 + r_{oi}^2}{r_{oo}^2 - r_{io}^2}\right] + \frac{\mu_o r_{oi}}{E_o}\right\}.$$

If inner cylinder and outer cylinder are of same material then

$$E_i = E_o = E$$

$$\mu_i = \mu_o = \mu$$

Also $r_{oi} = r_{io} = r$ (radius at the junction). Then

$$\Delta = \frac{P_s r_i}{E}\left\{\left(\frac{r_{oo}^2 + r_{ii}^2}{r_{io}^2 - r_{ii}^2}\right) + \left(\frac{r_{oo}^2 + r_{oi}^2}{r_{oo}^2 - r_{oi}^2}\right)\right\}.$$

PROBLEM 8.13

Objective 1, 3, 4

A compound cylinder is designed with a shrinkage pressure of 10 MPa. Estimate the shrinkage allowance if the outer diameter, diameter at the junction, and inner diameter are 200 mm, 150 mm, and 100 mm, respectively. Also determine the allowable internal pressure, if the maximum stress is not to exceed 50 MPa. Take $E = 200$ GPa.

SOLUTION

Given that,

$$r_{ii} = 50 \text{ mm}$$

$$r_{io} = r_{oi} = 75 \text{ mm} \qquad \text{(radius at the junction)}$$

$$r_{oo} = 100 \text{ mm}$$

Inner and outer tubes are of same material.

p_s = shrinkage pressure = 10 MPa

Shrinkage allowance $$\Delta = \frac{p_s \times r_{oi}}{E}\left[\left(\frac{r_{oo}^2 + r_{oi}^2}{r_{oo}^2 - r_{oi}^2}\right) + \left(\frac{r_{ii}^2 + r_{io}^2}{r_{io}^2 - r_{ii}^2}\right)\right]$$

$$\Delta = \frac{10 \times 75}{200 \times 10^3}\left(\frac{100^2 + 75^2}{100^2 - 75^2} + \frac{75^2 + 50^2}{75^2 - 50^2}\right)$$

$$\Delta = 0.231 \text{ mm}.$$

Let 'p' be the internal fluid pressure.

Stage 1: Hoop stress in the inner cylinder due to shrinkage pressure

$$\sigma_r = A_1 - \frac{B_1}{r^2}$$

At $r = 50$, $\sigma_r = 0$

At $r = 75$, $\sigma_r = -p_s = -10$

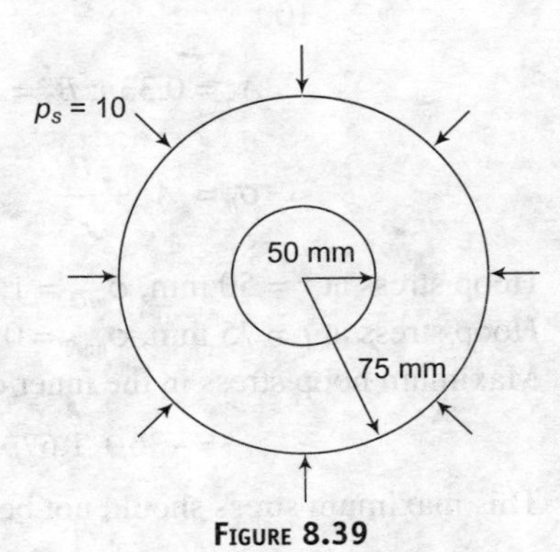

FIGURE 8.39

$$A_1 - \frac{B_1}{50^2} = 0$$

$$A_1 - \frac{B_2}{75^2} = -10$$

$$\Rightarrow \quad A_1 = 18 \text{ MPa}; B_1 = -45000 \text{ N}$$

$$\sigma_h = A_1 + \frac{B_1}{r^2}$$

For maximum hoop stress, $r = 50$ mm

$$\sigma_{hii1} = -18 + \frac{-45000}{50^2} = -36 \text{ MPa}$$

Stage 2: Hoop stress in the outer cylinder due to shrinkage pressure

$$\sigma_r = A_2 - \frac{B_2}{r^2}$$

At $r = 75$ mm, $\sigma_r = -p_s = -10$

At $r = 100$ mm, $\sigma_r = 0$

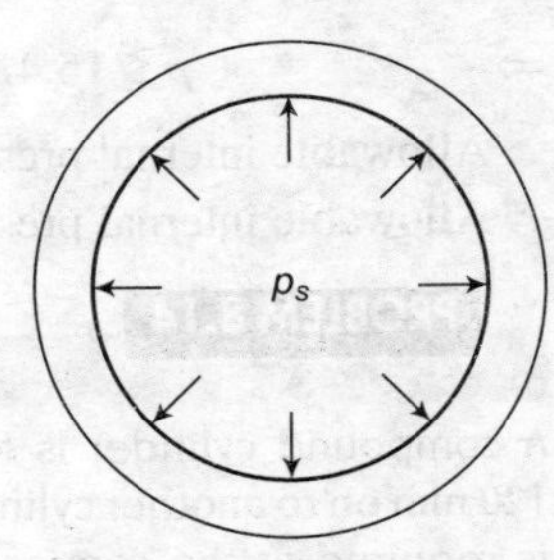

FIGURE 8.40

$$A_2 - \frac{B_2}{75^2} = -10$$

$$A_2 - \frac{B_2}{100^2} = 0$$

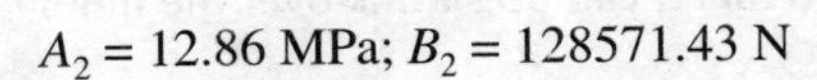

$A_2 = 12.86$ MPa; $B_2 = 128571.43$ N

Hoop stress $= \sigma_{ho} = A_2 + \dfrac{B_2}{r^2}$

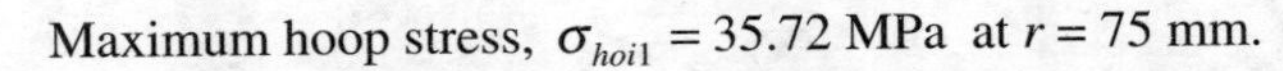

Maximum hoop stress, $\sigma_{hoi1} = 35.72$ MPa at $r = 75$ mm.

Stage 3: Hoop stress due to internal fluid pressure

$$\sigma_r = A_3 - \frac{B_3}{r^2}$$

At $r = 50$ mm, $\sigma_r = -p$

At $r = 100$ mm, $\sigma_r = 0$

$$A_3 - \frac{B_3}{50^2} = -p$$

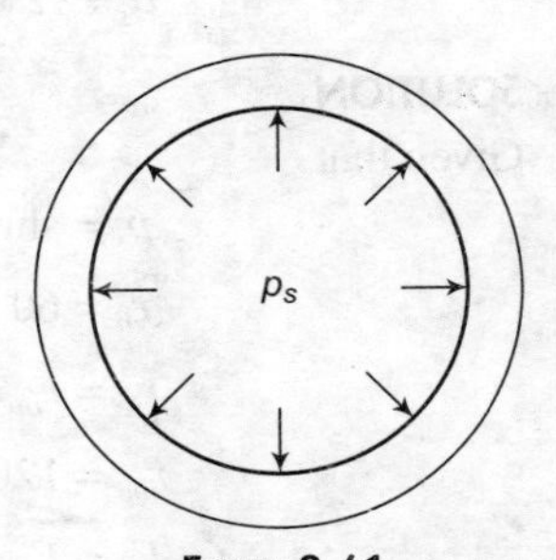

FIGURE 8.41

$$A_3 - \frac{B_3}{100^2} = 0$$

$$A_3 = 0.33p;\ B_3 = 3333.33p$$

$$\sigma_h = A_3 + \frac{B_3}{r^2}$$

Hoop stress at $r = 50$ mm, $\sigma_{hii2} = 1.67\ p$
Hoop stress at $r = 75$ mm, $\sigma_{hio2} = 0.926\ p$
Maximum hoop stress in the inner cylinder = $\sigma_{hii1} = \sigma_{hii2}$

$$= -36 + 1.67p$$

This maximum stress should not be more than 50.

$\therefore$ $$-36 + 1.67\ p \le 50$$

$\Rightarrow$ $$p \le 51.60 \text{ MPa}$$

Considering the hoop stress in the outer cylinder,
Maximum hoop stress in the outer cylinder $\sigma_{hoi} = \sigma_{hoi} = 0.926\ p + 35.72$.
Given that $\sigma_{hoi} \le 50$

$\Rightarrow$ $$0.926\ p + 35.72 \le 50$$

$\Rightarrow$ $$p \le 15.421 \text{ MPa}$$

$\therefore$ Allowable internal pressure is the lesser of the values 15.421 MPa and 51.60 MPa.
$\therefore$ Allowable internal pressure = 15.421 MPa.

PROBLEM 8.14

Objective 2, 4

A compound cylinder is formed by shrinking a cylinder of inner radius 100 mm and outer radius 120 mm on to another cylinder of inner radius 60 mm and outer radius 100 mm. The shrinkage pressure is required at the common surface. The inner cylinder is copper and the outer one is steel. Also estimate the temperature required to heat the outer cylinder, so that it can be shrunk over the inner cylinder.

$$E_c = 80 \text{ GPa};\ \mu_c = 0.32$$

$$E_s = 200 \text{ GPa};\ \mu_s = 0.30$$

$$\alpha_s = 12 \times 10^{-6}/^\circ\text{C rise}$$

SOLUTION

Given that

$$p_s = \text{shrinkage pressure} = 20 \text{ MPa}$$

$$r_{ii} = 60 \text{ mm}$$

$$r_{io} = r_{oi} = 100 \text{ mm}$$

$$r_{oo} = 120 \text{ mm}$$

$$E_o = E_s = E_{outer} = 200 \times 10^3 \text{ N/mm}^2$$

$$\mu_o = \mu_s = \mu_{\text{outer}} = 0.3$$

$$E_i = E_c = E_{\text{inner}} = 80 \times 10^3 \text{ N/mm}^2$$

$$\mu_i = \mu_c = \mu_{\text{inner}} = 0.32$$

Shrinkage allowance = Δ

$$\Delta = P_s \left\{ \frac{r_{oi}}{E_o} \left[\left(\frac{r_{oo}^2 + r_{oi}^2}{r_{oo}^2 - r_{oi}^2} \right) + \mu \right] + \frac{r_{io}}{E_C} \left[\left(\frac{r_{ii}^2 + r_{io}^2}{r_{io}^2 - r_{ii}^2} \right) - \mu \right] \right\}$$

$$\therefore \qquad \Delta = 20 \left\{ \frac{100}{200 \times 10^3} \left[\frac{120^2 + 100^2}{120^2 - 100^2} \right] + 0.3 + \frac{100}{80 \times 10^3} \left[\left(\frac{100^2 + 60^2}{100^2 - 60^2} \right) - 0.32 \right] \right\}$$

$$= 20\{0.00292 + 0.002226\}$$

$$= 0.1036 \text{ mm.}$$

Let T be the temperature required to match the allowance by heating the outer cylinder.

$$\Delta = r_{oi} \times \alpha \times T$$

Temperature required = T

$$T = \frac{\Delta}{r_{oi} \times \alpha}$$

$$= \frac{0.1036}{100 \times 12 \times 10^{-6}} = 86.3°$$

PROBLEM 8.15

Objective 1, 3

A vertical stream boiler is of 2 m internal diameter and 5 m high. It is constructed with 20 mm thick plates for a working pressure of 1 N/mm^2. The end plates are flat and are not stayed.

SOLUTION

Wall thickness of the cylinder = 20 mm

Ends of the cylinder are closed.

Internal pressure P_o = 1 N/mm^2

(a) Circumferential stress or hoop stress in the wall of the boiler

$$\sigma_h = \frac{P_0 d}{2t} = \frac{1 \times 2000}{2 \times 20} = 50 \text{ MPa}$$

(b) $\mu = 0.3$; $\quad E = 200$ GPa

$$= 2 \times 10^5 \text{ N/mm}^2$$

σ_l = Longitudinal stress

$$= \frac{P_0 d}{4t} = \frac{1 \times 2000}{4 \times 20}$$

$$= 25 \text{ MPa}$$

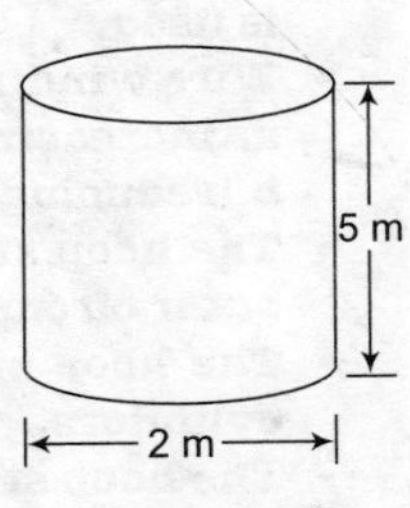

FIGURE 8.42

$$\varepsilon_l = \frac{\delta l}{L} = \text{Linear strain}$$

$$= \frac{\sigma_l}{E} \times \mu \times \frac{\sigma_h}{E}$$

$$L = 5000 \text{ mm}$$

Increase in length of the boiler $\delta L = \varepsilon_l \times L$

$$\varepsilon_l = \frac{25}{E} - 0.3 \times \frac{50}{E}$$

$$= \frac{10}{E} - \frac{10}{2\times10^5}$$

$$= 0.00005$$

$$\delta L = 0.00005 \times 5000$$

$$= 0.25 \text{ mm.}$$

SUMMARY

- **If the thickness of the wall of the cylinder vessel is less than 1/20 of its internal diameter, the cylindrical vessel is known as a thin cylinder. Or if $t/d < 1/20$, the cylinder is a thin cylinder.**
- **In case of thin cylinders, the stress distribution is assumed uniform over the thickness of wall.**
- **The circumferential stress (σ_1) is given by**

  $$\sigma_1 = \frac{p \times d}{2t}$$

  in which p = internal fluid pressure, d = internal diameter of the thin cylinder, and t = thickness of the wall of the cylinder.
- **The longitudinal stress (σ_2) is given by**

  $$\sigma_2 = \frac{p \times d}{4t}$$

- **The circumferential and longitudinal stresses are tensile stresses.**
- **If the thickness of the cylinder is to be determined, then circumferential stress is used.**
- **Wire winding of thin cylinder is necessary for**
  a. increasing the pressure-carrying capacity of the cylinder and
  b. reducing the chances of bursting of the cylinder in longitudinal direction.
- **The hoop stress is maximum at the inner circumference and minimum at the outer circumference of a thick cylinder.**
- **The hoop stress in case of thin cylinders is reduced by wire winding on the cylinders.**
- **The hoop stress in case of thick cylinders is reduced by shrinking one cylinder over another cylinder.**

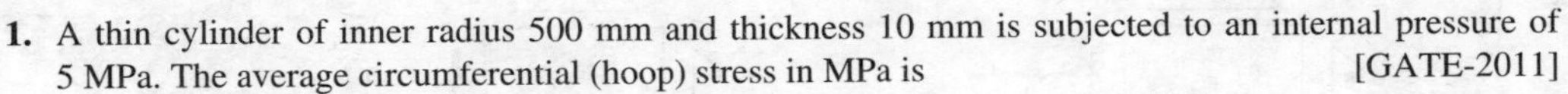

OBJECTIVE TYPE QUESTIONS

1. A thin cylinder of inner radius 500 mm and thickness 10 mm is subjected to an internal pressure of 5 MPa. The average circumferential (hoop) stress in MPa is [GATE-2011]

(a) 100 (b) 250 (c) 500 (d) 1000

2. The maximum principal strain in a thin cylindrical tank, having a radius of 25 cm and wall thickness of 5 mm when subjected to an internal pressure of 1 MPa, is (taking Young's modulus as 200 GPa and Poisson's ratio as 0.2) [GATE-1998]

(a) 2.25×10^{-4} (b) 2.25 (c) 2.25×10^{-6} (d) 22.5

3. A thin-walled spherical shell is subjected to an internal pressure. If the radius of the shell is increased by 1% and the thickness is reduced by 1%, with the internal pressure remaining the same, the percentage change in the circumferential (hoop) stress is [GATE-2012]

(a) 0 (b) 1 (c) 1.08 (d) 2.02

4. A thin cylindrical shell is subjected to internal pressure p. The Poisson's ratio of the material of the shell is 0.3. Because of internal pressure, the shell is subjected to circumferential strain and axial strain. The ratio of circumferential strain to axial strain is: [IES-2001]

(a) 0.425 (b) 2.25 (c) 0.225 (d) 4.25

5. A thin cylindrical shell of diameter d, length l, and thickness t is subjected to an internal pressure p. What is the ratio of longitudinal strain to hoop strain in terms of Poisson's ratio ($1/m$)? [IES-2004]

(a) $\frac{m-2}{2m+1}$ (b) $\frac{m-2}{2m-1}$ (c) $\frac{2m-1}{m-2}$ (d) $\frac{2m+2}{m-1}$

6. A thin cylinder contains fluid at a pressure of 500 N/m^2, the internal diameter of the shell is 0.6 m and the tensile stress in the material is to be limited to 9000 N/m^2. The shell must have a minimum wall thickness of nearly [IES-2000]

(a) 9 mm (b) 11 mm (c) 17 mm (d) 21 mm

7. A thin cylinder with closed lids is subjected to internal pressure and supported at the ends as shown in figure. The state of stress at point X is as represented in [IES-1999]

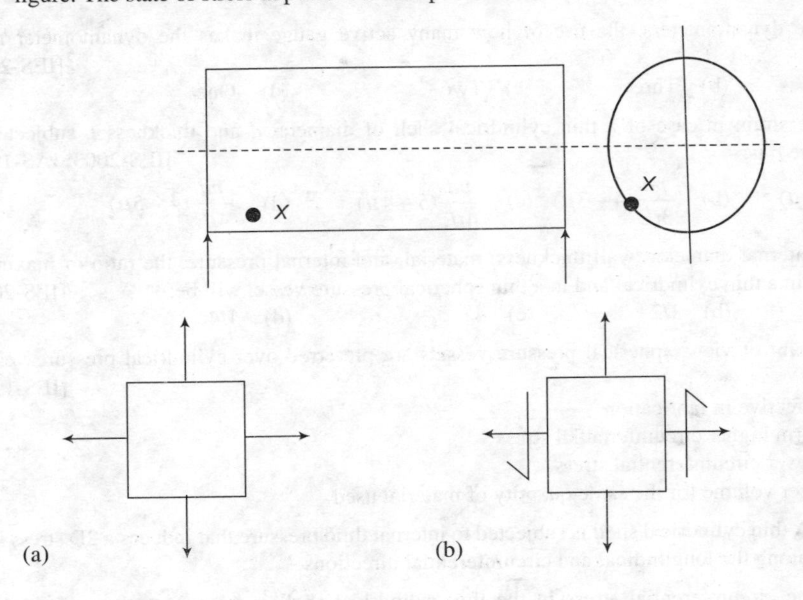

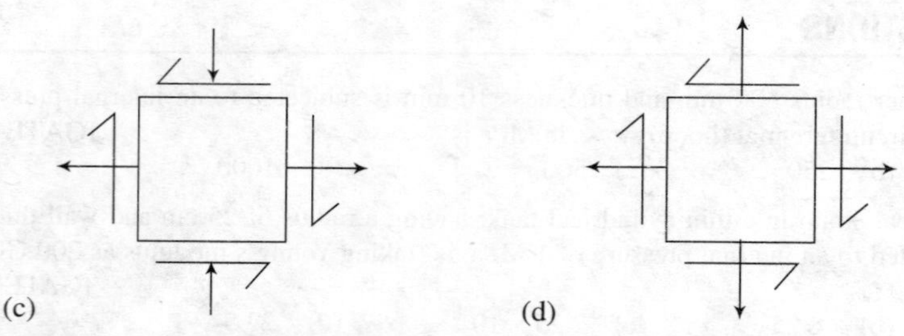

8. A metal pipe of 1 m diameter contains a fluid having a pressure of 10 kgf/cm^2. If the permissible tensile stress in the metal is 200 kgf/cm^2, then the thickness of the metal required for making the pipe would be: [IES-1993]
(a) 5 mm (b) 10 mm (c) 20 mm (d) 25 mm

9. Circumferential stress in a cylindrical steel boiler shell under internal pressure is 80 MPa. Young's modulus of elasticity and Poisson's ratio are respectively 2 × 105 MPa and 0.28. The magnitude of circumferential strain in the boiler shell will be: [IES-1999]
(a) 3.44×10^{-4} (b) 3.84×10^{-4} (c) 4×10^{-4} (d) 4.56×10^{-4}

10. A penstock pipe of 10 m diameter carries water under a pressure head of 100 m. If the wall thickness is 9 mm, what is the tensile stress in the pipe wall in MPa? [IES-2009]
(a) 2725 (b) 545.0 (c) 272.5 (d) 1090

11. A water main of 1 m diameter contains water at a pressure head of 100 m. The permissible tensile stress in the material of the water main is 25 MPa. What is the minimum thickness of the water main? (Take $g = 10$ m/s^2). [IES-2009]
(a) 10 mm (b) 20 mm (c) 50 mm (d) 60 mm

12. Hoop stress and longitudinal stress in a boiler shell under internal pressure are 100 MN/m^2 and 50 MN/m^2, respectively. Young's modulus of elasticity and Poisson's ratio of the shell material are 200 GN/m^2 and 0.3, respectively. The hoop strain in boiler shell is: [IES-1995]
(a) 0.425×10^{-3} (b) 0.5×10^{-3} (c) 0.585×10^{-3} (d) 0.75×10^{-3}

13. In strain gauge dynamometers, the use of how many active gauge makes the dynamometer more effective? [IES-2007]
(a) Four (b) Three (c) Two (d) One

14. The volumetric strain in case of a thin cylindrical shell of diameter d and thickness t, subjected to internal pressure p is: [IES-2003; IAS-1997]
(a) $\frac{pd}{2tE}(3-2\mu)$ (b) $\frac{pd}{3tE}(4-3\mu)$ (c) $\frac{pd}{4tE}(5-4\mu)$ (d) $\frac{pd}{4tE}(4-5\mu)$

15. For the same internal diameter, wall thickness, material, and internal pressure, the ratio of maximum stress, induced in a thin cylindrical and in a thin spherical pressure vessel will be: [IES-2001]
(a) 2 (b) 1/2 (c) 4 (d) 1/4

16. From design point of view, spherical pressure vessels are preferred over cylindrical pressure vessels because they [IES-1997]
(a) Are cost effective in fabrication
(b) Have uniform higher circumferential stress
(c) Uniform lower circumferential stress
(d) Have a larger volume for the same quantity of material used

17. Assertion (A): A thin cylindrical shell is subjected to internal fluid pressure that induces a 2D stress state in the material along the longitudinal and circumferential directions.

Reason (R): The circumferential stress in the thin cylindrical shell is two times the magnitude of longitudinal stress. [IAS-2000]

(a) Both A and R are individually true and R is the correct explanation of A
(b) Both A and R are individually true but R is NOT the correct explanation of A
(c) A is true but R is false
(d) A is false but R is true

18. A thick cylinder is subjected to an internal pressure of 60 MPa. If the hoop stress on the outer surface is 150 MPa, then the hoop stress on the internal surface is: [GATE-1996; IES-2001]
(a) 105 MPa (b) 180 MPa (c) 210 MPa (d) 135 MPa

19. If a thick cylindrical shell is subjected to internal pressure, then hoop stress, radial stress, and longitudinal stress at a point in the thickness will be: [IES-1999]
(a) Tensile, compressive, and compressive respectively
(b) All compressive
(c) All tensile
(d) Tensile, compressive, and tensile respectively

20. Where does the maximum hoop stress in a thick cylinder under external pressure occur? [IES-2008]
(a) At the outer surface (b) At the inner surface
(c) At the mid-thickness (d) At the 2/3rd outer radius

21. In a thick cylinder pressurized from inside, the hoop stress is maximum at [IES-1998]
(a) The center of the wall thickness (b) The outer radius
(c) The inner radius (d) Both the inner and the outer radii

22. Where does the maximum hoop stress in a thick cylinder under external pressure occur? [IES-2008]
(a) At the outer surface (b) At the inner surface
(c) At the mid-thickness (d) At the 2/3rd outer radius

23. A thick-walled hollow cylinder having outside and inside radii of 90 mm and 40 mm, respectively, is subjected to an external pressure of 800 MN/m^2. The maximum circumferential stress in the cylinder will occur at a radius of [IES-1998]
(a) 40 mm (b) 60 mm (c) 65 mm (d) 90 mm

24. A thick open-ended cylinder as shown in the figure is made of a material with permissible normal and shear stresses 200 MPa and 100 MPa, respectively. The ratio of permissible pressure based on the normal and shear stress is [IES-2002]

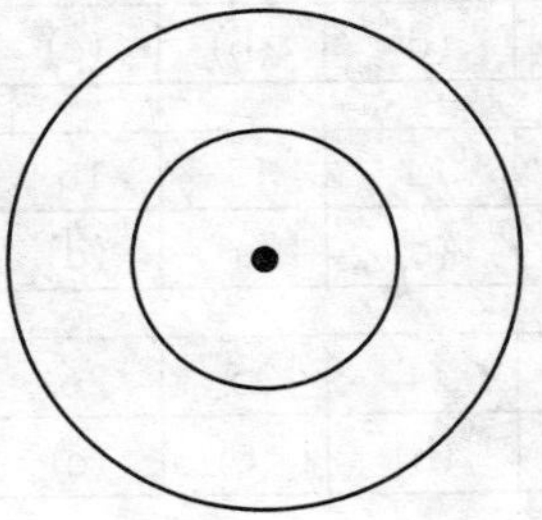

(a) 9/5 (b) 8/5 (c) 7/5 (d) 4/5

25 Consider the following statements: [IES-2007]

In a thick-walled cylindrical pressure vessel subjected to internal pressure, the tangential and radial stresses are:
1) Minimum at outer side
2) Minimum at inner side
3) Maximum at inner side and both reduce to zero at outer wall
4) Maximum at inner wall but the radial stress reduces to zero at outer wall

Which of the statements given above is/are correct?

(a) 1 and 2 (b) 1 and 3 (c) 1 and 4 (d) 4 only

26 Consider the following statements at given point in the case of thick cylinder subjected to fluid pressure: [IES-2006]

1) Radial stress is compressive
2) Hoop stress is tensile
3) Hoop stress is compressive
4) Longitudinal stress is tensile and it varies along the length
5) Longitudinal stress is tensile and remains constant along the length of the cylinder

Which of the statements given above are correct?

(a) Only 1, 2, and 4 (b) Only 3 and 4 (c) Only 1, 2, and 5 (d) Only 1, 3, and 5

27 Autofrettage is a method of [IES-1996; 2005; 2006]

(a) Joining thick cylinders (b) Relieving stresses from thick cylinders
(c) Prestressing thick cylinders (d) Increasing the life of thick cylinders

28 A solid thick cylinder is subjected to an external hydrostatic pressure p. The state of stress in the material of the cylinder is represented as: [IAS-1995]

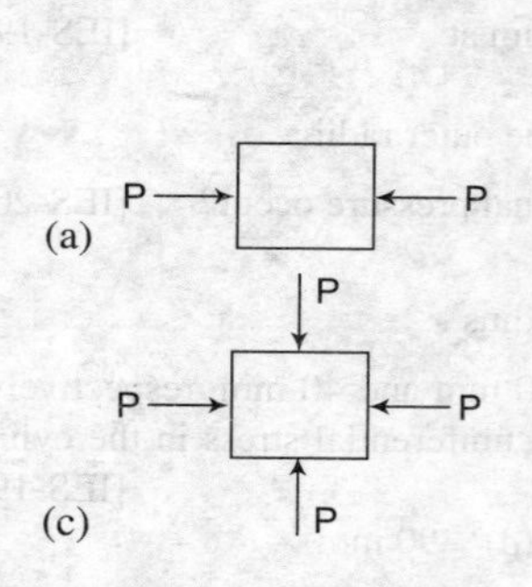

(a) (c)

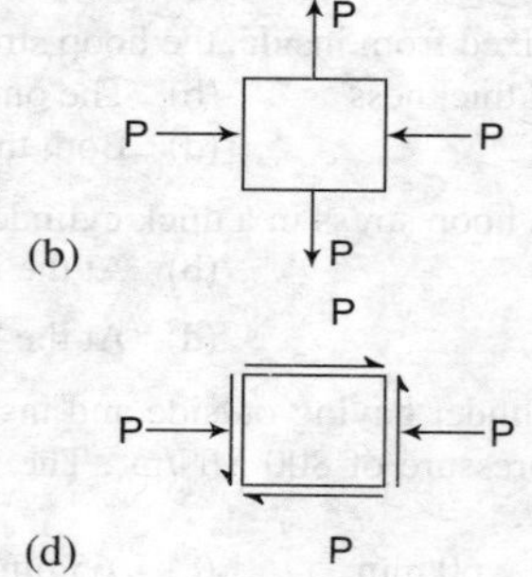

(b) (d)

Solutions for Objective Questions

Sl. No.	1.	2.	3.	4.	5.	6.	7.	8.	9.	10.
Answer	(b)	(a)	(d)	(d)	(b)	(c)	(a)	(d)	(a)	(b)

Sl.No.	11.	12.	13.	14.	15.	16.	17.	18.	19.	20.
Answer	(b)	(a)	(b)	(c)	(a)	(d)	(b)	(c)	(d)	(b)

Sl.No.	21.	22.	23.	24.	25.	26.	27.	28.
Answer	(c)	(a)	(a)	(b)	(c)	(c)	(c)	(c)

EXERCISE PROBLEMS

1. Derive an expression for the increase in the volume of a thin cylinder subjected to an internal fluid pressure.
2. A cast iron pipe of 1 m internal diameter is required to withstand a 200 m head of water. If the maximum tensile stress in the pipe is not to exceed 20 MPa, find the thickness of the pipe.

3. A copper tube 60 mm internal diameter and 1 mm thick is closely wound with 0.4 mm diameter steel wire under the tensile stress of 50 MPa. Determine the final stresses induced in wire and tube, if fluid is allowed into tube at 4 MPa. $\mu_c = 0.3$, $E_{steel} = 2 \times 10^5$ MPa, $E_{copper} = 1.25 \times 10^5$ MPa.

4. A steel tube of 300 mm external diameter is to be shrunk on to another tube of internal diameter 150 mm. After shrinking the diameter at the junction is 250 mm. If the shrinkage pressure is 28 MPa, determine the shrinkage allowance required. Take $E = 200$ GPa and Poisson's ratio is 0.25.

5. A cast iron cylinder of 200 mm inner diameter and 12.5 mm thick is closely wound with a layer of 4 mm diameter steel wire under a tensile stress of 55 MN/m^2. Determine the stress setup in the cylinder and steel wire, if water under a pressure of 3 MN/m^2 is admitted in the cylinder. EC = 100 GN/m^2; $E_S = 200$ GN/m^2, and Poisson's ratio is 0.25.

6. A copper tube 38 mm external diameter 35.5 mm internal diameter is closely wound with steel wire 0.75 mm diameter. Stating clearly the assumptions made, estimate the tension at which the wire must have been wound if an internal pressure of 2 N/mm^2 produces a tensile circumferential stress of 6.5 N/mm^2 in the tube. $E_S = 1.6 \times E_C$.

7. A cylinder shell 90 cm long and 20 cm internal diameter having thickness of metal as 8 mm is filled with fluid at atmosphere pressure. If an additional 20 cm^3 of fluid is pumped into the cylinder, find
   (i) The pressure exerted by the fluid on the cylinder
   (ii) The hoop stress induced.

   Take $E_C = 200$ GPa and $\mu_c = 0.25$; K_f (bulk modulus of the fluid) = 1.2 GPa

8. A steel tube is 18 mm internal diameter, 3 mm wall thickness, and 300 mm length. One end is closed and the other end is screwed into a pressure vessel. Find the safe internal pressure for the tube, if allowance stress is 150 MPa. Calculate the changes in dimensions and volume of tube. Take $E = 200$ GPa and $\mu = 0.25$.

9. A compound cylinder is formed by shrinking one tube of 240 mm outer diameter on to another tube of 120 mm inner diameter, with 200 mm as diameter at junction. If the radial pressure at the junction is 12 MPa due to shrinking, find the stresses in the two cylinders. If fluid is admitted into compound cylinder with 45 MPa, determine the final stresses in the compound cylinder. Sketch the variation of hoop stress and the radial stress across the thickness of the compound tube.

10. A compound cylinder of outer diameter 600 mm, diameter at the junction being 500 mm, and inner diameter 400 mm, is subjected to an internal pressure of 20 MPa. The contact pressure at the junction is 7 MPa. Sketch the variation of radial as well as hoop stress variation across the wall thickness of the tube.

11. A thickness is subjected to internal fluid pressure 'p'. If the maximum hoop stress developed is three times the internal fluid pressure find the ratio of outer diameter to inner diameter.

12. A thickness cylinder whose external diameter as 'k' times of its internal diameter is subjected to an internal fluid pressure 'p'. If the ratio to maximum to minimum hoop stress is 'n', find the relation between 'n' and 'k'. If the maximum hoop stress is 45 MPa and the value of 'n' is 3, find the internal fluid pressure and the necessary thickness of cylinder. The diameter of bore is 200 mm.

CHAPTER 9

TORSION

UNIT OBJECTIVE

This chapter provides information on the torsion of circular sections. The presentation attempts to help the student achieve the following:

Objective 1: Determine the maximum shear stress developed in a shaft and the angle of twist of the shaft subjected to a twisting moment.

Objective 2: Sketch the twisting moment diagram and angle of twist diagram for a loaded beam.

Objective 3: Determine the diameters of the shaft, estimate the rigidity modulus of the shaft, and calculate the angle of rotation and deflection of a coil subjected to load, torque, and shear stress.

Objective 4: Determine the strain energy stored in a shaft subjected to twisting moment.

9.1 INTRODUCTION

Torsion forms one of the basic structural actions besides axial force, bending, and shear. Rotating shafts, pulleys, and gears are the structural components, which are subjected to torsion/twisting moments. Torsion of circular shafts greatly differs from torsion of noncircular shafts. In this chapter, torsion of circular sections will be discussed. Torsion produces shearing stresses in the cross-section of the member. In case of circular cross-sections, the assumption that plane sections remain plane is valid, but the same is not valid in case of noncircular sections. However, solid noncircular sections are subjected to shearing stress alone. In case of open sections such as *T*, *I*, and *C*, the cross-section is subjected to shearing stresses and warping stresses. Warping stresses are normal stresses produced due to warping of the cross-section.

Assumption: The following assumptions were made to evaluate the shearing stress and tensional deformation (twist) in case of circular sections.

1. Material of the shaft is homogenous, elastic, and isotropic.
2. Plane sections remains plane, even after the application of the twisting moment.

Consider a circular shaft of length '*L*', subjected to twisting moment '*T*', as shown in Figure 9.1.

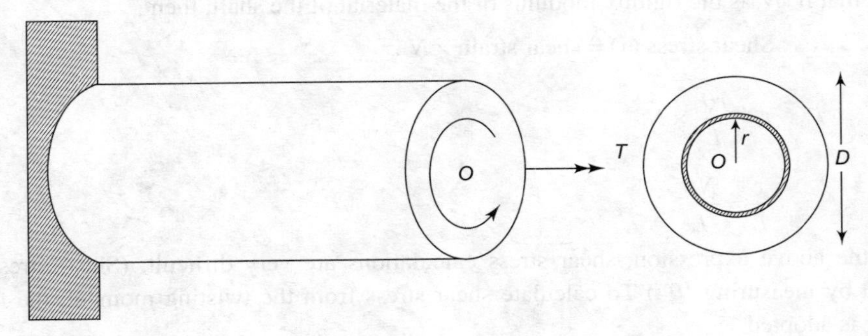

FIGURE 9.1

Let '*D*' be the diameter of the shaft. Consider an angular portion of the shaft. Let its radius be '*r*'. Fixidity at one of the supports does not allow the cross-section to rotate but free rotation is possible at the free end. AB, a fiber before the application of the twisting, takes the position AB′ after the application of the twisting moment. The inclination of the line segment AB′ with AB is same along the length of the shaft, as plane sections remain plane even after twisting.

The radial OB (shown in Figure 9.1) rotates by 'θ' and assumes the position OB′. 'θ' is called the angle of twist.

From geometrical considerations, assuming torsional deformations are very small.

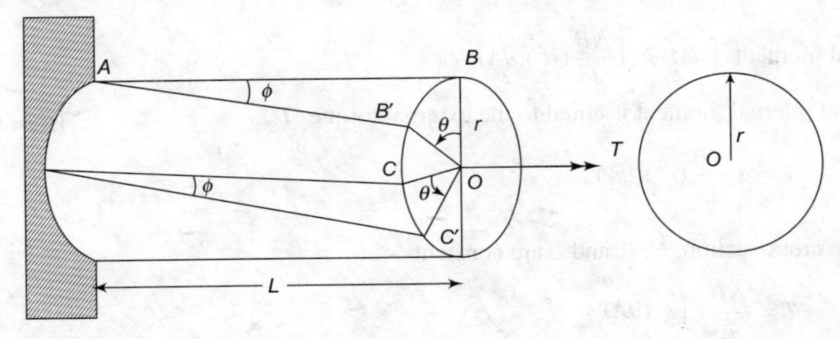

FIGURE 9.2

From Figure 9.2, elemental shaft of radius *r* is considered. *ABCD* is the curved plane before twisting *AB′C′D* is the curved plane after twisting. 'ϕ' is of the change of angle between *AB* and AD. Thus, ϕ can be taken as shear strain at a radial distance '*r*' from the center of the shaft cross-section.

$\therefore$ Shear strain = ϕ (constant along the length of the shaft)

$$\phi = \frac{BB'}{L}$$

$$BB' = r\theta$$

$$\therefore \quad \text{Shear strain} = \phi = \frac{r\theta}{L}.$$

We know that if '*N*' is the rigidity modulus of the material of the shaft, then

$$\text{Shear stress } (\tau) = \text{shear strain} \times N$$

$$\tau = \frac{r\theta}{L} \times N$$

$$\frac{\tau}{L} = \frac{N\theta}{L}. \qquad (9.1)$$

From the above expression, shear stress calculations are very difficult. (Shear stress can be calculated by measuring 'θ'.) To calculate shear stress from the twisting moment, the following procedure is adopted.

Shear stress 'τ' at a point is located 'r' distance from the center of the shaft.

$$\tau = \frac{N\theta}{L} \times r.$$

Consider an elemental area 'δA', located 'r' from the center of the shaft as shown in Figure 9.3.

Elemental shear force on the elemental area = $dF = \tau \times dA$.

$$dM = \tau\,(dA)\,(r)$$

$$= \frac{N\theta}{L}\,(r^2)\,(dA)$$

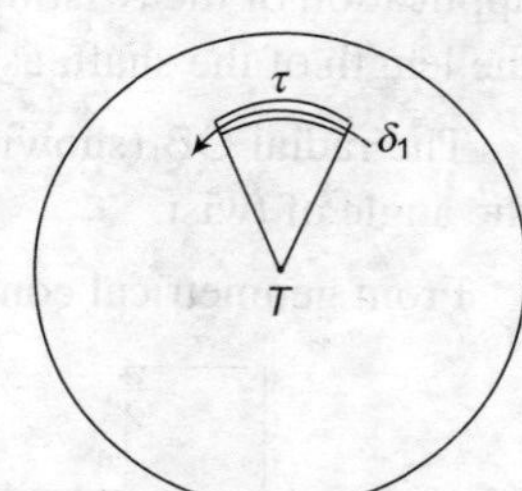

FIGURE 9.3

For total moment $\int dM = \int \frac{N\theta}{L}(r^2)(dA)$

This total internal moment is equal to the external torque '*T*'

$$T = \int_A \frac{N\theta}{L}(r^2)(dA)$$

Over the cross-section, *N*, θ, and *L* are constant

$$\therefore \quad T = \frac{N\theta}{L} \int_A r^2 (dA)$$

$\int r^2 (dA)$ is the polar moment of inertia (MI) of cross-section about center (*J*)

$$\therefore \quad T = \frac{N\theta}{L}(J) \qquad (9.2)$$

$$\frac{T}{J} = \frac{N\theta}{L}$$

in the above expression '*NJ*' is called torsional rigidity.

Equating Eqs. (9.1) and (9.2)

$$\frac{T}{J} = N\theta/L = \tau/r. \tag{9.3}$$

From Eq. (9.3), it is clear that τ shear stress linear varies from zero at center to a maximum (τ_{max}) value at the outer most fiber.

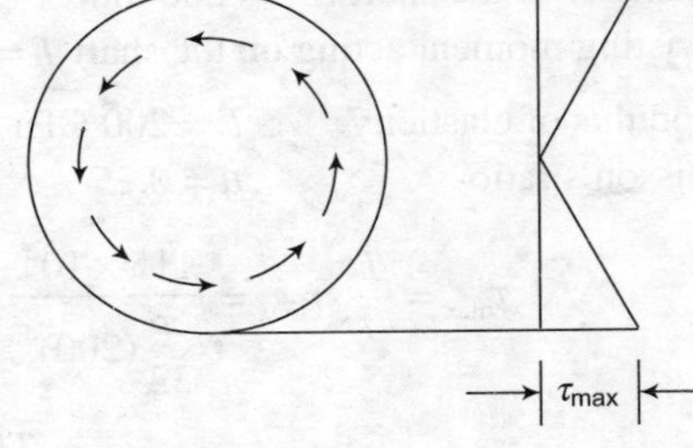

FIGURE 9.3

$$\therefore \qquad \tau_{max} = \frac{T}{J} r_{max}$$

$$\text{(or)} \qquad \tau_{max} = \frac{T}{J} \times (D/2)$$

For a solid circular cross-section, $J = \frac{\pi D^4}{32}$

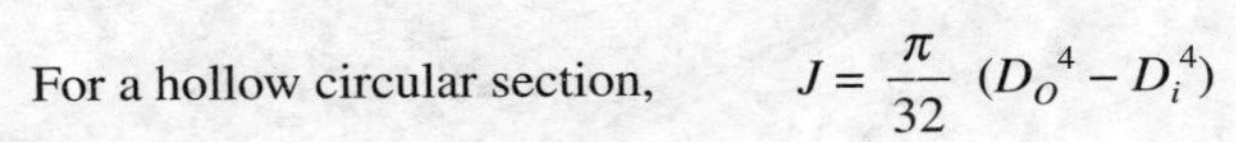

For a hollow circular section, $J = \frac{\pi}{32}(D_O^4 - D_i^4)$

For a thin tube, $J = \text{Lt}\ \frac{\pi}{32}(D_O^2 + D_i^2)(D_O + D_i)(D_O - D_i)$

In the case, a thin tube $D_O - D_i = 2t$, in which '*t*' is thickness of the tube.

For thin tube $J = \frac{\pi D^{3t}}{4}$

Hence for a solid circular cross-section $\tau_{max} = \left\{ \frac{T}{\frac{\pi}{32} D^4} \times \frac{D}{2} \right\}$

$$\tau_{max} = \frac{16T}{\pi D^3}$$

For a hollow shaft $\tau_{max} = \frac{T}{\frac{\pi}{32}\left(D_O^4 - D_i^4\right)} \times \left(\frac{D_O}{2}\right) = \frac{16T}{\pi D_O^3 \left[1 - \left[\frac{D_i}{D_O}\right]^4\right]}$

For a thin tube, $\tau_{max} = \frac{T}{\left(\frac{\pi D^3 t}{4}\right)} \times \left(\frac{D}{2}\right)$

$$\tau_{max} = \frac{2T}{\pi D^2 t}$$

It is to be remembered that polar moment of inertia (*J*) can be used only if the cross-section is circular.

PROBLEM 9.1

Objective 1

Determine the maximum shear stress developed in a solid circular shaft of a radius 100 mm, subjected to a twisting moment of 115 kN·m. Also determine the angle of twist per 1 m length of the shaft. Take $E = 200$ GPa; $\mu = 0.25$. If a hole of diameter 100 mm is bored at the center of the shaft along the length, estimate percentage increase of the maximum stress and angle of twist.

SOLUTION

Given that

Diameter of the shaft 'D' = 200 mm

Twisting moment acting on the shaft $T = 115$ kN·m

Modulus of elasticity $E = 200$ GPa

Poisson's ratio $\mu = 0.25$

$$\therefore \qquad \tau_{\max} = \frac{T}{J}\cdot r_{\max} = \frac{115\times10^6}{\frac{\pi}{32}(200)^4}\times 100 = 73.211 \text{ N/mm}^2$$

Let 'θ' be the angle of twist, then $\theta = \dfrac{TL}{GJ}$

$$\therefore \qquad \frac{\theta}{L} = \text{twist per unit length} = \frac{T}{GJ}$$

$$\therefore \quad \text{Twist per 1 mm (unit length)} = \frac{115\times10^6}{80\times10^3\times\frac{\pi}{32}(200)^4} = 9.15\times10^{-6} \text{ rad/mm}$$

$$\text{Twist per 1 mm length} = 9.15\times10^{-6}\times1000$$

$$= 0.00915 \text{ rad}$$

If a hole of 20 mm is bored into the shaft, at its center,

$$D_O = 200 \text{ mm}$$

$$D_i = 100 \text{ mm}$$

Then,

$$\tau_{\max} = \frac{T}{\frac{\pi}{32}(D_O^4 - D_i^4)}\times\left(\frac{D_O}{2}\right)$$

$$\tau_{\max} = \frac{16TD_O}{\pi(D_O^4 - D_i^4)} = 78.09 \text{ N/mm}^2$$

$$\text{Angle of twist per 1 m length} = \frac{T}{GJ} = \frac{115\times10^6\times1000}{80\times10^3\times\frac{\pi}{32}(200^4 - 100^4)}$$

$$= 0.0097 \text{ rad/m } \{0.5593°\text{/m}\}$$

∴ Percentage increase in the maximum shear stress due to boring the shaft

$$= 78.09 - \frac{73.211}{73.211}\times100 = 6.66\%$$

$$\text{Percentage increase in the twist} = \frac{0.00976 - 0.00915}{0.00915} \times 100 = 6.67\%$$

Percentage decrease in cross-sectional area

$$= \frac{\pi}{4(200)^2} - \frac{\pi / 4(200)^2 - 100^2}{\pi / 4(200)^2 \times 100} = 25\%$$

Note here that a decrease of 25% in cross-sectional area increased the maximum shear stress and twist by 6.67%. Thus, hollow sections are always preferable over solid sections.

PROBLEM 9.2 **Objective 1**

Determine shear stress and angle of twist at different sections, *A*, *B*, and *C* of the shaft shown is figure 9.5

$T_1 = 6$ kNm

$T_2 = 4$ kNm

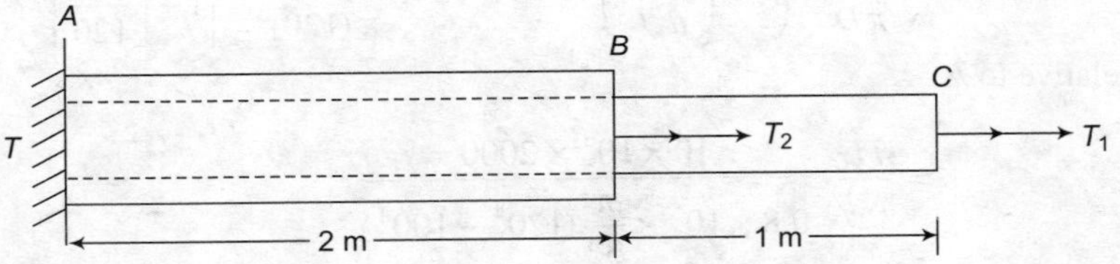

FIGURE 9.5

$N = 80 \times 10^3$ MPa in the portion *AB*, $d_O = 120$ mm;

$d_I = 100$ mm; in the region *BC*, $d = 60$ mm

SOLUTION

Draw the free body diagram (FBD) of individual parts of the member *ABC*

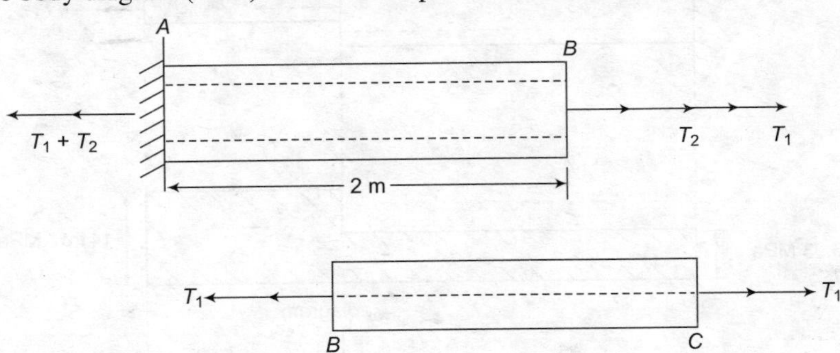

FIGURE 9.6

The stresses and deformations of the member *BC*

$$\tau_{max} = \frac{T_1}{J} \times r_{max}$$

$$= \frac{16\,T_1}{\pi d^3} = \frac{16 \times 6 \times 10^6}{\pi (60)^3} = 141.47 \text{ N/mm}^2$$

$$\text{Twist at } C \text{ relative to } B = \frac{TL}{NJ} = \frac{6 \times 10^6 \times 1000}{0.8 \times 10^5 \times \frac{\pi}{32}(60)^4}$$

Twist at 'C' relative to B = 0.0589 rad.

Shear stress and torsional deformation in the region AB

Twisting moment acting in the portion AB (from Figure 9.1.6) = $T_1 + T_2$

$= 6 + 4 = 10$ kN·m

Outer diameter = D = 120 mm

Inner diameter = d = 100 mm

Maximum shear stress in the portion $AB = \tau_{max}$

$$= \frac{16T}{\pi D^3}\left(1 - \left(\frac{D}{d}\right)^4\right) = 16 \times 10 \times \frac{10^6}{\pi(120)^3}\left\{1 - \left[\frac{100}{120}\right]^4\right\} = 56.93 \text{ MPa}$$

Twist at 'B' relative to A

$$= \frac{TL}{NJ} \quad \frac{10 \times 10^6 \times 2000}{0.8 \times 10^5 \times \frac{\pi}{32}(120^4 - 100^4)}$$

$= 0.0237$ rad

Twist at A is zero as the end 'A' is fixed end

$\therefore$ Angle of twist at B = 0.0237 rad

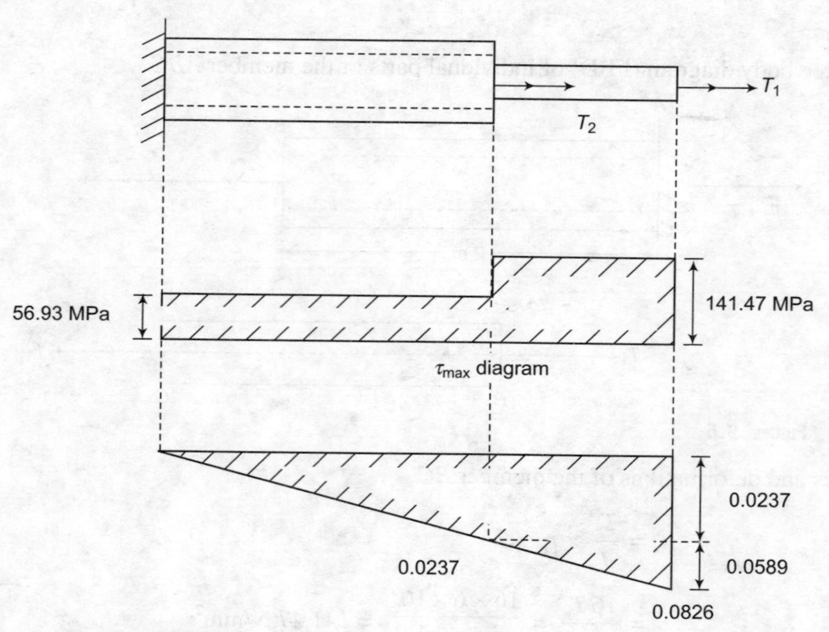

FIGURE 9.7 Angle of twist diagram

Angle of twist at C = angle of twist at B + angle of twist at C relative to B

$$= 0.0237 + 0.0589$$

$$= 0.0826 \text{ rad}$$

PROBLEM 9.3

Objective 1

Determine angle of twist and shear stress at different sections of the shaft shown in figure 9.8.

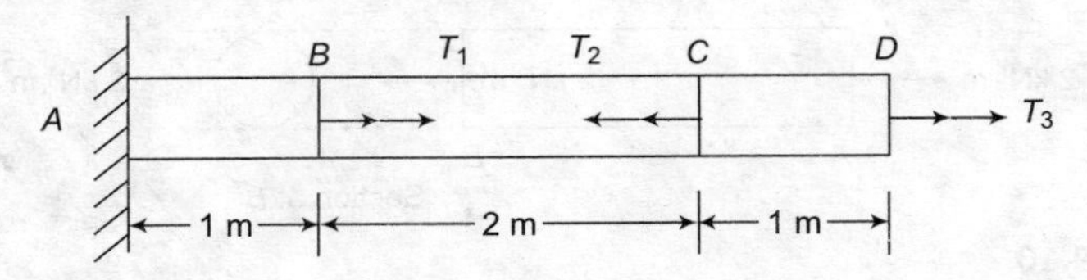

FIGURE 9.8

$$T_1 = 6 \text{ kN·m};\ T_2 = 10 \text{ kN·m};\ T_3 = 6 \text{ kN·m}$$

SOLUTION

Draw the *FBD* of the parts *AB*, *BC*, and *CD* of the shaft. The FBD of the different parts are shown in Figure 9.9.

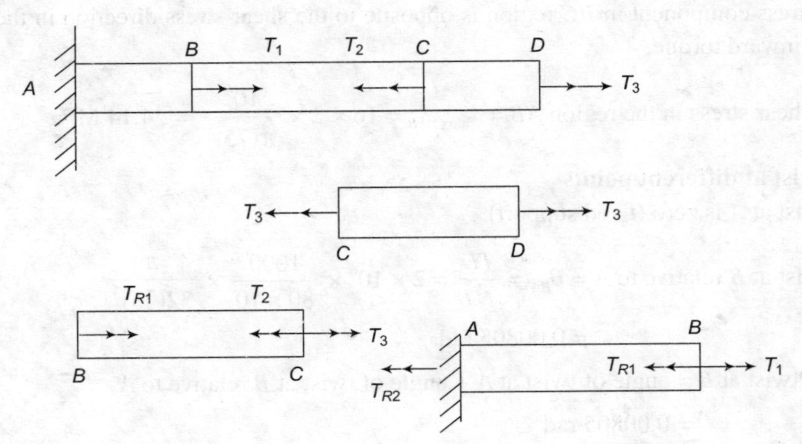

FIGURE 9.9

Given that

$$T_3 = 6 \text{ kN}\cdot\text{m};\ T_2 = 10 \text{ kN}\cdot\text{m}$$

$$\therefore \quad T_{R1} = T_2 - T_3 = 4 \text{ kN}\cdot\text{m};\ T_1 = 6 \text{ kN}\cdot\text{m}$$

$$T_{R2} = T_1 - T_{R1} = 6 - 4 = 2 \text{ kN}\cdot\text{m}$$

$\therefore$ Twisting moment acting in the portion $CD = T_3 = 6 \text{ kN}\cdot\text{m}$

Twisting moment acting in the portion $CB = T_{R1} = 4 \text{ kN}\cdot\text{m}$

Twisting moment in the region $BA = T_{R2} = 2 \text{ kN}\cdot\text{m}$

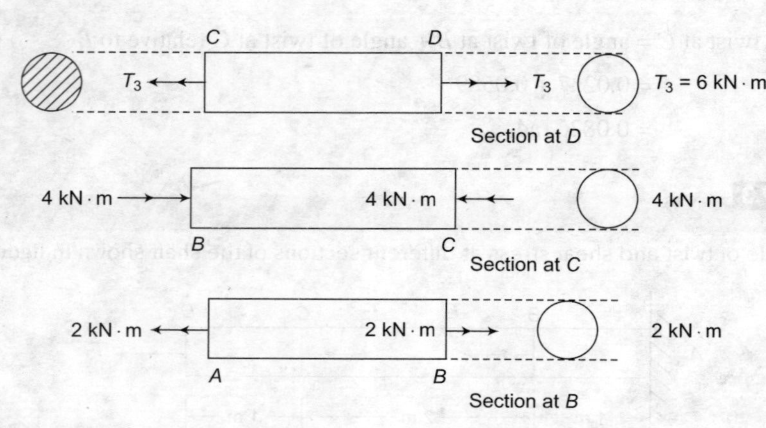

FIGURE 9.10

Maximum shear stress in the region $CD = \tau_{\max \cdot CD} = \dfrac{16T}{\pi d^3} = 16 \times 6 \times \dfrac{10^6}{\pi (75)^3} = 72.43$ MPa

Maximum shear stress in the region $BC = \tau_{\max \cdot BC} = 16 \times 4 \times \dfrac{10^6}{\pi (75)^3} = 48.29$ MPa

The shear stress component in BC region is opposite to the shear stress direction in the region CD as torque is inward torque.

Maximum shear stress in the region $AB = \tau_{\max \cdot AB} = 16 \times 2 \times \dfrac{10^6}{\pi (75)^3} = 24.14$ MPa

Angle of twist at different points

Angle of twist at A is zero (fixed support).

Angle of twist at B relative to $A = \theta_{BA} = \dfrac{TL}{NJ} = 2 \times 10^6 \times \dfrac{1000}{80 \times 10} \times \dfrac{\pi}{32(75)^4}$

$= 0.00805$ rad.

$\therefore$ Angle of twist at B = angle of twist at A + angle of twist at B relative to A

$= 0.00805$ rad

The twisting moment acting in region BC of the bars is $T_{R1} = 4$ kN·m (inward direction). Thus, angle of twist is opposite to the direction of angle of twist in the region AB.

$\therefore$ Angle of twist at C relative to $B = -T_{R1} \times \dfrac{L_{BC}}{NJ} = -4 \times 10^6 \times \dfrac{2000}{80 \times 10^3 \times \dfrac{\pi}{32}(75)^4}$

$= -0.0322$ rad

It implies that the cross-section at 'B' rotates in one direction, whereas the cross-section at C rotates in opposite direction.

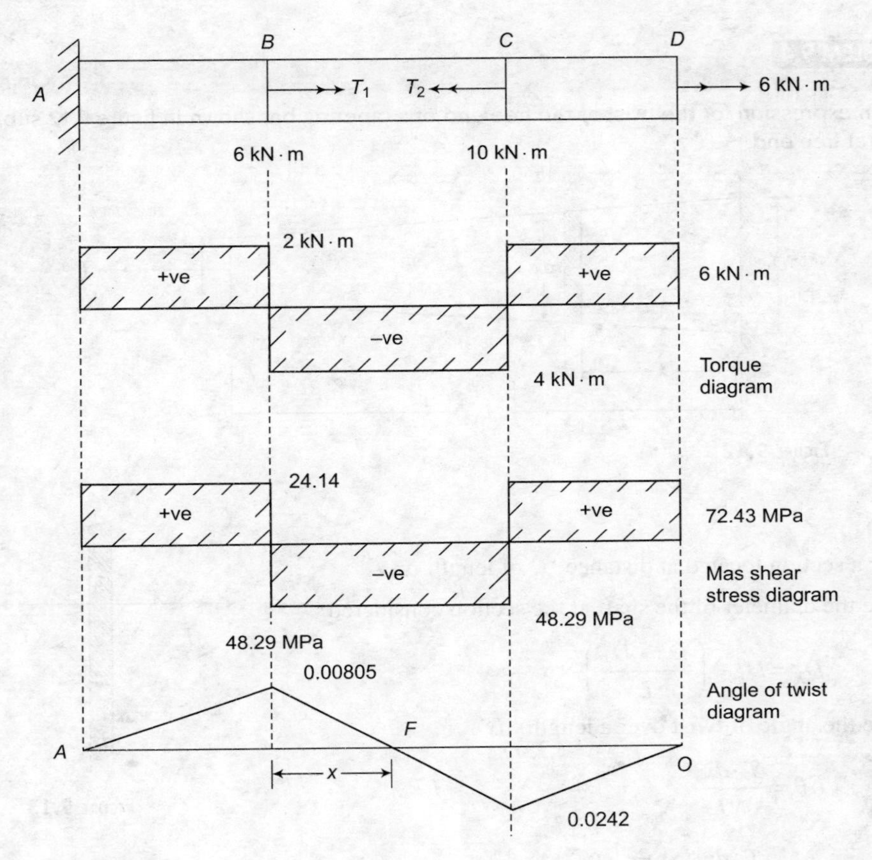

FIGURE 9.11

Angle of twist at D relative to $C = \theta_{DC} = 6 \times 10^6 \times 1000 \Big/ \left[80 \times 10^3 \times \dfrac{\pi}{32(75)^4} \right] = 0.0241$

Angle of twist at D = angle of twist at C + angle of twist at D relative to C

$$= -0.0242 + 0.0241$$

$$= 0$$

From the angle of twist diagram, it is clear that angle of twist is zero in the region BC.
Let the location of zero angle of twist be at a distance 'x' from B. Variation of angle of twist from B to C is linear.

$$\therefore \qquad \frac{x}{0.00805} = 2 - \frac{x}{0.0242} \Rightarrow x = 0.5 \text{ m}$$

$\therefore$ Angle of twist is zero at

(a) A, fixed support
(b) E, 1.5 m from the fixed support A
(c) D, free end.

PROBLEM 9.4

Objective 1

Obtain an expression for the twist at the free end of a tapering bar shown in figure 9.12 subjected to torque T at free end.

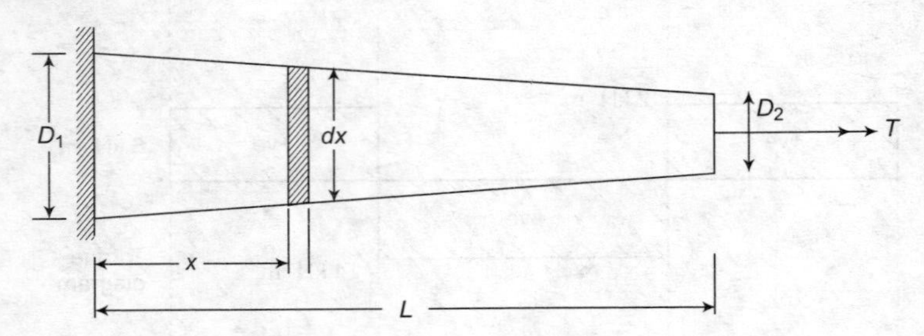

FIGURE 9.12

SOLUTION

Consider a section located at distance 'x' of length δx.

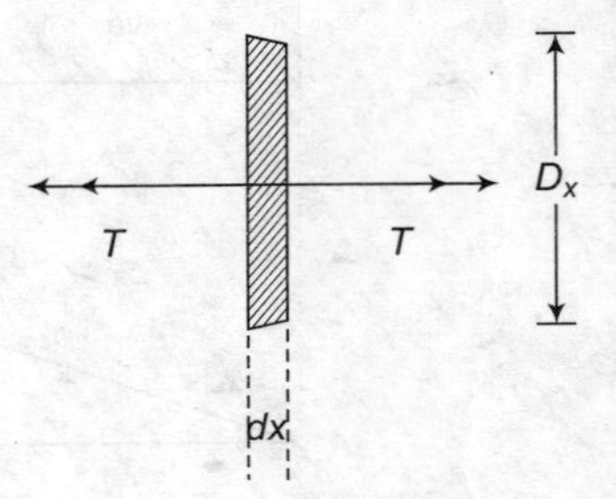

FIGURE 9.13

Let D_X be the diameter of the shaft at the section considered.

$$D_X = D_1 - \left(\frac{D_1 - D_2}{L}\right) \times x$$

Let $\delta\theta$ be the angle of twist over a length 'dx'.

Then, $$\delta\theta = \frac{T \cdot dx}{NJ}$$

$$d\theta = \frac{T \cdot dx}{N} \div \frac{\pi}{32}\left\{\frac{D_1 - D_2}{Lx}\right\}^4$$

Angle of twist at the free end relative to fixed end $= \delta = \int_0^L \frac{32T}{N \cdot \pi} \cdot \left\{D_1 - \frac{D_1 - D_2}{L} x\right\} dx$

$$= \frac{32T}{N\pi} \cdot \frac{1}{3} \left\{ \frac{\left[D_1 - \left(\frac{D_1 - D_2}{L}\right)x\right]^{-3}}{\left(\frac{D_1 - D_2}{L}\right)} \right\} L_0$$

$$\Rightarrow \qquad \theta = \frac{32TL}{NJ}\left\{D_1^2 + D_1 D_2 + \frac{D_2^2}{D_1^3 D_2^3}\right\}.$$

PROBLEM 9.5

Objective 2

Sketch the twisting moment diagram and angle of twist diagram for the beam loaded as shown in Figure 9.14.

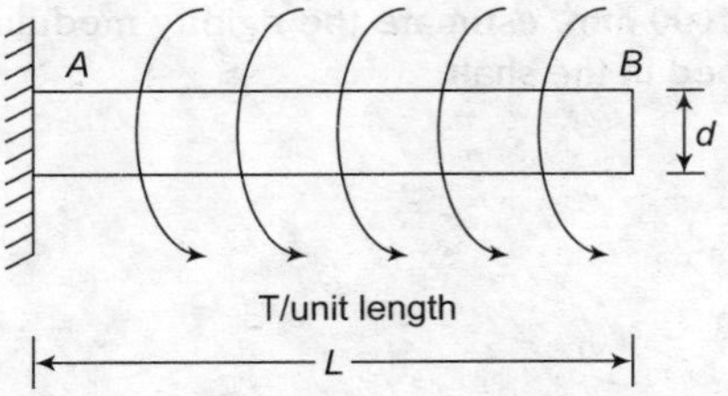

FIGURE 9.14

SOLUTION

Consider a section loaded x distant from free end. Let 'δx' be the length of the elemental strip.

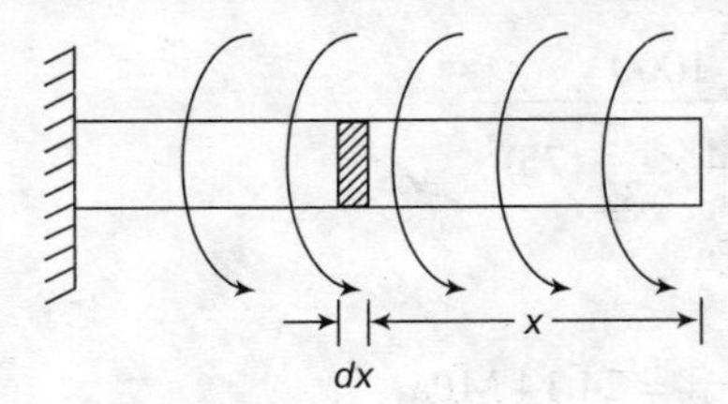

FIGURE 9.15

Twisting moment acting at the section considered is

$$T_X = Tx$$

At $x = 0$, $\quad T_B = 0$

At $x = L$, $\quad T_A = T \times L$

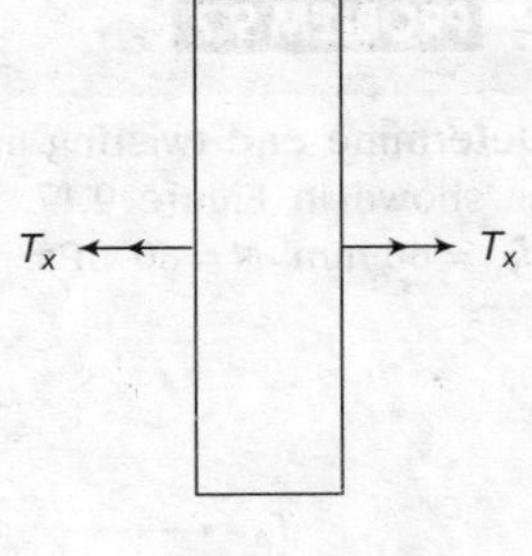

FIGURE 9.16

Consider the elemental length of the shaft, located 'x' distant from free end.

Angle of twist at the section considered $\delta\theta = \dfrac{T_X \cdot dx}{NJ}$

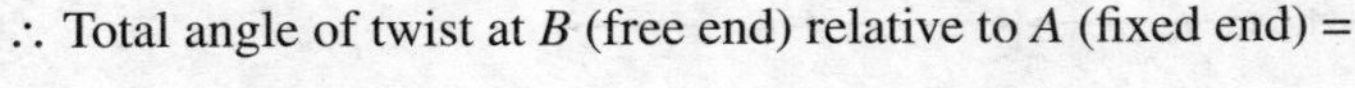

$\therefore$ Total angle of twist at B (free end) relative to A (fixed end) =

$$\theta = \int_0^L \theta$$

$$\therefore \qquad \theta = \int_0^L T(x)\frac{dx}{NJ}$$

$$\theta = \frac{T(L^2)}{2NJ}$$

PROBLEM 9.6 **Objective 1 and 3**

In an experiment, the relative angle of twist between two points A and B is found to be 0.0174 rad, for applied pure torque of 2 kN·m. If the diameter of the shaft is 75 mm and the distance between the twist measuring points is 1000 mm, estimate the rigidity modulus of the shaft material and the maximum shear stress developed in the shaft.

SOLUTION

Twisting moment = 2 kN·m

Diameter of the shaft = 75 mm

Guage length (L) = 1 m

Relative angle of twist = $\theta = 0.0174$ rad.

We know that $$\theta = \frac{TL}{NJ}$$

$$\therefore \quad N = \frac{TL}{\theta J} = 2 \times 10^6 \times \frac{1000}{0.0174 \times \frac{\pi}{32}(75)^4}$$

$$= 37{,}003 \text{ N/mm}^2 = 37 \text{ GPa}$$

$$\tau_{max} = \frac{16T}{\pi d^3} = 16 \times 2 \times \frac{10^6}{\pi (75)^3} = 24.14 \text{ MPa}$$

PROBLEM 9.7

Objective 1

Determine end twisting moments developed in the shaft subjected to twisting moment of 6 kN·m as shown in Figure 9.17. Also determine the maximum shear stress developed at d_{AB} = 40 mm; d_{BC} = 60 mm; N = 80 GPa.

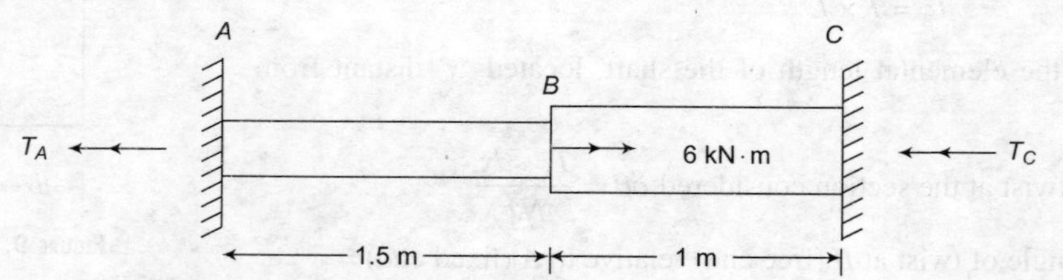

FIGURE 9.17

SOLUTION

Let T_A and T_C be the twisting moments developed at A and C, respectively.
Sum of the moments along the axis of the beam is zero.

$$\Rightarrow \quad T_A + T_C = 6 \qquad (1)$$

With the above single equilibrium equation, T_A and T_C cannot be evaluated. Thus, a compatibility condition is required for the analysis.

As the ends A and C are fixed, angle of twist at B relative to A is equal to the angle of twist at B relative to C.

Consider the portion AB,

FIGURE 9.18

Angle of twist at B relative to A = $\theta_{B/A} = \dfrac{T_A \cdot L_{AB}}{NJ_{AB}}$

$$\theta_{B/A} = \left[\frac{T_A \times 10^6 \times 1.5 \times 1000}{N \times \frac{\pi}{32}(40)^4}\right]$$

$$= 5968.31 T_A / N$$

Consider the portion BC,

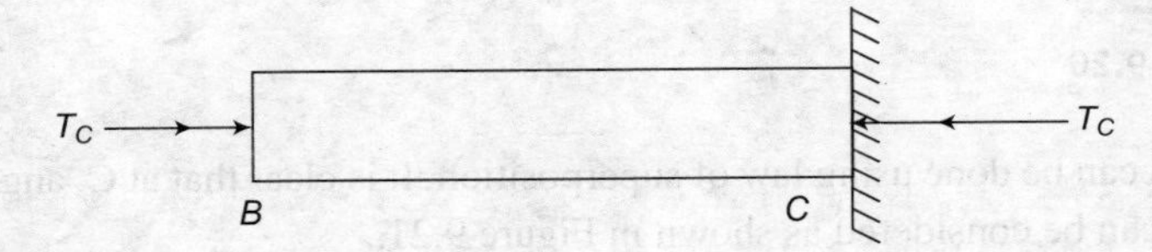

FIGURE 9.19

Angle of twist at B relative to C

$$\theta_{B/C} = \frac{T_C \times L_{BC}}{NJ_{BC}}$$

$$\theta_{B/C} = \left[\frac{T_C \times 10^6 \times 1.0 \times 1000}{N \times \frac{\pi}{32}(60)^4}\right] = \frac{785.95 T_C}{N}$$

$$\theta_{B/A} = \theta_{B/C} \text{ (compatibility condition)}$$

$$\Rightarrow \qquad \frac{5968.31 T_A}{N} = \frac{785.95 T_C}{N}$$

$$T_C = 7.594 T_A \qquad (2)$$

Substituting Eq. (2) in Eq. (1)

$$7.594 T_A + T_A = 6$$

$$T_A = 0.698 \text{ kN·m}$$

$$T_C = 5.302 \text{ kN·m}$$

If $N = 80$ GPa, $\qquad \theta_B = 0.0521$ rad

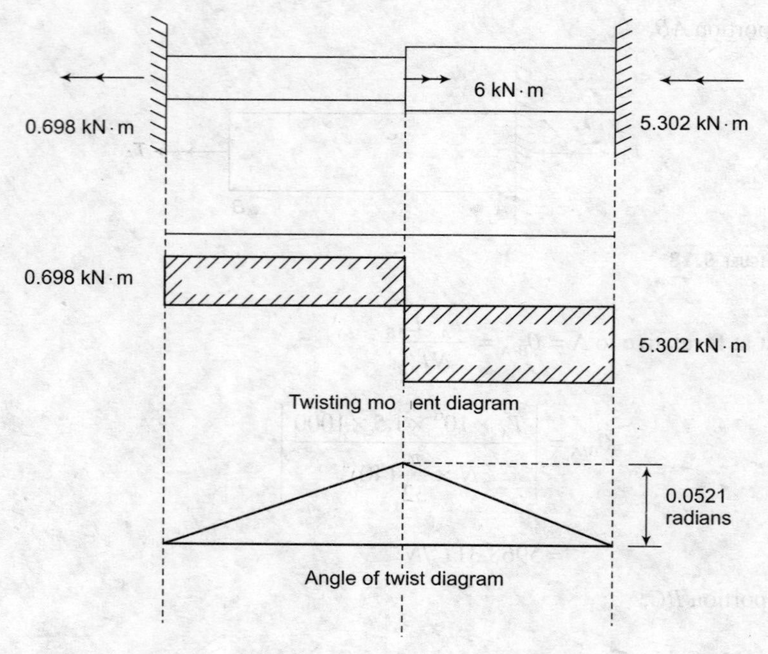

FIGURE 9.20

The same problem can be done using law of superposition. It is clear that at C, angle of twist is zero. Thus, given shaft can be considered as shown in Figure 9.21.

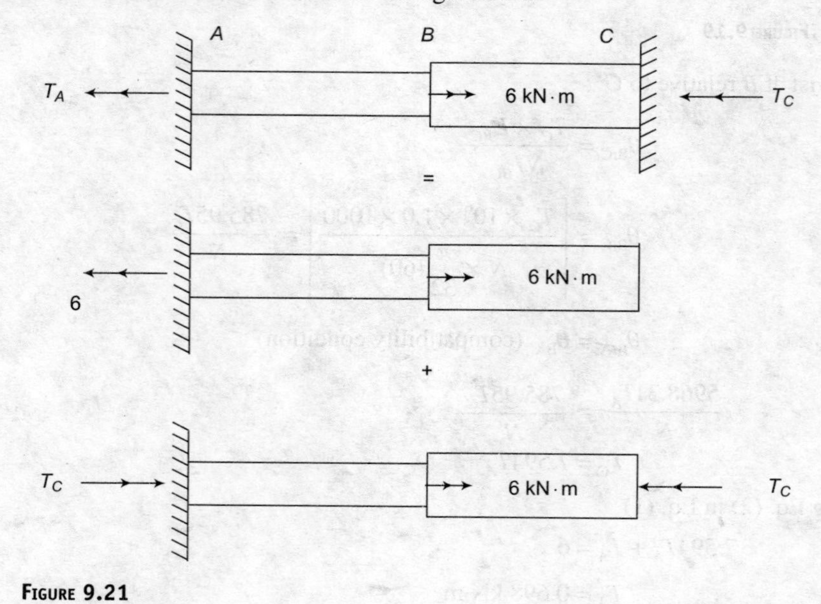

FIGURE 9.21

From Figure 9.20, it is clear that

$$T_A = 6 - T_C \qquad \text{(1) (equilibrium)}$$

Angle of twist in case (I), at C = θ_{C1}

θ_{C1} = angle of twist at B, relative to A + angle of twist at C, relative to B

$$= \frac{6 \times 10^6 \times 1500}{80 \times 10^3 \times \frac{\pi}{32}(40)^4} + 0 = 0.448 \text{ rad}$$

Angle of twist at C in case (II) = θ_{C2}

$$\theta_{C2} = -\left\{\frac{[T_C \times 10^6 \times 1500]}{\left[80 \times 10^3 \times \left(\frac{\pi}{32}\right)(40)^4\right]} + \frac{[T_C \times 10^6 \times 1000}{\left[80 \times 10^3 \times \left(\frac{\pi}{32}\right)(60)^4\right]}\right\}$$

$= -0.0844\ T_C$ rad (sign opposite to θ_{C1})

$$\theta_{C1} + \theta_{C2} = 0$$

$\Rightarrow \qquad 0.448 = 0.0844\ T_C$

$\Rightarrow \qquad T_C = 5.302$ kN·m

$\therefore \qquad T_A = 0.698$ kN·m

9.2 STRAIN ENERGY STORED IN SHAFTS SUBJECTED TO TWISTING MOMENT

Consider a shaft subjected to twisting moment T. Consider an elemental area 'dA' located 'r' distance from the center of the shaft.

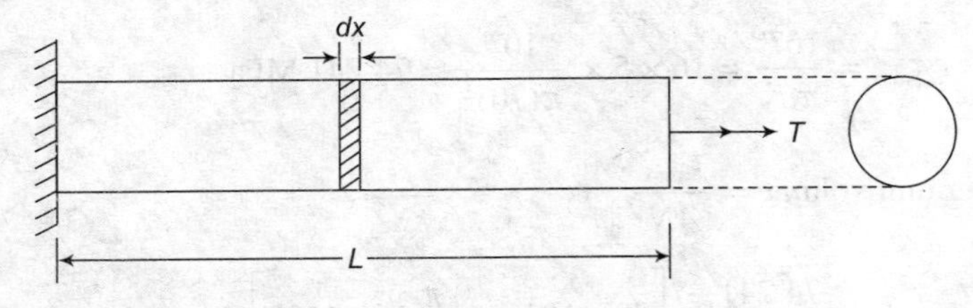

FIGURE 9.22

Let 'τ' be the shear stress developed over the elemental area. Let 'γ' be the shear strain.

Then, elemental strain energy (U) per unit volume $= \frac{1}{2} \cdot \tau \cdot \gamma$

$$= \frac{\tau^2}{2N}$$

Elemental strain energy per unit volume $= \frac{T^2}{J^2}(r^2)\frac{1}{2N}$

Elemental strain energy $= \frac{T^2}{J^2}(R^2)\frac{1}{2N} \cdot dA \cdot dx$

$$\therefore \text{ Total strain energy} = \iint\limits_{L\,A} \frac{T^2}{J^2}(R^2)\frac{1}{2N} \cdot dA \cdot dx$$

$$\int\limits_{L}\left(\frac{T^2}{J^2}\right)\frac{1}{2N}(J)\,dx = \int\limits_{L}\frac{T^2}{2NJ}dx$$

$$= \frac{T^2}{2NJ \cdot L} = \frac{1}{2T \cdot \theta_C}$$

$$U = \frac{T^2 L}{2NJ}$$

From a circular shift,

$$T_{\max} = \frac{16T}{\pi d^3}$$

$$U = \frac{\pi^2 d^6 \tau_{\max}^2}{16}(16)\frac{L}{2}(N)\frac{\pi}{32}d^4 = \frac{\tau_{\max}^2}{4N} \times \frac{\pi}{4}d^2 \times L$$

$$= \tau_{\max}^2/4N \times \text{volume of the shaft.}$$

PROBLEM 9.8

Objective 4

Determine the strain energy stored in a 70 mm diameter, shaft of length 1.2 m, subjected to twisting moment 5 kN·m, if $N = 80$ GPa.

SOLUTION

$$\tau_{\max} = \frac{16T}{\pi d^3} = 16 \times 5 \times \frac{10^6}{\pi(70)^3} = 74.241 \text{ MPa}$$

$$\therefore \text{ Strain energy per unit volume} = \frac{\tau_{\max}^2}{4N}$$

$$\text{Strain energy } U = \frac{74.241^2}{4} \Big/ \left[80 \times 10^3 \times \frac{\pi}{4(70)^2} \times 1200\right]$$

$$U = 0.0795 \text{ kN·m}$$

$$\text{Other expression } U = \frac{T^2 L}{2NJ} = (5 \times 10^6)^2 \times \frac{1200}{\frac{\pi}{32(70)^4}} \times \frac{1}{80 \times 10^3 \times 2}$$

$$= 79544.33 \text{ N·mm}$$

$$= 0.0795 \text{ kN·m}$$

PROBLEM 9.9

Objective 4

Determine strain energy stored in a thin tube of mean diameter 100 mm, thickness 10 mm, and length 1 m subjected to twisting moment of 2.5 kN·m. Take $G = 80$ GPa.

SOLUTION

$$T = 2.5\ \text{kN} \cdot \text{m}$$

$$\text{Strain energy } U = \frac{1}{2} T\theta$$

$$\text{Angle of twist } \theta = \frac{TL}{NJ}$$

$$J = \text{Polar MI}$$

$$= \frac{\pi d^3 t}{4} = 7.854 \times 10^6\ \text{mm}^4$$

$$\therefore \quad = \frac{(2.5 \times 10^6 \times 1000)}{(80 \times 10^3 \times 7.854 \times 10^6)}$$

$$\text{Strain energy} = \frac{1}{2} T\theta = 4973.59\ \text{N·mm.}$$

9.3 POWER TRANSMISSION THROUGH SHAFTS

Generally, shafts are used in mechanical appliances, where in power from generation point is to be transmitted to the point of its (power) application, through rotary motion. A best example for this is flour mills. The power transmission through shafts is shown in Figure 9.23.

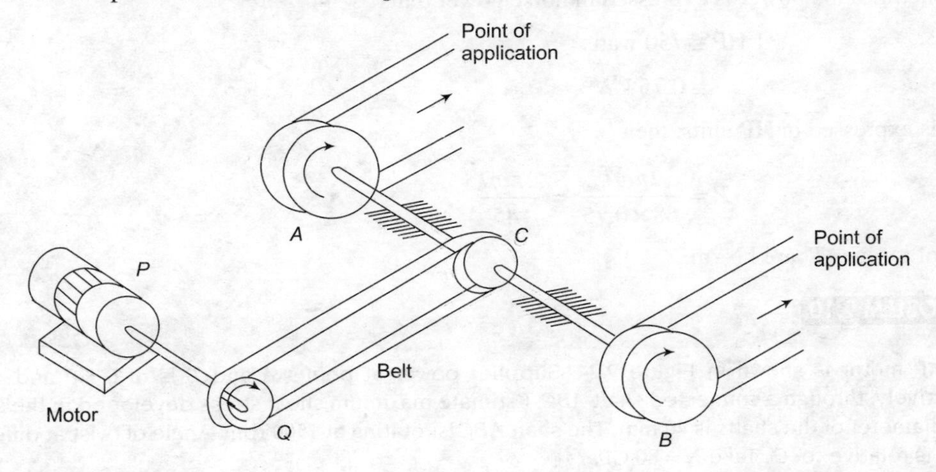

FIGURE 9.23

Power from the motor is transmitted to the shaft *ACB* through the belt. The shaft *ACB* rotates thereby rotating the fly wheels at *A* and *B*. The fly wheels at *A* and *B* transmit power to the point of application.

9.3.1 Relationship Between the Power and Torque of Rotating Shaft

Consider a shaft rotating at constant speed 'ω' under the action of torque 'T'. Let 'α' be the rotation of the shaft during a time interval of 't'.

Work done by the torque = $T \cdot \alpha$

Power required for rotating shaft = P = work done/time

$$\Rightarrow \qquad P = T \cdot \frac{\alpha}{t}$$

But we know that α/t is the angular speed of the rotating shaft

$$\therefore \qquad P = T \cdot \left\{\frac{d\alpha}{dt}\right\}$$

If 'ω' is expressed in revolution/min (rpm), then $\dfrac{d\alpha}{dt} = \dfrac{2\pi\omega}{60}$

$$\therefore \qquad P = \frac{2\pi\omega}{60} T$$

Units: If 'P' is expressed as kilowatts, (kW) then

$$P = \frac{2\pi\omega}{60} T$$

in which units for torque T are kN·m.

Sometimes the power is expressed in horse power units

$$1 \text{ HP} \cong 750 \text{ watts}$$

$$\cong 0.75 \text{ kW}.$$

If 'P' is expressed in HP units, then

$$P = \frac{2\pi\omega T}{68 \times 0.75} = \frac{2\pi\omega T}{45}$$

Units of torque 'T' are kN·m.

PROBLEM 9.10 **Objective 1**

A 10 HP motor is shown in Figure 9.24. Supplier power at points *A* and *B* is of 6 HP and 4 HP, respectively, through a connected shaft *ABC*. Estimate maximum shear stress developed in the shaft, if the diameter of the shafts is 40 mm. The shaft *ABC* is rotating at 1500 rpm. Angle of twist at different points is relative to '*C*'. Take $N = 80$ GPa.

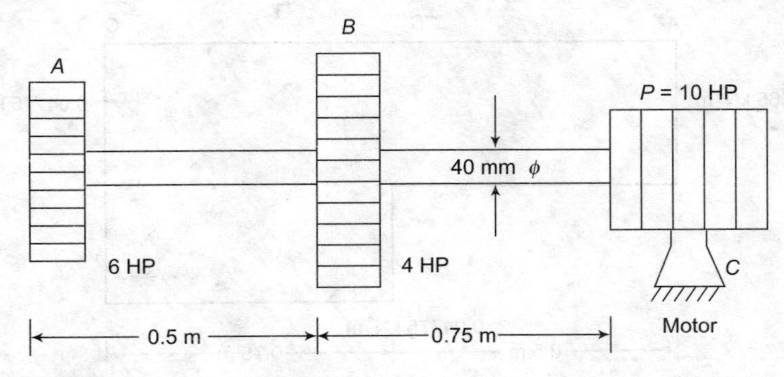

FIGURE 9.24

SOLUTION

Given that,

Angular speed of the shaft = ω = 1500 rpm

Let torque acting at 'A' be T_A.

Power delivered at A = 6 HP

$$P_A = \frac{2\pi\omega T_A}{45} \text{ (since power is given in HP units)}$$

$$\Rightarrow \qquad 6 = 2\pi \times 1500 \times \frac{T_A}{45}$$

$$\Rightarrow \qquad T_A = 0.02865 \text{ kN·m}$$

Power delivered at B = 4 HP.

Let the toque acting at B be T_B. Then

$$P_B = \frac{2\pi\omega T_B}{45}$$

$$\Rightarrow \qquad 4 = \frac{2\pi\omega \times 1500 \times T_B}{45}$$

$$\Rightarrow \qquad T_B = 0.01910.$$

Similarly torque at C,

$$T_C = 0.04775 \text{ kN·m}$$

Angle of twist at B relative to $C = \dfrac{T_{BC} \cdot L_{BC}}{NJ}$

$$\theta_{B/C} = \frac{[0.04775 \times 10^6 \times 750]}{\left[\dfrac{80 \times 10^3 \times \pi}{32(40)^4}\right]}$$

$$= 0.00178 \text{ rad}$$

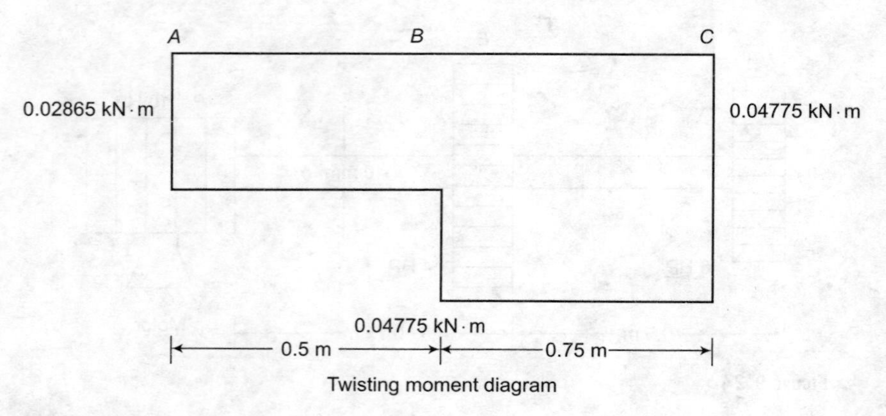

FIGURE 9.25

Angle of twist at A relative to B

$$\theta_{A/B} = \frac{T_{AB} \cdot L_{AB}}{NJ} = \frac{[0.02865 \times 10^6 \times 500]}{\left[\dfrac{80 \times 10^3 \times \pi}{32(40)^4}\right]}$$

$$= 0.00071 \text{ rad.}$$

$\therefore$ Angle of twist at A relative to $C = \theta_{A/B} + \theta_{B/C}$

$$\theta_{A/C} = 0.0249 \text{ rad.}$$

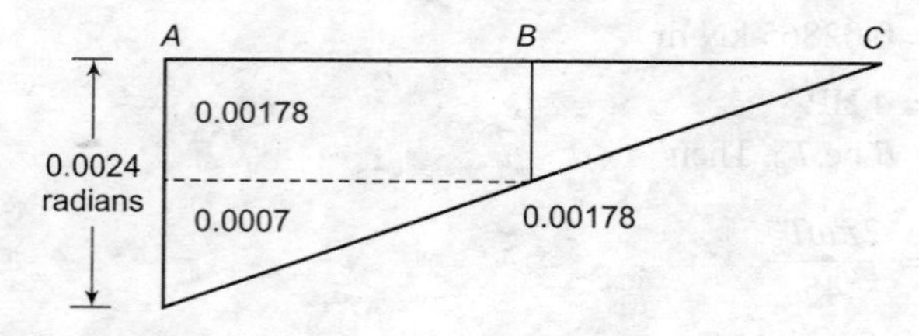

FIGURE 9.26

PROBLEM 9.11

Objective 3

A hollow circular shaft transmits 200 kW power rotating at 100 rpm. If the maximum torque developed is 1.3 times the mean torque, determine the outer diameter and inner diameter of the shaft. Allowable shear stress is limited to 80 MPa and angle of twist is limited to 1.5°/m. Take $N = 40$ GPa and diameter ratio of 0.6.

SOLUTION

Power = 100 kW

Angular speed of the shaft = 100 rpm

If mean torque acting on the shaft is T_{mean}, then

$$P = \frac{2\pi\omega T_{\text{mean}}}{60}$$

$$\Rightarrow \qquad 200 = 2\pi \times \frac{100}{60} T_{mean}$$

$$\rightarrow \qquad T_{mean} = 1.910 \text{ kN·m}.$$

Shaft is to be designed for the maximum torque.

$$T_{max} = 1.3 \times T_{mean}$$

$$= 1.3 \times 1.901$$

$$= 2.4328 \text{ kN·m}.$$

Maximum shear stress in shaft $\tau_{max} \le 80$ MPa (given)

$$\tau_{max} = \frac{(16T_{mean})}{\pi d_o^3 \left[1 - \left(\frac{d_i}{d_o}\right)^4\right]} = \frac{1 \times 2.483 \times 10^6}{\pi d_o^3 \{1 - (0.6)^4\}}$$

$$\Rightarrow \qquad \frac{1.4589 \times 10^7}{d_o} \le 80$$

$$\Rightarrow \qquad d_o \ge 56.63 \text{ mm}.$$

Consider the condition an angle of twist per meter length

$$\theta = 1.5° = \frac{1.5}{180} \times \pi = 0.0262 \text{ rad}$$

$$\theta \le \frac{TL}{NJ}$$

$$\Rightarrow \qquad 0.0262 \ge \frac{2.483 \times 10^6 \times 1000}{80 \times 10^3 \times \frac{\pi}{32} \times (d_o)^4 [1 - 0.6^4]}$$

$$\Rightarrow \qquad d_o^4 \ge 13873924.58$$

$$\Rightarrow \qquad d_o \ge 61.03 \text{ mm}.$$

$\therefore$ Outer diameter of shaft $d_O \ge 61.03$ mm

Provide $d_O = 70$ mm and $d_i = 42$ mm

PROBLEM 9.12 **Objective 5**

A solid shaft of running at 100 rpm is used to raise a weight of 1500 N by 3 m in the vertical direction in 2 s. Determine the diameter of the shaft if shear stress developed in the shaft is limited to 40 MPa. If the shaft is replaced by a thin tube of thickness 3 mm, determine the percentage saving in the material.

SOLUTION

The rotating shaft has to leave a weight of 1500 N through 3 m in 2 seconds.

$$\text{Power required} = \frac{1500 \times 3}{2} = 2250 \text{ N·m} = 2.25 \text{ kW}.$$

Shaft is rotating at 100 rpm.

Let the torque acting on the shaft be T.

$$P = \frac{2\pi\omega T}{60}$$

$$\Rightarrow \quad 2.25 = \frac{2\pi \times 100 \times T}{60}$$

$$\Rightarrow \quad T = 0.2149 \text{ kN·m}$$

When a solid shaft is used to resist the torque, then

$$\tau_{\max} = \leq 40 \text{ MPA}$$

$$\therefore \quad \frac{16 \times 0.2149 \times 10^6}{\pi d^3} \leq 40 \quad \Rightarrow \quad d \geq 30.13 \text{ mm}$$

$\therefore$ Cross-sectional area of the solid shaft $A_S = \frac{\pi}{4}(30.13)^2 = 713.07 \text{ mm}^2$

If the solid shaft is replaced by a thin tube of thickness 3 mm, then

$$\tau_{\max} = \frac{T}{J}\left(\frac{d}{2}\right) = \frac{T}{\frac{\pi d^3 t}{8}} \frac{d}{2} = \frac{4T}{\pi d^2 t}$$

$$\tau_{\max} \leq 40$$

$$\Rightarrow \quad \frac{4 \times 0.2149 \times 10^6}{\pi d^2 (3)} \leq 40$$

$$\Rightarrow \quad d \geq 47.75 \text{ mm}.$$

Cross-sectional area of the thin tube = $\pi dt = \pi \times 47.75 \times 3 = 450.04 \text{ mm}^2$

% Saving in adopting a thin tube replacing a solid one $= \frac{713.07 - 450.04}{713.07} \times 100$

$$= 36.89\%.$$

SUMMARY

- **A shaft is in torsion, when equal and opposite torques are applied at the two ends of a shaft.**
- **The relation of maximum shear stress induced in a shaft subjected to twisting moment is given by**

$$\frac{\tau}{R} = \frac{C\theta}{L}$$

in which, τ = maximum shear stress
R = radius of shaft
C = modulus of rigidity
θ = angle of twist in radian, and
L = length of the shaft
- **The shear stress is maximum on the surface of the shaft and is zero at the axis of the shaft.**
- **The power transmitted by a shaft is given by**

$$P = \frac{2\pi NT}{60}$$

- **Springs are the elastic bodies which absorb energy due to resilience. Two important types of springs are:**
  i. Laminated or leaf springs, and
  ii. Helical springs.

OBJECTIVE TYPE QUESTIONS

1 A solid circular shaft of 60 mm diameter transmits a torque of 1600 N·m. The value of maximum shear stress developed is: [GATE-2004]
(a) 37.72 MPa (b) 47.72 MPa (c) 57.72 MPa (d) 67.72 MPa

2 Maximum shear stress developed on the surface of a solid circular shaft under pure torsion is 240 MPa. If the shaft diameter is doubled then the maximum shear stress developed corresponding to the same torque will be: [GATE-2003]
(a) 120 MPa (b) 60 MPa (c) 30 MPa (d) 15 MPa

3 A steel shaft 'A' of diameter 'd' and length 'l' is subjected to a torque T. Another shaft 'B' made of aluminium of the same diameter 'd' and length 0.5l is also subjected to the same torque 'T'. The shear modulus of steel is 2.5 times the shear modulus of aluminium. The shear stress in the steel shaft is 100 MPa. The shear stress in the aluminium shaft, in MPa, is: [GATE-2000]
(a) 40 (b) 50 (c) 100 (d) 250

4 The diameter of shaft A is twice the diameter or shaft B and both are made of the same material. Assuming both the shafts to rotate at the same speed, the maximum power transmitted by B is: [IES-2001; GATE-1994]
(a) The same as that of A
(b) Half of A
(c) 1/8th of A
(d) 1/4th of A

5 A solid shaft can resist a bending moment (BM) of 3.0 kN·m and a twisting moment of 4.0 kN·m together, then the maximum torque that can be applied is: GATE-1996]
(a) 7.0 kN·m (b) 3.5 kN·m (c) 4.5 kN·m (d) 5.0 kN·m

6 A torque of 10 N·m is transmitted through a stepped shaft as shown in figure. The torsional stiffnesses of individual sections of lengths MN, NO, and OP are 20 N·m/rad, 30 N·m/rad, and 60 N·m/rad, respectively. The angular deflection between the ends M and P of the shaft is: [GATE-2004]

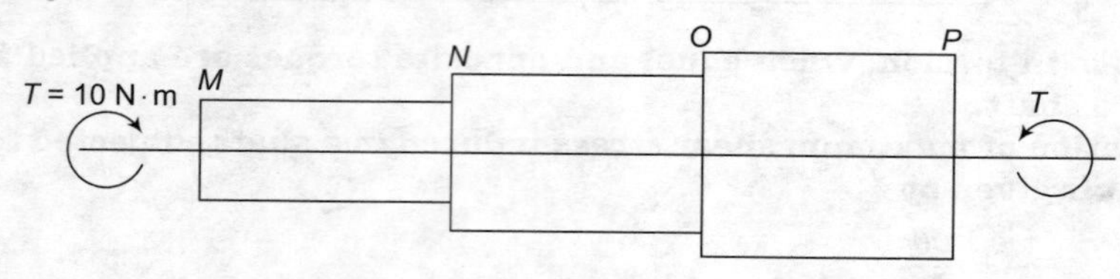

(a) 0.5 rad (b) 1.0 rad (c) 5.0 rad (d) 10.0 rad

7 The two shafts AB and BC, of equal length and diameters d and $2d$, are made of the same material. They are joined at B through a shaft coupling, whereas the ends A and C are built-in (cantilevered). A twisting moment T is applied to the coupling. If TA and TC represent the twisting moments at the ends A and C, respectively, then [GATE-2005]

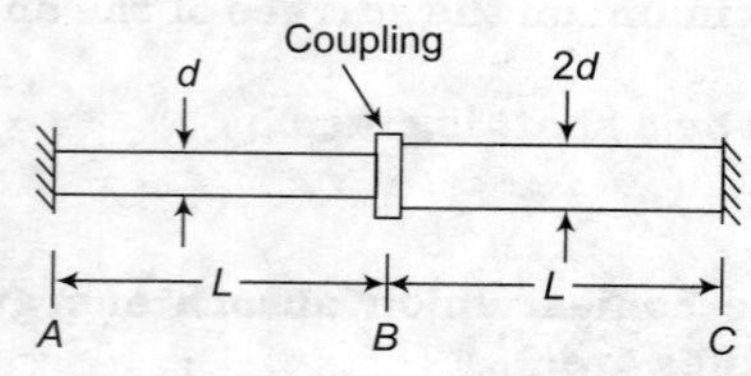

(a) $T_C = T_A$ (b) $T_C = 8\ T_A$ (c) $T_C = 16\ T_A$ (d) $T_A = 16\ T_C$

8 Consider the following statements: [IES-2008]

Maximum shear stress induced in a power transmitting shaft is:
1. Directly proportional to torque being transmitted.
2. Inversely proportional to the cube of its diameter.
3. Directly proportional to its polar MI. Which of the statements given above are correct?

(a) 1, 2, and 3 (b) 1 and 3 only (c) 2 and 3 only (d) 1 and 2 only

9 Maximum shear stress developed on the surface of a solid circular shaft under pure torsion is 240 MPa. If the shaft diameter is doubled, then what is the maximum shear stress developed corresponding to the same torque? [IES-2009]

(a) 120 MPa (b) 60 MPa (c) 30 MPa (d) 15 MPa

10 The diameter of a shaft is increased from 30 mm to 60 mm, all other conditions remaining unchanged. How many times is its torque carrying capacity increased? [IES-1995; 2004]

(a) 2 times (b) 4 times (c) 8 times (d) 16 times

11 A circular shaft subjected to twisting moment results in maximum shear stress of 60 MPa. Then, the maximum compressive stress in the material is: [IES-2003]

(a) 30 MPa (b) 60 MPa (c) 90 MPa (d) 120 MPa

12 The boring bar of a boring machine is 25 mm in diameter. During operation, the bar gets twisted though 0.01 rad and is subjected to a shear stress of 42 N/mm^2. The length of the bar is (taking $G = 0.84 \times 105$ N/mm^2) [IES-2012]

(a) 500 mm (b) 250 mm (c) 625 mm (d) 375 mm

13 The magnitude of stress induced in a shaft due to applied torque varies [IES-2012]

(a) From maximum at the center to zero at the circumference
(b) From zero at the center to maximum at the circumference
(c) From maximum at the center to minimum but not zero at the circumference
(d) From minimum but not zero at the center to maximum at the circumference

14 A solid circular shaft is subjected to pure torsion. The ratio of maximum shear to maximum normal stress at any point would be: [IES-1999]

(a) 1:1 (b) 1:2 (c) 2:1 (d) 2:3

15 Assertion (A): In a composite shaft having two concentric shafts of different materials, the torque shared by each shaft is directly proportional to its polar MI.

Reason (R): In a composite shaft having concentric shafts of different materials, the angle of twist for each shaft depends on its polar MI. [IES-1999]

(a) Both A and R are individually true and R is the correct explanation of A
(b) Both A and R are individually true but R is NOT the correct explanation of A
(c) A is true but R is false
(d) A is false but R is true

16 A round shaft of diameter 'd' and length 'l' fixed at both ends 'A' and 'B' is subjected to a twisting moment 'T' at 'C', at a distance of 1/4 from A (see figure). The torsional stresses in the parts AC and CB will be:

C
A
B
l/4
l

(a) Equal (b) In the ratio 1:3
(c) In the ratio 3 :1 (d) Indeterminate

17 In power transmission shafts, if the polar MI of a shaft is doubled, then what is the torque required to produce the same angle of twist? [IES-2006]

(a) 1/4 of the original value (b) 1/2 of the original value
(c) Same as the original value (d) Double the original value

18 A shaft can safely transmit 90 kW while rotating at a given speed. If this shaft is replaced by a shaft of diameter double of the previous one and rotated at half the speed of the previous, the power that can be transmitted by the new shaft is: [IES-2002]

(a) 90 kW (b) 180 kW (c) 360 kW (d) 720 kW

19 The diameter of shaft A is twice the diameter or shaft B and both are made of the same material. Assuming both the shafts to rotate at the same speed, the maximum power transmitted by B is: [IES-2001; GATE-1994]

(a) The same as that of A (b) Half of A
(c) 1/8th of A (d) 1/4th of A

20 When a shaft transmits power through gears, the shaft experiences [IES-1997]

(a) Torsional stresses alone
(b) Bending stresses alone
(c) Constant bending and varying torsional stresses
(d) Varying bending and constant torsional stresses

21 A shaft is subjected to fluctuating loads for which the normal torque (T) and BM (M) are 1000 N·m and 500 N·m, respectively. If the combined shock and fatigue factor for bending is 1.5 and combined shock and fatigue factor for torsion is 2, then the equivalent twisting moment for the shaft is: [IES-1994]

(a) 2000 N·m (b) 2050 N·m (c) 2100 N·m (d) 2136 N·m

22 A member is subjected to the combined action of BM 400 N·m and torque 300 N·m. What respectively are the equivalent BM and equivalent torque? [IES-1994]

(a) 450 N·m and 500 N·m (b) 900 N·m and 350 N·m
(c) 900 N·m and 500 N·m (d) 400 N·m and 500 N·m

23 A shaft was initially subjected to BM and then was subjected to torsion. If the magnitude of BM is found to be the same as that of the torque, then the ratio of maximum bending stress to shear stress would be: [IES-1993]

(a) 0.25 (b) 0.50 (c) 2.0 (d) 4.0

24 Which one of the following statements is correct? Shafts used in heavy duty speed reducers are generally subjected to: [IES-2004]

(a) Bending stress only
(b) Shearing stress only
(c) Combined bending and shearing stresses
(d) Bending, shearing, and axial thrust simultaneously

25 A hollow shaft of the same cross-sectional area and material as that of a solid shaft transmits: [IES-2005]

(a) Same torque
(b) Lesser torque
(c) More torque
(d) Cannot be predicted without more data

26 Assertion (A): A hollow shaft will transmit a greater torque than a solid shaft of the same weight and same material.

Reason (R): The average shear stress in the hollow shaft is smaller than the average shear stress in the solid shaft. [IES-1994]

(a) Both A and R are individually true and R is the correct explanation of A
(b) Both A and R are individually true but R is NOT the correct explanation of A
(c) A is true but R is false
(d) A is false but R is true

27 Assertion (A): In theory of torsion, shearing strains increase radically away from the longitudinal axis of the bar.

Reason (R): Plane transverse sections before loading remain plane after the torque is applied. [IAS-2001]

(a) Both A and R are individually true and R is the correct explanation of A
(b) Both A and R are individually true but R is NOT the correct explanation of A
(c) A is true but R is false
(d) A is false but R is true

28 The shear stress at a point in a shaft subjected to a torque is: [IAS-1995]

(a) Directly proportional to the polar MI and to the distance of the point form the axis
(b) Directly proportional to the applied torque and inversely proportional to the polar MI
(c) Directly proportional to the applied torque and polar MI
(d) inversely proportional to the applied torque and the polar MI

29 If two shafts of the same length, one of which is hollow, transmit equal torque and have equal maximum stress, then they should have equal. [IAS-1994]

(a) Polar MI (b) Polar modulus of section
(c) Polar MI (d) Angle of twist

30 Two shafts having the same length and material are joined in series. If the ratio of the diameter of the first shaft to that of the second shaft is 2, then the ratio of the angle of twist of the first shaft to that of the second shaft is: [IAS-1995; 2003]

(a) 16 (b) 8 (c) 4 (d) 2

Solutions for Objective Questions

Sl. No.	1.	2.	3.	4.	5.	6.	7.	8.	9.	10.
Answer	(a)	(c)	(c)	(c)	(d)	(b)	(c)	(d)	(c)	(c)

Sl.No.	11.	12.	13.	14.	15.	16.	17.	18.	19.	20.
Answer	(b)	(b)	(b)	(b)	(c)	(c)	(d)	(c)	(c)	(d)

Sl.No.	21.	22.	23.	24.	25.	26.	27.	28.	29.	30.
Answer	(d)	(a)	(c)	(c)	(c)	(a)	(b)	(b)	(b)	(a)

EXERCISE PROBLEMS

1. A hollow copper shaft of length is 200 mm. The diameter ratio is 0.65. Determine the outer and inner diameters of the shaft, if the angular twist is limited to 0.4°. The maximum shear stress is limited to 80 MPa. G = 80 GPa. [Objective 1]
2. A circular shaft has to transmit 700 kW power at 200 rpm. Determine the required diameter of solid shaft without exceeding the maximum shear stress of 100 MPa in it. Find the maximum angle of twist in a length of 3 m. $G = 80$ GPa. [Objective 1]
3. Determine the power transmitted by a solid shaft of 120 mm diameter at 100 rpm the maximum shear stress is not to exceed 80MPa. The maximum torque is 20% more than its mean value. Rigidity modulus of shaft material = 80 GPa. [Objective 3]
4. A hollow shaft of same material and same length to transmit the same power at same speed replaces a solid shaft of 120 mm diameter, what would be the percentage saving in material. The inner diameter is 75% of its outer diameter. Compare the torsional stiffness of both shafts. [Objective 5]
5. A circular shaft of 100 mm diameter is subjected to combined bending and twisting with BMs equal to three times of the twisting moment. If the safe stress in tension is 100 MPa and safe stress is 60 MPa, find the allowable twisting moment. [Objective 1]
6. Determine the power transmitted by a 120 mm diameter solid shaft at 100 rpm. The maximum shear stress is not to exceed 80 MPa. The maximum torque is 20% more than its mean value. [Objective 5]
7. Determine the power transmitted by a solid shaft of 120 mm diameter at 100 rpm. The maximum shear stress is not to exceed 80 MPa. The maximum torque is 20% more than its mean value. If the shaft is replaced by a hollow shaft of same weight with outer diameter to inner diameter ratio of 0.6, what will be percentage increase in the Power transmitted by this hollow shaft? [Objective 5]

CHAPTER 10

PRINCIPAL STRESSES

UNIT OBJECTIVE

This chapter provides information on the maximum or minimum normal stress and maximum shear stress in terms of σ_x, σ_y, and τ_{xy}. The presentation attempts to help the student achieve the following:

Objective 1: Determine the maximum or minimum normal and shear stress.

Objective 2: Determine the principal stresses on the shafts, inclined planes, and the inclination between the planes.

Objective 3: Draw Mohr circle for the given stress and determine the principal stress.

Objective 4: Determine the intensities of normal and shear stresses across a plane to the longitudinal axis of a steel bolt subjected to direct tension and shearing force.

10.1 INTRODUCTION

In generalized strained body, stress at a point can be expressed as a stress tensor given by

$$[\sigma] = \begin{bmatrix} \sigma_x & \tau_{yx} & \tau_{zx} \\ \tau_{xy} & \sigma_y & \tau_{zy} \\ \tau_{zx} & \tau_{zy} & \sigma_z \end{bmatrix}.$$

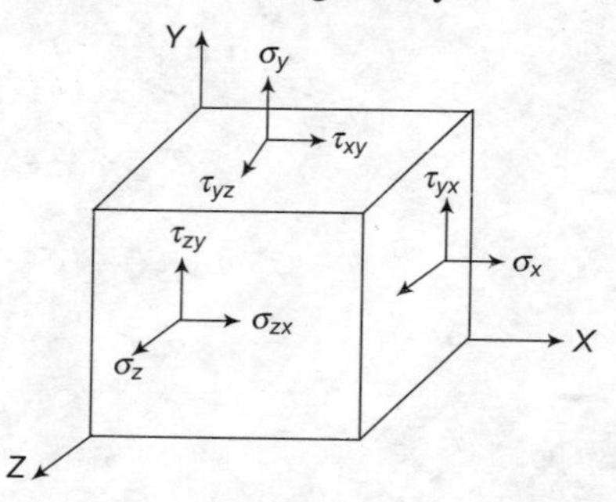

FIGURE 10.1

The stress components σ_x, τ_{yz}, τ_{zx}, etc. due to external force may be flexure, axial loads, torsion loads, or their combinations. In this chapter, we consider the stresses in a plane only. Consider stresses in X–Y plane.

σ_x and σ_y are the normal stresses τ_{yz} and τ_{xy} is the shearing or tangential stress $\tau_{xy} = \tau_{yx}$ as τ_{yx} is complimentary shear stress to τ_{xy}. Denoting $\tau_{xy} = \tau_{yx}$, the generalized stress in 2D stress system is given by

$$[\sigma] = \begin{bmatrix} \sigma_x & \tau_x \\ \tau_x & \sigma_y \end{bmatrix}$$

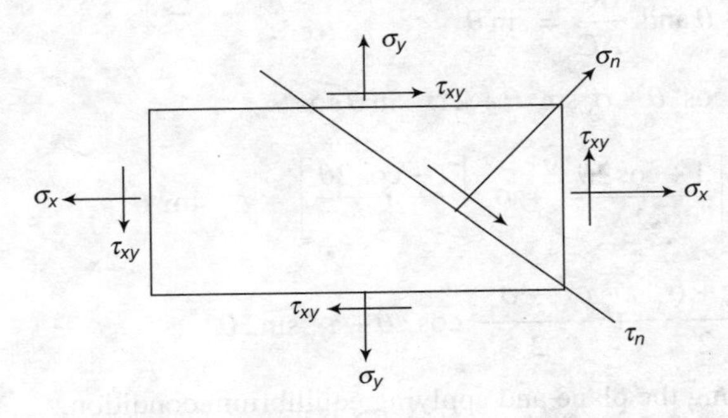

FIGURE 10.2

The effects σ_x, σ_y, and τ_{xy} on inclined plane shown in Figure 10.2 can be represented as σ_n and τ_n, that is, normal stress and shear stress, respectively.

Often in the design problems, the dimensions of the stained element are proportioned depending on the maximum normal stress criterion or maximum shear stress criterion. Thus, it is required to quantify the maximum or minimum normal stress and maximum shear stress in terms of σ_x, σ_y, and τ_{xy}.

10.2 EXPRESSION FOR σ_n AND τ_n

Consider a generalized 2D stress system, shown in Figure 10.3(a).

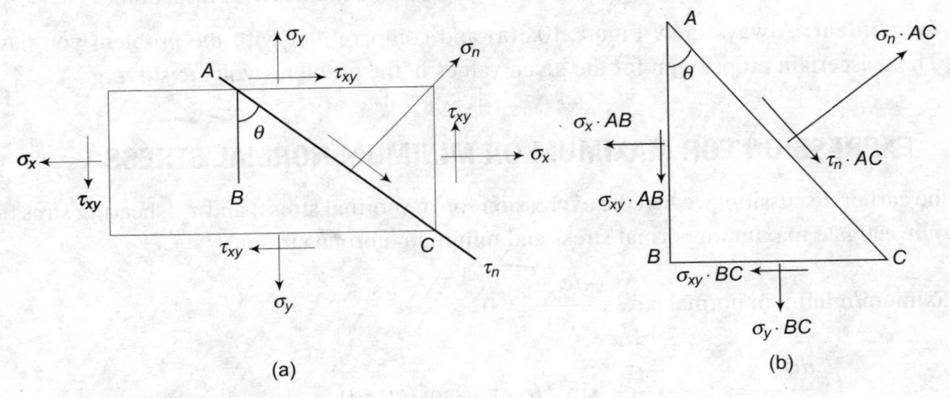

FIGURE 10.3

Consider the triangular portion ABC and convert the stresses into forces, as shown in Figure 10.3(b). Take unit width for force calculations, that is, normal forces on the plane AC is $\sigma_n \times \text{AC} \times 1$.

Resolving the forces along the normal to the plane AC and applying equilibrium condition,

$$\sigma_n \cdot AC - \sigma_x AB \cos\theta - \sigma_y \cdot BC \sin\theta - \tau_{xy} \cdot AB \sin\theta - \tau_{xy} \cdot BC \cos\theta = 0$$

$$\sigma_n = \sigma_x \cdot \frac{AB}{AC} \cos\theta + \sigma_y \cdot \frac{BC}{AC} \sin\theta + \tau_{xy} \cdot \frac{AB}{AC} \sin\theta + \tau_{xy} \cdot \frac{BC}{AC} \cos\theta$$

$$\frac{AB}{AC} = \cos\theta \text{ and } \frac{BC}{AC} = \sin\theta$$

$$\sigma_n = \sigma_x \cos^2\theta + \sigma_y \sin^2\theta + 2\tau_{xy} \sin\theta \cos\theta$$

$$\sigma_n = \sigma_x \left[\frac{1+\cos 2\theta}{2}\right] + \sigma_y \left[\frac{1-\cos 2\theta}{2}\right] + \tau_{xy} \sin 2\theta$$

$$\sigma_n = \frac{\sigma_x + \sigma_y}{2} + \frac{\sigma_x - \sigma_y}{2} \cos 2\theta + \tau_{xy} \sin 2\theta.$$

Resolving the forces along the plane and applying equilibrium condition,

$$\tau_n \cdot AC - \sigma_x \cdot AC \sin\theta + \sigma_y \cdot BC \cos\theta + \tau_{xy} \cdot BC \sin\theta + \tau_{xy} \cdot AB \cos\theta = 0$$

$$\tau_n = \sigma_x \sin\theta \cos\theta - \sigma_y \sin\theta \cos\theta + \tau_{xy} \sin^2\theta - \tau_{xy} \cos^2\theta$$

$$\tau_n = \frac{\sigma_x - \sigma_y}{2} \sin 2\theta - \tau_{xy} \cos 2\theta.$$

The sign convention to be followed:

$\sigma_n \rightarrow$ Normal +ve, if tensile and –ve, if compressive

$\tau_n \rightarrow$ Shearing stress +ve, if it rotates the block on which it acts

In clockwise direction, and –ve otherwise.

$\theta \rightarrow$ Measured from plane carrying σ_x, in anticlockwise direction.

A tip for student: Always draw Figure 10.3(a) and compare that with the problem you have to solve. Then, ascertain proper sign for the given values in the problems you are solving.

10.3 EXPRESSION FOR MAXIMUM OR MINIMUM NORMAL STRESS

From the earlier discussion, we had the expression for σ_n (normal stress) and τ_n (shearing stress). We have to investigate maximum normal stress and minimum normal stress.

For maximum/minimum normal stress $\frac{d\sigma_n}{d\theta} = 0$

$$\frac{d\sigma_n}{d\theta} = 2\frac{\sigma_x - \sigma_y}{2} \cdot \sin 2\theta + \tau_{xy} 2\cos 2\theta = 0$$

$$\frac{\sigma_x - \sigma_y}{2} \sin 2\theta - \tau_{xy} 2\cos 2\theta = 0$$

LHS of the above expression is an expression for τ_n.

Hence plane carrying maximum or minimum normal stress carry zero shear. These planes are called principal planes. The normal stresses on these principal planes are called principal stresses.

Let 'α' be the inclination of principal planes, then

$$\frac{\sigma_x - \sigma_y}{2} \cos 2\alpha - \tau_{xy} \sin 2\alpha = 0$$

$$\tan 2\alpha = \frac{\tau_{xy}}{\left[\frac{\sigma_x + \sigma_y}{2}\right]}.$$

Principal stresses is normal stress σ_n when $\theta = \alpha$.

$$\sigma_1 = \frac{\sigma_x + \sigma_y}{2} + \frac{\sigma_x - \sigma_y}{2} \cos 2\alpha + \tau_{xy} \sin 2\alpha$$

From the above triangle,

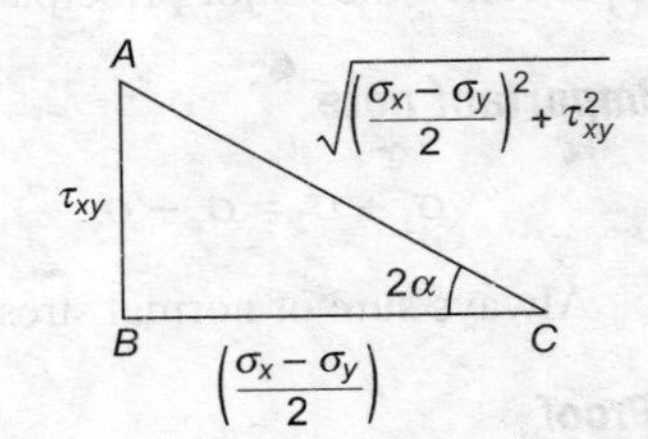

FIGURE 10.4

$$\sin 2\alpha = \frac{\tau_{xy}}{\sqrt{\left(\frac{(\sigma_x - \sigma_y)}{2}\right)^2 + \tau_{xy}^2}}$$

$$\cos 2\alpha = \frac{\left(\frac{\sigma_x - \sigma_y}{2}\right)}{\sqrt{\left(\frac{(\sigma_x - \sigma_y)}{2}\right)^2 + \tau_{xy}^2}}.$$

Substituting $\sin 2\alpha$ and $\cos 2\alpha$ values in the expression for σ_1

$$\sigma_1 = \frac{\sigma_x + \sigma_y}{2} + \frac{\sigma_x - \sigma_y}{2} \frac{\left(\frac{\sigma_x + \sigma_y}{2}\right)}{\sqrt{\left(\frac{(\sigma_x - \sigma_y)}{2}\right)^2 + \tau_{xy}^2}} + \tau_{xy} \frac{\tau_{xy}}{\sqrt{\left(\frac{(\sigma_x - \sigma_y)}{2}\right)^2 + \tau_{xy}^2}}$$

$$\sigma_1 = \frac{\sigma_x + \sigma_y}{2} + \frac{\left(\frac{\sigma_x - \sigma_y}{2}\right) + \tau_{xy}^2}{\sqrt{\left(\frac{(\sigma_x - \sigma_y)}{2}\right)^2 + \tau_{xy}^2}}.$$

The above expression yields two values as $\left(\frac{\sigma_x - \sigma_y}{2}\right)^2 + \tau_{xy}^2$ can be written as

$$\left\{\pm\left[\left(\frac{\sigma_x - \sigma_y}{2}\right)^2 + \tau_{xy}^2\right]^{1/2}\right\}^2.$$

Thus, two principal stresses are $\sigma_1 = \frac{\sigma_x + \sigma_y}{2} + \sqrt{\left(\frac{\sigma_x - \sigma_y}{2}\right)^2 + \tau_{xy}{}^2}$

$$\sigma_2 = \frac{\sigma_x + \sigma_y}{2} - \sqrt{\left(\frac{\sigma_x - \sigma_y}{2}\right)^2 + \tau_{xy}{}^2}$$

σ_1 is called the major principal stress and σ_2 is called the minor principal stress.

Important note

$$\sigma_1 + \sigma_2 = \sigma_x + \sigma_y.$$

Always sum of normal stresses on two orthogonal planes is constant.

Proof

Let θ be a plane on which σ_n be the normal stress and σ_n is given by

$$\sigma_n = \frac{\sigma_x + \sigma_y}{2} + \frac{\sigma_x - \sigma_y}{2}\cos 2\theta + \tau_{xy}\sin 2\theta.$$

Let σ_{n+90} be normal stress on a plane orthogonal to the plane carrying σ_n, that is, inclination of the plane $(\theta + 90°)$.

$$\sigma_{n+90} = \frac{\sigma_x + \sigma_y}{2} + \frac{\sigma_x - \sigma_y}{2}\cos 2(\theta + 90) + \tau_{xy}\sin 2(\theta + 90)$$

$$\sigma_{n+90} = \frac{\sigma_x + \sigma_y}{2} - \frac{\sigma_x - \sigma_y}{2}\cos 2\theta - \tau_{xy}\sin 2\theta$$

$$\sigma_n + \sigma_{n+90} = \frac{\sigma_x + \sigma_y}{2} + \frac{\sigma_x + \sigma_y}{2}$$

$$\sigma_n + \sigma_{n+90} = \sigma_x + \sigma_y = \sigma_1 + \sigma_2.$$

This implies that the sum of normal stress on two orthogonal planes is constant and equal to sum of principal stresses.

Maximum normal stress in a strained body is the maximum principal stress and the shear stress on the plane carrying maximum normal stress is zero.

In general, some materials fail when the shear stress exceeds a particular limit. Hence, in the strained body one need to find the expression for maximum shear stress also.

10.4 EXPRESSION FOR MAXIMUM SHEAR STRESS

For maximum shear stress $\frac{d\tau_n}{d\theta} = 0$.

$$\tau_n = \frac{\sigma_x - \sigma_y}{2}\sin 2\theta - \tau_{xy}\cos 2\theta$$

$$\frac{d\tau_n}{d\theta} = 2\frac{\sigma_x - \sigma_y}{2}\cos 2\theta - \tau_{xy} \cdot \sin 2\theta = 0$$

$$\tan 2\theta = \frac{\left(\frac{\sigma_x - \sigma_y}{2}\right)}{\tau_{xy}}.$$

For a plane carrying maximum shear stress $\theta = \beta$.

Thus,
$$\tan 2\beta = -\frac{\left(\frac{\sigma_x - \sigma_y}{2}\right)}{\tau_{xy}}.$$

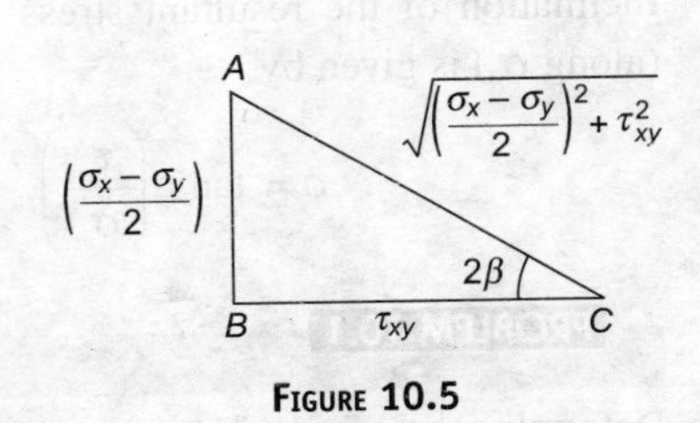

FIGURE 10.5

$$\cos 2\beta = \frac{\tau_{xy}}{\sqrt{\left(\frac{(\sigma_x - \sigma_y)}{2}\right)^2 + \tau_{xy}^2}}$$

$$\sin 2\beta = -\frac{\left(\frac{\sigma_x - \sigma_y}{2}\right)}{\sqrt{\left(\frac{(\sigma_x - \sigma_y)}{2}\right)^2 + \tau_{xy}^2}}.$$

Substituting the values of $\cos 2\beta$ and $\sin 2\beta$ in τ_n expression τ_n becomes $\tau_{\max}$.

Hence,
$$\tau_{\max} = \frac{\sigma_x - \sigma_y}{2}\sin 2\beta - \tau_{xy} \cdot \cos 2\beta$$

$$= \frac{\sigma_x - \sigma_y}{2} - \frac{\left(\frac{\sigma_x - \sigma_y}{2}\right)}{\sqrt{\left(\frac{(\sigma_x - \sigma_y)}{2}\right)^2 + \tau_{xy}^2}} + \tau_{xy}\frac{\tau_{xy}}{\sqrt{\left(\frac{(\sigma_x - \sigma_y)}{2}\right)^2 + \tau_{xy}^2}}$$

$$\tau_{max} = \sqrt{\left(\frac{\sigma_x - \sigma_y}{2}\right)^2 + \tau_{xy}^2}.$$

The same can be obtained as $\tau_{max} = \left(\frac{\sigma_1 - \sigma_2}{2}\right).$

Examine an important point 'α' is the inclination of the plane carrying principal stress and β is the plane carrying maximum shear stress

$$\tan 2\alpha \cdot \tan 2\beta = -1$$

This implies that the difference of 2α and 2β is 90°. The difference of α and β is 45°. The angle between the plane carrying principal planes and planes carrying maximum or minimum shear stress is $\pi/4$ or 45°.

Expression resultant stress on a plane:

The resultant stress on a plane due to σ_n and τ_n is given by.

$$\sigma_R = \sqrt{\sigma_n^2 + \tau_n^2}$$

Inclination of the resultant stress to the outward normal (along σ_n) is given by

$$\phi = \tan^{-1}\left(\frac{\tau_n}{\sigma_n}\right).$$

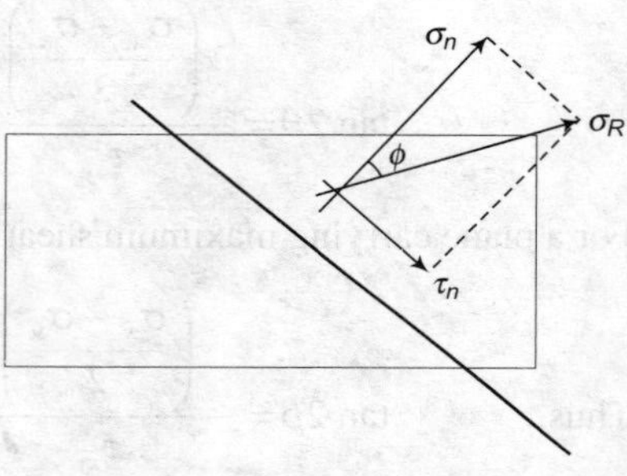

FIGURE 10.6

PROBLEM 10.1 **Objective 1**

Determine the principal stresses, corresponding planes, maximum shear stress and the corresponding planes of a stressed 2D plane as shown in Figure 10.7.
Also determine the resultant stress on a plane inclined 30 degrees to the major principal plane.

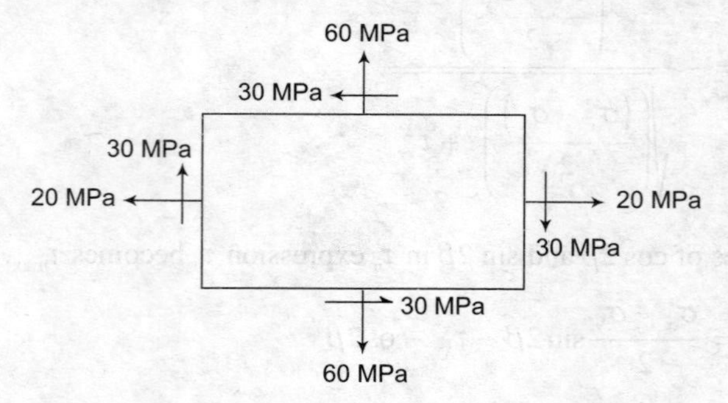

FIGURE 10.7

SOLUTION

Always draw a stress block for which derivation was done and compare that with the stress block given in the problem.

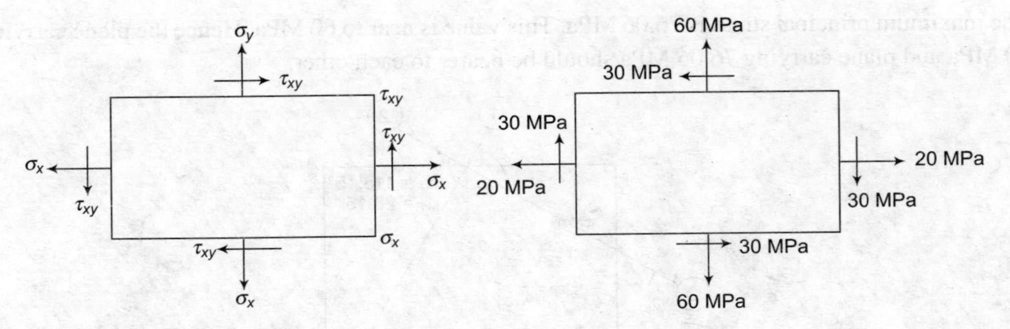

FIGURE 10.8

Now $\sigma_x = 20$ MPa

$\sigma_y = 60$ MPa

$\tau_{xy} = -30$ MPa

For principal stresses, $\sigma_1 = \dfrac{\sigma_x + \sigma_y}{2} + \sqrt{\left(\dfrac{\sigma_x - \sigma_y}{2}\right)^2 + (\tau_{xy})^2}$

$$= \frac{20 + 60}{2} + \sqrt{\left(\frac{20 - 60}{2}\right)^2 + (-30)^2}$$

= 76.06 MPa (tensile) (maximum principal stress)

$$\sigma_2 = \frac{\sigma_x + \sigma_y}{2} - \sqrt{\left(\frac{\sigma_x - \sigma_y}{2}\right)^2 + (\tau_{xy})^2}$$

or

$$\sigma_x + \sigma_y = \sigma_1 + \sigma_2$$

$\sigma_2 = 3.94$ MPa (tensile) (minimum principal stress)

Plane carrying principal stress,

$$\tan 2\alpha = \frac{\tau_{xy}}{\left(\dfrac{\sigma_x - \sigma_y}{2}\right)} = -\frac{30}{\left(\dfrac{20 - 60}{2}\right)}$$

$$\tan 2\alpha = \frac{3}{2}.$$

Satisfying the above condition 'α' will have two values

$\alpha_1 = 28.15°$

$\alpha_2 = 28.15 + 90 = 118.15°$.

Now it is clear that one of the two plane carrying maximum principal stress.

The maximum principal stress is 76.06 MPa. This value is near to 60 MPa. Hence the plane carrying 60 MPa and plane carrying 76.06 MPa should be nearer to each other.

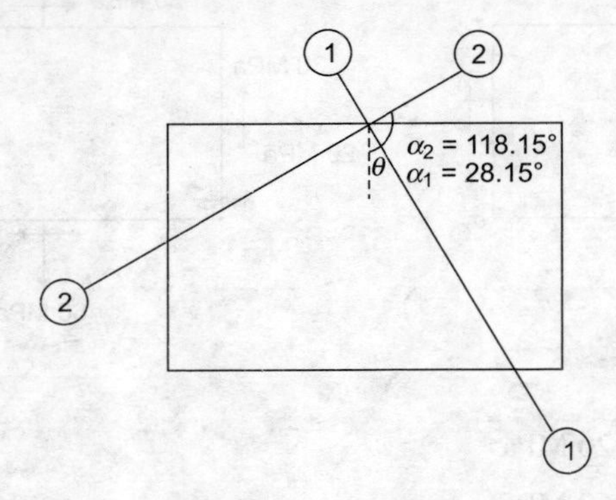

FIGURE 10.9

60 MPa is acting on the horizontal plane. It is clear that plane (2)-(2) is closer to the horizontal plane than (1)-(1).

(2)-(2) plane makes 28.15° with the plane carrying 60 MPa.

(1)-(1) plane makes 61.85° with the plane carrying 60 MPa.

Hence plane (2)-(2) carries the major principal stress, 76.06 MPa.

Plane (1)-(1) carries the minimum principal stress 3.94 MPa.

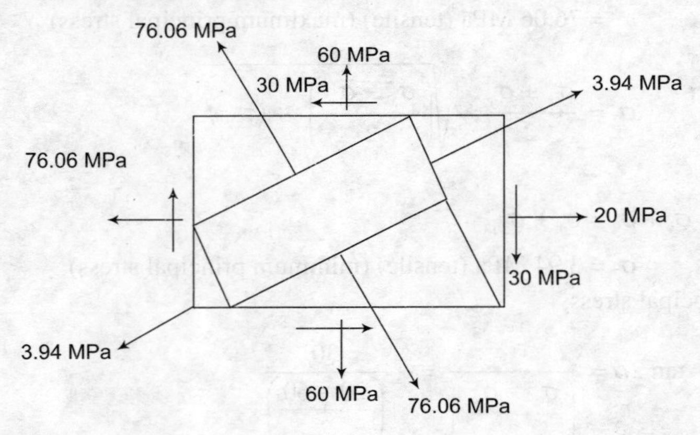

FIGURE 10.10

Inclination of major principal plane = 118.15°

Inclination of minor principal plane = 28.15°

Planes carrying maximum or minimum shear stress

$$\tan 2\beta = \frac{\left(\dfrac{\sigma_x - \sigma_y}{2}\right)}{\tau_{xy}} = \frac{\left(\dfrac{20 - 60}{2}\right)}{-30} = \frac{2}{3}$$

$\beta_1 = -16.85°$ $(28.15 - 45° = -16.85°)$

$\beta_2 = -73.15°$ $(118.15 - 45° = 73.15°)$

$$\text{Maximum shear stress} = \sqrt{\left(\frac{\sigma_x - \sigma_y}{2}\right)^2 + \tau_{xy}^2}$$

$$\text{(or)} \quad = \frac{\sigma_1 - \sigma_2}{2} = 36.06 \text{ MPa.}$$

$$\text{Minimum shear stress} = -\sqrt{\left(\frac{\sigma_x - \sigma_y}{2}\right)^2 + \tau_{xy}^2} = -36.06 \text{ MPa.}$$

Look at the plane carrying 20 MPa normal stress. This plane carries a shear stress of 30 MPa, which is rotating the block in clockwise direction (+ve). Maximum shear stress is 36.06 MPa. Hence plane carrying maximum shear stress must be near to vertical plane or plane carrying 20 MPa normal stress.

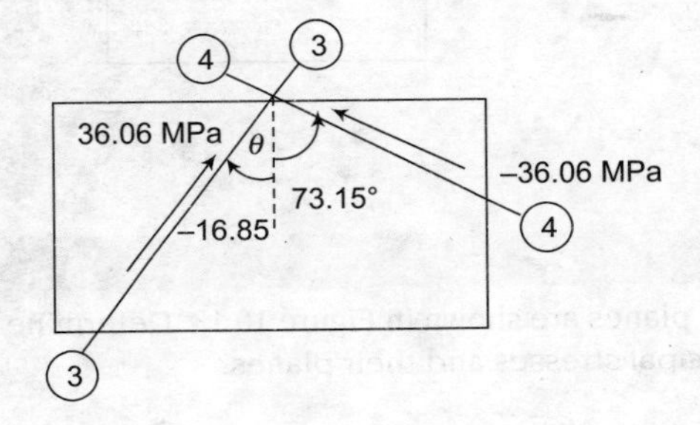

FIGURE 10.11

Plane (3)-(3) inclined −16.85° is nearer/closer to the vertical plane then plane (4)-(4). Hence, plane (3)-(3) carries maximum shear stress and plane (4)-(4) carries minimum shear stress.
Resultant stress on a plane 30° to major principal plane:

Inclination of the major principal plane $\alpha = 118.15°$

$$\therefore \quad \theta = 118.15° + 30° = 148.15°$$

$$2\theta = 296.3°.$$

σ_n = Normal stress on a plane inclined 148.15°

$$= \frac{\sigma_x + \sigma_y}{2} + \frac{\sigma_x - \sigma_y}{2}\cos 2\theta + \tau_{xy}\sin 2\theta$$

$$= 58.03 \text{ MPa (tensile)}$$

$$\tau_n = \frac{\sigma_x - \sigma_y}{2}\sin 2\theta - \tau_{xy}\cos 2\theta$$

$$= 31.22 \text{ MPa (tensile)}$$

Resultant stress $\sigma_R = \sqrt{\sigma_n^{\ 2} + \tau_n^{\ 2}} = 65.90$ MPa

Inclination of the resultant stress with the outward normal $= \phi = \tan^{-1}\left(\dfrac{\tau_n}{\sigma_n}\right) = 28.28°$

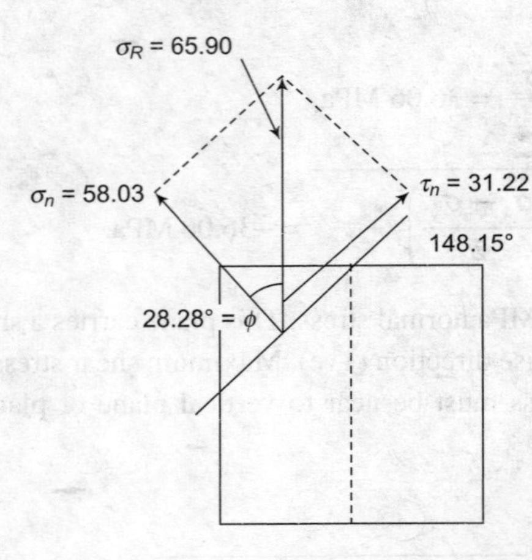

FIGURE 10.12

PROBLEM 10.2

Objective 2

The stresses on two inclined planes are shown in Figure 10.13. Determine inclination between planes and also determine the principal stresses and their planes.

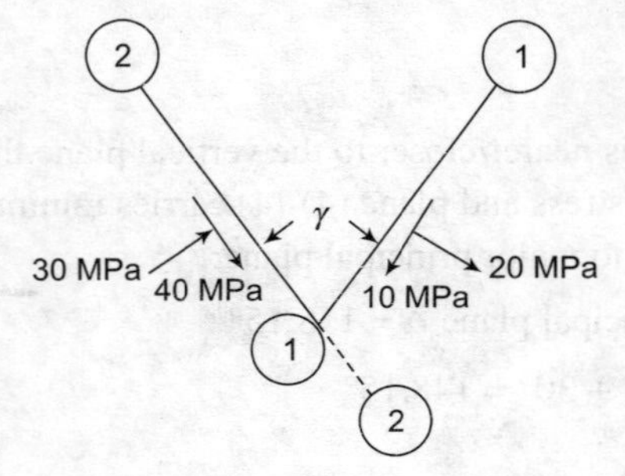

FIGURE 10.13

SOLUTION

Take the plane (1)-(1) as a vertical plane, then compare with the standard 2D stress block.

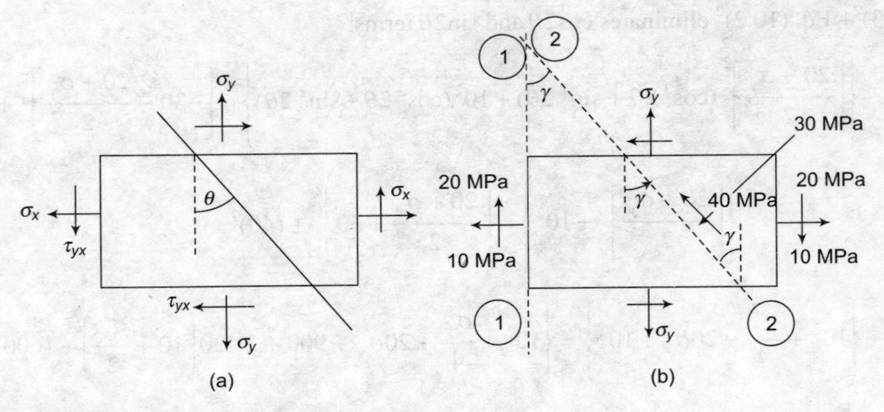

FIGURE 10.14

A comparison between Figure 10.14 (a) and Figure 10.14 (b) reveals that

$$\sigma_x = 20 \text{ MPa}$$

$$\sigma_y = \text{Unknown} = \sigma_y$$

$$\tau_{xy} = -10 \text{ MPa}$$

$$\sigma_n = -30 \text{ MPa}$$

$$\tau_n = -24 \text{ MPa}$$

$$\gamma = \theta$$

We know that equation for σ_n and τ_n. The unknowns are σ_y and θ.

Take
$$\sigma_n = \frac{\sigma_x + \sigma_y}{2} + \frac{\sigma_x - \sigma_y}{2}\cos 2\theta + \tau_{xy}\sin 2\theta$$

$$-30 = \frac{20 + \sigma_y}{2} + \frac{20 - \sigma_y}{2}\cos 2\theta - 10\sin 2\theta$$

$$\tau_n = \frac{\sigma_x - \sigma_y}{2}\sin 2\theta - \tau_{xy}\cos 2\theta$$

$$-40 = \frac{20 - \sigma_y}{2}\sin 2\theta + 10\cos 2\theta. \tag{10.1}$$

Rearranging Eq. (10.1)

$$\left(\frac{20 - \sigma_y}{2}\cos 2\theta - 10\sin 2\theta\right) = -30 - \frac{20 + \sigma_y}{2} \tag{10.2}$$

$$\frac{20 - \sigma_y}{2}\sin 2\theta + 10\cos 2\theta = -40 \tag{10.3}$$

Eq. $(10.3)^2$+ Eq. $(10.2)^2$ eliminates $\cos 2\theta$ and $\sin 2\theta$ terms,

$$\left\{\left[\frac{20-\sigma_y}{2}\right]^2(\cos^2 2\theta+\sin^2 2\theta)+10^2(\cos^2 2\theta+\sin^2 2\theta)\right\}=\left[-30-\frac{20+\sigma_y}{2}\right]^2+(-40)^2$$

$$\left\{\left[\frac{20-\sigma_y}{2}\right]^2+10^2\right\}=\left[\frac{20+\sigma_y}{2}+30\right]^2+(40)^2$$

$$\left\{10^2+\frac{\sigma_y^2}{4}-20\sigma_y+10^2\right\}=\left\{10^2+\frac{\sigma_y^2}{4}+20\sigma_y+900+2*30\left(10+\frac{\sigma_y}{2}\right)+1600\right\}$$

$$-70\sigma_y=3000$$

$$\sigma_y=-42.86 \text{ MPa.}$$

Using these values of σ_y in Eq. (10.2)

$$\frac{20-(-42.86)}{2}\sin 2\theta+10\cos 2\theta=-40$$

$$31.43 \sin 2\theta+10 \cos 2\theta+40=0$$

$$31.43\sin 2\theta+10\sqrt{1-\sin^2 2\theta}+40=0$$

$$31.43 \sin 2\theta+40= -10\sqrt{1-\sin^2 2\theta}.$$

Squaring on both side

$$987.85 \sin^2 2\theta+2514.4 \sin 2\theta+1600=-10+10 \sin^2 2\theta$$

$$977.85 \sin^2 2\theta+2514.4 \sin 2\theta+1610=0$$

$$\sin 2\theta=\frac{2514.4\pm 3552.40}{2\times 977.85}= -0.531 \text{ or } 3.102$$

$\sin 2\theta= 3.102$ is not possible

$$\sin 2\theta=-0.531$$

$$2\theta=327.92°$$

$$\theta=163.96°.$$

The angle between plane (1)-(1) and plane (2)-(2) is 163.96°.

Now, $\sigma_x=20$ MPa

$\sigma_y=-42.86$ MPa

$\tau_{xy}=-10$ MPa

$$\sigma_1 = \frac{\sigma_x + \sigma_y}{2} + \sqrt{\left(\frac{\sigma_x - \sigma_y}{2}\right)^2 + \tau_{xy}^2}$$

$$\sigma_1 = \frac{20 - 42.86}{2} + \sqrt{\left(\frac{20 - 42.86}{2}\right)^2 + (-10)^2}$$

$$= -11.43 + 32.98 = 21.55 \text{ MPa (tensile)}$$

$$\sigma_2 = -11.43 - 32.98$$

$$= -44.41 \text{ MPa (Compressive)}$$

$$\tan 2\alpha_2 = \frac{\tau_{xy}}{\left(\dfrac{\sigma_x - \sigma_y}{2}\right)} = \frac{-10}{\left(\dfrac{20 + 42.86}{2}\right)} = -0.138$$

$$\alpha_1 = -8.82°$$

$$\alpha_2 = 81.18°$$

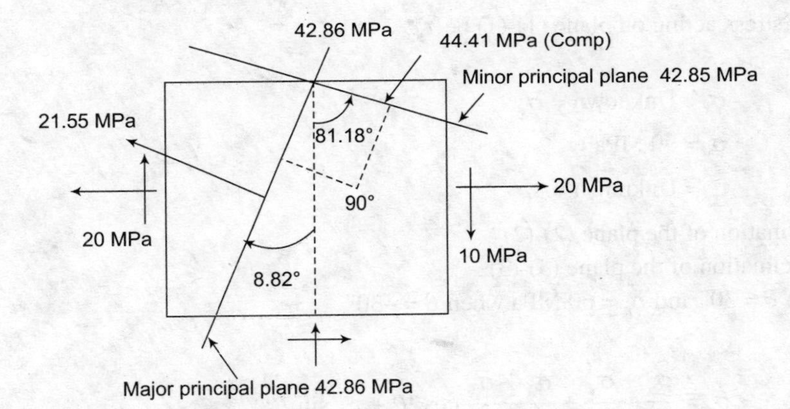

FIGURE 10.15

PROBLEM 10.3 **Objective 1**

Determine the principal stresses and maximum shear stress of a 2D plane body subjected to stresses as shown in Figure below.

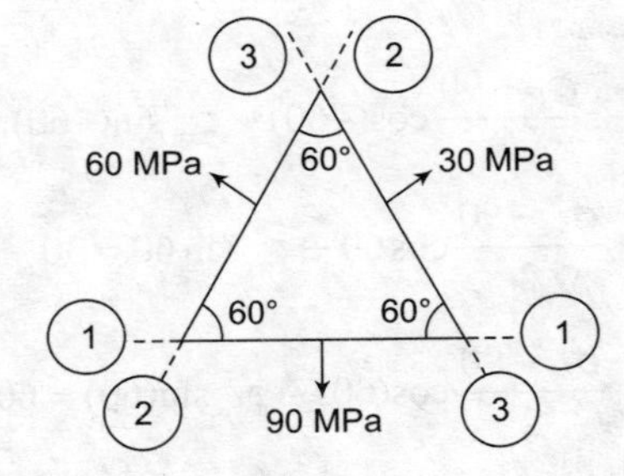

SOLUTION

Compare the given stress block with the standard stress block. Take plane (1)-(1) as horizontal plane, then

$$\sigma_y = 90 \text{ MPa.}$$

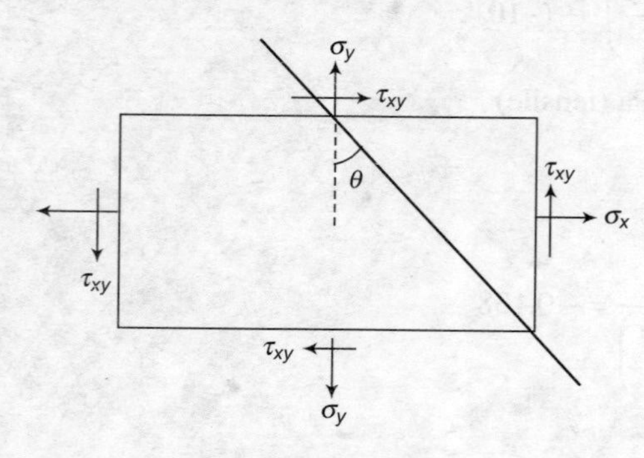

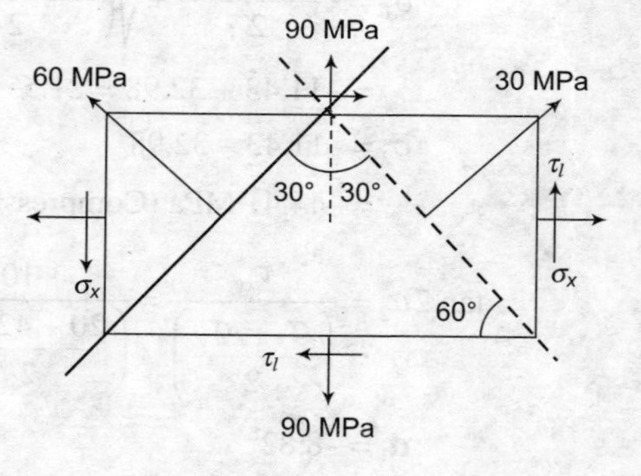

FIGURE 10.16

Let the shear stress acting on plane (1)-(1) be τ_1.
Then

$$\sigma_x = \text{Unknown} = \sigma_x$$

$$\sigma_y = 90 \text{ MPa}$$

$$\tau_{xy} = \text{Unknown} = \tau_1$$

$\theta_2 = 30°$, inclination of the plane (2)-(2)
$\theta_3 = -30°$, inclination of the plane (3)-(3)
$\sigma_n = 30$ when $\theta = 30°$ and $\sigma_n = 60$ MPa when $\theta = -30°$
If ($\theta = 30°$)

$$\sigma_n = \frac{\sigma_x + \sigma_y}{2} + \frac{\sigma_x - \sigma_y}{2}\cos 2\theta + \tau_{xy}\sin 2\theta$$

$$\sigma_n = \frac{\sigma_x + \sigma_y}{2} + \frac{\sigma_x - \sigma_y}{2}\cos 60 + \tau_{xy}\sin 60$$

$$30 = \frac{\sigma_x + 90}{2} + \frac{\sigma_x - 90}{2}\cos 60 + \tau_{xy}\sin 60$$

And

$$60 = \frac{\sigma_x + 90}{2} + \frac{\sigma_x - 90}{2}\cos(-60) + \tau_{xy}\sin(-60)$$

$$\frac{\sigma_x + 90}{2} + \frac{\sigma_x - 90}{2}\cos 60 + \tau_{xy}\sin 60 = 30$$

$$\frac{\sigma_x + 90}{2} + \frac{\sigma_x - 90}{2}\cos(60) - \tau_{xy}\sin(60) = 60$$

Eq. (10.2) – Eq. (10.1) given $-2\tau_1 \sin 60 = 30$

$$\tau_1 = -17.32 \text{ MPa}$$

Eq. (10.2) + Eq. (10.1) given $20 \times \dfrac{\sigma_x + 90}{2} + 2 \times \dfrac{\sigma_x - 90}{2} \cos 60 = 90$

$$\sigma_x + 90 + \frac{\sigma_x}{2} - 45 = 95$$

$$\sigma_x = 30 \text{ MPa}$$

Now, $\sigma_x = 3$ MPa, $\sigma_y = 90$ MPa, $\tau_{xy} = \tau_1 = -17.32$ MPa.
When $\theta = -30°$, $\tau_n = \tau_2$

$$\tau_2 = \frac{\sigma_x - \sigma_y}{2} \sin 2\theta - \tau_{xy} \cos 2\theta$$

$$= \frac{30 - 90}{2} \sin(-60) - (-17.32)\cos(-60)$$

$$= 25.98 + 8.66 = 34.64 \text{ MPa}$$

When $\theta = 30°$, $\tau_n = \tau_3$

$$\tau_3 = \frac{30 - 90}{2} \sin(60) - (-17.32)\cos(60)$$

$$= -25.98 + 8.66 = -17.32 \text{ MPa.}$$

Maximum principal stress

$$\sigma_1 = \frac{\sigma_x + \sigma_y}{2} + \sqrt{\left(\frac{\sigma_x - \sigma_y}{2}\right)^2 + \tau_{xy}^2}$$

$$\sigma_1 = \frac{30 + 90}{2} + \sqrt{\left(\frac{30 + 90}{2}\right)^2 + (-17.32)^2}$$

$$= 60 + 34.64 = 94.64 \text{ MPa}$$

Minimum principal stress = $\sigma_2 = 60 - 34.64 = 25.36$ MPa

Maximum shear stress $= -\sqrt{\left(\dfrac{\sigma_x - \sigma_y}{2}\right)^2 + \tau_{xy}^2} = -34.64$ MPa.

PROBLEM 10.4 **Objective 1**

The resultant stress and normal stress on two mutually perpendicular planes are 30 MPa and 5MPa as shown in Figure 10.17. Determine the principal stress and also find the maximum shear stress.

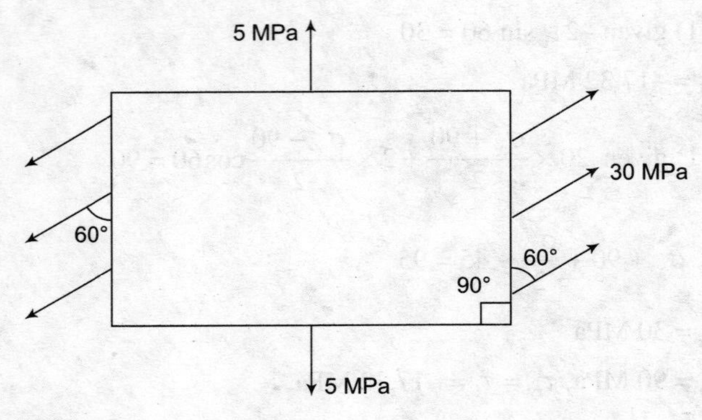

FIGURE 10.17

SOLUTION

Resolve the resultant stress of 30 MPa on the vertical plane.

On the vertical plane, normal stress = 30 sin 60 = 25.98 MPa

Shear stress on the vertical plane = 30 cos 60 = 15 MPa

Thus, the given stress block can be shown as

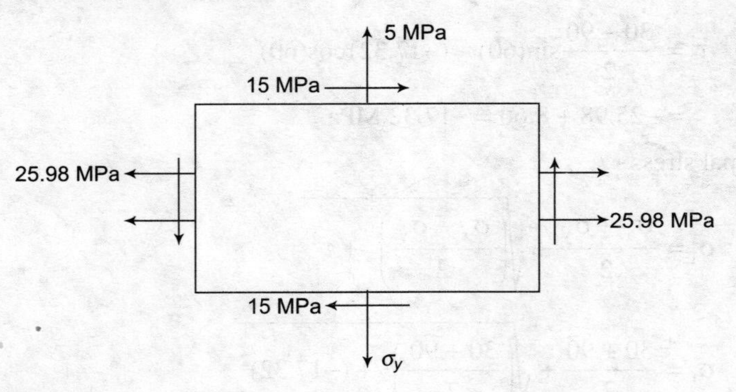

FIGURE 10.18

The shear stress on the horizontal plane is also 15 MPa. As sum of shearing stresses on two orthogonal planes is zero.

$$\sigma_x = 25.98 \text{ MPa}, \ \sigma_y = 5 \text{ MPa, and } \tau_{xy} = 15 \text{ MPa.}$$

Maximum principal stress:

$$\sigma_1 = \frac{\sigma_x + \sigma_y}{2} + \sqrt{\left(\frac{\sigma_x - \sigma_y}{2}\right)^2 + \tau_{xy}^2}$$

$$\sigma_1 = \frac{25.98 + 5}{2} + \sqrt{\left(\frac{25.98 - 5}{2}\right)^2 + 15^2}$$

$$= 15.49 + 18.30 = 33.79 \text{ MPa}$$

$$\sigma_1 = 15.49 - 18.30 = -2.81 \text{ MPa}$$

Maximum shear stress:

$$\tau_{max} = \frac{\sigma_1 - \sigma_2}{2} = 18.30 \text{ MPa.}$$

PROBLEM 10.5 **Objective 2**

Determine the principal stress and their plane for a pure shear state of stress.

SOLUTION

Put shear means only shear stress and normal stress.

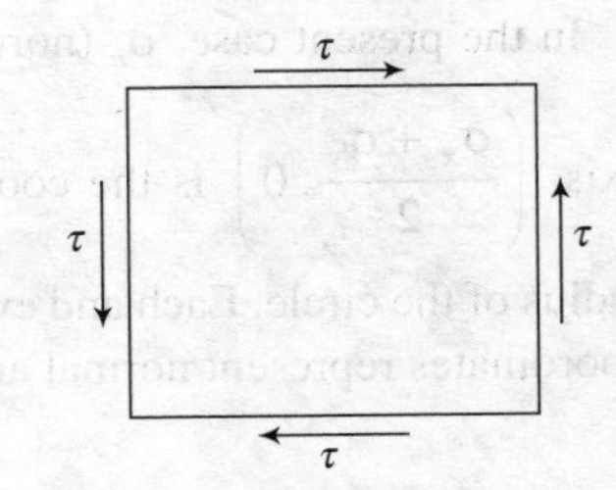

$$\sigma_x = 0,\ \sigma_y = 0, \text{ and } \tau_{xy} = \tau$$

$$\sigma_1 = \frac{\sigma_x + \sigma_y}{2} + \sqrt{\left(\frac{\sigma_x - \sigma_y}{2}\right)^2 + \tau_{xy}^2} = \tau$$

$$\sigma_2 = \frac{\sigma_x + \sigma_y}{2} - \sqrt{\left(\frac{\sigma_x - \sigma_y}{2}\right)^2 + \tau_{xy}^2} = -\tau$$

$$\tan 2\alpha = \frac{\tau_{xy}}{\left(\frac{\sigma_x - \sigma_y}{2}\right)} = \infty$$

$$2\alpha = 90°$$

$$\sigma_1 = 45°$$

$$\sigma_2 = 135°$$

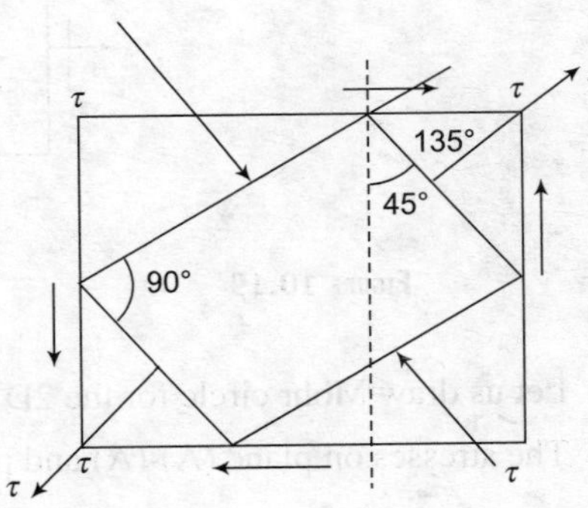

10.5 MOHR CIRCLE: GRAPHICAL METHOD

The stresses in a plane can be represented on a circle. The stresses on any plane can be found diagrammatically. We have,

$$\sigma_n = \frac{\sigma_x + \sigma_y}{2} + \frac{\sigma_x - \sigma_y}{2}\cos 2\theta + \tau_{xy}\sin 2\theta$$

$$\tau_n = \frac{\sigma_x - \sigma_y}{2}\sin 2\theta - \tau_{xy}\cos 2\theta$$

Rewriting the above equations,

$$\frac{\sigma_x - \sigma_y}{2}\cos 2\theta + \tau_{xy}\sin 2\theta = \sigma_n - \frac{\sigma_x + \sigma_y}{2} \tag{10.4}$$

$$\frac{\sigma_x - \sigma_y}{2} \sin 2\theta - \tau_{xy} \cos 2\theta = \tau_n \tag{10.5}$$

Squaring and adding the above two equations

$$\left(\frac{\sigma_x - \sigma_y}{2}\right)^2 + \tau_{xy}{}^2 = \left(\sigma_n - \frac{\sigma_x + \sigma_y}{2}\right)^2 + \tau_n^2$$

The above expression is in the form of $(x - a)^2 + (y - b)^2 = r^2$, which represents a circle with coordinating center as (a, b) and radius 'r'.

In the present case, σ_n (normal stress on any plane) is x axis, τ_n shear stress on any plane) is y axis, $\left(\frac{\sigma_x + \sigma_y}{2}, 0\right)$ is the coordinate of the center of the circle and $\sqrt{\left(\frac{\sigma_x - \sigma_y}{2}\right)^2 + \tau_{xy}{}^2}$ is the radius of the circle. Each and every point on the circle represents a plane in the stress block and their coordinates represent normal and shearing stress on the corresponding plane.

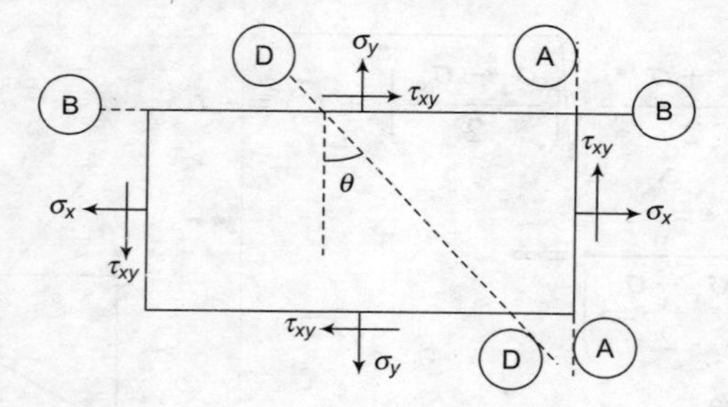

FIGURE 10.19

Let us draw Mohr circle for the 2D stress problem shown in Figure 10.19.

The stresses on plane (A)-(A) and plane (B)-(B) should be points on a circle.

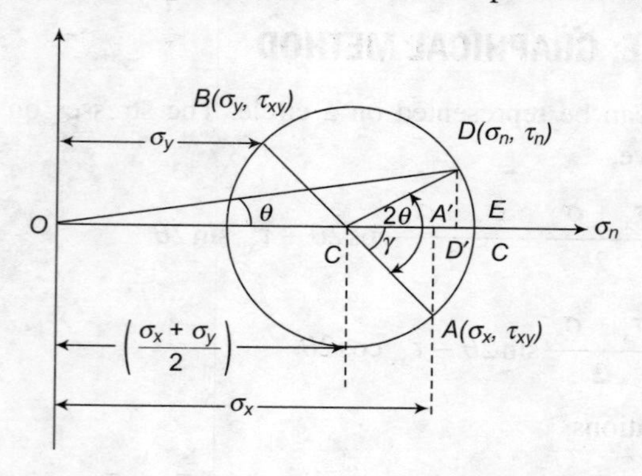

FIGURE 10.20

Stress on plane (A)-(A)

Normal stress is σ_x

Shear stress is $-\tau_{xy}$ (as this stress rotates the block in anticlockwise direction)

Plot normal stress on x axis and shear stress on y axis.

Stress on plane (B)-(B)

Normal stress is σ_y

Shear stress is τ_{xy} (This stress rotates the block in clockwise direction)

Midpoint of A and B is C whose coordinates are $\left(\frac{\sigma_x + \sigma_y}{2}, 0\right)$ center of Mohr circle.

Taking C as center, draw a circle passing through the points A and B. Take a point D on the Mohr circle, whose coordinates are σ_n and τ_n.

Let us have the expression for σ_n.

Draw a perpendicular DD′ from D onto the x axis.

$$\sigma_n = \text{OD}' \text{ and } \tau_n = \text{DD}'$$

$$\text{OD}' = \text{OC} + \text{CD}'$$

$$\text{OC} = \frac{\sigma_x + \sigma_y}{2} \quad (x \text{ coordinate of midpoint of AB})$$

$$\text{CD}' = \text{CD} \cos (2\theta - \gamma)$$

$$\text{CD} = \text{CA} = \text{radius of the circle}$$

$$\text{CD}' = \text{CA} \cos(2\theta) \cos \gamma + \text{CA} \sin \text{CD} \sin 2\theta \sin \gamma$$

$$\text{CA} \cos \gamma = \text{CA}'$$

$$= \text{OA}' - \text{OC} = \sigma_x - \frac{\sigma_x + \sigma_y}{2} = \frac{\sigma_x - \sigma_y}{2}$$

$$\text{CA} \sin \gamma = \text{AA}' = \tau_{xy}$$

Substituting CA cos γ and CA sin γ values in the expression for CD′

$$\text{CD}' = \frac{\sigma_x - \sigma_y}{2} \cos 2\theta + \tau_{xy} \sin 2\theta$$

Substituting CD′ in the expression for OD′

$$\text{OD}' = \frac{\sigma_x + \sigma_y}{2} + \frac{\sigma_x - \sigma_y}{2} \cos 2\theta + \tau_{xy} \sin 2\theta$$

OD′ is the X coordinate of D, i.e., σ_n

$$\sigma_n = \frac{\sigma_x + \sigma_y}{2} + \frac{\sigma_x - \sigma_y}{2} \cos 2\theta + \tau_{xy} \sin 2\theta$$

Similarly, $DD' = CD \sin(2\theta - \gamma) = AC \sin(2\theta - \gamma)$

$$DD' = CD \sin(2\theta - \gamma) = AC \sin 2\theta \cos\gamma - AC \cos 2\theta \sin\gamma$$

$$AC \cos\gamma = CA' = \frac{\sigma_X - \sigma_y}{2}$$

$$AC \sin\gamma = AA' = \tau_{xy}$$

$$DD' = \frac{\sigma_x - \sigma_y}{2} \sin 2\theta - \tau_{xy} \cos 2\theta$$

DD' is the y coordinate of D, i.e., τ_n

$$\tau_n = \frac{\sigma_x - \sigma_y}{2} \sin 2\theta - \tau_{xy} \cos 2\theta$$

As every point on the Mohr circle represents a plane, point E represents a plane on which shear stress is zero. The x coordinate of this point E is maximum normal stress (σ_1).

$$\sigma_1 = OE = OC + EC$$

$$= OC + CA$$

$$= \frac{\sigma_x + \sigma_y}{2} + \sqrt{\left\{\sigma_x - \frac{\sigma_x + \sigma_y}{2}\right\}^2 + \{\tau_{xy} - 0\}^2}$$

$$= \frac{\sigma_x + \sigma_y}{2} + \sqrt{\left\{\sigma_x - \frac{\sigma_x + \sigma_y}{2}\right\}^2 + \tau_{xy}^2}$$

Similarly, $\sigma_2 = OF = OC - CF$

CF = Radius of Mohr circle

$$OF = \sigma_2 = \frac{\sigma_x + \sigma_y}{2} - \sqrt{\left(\frac{\sigma_x + \sigma_y}{2}\right)^2 + \tau_{xy}^2}.$$

For maximum shear stress, Y coordinate should be maximum.

$$\tau_{max} = CG$$

= Radius of Mohr circle

$$= \sqrt{\left(\frac{\sigma_x + \sigma_y}{2}\right)^2 + \tau_{xy}^2}$$

OD = Resultant stress on the plane (D)-(D).

$$= \sqrt{(OD')^2 + (DD')^2}$$

$$= \sqrt{\sigma_n^2 + \tau_n^2}$$

ϕ = Inclination of resultant with outward normal

$$= \tan^{-1}\left[\frac{\tau_n}{\sigma_n}\right]$$

PROBLEM 10.6 **Objective 3**

Draw Mohr circle for the stresses shown in Figure 10.21. Determine principal stress, plane carrying principal stress and resultant stress on a plane inclined 30° with major principal plane.

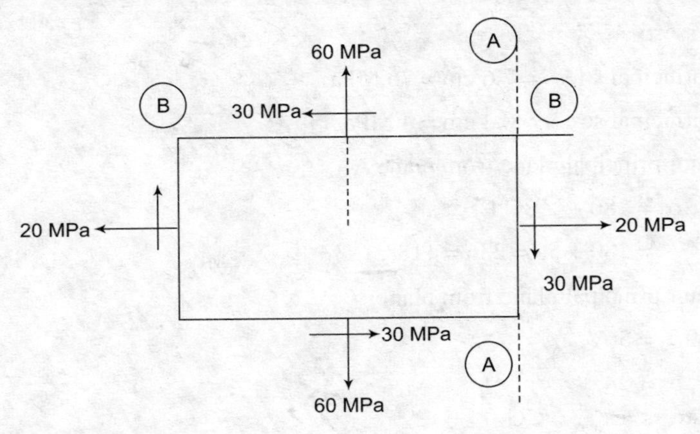

FIGURE 10.21

SOLUTION

Take a scale of 1 cm = 10 MPa on both X and Y axes. Plot normal stress on X axis and shear stress on Y axis.

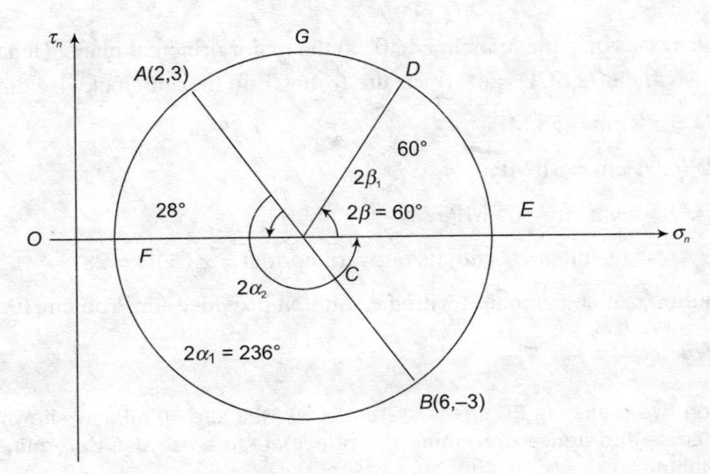

FIGURE 10.22

Locate the plane (A)-(A) as point A and on plane (A)-(A) normal stress = 20 MPa

Shear stress = 30 MPa (+ve as shear stress rotates the block in clockwise direction)

Locate the plane (B)-(B) as point B

Normal stress = 60 MPa

Shear stress = 30 MPa (–ve as shear stress rotates the block in anticlockwise direction)

Join A and B. The line AB cut the *X* axis at C. Taking C as center draw a circle passing through A and B.

For principal stresses:

OE is the major principal stress = 7.6 cm = 76 MPa

OF is the minor principal stress = 0.4 cm = 4 MPa

Inclination of major principal plane from plane A

$$2\alpha_1 = 180 + \angle BC\,E$$

$$= 180 + 56 = 236 = 118°$$

Inclination of minor principal plane from plan

$$2\alpha_2 = 56°$$

$$\alpha_2 = 28°$$

Maximum shear stress = τ_{max} = CG

$$= 3.6 \text{ cm} = 36 \text{ MPa}$$

Inclination of the plane that carries τ_{max}

$$2\beta_1 = 326°$$

$$\beta_1 = 163°$$

To determine the stresses on a plane inclined 30° to the major principal plane. Hence, locate a point 'D' on the circle, such that ∠ECD =60° (measure from CE in the anticlockwise direction).

X coordinate of D = 5.8 cm = 58 MPa

Y coordinate of D = 3.1 cm = 31 MPa

Resultant stress = OD = 6.5 cm = 65 MPa

Inclination of the stress resultant σ_R with its outward normal = ∠COD = 28°

This problem solution can be compared with the solution provided for Problem 10.1.

PROBLEM 10.7

Objective 3

Resultant stress on two planes in 2D stress system is 30 MPa and 40 MPa as shown in Figure 10.23. Sketch the Mohr circle and hence determine the principal stresses. Also determine the inclination between of the planes.

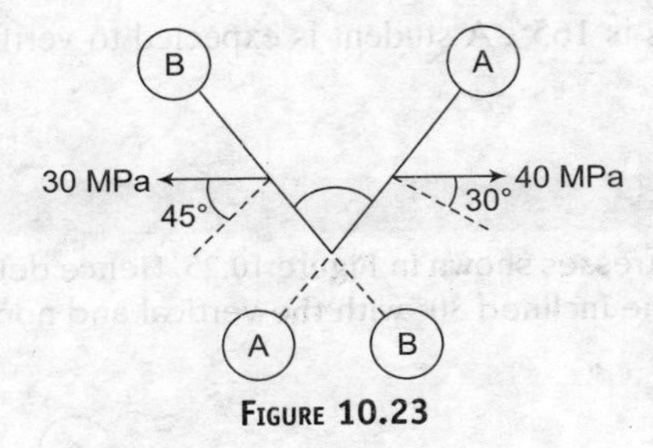

FIGURE 10.23

SOLUTION

Take 1 cm = 10 MPa

Location of point A: OA = 4 cm

φ_{OA} = 30° (–ve as the 40 MPa rotates in anticlockwise direction)

Location of point B:

OB = 5 cm

φ_{OB} = 45° (+ve as 30 MPa stress rotates the block in clockwise direction)

A and B are the points on the Mohr circle. To draw the circle, locate the center of the circle. To locate the center, join AB. Draw a perpendicular bisector to AB. This perpendicular bisector cuts the *X* axis at the center C.

OE = Major principal stress = 14.3 cm = 143 MPa

OF = Minor principal stress = 2.4 cm = 24 MPa

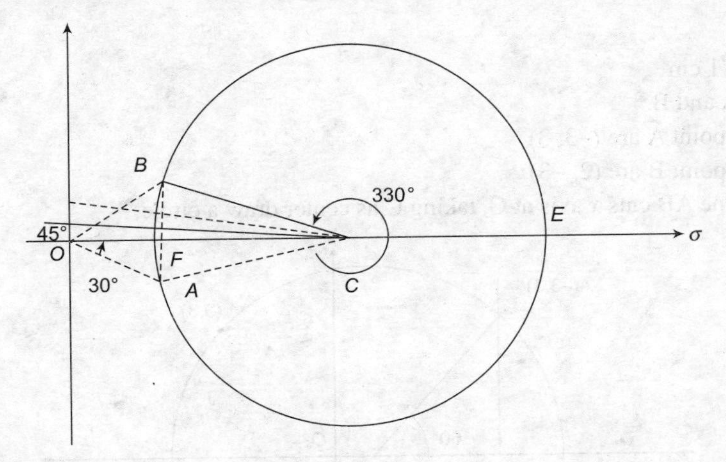

FIGURE 10.24

Inclination between planes (A)-(A) and (B)-(B) is given by 1/2 [∠ACB] measured in anticlockwise direction.

= 1/2 × ∠ACB

= 1/2 × 330° = 165°

Inclination between the planes is 165°. A student is expected to verify the solution of the problem analytically.

PROBLEM 10.8

Objective 3

Sketch the Mohr circle for the stresses shown in Figure 10.25. Hence determine the principal stresses. Estimate the stresses on a plane inclined 30° with the vertical and normal stress on a plane carrying maximum shear stress.

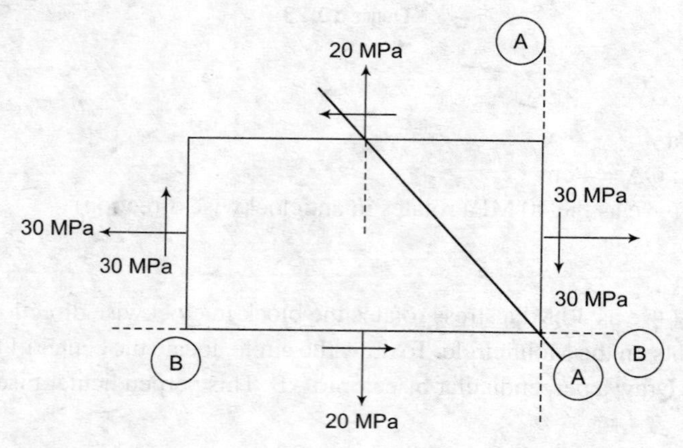

FIGURE 10.25

SOLUTION

Take 10 MPa = 1 cm

Locate points A and B.

Coordinates of point A are (−3, 3)

Coordinates of point B are (2, −3)

Join A and B, line AB cuts x axis at C, taking C as center draw a circle,

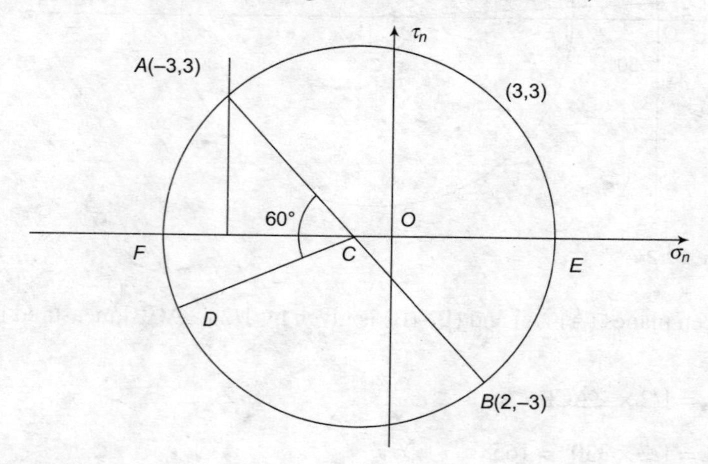

FIGURE 10.26

OE = Maximum principal stress = 3.5 cm = +35 MPa (tensile)

OF = Minimum principal stress = 4.5 cm = –4.5 MPa (compressive)

CG = Maximum shear stress = 4 cm = 40 MPa

Normal stress on plane carrying maximum shear stress

$$= x \text{ coordinate of G}$$

$$= \frac{(\sigma_x + \sigma_y)}{2} = \frac{(\sigma_1 + \sigma_2)}{2} = \frac{(-3+2)}{2} = -0.5 \text{ cm} = -5 \text{ MPa (compressive)}.$$

Plane (D)-(D) is inclined 30° with the plane (A)-(A).

To locate the point D, draw line CD such that, CD makes 60° with CA measured in anticlockwise direction.

Normal stress σ_D = 4.4 cm = 44 MPa (compressive)

Shear stress τ_D = 0.7 cm = 7 MPa

PROBLEM 10.9 **Objective 1**

Show that the sum of shear stresses acting on two mutually perpendicular planes is always zero.

SOLUTION

Consider a generalized 2D stress system

For the stress system shown in Figure 10.9.1,

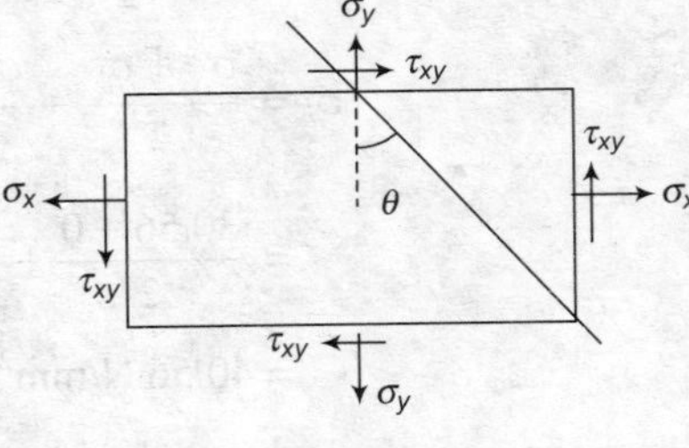

FIGURE 10.27

$$\sigma_n = \frac{\sigma_x + \sigma_y}{2} + \frac{\sigma_x - \sigma_y}{2}\cos 2\theta + \tau_{xy}\sin 2\theta$$

$$\tau_n = \frac{\sigma_x - \sigma_y}{2}\sin 2\theta - \tau_{xy}\cos 2\theta \qquad (1)$$

Let τ_{n+90} be the shear stress on a plane, orthogonal to the plane on which τ_n is acting

$$\tau_{n+90} = \tau_n \text{ at } \theta = (\theta + 90) = \frac{\sigma_x - \sigma_y}{2}\sin 2(\theta + 90) - \tau_{xy}\cos 2(\theta + 90)$$

$$\tau_{n+90} = -\frac{\sigma_x - \sigma_y}{2}\sin 2\theta + \tau_{xy}\cos 2\theta \qquad (2)$$

$$\tau_n + \tau_{n+90} = 0.$$

This implies that sum of shearing stresses on two mutually perpendicular planes is zero.

PROBLEM 10.10 **Objective 4**

A steel bolt of 25 mm diameter is subjected to a direct tension of 15 kN and a shearing force of 10 kN. Determine the intensities of normal and shear stresses across a plane at an angle of 60° to the longitudinal axis of the bolt. Also determine the resultant stress.

SOLUTION

Diameter of the bolt $d = 25\text{mm}$

Axial pull = 15 kN

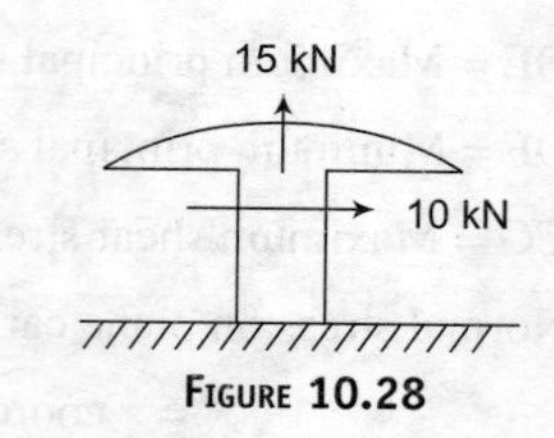

FIGURE 10.28

$$\text{Axial stress} = \frac{[(15 \times 10^3) \times 4]}{(\pi \times 25^2)}$$

$$= 30.56 \text{ N/mm}^2$$

$$\text{Shear stress} = \frac{[(10 \times 10^3) \times 4]}{(\pi \times 25^2)}$$

$$= 20.37 \text{ N/mm}^2$$

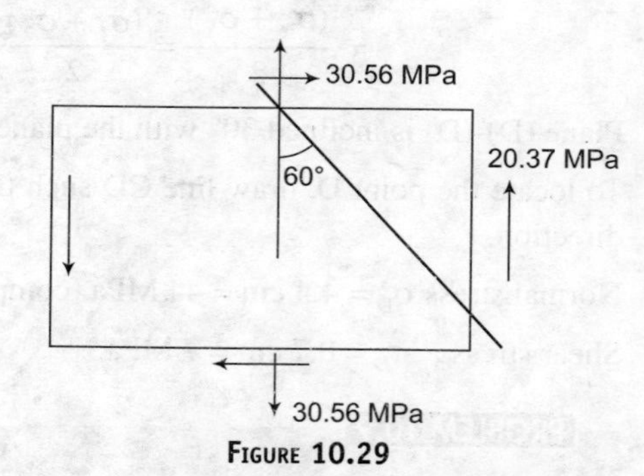

FIGURE 10.29

$$\sigma_x = 0$$

$$\sigma_y = 30.56 \text{ N/mm}^2$$

$$\tau_{xy} = 20.37 \text{ MPa}$$

$$\theta = 60°$$

Normal stress on the inclined plane,

$$\sigma_n = \frac{\sigma_x + \sigma_y}{2} + \frac{\sigma_x - \sigma_y}{2}\cos 2\theta + \tau_{xy}\sin 2\theta$$

$$= \frac{30.56 + 0}{2} + \frac{0 - 30.56}{2}\cos 120 + 20.37 \sin 120$$

$$= 40.56 \text{ N/mm}^2$$

Shear stress on the inclined plane,

$$\tau_2 = \frac{\sigma_x - \sigma_y}{2}\sin 2\theta - \tau_{xy}\cos 2\theta$$

$$= -13.23 + 10.185$$

$$= -3.045 \text{ N/mm}^2$$

$$\sigma_R = \text{Resultant stress} = \sqrt{\sigma_n^2 + \tau_n^2}$$

$$= 40.67 \text{ N/mm}^2$$

PROBLEM 10.11

Objective 2

A torque of 3.3 kN acts on the cross-section of a solid circular shaft, 80 mm diameter. What is the bending moment (BM) which can act on this section in addition to the given torque so that the maximum shear stress is 60 MPa and maximum normal stress is 100 MPa? Calculate the maximum and minimum principal stresses for this combination of torque and BM.

SOLUTION

T = Torque = 3.3 kN

BM acting on the shaft = M

σ = normal stress due to $M = \dfrac{32M}{\pi d^3}$

τ = shear stress due to $T = \dfrac{16T}{\pi d^3}$

$$\sigma_x = \sigma = \frac{32M}{\pi d^3}$$

$$\sigma_y = 0$$

$$\tau_{xy} = \tau = \frac{16T}{\pi d^3}$$

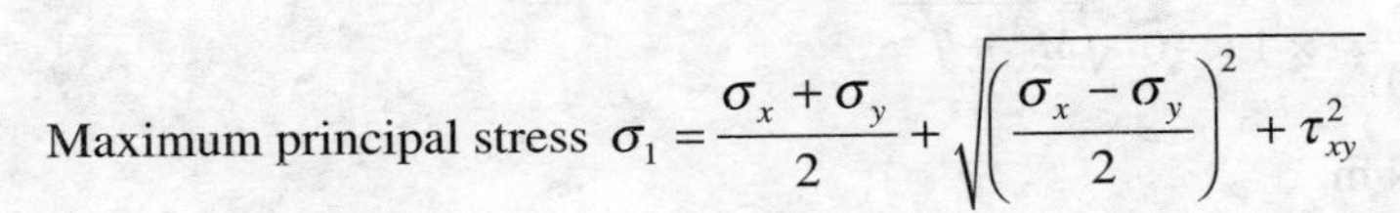

Maximum principal stress $\sigma_1 = \dfrac{\sigma_x + \sigma_y}{2} + \sqrt{\left(\dfrac{\sigma_x - \sigma_y}{2}\right)^2 + \tau_{xy}^2}$

$$\sigma_1 = \frac{16M}{\pi d} + \sqrt{\left(\frac{16M}{\pi d^3}\right)^2 + \left(\frac{16T}{\pi d^3}\right)^2} = \frac{16M}{\pi d^3}\{M + \sqrt{M^2 + T^2}\}$$

Minimum principal stress $\sigma_2 = \dfrac{16M}{\pi d^3}\{M - \sqrt{M^2 + T^2}\}$

Maximum shear stress

$$\tau_{\max} = \sqrt{\left(\frac{\sigma_x - \sigma_y}{2}\right)^2 + \tau_{xy}^2}$$

$$= \sqrt{\left(\frac{16\mu}{\pi d^3}\right)^2 + \left(\frac{16T}{\pi d^3}\right)^2}$$

$$= \frac{16}{\pi d^3}\{\sqrt{M^2 + T^2}\}.$$

Given that $d = 80$ mm; $T = 3.3 \times 10^6$ N·mm

$$\sigma_{\max} = \sigma_1 = 100 \text{ MPa}$$

$$\tau_{\max} = 60 \text{ MPa}$$

(a) If $\sigma_1 = 100$ MPa

Then $$100 = \frac{16}{\pi(80)^3}\{M + \sqrt{M^2 + [3.3*10^6]^2}\}$$

$$10.05 \times 10^6 = 1 \times 10^6\{M + \sqrt{M^2 + T^2}\}$$

If M and T are expressed in kN·m.

$$10.5 = M + \sqrt{M^2 + T^2} = M + \sqrt{M^2 + 3.3^2}$$

$$[10.05 - M]^2 = M^2 + 3.3^2$$

$$20.1M = 101 - 3.3^2 = 90.11$$

$$M = 4.48 \text{ kN·m.}$$

(b) If $\tau_{max} = 60$ MPa

Then $$60 = \frac{16}{\pi(80)^3} \times 1 \times 10^6 \sqrt{M^2 + T^2}$$

T and M are expressed in kN·m.

$$\sqrt{M^2 + 3.3^2} = 6.03$$

$$M = \sqrt{33.08} \text{ kN·m} = 5.75 \text{ kN·m}$$

Hence the allowable BM = 4.48 kN·m

Twisting moment = 3.3 kN·m

$$\sigma_1 = \frac{16}{\pi(80)^3} \times 1 \times 10^6\{4.48 + \sqrt{4.48^2 + 3.3^2}\} = 100 \text{ MPa}$$

$$\sigma_2 = \frac{16}{\pi(80)^3} \times 1 \times 10^6\{4.48 - \sqrt{4.48^2 + 3.3^2}\} = -10.78 \text{ MPa}$$

$$\tau_{max} = \frac{\sigma_1 - \sigma_2}{2} = \frac{100 + 10.78}{2} = 55.39 \text{ MPa}$$

PROBLEM 10.12 **Objective 1**

A propeller shaft, 160 mm external diameter, 80 mm internal diameter, transmits 450 kW at 4/3 Hz. There is, at the same time, a BM of 30 kN·m and end thrust of 250 kN. Find.

(a) The maximum principal stresses and their planes.
(b) The maximum shear stress and its plane.
(c) The stress, which acting alone, will produce the same maximum strain. Take Poisson's ratio = 0.3.

SOLUTION

External diameter of the shaft, $d_O = 160$ mm

Internal diameter of the shaft, $d_I = 80$ mm

Power to be transmitted is 450 kW.

$$\text{Speed} = \frac{4}{3}\text{ Hz} = \frac{4}{3}\text{ cycles/sec} = \frac{4}{3} \times 60 \text{ rotations per meter (rpm)}$$

$$\omega = 80 \text{ rpm}$$

Let torque acting on the shaft be T,

$$\frac{2\pi\omega T}{60} = P (T \text{ is in kN·m and } P \text{ is in kW})$$

$$450 = \frac{2\pi \times 80 \times T}{60}$$

$$T = 53.715 \text{ kN·m}$$

BM, $M = 30$ kN·m

Axial compression = 250 kN

τ = shear stress due to twisting moment T

$$= \frac{T}{J} r_{\max} = \frac{53.715 \times 10^6}{\frac{\pi}{32}(160^4 - 80^4)} \times 80$$

$$= 71.24 \text{ N/mm}^2$$

$$\text{Normal stress due to BM } \sigma_b = \frac{M}{I} y = \frac{30 \times 10^6}{\frac{\pi}{64}(160^4 - 80^4)} \times 80 = 79.58 \text{ N/mm}^2$$

$$\text{Normal thrust due to axial thrust } \sigma_a = \frac{P}{A} = \frac{250 \times 10^3}{\frac{\pi}{4}(160^2 - 80^2)} = 16.58 \text{ N/mm}^2$$

$$\text{Normal stress } \sigma = \sigma_a + \sigma_b = 79.58 + 16.58 = 96.16 \text{ N/mm}^2$$

Shear stress $\tau = 71.24$ N/mm^2

$$\sigma_x = -96.16 \text{ MPa}$$

$$\sigma_y = 0 \text{ and } \tau_{xy} = 71.24 \text{ MPa}$$

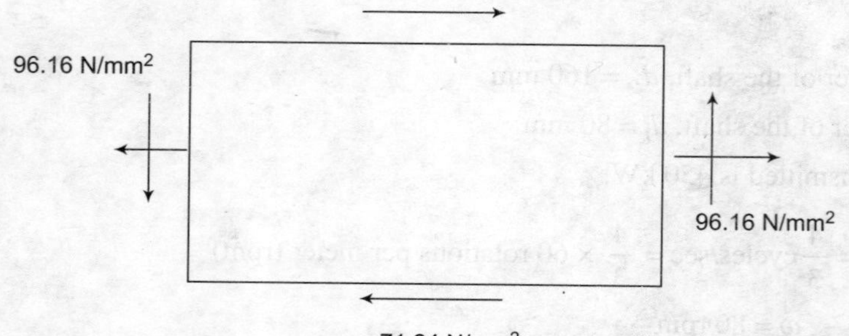

(a) $\sigma_1 = \dfrac{\sigma_x + \sigma_y}{2} \pm \sqrt{\left(\dfrac{\sigma_x - \sigma_y}{2}\right)^2 + \tau^2_{xy}}$

$= -48.08 - \sqrt{48.08^2 + 71.24^2} = -134.02 \text{ N/mm}^2$ (compressive)

Let α be inclination of principal plane

$$\tan 2\alpha = \frac{\tau_{xy}}{\left(\dfrac{\sigma_x - \sigma_y}{2}\right)} = \frac{71.24}{\left(\dfrac{-96.16}{2}\right)} = -1.48$$

$$\alpha_1 = -28°$$

$$\sigma_2 = -48.08 + \sqrt{48.08^2 + 71.24^2} = -48.08 + 85.95$$

$$= 37.87 \text{ N/mm}^2$$

(b) Maximum shear stress $\tau_{\max} = \sqrt{\left(\dfrac{\sigma_x - \sigma_y}{2}\right)^2 + \tau^2_{xy}} = 85.95 \text{ N/mm}^2$

(c) Minimum strain $\dfrac{\sigma_1}{E} - \mu \dfrac{\sigma_2}{E} = \in_{\max}$

$$\mu = 0.3$$

$$\in_{\max} = \frac{-134.02 - 0.3 \times (37.87)}{E} = -\frac{145.38}{E}$$

If $\in$ is strain due to normal stress (σ) acting alone, then

$$\in = \frac{\sigma}{E}$$

Given that, $\in = \in_{\max}$

$$\frac{\sigma}{E} = -\frac{145.38}{E}$$

$$\sigma = -145.38 \text{ N/mm}^2 \text{ (compressive stress)}$$

SUMMARY

- **The normal stress (σ_n) on a plane inclined 'θ' with the vertical in 2D stress system is given by**

$$\sigma_n = \frac{\sigma_x + \sigma_y}{2} + \frac{\sigma_x - \sigma_y}{2}\cos 2\theta - \tau_{xy}\sin 2\theta$$

- **The shear stress (τ_n) on a plane inclined 'θ' with the vertical in 2D stress system is given by**

$$\tau_n = \frac{\sigma_x + \sigma_y}{2}\sin 2\theta - \tau_{xy}\cos 2\theta$$

- **Principal stress is given by**

$$\sigma_{1,2} = \frac{\sigma_x + \sigma_y}{2} \pm \sqrt{\left(\frac{\sigma_x - \sigma_y}{2}\right)^2 + \tau_{xy}^2}$$

- **Maximum shear stress** $\tau_{max} = \dfrac{\sigma_1 - \sigma_2}{2}$ **or** $\sqrt{\left(\dfrac{\sigma_x - \sigma_y}{2}\right)^2 + \tau_{xy}^2}$

- **Sum of normal stress on two mutually perpendicular planes is constant and equal to sum of principal stress.**

$$\sigma_1 + \sigma_2 = \sigma_x + \sigma_y$$

- **Sum of shearing stress on two mutually perpendicular planes is zero.**
- **Normal stress on a plane carrying maximum shear stress is given by** $\dfrac{\sigma_1 + \sigma_2}{2}$.
- **Mohr circle reduces to point, if** $\sigma_x = \sigma_y * \sigma$ **and** $\tau_{xy} = 0$.
- **Center of Mohr circle is** $\dfrac{\sigma_x + \sigma_y}{2}, 0$.
- **Radius of Mohr circle is** $\sqrt{\left(\dfrac{\sigma_x - \sigma_y}{2}\right)^2 + \tau_{xy}^2}$.
- **Radius of Mohr circle is** $\dfrac{\sigma_1 - \sigma_2}{2}$.
- **In case of an axially loaded member with normal stress of σ, the maximum shear stress is $\sigma/2$. Principal stress for pure shear stress (τ) state is $\pm\tau$.**

OBJECTIVE TYPE QUESTIONS

1 A body is subjected to a pure tensile stress of 100 units. What is the maximum shear produced in the body at some oblique plane due to the above? [IES-2006]
(a) 100 units (b) 75 units (c) 50 units (d) 0 unit

2 A material element subjected to a plane state of stress such that the maximum shear stress is equal to the maximum tensile stress, would correspond to [IAS-1998]

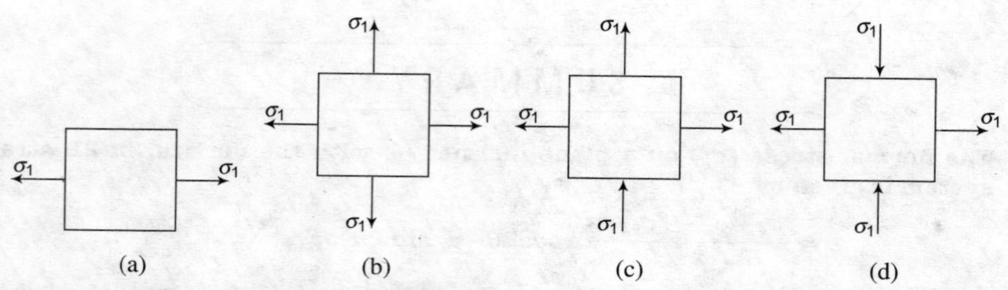

3 A solid circular shaft is subjected to a maximum shearing stress of 140 MPa. The magnitude of the maximum normal stress developed in the shaft is: [IAS-1995]

(a) 140 MPa (b) 80 MPa (c) 70 MPa (d) 60 MPa

4 The state of stress at a point in a loaded member is shown in the figure. The magnitude of maximum shear stress is [1 MPa = 10 kg/cm^2] [IAS 1994]

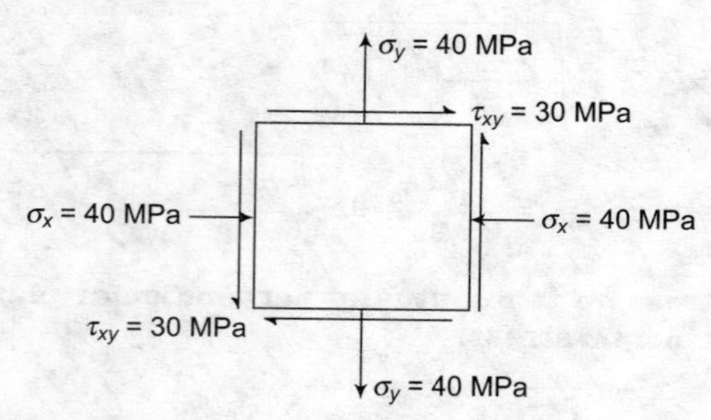

(a) 10 MPa (b) 30 MPa (c) 50 MPa (d) 100 MPa

5 A solid circular shaft of diameter 100 mm is subjected to an axial stress of 50 MPa. It is further subjected to a torque of 10 kN·m. The maximum principal stress experienced on the shaft is closest to [GATE-2008]

(a) 41 MPa (b) 82 MPa (c) 164 MPa (d) 204 MPa

6 In a biaxial stress problem, the stresses in x and y directions are (σ_x = 200 MPa and σ_y =100 MPa. The maximum principal stress in MPa is: [GATE-2000]

(a) 50 (b) 100 (c) 150 (d) 200

7 The maximum principle stress for the stress state shown in the figure is [GATE-2001]

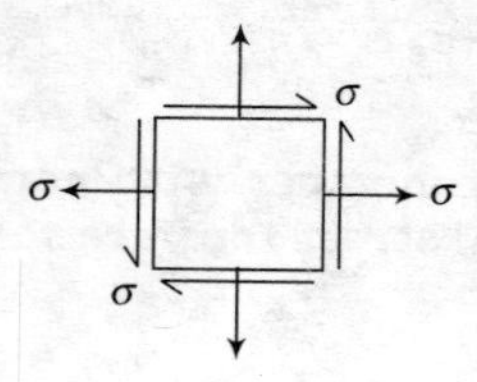

(a) σ (b) 2σ (c) 3σ (d) 1.5σ

8 The normal stresses at a point are σ_x = 10 MPa and, σ_y = 2 MPa; the shear stress at this point is 4 MPa. The maximum principal stress at this point is: [GATE-1998]

(a) 16 MPa (b) 14 MPa (c) 11 MPa (d) 10 MPa

9 Maximum shear stress in a Mohr's Circle [IES- 2008]

(a) Is equal to radius of Mohr's circle (b) Is greater than radius of Mohr's circle
(c) Is less than radius of Mohr's circle (d) Could be any of the above

10 A two-dimensional fluid element rotates like a rigid body. At a point within the element, the pressure is 1 unit. Radius of the Mohr's circle, characterizing the state of stress at that point, is: [GATE-2008]
(a) 0.5 unit (b) 0 unit (c) 1 unit (d) 2 units

11 The state of stress at a point under plane stress condition is σ_{xx} = 40 MPa, σ_{yy} = 100 MPa, and σ_{xy} = 40 MPa. The radius of the Mohr's circle representing the given state of stress in MPa is [GATE-2012]
(a) 40 (b) 50 (c) 60 (d) 100

12 Statement (I): Mohr's circle of stress can be related to Mohr's circle of strain by some constant of proportionality.
Statement (II): The relationship is a function of yield strength of the material. [IES-2012]
(a) Both Statement (I) and Statement (II) are individually true and Statement (II) is the correct explanation of Statement (I)
(b) Both Statement (I) and Statement (II) are individually true but Statement (II) is not the correct explanation of Statement (I)
(c) Statement (I) is true but Statement (II) is false
(d) Statement (I) is false but Statement (II) is true

13 The figure shows the state of stress at a certain point in a stressed body. The magnitudes of normal stresses in the *x* and *y* direction are 100 MPa and 20 MPa, respectively. The radius of Mohr's stress circle representing this state of stress is: [GATE-2004]
(a) 120 (b) 80 (c) 60 (d) 40

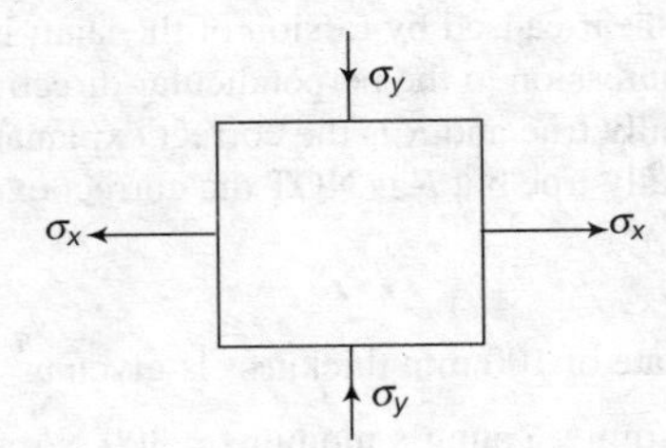

14 The state of stress at a point '*P*' in a two-dimensional loading is such that the Mohr's circle is a point located at 175 MPa on the positive normal stress axis. Determine the maximum and minimum principal stresses respectively from the Mohr's circle [GATE-2017]
(a) +175 MPa, –175 MPa (b) +175 MPa, +175 MPa
(c) 0, –175 MPa (d) 0, 0

15 A solid circular shaft is subjected to a maximum shearing stress of 140 MPa. The magnitude of the maximum normal stress developed in the shaft is: [IAS-1995]
(a) 140 MPa (b) 80 MPa (c) 70 MPa (d) 60 MPa

16 In the case of biaxial state of normal stresses, the normal stress on 45° plane is equal to [IES-1992]
(a) The sum of the normal stresses (b) Difference of the normal stresses
(c) Half the sum of the normal stresses (d) Half the difference of the normal stresses

17 Which one of the following Mohr's circles represents the state of pure shear? [IES-2000]

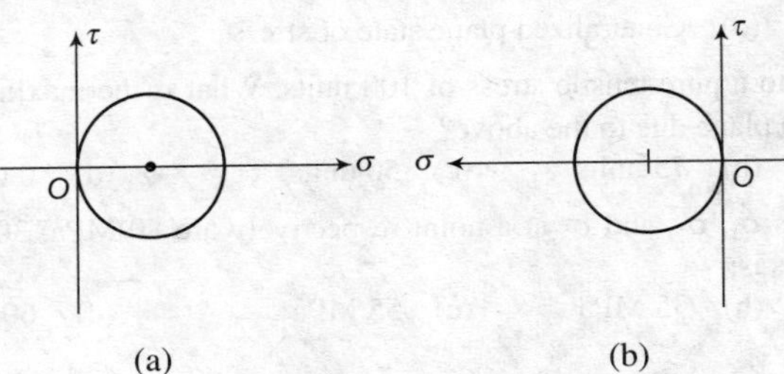

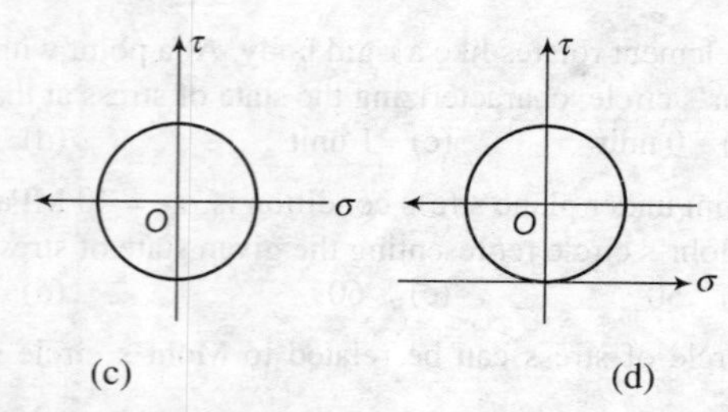

18 Assertion (*A*): If the state at a point is pure shear, then the principal planes through that point making an angle of 45° with plane of shearing stress carries principal stresses whose magnitude is equal to that of shearing stress.

Reason (*R*): Complementary shear stresses are equal in magnitude, but opposite in direction. [IES-1996]

(a) Both *A* and *R* are individually true and *R* is the correct explanation of A
(b) Both *A* and *R* are individually true but *R* is NOT the correct explanation of *A*
(c) *A* is true but *R* is false
(d) *A* is false but *R* is true

19 Assertion (*A*): Circular shafts made of brittle material fail along a helicoidally surface inclined at 45° to the axis (artery point) when subjected to twisting moment.

Reason (*R*): The state of pure shear caused by torsion of the shaft is equivalent to one of tension at 45° to the shaft axis and equal compression in the perpendicular direction. [IES-1995]

(a) Both *A* and *R* are individually true and *R* is the correct explanation of *A*
(b) Both *A* and *R* are individually true but *R* is NOT the correct explanation of *A*
(c) *A* is true but *R* is false
(d) *A* is false but *R* is true

20 The state of plane stress in a plate of 100 mm thickness is given as [IES-2000]

σ_{xx} = 100 N/mm^2, σ_{yy} = 200 N/mm^2, Young's modulus = 300 N/mm^2, Poisson's ratio = 0.3. The stress developed in the direction of thickness is:

(a) Zero (b) 90 N/mm^2 (c) 100 N/mm^2 (d) 200 N/mm^2

21 Consider the following statements: [IES-1996, 1998]

State of stress in two dimensions at a point in a loaded component can be completely specified by indicating the normal and shear stresses on

1. A plane containing the point
2. Any two planes passing through the point
3. Two mutually perpendicular planes passing through the point of these statements

(a) 1 and 3 are correct (b) 2 alone is correct
(c) 1 alone is correct (d) 3 alone is correct

22 If the principal stresses and maximum shearing stresses are of equal numerical value at a point in a stressed body, the state of stress can be termed as [IES-2010]

(a) Isotropic (b) Uniaxial
(c) Pure shear (d) Generalized plane state of stress

23 A body is subjected to a pure tensile stress of 100 units. What is the maximum shear produced in the body at some oblique plane due to the above? [IES-2006]

(a) 100 units (b) 75 units (c) 50 units (d) 0 unit

24 The principal stresses σ_1, σ_2, and σ_3 at a point respectively are 80 MPa, 30 MPa, and –40 MPa. The maximum shear stress is: [IES-2001]

(a) 25 MPa (b) 35 MPa (c) 55 MPa (d) 60 MPa

25 A piece of material is subjected to two perpendicular tensile stresses of 70 MPa and 10 MPa. The magnitude of the resultant stress on a plane in which the maximum shear stress occurs is [IES-2012]
(a) 70 MPa (b) 60 MPa (c) 50 MPa (d) 10 MPa

26 Plane stress at a point in a body is defined by principal stresses 3σ and σ. The ratio of the normal stress to the maximum shear stresses on the plane of maximum shear stress is: [IES- 2000]
(a) 1 (b) 2 (c) 3 (d) 4

27 For the state of plane stress shown, the maximum and minimum principal stresses are:

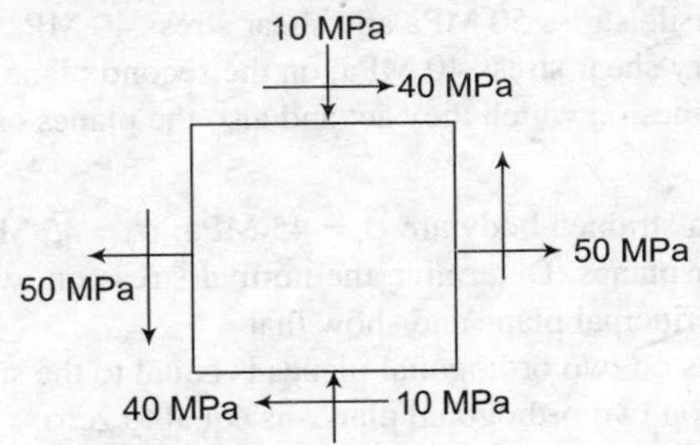

(a) 60 MPa and 30 MPa (b) 50 MPa and 10 MPa
(c) 40 MPa and 20 MPa (d) 70 MPa and 30 MPa

28 Normal stresses of equal magnitude *p*, but of opposite signs, act at a point of a strained material in perpendicular direction. What is the magnitude of the resultant normal stress on a plane inclined at 45° to the applied stresses? [IES-2005]
(a) $2p$ (b) $p/2$ (c) $p/4$ (d) Zero

Solutions for Objective Questions

Sl. No.	1.	2.	3.	4.	5.	6.	7.	8.	9.	10.
Answer	(c)	(d)	(a)	(c)	(b)	(d)	(b)	(c)	(a)	(b)

Sl.No.	11.	12.	13.	14.	15.	16.	17.	18.	19.	20.
Answer	(b)	(c)	(c)	(b)	(a)	(c)	(c)	(b)	(a)	(a)

Sl.No.	21.	22.	23.	24.	25.	26.	27.	28.
Answer	(d)	(c)	(c)	(d)	(c)	(b)	(d)	(d)

EXERCISE PROBLEMS

1. Normal stresses of 100 MPa tension and 100 N/mm^2 compression are applied to an elastic body at a certain point, on planes orthogonal to each other. The greater principal stress is limited to 150 MPa. What shearing stress may be applied to the given planes, and what will be the maximum shearing stress at a point? **[Objective 1]**
2. Sketch Mohr circle for 2D stress: $\sigma_x = 20$ MPa; $\sigma_y = 40$ MPa; $\tau_{xy} = 20$ MPa; Hence determine the planes carrying principal stress, principal planes, planes carrying maximum shear stress and minimum shear stress. **[Objective 3]**
3. Normal stresses of 120 MPa tension and 50 N/mm^2 compression are applied to an elastic body at a certain point, on planes orthogonal to each other. If the maximum shear stress is limited to 150 MPa, what shearing stress may be applied to the given planes, and what will be the principal stresses at the point? **[Objective 2]**

4. At a point in a strained body, the minimum and maximum principal stresses are 20 MPa and 40 MPa, both tension, find the shear stress and normal stress on a plane through this point making an angle of 30° with the major principal. **[Objective 1]**

5. A bar of circular cross-section is in tension under an axial stress of 120 MPa. If $\mu = 1/3$ for the material, what stresses must be applied to the side faces to prevent any change in cross-sectional dimensions? Also determine the maximum shear stress for this case. **[Objective 1]**

6. At a point in a piece of elastic material, there are two mutually perpendicular planes on which the stresses are as follows: tensile stress 50 MPa and shear stress 40 MPa on one plane, compressive stress 20 MPa and complementary shear stress 40 MPa, on the second plane. Determine the principal stresses and the positions of the planes on which they act and also the planes on which there is no normal stress. **[Objective 2]**

7. The stress components in a strained body are $\sigma_x = 45$ MPa; $\sigma_y = 35$ MPa: $\tau_{xy} = 15$ MPa; Determine the principal stresses and their planes. Determine the normal stress on two mutually perpendicular planes inclined 30° to the major principal plane and show that
   (a) Sum of normal stresses on two orthogonal planes is equal to the sum of principal stresses
   (b) Sum of shear stresses on two orthogonal planes is equal to zero. **[Objective 1]**

8. The stress components in a strained body are $\sigma_x = -25$ MPa; $\sigma_y = -25$ MPa: $\tau_{xy} = 25$ MPa: Find the resultant normal and shear stress on plane equally inclined to the principal axes. **[Objective 1]**

9. In a strained body, the normal and shear stresses were observed on two planes AB and AC as shown below. Determine the angle between the planes AB and AC. Also find the principal stresses and maximum shear stress. **[Objective 1 and 2]**

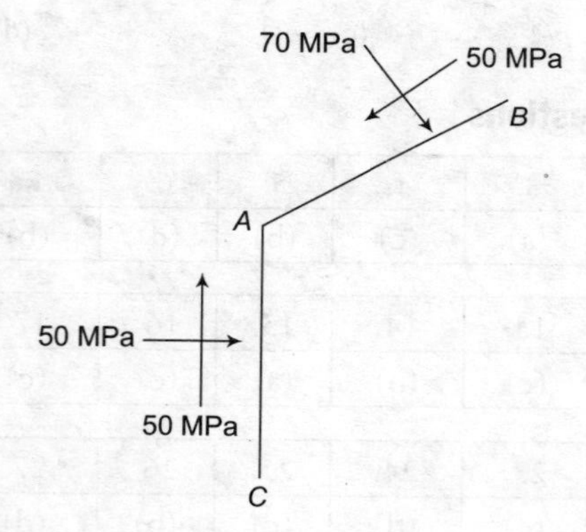

FIGURE 1

10. Find the principal stresses and their planes in an element subjected to pure shear stress. **[Objective 2]**

11. The normal stresses on three planes shown in Figure 2 carry 60 MPa (tensile) on plane OA, 30 MPa (tensile) on plane OC and 45 MPa (compressive) on plane OB. Determine the principal stresses and shear stress acting on these planes. **[Objective 2]**

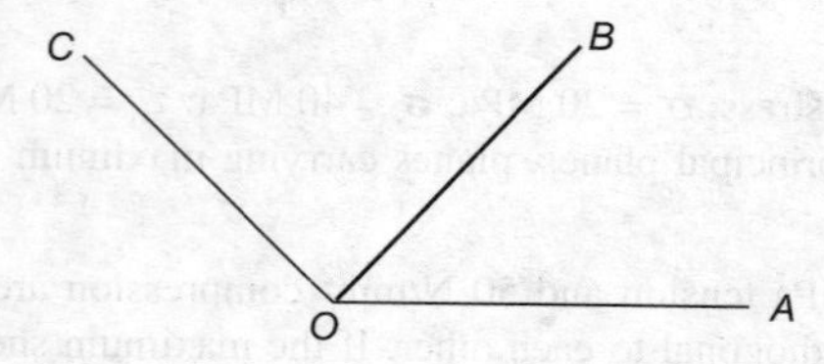

FIGURE 2

12. At a point in a strained body subjected to two-dimensional stress system, one of the principal stresses is 60 MPa (tensile). On a plane at 60° to this principal plane, the normal stress is zero. Determine (i) the other principal stress, (ii) the shear stress on the plane of zero normal stress, and (iii) the planes on which the normal and shear stresses are equal in magnitude. **[Objective 2]**

CHAPTER 11

COILED SPRINGS

UNIT OBJECTIVE

This chapter provides information about the theory and derivation of formulae for open- and close-coiled helical springs. The presentation attempts to help the student achieve the following:

Objective 1: Determine the extension of springs corresponding to the load.

Objective 2: Analysis of stress in the springs.

Objective 3: Determine the equivalent stiffness, load, etc., of springs.

Objective 4: Determine the angular rotation and dimensional parameters of springs.

11.1 INTRODUCTION

Springs are the structural elements used to absorb energy or impact. Energy absorption in these elements is due to their ability to deform. Members deform more under loads and absorb more energy. For example, a body falling on a sand heap receives less impact as the falling body moves the sand particles, thereby dissipating energy. If the same body falls on a concrete floor, the body receives more impact and very less energy is dissipated. Here, concrete floor could not absorb the energy as the particles of concrete are bound together, whereas in case of sand, particles are not bound together allowing energy dissipation. Thus, energy absorbing elements should be capable of deforming without getting ruptured. Coiled springs are the best examples for the energy absorption. Coil springs are of two types:

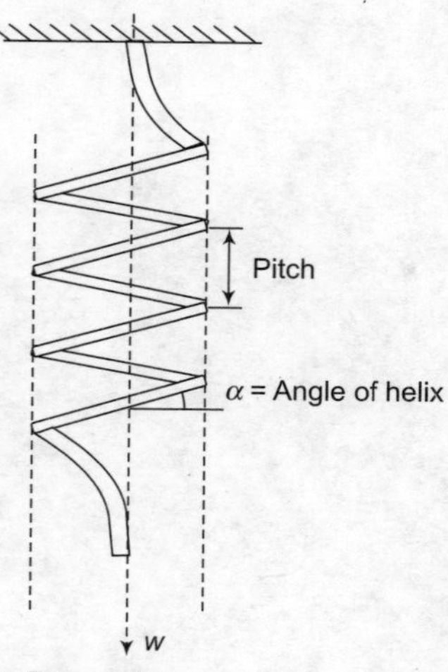

FIGURE 11.1

1. Close-coiled helical springs
2. Open-coiled helical springs.

In case of a close-coiled helical spring, the pitch (distance between two coils) is very less and angle of helix is zero.

In case of open-coiled helical springs, the gap between the coils (pitch) is considerable and the angle of helix is also finite value. The components of helical spring are

d = wire diameter

$D = 2R$ = diameter of the coil

R = mean radius of the coil

n = number of turns/coils present in the spring.

11.2 EXPRESSION FOR THE EXTENSION OF A CLOSE-COILED HELICAL SPRING

Consider a close-coiled helical spring subjected to axial load.

The load 'W' passes through the center of coil.

The axial load W is transformed on to the cross-section of the coil. Shear force acting on the wire is W, twisting moment acting on the wire $T = WR$. Compared to wire diameter the radius of the coil is generally high ($R > d$).

Thus, the shear stress due to twisting moment (WR) is considerable.

The maximum shear stress in the wire of the coil = $\tau_{max} = \frac{T}{J} r$.

$$\tau_{max} = \frac{WR}{\frac{\pi d^4}{32}} \times \left(\frac{d}{2}\right) \Rightarrow \tau_{max} = \frac{16WR}{\pi d^3}.$$

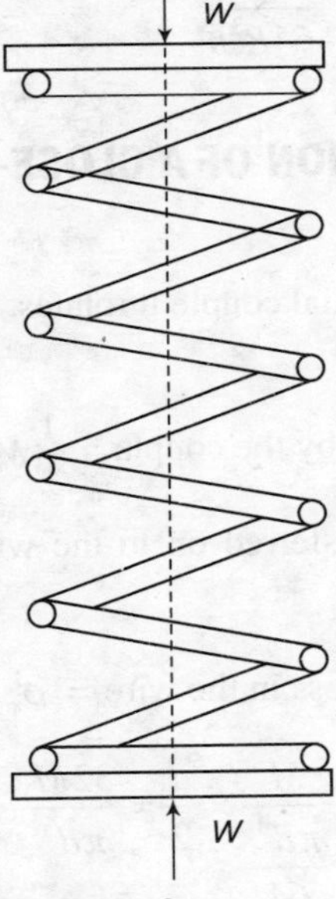

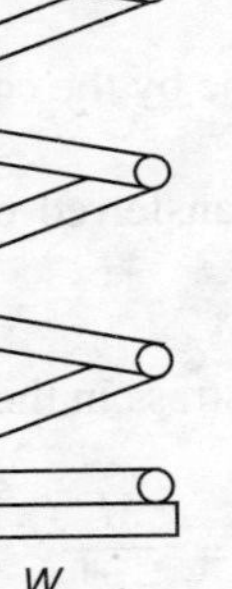

FIGURE 11.2

U = strain energy stored in the coil = $\frac{\tau_{max}^2}{4N} \times$ volume, in which N is the rigidity modulus of the material of the wire.

Volume of the spring = $\frac{\pi}{4} d^2 \cdot L$, in which '$L$' is the length of the wire of the spring. If 'n' is the number of turns present in a coil, wire length = $L = 2\pi R \times n = 2\pi Rn$.

$\therefore$ Strain energy stored in spring =

$$U = \left[\frac{16WR}{\pi d^3}\right]^2 \times \frac{2\pi Rn}{4N} \times \frac{\pi}{4} d^2$$

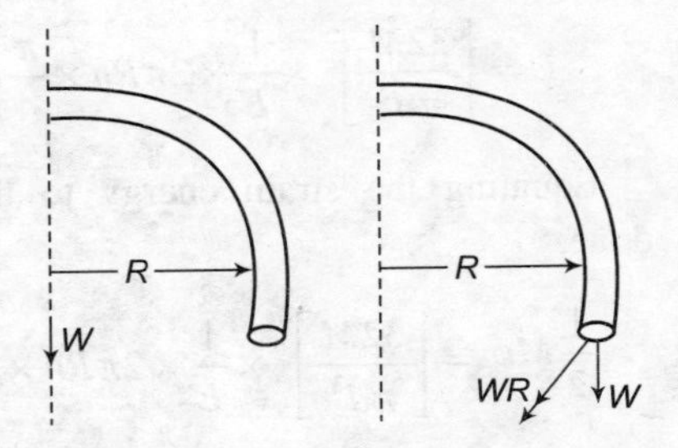

FIGURE 11.3

This strain energy stored in the opening is equal to the work done by the load W in undergoing axial compression of D

Work done by $W = \frac{1}{2} W\Delta$, as W is gradually applied load.

Equating the work done to the strain energy stored in the coil,

$$\frac{1}{2} W\Delta = \left[\frac{16WR}{\pi d^3}\right]^2 \times \frac{2\pi Rn}{4N} \times \frac{\pi}{4} d^2$$

$$\Rightarrow \qquad \Delta = \frac{64WR^3 n}{Nd^4}$$

The stiffness of the spring $= K = \frac{W}{\Delta}$

$$\Rightarrow \qquad K = \frac{Nd^4}{64R^3 n}.$$

11.3 ANGULAR ROTATION OF A CLOSE-COILED SPRING DUE TO AXIAL COUPLE M

When a coil is subjected to axial couple it rotates. Let M_c be the couple acting on the coil and 'ϕ' be the rotation of the coil.

$$\text{Work done by the couple} = \frac{1}{2} M\phi$$

If the axial couple is transferred on to the wire of the coil, the wire is subjected to bending moment (BM) of 'M'.

The maximum bending stress in the wire $= \sigma_{\max} = \frac{M}{I} y_{\max}$

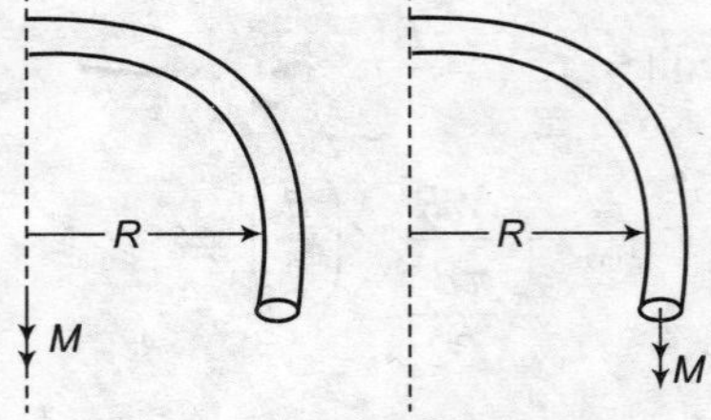

$$\sigma_{\max} = \frac{M}{\frac{\pi d^4}{64}} \times \frac{d}{2} = \frac{32M}{\pi d^3}$$

FIGURE 11.4

Energy stored in the coil $= \frac{\sigma_{\max}{}^2}{E} \times$ volume

$$U = \left[\frac{32M}{\pi d^3}\right]^2 \times \frac{1}{E} \times 2\pi Rn \times \frac{\pi}{4} d^2.$$

Equating the strain energy to the work done,

$$\frac{1}{2} M\phi = \left[\frac{32M}{\pi d^3}\right]^2 \times \frac{1}{E} \times 2\pi Rn \times \frac{\pi}{4} d^2$$

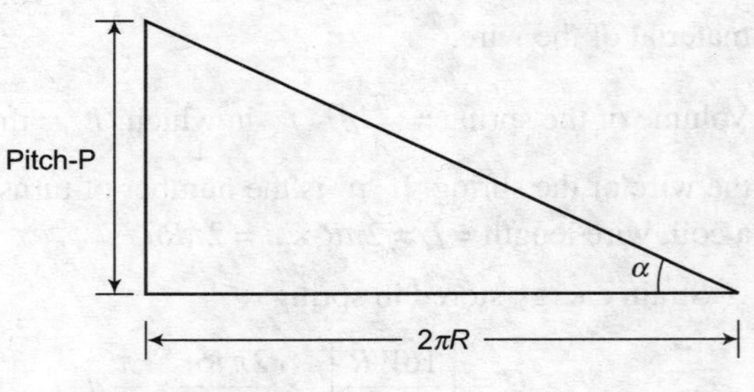

FIGURE 11.5

$$\Rightarrow \qquad \phi = \frac{128MRn}{Ed^4}$$

Bending the stiffness of the spring $K_b = \dfrac{Ed^4}{128Rn}$.

11.4 OPEN-COILED HELICAL SPRING

In an open-coiled helical spring, the pitch distance between the coils is considerable and angle of helix is also considerable.

Relation between angle of helix and pitch

Pitch is the vertical movement of the coil for one complete rotation. It can be diametrically represented as α shown in Figure 11.6.

Thus, $\qquad \tan\alpha = \dfrac{p}{2\pi R}$

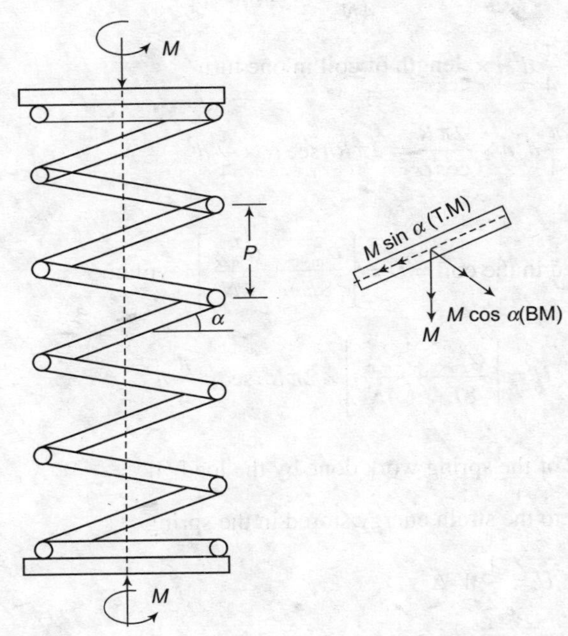

FIGURE 11.6

Expression for extension of open-coiled helical spring due to axial load *W*

Consider an open-coiled helical spring of angle of helix 'α' mean coil radius 'R', and wire diameter 'd'.

When the load is transferred to wire of the coil shown in Figure 11.6. WR the moment due to 'W' can be resolved into two components.

Along the wire the components of moment (twisting moment) = $T = WR\cos\alpha$.

Component of moment across the wire (BM) = $M_b = WR\sin\alpha$.

Maximum normal stress due to BM $M_b = \sigma_{max} = \dfrac{32M_b}{\pi d^3}$

$$\sigma_{max} = \frac{32WR\sin\alpha}{\pi d^3}$$

Strain energy due to BM $M_b = U_b = \dfrac{\alpha_{max}^2}{8E} \times$ volume

Similarly,

Maximum shear stress due to twisting moment $(T) = \tau_{max} = \dfrac{16T}{\pi d^3}$

$$\tau_{max} = \frac{16WR\cos\alpha}{\pi d^3}$$

Strain energy due to twisting moment = $\dfrac{\tau_{max}^2}{4N} \times$ volume.

Volume of the spring = $\dfrac{\pi}{4} d^2 n \times$ length of coil in one turn

$$= \frac{\pi}{4} d^2 n \times \frac{2\pi R}{\cos\alpha} = 2\pi Rn\sec\alpha \times \frac{\pi}{4} d^2$$

in which α = helix angle.

Total strain energy stored in the coil = $U = \left[\dfrac{\alpha_{max}^2}{8E} + \dfrac{\alpha_{max}^2}{4N}\right] \times$ volume

$$U = \left[\frac{\alpha_{max}^2}{8E} + \frac{\tau_{max}^2}{4N}\right] \times 2\pi Rn\sec\alpha \frac{\pi}{4} d^2.$$

Let 'Δ' be the extension of the spring work done by the load 'W' = $\dfrac{1}{2} W\,\Delta$.

Equating the work done to the strain energy stored in the spring

$$U = \frac{1}{2} W\,\Delta$$

$$= \left\{\left[\frac{32WR\sin\alpha}{\pi d^3}\right]^2 \times \frac{1}{8E} + \left[\frac{16WR\cos\alpha}{\pi d^3}\right]^2 \times \frac{1}{4N}\right\} 2\pi Rn\sec\alpha \times \frac{\pi}{4} d^2$$

$$\Delta = \left\{\left[\frac{128W\sin^2\alpha}{Ed^4}\right] + \left[\frac{64W\cos^2\alpha}{Nd^4}\right]\right\} R^3 n\sec\alpha$$

$$\Rightarrow \qquad \Delta = \frac{64WR^3 n\sec\alpha}{d^4}\left\{\frac{\cos^2\alpha}{N} + \frac{2\sin^2\alpha}{E}\right\}$$

The above expression reduces to the expression for close-coiled helical spring, if α is equal to zero.

Stiffness of an open-coiled helical spring is $K = \frac{W}{\Delta}$.

$$K = \frac{W}{\Delta} = \frac{d^4}{64R^3 n \left\{ \frac{\cos^2 \alpha}{N} + \frac{2 \sin^2 \alpha}{E} \right\} \sec \alpha}$$

$$K = \frac{N \cdot E \cdot d^4}{64R^3 n \sec \alpha \{E \cos^2 \alpha + 2N \sin^2 \alpha\}}.$$

11.5 ROTATION OF AN OPEN-COILED HELICAL SPRING DUE TO AXIAL COUPLE

Consider an open-coiled helical spring subjected to an axial couple M

Let ϕ be the rotation of the spring work done by couple = ½ $M\phi$.

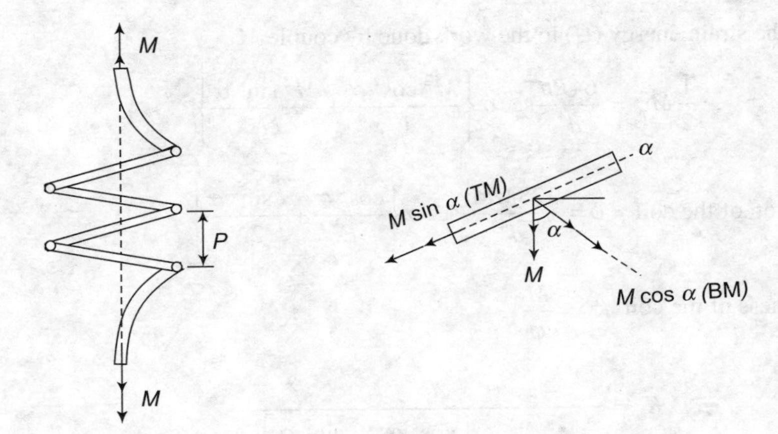

FIGURE 11.7

If the axial couple is transferred to the wire of the coil, the couple 'M' can be resolved into two components namely.

Torsional moment (T) = $M \sin \alpha$

BM (M_b) = $M \cos \alpha$

Maximum bending stress due to $M_b = \sigma_{max} = \dfrac{32M\cos\alpha}{\pi d^3}$

Maximum shear stress due to $T = \tau_{max} = \dfrac{16M\sin\alpha}{\pi d^3}$

Total strain energy U = strain energy due to bending + strain energy due to twisting

$$U = \frac{\sigma_{max}^2}{8E} \times \text{volume} + \frac{\tau_{max}^2}{4N} \times \text{volume}$$

$$= \left\{\left[\frac{32M\cos\alpha}{\pi d^3}\right]^2 \frac{1}{8E} + \left[\frac{16M\sin\alpha}{\pi d^3}\right]^2 \times \frac{1}{4N}\right\} \times \text{volume}$$

$$= \left[\frac{M^2\cos^2\alpha}{E} + \frac{M^2\sin^2\alpha}{2N}\right]\frac{128}{\pi^2 d^6} \times 2\pi Rn\sec\alpha \frac{\pi}{4} d^2$$

$$= \frac{64Rn}{d^4}\sec\alpha\left\{\frac{M^2\cos^2\alpha}{E} + \frac{M^2\sin^2\alpha}{2N}\right\}$$

Equating the strain energy (U) to the work done by couple M

$$\frac{1}{2}M_\varphi = \frac{64Rn}{d^4}\sec\alpha\left\{\frac{M^2\cos^2\alpha}{E} + \frac{M^2\sin^2\alpha}{2N}\right\}$$

$$\Rightarrow \quad \text{Rotation of the coil} = \phi = \frac{128MRn}{d^4}\sec\alpha\left\{\frac{\cos^2\alpha}{E} + \frac{\sin^2\alpha}{2N}\right\}$$

Bending stiffness of the coil $K_b = \dfrac{M}{\varphi}$

$$K_b = \frac{d^4}{128Rn\sec\alpha\left[\dfrac{\cos^2\alpha}{E} + \dfrac{\sin^2\alpha}{2N}\right]}$$

If $\alpha = 0$, the above expression reduces to close-coiled spring as

$$\phi = \frac{128MRn}{Ed^4}$$

11.6 SPRINGS IN SERIES

If the springs are connected in series, each spring is subjected to same axial load and total extension is equal to the sum of extensions of all springs.

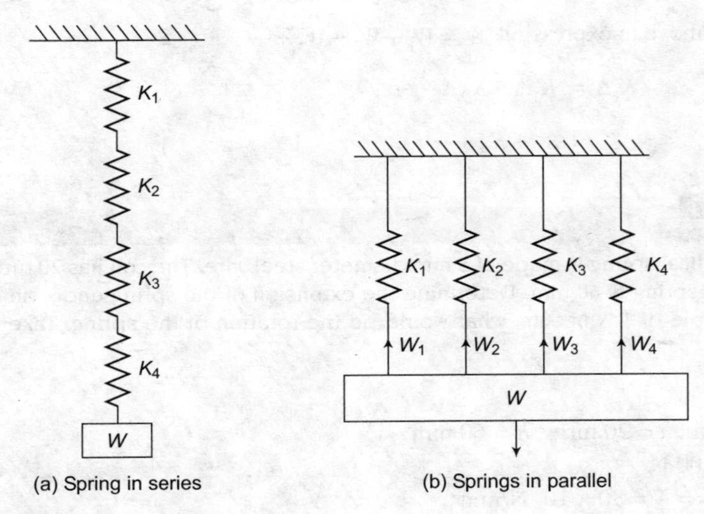

(a) Spring in series (b) Springs in parallel

FIGURE 11.8

An equivalent spring is a single spring, which gives the same extension under the same load.

Let k_e be the stiffness of the equivalent spring.

For springs connected in series,

Total extension $= \Delta = \Delta_1 + \Delta_2 + \Delta_3 + \cdots$

By definition, $\Delta = \dfrac{W}{K_e}$; $\Delta_1 = \dfrac{W}{K_1}$

$$\therefore \quad \frac{W}{K_e} = \frac{W}{K_1} + \frac{W}{K_2} + \cdots$$

$$\Rightarrow \quad \frac{1}{K_e} = \frac{1}{K_1} + \frac{1}{K_2} + \cdots$$

K_e is the equivalent spring stiffness, when springs are connected in series.

11.7 SPRINGS IN PARALLEL

Springs are connected parallel as shown in Figure 11.6.1. If K_1, K_2, K_3, etc., are the stiffness of springs connected in parallel, then the load is shared by all the springs, and each and every spring undergoes same extension.

$$W = W_1 + W_2 + W_3 + \cdots$$

$$K_1 = \frac{W_1}{\Delta}; \quad K_2 = \frac{W_2}{\Delta}; \quad K_3 = \frac{W_3}{\Delta}.$$

If K_e be the equivalent spring constant or stiffness,

$$K_e = \frac{W}{\Delta}; \quad W = K_e\Delta; \quad W_1 = K_1\Delta; \quad W_2 = K_2\Delta.$$

Substituting the above in expression $W = W_1 + W_2 + W_3 + \cdots$

$$K_e\Delta = K_1\Delta + K_2\Delta + \cdots$$

$$\Rightarrow \quad K_e = K_1 + K_2 + \cdots$$

PROBLEM 11.1 **Objective 1**

A close-coiled helical spring is made of 8 mm diameter steel wire. The coil has 20 turns and the mean coil radius of the spring is 60 mm. Determine the extension of the spring under an axial load of 100 N. If an axial couple of 1 N·m acts, what would be the rotation of the spring. Take $N = 80$ GPa and $E = 200$ GPa.

SOLUTION

$d = 8$ mm, $n = 20$ turns, $R = 60$ mm

Axial load $W = 100$ N

Rigidity modulus $= N = 80 \times 10^3$ N/mm^2

Extension of the spring $\Delta = \dfrac{64WR^3n}{Nd^4}$

$$\Delta = \frac{64 \times 100 \times 60^3 \times 20}{80 \times 10 \times 10 \times 10 \times 8^4} = 0.1875 \text{ mm}$$

PROBLEM 11.2 **Objective 1 and 2**

A wagon of weight 1200 N is moving with a speed of 12 kmph, brought to rest with the help of eight parallel springs. Each spring is made with 8 mm diameter steel wire and a mean coil radius of 65 mm. If the number of turns in each coil is 20, determine the compression of the springs. Take $N = 80 \times 10^3$ MPa. Determine the maximum shear stress induced in the wire of the coil.

SOLUTION

Mass of the wagon $= \dfrac{1200}{9.81}$ kg

Velocity of the wagon = 120 kmph $= \dfrac{10}{3}$ m/sec

Energy associated with the wagon $= \dfrac{1}{2} \times mV^2$

$$= \frac{1}{2} \times \frac{1200}{9.81} \times \left(\frac{10}{3}\right)^2 = 679.58 \text{ N·m}$$

Number of springs connected in parallel are eight.

Energy to be absorbed by each spring = 84.95 N·m.

Let W be the statically equivalent load, which produces the same effect of impact load.

Energy absorbed by each spring $= U = \dfrac{1}{2} W \times \delta$

in which δ is the compression of the spring.

$$U = \frac{1}{2}W \times \left[\frac{64WR^3n}{Nd^4}\right] = 84.95 \times 10^3$$

$R = 65$ mm, $d = 8$ mm, $n = 20$

$$\frac{1}{2}W^2 \times \frac{64 \times 65^3 \times 20}{80 \times 10^3 \times 8^4} = 84950 \Rightarrow W = 398 \text{ N}$$

Compression of the coil $\delta = \dfrac{64W \times R^3n}{Nd^4} = \dfrac{64 \times 398 \times 65^3 \times 20}{80 \times 10^3 \times 8^4} = 426.96$ mm

Shear stress produced in the wire $= \tau_{max} = \dfrac{16W \times R}{\pi d^3} = \dfrac{16 \times 398 \times 65}{\pi(8)^3} = 257.33$ MPa.

PROBLEM 11.3 — **Objective 1 and 2**

A close-coiled helical spring of radius 50 mm, made of 100 mm diameter rod has 20 turns. A weight of 150 N dropped on the spring from a height of 0.5 m. Determine compression if the spring and maximum shear stress are produced in the spring. Take $N = 80$ GPa.

SOLUTION

Work done by the falling weight = $W(h + \delta)$. 'δ' is the compression of the spring.

Let W be the statically equivalent load which produces the same effect of impact load.

Then, $$\delta = \left[\frac{64W \times R^3n}{Nd^4}\right]$$

Energy absorbed by the spring $= \dfrac{1}{2}W\delta$

Equating the work done to energy absorbed by the spring.

$$W\left[h + \frac{64W \times R^3n}{Nd^4}\right] = \frac{1}{2}W \times \delta \frac{64W \times R^3n}{Nd^4}$$

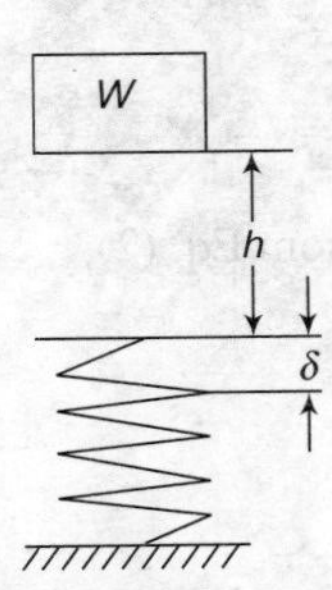

FIGURE 11.9

$W = 150$ N, $R = 50$ mm, $d = 10$ mm, $n = 20$

$N = 80$ GPa, $h = 0.5$ m $= 500$ mm

$$\frac{64R^3n}{Nd^4} = \frac{50^3 \times 20 \times 64}{80 \times 10^3 \times 10^4} = 0.2$$

$$150(500 + 0.2\,W) = 0.5 \times W^2 \times 0.2$$

$$0.1W^2 - 30W - 75000 = 0$$

$$W = 1029 \text{ N}$$

Compression of the spring = $\dfrac{64W \times R^3 n}{Nd^4} = 0.2 \times W = 0.2 \times 1029 = 205.78$ mm

Maximum shear stress induced in the wire of the coil = $\tau_{max} = \dfrac{16W \times R}{\pi d^3}$

$$= \frac{16 \times 1029 \times 50}{\pi \times 10^3} = 262.03\ \text{N/mm}^2$$

PROBLEM 11.4

Objective 4

The close-coiled helical spring stiffness is 2 N/mm. Under axial tensile load of 100 N, the maximum shear stress is 150 MPa. When the coil is touching each other, height of the spring is 60 mm. Determine the wire diameter, mean coil radius, and number of turns present in the spring. Take $N = 80$ GPa.

SOLUTION

Stiffness of the spring $R = 2$ N/mm, $W = 100$ N, $\tau_{max} = 150$ N/mm^2

Height of the spring when coils touch each other = $nd = 60$ mm

$$\tau_{max} = \frac{16WR}{\pi d^3} = 150\ \text{N/mm}^2 \tag{11.1}$$

$$K = \frac{Nd^4}{64R^3 n} = 2\ \text{N/mm} \tag{11.2}$$

$$nd = 60\ \text{mm}$$

From Eq. (1),

$$\frac{150\pi d^3}{16W} = \frac{150\pi d^3}{16 \times 100} = 0.2945d^3$$

From Eq. (2),

$$d^4 = \frac{2 \times 64 \times R^3 n}{N} = \frac{2 \times 64 \times (0.2945d^3)^3 n}{80 \times 10^3}$$

$$d^4 = \frac{2 \times 64 \times 0.2945^3 d^9 n}{80 \times 10^3}$$

$$nd^5 = \frac{80 \times 10^3}{2 \times 64 \times 0.2945^3} = 24469.44$$

$$nd = 60\ \text{mm}$$

$$nd^5 = nd \cdot d^4 = 24469.44$$

$$60\ d^4 = 24469.44$$

$$d^4 = 407.82$$

$$d = 4.49\ \text{mm}$$

Number of turns $= \dfrac{60}{4.49} = 13.36 = 14$

Radius of the coil $= R = 0.2945d^3 = 0.2945(4.49)^3 = 26.66$ mm.

PROBLEM 11.5

Objective 1

Determine the extension of an open-coiled helical spring of mean coil radius 65 mm, wire diameter 8 mm, and 10 turns. The helix angle is 10° and the axial load on the spring is 200 N. Take $E = 200$ GPa and $N = 80$ GPa. If the coil is considered as a close-coiled one, what is percentage error in the estimation of the extension?

SOLUTION

$$\alpha = 10°;\ R = 65\,\text{mm};\ d = 8\,\text{mm};\ n = 10;\ W = 200\,\text{N}$$
$$E = 200\ \text{GPa};\ N = 80000\ \text{N/mm}^2$$

Considering the coil as open-coiled one,

$$\Delta_0 = \frac{64WR^3 n \sec\alpha}{d^4}\left\{\frac{\cos^2\alpha}{N} + \frac{2\sin^2\alpha}{E}\right\}$$

$$\Delta_0 = \frac{64 \times 200 \times 65^3 \times 10 \sec 10}{8^4}\left\{\frac{\cos^2 10}{80 \times 10^3} + \frac{2\sin^2 10}{200 \times 10^3}\right\}$$

$$= 108.27\ \text{mm}.$$

If the coil is treated as a close-coiled one (assuming $\alpha = 0$)

$$\Delta_c = \frac{64WR^3 n}{Nd^4} = \frac{64 \times 200 \times 65^3 \times 10}{80 \times 10^3 \times 8^4} = 107.28\,\text{mm}.$$

$$\%\ \text{error in estimation of extension} = \frac{\Delta_o - \Delta_c}{\Delta_0} \times 100 = \frac{108.27 - 107.28}{108.27} \times 100 = 0.91\%.$$

The point to be noted here is that even if the helix angle is 10°, the error of estimating the extension is only 0.91%.

PROBLEM 11.6

Objective 3

A close-coiled spring '*A*' is placed coaxially in an another spring *B*. The height of spring *A* is 10 mm more than that of *B*; stiffness of spring *A* is 10 N/mm. If a load of 2000 N is placed on A, determine the load shared by the springs *A* and *B*. Mean diameter of the spring *B* is 200 mm, wire diameter is 18 mm, and the number of turns present in it is 16. Take $N = 80$ GPa. Determine the final compression of the combined spring.

SOLUTION

Outer spring is *B* and K_b = stiffness = 10 N/mm

Details of inner spring 'A'

$$D_a = 18\ \text{mm}$$

$$R_A = \frac{200}{2} = 100\,\text{mm}$$

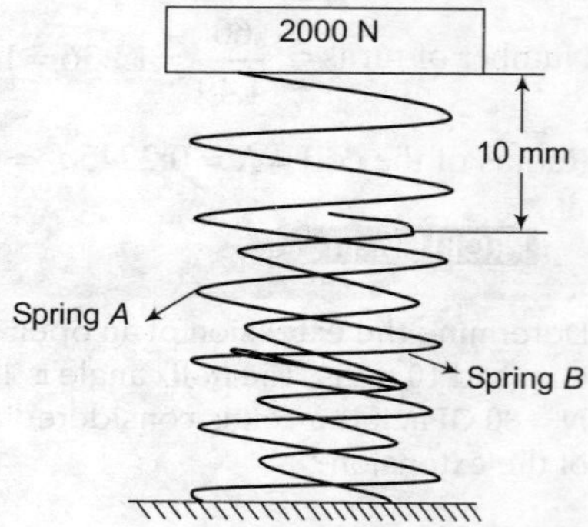

$$n_a = 16$$

$$N = 80 \text{ GPa}$$

Stiffness of the spring $A = K_a = \dfrac{Nd_a^4}{64R_a^3 n_a}$

$$K_a = \frac{80 \times 10^3 \times 18^4}{64 \times 100^3 \times 16} = 8.20 \text{ N/mm}$$

FIGURE 11.10

Let δ be the common compression of the springs A and B.

Compression of spring $A = \delta_a = \delta + 100$

Compression of spring $B = \delta_b = \delta$

$$W_a = \text{Force required to cause } \delta_a \text{ in spring } A = K_a\delta_a$$

$$W_b = \text{Force required to cause } \delta_b \text{ in spring } B = K_b\delta_b$$

We know that the total load of 2000 N is shared by two springs.

$\therefore \quad W_a + W_b = 2000$ N

$\Rightarrow \quad 8.2(\delta + 100) + 10\delta = 2000$

$\Rightarrow \quad \delta = 64.8$ mm

$\therefore$ Compression of spring $A = 164.84$ mm

Compression of spring $B = 64.84$ mm

Common compression of two springs $\delta = 64.84$ mm

$$W_a = \text{load shared by spring } A = 168.84 \times 8.2 = 1351.6 \text{ N}$$

$$W_b = \text{load shared by spring } B = 64.84 \times 10 = 648.4 \text{ N.}$$

PROBLEM 11.7

Objective 4

An open-coiled helical spring is made of 10 mm diameter wire. It is having 5 number of coils with mean coil radius of 100 mm. The angle of helix is 20°. This coil showed an extension of 40 mm under an axial load of 100 N and showed a rotation of 0.032 rad under an axial couple of 1 N·m. Determine the Poisson's ratio of the wire of the coil.

SOLUTION

Given that,

Wire diameter $= d = 10$ mm

Radius of coil $= R = 100$ mm

Number of coils $= n = 5$

Angle of helix $= \alpha = 20°$

$\Delta = 40$ mm due to load $W = 100$ N

$\phi = 0.032$ rad due to couple $M = 1$ N·m $= 1000$ N·mm

To determine Poisson's ratio μ, use expression $E = 2N(1 + \mu)$

To get N, the rigidity modulus, use data for Δ and W.

To get E, the modulus of elasticity, use data of ϕ and M.

We know that,

$$\Delta = \frac{64WR^3 N \sec\alpha}{d^4}\left(\frac{\cos^2\alpha}{N} + \frac{2\sin^2\alpha}{E}\right)$$

$$40 = \frac{64 \times 100 \times 100^3 \times 5 \sec 20°}{10^4}\left(\frac{\cos^2 20°}{N} + \frac{2\sin^2 20°}{E}\right)$$

$$\frac{0.883}{N} + \frac{0.23}{E} = 11.746 \times 10^{-6}$$

$$\phi = \frac{128MRn\sec}{d^4}\left(\frac{\cos^2\alpha}{E} + \frac{\sin^2\alpha}{2N}\right)$$

$$\Rightarrow \quad 0.032 = \frac{128 \times 1000 \times 100 \times 5 \times \sec 20}{10^4}\left(\frac{\cos^2\alpha}{E} + \frac{\sin^2\alpha}{2N}\right)$$

$$\Rightarrow \quad \frac{0.883}{E} + \frac{0.0585}{N} = 4.698 \times 10^{-6}$$

Solving Eqs. (1) and (2)

$$\frac{0.0585 \times 0.234}{E - 0.883^2 / E} = 0.0585 \times 11.746 \times 10^{-6} - 0.883 \times 4.698 \times 10^{-6}$$

$$\Rightarrow \quad \frac{1}{E(0.013689 - 0.789077)} = -3.461 \times 10^{-6}$$

$$\Rightarrow \quad E = 2.24 \times 10^5 \text{ MPa}$$

$$\Rightarrow \quad N = 0.773 \times 10^5 \text{ MPa}$$

$$E = 2N(1 + \mu)$$

$$\Rightarrow \quad \mu = 0.45$$

PROBLEM 11.8 **Objective 3**

An open coiled helical spring is made out of 10 mm diameter steel rod, the coils having 10 complete turns, and a mean diameter 80 mm; the angle of helix 15° calculates the deflection under an axial load of 250 N and the maximum intensities of direct and shear stresses induce an axial load of 250 N and the maximum intensities of direct and shear stresses are induced in the section of the wire. If the axial load of 250 N is replaced by an axial torque of 6 N·m, calculate the angle of rotation about axis of the coil and actual deflection. $N = 0.85 \times 10^5$ N/mm² and $E = 2.5 \times 10^5$ N/mm².

SOLUTION

Given Data: Number of turns $n = 10$

Mean coil diameter $D = 2R = 80$ mm

$\Rightarrow$ $R = 40$ mm

Diameter of the wire $d = 10$ mm

Angle of helix $\theta = 15°$

Axial load $W = 250$ N
Axial torque or axial couple $M = 6$ N·m $= 6000$ N·mm
Modulus of elasticity $E = 2.5 \times 10^5$ N/mm^2
Rigidity modulus $N = 0.85 \times 10^5$ N/mm^2
Deflection or extension due to axial load W:

$$\delta = \frac{64WR^3 n \sec\alpha}{d^4}\left\{\frac{\cos^2\alpha}{N} + \frac{2\sin^2\alpha}{E}\right\}$$

$$\delta = \frac{64 \times 250 \times 40^3 \times 10 \times \sec 15^0}{(10)^4}\left\{\frac{\cos^2 15}{0.85 \times 10^5} + \frac{2 \times \sin^2 15}{2.5 \times 10^5}\right\} = 12.2 \text{ mm}$$

Maximum direct stress/normal stress $\sigma_b = \dfrac{32WR\sin\alpha}{\Pi d^3}$

$$= \frac{32 \times 250 \times 40 \times \sin 15}{\pi (10)^3} = 26.36 \text{ N/mm}^2$$

Maximum shear stress $\tau_{\max} = \dfrac{16WR\cos\alpha}{\pi d^3} = \dfrac{16 \times 250 \times 40 \cos 15}{\pi (10)^3} = 49.19$ N/mm^2

Angle of rotation of the coil due to axial couple of 6000 N·mm

$$\phi = \frac{128MRn \sec\alpha}{d^4}\left\{\frac{\cos^2\alpha}{E} + \frac{\sin^2\alpha}{2N}\right\}$$

$$= \frac{128 \times 6000 \times 40 \times 10 \sec 15}{(10)^4}\left\{\frac{\cos^2 15}{2.5 \times 10^5} + \frac{\sin^2 15}{2 \times 0.85 \times 10^5}\right\}$$

$= 0.131$ rad.

11.8 LEAF SPRINGS OR CARRIAGE SPRINGS

Leaf springs are also called carriage springs. These springs are used in vehicles for absorbing shocks or impacts. In a carriage spring, a number of plates are attached one over the other without any bonding within them and are allowed to bend individually. A typical sketch of carriage spring is shown in Figure 11.11.

FIGURE 11.11 Carriage spring

This arrangement of plates structurally allows the member to undergo large deformations as moment of inertia (MI) reduces from midspan to end supports uniformly. How this reduction in MI is achieved is presented here in a systematic manner.

11.8.1 Structural Aspects of Leaf Spring

Consider a simply supported beam of span 'L' having number of detached plates just placed one over the other, subjected to a concentrated load 'W'.

Let 'n' be the number of plates with breadth as 'b' and thickness 't'.

The maximum BM is at midspan and is equal to $\frac{WL}{4}$.

As the plates are not bonded together, each plate deforms individually under the application of load. It means that top of the plate bends with compressive stress at its top fiber and tensile stress at its bottom fiber and similarly the plates underneath it are also subjected to similar stress distribution. The variation of bending stress across the depth of the spring at midspan is shown in Figure 11.12.

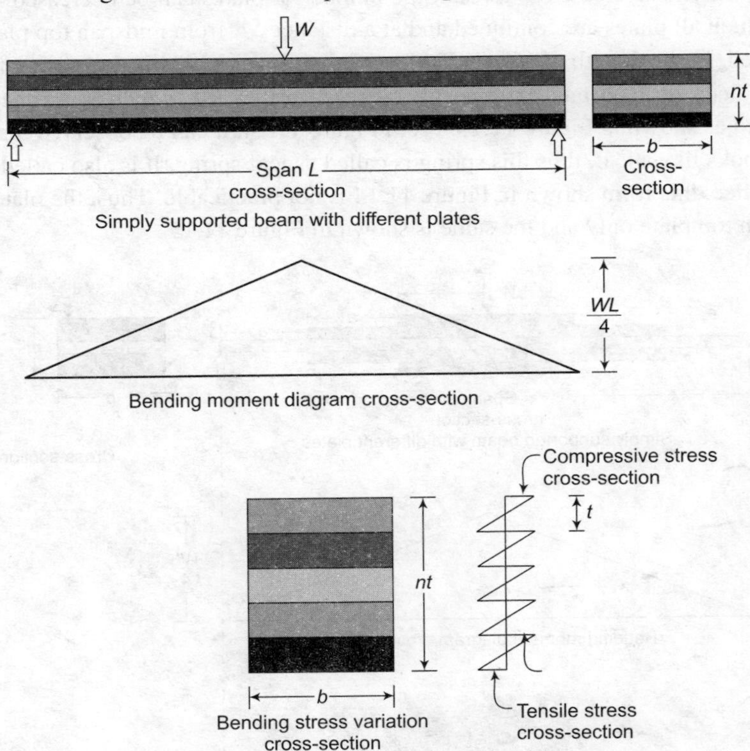

FIGURE 11.12 Carriage spring model

The BM is shared by the plates equally. Thus, the maximum bending stress in each plate is given by

$$\sigma = \frac{\left(\frac{M}{n}\right) y_{\max}}{I} = \frac{\left(\frac{WL}{4n}\right)\left(\frac{t}{2}\right)}{\left(\frac{bt^3}{12}\right)} = \frac{3WL}{2nbt^2}.$$

The requirement of a carriage spring is to undergo large deformation for a given load. This requirement can be achieved by proportioning the cross-section in such a way that everywhere along the length of the beam in every plate maximum stress shall be same. It means that the material in excess to resist the stresses shall be removed.

Maximum stress in a plate is the function of number of plates (n), breadth of each plate (b), and BM (M), when the thickness of the plate is maintained constant.

Theoretically, constant strength along the length of the spring and in each plate can be achieved as

(a) Decreasing the number of plates along the length of the beam from midspan considering BM variation.

(b) Decreasing the width of the outer most plates considering the BM variation.

The above two operations are shown in Figure 11.13.

At the midspan, BM is maximum; thus the required number of plates 'n' with full breadth 'b' shall be provided for a given allowable stress (σ) in the plate. As we move towards the support from midspan, the BM decreases as a result the number of plates can be decreased. Let 'a' be the distance for which all plates are continued and at a distance 'a' from midspan top plate and bottom plate is removed. In this length of 'a', the BM is varying linearly and is decreasing from midspan. Thus, even to keep the constant extreme fiber stress, further MI, which is a function of 'b', is linearly reduced as shown in Figure 11.14. From Figure 11.14, it can be observed that the top view of the spring looks like a leaf, thus this spring is called as leaf spring. It is also called semielliptical spring. In practice, this form shown in Figure 11.14 is not practicable. Thus, the plates are reduced in number from top plate only and the same is shown in Figure 11.15.

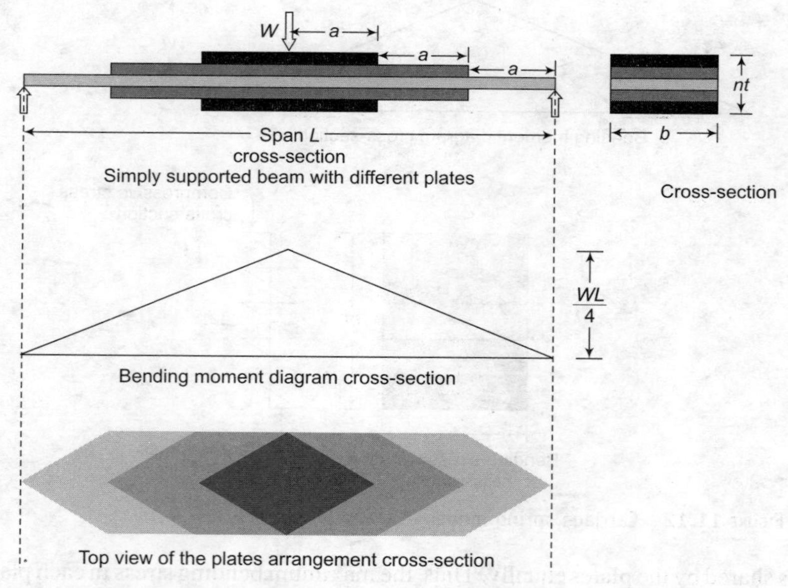

FIGURE 11.13 Curtailment of plates and shaping of plates

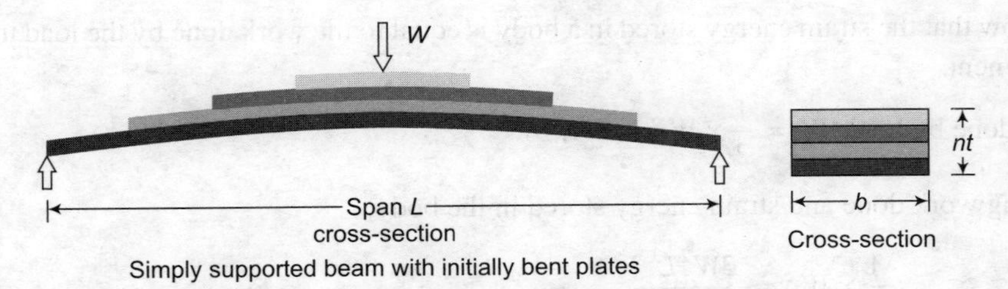

FIGURE 11.14 A typical sketch of carriage spring with initially bent plates

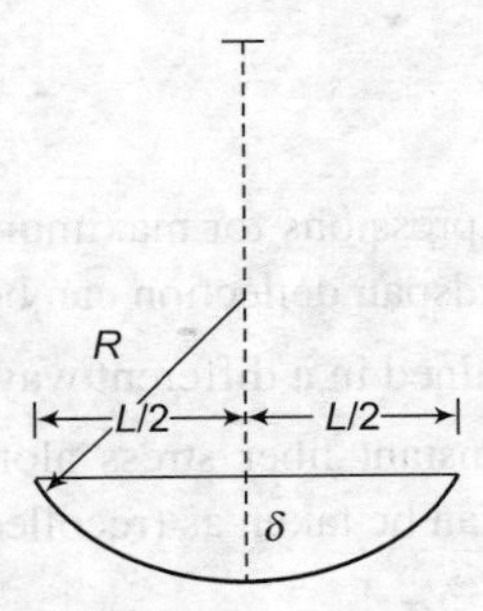

FIGURE 11.15 Bent profile of the leaf spring

The Deformation of Spring

Let 'δ' be the midspan deflection of the spring. In the previous chapters, it is learnt that the strain energy stored in the spring is given by

$$U = \frac{1}{6E}\sigma_{\max}^{2} \times \text{Volume of the spring.}$$

The spring is modeled like a triangle of height nt, breadth b, and length L. This modeling is acceptable as the number of plates and the width of the plate are linearly varied from zero to a maximum value.

Thus, volume of the total spring is given by

$$V = \frac{1}{2} \times ntbL.$$

Maximum stress in the plate material is given by

$$\sigma_{\max} = \frac{3WL}{2nbt^2}.$$

Substituting the values in strain energy expression we get,

$$U = \frac{1}{6E} \times \left\{\frac{3WL}{2nbt^2}\right\}^2 \times \frac{1}{2} \times ntbL$$

$$U = \frac{3W^2L^3}{16Enbt^3}.$$

We know that the strain energy stored in a body is equal to the work done by the load undergoing a displacement.

Work done by load 'W' $= \frac{1}{2} \times W\delta$

Equating work done and strain energy stored in the body,

$$\frac{1}{2} \times W\delta = \frac{3W^2L^3}{16Enbt^3}$$

$$\delta = \frac{3WL^3}{8Enbt^3}$$

Thus, in case of a leaf spring the expressions for maximum stress developed in the spring, strain energy stored in the spring, and the midspan deflection can be evaluated.

Deflection of the spring can be obtained in a different way also.

The leaf spring is subjected to constant fiber stress along the length of the beam. Thus, the expression for curvature of the beam can be taken as (recollecting flexure formula)

$$\frac{1}{R} = \frac{\sigma_{max}}{E\left(\frac{t}{2}\right)} = \frac{2\sigma_{max}}{Et}.$$

In the above expression, substituting value of $\sigma_{max} = \frac{3WL}{2nbt^2}$, we get

$$\frac{1}{R} = \frac{3WL}{Enbt^3}$$

In the above expression, all terms are constants; thus, the curvature of the beam does not change with length, and it is constant. It implies that the deflection profile of a leaf spring is arc of a circle. The same is shown in Figure 11.8.5.

Mathematically, it can be written as

$$(2R - \delta)\delta = \left(\frac{L}{2}\right)^2$$

$$\delta = \frac{L^2}{8R} \text{ (neglecting } \delta^2 \text{ term).}$$

In the above expression, substituting the value of curvature ($1/R$)

$$\delta = \frac{L^2}{8} \times \frac{3WL}{Enbt^3} = \frac{3WL^3}{8Enbt^3}.$$

A carriage spring deforms under the application of load. However, in practice the plates of the carriage spring are initially bent to the shape of deflected profile of the beam under external load. As and when the external load comes on to the member, the member becomes straight.

Stiffness of leaf spring is given by

$$k = \frac{W}{\delta} = \frac{8Enbt^3}{3L^3}.$$

11.8.2 Quarter Elliptical Spring

The quarter elliptical spring is a cantilever type spring. A typical sketch of quarter elliptical spring is shown in Figure 11.16.

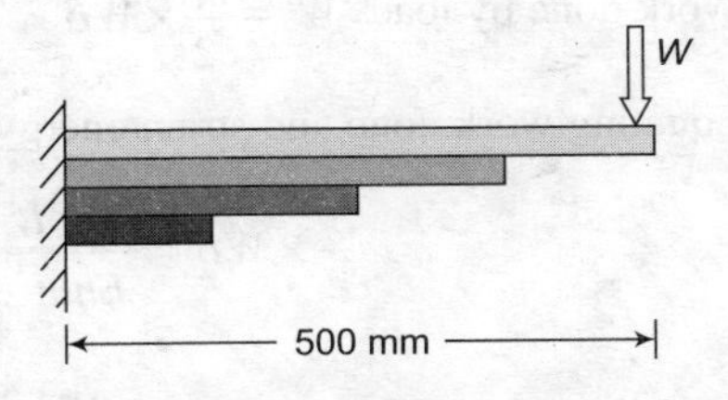

FIGURE 11.16 Quarter elliptical spring

Consider a quarter elliptical spring of span L having number of detached plates just placed one over the other, subjected to a concentrated load W at free end.

Let n be the number of plates with breadth as b and thickness t.

The maximum BM is at midspan and is equal to WL.

As the plates are not bonded together, each plate deforms individually under the application of load. The BM is shared by the plates equally. Thus, the maximum bending stress in each plate is given by

$$\sigma_{max} = \frac{\left(\frac{M}{n}\right) y_{max}}{I} = \frac{\left(\frac{WL}{n}\right)\left(\frac{t}{2}\right)}{\frac{bt^3}{12}} = \frac{6WL}{nbt^2}.$$

The Deformation of Spring

Let 'δ' be the midspan deflection of the spring. In the previous chapters, it is learnt that the strain energy stored in the spring is given by

$$U = \frac{1}{6E}\sigma_{max}^2 \times \text{volume of the spring.}$$

The spring is modeled like a triangle of height nt, breadth b, and length L. This modeling is acceptable as the number of plates and the width of the plate are linearly varied from zero to a maximum value.

Thus, volume of the total spring is given by

$$V = \frac{1}{2} \times ntbL.$$

Maximum stress in the plate material is given by

$$\sigma_{max} = \frac{6WL}{nbt^2}.$$

Substituting the values in strain energy expression we get,

$$U = \frac{1}{6E} \times \left\{\frac{6WL}{nbt^2}\right\}^2 \times \frac{1}{2} \times ntbL$$

$$U = \frac{3W^2L^3}{Enbt^3}$$

We know that the strain energy stored in a body is equal to the work done by the load undergoing a displacement.

Work done by load 'W' $= \frac{1}{2} \times W\delta$

Equating work done and strain energy stored in the body,

$$\frac{1}{2} \times W\delta = \frac{3W^2L^3}{Enbt^3}$$

$$\delta = \frac{6WL^3}{Enbt^3}$$

Thus, in case of a cantilever type leaf spring the expressions for maximum stress developed in the spring, strain energy stored in the spring, and the deflection at free end can be evaluated.

PROBLEM 11.9

Objective 4

Determine the width and thickness of a semielliptical or carriage spring carrying a central load of 200 N. The deflection is limited to 120 mm. The span of the simply supported spring is 0.6 m. The allowable stress is 150 MPa and modulus of elasticity is 200 GPa. If eight plates are to be adopted what shall be the thickness and width of each plate.

SOLUTION

In the present problem, limits on deflection and maximum stress are given.

$$\sigma_{max} \leq 150\,\text{MPa and } \delta \leq 120\,\text{mm.}$$

Given that $W = 200$ N

Span of the spring $= L = 600$ mm.

$$\sigma_{max} = \frac{3WL}{2nbt^2} \text{ and } \delta = \frac{3WL^3}{8Enbt^3}$$

$$\frac{\sigma_{max}}{\delta} = \left\{ \frac{\frac{3WL}{2nbt^2}}{\frac{3WL^3}{8Enbt^3}} \right\}$$

$$\frac{\sigma_{max}}{\delta} = \left\{ \frac{4Et^2}{L^2} \right\}$$

$$\frac{150}{120} = \left\{ \frac{4 \times 2 \times 10^5 t^2}{600^2} \right\}$$

$$\therefore \qquad t = 0.5625\text{ mm.}$$

0.5625-mm-thick plates are not available, so adopt 1-mm-thick plates.

If a single plate is adopted, then breadth of the plate is obtained by any of the above two conditions of stress or deflection whichever gives a higher value.

$$\sigma_{\max} \le 150$$

$$150 \ge \frac{3WL}{2nbt^2} \quad \text{and} \quad nb \ge \frac{3 \times 200 \times 600}{150 \times 2 \times 1 \times 1^2} = 1200\,\text{mm}$$

$$nb \ge \frac{3WL^3}{8\delta Et^3} = \frac{3 \times 200 \times 600^3}{8 \times 120 \times 2 \times 10^5 \times 1^3} = 675\,\text{mm}$$

Breadth $b \ge 1200$ mm. It is practically not possible as the span itself is 600 mm. Thus, provide eight plates as mentioned in the problem.

$$nb \ge \frac{3WL}{\sigma_{\max} 2t^2} \quad \text{hence } b \ge \frac{3 \times 200 \times 600}{8 \times 150 \times 2 \times 1^2} = 150\ \text{mm.}$$

Finally, provide eight plates of thickness 1 mm each and breadth 150 mm.

PROBLEM 11.10 **Objective 4**

Determine the width and thickness of a quarter elliptical spring (cantilever type) carrying an end concentrated load of 100 N. The deflection is limited to 100 mm. The span of the simply supported spring is 0.5 m. The allowable stress is 400 MPa and modulus of elasticity 200 GPa. Four plates are to be adopted. Also find the stiffness of the spring.

SOLUTION

In the present problem, limits on deflection and maximum stress are given.

$$\sigma_{\max} \le 400\,\text{MPa and } \delta \le 100\,\text{mm.}$$

A typical sketch of quarter elliptical leaf spring (cantilever type) is shown in Figure 11.17.

Given that $W = 100$ N

Span of the spring $= L = 500$ mm.

$$\sigma_{\max} = \frac{6WL}{nbt^2} \quad \text{and} \quad \delta = \frac{3WL^3}{Enbt^3}$$

$$\frac{\sigma_{\max}}{\delta} = \left\{ \frac{\dfrac{6WL}{nbt^2}}{\dfrac{3WL^3}{Enbt^3}} \right\}$$

$$\frac{\sigma_{\max}}{\delta} = \left\{ \frac{2Et^2}{L^2} \right\}$$

$$\frac{400}{100} = \left\{ \frac{2 \times 2 \times 10^5 t^2}{500^2} \right\}$$

$$\therefore \qquad t = 1.581\ \text{mm.}$$

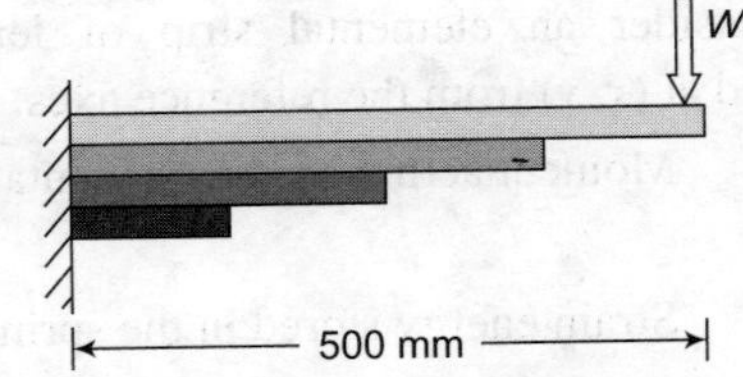

FIGURE 11.17 A typical quarter elliptical spring

1.581-mm-thick plates are not available, so adopt 2-mm-thick plates.

$$\sigma_{max} \leq 400$$

$$400 \geq \frac{6WL}{nbt^2} \quad \text{and} \quad b \geq \frac{6 \times 100 \times 500}{400 \times 4 \times 2^2} = 46.875 \text{ mm}$$

$$\delta \leq 100$$

$$b \geq \frac{3WL^3}{n\delta Et^3} = \frac{3 \times 100 \times 500^3}{4 \times 100 \times 2 \times 10^5 \times 2^3} = 58.59 \text{ mm}$$

Higher value of breadth is to be adopted, which is determined from above two considerations. Breadth $b \geq 58.59$ mm.

Adopt 60-mm-wide plates.

$$nb \geq \frac{3WL}{\sigma_{max} 2t^2} \text{ hence } b \geq \frac{3 \times 200 \times 600}{8 \times 150 \times 2 \times 1^2} = 150 \text{ mm.}$$

11.9 FLAT SPRINGS

Flat springs are energy-storing structural element, generally used in toys or clock machines, wherein energy is supplied to this spring to do a needful work. Thin steel flats are hinged at one end and wound in a spiral fashion with a spindle connected at the other end. Through the spindle, couple is applied. The spring stores the energy undergoing rotation due to the couple. A typical flat spring is shown in Figure 11.18.

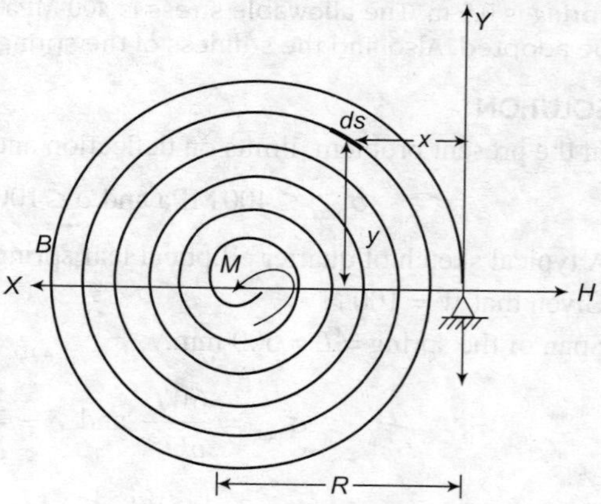

Let R be the radius of the spring. Let breadth and thickness of the flat of the spring be b and t, respectively. M is the couple applied at the spindle and the reactions developed at the other end are T and H, respectively.

FIGURE 11.18 A typical flat spring

Consider an elemental strip of length ds, located at (x, y) from the reference axes.

Moment acting on the elemental strip $M_s = Tx - Hy$

Strain energy stored in the spring is $U = \int \frac{M_s^2}{2EI} ds$.

Horizontal displacement of the spring at the hinged can be found from energy principle that partial derivative of strain energy with respect to force H gives the displacement in the direction of H. In the present flat spring, the horizontal displacement at the hinged end is zero.

$$\therefore \qquad \frac{dU}{dH} = 0$$

$$U = \int \frac{M_s^2}{2EI} ds = \int \frac{(Tx - Hy)^2}{2EI} ds$$

$$\frac{dU}{dH} = \int \frac{M_s}{EI} \frac{dM_s}{dH} ds = \int \frac{(Tx - Hy)}{EI} (-y) ds$$

$$\int \frac{(Tx - Hy)}{EI} (-y) ds = \int -\frac{Txy}{EI} ds + \int \frac{Hy^2}{EI} ds.$$

Here, an assumption that the wire of the spring is closely wound, then ds in the above integral can be taken as dA.

$$\int \frac{Txy}{EI} dA = \frac{TI_{xy}}{EI} = 0 \text{ (as } I_{xy} = 0\text{, for symmetrical section)}$$

$$\int \frac{Hy^2}{EI} dA = \frac{HI_{xx}}{EI};$$

As $\frac{dU}{dH} = 0$, this implies that $\int -\frac{Txy}{EI} ds + \int \frac{Hy^2}{EI} ds = 0.$. The first term is zero; hence second term should be equal to zero. Hence

$$\int \frac{Hy^2}{EI} dA = \frac{HI_x}{EI} = 0; \text{ or } H = 0.$$

To get the value of T, take moments about the spindle

$$M - TR = 0 \text{ or } T = \frac{M}{R}$$

Thus, maximum stress, at B, in the wire is given by $\sigma = \frac{6 \times T(2R)}{bt^2} = \frac{12M}{bt^2}$

To find the rotation '$\varnothing$' of the spindle due to couple M

$$\varnothing = \frac{\partial U}{\partial M};$$

$$M_s = (Tx - Hy) = \frac{M}{R} x \quad \text{as} \quad H = 0 \text{ and } T = \frac{M}{R}$$

$$\frac{dM_s}{dM} = \frac{x}{R} \; \varnothing = \frac{dU}{dM} = \int \frac{M_s}{EI} \frac{dM_s}{dM} ds = \int \frac{\frac{M}{R} x}{EI} \left(\frac{x}{R} \right) ds$$

$$\varnothing = \int \frac{M}{EI} \left(\frac{x}{R} \right)^2 ds = \int \frac{M}{EI} \left(\frac{x}{R} \right)^2 dA = \frac{M}{EI} \frac{1}{R^2} (I_y + AR^2)$$ in which A is area of the disc.

Here,
$$I_y = \frac{\pi R^4}{4} = \frac{AR^2}{4}$$

$$\varnothing = \frac{M}{EIR^2} \times \left\{ \frac{AR^2}{4} + AR^2 \right\} = 1.25 \frac{AM}{EI}$$

If the wire of the spring is closely spaced area of the spring A = length of the spring l.

Hence, the rotation of the spindle $\varnothing = 1.25 \frac{Ml}{EI}$

Thus, energy stored in the spring $U = \frac{1}{2} M\varnothing = \frac{5M^2 l}{8EI}$

Maximum stress in the wire of the spring $\sigma_{max} = \frac{12M}{bt^2}$.

PROBLEM 11.11 **Objective 4**

A flat spiral of radius 200 mm, length 1 m, 25 mm wide, and 1.2 mm thick is formed with steel, whose modulus of elasticity is 200 GPa. If the allowable stress is 800 MPa, determine the safe couple that can be applied on the spring. Determine the rotation of the spindle of the flat spring.

SOLUTION

Radius of the flat spring = R = 200 mm
Width of the flat = b = 25 mm
Thickness of the flat = t = 1.2 mm
Allowable bending stress in the flat = $\sigma_{max} \le 800\,\text{MPa}$
Let M be the couple that can be applied at the spindle.

We know that $\sigma_{max} = \frac{12M}{bt^2}$ and it is given that $\sigma_{max} = \frac{12M}{bt^2} \le 800$.

$$\frac{12M}{25 \times 1.2^2} \le 800 \Rightarrow M \le 2400\,\text{N.·mm}.$$

Rotation of the spindle = $\varnothing = 1.25 \frac{Ml}{EI}$

Length of the flat is 1 m.

$$\varnothing = 1.25 \frac{Ml}{EI} = \frac{1.25 \times 2400 \times 1000}{2 \times 10^5 \times \left(\frac{25 \times 1.2^3}{12} \right)\left(\frac{25 \times 1.2^3}{12} \right)} = 4.167\,\text{radians}$$

Energy stored in the spring $U = \frac{1}{2} M\varnothing = \frac{1}{2} \times 2400 \times 4.167 = 5000\,\text{N·mm}.$

SUMMARY

- **Helical springs are the thick spring wires coiled into a helix. They are of two types:**
  Close-coiled helical springs, and
  Open-coiled helical springs.
- **For springs in series, the effective stiffness,** $\frac{1}{K_e} = \frac{1}{K_1} + \frac{1}{K_2} + \cdots$.
- **For springs in parallel, the effective stiffness,** $K_e = K_1 + K_2 + \cdots$.
- **Extension of close-coiled spring,** $\Delta = \frac{64WR^3n}{Nd^4}$.
- **Extension of open-coiled spring,** $\Delta_0 = \frac{64WR^3n \sec\alpha}{d^4}\left\{\frac{\cos^2\alpha}{N} + \frac{2\sin^2\alpha}{E}\right\}$.
- **Maximum stress in a carriage spring is given by** $\sigma_{max} = \frac{6WL}{nbt^2}$.
- **Deflection at the midspan of a carriage spring** $\delta = \frac{3WL^3}{8Enbt^3}$.
- **Maximum stress in a flat spring is given by** $\sigma_{max} = \frac{12M}{bt^2}$.
- **Rotation of the spindle of a flat spring is given by** $\varnothing = 1.25\frac{Ml}{EI}$.

OBJECTIVE TYPE QUESTIONS

1. If the wire diameter of a close-coiled helical spring subjected to compressive load is increased from 1 cm to 2 cm, other parameters remaining same, then deflection will decrease by a factor of [GATE-2002]
   (a) 16 (b) 8 (c) 4 (d) 2
2. A compression spring is made of music wire of 2 mm diameter having a shear strength and shear modulus of 800 MPa and 80 GPa, respectively. The mean coil diameter is 20 mm, free length is 40 mm, and the number of active coils is 10. If the mean coil diameter is reduced to 10 mm, the stiffness of the spring is approximately [GATE-2008]
   (a) Decreased by 8 times (b) Decreased by 2 times
   (c) Increased by 2 times (d) Increased by 8 times
3. Two helical tensile springs of the same material and also having identical mean coil diameter and weight have wire diameters d and $d/2$. The ratio of their stiffness is [GATE-2001]
   (a) 1 (b) 4 (c) 64 (d) 128
4. A weighing machine consists of a 2 kg pan resting on spring. In this condition, with the pan resting on the spring, the length of the spring is 200 mm. When a mass of 20 kg is placed on the pan, the length

of the spring becomes 100 mm. For the spring, the undeformed length l_o and the spring constant k (stiffness) are [GATE-2005]

(a) $l_o = 220$ mm, $k = 1862$ N/m (b) $l_o = 210$ mm, $k = 1960$ N/m
(c) $l_o = 200$ mm, $k = 1960$ N/m (d) $l_o = 200$ mm, $k = 2156$ N/m

5. The deflection of a spring with 20 active turns under a load of 1000 N is 10 mm. The spring is made into two pieces each of 10 active coils and placed in parallel under the same load. The deflection of this system is [GATE-1995]

(a) 20 mm (b) 10 mm (c) 5 mm (d) 2.5 mm

6. Assertion (*A*): Concentric cylindrical helical springs are used to have greater spring force in a limited space.

Reason (*R*): Concentric helical springs are wound in opposite directions to prevent locking of coils under heavy dynamic loading. [IES-2006]

(a) Both *A* and *R* are individually true and *R* is the correct explanation of *A*
(b) Both *A* and *R* are individually true but *R* is NOT the correct explanation of *A*
(c) A is true but *R* is false
(d) A is false but *R* is true

7. Assertion (*A*): Two concentric helical springs used to provide greater spring force are wound in opposite directions.

Reason (*R*): The winding in opposite directions in the case of helical springs prevents buckling. [IES-1995; IAS-2004]

(a) Both *A* and *R* are individually true and *R* is the correct explanation of *A*
(b) Both *A* and *R* are individually true but *R* is NOT the correct explanation of *A*
(c) *A* is true but *R* is false
(d) *A* is false but *R* is true

8. Which one of the following statements is correct? [IES-1996; 2007; IAS-1997]

If a helical spring is halved in length, its spring stiffness

(a) Remains same (b) Halves (c) Doubles (d) Triples

9. A body having weight of 1000 N is dropped from a height of 10 cm over a close-coiled helical spring of stiffness 200 N/cm. The resulting deflection of spring is nearly [IES-2001]

(a) 5 cm (b) 16 cm (c) 35 cm (D) 100 cm

10. A close-coiled helical spring is made of 5 mm diameter wire coiled to 50 mm mean diameter. Maximum shear stress in the spring under the action of an axial force is 20 N/mm2. The maximum shear stress in a spring made of 3 mm diameter wire coiled to 30 mm mean diameter, under the action of the same force will be nearly [IES-2001]

(a) 20 N/mm^2 (b) 33.3 N/mm^2 (c) 55.6 N/mm^2 (d) 92.6 N/mm^2

11. A close-coiled helical spring is acted upon by an axial force. The maximum shear stress developed in the spring is t. Half of the length of the spring is cut off and the remaining spring is acted upon by the same axial force. The maximum shear stress in the spring of the new condition will be: [IES-1995]

(a) ½ *t* (B) *t* (c) 2 *t* (d) 4 *t*

12. The maximum shear stress occurs on the outermost fibers of a circular shaft under torsion. In a close-coiled helical spring, the maximum shear stress occurs on the [IES-1999]

(a) Outermost fibers (b) Fibers at mean diameter
(c) Innermost fibers (d) End coils

13. A helical spring has *N* turns of coil of diameter D, and a second spring, made of same wire diameter and of same material, has *N*/2 turns of coil of diameter 2D. If the stiffness of the first spring is *k*, then the stiffness of the second spring will be: [IES-1999]

(a) *k*/4 (b) *k*/2 (c) 2*k* (d) 4*k*

14. A close-coiled helical spring is subjected to a torque about its axis. The spring wire would experience a [IES-1996; 1998]

(a) Bending stress
(b) Direct tensile stress of uniform intensity at its cross-section
(c) Direct shear stress
(d) Torsional shearing stress

15. A close-coiled helical spring of 20 cm mean diameter is having 25 coils of 2 cm diameter rod. The modulus of rigidity of the material is 107 N/cm^2. What is the stiffness for the spring in N/cm? [IES-2004]

(a) 50 (b) 100 (c) 250 (d) 500

16. In the calculation of induced shear stress in helical springs, the Wahl's correction factor is used to take care of [IES-1995; 1997]

(a) Combined effect of transverse shear stress and bending stresses in the wire
(b) Combined effect of bending stress and curvature of the wire
(c) Combined effect of transverse shear stress and curvature of the wire
(d) Combined effect of torsional shear stress and transverse shear stress in the wire

17. While calculating the stress induced in a close-coiled helical spring, Wahl's factor must be considered to account for [IES-2002]

(a) The curvature and stress concentration effect
(b) Shock loading
(c) Poor service conditions
(d) Fatigue loading

18. Cracks in helical springs used in Railway carriages usually start on the inner side of the coil because of the fact that [IES-1994]

(a) It is subjected to the higher stress than the outer side
(b) It is subjected to a higher cyclic loading than the outer side
(c) It is more stretched than the outer side during the manufacturing process
(d) It has a lower curvature than the outer side

19. Wire diameter, mean coil diameter, and number of turns of a close-coiled steel spring are *d*, *D*, and N, respectively, and stiffness of the spring is *K*. A second spring is made of same steel but with wire diameter, mean coil diameter, and number of turns 2d, 2D, and 2N, respectively. The stiffness of the new spring is: [IES-1998; 2001]

(a) *K* (b) 2*K* (c) 4*K* (d) 8*K*

20. When a weight of 100 N falls on a spring of stiffness 1 kN/m from a height of 2 m, the deflection caused in the first fall is: [IES-2000]

(a) Equal to 0.1 m (b) Between 0.1 and 0.2 m
(c) Equal to 0.2 m (d) More than 0.2 m

21. When a helical compression spring is cut into two equal halves, the stiffness of each of the result in springs will be: [IES-2002; IAS-2002]

(a) Unaltered (b) Double (c) One-half (d) One-fourth

22. If a compression coil spring is cut into two equal parts and the parts are then used in parallel, the ratio of the spring rate to its initial value will be [IES-1999]

(a) 1 (b) 2
(c) 4 (d) Indeterminable for want of sufficient data

23. Assertion (*A*): Concentric cylindrical helical springs which are used to have greater spring force in a limited space are wound in opposite directions.

Reason (*R*): Winding in opposite directions prevents locking of the two coils in case of misalignment or buckling. [IAS-1996]

(a) Both A and R are individually true and R is the correct explanation of A
(b) Both A and R are individually true but R is NOT the correct explanation of A
(c) A is true but R is false
(d) A is false but R is true

24. Which one of the following statements is correct? [IES-1996; 2007; IAS-1997]

If a helical spring is halved in length, its spring stiffness
(a) Remains same (b) Halves (c) Doubles (d) Triples

25. A close-coiled helical spring has wire diameter 10 mm and spring index 5. If the spring contains 10 turns, then the length of the spring wire would be: [IAS-2000]
(a) 100 mm (b) 157 mm (c) 500 mm (d) 1570 mm

26. The springs of a chest expander are 60 cm long when unstretched. Their stiffness is 10 N/mm. The work done in stretching them to 100 cm is [IAS-1996]
(a) 600 N·m (b) 800 N·m (c) 1000 N·m (d) 1600 N·m

27. A block of weight 2 N falls from a height of 1 m on the top of a spring. If the spring gets compressed by 0.1 m to bring the weight momentarily to rest, then the spring constant would be: [IAS-2000]
(a) 50 N/m (b) 100 N/m (c) 200 N/m (d) 400N/m

28. The equivalent spring stiffness for the system shown in the given figure (S is the spring stiffness of each of the three springs) is: [IES-1997; IAS-2001]

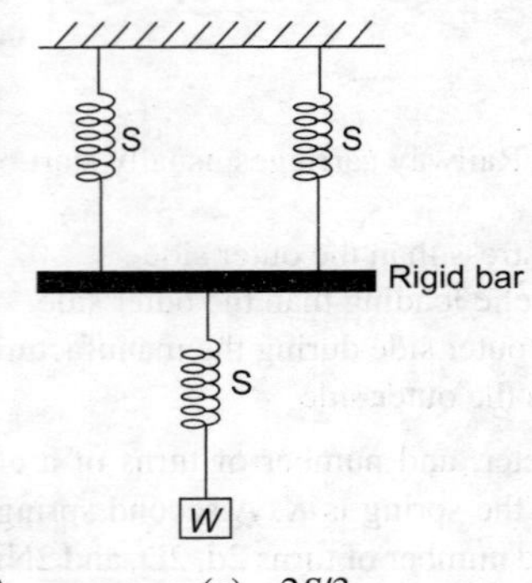

(a) $S/2$ (b) $S/3$ (c) $2S/3$ (d) S

Solutions for Objective Questions

Sl. No.	1.	2.	3.	4.	5.	6.	7.	8.	9.	10.
Answer	(a)	(d)	(c)	(b)	(d)	(b)	(c)	(c)	(b)	(c)

Sl.No.	11.	12.	13.	14.	15.	16.	17.	18.	19.	20.
Answer	(b)	(c)	(c)	(a)	(b)	(c)	(a)	(a)	(a)	(d)

Sl.No.	21.	22.	23.	24.	25.	26.	27.	28.
Answer	(b)	(c)	(a)	(c)	(d)	(b)	(d)	(c)

EXERCISE PROBLEMS

1. A close-coiled helical spring is to have a stiffness of 1000 N/m in compression with a max load of 50 N and a max shearing stress of 100 MPa. The length of the spring when coils are touching each other is 45 mm. Find the wire diameter, mean coil radius, and the number of coils. $N = 80$ GPa.
2. In a compound helical spring, the inner spring is arranged within and concentric with the outer one, but is 12 mm shorter. The outer spring has nine coils of mean diameter 26 mm and the wire diameter is 4 mm. Find the stiffness of the spring, if an axial load of 200 N causes the outer one to compress 20 mm. If the radial clearance between the springs is 1.5 mm, find the wire diameter of the inner spring, when it has eight coils. $N = 80$ GPa.
3. A composite spring has two close-coiled helical springs connected in series; each spring has 15 coils at a mean diameter of 40 mm. Find the wire diameter in one, if the other is 2.5 mm and the stiffness of the composite spring is 900 N/m. Estimate the greatest load that can be carried by the composite spring, and the corresponding extension, for a maximum shear stress of 200 MPa. $N = 80$ GPa.
4. A close-coiled helical spring of circular section extends 15 mm, when subjected to an axial load W, and there is an angular rotation of 1 rad, when torque T is independently applied about the axis. If R is the mean coil radius, obtain a relationship between the ratio of torque to moment in terms of mean coil, radius, and Poisson's ratio.
5. In a compound helical spring, the inner spring is arranged within and concentric with the outer one, but short by 10 mm. The outer spring has 10 coils of 30 mm mean diameter and 3 mm as the diameter of wire. Find the stiffness of inner spring, if an axial load of 100 N on the compound spring causes the outer spring to compress by 20 mm. If the radial clearance between the springs is 2 mm, find the wire diameter of inner spring, when it has eight coils.
6. Determine the maximum angle of helix for which the error in calculating the extension of a helical spring under axial load by the closely formula is less than 1%.
7. A coiled spring consists of 12 number of turns with mean coil radius 28 mm, wire diameter 10 mm, and helix angle 6°. This spring is subjected to an axial load of 300 N. Estimate the elongation of the spring under this load considering the spring as (i) close-coiled helical spring and (ii) open-coiled helical spring. Also estimate the percentage error in assuming the coil as close-coiled spring. $E = 200$ GPa; $\mu = 0.25$.

CHAPTER 12

COLUMNS AND STRUTS

UNIT OBJECTIVE

This chapter provides the information regarding types of columns, and its behavior in detail with neat sketches. It also provides theories to calculate the failure load of columns along with formulae and derivations. The presentation attempts to help the student achieve the following:

Objective 1: Find the stresses and applied load quantities of the columns.

Objective 2: Find the crippling and buckling for both long and short columns.

Objective 3: Apply Euler's and Rankine's formula for finding the failure loads.

Objective 4: Find the position of load of the given column.

12.1 INTRODUCTION

Columns are the structural elements that are subjected to axial compressive loads. Columns are referred as struts and stanchions also. Long columns (slender columns) when subjected to axial compressive load are failed by deflecting laterally undergoing bending rather than failed by direct compression; this type of column failure is called buckling. The phenomenon of buckling can be demonstrated by applying axial compressive load on slender bars (such as plastic rule or wooden rule). Buckling occurs in structures, which are subjected to compressive stress. The compression flange of a girder also buckles under bending load.

The axial load on a column that is corresponding to the buckling phenomenon is called critical load or buckling load or crippling load. If P_{cr} is the critical load of an axial loaded column, then the axial load on the column P can be

(a) $P < P_{cr}$—Column is in stable equilibrium condition and even if the column is slightly displaced in lateral direction the column comes backs to its stable configuration.

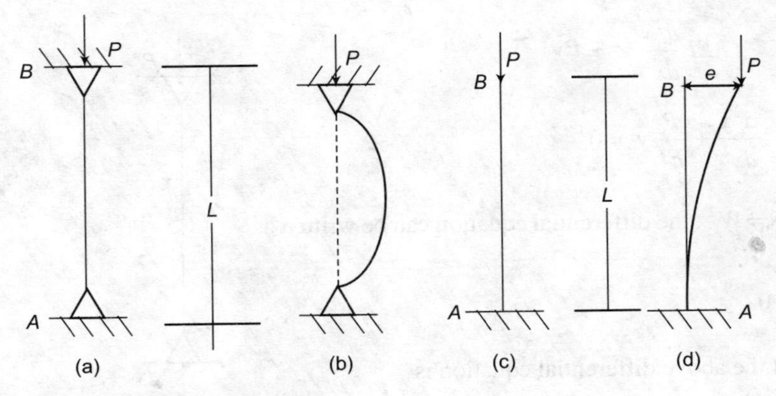

FIGURE 12.1 Buckling of columns

(b) $P = P_{cr}$—Column is in neutral equilibrium condition. There is a slight displacement in the lateral direction of the column in the new configuration. Column does not come back to original configuration.

(c) $P > P_{cr}$—Column is in unstable equilibrium; column continuously bends and becomes unserviceable.

The above three equilibrium configurations are analogous to the equilibrium of a bar placed on smooth surfaces shown in Figure 12.2.

The basic reason for buckling of a column is due to the in perfection inherently present in the column.

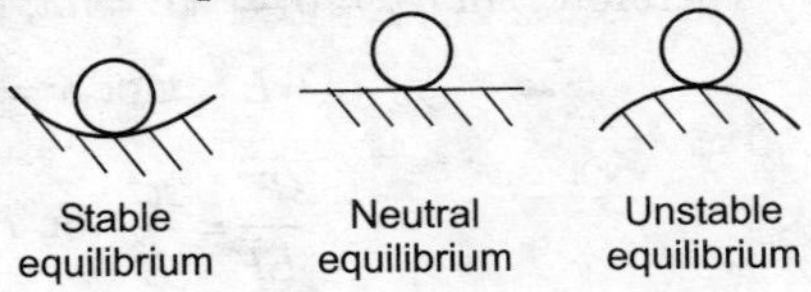

Stable equilibrium | Neutral equilibrium | Unstable equilibrium

FIGURE 12.2

12.2 EULER'S THEORY OF BUCKLING

Buckling phenomenon of slender columns was first investigated by Euler in 1744. The assumptions made for the analysis of column are:

(i) The material of the column is homogenous, isotropic, and elastic, and obeys Hook's law.
(ii) Axial load passes through the centroid of the column.
(iii) Column is perfectly straight.

12.2.1 Buckling Load of a Simply Supported Column (end conditions are hinged)

Consider a column AB of length 'L' subjected to axial load 'P' as shown in Figure 12.3 (a). Let EI be the flexural rigidity of the cross-section of the column. Let the laterally deformed configuration of the column as shown in Figure 12.3 (b) along length AB at any location 'x' distant from A and the lateral displacement of the column be y.

Recalling bending equation of the column in its deflected configuration, we have

$$EI\frac{d^2y}{dx^2} = -M_x \text{ and } M_x = Py$$

$$\Rightarrow \qquad EI\frac{d^2y}{dx^2} = -Py$$

$$\Rightarrow \qquad \frac{d^2y}{dx^2} + \frac{P}{EI}y = 0$$

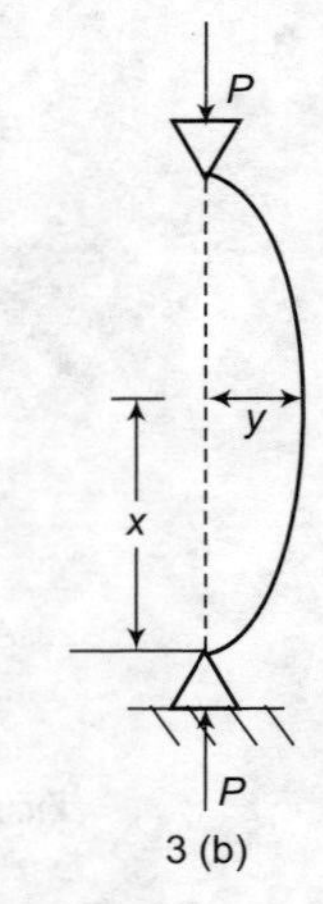

FIGURE 12.3

Taking $\frac{P}{EI} = W^2$, the differential equation can be written as

$$\frac{d^2y}{dx^2} + W^2y = 0.$$

Solution of the above differential equation is

$$y = C_1\cos wx + C_2\sin wx$$

Constants C_1 and C_2 can be obtained from the boundary conditions that at $x = 0$, $y = 0$ and at $x = L$, $y = 0$.

At $x = 0$, $y = 0 \Rightarrow C_1 = 0$

At $x = L$, $y = 0 \Rightarrow C_2\sin wL = 0$

In the above condition, $C_2 \neq 0$ (if $C_2 = 0$, $y = 0$ for all values of x which is not true).

Therefore, $\sin wL = 0 \Rightarrow wL = n\pi$, in which n is an integer taking $n = 1$.

$$wL = \pi \text{ or } w = \pi/L$$

$$\frac{P}{EI} = \frac{\pi^2}{L^2} \text{ or } P = EI\frac{\pi^2}{L^2}.$$

The above equation implies that at $P = EI\frac{\pi^2}{L^2}$ buckling of the column takes place.

Thus, the critical load $P_{cr} = EI\frac{\pi^2}{L^2}$. In the above expression, 'I' is the second moment area of cross-section. Therefore, for critical 'I' the second moment area shall be the least value and bending of the cross-section takes place about the axis about which second moment area is the least.

$$P_{cr} = \frac{\pi^2}{L^2}EI_{min}$$

$I_{min} = AK^2$ in which 'K' is the least radius of gyration of the cross-section.

$$\Rightarrow \qquad P_{cr} = \frac{\pi^2}{L^2}EAK^2 \text{ or } \frac{P_{cr}}{A} = \frac{\pi^2 E}{\left(\frac{L}{K}\right)^2}$$

L/K is called slenderness ratio denoted by λ.

Therefore, $\frac{P_{cr}}{A}$ = axial stress in the column that corresponds to the buckling. It is called critical stress denoted by σ_c.

$$\sigma_c = \frac{P_{cr}}{A} = \frac{\pi^2 E}{\lambda^2} \quad \Rightarrow \quad \sigma_c = \frac{\pi^2 E}{\lambda^2}.$$

The above equation indicates that for low values of slenderness ratio (λ), the critical stress will be high. A typical $\sigma_c - \lambda$ variation is shown in Figure 12.4 (taking $E = 200$ GPa).

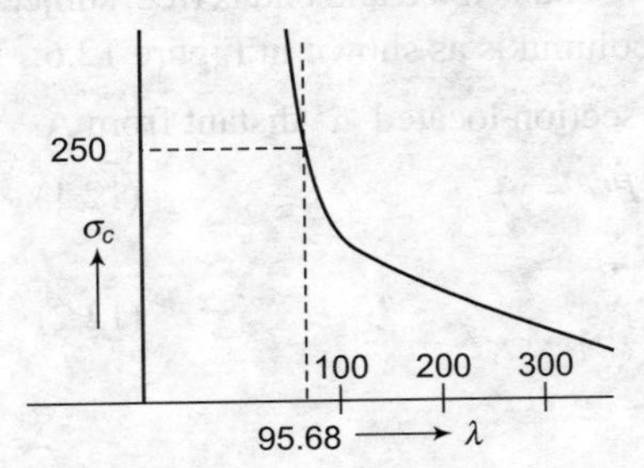

FIGURE 12.4 Typical critical stress–slenderness ratio variation

12.3 MODES OF FAILURE

In solving the differential equation, the solution of the column problem is given by

$$wL = n\pi$$

$$w = \frac{n\pi}{L} \text{ for } n = 1$$

$$w = \frac{\pi}{L} \text{ and } w^2 = \frac{P}{EI}$$

Therefore, $\frac{P}{EI} = \frac{\pi^2}{L^2}$ or $P = EI\frac{\pi^2}{L^2}$

If $n = 2$, $W = \frac{2\pi}{L}$

$$\Rightarrow \qquad P = \frac{4\pi^2 EI}{L^2}$$

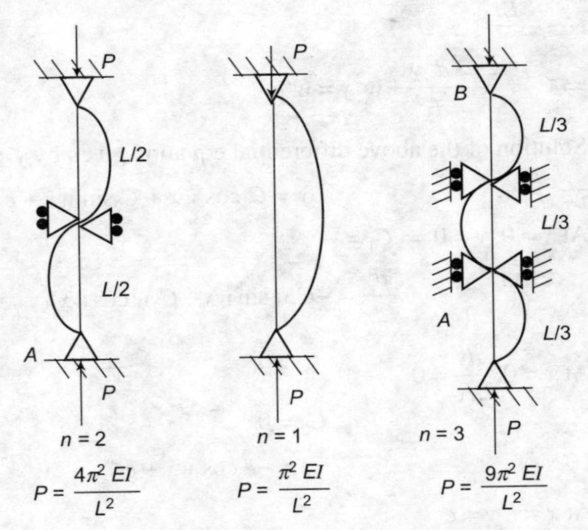

FIGURE 12.5 Different modes of buckling failure of columns

The above expression means that at $P = \frac{4\pi^2 EI}{L^2}$ also the column buckles. This is possible if the buckling possibility at prevented for $n = 1$ $\left(\text{i.e., } P = \frac{\pi^2 EI}{L^2}\right)$

From the above discussion, it can be stated that if $n = \infty$ and indicates that $P = \infty$ for buckling. This is true if the buckling is avoided by putting several supports along the length of the column. When several supports are present along length of the column, the column never buckles.

12.4 BUCKLING LOAD OF A CANTILEVER COLUMN

Consider a cantilever column one end is fixed and end is free, subjected to axial load 'P'. At buckling stage, the configuration of the column is as shown in Figure 12.6.

Bending moment (BM) at a section located 'x' distant from A.

$$M_x = -P(e - y) \tag{12.1}$$

When $M_x = -EI\dfrac{d^2y}{dx^2}$ (12.2)

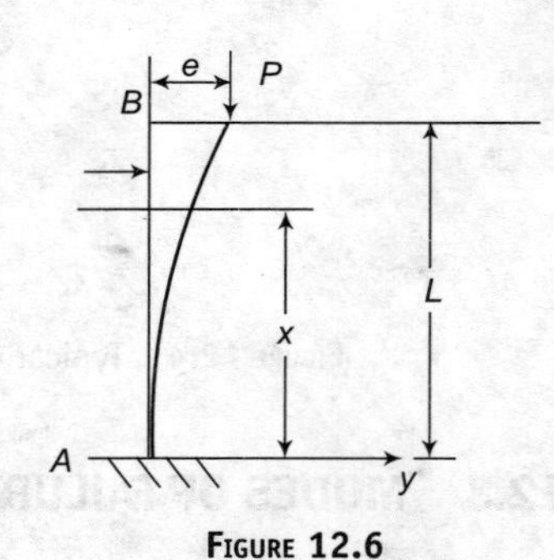

FIGURE 12.6

Equating Eqs. (1) and (2)

$$EI\frac{d^2y}{dx^2} = Pe - py$$

Take $\dfrac{P}{EI} = w^2$

$$\Rightarrow \qquad \frac{d^2y}{dx^2} + w^2y = w^2e \tag{12.3}$$

Solution of the above differential equation given by $y = C_1\cos wx + C_2\sin wx + e$.

$$y = C_1\cos wx + C_2\sin wx + e$$

At $x = 0$, $y = 0 \Rightarrow C_1 = -e$

$$\frac{dy}{dx} = -C_1 w\sin wx + C_2 w\cos wx$$

At $x = 0$, $\dfrac{dy}{dx} = 0$

$$C_2 = 0$$

$$y = -e\cos wx + e$$

At $x = l$, $y = e$

$$\Rightarrow \qquad e = -e\cos wL + e \Rightarrow \cos wL = 0$$

$$L = \frac{(2n+1)\pi}{2w}$$

For $n = 0$, $w = \dfrac{\pi}{2L}$

$$w^2 = \frac{\pi^2}{4L^2} = \frac{P}{EI} \text{ or } P = \frac{\pi^2 EI}{4L^2}$$

Thus, the buckling load of a column with one end fixed and other end free is given by $\dfrac{\pi^2 EI}{4l^2}$.

12.5 BUCKLING LOAD OF A FIXED COLUMN

Consider a column with both the ends fixed as shown in Figure 12.7.

Let 'M' be the fixed end moment developed at the fixed ends. Bending at a section located 'x' distant from A is given by

$$M_x = Py - M \tag{12.4}$$

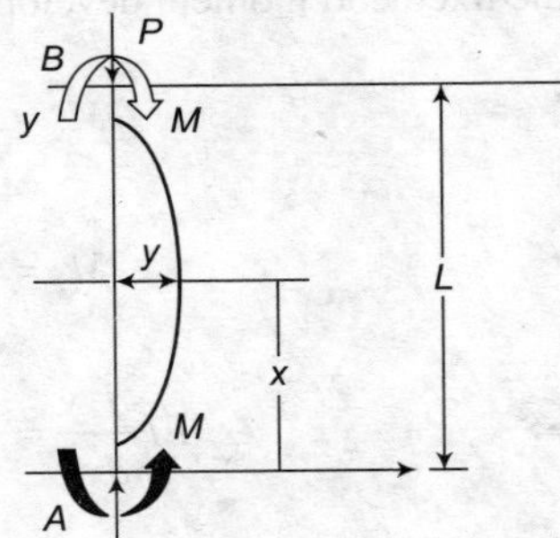

FIGURE 12.7

When $M_x = -EI\dfrac{d^2y}{dx^2}$ (12.5)

Equating Eqs. (1) and (2)

$$EI\frac{d^2y}{dx^2} = -Py + M \tag{12.6}$$

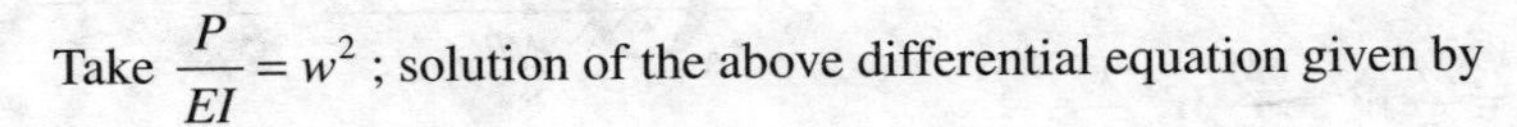

Take $\dfrac{P}{EI} = w^2$; solution of the above differential equation given by

$$\frac{d^2y}{dx^2} + w^2y = \frac{M}{P}w \tag{12.7}$$

$$y = C_1\cos wx + C_2\sin wx + \frac{M}{P}$$

$$\frac{dy}{dx} = -C_1w\sin wx + C_2w\cos wx$$

Substituting the boundary conditions that at $x = 0$, $y = 0$, at $x = 0$, $\dfrac{dy}{dx} = 0$, and at $x = L$, $y = 0$.

At $x = 0$, $y = 0 \Rightarrow C_1 + \dfrac{M}{P} = 0 \Rightarrow C_1 = -\dfrac{M}{P}$

At $x = 0$, $\dfrac{dy}{dx} = 0 \Rightarrow C_2 = 0$

$$y = -\frac{M}{P}\cos wx + \frac{M}{P}$$

At $x = L$, $y = 0 \Rightarrow 0 = -\dfrac{M}{P}\cos wL + \dfrac{M}{P}$

$\Rightarrow \qquad \cos wL = 1$

$\Rightarrow \qquad wL = 2n\pi$

$\Rightarrow$ for $n = 1$, $\qquad wL = 2\pi$

$\Rightarrow \qquad w = \dfrac{2\pi}{L}$

We know that $w^2 = \dfrac{P}{EI}$

$\Rightarrow \qquad P = \dfrac{4\pi^2 EI}{L^2}$

Thus, the crippling or buckling load of a column with fixed ends is given by $\dfrac{4\pi^2 EI}{L^2}$.

12.6 BUCKLING LOAD OF A COLUMN WITH ONE END FIXED AND THE OTHER HINGED

Consider the BM section located 'x' distant from A of a column shown in Figure 12.8. Let 'M' be the fixed end moment developed at A.

$$M_x = Py - M + \frac{M}{L}x \qquad (12.8)$$

$$M_x = -EI\frac{d^2y}{dx^2} \qquad (12.9)$$

$$\Rightarrow \qquad EI\frac{d^2y}{dx^2} = -Py + M - \frac{M}{L}x$$

Take $\frac{P}{EI} = w^2 \Rightarrow \frac{d^2y}{dx^2} + W^2y = \frac{M}{P}W^2 - \frac{M}{P}W^2\frac{x}{L}$ (12.10)

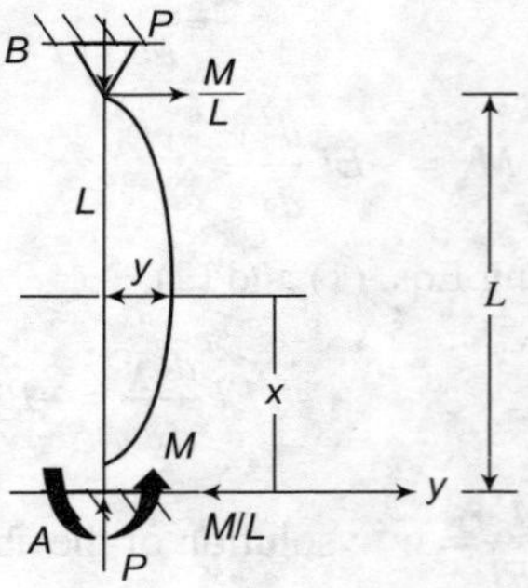

FIGURE 12.8 Buckling of a column with one end fixed and the other hinged

Solution of the above differential equation is given by

$$y = C_1\cos wx + C_2\sin wx + \left(\frac{M}{P} - \frac{M}{P}\frac{x}{L}\right).$$

Applying the boundary conditions

At $x = 0$, $y = 0 \Rightarrow C_1 = -\frac{M}{P}$

At $x = 0$, $\frac{dy}{dx} = 0 \Rightarrow \frac{dy}{dx} = -C_1w\sin wx + C_2w\cos wx - \frac{M}{PL}$

At $x = 0$, $\frac{dy}{dx} = 0 \Rightarrow C_2w = \frac{M}{PL}$ or $C_2 = \frac{M}{PLw}$

Therefore, $y = -\frac{M}{P}\cos wx + \frac{M}{PLw}\sin wx + \left(\frac{M}{P} - \frac{M}{P}\frac{x}{L}\right)$

At $x = l$, $y = 0 \Rightarrow -\frac{M}{P}\cos wL + \frac{M}{PLw}\sin wL = 0$

$$\Rightarrow \qquad \tan wL = wL \qquad (12.11)$$

Solution of the above equation yields the buckling load of the column.

Solution of the above equation is given by

$$wL = 4.49\ (wL \neq 0) \text{ or } w^2 = \frac{P}{EI} = \frac{20.16}{L^2}$$

$$\Rightarrow \qquad P = \frac{20.16EI}{L^2} = \frac{2\pi^2 EI}{L^2} \text{ or } P = \frac{\pi^2 EI}{\left(\frac{L}{\sqrt{2}}\right)^2}.$$

From the above four sections, it is clear that the critical or buckling load of a column depends on the end conditions for constant length and flexural rigidity (*EI*) of the column.

In general, the buckling load of a column is expressed as $P = \dfrac{\pi^2 EI}{l_e^{\,2}}$.

In the above expression, 'l_e' is effective length and it depends on the end conditions of the column. Table 12.1 gives the effective length of different columns.

Table 12.1 Effective length of columns with different end conditions

End Conditions	l_e (Effective Length)
(i) Both ends are hinged	 L
(ii) One end is fixed and other end is free	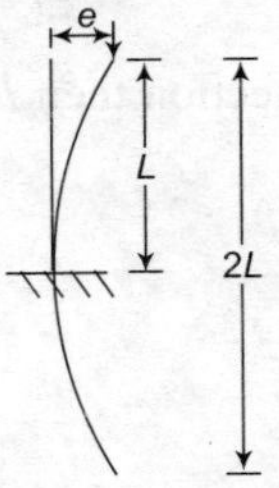 $2L$
(iii) Both ends are fixed	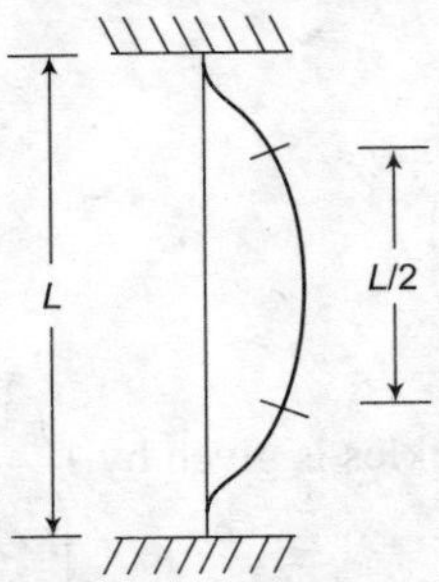 $\dfrac{L}{2}$
(iv) One end is fixed and other end is hinged	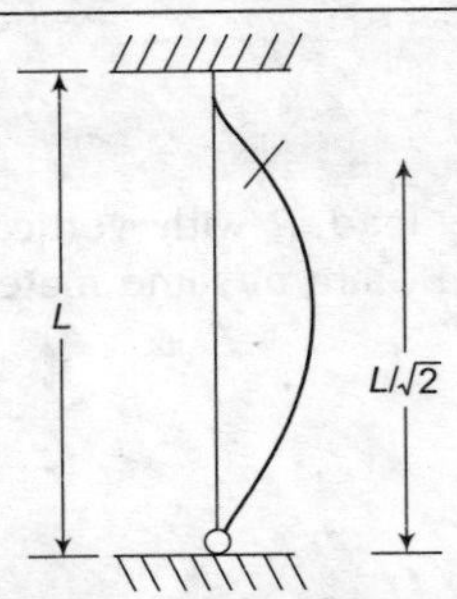 $\dfrac{L}{\sqrt{2}}$

For a general cross-section of a column, several values of radius of gyration exist and these values depend on the axis of bending. Thus, for least of $\frac{P}{A}$, for which buckling occurs, it depends only on (r_{min}) minimum radius of gyration. Thus, the buckling load of a column is given by

$$\frac{P}{A} = \frac{\pi^2 E}{\left(\frac{l_e}{r_{min}}\right)^2}.$$

Always a column buckles about an axis about which radius of gyration is minimum. The above expression is represented as if $\sigma_{cc} = \frac{\pi^2 E}{\lambda^2}$, in which λ = ratio of effective length of column to the least radius of gyration. λ is called slenderness ratio.

PROBLEM 12.1 **Objective 1**

Determine the stress at which a column with hinged ends buckles.

SOLUTION

The buckling load of a column is given by $P = \frac{\pi^2 EI}{l_e^2}$.

If r = radius of gyration of the cross-section then $I = Ar^2$, in which A is the cross-sectional area of the column

$$P = \frac{\pi^2 EAr^2}{l_e^2}$$

$$\frac{P}{A} = \frac{\pi^2 Er^2}{l_e^2}$$

or

$$\frac{P}{A} = \frac{\pi^2 E}{\left(\frac{l_e}{r}\right)^2}.$$

The stress at which the column buckles is given by $\frac{\pi^2 E}{\left(\frac{l_e}{r}\right)^2}$.

PROBLEM 12.2 **Objective 1**

Determine the inclination of the load P with vertical, such that columns AB and BC buckle simultaneously. Members AB and BC are of same material and same cross-section. AB = 3 m and BC = 4 m.

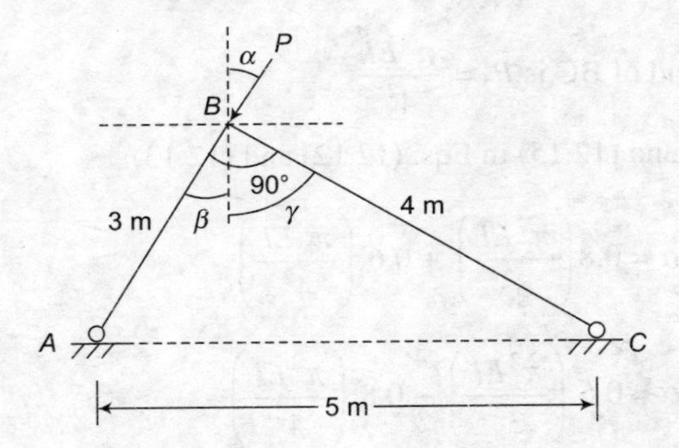

FIGURE 12.9

SOLUTION

Let α be the inclination of the load P with vertical. Load P causes compressive force in members AB and BC; let P_1 and P_2 be the compressive forces developed in AB and BC, respectively. Inclination of AB with vertical: angle $B = 90°$, angle A = $\tan^{-1}(4/3) = 53.13°$ and angle C = 36.87°; $\beta = 36.87°$ and $\gamma = 53.13°$.

To find the force in the members AB and BC resolve the forces vertically

$$P \cos \alpha = P_1 \cos \beta + P_2 \cos \gamma$$

$$P \cos a = P_1\, 0.8 + P_2\, 0.6 \tag{12.12}$$

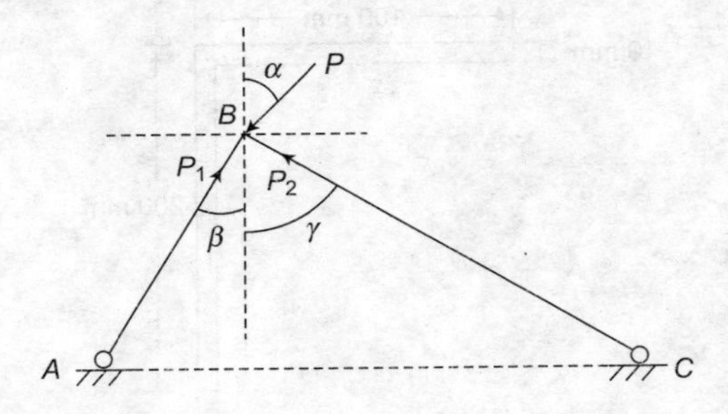

FIGURE 12.10

Resolving forces horizontally

$$P \sin \alpha = P_1 \sin \beta - P_2 \sin \gamma = P_1\, 0.6 - P_2\, 0.8 \tag{12.13}$$

Above two equations are not sufficient to find the value of α. The other condition is that the members AB and BC should buckle simultaneously.

Buckling load of $AB = \dfrac{\pi^2 EI}{L_{AB}^{\;2}}$ (ends of the column are hinged)

$$P_1 = \frac{\pi^2 EI}{3^2} \tag{12.14}$$

Similarly, the buckling load of BC is $P_2 = \dfrac{\pi^2 EI}{4^2}$ (12.15)

Substituting Eqs. (12.14) and (12.15) in Eqs. (12.12) and (12.13)

$$P \cos \alpha = 0.8\left(\frac{\pi^2 EI}{3^2}\right) + 0.6\left(\frac{\pi^2 EI}{4^2}\right)$$

$$P \sin \alpha = 0.6\left(\frac{\pi^2 EI}{3^2}\right) - 0.8\left(\frac{\pi^2 EI}{4^2}\right)$$

$$\tan \alpha = \frac{\dfrac{0.6}{3^2} - \dfrac{0.8}{4^2}}{\dfrac{0.8}{3^2} + \dfrac{0.6}{4^2}} = \frac{0.0167}{0.1264} = 0.132$$

$$\Rightarrow \quad \alpha = 7°30'$$

PROBLEM 12.3 **Objective 2**

Determine the buckling load of an angular column cross-section shown in Figure 12.11. Length of the column is 4 m and ends of the column are hinged. Take $E = 200$ GPa.

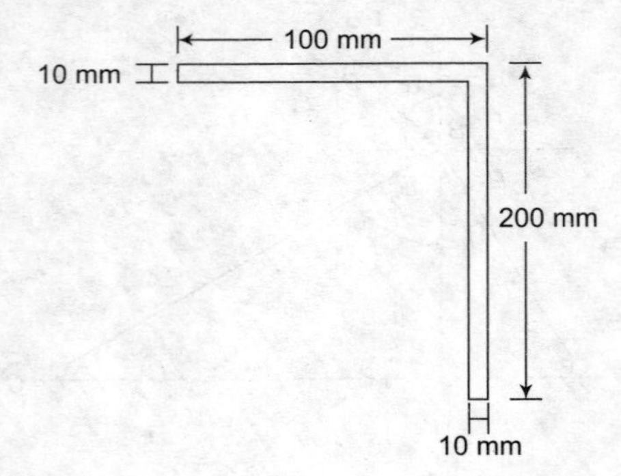

FIGURE 12.11

SOLUTION

Buckling load of the column $P = \dfrac{\pi^2 EI}{l_e^2}$

As ends of the column are hinged $l_e = L = 4$ m $= 4000$ mm

In the above expression, 'I' is the least second moment of the area.

Determination of Least Second Moment of the Area

Cross-sectional area of the column = (100 × 10) + (190 × 10) = 2900 mm^2

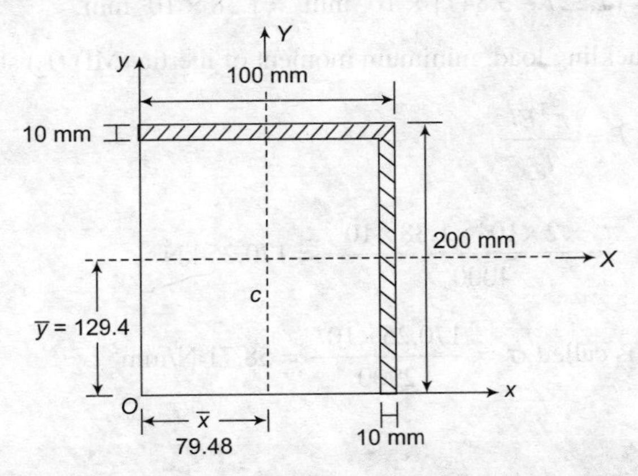

FIGURE 12.12

Coordinates of the centroid with reference to *ox–oy* axis

$$\overline{x} = \frac{(100\times10)50+(190\times10)95}{(100\times10)+(190\times10)} = 79.48 \text{ mm}$$

$$\overline{y} = \frac{(100\times10)195+(190\times10)95}{(100\times10)+(190\times10)} = 129.48 \text{ mm}$$

I_{cx} = second moment of area about *x* axis passing through centroid

$$= 100\times\frac{10^3}{12}+100\times10(195-129.48)^2+10\times\frac{190^3}{12}+10\times190\times(129.48-95)^2$$

$$= 12.276 \times 10^6 \text{ mm}^4$$

$$I_{cy} = 10\times\frac{100^3}{12}+100\times10(79.48-50)^2+190\times\frac{10^3}{12}+10\times190\times(95-79.48)^2$$

$$= 2.177 \times 10^6 \text{ mm}^4$$

I_{cxy} = Product of inertia about centroidal *X* and *Y* axis

$$= 100\times10(50-79.48)(195-129.48)+190\times10(95-79.48)(95-129.48)$$

$$= -2.9482 \times 10^6 \text{ mm}^4$$

I_1 and I_2 are the principal moments of inertia

$$I_{1,2} = \frac{I_{cx}+I_{cy}}{2} \pm \sqrt{(\frac{I_{cx}-I_{cy}}{2})^2 + I_{xy}{}^2}$$

$$I_1 = \left(\frac{12.276+2.177}{2}\right)\times10^6+10^6\sqrt{\left(\frac{12.276-2.177}{2}\right)^2+(-2.9482)^2}$$

$$= \{7.227 + 5.847\} \times 10^6 \text{ mm}^4 = 13.073 \times 10^6 \text{ mm}^4$$

$$I_2 = \{7.227 - 5.847\} \times 10^6 \text{ mm}^4 = 1.38 \times 10^6 \text{ mm}^4$$

For determination of buckling load, minimum moment of inertia (MI) (I_2) should be used.

$$\text{Buckling load } P = \frac{\pi^2 EI}{l_e^2}$$

$$P = \frac{\pi^2 \times 2\times 10^5 \times 1.38\times 10^6}{4000^2} = 170.25 \text{ kN}$$

$$\text{Stress at buckling load is called } \sigma_c = \frac{170.25\times 10^6}{2900} = 58.71 \text{ N/mm}^2$$

PROBLEM 12.4

Objective 2

Determine the buckling load of a hollow circular tube of mean diameter 100 mm, thickness 8 mm, and length 5 m. The tube is hinged at both ends take E = 200 GPa if four similar tubes are arranged as shown in Figure 12.13. What is the buckling load? Comment on the increase in the buckling load with this type arrangement.

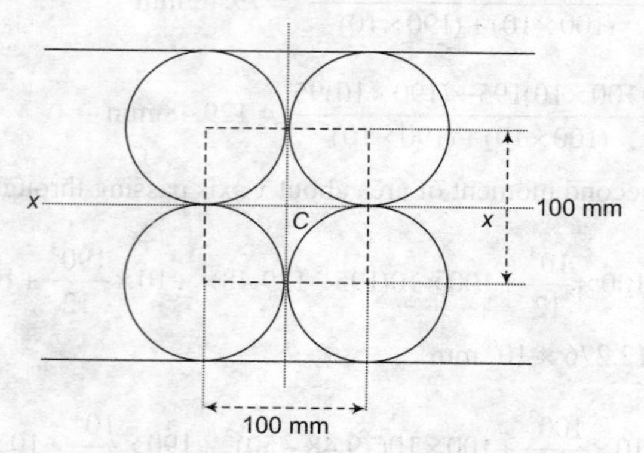

FIGURE 12.13

SOLUTION

Both ends of the column are hinged. Hence the effective length of the column is 5 m.

$$\text{For a single tubular column, } P = \frac{\pi^2 EI}{l_e^2}$$

$$I = \frac{\pi d^3 t}{8} = \frac{\pi \times 100^3 \times 8}{8} = 3.141\times 10^6 \text{ mm}^4$$

$$P = \frac{\pi^2 \times 2\times 10^5 \times 3.141\times 10^6}{5000^2} = 248 \text{ kN}$$

Stress at which the column buckles $\sigma_{cc} = \dfrac{248 \times 10^3}{\pi \times 100 \times 8} = 98.67\ \text{N/mm}^2$

Buckling load of four columns is arranged as shown in Figure 12.13.

$$I_{cx} = I_{cy} = \frac{\pi d^3 t}{8} \times 4 + 4 \times \pi d t \times \left(\frac{d}{2}\right)^2 = 3.141 \times 10^6 \times 4 + 4 \times \pi \times 8 \times 100 \times 50^2 = 37.699 \times 10^6\ \text{mm}^4$$

$$\text{Buckling load } P = \frac{\pi^2 EI}{l_e^2} = \frac{\pi^2 \times 2 \times 10^5 \times 37.699 \times 10^6}{5000^2} = 2976.59\ \text{kN.}$$

Points to be observed:

Buckling load of single column: 248 kN
Buckling load of four columns arranged as mentioned: 2976.59 kN
This implies that the hollow tubes when arranged properly increased the buckling load by 12 times, resulting in a great improvement in the load carrying capacity. For increasing the column strength, material of the column cross-section shall be spread away from the center of cross-section so that MI increases. But to achieve this increase in buckling load, four columns shall be connected by welding along the length of the column.

PROBLEM 12.5 **Objective 2**

Two channel sections are placed back to back as shown in Figure 12.14. Determine the spacing between the channels so that the buckling load is maximum. Determine the buckling loads if $E = 200$ GPa. Length of the column is 8 m. One end of the column is hinged while the other is fixed.

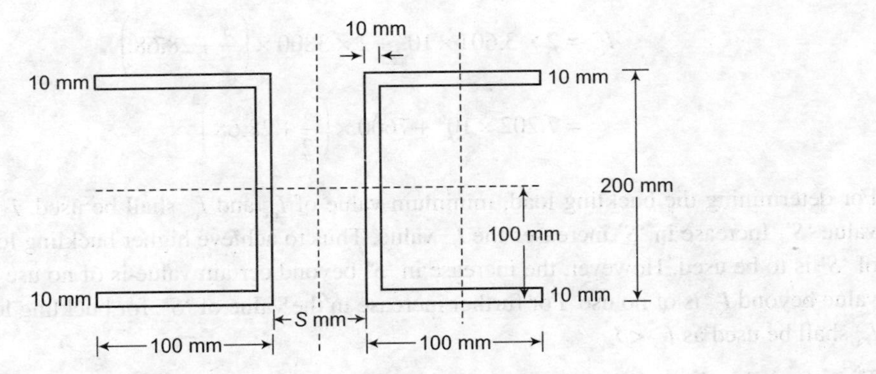

FIGURE 12.14

SOLUTION

Let $\bar{x}$ be the centroid of the channel section

$$\bar{x} = \frac{100 \times 10 \times 50 \times 2 + 180 \times 10 \times 5}{2 \times 100 \times 10 + 180 \times 10} = 28.68\ \text{mm}$$

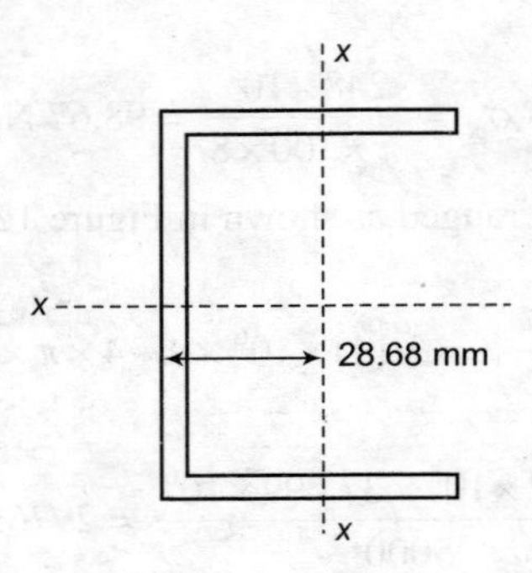

FIGURE 12.15

$$A = 100 \times 10 \times 2 + 180 \times 10 = 3800\,\text{mm}^2$$

$$I_x = \frac{200^3 \times 100}{12} - \frac{180^3 \times 90}{12}$$
$$= 22.9267 \times 10^6\ \text{mm}^4$$

$$I_y = \frac{2 \times 10 \times 100^3}{3} + \frac{10^3 \times 180}{3} - 3800 \times 28.68^2$$
$$= 3.601 \times 10^6\ \text{mm}^4$$

For the total combined section,

$$I_{cx} = 2 \times 22.9267 \times 10^6\ \text{mm}^4$$
$$= 45.8534 \times 10^6\ \text{mm}^4$$

$$I_{cy} = 2 \times 3.601 \times 10^6 + 2 \times 3800 \times \left(\frac{s}{2} + 28.68\right)^2$$
$$= 7.202 \times 10^6 + 7600 \times \left(\frac{s}{2} + 28.68\right)^2$$

For determining the buckling load, minimum value of I_{cx} and I_{cy} shall be used. I_{cy} depends on the value 'S'. Increase in 'S' increases the I_{cy} value. Thus to achieve higher buckling load, higher value of 'S' is to be used. However, the increase in 'S' beyond certain value is of no use as increase is I_{cy} value beyond I_{cx} is of no use. For further increase in the value of 'S', for buckling load calculations, I_{cx} shall be used as $I_{cx} < I_{cy}$.

Thus, for achieving maximum value of buckling load

$$I_{cx} = I_{cy}$$

$$\Rightarrow \quad 45.8534 \times 10^6 = 7.202 \times 10^6 + 7600 \times \left(\frac{S}{2} + 28.68\right)^2$$

$$\Rightarrow \quad s = 85.26\ \text{mm}$$

Adopt 90 mm distance between the channels. Use I_{cx} for calculation of buckling load.

Therefore, buckling load $P = \dfrac{\pi^2 EI_{cx}}{L_e^2}$

$$L_e = 0.707\,L$$
$$= 0.707 \times 8$$
$$= 5.656 \text{ m}$$
$$P = \frac{\pi^2 \times 2 \times 10^5 \times 45.8534 \times 10^6}{(5.656 \times 1000)^2}$$
$$= 2829.3 \text{ kN}$$

For keeping the two channels together, plates or bars are welded or bolted to the channels along the length of the column. Such a system is called lacing and battening, the details of which can be in any text book of design of steel structures.

12.7 LIMITATIONS OF EULER'S THEORY

Euler's buckling theory has got certain limitations. For example in the expression for buckling load $P = \dfrac{\pi^2 EI}{l_e^2}$, if effective length of the column is reduced the value of 'P' increases, thus the stress at buckling load P/A also increases. If effective length is reduced beyond certain value, the stress in the member (P/A) crosses the yield stress of the material. Thus, the member fails by yield before it buckles. This is the limitation of Euler's theory. Thus, Euler's buckling theory can explain the failure of columns of short length. The following are the limitations of Euler's theory.

1. Euler's theory is applicable for columns which are perfectly straight and the axial load should pass through centroid of the cross-section. In most of the practical case, this condition does not exist.
2. Euler's theory cannot explain the yielding failure of short column. This theory is applicable to only long columns.

   Stress at which column buckles $(\sigma_c) = \dfrac{\pi^2 EA}{\lambda^2}$, in which λ is slenderness ratio.

A typical variation of buckling stress σ_c and slenderness ratio of column is shown in Figure 12.16.

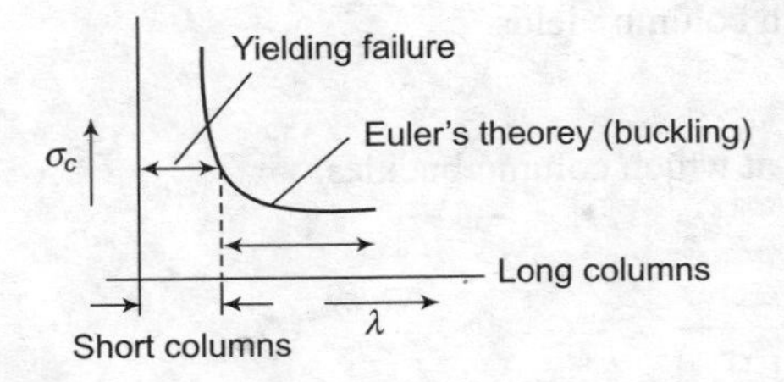

FIGURE 12.16

From Figure 12.16, it is clear that short columns fail by yielding, whereas the long columns fail by buckling. This aspect is taken care in Rankine–Gordon formula.

12.8 RANKINE–GORDON FORMULA

This formula takes into account the buckling failure as well as yielding failure of a column into a single expression of 'P' is the failure load of a column. If P_c is the buckling load of the column and P_y be the yielding load, then the failure load 'P' is given by

$$1/P = (1/P_c) + (1/P_y) \tag{12.16}$$

$$P_c = \text{buckling load} = \frac{\pi^2 EI}{l_e^2}$$

$$P_y = \text{yielding load} = \sigma_y \cdot A$$

σ_y = yield stress of the material of column

Eq. (12.16) can be further expanded as

$$P_c = \frac{\pi^2 EI}{l_e^2} = \frac{\pi^2 EA}{\left(\frac{l_e}{k}\right)^2} = \frac{\pi^2 EA}{\lambda^2}$$

$$P_y = \sigma_y . A$$

Substituting P_y and P_c values in Eq. (12.9.1)

$$P = \frac{P_C \times P_y}{P_C + P_v}$$

$$\Rightarrow \quad \frac{P}{A} = \frac{\frac{P_c}{A} \times \frac{P_y}{A}}{\frac{P_c}{A} + \frac{P_y}{A}}$$

$\sigma = \frac{P}{A}$ = stress at which column fails

$\sigma_y = \frac{P_y}{A}$ = stress at which column yields

$\sigma_c = \frac{P_c}{A} = \frac{\Pi^2 E}{\lambda^2}$ = stress at which column buckles.

Therefore, $\frac{\sigma_c \times \sigma_y}{\sigma_c + \sigma_y} = \frac{\sigma_y}{1 + \left(\frac{\sigma_y}{\sigma_c}\right)}$

$$= \frac{\sigma_y}{1+\left(\frac{\sigma_y}{\Pi^2 E}\right)\lambda^2}$$

For a material, $\frac{\sigma_y}{\Pi^2 E}$ is constant.

Hence, $\sigma = \frac{\sigma_y}{1+\alpha\lambda^2}$.

or $P = \frac{\sigma_y \cdot A}{1+\alpha\lambda^2}$.

In the above formula, if λ is very small, that is, for a short column, the denominator approaches 1 and hence the failure load 'P' approaches $P_c = (\sigma_y A)$.

If λ is high, then the denominator increases and $\sigma_y A$ reduces to P_c.

For predicting the buckling load of a column, few empirical formula available in the literature are given below.

1. Straight line formula

$$P = \sigma_y A\{1-\alpha_1\lambda\}$$

$$\alpha_1 = \frac{\sigma_y}{4\pi^2 E}$$

2. Johnson's parabolic formula

$$P = \sigma_y\ A\{1-\alpha_2\lambda^2\}$$

α_1 = constant.

PROBLEM 12.6 **Objective 3**

Determine the Rankine's failure load and failure stress of a hollow circular column of outer diameter 50 mm and inner diameter 38 mm. Column length is 3 m. Ends of the column are fixed. Yield stress and elastic modulus of the material of the column is 320 MPa and 200 GPa. Also find the length of the column, for which stress at Euler's load is equal to yield stress.

SOLUTION

Length of the column = 3 m

Ends of the column are fixed

$$L_e = L/2 = 1.5 \text{ m}$$

$$\text{Euler's load} = P_e = \frac{\pi^2 EI}{l_e^2}$$

$$I = \frac{\pi}{64}\{d_o^4 - d_i^4\}$$

$$= \frac{\pi}{64}\{50^4 - 38^4\} = 0.204 \times 10^6 \text{ mm}^4$$

$$P_c = \frac{\pi^2 \times 2 \times 10^5 \times 0.204 \times 10^6}{1500^2} = 178.97 \text{ kN}$$

The stress at which the column yields = P_y

$$P_y = \sigma_y A$$

$$= \sigma_y \times \frac{\pi}{4}\{d_o{}^4 - d_i{}^4\}$$

$$= 320 \times \frac{\pi}{4}\{50^4 - 38^4\} = 265.40 \text{ kN}$$

The failure load

$$\frac{1}{P} = \frac{1}{P_c} + \frac{1}{P_y}$$

$$\Rightarrow \quad \frac{1}{P} = \frac{1}{265.4} + \frac{1}{178.94}$$

$$\Rightarrow \quad P = 106.88 \text{ kN}$$

$$\text{Failure stress} = \frac{106.88 \times 10^3}{\frac{\pi}{4}(50^2 - 38^2)} = 128.86 \text{ N/mm}^2$$

Let 'l' be the effective length of the column for which yield load and buckling load are same.

$$\Rightarrow \quad \sigma_A A = \frac{\pi^2 EI}{l_e^2}$$

Or

$$l_e^2 = \frac{\pi^2 E}{\sigma_A}\left\{\frac{I}{A}\right\}$$

$$l_e^2 = \frac{\pi^2 \times 2 \times 10^5}{320}\left\{\frac{\frac{\pi}{64}(d_o{}^4 - d_i{}^4)}{\frac{\pi}{4}(d_o{}^2 - d_i{}^2)}\right\}$$

$$= \frac{\pi^2 \times 2 \times 10^5}{320}\left\{\frac{d_o{}^2 + d_i^2}{16}\right\}$$

$$= 1.5205 \times 10^5 \text{ mm}^2$$

$$\Rightarrow \quad l_e = 1233 \text{ mm}$$

$\Rightarrow$ Length of the column = $2 \times l_e$

$$= 2.466 \text{ m}$$

PROBLEM 12.7 **Objective 3**

A straight column of length 6 m, with hinged ends, is allowed to buckle due to the application of axial load. Cross-section of the column is 200 mm wide, 100 mm thick, and wall thickness 10 mm. Considering the buckling to take place as per Euler's formula, determine the maximum central deflection for maximum stress in the highly compressed fiber to be 300 MPa. Take $E = 200$ GPa.

SOLUTION

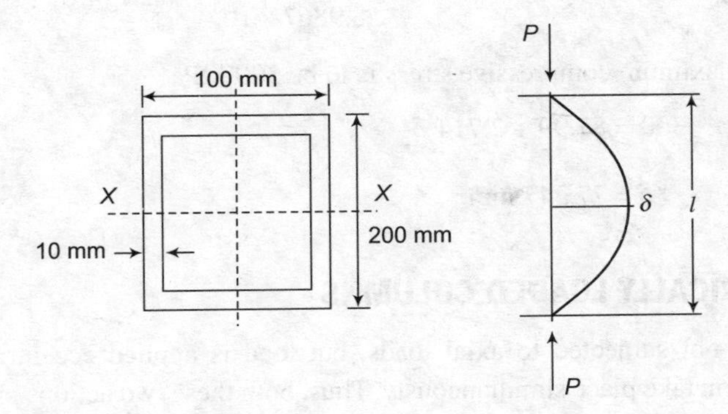

FIGURE 12.17

$L = 6$ m

$l_e = 6$ m as both the ends of the column are hinged

If 'δ' be the maximum deflection of the column at mid height, the midspan is subjected to axial load 'P' and BM, $M = P\delta$.

$$\text{Axial stress} = P/A$$

$$\text{Bending stress} = \frac{M}{I_{\min}} \cdot x_{\max}.$$

(I_{yy} should be used, as buckling takes place about YY axis of the cross-section since $I_{yy} < I_{xx}$)

$$I_{xx} = \frac{(200 \times 100^3 - 180 \times 80^3)}{12}$$

$$= 27.7867 \times 10^6 \text{ mm}^4$$

$$I_{yy} = \frac{(200 \times 100^3 - 180 \times 80^3)}{12}$$

$$= 8.9867 \times 10^6 \text{ mm}^4$$

Buckling load P is given by

$$P = \frac{\pi^2 EI_{YY}}{l_e^2}$$

$$P = \frac{(\pi^2 \times 2 \times 10^5 \times 8.9867 \times 10^6)}{(6000^2)}$$

$$= 492.75 \text{ kN}$$

$$\text{Axial compressive stress} = \frac{P}{A} = \frac{492750}{200 \times 100 - 180 \times 80} = 87.99 \text{ N/mm}^2$$

$$\text{Maximum compressive stress due to bending} = \frac{P\delta}{I_{yy}} \times 50$$

$$= \frac{492750 \times 50\delta}{8.9867 \times 10^6} = 2.714\,\delta$$

It is given that the maximum compressive stress is to be 300 MPa

$$300 = 87.99 + 2.714\,\delta$$

$$\delta = 77.347 \text{ mm}$$

12.9 ECCENTRICALLY LOADED COLUMNS

When columns are not subjected to axial loads, but load is applied eccentrically buckling and bending of the column take place simultaneously. Thus, both these two actions produce compressive stress in the column. Failure of column takes place when the maximum compressive stress reaches the yield stress of the material. Based on this hypothesis, secant formula for failure load of columns is derived.

Secant Formula

Consider a column AB, subjected to eccentric load 'P' acting at a distant 'e' from the centroid of the cross-section of the column.

BM at any section 'x' distant from A is given by

$$M_x = P(e + y)$$

We know that $M_x = -EI\dfrac{d^2y}{dx^2}$

$$\Rightarrow \qquad EI\frac{d^2y}{dx^2} = -Pe - Py$$

Taking $\dfrac{P}{EI} = w^2$

The above differential equation can be written as

$$\frac{d^2y}{dx^2} + w^2 y = w^2(-e)$$

Solution of this differential equation is $y = C_1 \cos wx + C_2 \sin wx - e$.

We know that,

At $x = 0$; $y = 0$

$$\Rightarrow \qquad C_1 = e$$

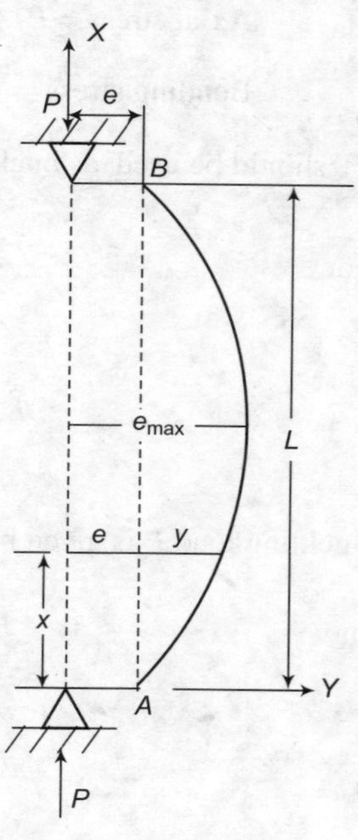

FIGURE 12.18

At $x = l$; $\quad y = 0$

$\Rightarrow e \cos wl + C_2 \sin wl - e = 0$

$$\Rightarrow \quad C_2 = \frac{e - e\cos wl}{\sin wl}$$

$$\therefore \quad y = e\cos wx + \frac{e(1-\cos wl)}{\sin wl}\sin wx - e$$

At $x = \dfrac{l}{2}$; $\quad y = y_{\max}$

$$\Rightarrow \quad y_{\max} = e\cos\frac{wl}{2} + \frac{e(1-\cos wl)}{\sin wl}\sin\frac{wl}{2} - e$$

$$\Rightarrow \quad y_{\max} = e\cos\frac{wl}{2} + e\frac{2\sin^2\dfrac{wl}{2}}{2\sin\dfrac{wl}{2}\cos\dfrac{wl}{2}}\sin\frac{wl}{2} - e$$

$$= e\left\{\frac{\cos^2\dfrac{wl}{2} + \sin^2\dfrac{wl}{2}}{\cos\dfrac{wl}{2}}\right\} - e$$

$$\Rightarrow \quad y_{\max} = e\left\{\sec\frac{wl}{2}\right\} - e$$

Maximum eccentricity occurs at mid height

$$e_{\max} = e + y_{\max}$$

$$= e\sec\frac{wl}{2}.$$

For this maximum eccentricity, the total stress in the column cross-section is given by

$$\sigma_{\max} = \frac{P}{A} + \frac{Pe_{\max}}{I}y_c$$

y_c is the location of highly compressed fiber from centroidal axis about axis of bending.

$$\sigma_{\max} = \frac{P}{A} + \frac{Pe\sec\dfrac{wl}{2}}{Ar^2}y_c$$

in which 'r' is radius of gyration about axis of bending.

$$\Rightarrow \quad \sigma_{\max} = \frac{P}{A}\left\{1 + \frac{e\sec\dfrac{wl}{2}}{r^2}y_c\right\}$$

Equating σ_{max} to yield stress σ_y of the column, the buckling load 'P' can be obtained by trial and error procedure.

$$\therefore \quad \sigma_y = \frac{P}{A}\left\{1+\frac{ey_c}{r^2}\cdot \sec\left(\sqrt{\frac{P}{EI}}\right)\frac{l}{2}\right\}$$

in which l = effective length of the column.
The above formula is called secant's formula.

Perry's Formula

The secant formula is modified for the convenience of calculations.

$$\sigma_y = \frac{P}{A}\left\{1+\frac{ey_c}{r^2}\sec\sqrt{\frac{P}{EI}}\cdot\frac{l}{2}\right\}$$

$$\sigma_y = \frac{P}{A}\left\{1+\frac{ey_c}{r^2}\cdot \sec\left[\sqrt{\frac{Pl^2}{EI\pi^2}}\cdot\frac{\pi}{2}\right]\right\}$$

$$\sigma_y = \frac{P}{A}\left\{1+\frac{ey_c}{r^2}\sec\frac{\pi}{2}\left(\sqrt{\frac{P}{P_e}}\right)\right\} \qquad \left(\because \frac{\pi^2 EI}{l^2} = P_e\right)$$

in which P_e is Euler's buckling load.

Perry has simplified the secant term as $\sec\frac{\pi}{2}\sqrt{\frac{P}{P_e}} \cong \frac{1.2P_e}{P_e - P}$

$$\therefore \quad \sigma_y = \frac{P}{A}\left\{1+\frac{ey_c}{r^2}\left(\frac{1.2P_e}{P_e - P}\right)\right\}$$

Knowing the value of σ_y, A, e, y_c, r^2, and P_e, the buckling load can be found easily. The above formula can be further simplified. Take

$$\frac{P}{A} = \sigma\,;\ \frac{P_e}{A} = \sigma_e\text{, then}$$

$$\sigma_y = \sigma\left\{1+\frac{ey_c}{r^2}\times 1.2\left(\frac{\sigma_e}{\sigma_e - \sigma}\right)\right\}$$

$$\left\{\frac{\sigma_y}{\sigma} - 1\right\} = \frac{1.2ey_c}{r^2}\left\{\frac{1}{1-\dfrac{\sigma}{\sigma_e}}\right\}.$$

$$\Rightarrow \quad \left[\frac{\sigma_Y}{\sigma} - 1\right]\left[1-\frac{\sigma_Y}{\sigma}\right] = \frac{1.2eY_c}{\pi^2}$$

This formula is called Perry's formula.

PROBLEM 12.8

Objective 4

In brittle structures like concrete, allowable axial compressive stress (σ) is taken as 0.8 times allowable compressive stress in bending (σ_y). A column of length 12 times the least lateral dimension is treated as short column. Taking these two into consideration, determine the eccentricity (e/D) for which the column fails under allowable axial compressive (0.8 σ_y) taking the column cross-section as square and $E = 5000\sigma$.

SOLUTION

Let 'σ' be the failure stress which corresponds to the allowable stress in axial compression (treating the column as axially loaded one)

$$\sigma_y = \sigma/0.8$$

$$\sigma_y = 1.25\sigma$$

σ_e = bulking stress

Let 'D' be the size of the square column.

Then, $I = \dfrac{D^4}{12}$; $A = D^2$

$$r^2 = \frac{I}{A} = \frac{D^2}{12}; r = \frac{D}{\sqrt{12}}$$

$$y_c = \frac{D}{2} \text{ (highly stress fiber)}$$

Effective length of the column = $12D$

Slenderness ratio = $\left(\dfrac{12D}{r}\right) = \lambda = 41.57$

$$\sigma_e = \frac{\pi^2 \times 5000\sigma}{41.57^2} = 28.55\sigma$$

Using Perry's formula

$$\left\{\frac{\sigma_y}{\sigma} - 1\right\}\left\{1 - \frac{\sigma}{\sigma_e}\right\} = \frac{1.2 e\, y_c}{r^2}$$

$$\left\{\frac{1.25\sigma}{\sigma} - 1\right\}\left\{1 - \frac{\sigma}{28.55\sigma}\right\} = \frac{1.2 e D}{2 \times \dfrac{D^2}{12}}$$

$$0.2412 = 7.2\frac{e}{D}$$

$\dfrac{e}{D} < 0.0335$ {This value is ~$L/30$}.

In most of the concrete like structure, if eccentricity is less than $D/30$, the column may be designed for axial load only neglecting eccentricity effect.

PROBLEM 12.9

Objective 3

Determine the failure load of an eccentrically loaded column of effective length 4 m. The cross-section of the column is 140 mm × 60 mm. Eccentricity is about major axis of bending and is 8 mm. Take $\sigma_y = 300$ MPa and $E = 200$ GPa.

SOLUTION

For the cross-section, the major axis of bending is *xx* axis. Thus, while using Perry's formula, bending about *xx* axis shall be considered. Thus in the expression σ_e, I_{xx} should be considered not I_{yy} through $I_{yy} < I_{xx}$.

$$\sigma_y = 300 \text{ MPa}$$

$$I_{xx} = \frac{60 \times 140^3}{12} = 13.72 \times 10^6 \text{ mm}^4$$

$$A = 60 \times 140 = 8400 \text{ mm}^2$$

$$l_e = 4 \text{ m}$$

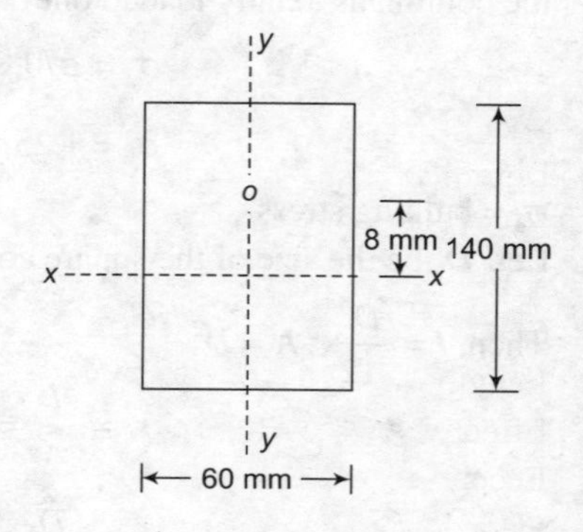

FIGURE 12.19

$$r = \text{radius of gyration} = \sqrt{\frac{13.72 \times 10^6}{8400}}$$

$$r = 40.41;\ \lambda = \frac{4000}{40.41} = 98.98$$

$$\sigma_e = \frac{\Pi^2 \times 2 \times 10^5}{98.98^2} = 201.48$$

$$e = 8 \text{ mm};\ y_c = \frac{140}{2} = 70 \text{ mm}$$

Using Perry's formula,

$$\left\{\frac{\sigma_y}{\sigma} - 1\right\}\left\{1 - \frac{\sigma}{\sigma_e}\right\} = \frac{1.2\,e\,y_c}{r^2}$$

$$\Rightarrow \quad \left\{\frac{300}{\sigma} - 1\right\}\left\{1 - \frac{\sigma}{201.48}\right\} = \frac{1.2 \times 8 \times 70}{40.41^2} = 0.412$$

By trial and error, $\sigma = 134.22$ N/mm^2.

Actress of 134.22 MPa is needed for the column to bend about *x–x* axis, that is, about axis of bending.

But the column may buckle about weaker axis I_{yy}. When the column buckles about weaker axis I_{yy}, it behaves like an axially loaded column only as eccentricity about this axis is absent.

The stress at which column buckles about weaker axis

$$I_{yy} = \frac{60^3 \times 140}{12} = 2.52 \times 10^6 \text{ mm}^4;$$

$$\sigma = \frac{\pi^2 EI_{yy}}{Al_e^2} = \frac{\pi^2 \times 2 \times 10^5 \times 2.52 \times 10^6}{4000^2 \times 140 \times 60} = 37.01 \text{ N/mm}^2.$$

As buckling stress about *YY* axis (37.01 MPa) is less than the buckling stress about *x–x* axis due to eccentric load (134.22 MPa), the failure stress is to be taken as 37.01 MPa.

Failure load = P = 37.01 × 140 × 60 = 310.86 kN.

PROBLEM 12.10 **Objective 4**

Determine the eccentricity that can be allowed in a square column of size 80 mm × 80 mm and length 2 m. The eccentricity is along the diagonal of the cross-section. Take σ_y = 320 MPa and E = 200 GPa. One end of the column is fixed while the other is free.

SOLUTION

Length of the column = 2.0 m

Effective length of the column = $l_e = 2l = 4$ m

Eccentricity is along the diagonal. Thus, buckling takes place along the diagonal perpendicular to the axis along which eccentricity acts.

$$\therefore \quad e = \frac{1}{2}\{80^2 + 80^2\}^{1/2} = 56.57 \text{ mm}$$

Using Perry's formula

$$\left\{\frac{\sigma_y}{\sigma} - 1\right\}\left\{1 - \frac{\sigma}{\sigma_e}\right\} = \frac{1.2ey_c}{r^2}$$

$$I_{xx} = \frac{80^4}{12} = 3.413 \times 10^6 \text{ mm}^4.$$

y_c = location of highly compressed fiber from axis of bending 'B' = 56.57 mm

$$r^2 = \frac{I_{xx}}{A} = \frac{\frac{80^4}{12}}{80 \times 80} = \frac{80^2}{12}$$

$$\sigma_y = 320 \text{ MPa}$$

$$\sigma_e = \frac{P_e}{A} = \frac{\pi^2 E}{\lambda^2}$$

$$\lambda = \frac{l_e}{r} = \frac{4000}{\sqrt{\frac{80^2}{12}}}$$

$$\lambda^2 = 30{,}000 \text{ mm}^2$$

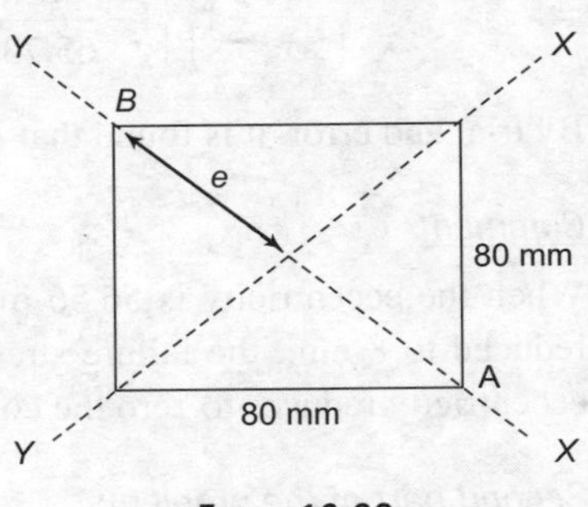

FIGURE 12.20

$$\therefore \qquad \sigma_e = \frac{\pi^2 \times 2 \times 10^5}{30000} = 65.79 \text{ N/mm}^2$$

Let 'σ' be the stress at which the column fails. Then, $\left(\frac{\sigma_y}{\sigma} - 1\right)\left(1 - \frac{\sigma}{\sigma_e}\right) = \frac{1.2ey_c}{r^2}$

$$\Rightarrow \qquad \left\{\frac{320}{\sigma} - 1\right\}\left\{1 - \frac{\sigma}{65.79}\right\} = \frac{1.2 \times 56.57 \times 56.57}{\left(\frac{80^2}{12}\right)}$$

$$\left\{\frac{320}{\sigma} - 1\right\}\left\{1 - \frac{\sigma}{65.79}\right\} = 7.2$$

By trial and error, $\sigma = 25.235$ N/mm^2.
Thus, the failure load on the column is given by $P = 25.235 * 80^2 = 161.5$ kN.

PROBLEM 12.11

Objective 4

Repeat the Problem 12.7, eccentricity is reduced to 8 mm. Compare solution and comment. Also find the failure stress when effective length of the column is reduced to 1 m.

SOLUTION

Eccentricity = 8 mm

$$\therefore \qquad \left(\frac{\sigma_y}{\sigma} - 1\right)\left(1 - \frac{\sigma}{\sigma_e}\right) = \frac{1.2ey_c}{r^2}$$

$$= \frac{1.2 \times 8 \times 56.57}{\left(\frac{80^2}{12}\right)}$$

$$= 1.018$$

$$\Rightarrow \qquad \left\{\frac{320}{\sigma} - 1\right\}\left\{1 - \frac{\sigma}{65.73}\right\} = 1.018$$

By trial and error, it is found that $\sigma = 52.60$ N/mm^2

Comment

When the eccentricity is 56.56 mm, the failure stress is 25.235 N/mm^2. When the eccentricity is reduced to 8 mm, the failure stress is observed to be 52.60 N/mm^2. This indicates that when the eccentricity reduces to zero the column fails by buckling stress (σ_e) as σ is approaching σ_e.

Second part of the problem:

Effective length of the column = 1 m

$$\therefore \qquad \sigma_e = \frac{\pi^2 E}{\lambda^2};\ \lambda = \frac{l_e}{r} = \frac{1000}{\sqrt{\dfrac{80^2}{12}}} = 43.301$$

$$\Rightarrow \qquad \sigma_e = \frac{\pi^2 \times 2\times 10^5}{43.301^2} = 1052.75\,\text{N/mm}^2$$

$$\lambda = 43.301$$

$$\sigma_e = 1052.77\ \text{N/mm}^2$$

$$\sigma = ?$$

$$\sigma_y = 320$$

$$\left\{\frac{320}{\sigma}-1\right\}\left\{1-\frac{\sigma}{1052.77}\right\} = \frac{1.2ey_c}{r^2}$$

$$e = 8\ \text{mm};\ y_c = 56.57;\ r = 23.09$$

$$\Rightarrow \qquad \left\{\frac{320}{\sigma}-1\right\}\left\{1-\frac{\sigma}{1052.77}\right\} = \frac{1.2\times 8\times 56.57}{23.09^2} = 1.018.$$

By trial and error, $\sigma = 146.60\ \text{N/mm}^2$.

In the present case, when effective length of the column is decreased the failure stress increased from 52.6 N/mm^2 to 146.6 N/mm^2.

Comment

This indicates that when the eccentricity is very less and length of the column is decreased to a short column, the failure stress approaches maximum fiber due to axial compression and bending.

$$\text{Maximum fiber stress due to bending and axial load} = \sigma\left\{1+\frac{6\sqrt{2}e}{D}\right\}$$

$$= 146.6\left\{1+\frac{6\sqrt{2}\times 8}{80}\right\} = 271$$

12.10 INELASTIC BUCKLING OF COLUMNS

In the earlier sections, it is observed that though the yield stress of the material is high, long columns buckle at a stress less than the yield stress. Most of the material exhibits stress–strain variation linear to the proportionality and nonlinear behavior prior to plastic yielding as shown in Figure 12.21.

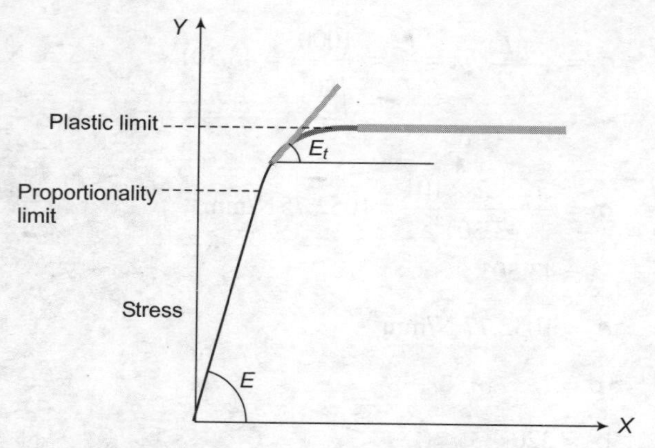

FIGURE 12.21 Inelastic behavior

The failure stress in respect of buckling in a column is given by

$$\sigma = \frac{\pi^2 E}{\left(\frac{l_e}{r}\right)^2}$$

In the above expression, as the length of the column is large enough, the stress in the column at the instant of buckling will be below proportionality limit. When the length of the column is very low, the stress in the column will be limited to plastic yield stress. For column of the length between very long and very short, inelastic bucking where stress in member is less than plastic stress and more than limit of proportionality. For this type of column, inelastic buckling theory is to be adopted.

Two theories are in vogue to predict the inelastic buckling load of columns. They are

1. Tangent modulus theory—Engesser's formula
2. Reduced modulus or double modulus theory

12.10.1 Tangent Modulus Theory

1. The strains in the cross-section of the column even at the instant of buckling continue to increase and the variation is not uniform.
2. At the instant of buckling, the strains on the convex side decrease while the same increase on the concave side.

The second assumption clearly indicates that the fibers on the concave face of the column are highly stressed, that is, beyond elastic limit, whereas the fibers on the convex side would be within the elastic limit. Thus to estimate the stress decrease on convex side, modulus of elasticity (E) is to be used and to estimate increase on the concave side tangent modulus (E_t) is to be used. The value of E_t depends on the stress level in the cross-section due to load P. The critical load using tangent modulus theory can be obtained as

$$P = \frac{\pi^2 E_t I}{l_e^2}$$

In the above expression, the value of E_t is obtained as the tangent to the stress–strain variation of the material of the column as shown in Figure 12.21.

12.10.2 Double Modulus Theory

In double modulus theory, the basic assumption is that the fibers that are highly stressed on concave side of the column cross-section are governed by tangent modulus E_t and the fibers on the convex side of the cross-section of the column are governed by elastic modulus E.

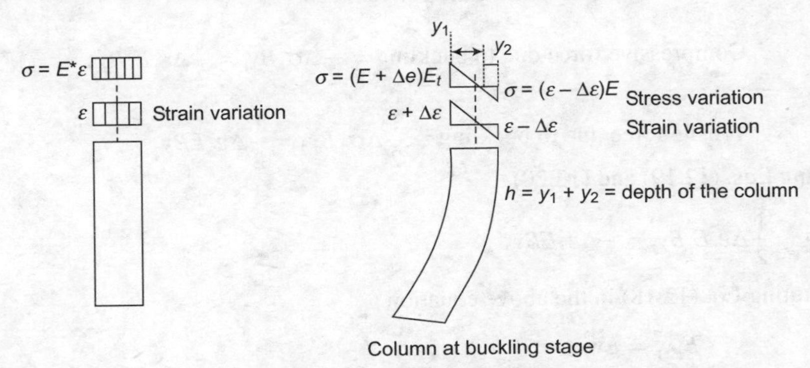

FIGURE 12.22 Column at prior to bucking stage and at buckling stage.

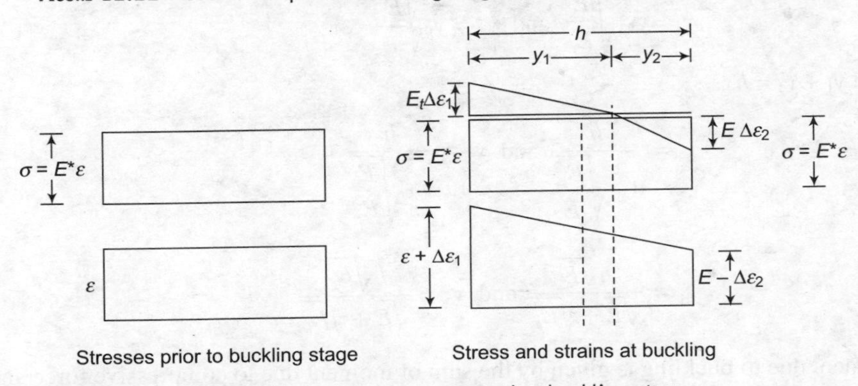

FIGURE 12.23 Stresses in the column cross-section buckling stage.

At the buckling stage, let P_c be the critical load that causes buckling. This buckling induces high stress on the concave face and low stress on the convex side same as bending stresses. The variation of these stresses is linear across the depth of the cross-section. Prior to buckling the stress in the column cross-section is uniform and is given by $\sigma = \varepsilon \times E$ and $P_c = A \times \sigma$. After buckling the stress in the cross-section of the column is only altered keeping the resultant axial load P_c constant. Thus, total moment due to internal stresses must be equal to $P_c \times y$, in which 'y' is the eccentricity of this load from centroid of the cross-section. To combine the effect of two modules, the following conditions are to be used.

1. Strain variation is linear about the axis of buckling. Thus, the curvature is same on both convex and concave faces.

$$\text{Curvature} = \frac{1}{R} = \frac{1}{E_d I} = \frac{\Delta\varepsilon_1}{y_1} = \frac{\Delta\varepsilon_2}{y_2} \quad (12.17)$$

$$\Delta\varepsilon_1 = \frac{y_1}{R} \text{ and } \Delta\varepsilon_2 = \frac{y_2}{R} \quad (12.18)$$

2. Compressive force developed due to buckling (over and above $\sigma = P/A$) on the concave face is equal to tensile force developed on the convex face.

$$\text{Compressive force due to buckling} = \frac{1}{2}\Delta\sigma_1 By_1 = \frac{1}{2}\Delta\varepsilon_1 E_t By_1 \quad (12.19)$$

$$\text{Tensile force due to buckling} = \frac{1}{2}\Delta\sigma_2 By_2 = \frac{1}{2}\Delta\varepsilon_2 EBy_2 \quad (12.20)$$

Equating Eqs. (12.19) and (12.20)

$$\frac{1}{2}\Delta\varepsilon_1 E_t By_1 = \frac{1}{2}\Delta\varepsilon_2 EBy_2$$

Substituting Eq. (12.18) in the above equation

$$E_t y_1^2 = Ey_2^2 \quad (12.21)$$

$$y_1 = y_2\sqrt{\frac{E}{E_t}} \text{ and } y_2 = y_1\sqrt{\frac{E_t}{E}}$$

And $y_1 + y_2 = h$

$$y_1 = \frac{h}{1+\sqrt{\frac{E_t}{E}}} \text{ and } y_2 = \frac{h}{1+\sqrt{\frac{E}{E_t}}}$$

$$y_1 = \frac{h\sqrt{E}}{\sqrt{E}+\sqrt{E_t}} \text{ and } y_2 = \frac{h\sqrt{E_t}}{\sqrt{E}+\sqrt{E_t}}$$

3. Moment due to buckling is given by the sum of moment due to compressive force and tensile force.

Moment due to Compressive Force

$$M_1 = \frac{1}{2}E_t\Delta\varepsilon_1 By_1\frac{2y_1}{3} = \frac{1}{3}E_t B\frac{y_1^3}{R} = \frac{1}{3}E_t Bh^3\frac{1}{R}\frac{E\sqrt{E}}{\{\sqrt{E}+\sqrt{E_t}\}^3}$$

Moment due to Tensile Force

$$M_2 = \frac{1}{2}E\Delta\varepsilon_2 By_2\frac{2y_2}{3} = \frac{1}{3}EB\frac{y_2^3}{R} = \frac{1}{3}EBh^3\frac{1}{R}\frac{E_t\sqrt{E_t}}{\{\sqrt{E}+\sqrt{E_t}\}^3}$$

Total moment $= M = M_1 + M_2$

$$M = \frac{1}{3}E_t Bh^3 \frac{1}{R}\frac{E\sqrt{E}}{\{\sqrt{E}+\sqrt{E_t}\}^3} + \frac{1}{3}EBh^3 \frac{1}{R}\frac{E_t\sqrt{E_t}}{\{\sqrt{E}+\sqrt{E_t}\}^3}$$

$$= \frac{1}{3}Bh^3 \frac{1}{R}\frac{EE_t\{\sqrt{E_t}+\sqrt{E}\}}{\{\sqrt{E}+\sqrt{E_t}\}^3}$$

$$M = \frac{1}{12}Bh^3 \frac{1}{R}\frac{4EE_t}{\{\sqrt{E}+\sqrt{E_t}\}^2}$$

$$M = \frac{1}{12}Bh^3 \frac{1}{R}\frac{4EE_t}{\{\sqrt{E}+\sqrt{E_t}\}^2}$$

$$M = \frac{I}{R}\frac{4EE_t}{\{\sqrt{E}+\sqrt{E_t}\}^2}$$

Taking $\frac{4EE_t}{\{\sqrt{E}+\sqrt{E_t}\}^2} = E_D$ = Double Modulus

$$M = \frac{E_D I}{R} \quad \text{or} \quad \frac{1}{R} = \frac{M}{E_D I}$$

This indicates that

$$\frac{1}{R} = \frac{d^2y}{dx^2} = \frac{M}{E_D I}$$

Thus

$$E_D I \frac{d^2y}{dx^2} = -Py$$

And hence buckling load is given by

$$P = \frac{\pi^2 E_D I}{l_e^2}$$

in which E_D is double modulus.

12.11 BUCKLING OF INITIALLY CURVED COLUMNS

Euler's assumption for the buckling load of a column that the column is initially straight may not be applicable in all cases. Sometimes, the column may have initial curvature. When the columns have initial curvature the buckling load of the column is to be obtained with the following procedure.

Consider a column AB initially curved, having maximum offset of δ_1 at mid height of the column. The initial bent portion is defined by y_1 at a distant x from A.

Let the buckled profile of the column be defined as y offset at a distant x from A.

Thus, the buckled profile of the column from its initial position is defined by $y - y_1$ at x.

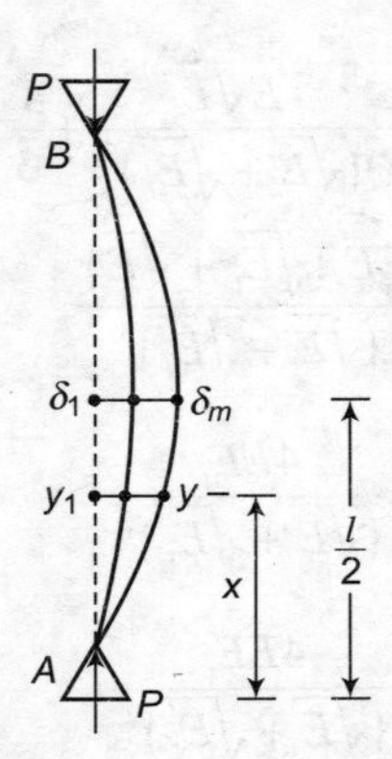

FIGURE 12.24 Buckled profile of the initially curved column.

BM in the column at x distant from A is given by $M = Py$ (12.22)

The column buckled from initially bent position to a new buckled position. It indicates that the column deviated by $(y - y_1)$.

Thus, $$EI\frac{d^2(y-y_1)}{dx^2} = -M \quad (12.23)$$

Hence, $$EI\frac{d^2(y-y_1)}{dx^2} = -Py \quad (12.24)$$

$$EI\left\{\frac{d^2y}{dx^2} - \frac{d^2y_1}{dx^2}\right\} = -Py$$

Assuming the initial curved portion of the column as

$$y_1 = \delta_1 \sin\frac{\pi x}{l}$$

$$EI\frac{d^2y_1}{dx^2} = -\frac{EI\pi^2\delta_1}{l^2}\sin\frac{\pi x}{l}$$

Taking $\dfrac{\pi^2 EI}{l^2} = P_e =$ Euler's Buckling load

$$EI\frac{d^2y}{dx^2} + Py = -P_e\delta_1 \sin\frac{\pi x}{l} \quad (12.25)$$

The solution of the above differential equation may be taken as

$$y = A\sin\frac{\pi x}{l}$$

Substituting the above expression in Eq. (12.25)

$$EI\left(-A\frac{\pi^2}{l^2}\sin\frac{\pi x}{l}\right)\left(-A\frac{\pi^2}{l^2}\sin\frac{\pi x}{l}\right) + P\left(A\sin\frac{\pi x}{l}\right)\left(A\sin\frac{\pi x}{l}\right) = -P_e\delta_1\sin\frac{\pi x}{l}$$

$$\sin\frac{\pi x}{l} \neq 0$$

Hence, $A = -\dfrac{P_e\delta_1}{\left\{P - \dfrac{\pi^2 EI}{l^2}\right\}}$ taking $\dfrac{\pi^2 EI}{l^2} = P_e$

We get

$$A = -\frac{-P_e\delta_1}{\{P - P_e\}} = \frac{P_e\delta_1}{\{P_e - P\}} = \frac{\delta_1}{\left\{1 - \dfrac{P}{P_e}\right\}} \tag{12.26}$$

Thus

$$y = \frac{\delta_1}{\left\{1 - \dfrac{P}{P_e}\right\}}\sin\frac{\pi x}{l}$$

At $x = l/2$,

$$y_{\max} = \frac{\delta_1}{\left\{1 - \dfrac{P}{P_e}\right\}}$$

Maximum bending stress is given by

$\sigma_{\max} = \dfrac{P}{A} + \dfrac{Py_{\max}}{I}y_c$ in which y_c is the high stressed fiber in the cross-section

$\sigma_{\max} = \dfrac{P}{A} + \dfrac{P}{I}\dfrac{\delta_1}{\left\{1 - \dfrac{P}{P_e}\right\}}y_c$ taking $I = Ar^2$, in which r is radius of gyration about axis of bending

$$\sigma_{\max} = \frac{P}{A} + \frac{P}{Ar^2}\frac{\delta_1}{\left\{1 - \dfrac{P}{P_e}\right\}}y_c$$

$$\sigma_{\max} = \frac{P}{A}\left\{1 + \frac{\delta_1 y_c}{r^2}\frac{1}{\left\{1 - \dfrac{\frac{P}{A}}{\frac{P_e}{A}}\right\}}\right\}$$

Taking $\frac{P}{A}$ as σ and $\frac{P_e}{A}$ as σ_e, we get

$$\sigma_{\max} = \sigma\left\{1+\frac{\delta_1 y_c}{r^2}\frac{1}{\left\{1-\frac{\sigma}{\sigma_e}\right\}}\right\}$$

$$\left\{\frac{\sigma_{max}}{\sigma}-1\right\}\left\{1-\frac{\sigma}{\sigma_e}\right\}=\frac{\delta_1 y_c}{r^2}$$

The above expression is similar to expression of an eccentrically loaded column.

SUMMARY

- **The buckling load of a column is given by $P_{cr}=\frac{\pi^2 EI}{l_e^2}$, in which l_e is the effective length of the column and it depends on the end conditions of the column.**
- **Radius of gyration of a column is given by $k=\sqrt{\frac{I}{A}}$, in which A is cross-sectional area and I is MI or second moment area.**
- **The load carrying capacity of a column is given by Rankine's formula as $\frac{1}{P}=\frac{1}{P_e}+\frac{1}{P_y}$, in which P_e is the Euler's critical load and P_y is the yield load.**
- **The failure stress in eccentrically loaded column is given by $\sigma_y=\frac{P}{A}\left\{1+\frac{ey_c}{k^2}\sec\frac{\pi}{2}\left(\sqrt{\frac{P}{P_e}}\right)\right\}$.**

OBJECTIVE TYPE QUESTIONS

1. A dam under axial and transverse load is a case of
   (a) Buckling (b) Eccentric loading
   (c) Bending (d) None
2. A column that fails due to direct stress is called
   (a) Short column (b) Long column (c) Medium column (d) Slender column
3. The direct stress included in a long column is……as compared to bending stress
   (a) More (b) Less (c) Same (d) Negligible
4. For long columns, the value of buckling load is……………..crushing load.
   (a) Less than (b) More than (c) Equal to (d) None of these

5. The slenderness ratio is the ratio of
(a) Length of column to least radius of gyration
(b) MI to area of cross-section
(c) Area of cross-section to MI
(d) Least radius of gyration to length of the column

6. Compression members always tend to buckle in the direction of
(a) Vertical axis (b) Horizontal axis
(c) Minimum cross-section (d) Least radius of gyration

7. A column has MI about X–X and Y–Y axis as $I_{XX} = 4234.4$ mm^4 and $I_{YY} = 236.3$ mm^4. This column will buckle about
(a) X–X axis (b) Y–Y axis
(c) It depends on the applied load (d) None of these

8. A column of length 4 m with both ends fixed may be considered as equivalent to a column of length ………… with both ends hinged
(a) 2 m (b) 1 m (c) 3 m (d) 6 m

9. The ratio of effective length and least lateral dimension for short column is…
(a) >12 (b) <12 (c) ≥12 (d) None of the above

10. In Euler's theory, long columns having the ratio of (Le/LLD) ≥ 12 fail due to
(a) Crushing (b) Buckling (c) Both a and b (d) None of the above

11. Euler's formula is applicable only ___________

1. for short columns
2. for long columns
3. if slenderness ratio is greater than $\sqrt{(\pi^2 E/\sigma_c)}$
4. if crushing stress < buckling stress
5. if crushing stress ≥ buckling stress

(a) 1, 2, and 3 (b) 2, 3, and 5 (c) 3 and 4 (d) All of the above

12. What is the safe load acting on a long column of 2 m having diameter of 40 mm. The column is fixed at both the ends and modulus of elasticity is 2×10^5 N/mm^2? (F.O.S = 2)
(a) 120 kN (b) 124 kN (c) 130 kN (d) 150 kN

13. Keeping loading same but increasing the length, shear stresses in a beam will
(a) Increase (b) Decrease (c) No change (d) None

14. A long column with fixed ends can carry load as compared to both ends hinged
(a) 4 times (b) 8 times (c) 16 times (d) None

15. A strut of length 'L' and a flexural rigidity 'EI' are fixed at both the ends. The buckling load under compression is
(a) $\dfrac{\pi^2 EI}{4L^2}$ (b) $\dfrac{\pi^2 EI}{L^2}$ (c) $\dfrac{2\pi^2 EI}{L^2}$ (d) $\dfrac{4\pi^2 EI}{L^2}$

16. In case of eccentrically loaded struts
(a) Hollow section is preferred (b) Solid section is preferred
(c) Composite section is preferred (d) Reinforced section is preferred

17. A 10 mm mild steel rod 40 mm long lying on ground will be known as
(a) Short column (b) Long column (c) Strut (d) None of the above

Solutions for Objective Questions

Sl. No.	1.	2.	3.	4.	5.	6.	7.	8.	9.	10.
Answer	(b)	(a)	(d)	(a)	(a)	(d)	(b)	(a)	(b)	(b)

Sl.No.	11.	12.	13.	14.	15.	16.	17.
Answer	(b)	(b)	(c)	(a)	(b)	(a)	(d)

EXERCISE PROBLEMS

1. A hollow mild steel tube 6 m long, 4 cm internal diameter, and 6 mm thick is used as a strut with both ends hinged. Find the crippling load and safe load taking factor of safety as 3. Take $E = 2 \times 10^5$ N/mm^2.

2. A hollow circular column of internal diameter 20 mm and external diameter 40 mm has a total length of 5 m. One end of the column is fixed and the other end is hinged. Find out the crippling stress of the column if $E = 2 \times 105$ N/mm^2. Also find out the shortest length of this column.

3. A T section 150 mm × 120 mm × 20 mm is used as a strut of 4 m long with hinged at its both ends. Calculate the crippling load if modulus of elasticity for the material be 2.0×10^5 N/mm^2. Column for which Euler's formula is valid taking the yield stress equal to 250 N/mm^2.

4. A short length of tube having internal diameter and external diameter are 4 cm and 5 cm, respectively, which failed in compression at a load of 250 kN. When a 1.8 m length of the same tube was tested as a strut with fixed ends, the load failure was 160 kN. Assuming that σ_c in Rankine's formula is given by the first test, find the value of the constant α in the same formula. What will be the crippling load of this tube if it is used as a strut 2.8 m long with one end fixed and the other hinged?

5. A hollow cast iron column 4.5 m long with both ends fixed is to carry an axial load of 250 kN under working conditions. The internal diameter is 0.8 times the outer diameter of the column. Using Rankine–Gordon's formula, determine the diameters of the column adopting a factor of safety of 4. Assume f_c, the compressive strength to be 550 N/mm^2 and Rankine's constant $a = 1/1600$.

6. A round steel rod of diameter 15 mm and length 2 m is subjected to a gradual increasing axial compressive load. Using Euler's formula, find the buckling load. Find also the maximum lateral deflection corresponding to the buckling condition. Both ends of the rod may be taken as hinged. Take $E = 2.1 \times 10^5$ N/mm^2 and the yield stress of steel = 240 N/mm^2.

7. A simply supported beam of length 4 m is subjected to a uniformly distributed load of 30 kN/m over the whole span and deflects 15 mm at the center. Determine the crippling load if the beam is used as a column with the following conditions:
   (i) One end fixed and another end hinged (ii) Both the ends pin jointed.

8. A hollow cylindrical cast iron column is 4 m long with both ends fixed. Determine the minimum diameter of the column if it has to carry a safe load of 250 kN with a factor of safety of 5. Take the internal diameter as 0.8 times the external diameter. Take $\sigma_c = 550$ N/mm^2 and $a = 1/1600$ in Rankine's formula.

CHAPTER 13

ANALYSIS OF MEMBERS UNDER COMBINED LOADING

UNIT OBJECTIVE

This chapter helps in the estimation of stresses for combined actions, namely axial compression, bending, shear, and torsion. The presentation attempts to help the student achieve the following:

Objective 1: Determine the resulting stress due to combined bending and axial load.

Objective 2: Plot kern of a given cross-section.

Objective 3: Determine the principal stresses when a shaft is subjected to combined bending and torsion.

Objective 4: Find principal stresses when a shaft is subjected to combined torsion and shear.

Objective 5: Evaluate principal stresses when a shaft is subjected to combined bending, torsion, and shear.

Objective 6: Apply different failure theories in deciding the dimensions of a structural element or safety in a structural element.

13.1 INTRODUCTION

In the previous chapters, we have learnt the procedure to determine the stress in structural elements when they are subjected to forces such as axial compression, axial tension, bending, shear, and torsion. These forces rarely occur alone. Frequently, structural elements are subjected to combination of forces such as combined bending and axial load; axial load and torsion; torsion and bending; and torsion, bending, and axial load. In this chapter, we determine the resulting stresses due to combined loads. Different types of stresses developed due to combined loads lead to stresses such as normal stress and shear stress. These normal and shear stresses result in principal stresses. This chapter provides the information regarding such combined stresses.

13.2 SHORT COLUMNS SUBJECTED TO AXIAL COMPRESSION AND BENDING

Short column are axially loaded elements which are very often subjected to combined bending and axial compression. Axial force on a member produces normal stress. Bending moment (BM) produces bending stress, which is also normal to the cross-section. As these two stresses are normal to the cross-section their algebraic addition results in the final stress.

Consider a cross-section subjected to axial load P and BM M. Let A be the cross-sectional area of the cross-section and I be the moment of inertia (MI) of the cross-section. Axial stress is given by $\sigma_a = \frac{P}{A}$. The total cross-section is subjected to axial compressive stress of uniform intensity. BM produces bending stress.

$$\sigma_b = \frac{M}{I} y,,$$

in which y is the location of the fiber from neutral axis (NA) where the stress is to be found. Stress in the extreme fiber A is given by

$$\sigma_b = \frac{P}{A} - \frac{M}{I} y_2.$$

Stress in the extreme fiber B is given by

$$\sigma_b = \frac{P}{A} + \frac{M}{I} y_1.$$

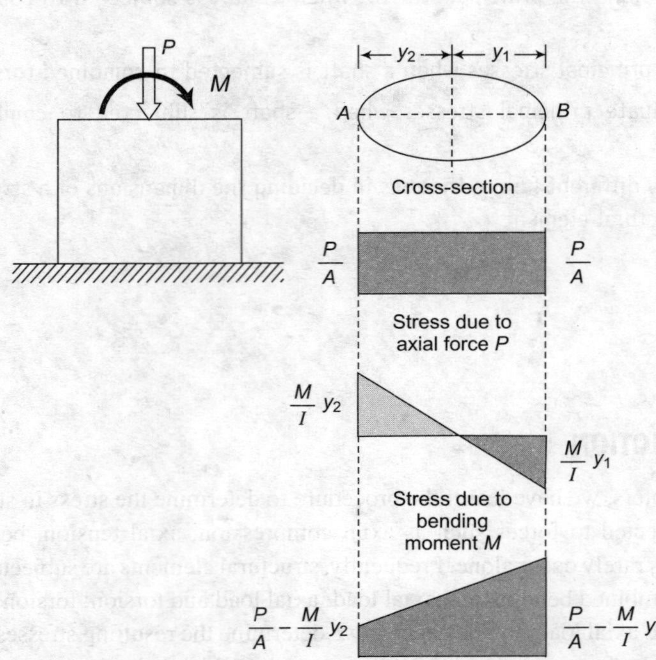

FIGURE 13.1 Resulting stresses due to combined bending and axial load

A column subjected to eccentric longitudinal load falls under the category of combined bending and axial load case. This case is more common in many structural elements.

The stress analysis of an eccentrically loaded short column is presented here as atypical example of combined bending and axial load.

13.3 KERN OF DIFFERENT SECTIONS AND APPLICATIONS

Consider a cross-section that is having at least one axis of symmetry subjected to eccentric loading. Let the eccentricity of load P is e_x about Y axis and e_y about X axis. The eccentric loading can be taken as the combination of axial load, bending about X axis, and bending about Y axis.

Let σ be the stress at a point whose coordinates are $S(x, y)$ with reference to the coordinate axes X and Y.

Axial load produces axial stress compressive in nature $\sigma_a = \dfrac{P}{A}$. This stress is same everywhere in the cross-section.

Stress at point $S(x,y)$ due to axial load P is $\sigma_a = \dfrac{P}{A}$.

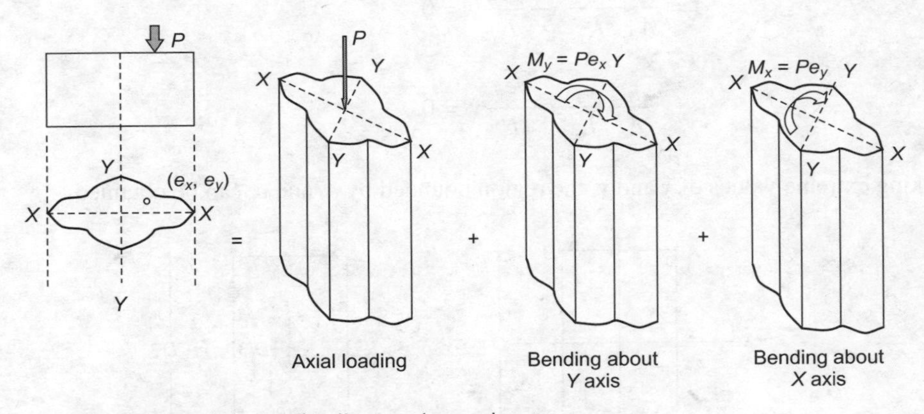

FIGURE 13.2 Eccentric loading on short column.

Eccentricity of load about Y axis produces bending about Y axis = $M_y = Pe_x$. Stress at $S(x,y)$ due to bending about Y axis

$$\sigma_{by} = \frac{M_y}{I_{yy}} x = \frac{Pe_x}{I_{yy}} x.$$

Eccentricity of load about X axis produces bending about X axis = $M_x = Pe_y$. Stress at $S(x,y)$ due to bending about X axis

$$\sigma_{bx} = \frac{M_x}{I_{xx}} y = \frac{Pe_y}{I_{xx}} y.$$

Stress at $S(x,y)$ due to load P acting (e_x,e_y) is given by the addition of above three quantities. Thus

$$\sigma = \sigma_a + \sigma_{by} + \sigma_{bx}$$

$$\sigma = \frac{P}{A} + \frac{Pe_x}{I_{yy}} x + \frac{Pe_y}{I_{xx}} y.$$

In the above expression, positive value of stress indicates compressive stress. Location of load P and the point where the stress is to be determined decides the nature of the stress. Negative value of stress indicates that the nature of stress is tensile.

In many brittle structures such as plain concrete, rock, or brick masonry structures, tensile stress in the structural member is not permitted as the tensile resistance of these materials is very low. Thus, it is very important to see that the eccentricity is within the certain region nearer to the centroid of the cross-section, which avoids tensile stress in the entire cross-section.

The region within which application of longitudinal load avoids the tensile stress in the entire cross-section is called kern or core of the cross-section. This region does not depend on the intensity of the load.

Thus, the kern or core of a cross-section can be obtained by equating the term σ to zero.

$$\sigma = \frac{P}{A} + \frac{Pe_x}{I_{yy}} x + \frac{Pe_y}{I_{xx}} y = 0$$

$$\frac{1}{A} + \frac{e_x}{I_{yy}} x + \frac{e_y}{I_{xx}} y = 0.$$

Taking extreme values of x and y, the region bounded by e_x and e_y can be obtained.

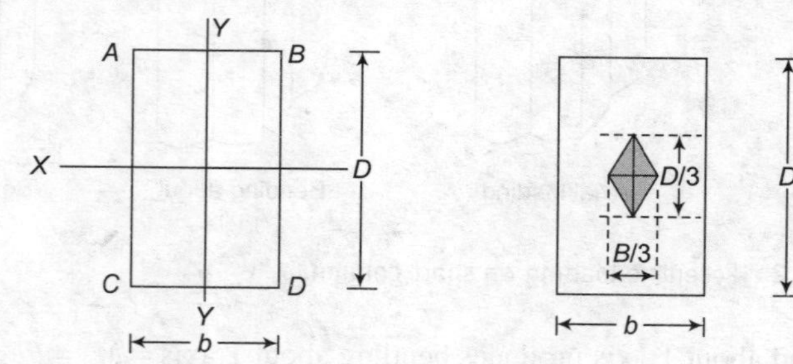

FIGURE 13.3

13.3.1 Kern of a Rectangular Section

Consider a rectangular section of width 'b' and height 'D'.

$$A = bD$$

$$I_{xx} = \frac{bD^3}{12}$$

$$I_{yy} = \frac{Db^3}{12}.$$

If the load is in the positive X and positive Y region, that is, e_x and e_y are positive, then the extreme point would be $C\left(-\frac{b}{2},-\frac{D}{2}\right)$.

Hence for the kern $\sigma \geq 0$

$$\frac{P}{A}+\frac{Pe_x}{I_{yy}}x+\frac{Pe_y}{I_{xx}}y \geq 0$$

$$\frac{1}{bD}+\frac{Pe_x}{\frac{b^3D}{12}}\left(-\frac{b}{2}\right)+\frac{Pe_y}{\frac{D^3b}{12}}\left(-\frac{D}{2}\right) \geq 0$$

$$1-\frac{6e_x}{b}-\frac{6e_y}{D} \geq 0$$

$$\frac{e_x}{\frac{b}{6}}+\frac{e_y}{\frac{D}{6}} \leq 1$$

The location of load that satisfies the above equation avoids tensile anywhere in the cross-section. It is considered in the above derivation that e_x and e_y are positive.

Thus, the kern region is $\left(\frac{b}{6},0\right)$, $\left(0,\frac{D}{6}\right)$, $\left(-\frac{b}{6},0\right)$, and $\left(0,-\frac{D}{6}\right)$.

The kern region along X axis is $\frac{b}{3}$ and along Y axis is $\frac{D}{3}$.

This is important to understand that within middle third of the cross-section in breadth direction, application of load does not create tension in the cross-section. This is called **middle third rule**.

13.3.2 Kern of a Circular Section

Consider a circular section of diameter 'D'.

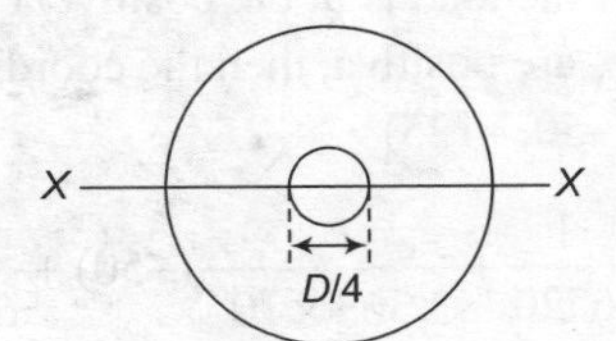

FIGURE 13.4 Kern of a circular cross-section

$$A = \frac{\pi D^2}{4}$$

$$I_{xx} = \frac{\pi D^4}{64}.$$

As the cross-section is circular, eccentricity about two axes need not be considered and eccentricity about a single axis would be sufficient. Thus, let 'e' be the eccentricity about axis of bending.

For the kern $\sigma \geq 0$

$$\frac{P}{A}+\frac{Pe}{I_{xx}}r \geq 0$$

$$\frac{1}{\frac{\pi D^2}{4}}+\frac{Pe}{\frac{\pi D^4}{64}}\left(-\frac{D}{2}\right)\geq 0$$

$$1-\frac{8e}{D}\geq 0$$

$$e\leq\frac{D}{8}$$

Thus, kern of circular cross-section of diameter D is a circle of diameter $\frac{D}{4}$.

PROBLEM 13.1

Objective 2

Sketch the kern of an I section of flange width 100 mm, flange thickness 20 mm, and web thickness 12 mm. Overall depth of the cross-section is 250 mm.

SOLUTION

Cross-sectional area of the I section = $A = (100 \times 20 \times 2) + (210 \times 12) = 6520\ \text{mm}^2$

MI about X axis

$$I_{xx}=\frac{1}{12}\{100\times 250^3-88\times 210^3\}=62.294\times 10^6\ \text{mm}^4$$

MI about Y axis

$$I_{yy}=\frac{1}{12}\times 210\times 12^3+2\times\frac{1}{12}\times 20\times 100^3=3.363\times 10^6\ \text{mm}^4$$

For determining the kern region,

$$\frac{P}{A}+\frac{Pe_x}{I_{yy}}x+\frac{Pe_y}{I_{xx}}y\geq 0.$$

If the load is in the positive X and positive Y region, that is, e_x and e_y are positive, then the coordinates of the extreme point would be (–50, –125)

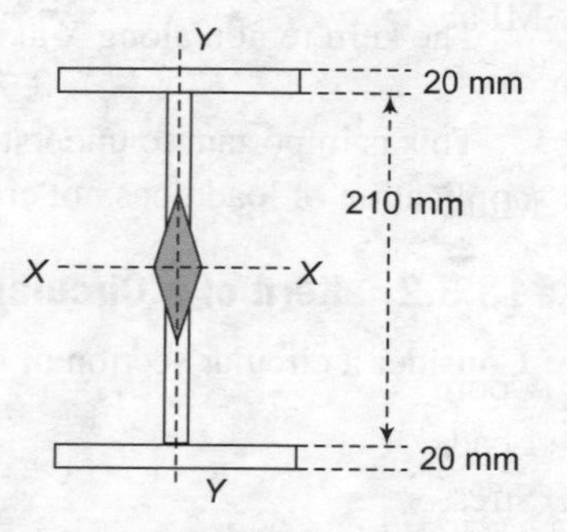

FIGURE 13.5 Kern of I section

$$\frac{1}{6520}+\frac{e_x}{3.363\times 10^6}(-50)+\frac{e_y}{62.294\times 10^6}(-125)\geq 0$$

$$1-\frac{e_x}{10.32}-\frac{e_y}{75.43}\geq 0$$

$$\frac{e_x}{10.32}+\frac{e_y}{75.43}\leq 1$$

From the above equation, it can be concluded that if the load is applied within the region bounded by (10.32,0) and (0,75.43), there will be no tension in the cross-section. Substituting the same region on the other quadrants, the kern region can be obtained.

The region bounded by (10.32,0), (0,75.43), (–10.32,0), and (0,–75.43) is the kern region for the I section shown.

PROBLEM 13.2 **Objective 1**

An eccentric load of 50 kN acts at 25 mm from *X* axis and 25 mm from *Y* axis of a *T* cross-section shown in Figure 13.6. Flange width is 100 mm, flange thickness is 20 mm, and web thickness is 10 mm. Overall depth of the cross-section is 200 mm. Determine the stress at the four corners and sketch the NA.

SOLUTION

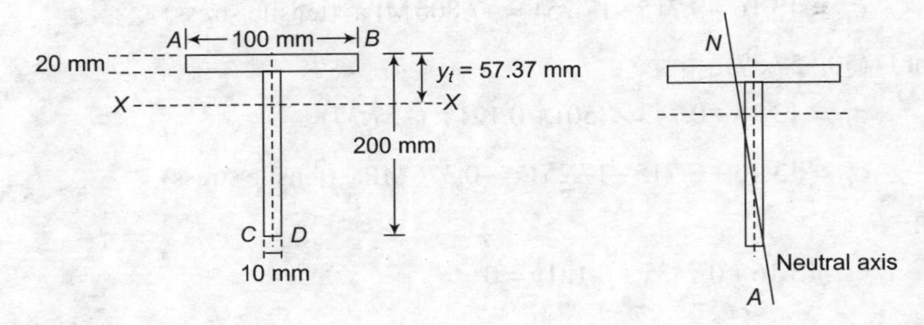

FIGURE 13.6

Cross-sectional area of the I section = $A = (100 \times 20) + (180 \times 10) = 3800\ \text{mm}^2$
Centroid of the cross-section from top fiber

$$\bar{y} = \frac{100 \times 20 \times 10 + 180 \times 10 \times (90 + 20)}{3800} = 57.37\,\text{mm}$$

MI about *X* axis

$$I_{xx} = \frac{1}{3}\{(100 \times 57.37^3) - (90 \times 57.37^3) + (10 \times (200 - 57.37)^3)\} = 10.301 \times 10^6\ \text{mm}^4$$

MI about *Y* axis

$$I_{yy} = \frac{1}{12}\{20 \times 100^3 + 180 \times 10^3\} = 1.682 \times 10^6\ \text{mm}^4$$

Coordinates of the eccentricity = (25, 25)
Load = 50 kN
Stress at any point $S(x,y)$ is given by

$$\sigma = \frac{P}{A} + \frac{Pe_x}{I_{yy}}x + \frac{Pe_y}{I_{xx}}y$$

$$\sigma = \frac{50 \times 1000}{3800} + \frac{50 \times 1000 \times 25}{1.682 \times 10^6}x + \frac{50 \times 1000 \times 25}{10.301 \times 10^6}y = 13.16 + 0.743x + 0.121y$$

$$\sigma = 13.16 + 0.743x + 0.121y$$

Stress at point A (–50,57.37)

$$\sigma_A = 13.16 + 0.743 \times (-50) + 0.121 \times (57.37)$$

$$\sigma_A = 13.16 - 37.15 + 6.94 = -17.05\,\text{MPa} \text{ (tensile stress)}$$

Stress at point B (50,57.37)

$$\sigma_B = 13.16 + 0.743 \times (50) + 0.121 \times (57.37)$$

$$\sigma_B = 13.16 + 3.715 + 6.94 = 57.25\,\text{MPa} \text{ (compressive stress)}$$

Stress at point C (−5,−142.63)

$$\sigma_C = 13.16 + 0.743 \times (-5) + 0.121 \times (-142.63)$$

$$\sigma_C = 13.16 - 3.715 - 17.251 = -7.806\,\text{MPa} \text{ (tensile stress)}$$

Stress at point D (50,−57.37)

$$\sigma_D = 13.16 + 0.743 \times (50) + 0.121 \times (-57.37)$$

$$\sigma_D = 13.16 + 3.715 - 17.251 = -0.376\,\text{MPa} \text{ (tensile stress)}$$

For NA $\sigma = 0$

$$13.16 + 0.743x + 0.121y = 0$$

Equation of the NA is rewritten as

$$\frac{x}{(-17.71)} + \frac{y}{(-108.76)} = 1$$

For the NA, x intercept is −17.71 mm and y intercept is −108.76 mm.

PROBLEM 13.3

Objective 1

Determine the position and intensity of maximum normal stress developed at top fiber of the tapering bar subjected to axial load as shown in Figure 13.7. A Load of 360 kN acts at the centroid at one end where the cross-section is 60 mm × 60 mm and eccentric at the other end, where the cross-section is 60 mm wide and 180 mm deep. Length of the bar is 900 mm. Also plot the variation of top fiber stress along the length of the bar.

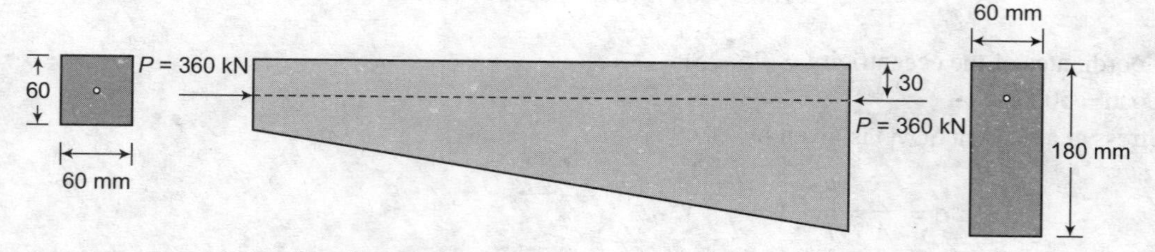

FIGURE 13.7 Tapering bar with longitudinal load

SOLUTION

At the left end, the cross-section is square and the load is axial load.

Hence normal stress at left end is $\sigma = \dfrac{P}{A} = \dfrac{360 \times 1000}{60 \times 60} = 100\text{ MPa}$.

Consider a section located x distant from left end.

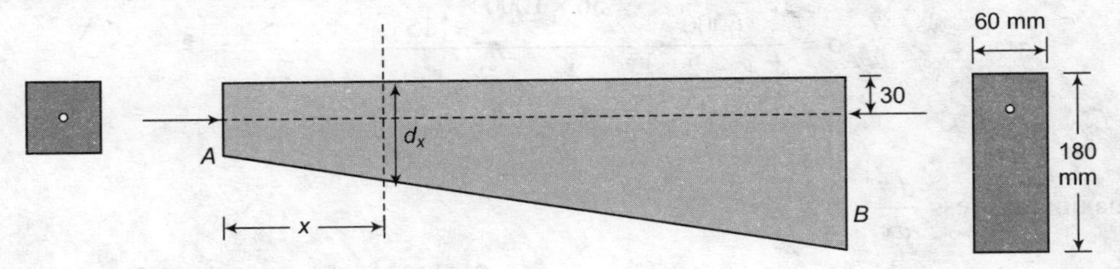

FIGURE 13.8

Let d be the depth of the member at a distant x from left end.

Then $$d_x = 60 + \frac{180-60}{900}x = 60 + \frac{2}{15}x$$

Eccentricity of load $e = \dfrac{d_x}{2} - 30 = \dfrac{1}{15}x$

At the section under consideration:

Cross-sectional area $A = 60 \times \left\{60 + \dfrac{2}{15}x\right\}$

$$\text{MI} = I = \frac{60 \times \left\{60 + \frac{2}{15}x\right\}^3}{12}$$

Axial stress due to P is $\sigma_a = \dfrac{P}{60 \times \left\{60 + \frac{2}{15}x\right\}} = \dfrac{360 \times 1000}{60 \times \left\{60 + \frac{2}{15}x\right\}} = \dfrac{6000}{\left\{60 + \frac{2}{15}x\right\}}$

Maximum bending stress $\sigma_b = \dfrac{Pe}{I}y_t = \dfrac{360 \times 1000 \times \left\{\frac{x}{15}\right\} \times \frac{d}{2}}{\frac{bd^3}{12}}$

$$\sigma_b = \frac{360 \times 1000 \times \left\{\frac{x}{15}\right\}}{\frac{60 \times \left\{60 + \frac{2}{15}x\right\}^2}{6}} = \frac{36 \times 1000 \times \left\{\frac{x}{15}\right\}}{\left\{60 + \frac{2}{15}x\right\}^2}$$

Maximum stress due to axial load as well as bending due to eccentricity $\sigma = \sigma_a + \sigma_b$

$$\sigma = \frac{6000}{\left\{60+\frac{2}{15}x\right\}} + \frac{36\times 1000\times\left\{\frac{x}{15}\right\}}{\left\{60+\frac{2}{15}x\right\}^2}$$

For maximum stress $\frac{d\sigma}{dx}=0$

$$\frac{d\sigma}{dx} = -\frac{6000}{\left\{60+\frac{2}{15}x\right\}^2}\left(\frac{2}{15}\right) + \frac{\left\{36000\left\{\frac{1}{15}\right\}\left\{60+\frac{2}{15}x\right\}^2 - 2\times 36000\left\{\frac{x}{15}\right\}\left\{60+\frac{2}{15}x\right\}\left\{\frac{2}{15}\right\}\right\}}{\left\{60+\frac{2}{15}x\right\}^4} = 0$$

This indicates

$$6\times 2\times\left\{60+\frac{2}{15}x\right\} = \left\{36\left\{60+\frac{2}{15}x\right\} - 2\times 36\{x\}\left\{\frac{2}{15}\right\}\right\}$$

$$\left\{60+\frac{2}{15}x\right\} = \left\{180+\frac{6}{15}x\right\} - \left\{\frac{12x}{15}\right\}$$

$$x\left(\frac{8}{15}\right) = 120$$

$$x = 225.00 \text{ mm.}$$

At a distant 225 mm from left end, the maximum stress at extreme fiber (at top) occurs.

$$\sigma_{\max} = \frac{6000}{\left\{60+\frac{2}{15}\times 225\right\}} + \frac{36\times 1000\times\left\{\frac{225}{15}\right\}}{\left\{60+\frac{2}{15}\times 225\right\}^2} = 66.67+66.67 = 133.34 \text{ MPa.}$$

Top fiber stress at right end (B):

The eccentricity is $e = \frac{180}{2} - \frac{60}{2} = 60$ mm.

Dimensions are 60 mm wide and 180 mm deep.
Load $P = 360$ kN.

Axial stress $\sigma_a = \frac{360\times 1000}{60\times 180} = 33.33$ MPa

Maximum bending stress $\sigma_b = \frac{Pe}{I}y_t = \frac{360\times 1000\times 60\times\frac{180}{2}}{\frac{60\times 180^3}{12}} = 66.67$

Maximum stress at top fiber at $B = \sigma = \sigma_a + \sigma_b = 33.33 + 66.67 = 100$ MPa

Variation of bending stress at top fiber along the length of the bar:

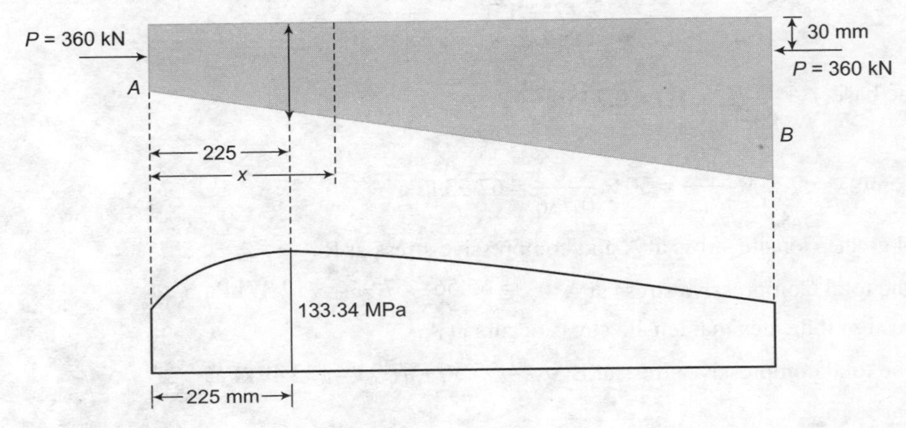

FIGURE 13.9

PROBLEM 13.4

Objective 1

Determine the stresses at the base of a hollow circular chimney of height 10 m is subjected to lateral load of 5 kN at the top as shown in Figure 13.10. Density of the masonry material is 25 kN/m^3. Diameter of the central hollow portion is 1.0 m. Outer diameter of the base portion is 2.0 m.

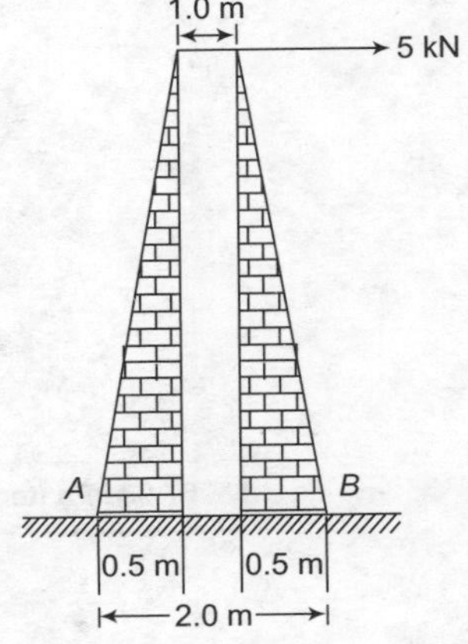

FIGURE 13.10

SOLUTION

Base section of the chimney is subjected to axial load due to self weight NA BM due to lateral load 25 kN.

Dead weight of the chimney (*W*):

$$W = \left\{\frac{1}{2}\times 0.5\times 10\right\}\left\{\pi\times\left(\frac{1}{2}+\frac{1}{3}\times 0.5\right)\right\}\times 25 = 130.90\,\text{kN}$$

In the above expression, volume of the masonry portion can be obtained from the Pappu's theorem. The volume generated by a surface about a nonintersecting axis is given by the product of the area of the surface and the distance travelled by the centroid of the surface in one revolution. First term is area of the triangular portion. Second term is the distance travelled by the centroid of the triangular portion about the center of the chimney (axis of rotation). The third term is the density of the masonry material.

BM at the base

$$M = 5\times 10 = 50\,\text{kN·m}$$

The cross-section at the base is hollow circular.

Axial stress due to self weight of the column

$$\sigma_a = \frac{W}{A} = \frac{130.90}{\frac{\pi}{4}(2^2 - 1^2)} = 55.56 \text{ kPa (compressive stress)}$$

MI at the base $I = \frac{\pi}{64}(2^4 - 1^4) = 0.736\,\text{m}^4$

Bending stress $\sigma_b = M\frac{y_{\max}}{I} = 50 \times \frac{1}{0.736} = 67.93$ kPa.

The BM creates tensile stress at A and compressive stress at B.

Hence the total Compressive stress at A $\sigma_A = 55.56 - 67.93 = -12.37\,\text{kPa}$

Negative sign indicates that tensile stress occurs at A.

Hence the total compressive stress at B $\sigma_B = 55.56 + 67.93 = 123.49\,\text{kPa}$

Location of NA

Let the NA is located x distant from A. At NA, final stress is zero.

$$\sigma = 55.56 - 50 \times \frac{(1-x)}{0.736} = 0$$

$$x = 0.182 \text{ m}$$

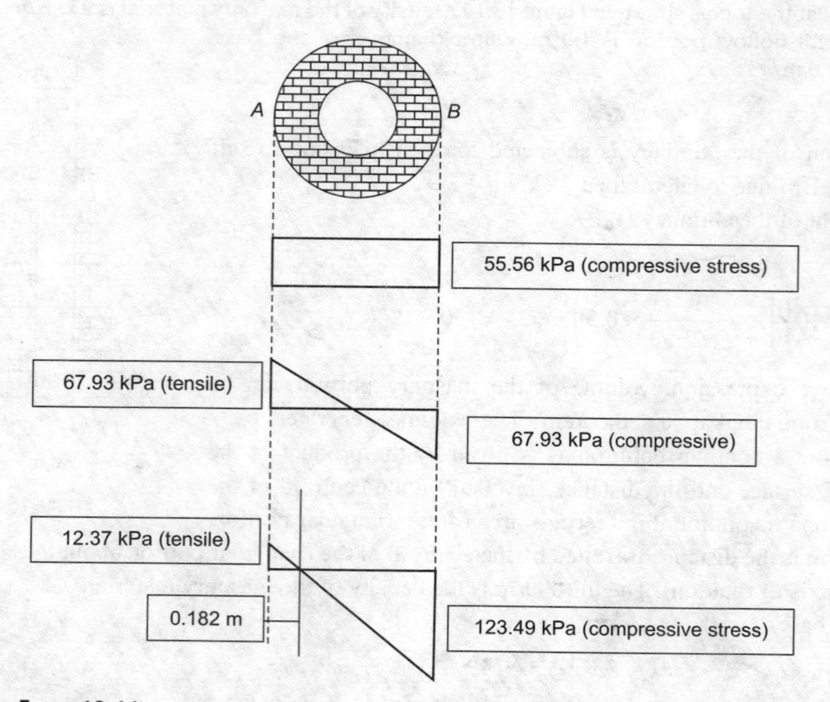

FIGURE 13.11

PROBLEM 13.5 **Objective 1**

A concrete dam 8 m high, 1.5 m wide at the top, and 4 m wide at the base has its front face vertical and retains water to a depth of 6 m. Find the maximum and minimum stress intensities at the base. The density of water is 10 kN/m^3 and that of masonry is 24 kN/m^3.

SOLUTION

Consider 1 m width of the dam.

To find stresses at the base of the dam, the base is acted upon the vertical load, eccentric load, and moment due to horizontal water pressure.

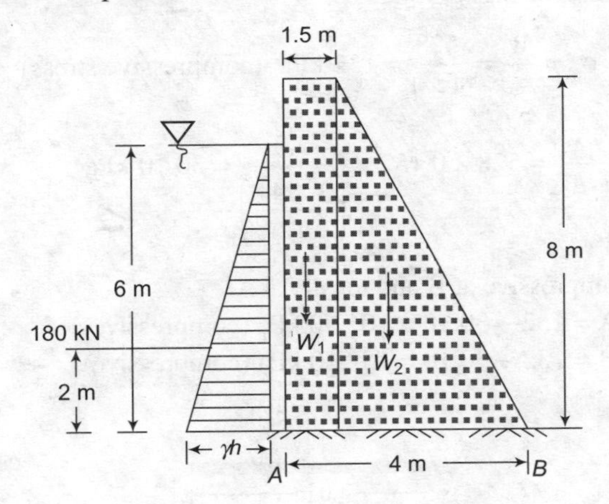

FIGURE 13.12.

Pressure at the base of the dam due to water $= P_h = \gamma h = 10 \times 6 = 60\,\text{kN/m}^2$

Total horizontal force $= H = \frac{1}{2}\gamma h \times h = \frac{1}{2} \times 60 \times 6 = \frac{180\,\text{kN}}{\text{m}}$ acts at $\frac{6}{3} = 2\,\text{m}$ (y)rom base

Vertical load of the dam portion is divided into two parts.

Weight of the rectangular portion $W_1 = 1.5 \times 8 \times 24 = 288\,\text{kN}$ acts at $\frac{1.5}{2} = 0.75$ m from A

Weight of the rectangular portion $W_2 = \frac{1}{2} \times 2.5 \times 8 \times 24 = 240\,\text{kN}$

W_2 acts at $1.5 + \frac{2.5}{3} = \frac{7}{3}$ m from A.

The three forces H, W_1, and W_2 acting at the base of the dam reduces to a single force and moment. The $\bar{x}$ be the location of the force from resultant force from A.

Taking moment about A, $\bar{x} = \dfrac{W_1 x_1 + W_2 x_2 + Hy}{W_1 + W_2}$

$$\bar{x} = \frac{288 \times 0.75 + 240 \times \frac{7}{3} + 180 \times 2}{288 + 240} = 2.152 \text{ m from A}$$

Total downward load = 288 + 240 = 528 kN.

The net effect of horizontal water force and weight of the dam can be replaced by a single vertical force 528 kN acting at 2.152 m from A.

This indicates that 528 kN load acts at an eccentricity of 2.152 – (4/2) = 0.152 m from the center of the base.

Axial compressive stress at the base

$$\sigma_a = \frac{W}{A} = \frac{528}{4 \times 1} = 132 \text{ kPa (compressive stress)}$$

$$\text{Bending stress } \sigma_b = M\frac{y_{max}}{I} = 528 \times 0.152 \times \frac{2.0}{\left(\frac{1 \times 4^3}{12}\right)} = 30.10 \text{ kPa}$$

This bending stress is compressive at B and tensile at A.

Final stress at A = 132 – 30.10 = 101.90 kPa (compressive)

Final stress at B = 132 + 30.10 = 152.10 kPa (compressive).

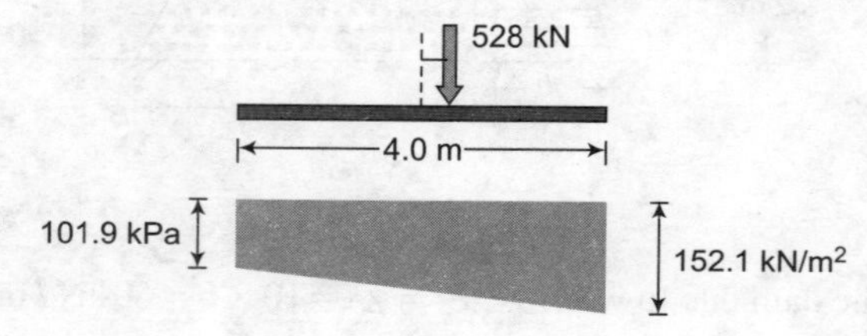

FIGURE 13.13

PROBLEM 13.6 — Objective 1

A masonry retaining wall is 10 m high and retains earth weighting 2000 kg/m^3. The top width of the retaining wall is 2 m and bottom width 6 m. The angle of repose is 30°. Weight of masonry is 2400 kg/m^3. Determine the maximum and minimum stresses in the wall at base.

SOLUTION

Considering 1 m length of the wall, weight of masonry

$$W = \frac{1}{2}(2 + 6) \times 10 \times 2400 \times 9.81 = 941.76 \text{ kN}$$

Distance of C.G (Center of Gravity) from vertical face

$$\bar{x} = \frac{10 \times 2 \times 1 + \frac{1}{2} \times 10 \times 4 \times \left(2 + \frac{4}{3}\right)}{60} = \frac{20 + 66.67}{60} = \frac{86.667}{60} = 1.44 \text{ m}$$

Earth pressure, $P = \frac{wH^2}{2}\left(\frac{1-\sin\phi}{1+\sin\phi}\right) = 9.81 \times 2000 \times \frac{100}{2}\left(\frac{1-\sin 30°}{1+\sin 30°}\right)$

$= 327 \text{ kN}$

Eccentricity, $e = \bar{x} + \frac{P}{W}\frac{H}{3} - \frac{B}{2} = 1.44 + \frac{327}{941.76} \times \frac{10}{3} - 3 = 1.44 - 1.157 - 3 = -0.403 \text{ m}$

Axial stress, $\sigma_a = \frac{W}{B} = \frac{941.76}{6} = 156.96 \text{ kN/m}^2$

Bending stress, $\sigma_b = \frac{We}{\frac{B^2}{6}} = \frac{941.76 \times 0.403}{36} = 63.25 \text{ kN/m}^2$

Maximum compressive stress = $156.96 + 63.25 = 220.21 \text{ kN/m}^2$
Minimum compressive stress = $156.96 - 63.25 = 93.71 \text{ kN/m}^2$

PROBLEM 13.7 **Objective 1**

A 150 mm × 20 mm steel plate is subjected to a pull of 150 kN along its longitudinal censorial axis. A hole of 40 mm diameter is drilled through the plate whose center is 50 mm from the original longitudinal axis of the plate as shown in Figure 13.14. Calculate the maximum stresses induced in the plate.

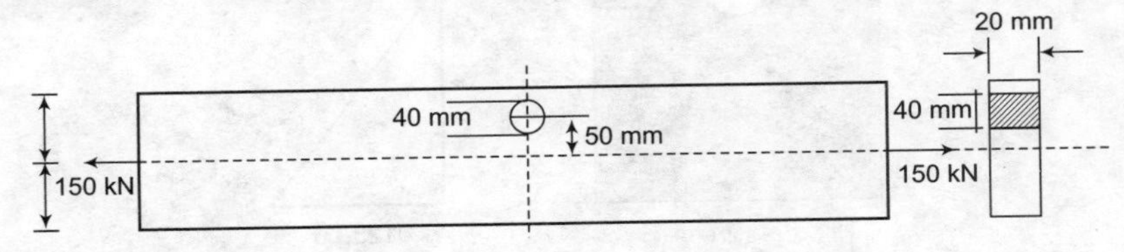

FIGURE 13.14

SOLUTION

Net area of plate, A = $150 \times 20 - 20 \times 40 = 2200 \text{ mm}^2$

$$2200 \times \bar{y} = 150 \times 20 \times 75 - 40 \times 20 \times 125$$

$$\bar{y} = 56.82 \text{ mm}$$

$$I_{NA} = \left[\frac{20 \times 150^3}{12} + 150 \times 20 \times (75 - 56.82)^2\right] - \left[\frac{20 \times 40^3}{12} + 20 \times 40 \times (125 - 56.82)^2\right]$$

$$= (5.625 \times 10^6 + 0.99154) - (0.1067 + 3.71881) = 2.791 \times 10^6 \text{ mm}^4$$

Direct stress, $\sigma_a = \dfrac{P}{A} = \dfrac{150 \times 10^3}{2200} = +\,68.18$ N/mm^2 (tensile stress)

Eccentricity of load, $e = 75 - 56.82 = 18.18$ mm

BM, M = Pe = $150 \times 10^3 \times 18.18 = 2727 \times 10^3$ N·mm (hogging BM)

$$\text{Bending stress at top of plate} = \frac{2727 \times 10^3 \times (150 - 56.82)}{2.791 \times 10^6}$$

$$= -91.04 \text{ N/mm}^2 \text{ (compressive stress)}$$

$$\text{Bending stress at bottom of plate} = \frac{+2727 \times 10^3 \times 56.82}{2.791 \times 10^6}$$

$$= 55.52 \text{ N/mm}^2 \text{ (tensile stress)}$$

Resultant stress at top = $68.18 - 91.04 = -22.86$ N/mm^2 (compressive stress)

Resultant stress at bottom = $68.18 + 55.52 = 123.70$ N/mm^2 (tensile stress)

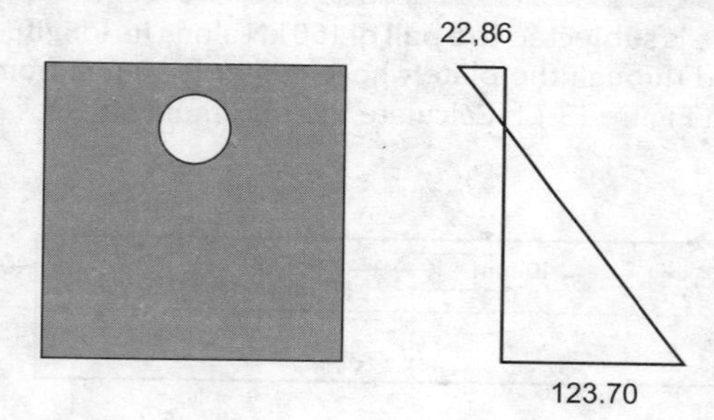

FIGURE 13.15

PROBLEM 13.8

Objective 1

A cracked rod of rectangular cross-section as shown in Figure 13.16 is loaded by a force *P*, the direction of which passes through the centroid *A* and *B* of the cross-section. At what point will the normal stress be the greatest? Find its value, if $P = 35$ kN, $h = 15$ cm, $b = 8$ cm, $a = 100$ cm, $c = 85$ cm, and $l = 2$ m.

SOLUTION

$$\tan\theta = \frac{\text{BC}}{\text{AC}} = \frac{115}{100} = 1.15$$

$$\theta = 49°$$

Vertical component of load,

$$P_v = P\cos\theta = 35\cos 49° = 22.96 \text{ kN}$$

Horizontal component of load,

$$P_H = P\sin\theta = 35\sin 49° = 26.42 \text{ kN}$$

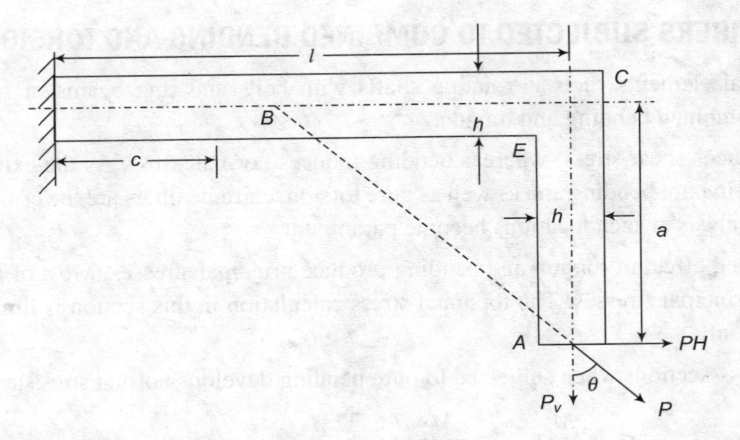

FIGURE 13.16

BM in member CD,

$$M = P_v \times l - P_H \times a$$
$$= 22.96 \times 2 - 26.42 \times 1$$
$$= 19.5 \text{ kN·m}$$

$$Z = \frac{bh^2}{6} = \frac{8 \times 15^2}{6} = 300 \text{ cm}^3$$

Bending stress, $\sigma_b = \dfrac{M}{z} = \dfrac{19.5 \times 10^3}{300 \times 10^{-6}} = 65 \text{ MPa}$

Direct stress, $\sigma_d = \dfrac{P_H}{A} = \dfrac{26.42 \times 10^3}{8 \times 15 \times 10^{-4}} = +2.201 \text{ MPa}$

Maximum stress at point D

$$= \sigma_a + \sigma_b = 65 + 2.201 = 67.201 \text{ MPa}$$

BM in member AC,

$$M_E = P_H \times a = 26.42 \times 1 = 26.42 \text{ kN·m}$$

BM at point E,

$$\sigma_b = \frac{26.42 \times 10^3}{300 \times 10^{-6}} = 88.06 \text{ MPa}$$

$$\sigma_b = \frac{P_V}{A} = \frac{22.96 \times 10^3}{(8 \times 15 \times 10^{-4})} = 1.913 \text{ MPa}$$

Maximum stress at point $E = \sigma_b + \sigma_d = 88.06 + 1.913 = 89.973$ MPa

Total stress at point $E = 89.973 + 2.201 = 92.174$ MPa.

13.4 MEMBERS SUBJECTED TO COMBINED BENDING AND TORSION

Many structural elements such as rotating shafts with belts and ring beams of water tanks are subjected to combined bending and torsion.

Torsion induces shear stress, whereas bending induces normal stress. As the extreme fibers are highly stressed in pure bending and as well as pure torsion, extreme fibers are the critical section and hence stress analysis at such locations become paramount.

The combined effect of torsion and bending produce principal stresses. Most of the designs are based on the principal stresses. The torsional stress calculation in this section is limited to circular cross-sections only.

Circular cross-sections when subjected to pure bending develops normal stress at extreme fibers

$$\sigma_b = \frac{M}{I} y_{\max} = \frac{M}{\frac{\pi d^4}{64}} \frac{d}{2} = \frac{32M}{\pi d^3}$$

Circular cross-sections when subjected to pure twisting moment develops shear stress at extreme fibers

$$\tau_t = \frac{T}{J} r_{\max} = \frac{T}{\frac{\pi d^4}{32}} \frac{d}{2} = \frac{16T}{\pi d^3}$$

Thus, at extreme fiber both shear stress due to torsion and normal stress due to bending occur. These two stresses are orthogonal to each other.

The principal stresses thus induced are

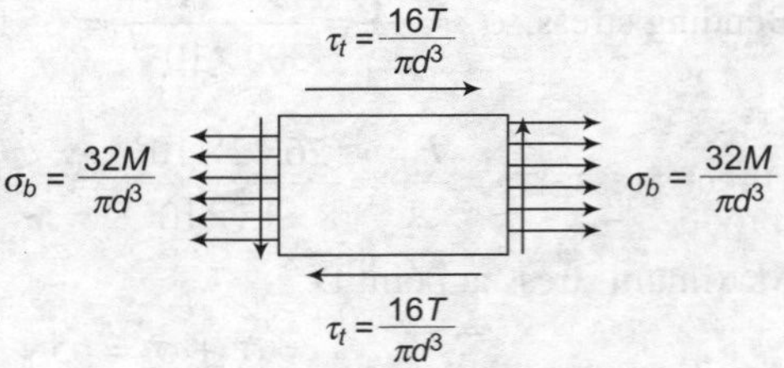

$$\sigma_{1.2} = \left\{\frac{\sigma_x + \sigma_y}{2}\right\} \pm \sqrt{\left\{\frac{\sigma_x - \sigma_y}{2}\right\}^2 + \{\tau\}^2}$$

$$\sigma_1 = \left\{\frac{\frac{32M}{\pi d^3}}{2}\right\} \pm \sqrt{\left\{\frac{\frac{32M}{\pi d^3}}{2}\right\}^2 + \left\{\frac{16T}{\pi d^3}\right\}^2}$$

FIGURE 13.17

$$\sigma_1 = \frac{16}{\pi d^3}\{M + \sqrt{\{M\}^2 + \{T\}^2}$$

$$\sigma_2 = \frac{16}{\pi d^3}\{\mathrm{M} - \sqrt{\{M\}^2 + \{T\}^2}\}$$

Maximum shear stress

$$\tau_{\max} = \sqrt{\left\{\frac{\sigma_x - \sigma_y}{2}\right\}^2 + \{\tau\}^2} \quad \text{or} \quad \frac{\sigma_1 - \sigma_2}{2}$$

$$\tau_{\max} = \frac{16}{\pi d^3}\{\sqrt{\{M\}^2 + \{T\}^2}\}.$$

Most of the structural elements design is based on either maximum normal stress theory or maximum shear stress theory. For maximum normal stress theory, σ_1 governs the design and in case of maximum shear stress theory τ_{max} governs the design.

The effect of torsion and bending can be combined in a such a way that single action either bending or torsion may result in normal stress or shear stress due to the combined torsion and bending.

In this direction, the equivalent twisting moment and equivalent bending moment are defined.

Equivalent BM Equivalent BM is the BM that alone produces a maximum normal stress equal to the maximum normal stress developed due to combined bending and torsion.

Let M_{eq} be the equivalent BM. Maximum normal stress due to M_{eq} is given by

$$\sigma = \frac{32M_{eq}}{\pi d^3}$$

Maximum normal stress due to combined bending and twisting is given by

$$\sigma_1 = \frac{16}{\pi d^3}\{M + \sqrt{\{M\}^2 + \{T\}^2}\}$$

By definition of equivalent BM $\sigma = \sigma_1$.

$$\frac{32M_{eq}}{\pi d^3} = \frac{16}{\pi d^3}\{M + \sqrt{\{M\}^2 + \{T\}^2}\}$$

Thus

$$M_{eq} = \frac{1}{2}\{M + \sqrt{\{M\}^2 + \{T\}^2}\}.$$

Equivalent twisting moment Equivalent twisting moment is the twisting moment that alone produces a maximum shear stress equal to the maximum shear stress developed due to combined bending and torsion.

Let T_{eq} be the equivalent twisting moment. Maximum shear stress due to T_{eq} is given by

$$\tau = \frac{16T_{eq}}{\pi d^3}$$

Maximum shear stress due to combined bending and twisting is given by

$$\tau_{max} = \frac{16}{\pi d^3}\{\sqrt{\{M\}^2 + \{T\}^2}\}$$

By definition of equivalent BM $\tau = \tau_{max}$.

$$\frac{16T_{eq}}{\pi d^3} = \frac{16}{\pi d^3}\{\sqrt{\{M\}^2 + \{T\}^2}\}$$

Thus, $T_{eq} = \sqrt{\{M\}^2 + \{T\}^2}$.

PROBLEM 13.9

Objective 3

A shaft of hollow shaft of outer diameter 100 mm and inner diameter 50 mm is subjected to a BM of 10 kN·m and twisting moment of 6 kN·m. Determine the maximum normal stress and maximum shear stress induced in the shaft at *A* and *B* shown Figure 13.18.

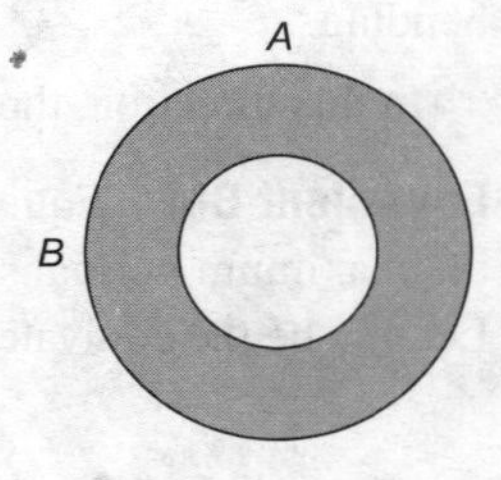

FIGURE 13.18

SOLUTION

BM = 10 kN·m

Twisting moment = 6 kN·m

Outer diameter = D_o = 100 mm

Inner diameter = D_i = 50 mm

$$\text{MI} = I = \frac{\pi}{64}(100^4 - 50^4) = 4.601 \times 10^6 \text{ mm}^4$$

$$\text{Polar MI} = J = \frac{\pi}{32}(100^4 - 50^4) = 9.202 \times 10^6 \text{ mm}^4$$

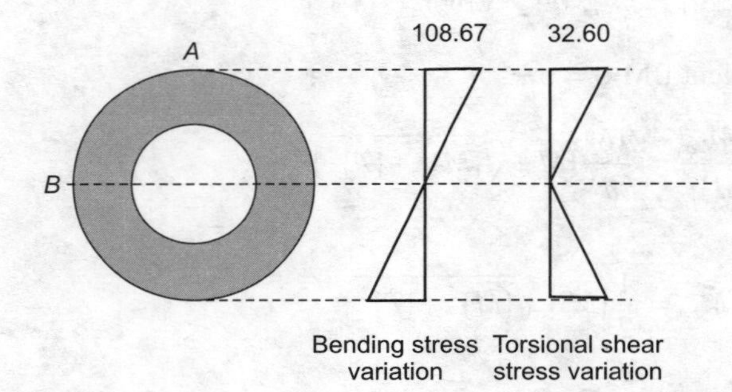

FIGURE 13.19

Maximum bending normal stress at the extreme fibers

$$\sigma_{\text{bending}} = \sigma = \frac{M}{I} y_{\max} = \frac{10 \times 10^6}{4.601 \times 10^6} \times 50 = 108.67 \text{ MPa}$$

Maximum torsional shear stress at the extreme fibers

$$\tau_{\text{torisional}} = \tau = \frac{T}{J} r_{\max} = \frac{6 \times 10^6}{9.202 \times 10^6} \times 50 = 32.60 \text{ MPa}$$

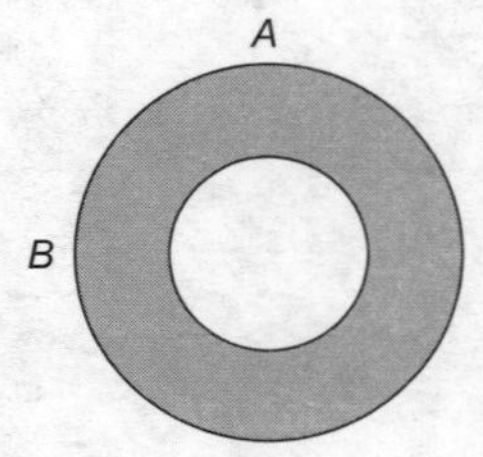

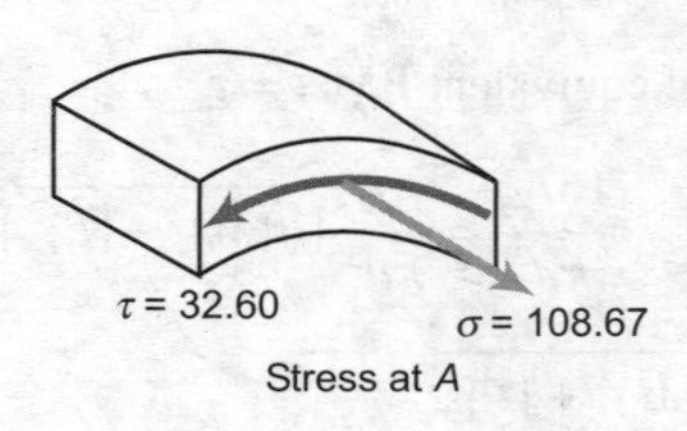

FIGURE 13.20

Maximum normal stress (maximum principal stress) at A due to combined bending and torsion is to be obtained.

$$\sigma_{1,2} = \left\{\frac{\sigma_x + \sigma_y}{2}\right\} \pm \sqrt{\left\{\frac{\sigma_x - \sigma_y}{2}\right\}^2 + \{\tau\}^2}$$

$$\sigma_x = 108.67 \text{ MPa}; \ \sigma_y = 0; \text{ and } \tau = 32.60 \text{ MPa}$$

$$\sigma_1 = \left\{\frac{108.67}{2}\right\} + \sqrt{\left\{\frac{108.67}{2}\right\}^2 + \{32.60\}^2} = 117.70 \text{ MPa}$$

Maximum shear stress due to combined bending and torsion

$$\tau_{max} = \sqrt{\left\{\frac{\sigma_x - \sigma_y}{2}\right\}^2 + \{\tau\}^2} \quad \text{or} \quad \frac{\sigma_1 - \sigma_2}{2}$$

$$\tau_{max} = \sqrt{\left\{\frac{108.67}{2}\right\}^2 + \{32.60\}^2} = 63.36 \text{ MPa}$$

Stresses at B:

Maximum bending normal stress at outer most fiber at neutral axis = 0

Maximum torsional shear stress at outer most fiber = τ = 32.60 MPa

Maximum normal stress (maximum principal stress) at *B* due to combined bending and torsion is to be obtained.

$$\sigma_{1,2} = \left\{\frac{\sigma_x + \sigma_y}{2}\right\} \pm \sqrt{\left\{\frac{\sigma_x - \sigma_y}{2}\right\}^2 + \{\tau\}^2}$$

$$\sigma_x = 0 \text{ MPa}; \ \sigma_y = 0; \text{ and } \tau = 32.60 \text{ MPa}$$

$$\sigma_1 = 32.60 \text{ MPa}$$

Maximum shear stress due to combined bending and torsion

$$\tau_{max} = 32.60$$

PROBLEM 13.10 **Objective 4 and 5**

A hollow shaft of outer diameter 150 mm and inner diameter 80 mm is rotating at 200 rpm, transmitting a power of 132 kW. The axial compressive load acting on the shaft is 800 kN. Determine the principal stresses and maximum shear stress due to this combined action of loads.

SOLUTION

Hollow shaft is rotating (ω) at 200 rpm

Power transmitted by the shaft = P = 132 kW

If torque developed in the shaft is T, then

$$P = \frac{2\pi\omega T}{60}$$

$$T = \frac{60P}{2\pi\omega} = \frac{60 \times 132}{2\pi \times 200} = 6.303 \text{ kN·m}$$

Axial compressive load on the shaft = P_a = 800 kN.

Axial compressive stress on the shaft = $\sigma = \dfrac{P_a}{A} = \dfrac{800 \times 1000}{\dfrac{\pi}{4}(150^2 - 80^2)} = 63.26$ MPa

This stress is uniform within the cross-section.

Polar MI = $J = \dfrac{\pi}{32}(150^4 - 80^4) = 45.68 \times 10^6 \text{ mm}^4$

Maximum torsional shear stress at the extreme fibers

$$\frac{T}{J} r_{\max} = \frac{6.303 \times 10^6}{45.68 \times 10^6} \times \left(\frac{150}{2}\right) = 10.35 \text{ MPa}$$

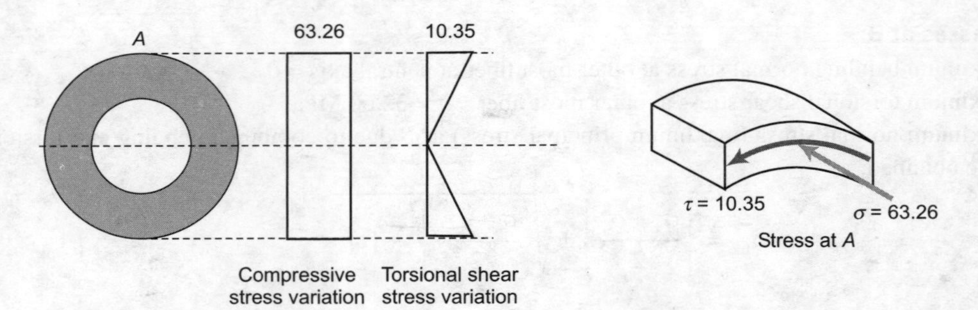

FIGURE 13.21

Maximum normal stress (maximum principal stress) at A due to combined bending and torsion is to be obtained.

$$\sigma_{1,2} = \left\{\frac{\sigma_x + \sigma_y}{2}\right\} \pm \sqrt{\left\{\frac{\sigma_x - \sigma_y}{2}\right\}^2 + \{\tau\}^2}$$

$$\sigma_x = 63.26 \text{ MPa}; \; \sigma_y = 0; \text{ and } \tau = 10.35 \text{ MPa}$$

$$\sigma_1 = \left\{\frac{63.26}{2}\right\} + \sqrt{\left\{\frac{63.26}{2}\right\}^2 + \{10.35\}^2} = 64.91 \text{ MPa}$$

Maximum principal stress is compressive in nature.

$$\sigma_2 = \left\{\frac{63.26}{2}\right\} - \sqrt{\left\{\frac{63.26}{2}\right\}^2 + \{10.35\}^2} = -1.65 \text{ MPa}$$

Minimum principal stress is tensile in nature.

Maximum shear stress due to combined bending and torsion

$$\tau_{max} = \sqrt{\left\{\frac{\sigma_x - \sigma_y}{2}\right\}^2 + \{\tau\}^2} \quad \text{or} \quad \frac{\sigma_1 - \sigma_2}{2}$$

$$\tau_{max} = \frac{\sigma_1 - \sigma_2}{2} = \frac{64.91 - (-1.65)}{2} = 33.28 \text{ MPa.}$$

13.5 COMBINED TORSION AND SHEAR

Combined shear and torsion occurs in heavy springs. Heavy spring under axial develop shear as well as torsion in the cross-section. As the stress developed due to these actions are shearing stress, the stresses can be added algebraically.

Shear stress due to shear force (SF) in the absence of bending = $\tau = \frac{V}{A}$, in which V is the direct SF and A is the cross-sectional area.

$$\tau = \frac{4V}{\pi d^2}$$

Shear stress due to torsion $\tau_t = \frac{T}{J} r_{max}$. The stress components are normal to the radial drawn from center to the point where shear stress is acting.

A diagrammatic presentation of this combined action is shown in Figure 13.22.

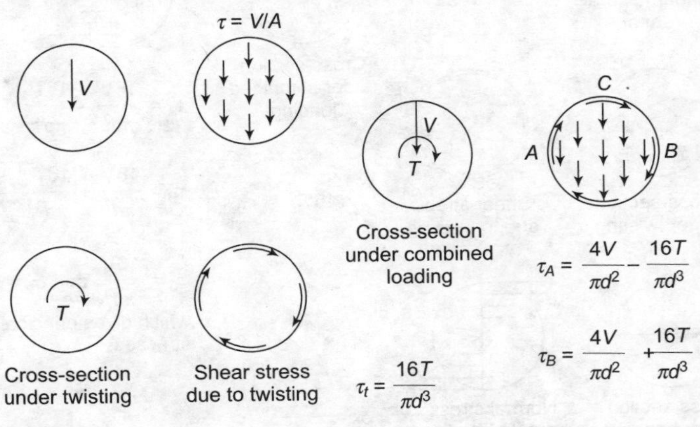

FIGURE 13.22

At A, shear stress is given by $\tau_A = \frac{4V}{\pi d^2} - \frac{16T}{\pi d^3}$

At B, shear stress is given by $\tau_B = \frac{4V}{\pi d^2} + \frac{16T}{\pi d^3}$

At C, shear stress due to torsion and direct shear are in one plane and orthogonal to each other. Hence the shear stress at C is given by their vectorial addition

$$\tau_c = \sqrt{\left(\frac{4V}{\pi d^2}\right)^2 + \left(\frac{16T}{\pi d^3}\right)^2}$$

13.6 COMBINED BENDING, TORSION, AND SHEAR

While considering the combined effect of torsion, bending, and shear, the shear is associated with bending. Thus, the expression for shear stress is given by

$$\tau_v = \frac{V}{Ib} A\bar{y}$$

This variation is parabolic with zero intensity at top fiber and bottom fiber.

The normal stress due to bending and shear stress due to torsion are to be evaluated as per the procedure mentioned earlier.

The stress due to these actions is presented in Figure 13.23. Circular cross-section is taken for convenience of explanation.

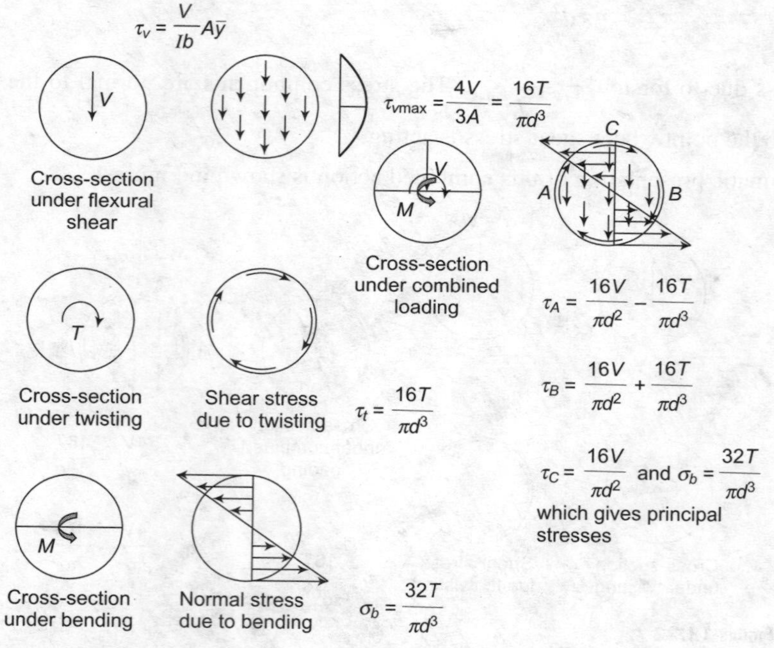

FIGURE 13.23

PROBLEM 13.11

Objective 5

A shaft of diameter 100 mm is subjected to a BM of 20 kN·m, SF of 50 kN due to varying bending, and a twisting moment of 10 kN·m. Determine the principal stresses and maximum shear stress in the cross-section at three points shown in Figure 13.24.

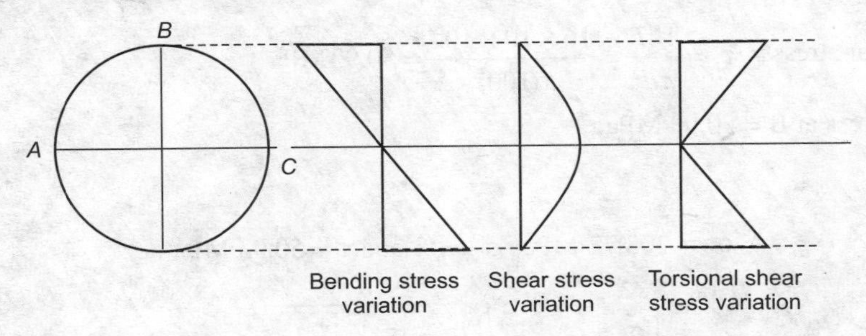

FIGURE 13.24

SOLUTION

BM = 20 kN·m

SF = 50 kN

Twisting moment = 10 kN·m

Stresses at A

Bending stress = 0 being NA

Shear stress due shear = $\tau_v = \frac{4}{3} \times \frac{V}{A} = \frac{16V}{3\pi d^2} = \frac{16 \times 50 \times 1000}{3\pi (100)^2} = 8.49\,\text{MPa}$

Torsional shear stress = $\tau_t = \frac{16T}{\pi d^3} = \frac{16 \times 10 \times 10^6}{\pi (100)^3} = 50.93$

Total shear stress at A = 50.93 – 8.49 = 42.44 MPa

Normal stress at A = 0

Thus at A,

$$\sigma_x = 0 \text{ MPa};\ \sigma_y = 0;\ \text{and}\ \tau = 42.44 \text{ MPa}$$

$$\sigma_{1,2} = \left\{\frac{\sigma_x + \sigma_y}{2}\right\} \pm \sqrt{\left\{\frac{\sigma_x - \sigma_y}{2}\right\}^2 + \{\tau\}^2}$$

$$\sigma_1 = 42.44 \text{ MPa}$$

$$\sigma_2 = -42.44 \text{ MPa}$$

$$\tau_{max} = \frac{\sigma_1 - \sigma_2}{2} = 42.44 \text{ MPa}$$

Stresses at B

Bending stress = $\sigma_b = \dfrac{32M}{\pi d^3} = \dfrac{32 \times 20 \times 10^6}{\pi(100)^3} = 203.72\,\text{MPa}$

Shear stress due shear = 0 as flexural shear at top fiber and bottom fiber is zero.

Torsional shear stress = $\tau_t = \dfrac{16T}{\pi d^3} = \dfrac{16 \times 10 \times 10^6}{\pi(100)^3} = 50.93\,\text{MPa}$

Total shear stress at B = 50.93 MPa.

Thus at B,

$$\sigma_x = 203.72 \text{ MPa; } \sigma_y = 0; \text{ and } \tau = 50.93 \text{ MPa}$$

$$\sigma_{1,2} = \left\{\frac{\sigma_x + \sigma_y}{2}\right\} \pm \sqrt{\left\{\frac{\sigma_x - \sigma_y}{2}\right\}^2 + \{\tau\}^2}$$

$$\sigma_1 = 101.86 + 113.88 = 2.15.74 \text{ MPa}$$

$$\sigma_2 = 101.86 - 113.88 = -12.02 \text{ MPa}$$

$$\tau_{max} = \frac{\sigma_1 - \sigma_2}{2} = 113.88 \text{ MPa.}$$

13.7 THEORIES OF FAILURE

In the previous sections, the stresses in the structural elements are determined for individual and combined actions such as axial compression, axial tension, bending, shear, and torsion. For the design of structural elements, we need to present a criterion for the failure of an element. Once we adopt a criterion for the failure of an element, the design can be done by suitably adopting factor of safety (FOS). It is mandatory to say that any criterion proposed for the failure of a structural element should be applicable irrespective of combination of loads. For example, a criterion proposed for failure of a member under axial tension should be same when the body is subjected to compression on all three sides. A member subjected to stress in all three directions is referred as member under hydrostatic state of stress. In practice, it is evident that under hydrostatic state of stress (compression) no material fails irrespective of the intensity of stress. Example for this is the existence of aquatic life underneath sea at large depths.

Let us consider a body acted upon by several forces. At a point within the body, the stresses can be represented by stress tensor explained in the first chapter.

$$[\sigma] = \begin{bmatrix} \sigma_{xx} & \tau_{yx} & \tau_{zx} \\ \tau_{xy} & \sigma_{yy} & \tau_{zy} \\ \tau_{xz} & \tau_{yz} & \sigma_{zz} \end{bmatrix}$$

The corresponding principal stress tensor can be presented as

$$[\sigma] = \begin{bmatrix} \sigma_1 & 0 & 0 \\ 0 & \sigma_2 & 0 \\ 0 & 0 & \sigma_3 \end{bmatrix}$$

It is already learnt in the principal stresses chapter that the effect above two stress tensors are the same. The only difference is that the stresses in the first tensor are with respect to *x*, *y*, and *z* Cartesian coordinate system, whereas the other stress tensor is with respect to principal axes. In fact, both the stress tensors are due to same loading only. The failure theories are generally proposed in terms of the principal stresses only.

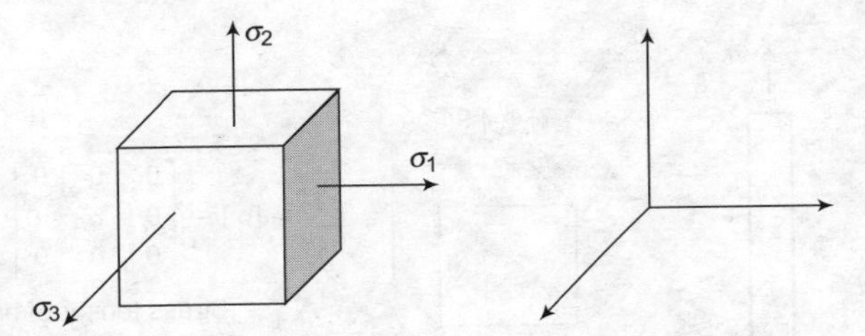

FIGURE 13.25 Principal stresses

13.8 MAXIMUM PRINCIPAL STRESS THEORY

This theory is proposed by Rankine. Often this theory is called Rankine's theory or maximum normal stress theory. This theory states that 'When maximum normal stress or maximum principal stress in a strained body reaches a particular value then failure of that element takes place'.

Consider a bar subjected to axial compressive stress. Let σ_f (yield stress) be the stress at which the bar has failed. The stress tensor for this loading case is given by

$$[\sigma] = \begin{bmatrix} 0 & 0 & 0 \\ 0 & \sigma_2 & 0 \\ 0 & 0 & 0 \end{bmatrix}$$

Maximum normal stress is σ_2.

As per Rakine's theory, when σ_2 reaches σ_f the bar fails. Let us adopt this theory for a body subjected to hydrostatic state of stress.

The stress tensor for hydrostatic state of stress is

$$[\sigma] = \begin{bmatrix} \sigma & 0 & 0 \\ 0 & \sigma & 0 \\ 0 & 0 & \sigma \end{bmatrix}$$

The maximum principal stress is $\sigma_1 = \sigma_2 = \sigma_3 = \sigma$.

As per maximum normal stress theory or maximum principal stress theory, when σ reaches σ_f the body should fail. That means a body fails, when hydrostatic stress in a body reaches the yield stress of the same body obtained in an uniaxial test. This means that if a concrete block fails at a stress of 20 MPa under uniaxial compression, then the same concrete element fails under 20 MPa. But any body subjected to hydrostatic stress never fails irrespective of the intensity of stress. This is the drawback of this theory.

PROBLEM 13.12 **Objective 6**

Determine the diameter of a solid circular shaft as per maximum normal stress theory subjected to a BM of 20 kN·m and twisting moment of 12 kN·m. The material of the shaft is yielded under axial tensile stress of 250 MPa. Take E = 200 GPa and Poisson's ratio $\mu = 0.25$. Apply an FOS as 2.5.

SOLUTION

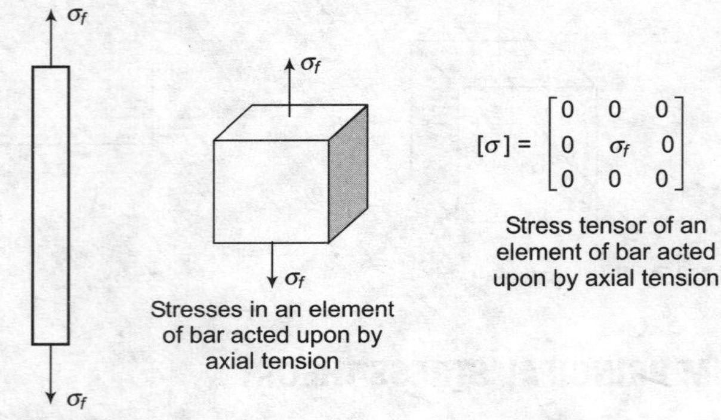

FIGURE 13.26

In case of a member subjected to axial tension, from the stress tensor, it is clear that $\sigma_1 = \sigma_f$ and $\sigma_2 = \sigma_3 = 0$.

Maximum principal stress = $\sigma_1 = \sigma_f = 250$ MPa.

Consider the case of a shaft subjected to torsion and bending.

$$\text{BM} = 20 \text{ kN·m}$$

Twisting moment = 12 kN·m

Shaft diameter = d

$$\text{MI} = I = \frac{\pi}{64} d^4$$

$$\text{Polar MI} = J = \frac{\pi}{32} d^4$$

Maximum bending normal stress at the extreme fibers

$$\sigma_{\text{bending}} = \sigma = \frac{M}{I} y_{\max} = \frac{M}{\frac{\pi}{64} d^4} \times \frac{d}{2} = \frac{32M}{\pi d^3} = \frac{32 \times 20 \times 10^6}{\pi d^3} = \frac{6.4 \times 10^8}{\pi d^3}$$

Maximum torsional shear stress at the extreme fibers

$$\tau_{\text{torisional}} = \tau = \frac{T}{J} r_{\max} = \frac{T}{\frac{\pi}{32} d^4} \times \frac{d}{2} = \frac{16T}{\pi d^3} = \frac{16 \times 12 \times 10^6}{\pi d^3} = \frac{1.92 \times 10^8}{\pi d^3}$$

Maximum normal stress (maximum principal stress) due to combined bending and torsion is to be obtained.

$$\sigma_{1,2} = \left\{\frac{\sigma_x + \sigma_y}{2}\right\} \pm \sqrt{\left\{\frac{\sigma_x - \sigma_y}{2}\right\}^2 + \{\tau\}^2}$$

$$\sigma_x = \frac{6.4 \times 10^8}{\pi d^3} \text{MPa}; \sigma_y = 0; \text{ and } \tau = \frac{1.92 \times 10^8}{\pi d^3} \text{ MPa}$$

$$\sigma_1 = \left\{\frac{\frac{6.4 \times 10^8}{\pi d^3}}{2}\right\} + \sqrt{\left\{\frac{\frac{6.4 \times 10^8}{\pi d^3}}{2}\right\}^2 + \left\{\frac{1.92 \times 10^8}{\pi d^3}\right\}^2} = \frac{6.93 \times 10^8}{\pi d^3}$$

$$\sigma_2 = \left\{\frac{\frac{6.4 \times 10^8}{\pi d^3}}{2}\right\} - \sqrt{\left\{\frac{\frac{6.4 \times 10^8}{\pi d^3}}{2}\right\}^2 + \left\{\frac{1.92 \times 10^8}{\pi d^3}\right\}^2} = -\frac{0.53 \times 10^8}{\pi d^3}$$

The stress tensor for this combined bending and torsion is

$$[\sigma] = \frac{10^8}{\pi d^3} \begin{bmatrix} 6.93 & 0 & 0 \\ 0 & 0 & 0 \\ 0 & 0 & -0.53 \end{bmatrix}$$

As per Rankine's theory, failure of the shaft takes place when maximum normal stress (in the above case $\sigma_1 = \frac{6.93 \times 10^8}{\pi d^3}$) reaches the maximum normal stress in axial tension case $\sigma_f = 250$ MPa.

Applying an FOS 2.5

Allowable stress will be $\frac{\text{failure stress}}{\text{FOS}} = \frac{250}{2.5} = 100$ MPa

Now, for design of the shaft $\sigma_1 = \frac{\sigma_f}{\text{FOS}} = \frac{250}{2.5} = 100$ MPa

$$\frac{6.93 \times 10^8}{\pi d^3} = 100$$

$$d = 130.2 \text{ mm}$$

Provide a shaft of diameter more than 130.2 mm.

13.9 MAXIMUM PRINCIPAL STRAIN THEORY

This theory is proposed by Saint Venant. Often this theory is called maximum normal strain theory. This theory states that 'When maximum normal strain or maximum principal strain in a strained body reaches a particular value then failure of that element takes place'.

Maximum normal strain = $\in_{max} = \frac{1}{E}\{\sigma_1 - \mu(\sigma_2 + \sigma_3)\}$, in which σ_1, σ_2, and σ_3 are the principal stresses.

Consider a bar subjected to axial compressive stress. Let σ_f (yield stress) be the stress at which the bar has failed. The stress tensor for this loading case is given by

$$[\sigma] = \begin{bmatrix} 0 & 0 & 0 \\ 0 & \sigma_f & 0 \\ 0 & 0 & 0 \end{bmatrix}$$

Maximum normal strain is $\in_{max} = \frac{\sigma_f}{E}$, in which E is modulus of elasticity of the material.

As per maximum normal strain theory, when $\in_{max}$ reaches $\in_f$ the bar fails. Let us adopt this theory for a body subjected to hydrostatic state of stress.

The stress tensor for hydrostatic state of stress is

$$[\sigma] = \begin{bmatrix} \sigma & 0 & 0 \\ 0 & \sigma & 0 \\ 0 & 0 & \sigma \end{bmatrix}$$

The principal stresses are $\sigma_1 = \sigma_2 = \sigma_3 = \sigma$.

Maximum normal strain = $\in_{max} = \frac{1}{E}\{\sigma_1 - \mu(\sigma_2 + \sigma_3)\}$

In case of hydrostatic stress,

$$\in_{max} = \frac{1}{E}\{\sigma - \mu(\sigma + \sigma)\} = \sigma\left\{\frac{1-2\mu}{E}\right\}.$$

As per maximum normal strain theory or maximum principal strain theory, when $\in_{max}$ reaches $\in_f$ the body should fail. This indicates that

$$\sigma\left\{\frac{1-2\mu}{E}\right\} = \frac{\sigma_f}{E}$$

$$\sigma = \frac{\sigma_f}{1-2\mu}$$

Let us consider a case of concrete block that fails under an axial compressive stress of 25 MPa. Taking $E = 25000$ MPa and $\mu = 0.15$.

$$\sigma = \frac{\sigma_f}{1-2\mu} = \frac{25}{1-2\times 0.15} = 35.71\,\text{MPa}$$

As per this theory of failure, concrete block failed at an axial compressive stress of 25 MPa fails at a stress of 35.71 MPa under hydrostatic stress.

However, this not true. Any body subjected to hydrostatic stress never fails irrespective of the intensity of stress.

This theory also cannot explain the failure mechanism of a body subjected to hydrostatic stress.

PROBLEM 13.13

Objective 6

Determine the diameter of a solid circular shaft as per maximum normal stress theory subjected to a BM of 20 kN·m and twisting moment of 12 kN·m. The material of the shaft is yielded under axial tensile stress of 250 MPa. Take $E = 200$ GPa and Poisson's ratio $\mu = 0.25$. Apply an FOS as 2.5.

SOLUTION

In case of a member subjected to axial tension, from the stress tensor, it is cleat that $\sigma_1 = \sigma_f$ and $\sigma_2 = \sigma_3 = 0$.

Maximum principal stress = $\sigma_1 = \sigma_f = 250\,\text{MPa}$.

Consider the case of a shaft subjected to torsion and bending

BM = 20 kN·m

Twisting moment = 12 kN·m

$E = 200$ GPa

Poisson's ratio $\mu = 0.25$

Shaft diameter = d

$$\text{MI} = I = \frac{\pi}{64}d^4$$

$$\text{Polar MI} = J = \frac{\pi}{32}d^4$$

Maximum bending normal stress at the extreme fibers

$$\sigma_{\text{bending}} = \sigma = \frac{M}{I}y_{\max} = \frac{M}{\frac{\pi}{64}d^4}\times\frac{d}{2} = \frac{32M}{\pi d^3} = \frac{32\times 20\times 10^6}{\pi d^3} = \frac{6.4\times 10^8}{\pi d^3}$$

Maximum torsional shear stress at the extreme fibers

$$\tau_{\text{torisional}} = \tau = \frac{T}{J}r_{\max} = \frac{T}{\frac{\pi}{32}d^4}\times\frac{d}{2} = \frac{16T}{\pi d^3} = \frac{16\times 12\times 10^6}{\pi d^3} = \frac{1.92\times 10^8}{\pi d^3}$$

Maximum normal stress (maximum principal stress) due to combined bending and torsion is to be obtained.

$$\sigma_{1,2} = \left\{\frac{\sigma_x + \sigma_y}{2}\right\} \pm \sqrt{\left\{\frac{\sigma_x - \sigma_y}{2}\right\}^2 + \{\tau\}^2}$$

$$\sigma_x = \frac{6.4 \times 10^8}{\pi d^3}\text{MPa};\ \sigma_y = 0;\ \text{and}\ \tau = \frac{1.92 \times 10^8}{\pi d^3}\text{MPa}$$

$$\sigma_1 = \left\{\frac{\frac{6.4 \times 10^8}{\pi d^3}}{2}\right\} + \sqrt{\left\{\frac{\frac{6.4 \times 10^8}{\pi d^3}}{2}\right\}^2 + \left\{\frac{1.92 \times 10^8}{\pi d^3}\right\}^2} = \frac{6.93 \times 10^8}{\pi d^3}$$

$$\sigma_{1,2} = \left\{\frac{\frac{6.4 \times 10^8}{\pi d^3}}{2}\right\} - \sqrt{\left\{\frac{\frac{6.4 \times 10^8}{\pi d^3}}{2}\right\}^2 + \left\{\frac{1.92 \times 10^8}{\pi d^3}\right\}^2} = -\frac{0.53 \times 10^8}{\pi d^3}$$

The stress tensor for this combined bending and torsion is

$$[\sigma] = \frac{10^8}{\pi d^3}\begin{bmatrix} 6.93 & 0 & 0 \\ 0 & 0 & 0 \\ 0 & 0 & -0.53 \end{bmatrix}$$

In this case, maximum normal strain = $\epsilon_{\max} = \frac{1}{E}\{\sigma_1 - \mu(\sigma_2 + \sigma_3)\}$

$$\epsilon_{\max} = \frac{\frac{10^8}{\pi d^3}}{200 \times 1000}\{6.93 - 0.25(0 - 0.53)\} = \frac{3.53 \times 10^3}{\pi d^3}$$

In case of axial tension, $\sigma_1 = \sigma_f$; and $\sigma_2 = \sigma_3 = 0$.

Applying an FOS 2.5

Allowable stress will be $\frac{\text{failure stress}}{\text{FOS}} = \frac{250}{2.5} = 100$ MPa

Now, for design of the shaft $\sigma_1 = \frac{\sigma_f}{\text{FOS}} = \frac{250}{2.5} = 100$ MPa

Maximum normal strain is $\epsilon_{\max} = \frac{1}{E}\{\sigma_1 - \mu(\sigma_2 + \sigma_3)\} = \frac{\sigma_f}{E} = \frac{100}{200 \times 1000} = 5 \times 10^{-4}$

For design, maximum normal strain in the above cases shall be equated.

$$\in_{max} = \frac{3.53 \times 10^3}{\pi d^3} = 5 \times 10^{-4}$$

$$\frac{3.53 \times 10^3}{\pi d^3} = 5 \times 10^{-4}$$

$$d = 130.98 \text{ mm}$$

Provide a shaft of diameter more than 130.98 mm.

13.10 TOTAL STRAIN ENERGY DENSITY THEORY

Maximum strain energy density theory is proposed by Haigh. This theory states that 'When total strain energy density in a strained body reaches a particular value then failure of that element takes place'.

Strain energy density in a strained body is the area bounded by the stress–strain curve of the body.

When the stress at a point are considered, the three principal stresses are σ_1, σ_2 and σ_3 and the corresponding principal strains are ε_1, ε_2 and ε_3, then the total strain energy density is given by

$$U = \frac{1}{2}\{\sigma_1\varepsilon_1 + \sigma_2\varepsilon_2 + \sigma_3\varepsilon_3\}$$

We know that

$$\varepsilon_1 = \frac{1}{E}\{\sigma_1 - \mu(\sigma_2 + \sigma_3)\}$$

$$\varepsilon_2 = \frac{1}{E}\{\sigma_2 - \mu(\sigma_1 + \sigma_3)\}$$

$$\varepsilon_2 = \frac{1}{E}\{\sigma_3 - \mu(\sigma_1 + \sigma_2)\}$$

Thus, the total strain energy density

$$U = \frac{1}{2}\left\{\sigma_1\frac{1}{E}\{\sigma_1 - \mu(\sigma_2 + \sigma_3)\} + \sigma_2\frac{1}{E}\{\sigma_2 - \mu(\sigma_1 + \sigma_3)\} + \sigma_3\frac{1}{E}\{\sigma_3 - \mu(\sigma_1 + \sigma_2)\}\right\}$$

$$U = \frac{1}{2E}\{\{\sigma_1^2 - \mu(\sigma_1\sigma_2 + \sigma_1\sigma_3)\} + \{\sigma_2^2 - \mu(\sigma_1\sigma_2 + \sigma_2\sigma_3)\} + \{\sigma_3^2 - \mu(\sigma_1\sigma_3 + \sigma_2\sigma_3)\}\}$$

$$U = \frac{1}{2E}\{\sigma_1^2 + \sigma_2^2 + \sigma_3^2 - 2\mu\sigma_1\sigma_2 - 2\mu\sigma_2\sigma_3 - 2\mu\sigma_3\sigma_1\}$$

Consider a bar subjected to axial compressive stress. Let σ_f (yield stress) be the stress at which the bar has failed. The stress tensor for this loading case is given by

$$[\sigma] = \begin{bmatrix} 0 & 0 & 0 \\ 0 & \sigma_f & 0 \\ 0 & 0 & 0 \end{bmatrix}$$

$$\sigma_1 = \sigma_f;\ \sigma_2 = \sigma_3 = 0$$

Total strain energy density $= U_f = \dfrac{\sigma_1^2}{2E} = \dfrac{\sigma_f^2}{2E}$

As per total strain energy density theory, when U reaches U_f the bar fails. Let us adopt this theory for a body subjected to hydrostatic state of stress.

The stress tensor for hydrostatic state of stress is

$$[\sigma] = \begin{bmatrix} \sigma & 0 & 0 \\ 0 & \sigma & 0 \\ 0 & 0 & \sigma \end{bmatrix}$$

The principal stresses are $\sigma_1 = \sigma_2 = \sigma_3 = 0$.

Total strain energy density $= U = \dfrac{1}{2E}\{\sigma_1^2 + \sigma_2^2 + \sigma_3^2 - 2\mu\sigma_1\sigma_2 - 2\mu\sigma_2\sigma_3 - 2\mu\sigma_3\sigma_1\}$

$$U = \frac{3\sigma^2}{2E}\{1 - 2\mu\}$$

As per total strain energy density theory or when U reaches U_f the body should fail. This indicates that

$$\frac{3\sigma^2}{2E}\{1 - 2\mu\} = \frac{\sigma_f^2}{2E}$$

This yields the condition that

$$\sigma = \frac{\sigma_f}{\sqrt{3(1 - 2\mu)}}.$$

Let us consider a case of concrete block that fails under an axial compressive stress of 25 MPa. Taking $E = 25000$ MPa and $\mu = 0.15$.

$$\sigma = \frac{\sigma_f}{\sqrt{3(1 - 2\mu)}} = \frac{25}{\sqrt{3(1 - 2 \times 0.15)}} = 11.228\ \text{MPa}$$

As per this theory of failure, concrete block failed at an axial compressive stress of 25 MPa fails at a stress of 11.228 MPa under hydrostatic stress.

However, this is not true. Any body subjected to hydrostatic stress never fails irrespective of the intensity of stress.

This theory also cannot explain the failure mechanism of a body subjected to hydrostatic stress.

PROBLEM 13.14

Objective 6

Determine the diameter of a solid circular shaft as per total strain energy theory subjected to a BM of 20 kN·m and twisting moment of 12 kN·m. The material of the shaft is yielded under axial tensile stress of 250 MPa. Take $E = 200$ GPa and Poisson's ratio $\mu = 0.25$. Apply a FOS of 2.5.

SOLUTION

In case of a member subjected to axial tension, from the stress tensor, it is cleat that $\sigma_1 = \sigma_f$ and $\sigma_2 = \sigma_3 = 0$.

Total strain energy density

$$U_f = \frac{1}{2E}\{\sigma_f^2\}.$$

Consider the case of a shaft subjected to torsion and bending

$$\text{BM} = 20 \text{ kN·m}$$

Twisting moment = 12 kN·m

$$E = 200 \text{ GPa}$$

Poisson's ratio $\mu = 0.25$

Shaft diameter = d

$$\text{MI} = I = \frac{\pi}{64}d^4$$

$$\text{Polar MI} = J = \frac{\pi}{32}d^4$$

Maximum bending normal stress at the extreme fibers

$$\sigma_{\text{bending}} = \sigma = \frac{M}{I}y_{\max} = \frac{M}{\frac{\pi}{64}d^4} \times \frac{d}{2} = \frac{32M}{\pi d^3} = \frac{32 \times 20 \times 10^6}{\pi d^3} = \frac{6.4 \times 10^8}{\pi d^3}$$

Maximum torsional shear stress at the extreme fibers

$$\tau_{\text{torisional}} = \tau = \frac{T}{J}r_{\max} = \frac{T}{\frac{\pi}{32}d^4} \times \frac{d}{2} = \frac{16T}{\pi d^3} = \frac{16 \times 12 \times 10^6}{\pi d^3} = \frac{1.92 \times 10^8}{\pi d^3}$$

Maximum normal stress (maximum principal stress) due to combined bending and torsion is to be obtained.

$$\sigma_{1,2} = \left\{\frac{\sigma_x + \sigma_y}{2}\right\} \pm \sqrt{\left\{\frac{\sigma_x - \sigma_y}{2}\right\}^2 + \{\tau\}^2}$$

$$\sigma_x = \frac{6.4 \times 10^8}{\pi d^3}\text{MPa}; \sigma_y = 0; \text{ and } \tau = \frac{1.92 \times 10^8}{\pi d^3}\text{MPa}$$

$$\sigma_1 = \left\{\frac{\frac{6.4 \times 10^8}{\pi d^3}}{2}\right\} + \sqrt{\left\{\frac{\frac{6.4 \times 10^8}{\pi d^3}}{2}\right\}^2 + \left\{\frac{1.92 \times 10^8}{\pi d^3}\right\}^2} = \frac{6.93 \times 10^8}{\pi d^3}$$

$$\sigma_2 = \left\{\frac{\frac{6.4\times10^8}{\pi d^3}}{2}\right\} - \sqrt{\left\{\frac{\frac{6.4\times10^8}{\pi d^3}}{2}\right\}^2 + \left\{\frac{1.92\times10^8}{\pi d^3}\right\}^2} = -\frac{0.53\times10^8}{\pi d^3}$$

The stress tensor for this combined bending and torsion is

$$[\sigma] = \frac{10^8}{\pi d^3}\begin{bmatrix} 6.93 & 0 & 0 \\ 0 & 0 & 0 \\ 0 & 0 & -0.53 \end{bmatrix}$$

In this case, total strain energy density

$$U = \frac{1}{2E}\{\sigma_1^{\,2} + \sigma_2^{\,2} + \sigma_3^{\,2} - 2\mu\sigma_1\sigma_2 - 2\mu\sigma_2\sigma_3 - 2\mu\sigma_3\sigma_1\}$$

$$U = \frac{10^{16}}{2E\pi^2 d^6}\{6.93^2 + 0^2 + (-0.53)^2 + 2\times0.25\times6.93\times0.53\} = \frac{50.16\times10^{16}}{2E\pi^2 d^6}$$

In case of axial tension, $\sigma_1 = \sigma_f;\ \sigma_2 = \sigma_3 = 0$.

Applying an FOS 2.5

Allowable stress will be $\sigma_f = \dfrac{\text{failure stress}}{\text{FOS}} = \dfrac{250}{2.5} = 100$ MPa

$$U_f = \frac{1}{2E}\{\sigma_f^2\}.$$

Now, for design of the shaft

$$U_f = \frac{1}{2E}\{\sigma_f^2\} = \frac{100\times100}{2E} = \frac{10^4}{2E}$$

For design equating total strain energy density from two cases

$$\frac{10^4}{2E} = \frac{50.16\times10^{16}}{2E\pi^2 d^6}$$

$$d = 131.12 \text{ mm}$$

Provide a shaft of diameter more than 131.12 mm.

13.11 MAXIMUM SHEAR STRESS THEORY

Maximum shear stress theory is proposed by Tresca. This theory states that 'When maximum shear stress in a strained body reaches a particular value then failure of that element takes place'.
Maximum shear stress in terms of principal stress is given by

$$\tau_{max} = \frac{1}{2}\{\sigma_{max} - \sigma_{min}\}$$

in which σ_{max} and σ_{min} are maximum value and minimum values of principal stresses.

Consider a bar subjected to axial compressive stress. Let σ_f (yield stress) be the stress at which the bar has failed. The stress tensor for this loading case is given by

$$[\sigma] = \begin{bmatrix} 0 & 0 & 0 \\ 0 & \sigma_f & 0 \\ 0 & 0 & 0 \end{bmatrix}$$

$$\sigma_1 = \sigma_f;\ \sigma_2 = \sigma_3 = 0$$

Maximum shear stress is $\tau_{max\,f} = \frac{1}{2}\{\sigma_{max} - \sigma_{min}\} = \frac{\sigma_f}{2}$

As per maximum shear stress theory when τ_{max} reaches $\tau_{max\,f}$ the bar fails. Let us adopt this theory for a body subjected to hydrostatic state of stress.

The stress tensor for hydrostatic state of stress is

$$[\sigma] = \begin{bmatrix} \sigma & 0 & 0 \\ 0 & \sigma & 0 \\ 0 & 0 & \sigma \end{bmatrix}$$

The principal stresses are $\sigma_1 = \sigma_2 = \sigma_3 = \sigma$.

Maximum shear stress $\tau_{max} = \frac{1}{2}\{\sigma_{max} - \sigma_{min}\} = 0$

A body under hydrostatic state of stress fails if maximum shear stress reaches a value $\frac{\sigma_f}{2}$, in which σ_f is the yield stress of the material under uniaxial loading.

This never happens in the case of a block subjected to hydrostatic state of stress, where in the maximum shear stress $\tau_{max} = \frac{1}{2}\{\sigma_{max} - \sigma_{min}\} = 0$ never reach the value $\frac{\sigma_f}{2}$, irrespective of the intensity of hydrostatic stress.

This theory can explain the failure mechanism of a body subjected to hydrostatic stress.

This theory is applicable for brittle materials such as glass, cast iron, and concrete. However, this theory cannot be applied to ductile materials.

A theory applicable for ductile materials is maximum shear strain energy theory or distorsional strain energy theory.

PROBLEM 13.15 **Objective 6**

Determine the diameter of a solid circular shaft as per total strain energy theory subjected to a BM of 20 kN·m and twisting moment of 12 kN·m. The material of the shaft is yielded under axial tensile stress of 250 MPa. Take $E = 200$ GPa and Poisson's ratio $\mu = 0.25$. Apply a FOS of 2.5.

SOLUTION

In case of a member subjected to axial tension, from the stress tensor, it is cleat that $\sigma_1 = \sigma_f$ and $\sigma_2 = \sigma_3 = 0$.

Total strain energy density

$$U_f = \frac{1}{2E}\{\sigma_f^2\}.$$

Consider the case of a shaft subjected to torsion and bending

BM = 20 kN·m

Twisting moment = 12 kN·m

E = 200 GPa

Poisson's ratio $\mu = 0.25$

Shaft diameter = d

$$\text{MI} = I = \frac{\pi}{64}d^4$$

$$\text{Polar MI} = J = \frac{\pi}{32}d^4$$

Maximum bending normal stress at the extreme fibers

$$\sigma_{\text{bending}} = \sigma = \frac{M}{I}y_{\max} = \frac{M}{\frac{\pi}{64}d^4} \times \frac{d}{2} = \frac{32M}{\pi d^3} = \frac{32 \times 20 \times 10^6}{\pi d^3} = \frac{6.4 \times 10^8}{\pi d^3}.$$

Maximum torsional shear stress at the extreme fibers

$$\tau_{\text{torsional}} = \tau = \frac{T}{J}r_{\max} = \frac{T}{\frac{\pi}{32}d^4} \times \frac{d}{2} = \frac{16T}{\pi d^3} = \frac{16 \times 12 \times 10^6}{\pi d^3} = \frac{1.92 \times 10^8}{\pi d^3}.$$

Maximum normal stress (maximum principal stress) due to combined bending and torsion is to be obtained.

$$\sigma_{1,2} = \left\{\frac{\sigma_x + \sigma_y}{2}\right\} \pm \sqrt{\left\{\frac{\sigma_x - \sigma_y}{2}\right\}^2 + \{\tau\}^2}$$

$$\sigma_x = \frac{6.4 \times 10^8}{\pi d^3} \text{ MPa; } \sigma_y = 0; \text{ and } \tau = \frac{1.92 \times 10^8}{\pi d^3} \text{ MPa}$$

$$\sigma_1 = \left\{\frac{\frac{6.4 \times 10^8}{\pi d^3}}{2}\right\} + \sqrt{\left\{\frac{\frac{6.4 \times 10^8}{\pi d^3}}{2}\right\}^2 + \left\{\frac{1.92 \times 10^8}{\pi d^3}\right\}^2} = \frac{6.93 \times 10^8}{\pi d^3}$$

$$\sigma_2 = \left\{\frac{\frac{6.4\times10^8}{\pi d^3}}{2}\right\} - \sqrt{\left\{\frac{\frac{6.4\times10^8}{\pi d^3}}{2}\right\}^2 + \left\{\frac{1.92\times10^8}{\pi d^3}\right\}^2} = -\frac{0.53\times10^8}{\pi d^3}$$

For this case, $\tau_{\max} = \frac{1}{2}\{\sigma_{\max} - \sigma_{\min}\} = \frac{1\times10^8}{2\pi d^3}(6.93-(-0.53)) = \frac{7.46\times10^8}{2\pi d^3}$

In case of axial tension, $\sigma_1 = \sigma_f$; $\sigma_2 = \sigma_3 = 0$.

Applying an FOS 2.5

Allowable stress will be $\sigma_f = \frac{\text{failure stress}}{\text{FOS}} = \frac{250}{2.5} = 100$ MPa

$$\tau_{\max f} = \frac{1}{2}\{\sigma_{\max} - \sigma_{\min}\} = \frac{\sigma_f}{2} = \frac{100}{2}$$

For design equating maximum shear stress from two cases

$$\frac{100}{2} = \frac{7.46\times10^8}{2\pi d^3}$$

$$d = 133.41 \text{ mm}$$

Provide a shaft of diameter more than 133.41 mm.

13.12 MAXIMUM SHEAR STRAIN ENERGY THEORY

Maximum shear strain energy density theory is proposed by Von-Mises. This theory states that 'When total shear strain energy density in a strained body reaches a particular value then failure of that element takes place'. This theory is also called distorsional strain energy theory.

The stresses at point in a strained body are represented as σ_1, σ_2 and σ_3. Action of these stresses result in dialation (change in the volume of the element) and distorsion (change in the shape of the element). Dialation is due to average stress $\left(\sigma_v = \frac{\sigma_1+\sigma_2+\sigma_3}{3}\right)$. The stress tensor responsible for dialation and distorsion are shown in Figure 13.27.

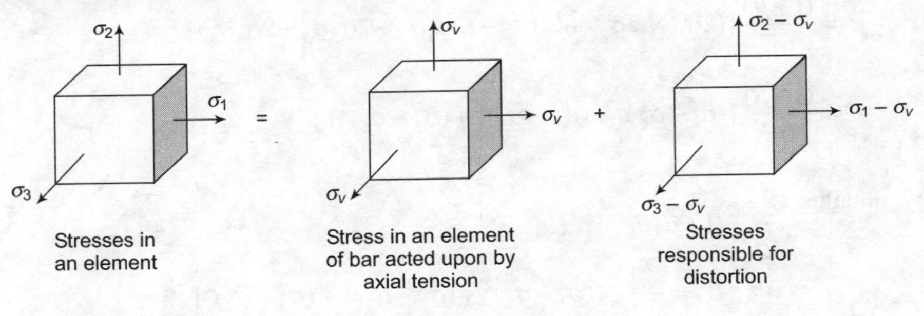

FIGURE 13.27

Shear strain energy is the strain energy stored in a body due the stresses that are responsible for distorsion. Thus, this shear strain energy density is referred as distorsional strain energy density. Distorsional strain energy = total strain energy – strain energy due to stresses that are responsible for dilation.

Distorsional strain energy $U_* = U - U_D$

Strain energy due to stresses that are responsible for dilation U_D.

$$U_D = \frac{(1-2\mu)\sigma_v^{\ 2}}{2E} = \frac{3(1-2\mu)\left(\dfrac{\sigma_1+\sigma_2+\sigma_3}{3}\right)^2}{2E} \tag{13.1}$$

Strain energy density due to three principal stresses is given by

$$U = \frac{1}{2E}\{\sigma_1^{\ 2}+\sigma_2^{\ 2}+\sigma_3^{\ 2}-2\mu\sigma_1\sigma_2-2\mu\sigma_2\sigma_3-2\mu\sigma_3\sigma_1\} \tag{13.2}$$

Distorsional strain energy $U_* = U - U_D$

$$U_* = \frac{1}{2E}\left\{(\sigma_1^{\ 2}+\sigma_2^{\ 2}+\sigma_3^{\ 2}-2\mu\sigma_1\sigma_2-2\mu\sigma_2\sigma_3-2\mu\sigma_3\sigma_1)-3(1-2\mu)\left(\frac{\sigma_1+\sigma_2+\sigma_3}{3}\right)^2\right\}$$

$$= \frac{1}{6E}\{(3\sigma_1^{\ 2}+3\sigma_2^{\ 2}+3\sigma_3^{\ 2}-6\mu\sigma_1\sigma_2-6\mu\sigma_2\sigma_3-6\mu\sigma_3\sigma_1)-(1-2\mu)(\sigma_1+\sigma_2+\sigma_3)^2\}$$

$$= \frac{1}{6E}\{(3\sigma_1^{\ 2}+3\sigma_2^{\ 2}+3\sigma_3^{\ 2}-6\mu\sigma_1\sigma_2-6\mu\sigma_2\sigma_3-6\mu\sigma_3\sigma_1)$$
$$-(\sigma_1^{\ 2}+\sigma_2^{\ 2}+\sigma_3^{\ 2}+2\sigma_1\sigma_2+2\sigma_2\sigma_3+2\sigma_3\sigma_1)$$
$$+2\text{ì}\,(\sigma_1^{\ 2}+\sigma_2^{\ 2}+\sigma_3^{\ 2}+2\sigma_1\sigma_2+2\sigma_2\sigma_3+2\sigma_3\sigma_1)\}$$

$$= \frac{1}{6E}\{(2\sigma_1^{\ 2}+2\sigma_2^{\ 2}+2\sigma_3^{\ 2}-2\mu\sigma_1\sigma_2-2\mu\sigma_2\sigma_3-2\mu\sigma_3\sigma_1)$$
$$-(2\sigma_1\sigma_2+2\sigma_2\sigma_3+2\sigma_3\sigma_1)+2\text{ì}\,(\sigma_1^{\ 2}+\sigma_2^{\ 2}+\sigma_3^{\ 2})\}$$

$$= \frac{1}{6E}\{(2\sigma_1^{\ 2}+2\sigma_2^{\ 2}+2\sigma_3^{\ 2})(1+\mu)-(2\sigma_1\sigma_2+2\sigma_2\sigma_3+2\sigma_3\sigma_1)(1+\mu)\}$$

$$= \frac{(1+\mu)}{6E}\{(2\sigma_1^{\ 2}+2\sigma_2^{\ 2}+2\sigma_3^{\ 2})-(2\sigma_1\sigma_2+2\sigma_2\sigma_3+2\sigma_3\sigma_1)\}$$

$$= \frac{(1+\mu)}{6E}\{(\sigma_1-\sigma_2)^2+(\sigma_2-\sigma_3)^2+(\sigma_3-\sigma_1)^2\}$$

Rigidity modulus $G = \dfrac{E}{2(1+\mu)}$

$$U_* = \frac{1}{12G}\{(\sigma_1-\sigma_2)^2+(\sigma_2-\sigma_3)^2+(\sigma_3-\sigma_1)^2\}.$$

When the stresses at a point are considered, the three principal stresses are σ_1, σ_2, and σ_3 then the shear strain energy density or distorsional strain energy density is given by

$$U_* = \frac{1}{12G}\{(\sigma_1-\sigma_2)^2+(\sigma_2-\sigma_3)^2+(\sigma_3-\sigma_1)^2\}$$

Consider a bar subjected to axial compressive stress. Let σ_f (yield stress) be the stress at which the bar has failed. The stress tensor for this loading case is given by

$$[\sigma] = \begin{bmatrix} 0 & 0 & 0 \\ 0 & \sigma_f & 0 \\ 0 & 0 & 0 \end{bmatrix}$$

$$\sigma_1 = \sigma_f;\ \sigma_2 = \sigma_3 = 0$$

$$\text{Total shear strain energy density} = U_{*f} = \frac{2\sigma_1^2}{12G} = \frac{2\sigma_f^2}{12G}$$

As per shear strain energy density theory, when U_* reaches U_{*f} the bar fails. Let us adopt this theory for a body subjected to hydrostatic state of stress.

The stress tensor for hydrostatic state of stress is

$$[\sigma] = \begin{bmatrix} \sigma & 0 & 0 \\ 0 & \sigma & 0 \\ 0 & 0 & \sigma \end{bmatrix}$$

The principal stresses are $\sigma_1 = \sigma_2 = \sigma_3 = \sigma$.

$$\text{Shear strain energy density} = U_* = \frac{1}{12G}\{(\sigma-\sigma)^2+(\sigma-\sigma)^2+(\sigma-\sigma)^2\}$$

$$U_* = 0.$$

This indicates that shear strain energy density in case of hydrostatic stress will never reach the shear strain energy density in an axially loaded element U_{*f}

Thus, this distorsional strain energy theory is able to explain why a body subjected to hydrostatic stress never fails irrespective of the intensity of stress.

PROBLEM 13.16

Objective 6

Determine the diameter of a solid circular shaft as per shear strain energy theory subjected to a BM of 20 kN·m and twisting moment of 12 kN·m. The material of the shaft is yielded under axial tensile stress of 250 MPa. Take $E = 200$ GPa and Poisson's ratio $\mu = 0.25$. Apply a FOS of 2.5.

SOLUTION

In case of a member subjected to axial tension, from the stress tensor, it is cleat that $\sigma_1 = \sigma_f$ and $\sigma_2 = \sigma_3 = 0$.

$$\text{Shear strain energy density} = U_* = \frac{1}{12G}\{(\sigma_1-\sigma_2)^2+(\sigma_2-\sigma_3)^2+(\sigma_3-\sigma_1)^2\}$$

$$U_{*f} = \frac{1}{12G}\{2\sigma_f{}^2\}$$

Consider the case of a shaft subjected to torsion and bending

BM = 20 kN·m

Twisting moment = 12 kN·m

E = 200 GPa

Poisson's ratio $\mu = 0.25$

Shaft diameter = d

$$\text{MI} = I = \frac{\pi}{64}d^4$$

$$\text{Polar MI} = J = \frac{\pi}{32}d^4$$

Maximum bending normal stress at the extreme fibers

$$\sigma_{\text{bending}} = \sigma = \frac{M}{I}y_{\max} = \frac{M}{\frac{\pi}{64}d^4}\times\frac{d}{2} = \frac{32M}{\pi d^3} = \frac{32\times 20\times 10^6}{\pi d^3} = \frac{6.4\times 10^8}{\pi d^3}$$

Maximum torsional shear stress at the extreme fibers

$$\tau_{\text{torisional}} = \tau = \frac{T}{J}r_{\max} = \frac{T}{\frac{\pi}{32}d^4}\times\frac{d}{2} = \frac{16T}{\pi d^3} = \frac{16\times 12\times 10^6}{\pi d^3} = \frac{1.92\times 10^8}{\pi d^3}$$

Maximum normal stress (maximum principal stress) due to combined bending and torsion is to be obtained.

$$\sigma_{1,2} = \left\{\frac{\sigma_x+\sigma_y}{2}\right\} \pm \sqrt{\left\{\frac{\sigma_x-\sigma_y}{2}\right\}^2 + \{\tau\}^2}$$

$$\sigma_x = \frac{6.4\times 10^8}{\pi d^3}\text{MPa};\ \sigma_y = 0;\ \text{and}\ \tau = \frac{1.92\times 10^8}{\pi d^3}\text{MPa}$$

$$\sigma_1 = \left\{\frac{\frac{6.4\times 10^8}{\pi d^3}}{2}\right\} + \sqrt{\left\{\frac{\frac{6.4\times 10^8}{\pi d^3}}{2}\right\}^2 + \left\{\frac{1.92\times 10^8}{\pi d^3}\right\}^2} = \frac{6.93\times 10^8}{\pi d^3}$$

$$\sigma_2 = \left\{\frac{\frac{6.4\times 10^8}{\pi d^3}}{2}\right\} - \sqrt{\left\{\frac{\frac{6.4\times 10^8}{\pi d^3}}{2}\right\}^2 + \left\{\frac{1.92\times 10^8}{\pi d^3}\right\}^2} = -\frac{0.53\times 10^8}{\pi d^3}$$

The stress tensor for this combined bending and torsion is

$$[\sigma] = \frac{10^8}{\pi d^3}\begin{bmatrix} 6.93 & 0 & 0 \\ 0 & 0 & 0 \\ 0 & 0 & -0.53 \end{bmatrix}$$

In this case, shear strain energy density $= U_* = \frac{1}{12G}\{(\sigma_1-\sigma_2)^2+(\sigma_2-\sigma_3)^2+(\sigma_3-\sigma_1)^2\}$

$$U_* = \frac{10^{16}}{12G\pi^2 d^6}\{(6.93-0)^2+(0-(-0.53))^2+((-0.53)-6.93)^2\} = \frac{103.96\times10^{16}}{12G\pi^2 d^6}$$

In case of axial tension, $\sigma_1 = \sigma_f;\ \sigma_2 = \sigma_3 = 0$.

Applying an FOS 2.5

Allowable stress will be $\sigma_f = \frac{\text{failure stress}}{\text{FOS}} = \frac{250}{2.5} = 100\text{ MPa}$

$$U_{*f} = \frac{1}{12G}\{2\sigma_f^{\,2}\}$$

Now, for design of the shaft

$$U_{*f} = \frac{1}{12G}\{2\sigma_f^{\,2}\} = \frac{2\times100\times100}{12G} = \frac{2\times10^4}{12G}$$

For design equating shear strain energy density from two cases

$$\frac{2\times10^4}{12G} = \frac{103.96\times10^{16}}{12G\pi^2 d^6}$$

$$d = 131.91\text{ mm}$$

Provide a shaft of diameter more than 131.91 mm.

PROBLEM 13.17

Objective 6

Determine the FOS as different failure theories of a body subjected to the following principal stresses. $\sigma_1 = 100$ MPa, $\sigma_2 = 20$ MPa, and $\sigma_3 = -20$ MPa. The material of the body yielded under uniaxial tensile stress of $\sigma_y = 300$ MPa.

SOLUTION

Let '*f*' be the FOS.

Then stress in the member is to be limited to $\sigma_{all} = \frac{\sigma_y}{f} = \frac{300}{f}$

The principal stresses in the strained body are given as

$\sigma_1 = 100$ MPa, $\sigma_2 = 20$ MPa, and $\sigma_3 = -20$ MPa

(a) Rankine's theory:

The element fails when maximum principal stress reaches $\sigma_{all} = \dfrac{\sigma_y}{f} = \dfrac{300}{f}$

Maximum principal stress = σ_1 = 100 MPa

$$\sigma_{all} = \sigma_1$$

$$\frac{300}{f} = 100$$

$$f = 3.0$$

(b) Maximum normal strain theory:

Maximum normal strain in case of uniaxial tension after allowing for safety $\varepsilon_f = \dfrac{\dfrac{\sigma_y}{f}}{E} = \dfrac{\sigma_y}{Ef}$

The element fails when maximum principal strain reaches $\varepsilon_f = \dfrac{\sigma_y}{Ef} = \dfrac{300}{f \times 2 \times 10^5}$

Maximum normal strain in case of the strained body = $\in_{max} = \in_1 = \dfrac{1}{E}\{\sigma_1 - \mu(\sigma_2 + \sigma_3)\}$

$$\sigma_1 = 100 \text{ MPa}, \ \sigma_2 = 20 \text{ MPa}, \text{ and } \sigma_3 = -20 \text{ MPa}$$

$$\in_{max} = \frac{1}{2 \times 10^5}\{100 - \mu(20 - 20)\} = \frac{100}{2 \times 10^5}$$

For assessing the safety in the element equate $\in_{max}$ to $\in_f$

$$\frac{100}{2 \times 10^5} = \frac{300}{f \times 2 \times 10^5}$$

$$f = 3.0$$

(c) Strain energy density theory:

In case of uniaxial tension case

$$\sigma_1 = \frac{\sigma_y}{f}; \ \sigma_2 = \sigma_3 = 0$$

Strain energy density in case of uniaxial tension after allowing for safety

$$U_f = \frac{\left(\dfrac{\sigma_y}{f}\right)^2}{2E}$$

The strained element fails when strain energy density reaches $U_f = \dfrac{\sigma_y}{Ef} = \dfrac{300^2}{f^2 \times 2 \times 2 \times 10^5}$

Strain energy density in case of the strained body

$$U = \frac{1}{2E}\{\sigma_1^{\,2} + \sigma_2^{\,2} + \sigma_3^{\,2} - 2\mu\sigma_1\sigma_2 - 2\mu\sigma_2\sigma_3 - 2\mu\sigma_3\sigma_1\}$$

$$\sigma_1 = 100\,\text{MPa}, \sigma_2 = 20\,\text{MPa}, \text{ and } \sigma_3 = -20\,\text{MPa}$$

$$U = \frac{1}{2\times 2\times 10^5}\{100^2 + 20^2 + (-20)^2 - 2\times 0.25(100\times 20 - 100\times 20 - 20\times 20)\} = \frac{11000}{2\times 2\times 10^5}$$

For assessing the safety in the element equate U to U_f

$$\frac{11000}{2\times 2\times 10^5} = \frac{300^2}{f^2\times 2\times 2\times 10^5}$$

$$f = 2.86$$

(d) Maximum shear stress theory:

Maximum shear stress is given by

$$\tau_{max} = \frac{\sigma_{max} - \sigma_{min}}{2}$$

in which σ_{max} and σ_{min} are the maximum and minimum values of three principal stresses.

In case of uniaxial tension case

$$\sigma_1 = \frac{\sigma_y}{f};\ \sigma_2 = \sigma_3 = 0$$

$$\sigma_{max} = \frac{\sigma_y}{f}; \sigma_{min} = 0$$

Maximum shear stress in case of uniaxial tension after allowing for safety $\tau_{max\,f} = \dfrac{\frac{\sigma_y}{f}}{2} = \dfrac{\sigma_y}{2f}$

As per maximum shear stress theory, the element fails when maximum shear stress reaches $\tau_{max\,f} = \dfrac{\sigma_y}{2f} = \dfrac{300}{2f}$

Maximum shear stress in case of the strained body $= \tau_{max} = \dfrac{\sigma_{max} - \sigma_{min}}{2}$

$$\sigma_1 = 100\text{ MPa},\ \sigma_2 = 20\text{ MPa, and } \sigma_3 = -20\text{ MPa}$$

$$\tau_{max} = \frac{\{100 - (-20)\}}{2} = \frac{120}{2}$$

For assessing the safety in the element equate τ_{max} to $\tau_{max\,f}$

$$\frac{120}{2} = \frac{300}{f\times 2}$$

$$f = 2.50$$

(e) Shear strain energy density theory:

Shear strain energy density is given by

$$U_* = \frac{1}{12G}\{(\sigma_1-\sigma_2)^2+(\sigma_2-\sigma_3)^2+(\sigma_3-\sigma_1)^2\}$$

In case of uniaxial tension

$$\sigma_1 = \frac{\sigma_y}{f};\ \sigma_2=\sigma_3=0$$

Shear strain energy density in case of uniaxial tension after allowing for safety

$$U_{*f} = \frac{2\sigma_f^2}{(12G)f^2} = \frac{180000}{(12G)f^2}$$

As per shear strain energy theory, the element fails when shear strain energy density reaches $U_{*f} = \frac{180000}{(12G)f^2}$.

Shear strain energy density in case of the strained body

$$\sigma_1 = 100\text{ MPa},\ \sigma_2 = 20\text{ MPa, and } \sigma_3 = -20\text{ MPa}$$

$$U_* = \frac{1}{12G}\{(\sigma_1-\sigma_2)^2+(\sigma_2-\sigma_3)^2+(\sigma_3-\sigma_1)^2\}$$

$$U_* = \frac{1}{12G}\{(100-20)^2+(20+20)^2+(-20-100)^2\} = \frac{22400}{12G}$$

For assessing the safety in the element equate U_* to U_{*f}

$$\frac{22400}{12G} = \frac{180000}{f^2\times 12G}$$

$$f = 2.83$$

Based on the five theories of failure, the factors of safety in the element are tabled below.

Table 13.1

Failure Theory	Factor of Safety
Maximum normal stress theory	3.00
Maximum normal strain theory	3.00
Strain energy density theory	2.86
Shear stress theory	2.50
Shear strain energy density theory	2.83

PROBLEM 13.18

Objective 6

A sign board of size 1 m × 1.2 m is rigidly attached to hollow steel pole of outer diameter 100 mm and inner diameter 50 mm as shown in Figure 13.28. Determine the safety present in the structure if the material of the shaft yielded at an axial stress of 150 MPa. Adopt a maximum normal stress theory and shear strain energy theory. The wind pressure normal to the sign board is 2.5 kN/m^2. Centroid of the sign board is 1.1 m from the center of the pole.

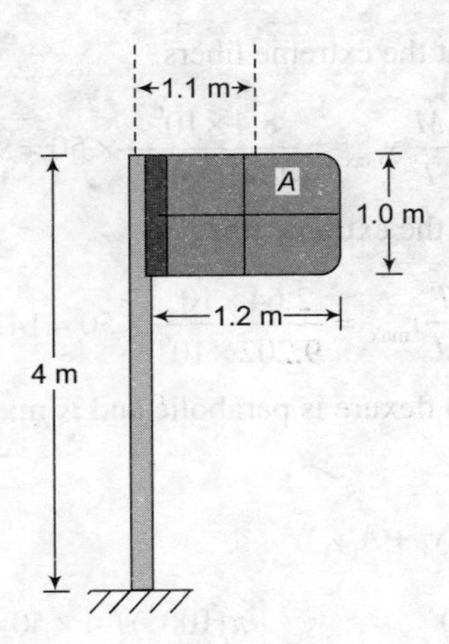

Figure 13.28

SOLUTION

Wind pressure acts normal to the plane of the sign board.

Total wind force = $F = 2.0 \times 1.2 \times 1.0 = 2.4$ kN acts at 1.1 m from the center of the pole and 3.5 m from base at A.

Wind force creates maximum BM, twisting moment, and SF at the base of the pole.

$$\text{BM at base} = M = 2.4 \times 3.5 = 8.4 \text{ kN}\cdot\text{m}$$

$$\text{Twisting moment at base} = T = 2.4 \times 1.1 = 2.64 \text{ kN}\cdot\text{m}$$

$$\text{SF at the base } F = 2.4 \text{ kN}$$

At the base of the pole

$$\text{Outer diameter} = D_o = 100 \text{ mm}$$

$$\text{Inner diameter} = D_i = 50 \text{ mm}$$

$$\text{MI} = I = \frac{\pi}{64}(100^4 - 50^4) = 4.601 \times 10^6 \text{ mm}^4$$

$$\text{Polar MI} = J = \frac{\pi}{32}(100^4 - 50^4) = 9.202 \times 10^6 \text{ mm}^4$$

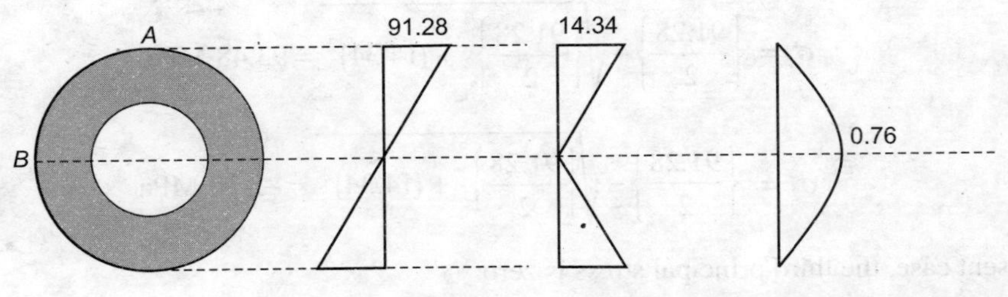

Figure 13.29

Maximum bending normal stress at the extreme fibers

$$\sigma_{\text{bending}} = \sigma = \frac{M}{I} y_{\max} = \frac{8.4 \times 10^6}{4.601 \times 10^6} \times 50 = 91.28\,\text{MPa}$$

Maximum torsional shear stress at the extreme fibers

$$\tau_{\text{torisional}} = \tau = \frac{T}{J} r_{\max} = \frac{2.64 \times 10^6}{9.202 \times 10^6} \times 50 = 14.34\,\text{MPa}$$

The variation of shear stress due to flexure is parabolic and is maximum at NA at B.
Maximum flexural shear stress =

$$\tau_v = \frac{V}{Ib} A\bar{y} = \frac{V}{Ib}(A_1 y_1 + A_1 y_1)$$

$$= \frac{2.4 \times 10^3}{4.601 \times 10^6 \times (100 - 50)} \times \left\{ \frac{\pi(100^2)}{4 \times 2}\left(\frac{4 \times 50}{3\pi}\right) - \frac{\pi(50^2)}{4 \times 2}\left(\frac{4 \times 25}{3\pi}\right) \right\} = 0.76\,\text{MPa}$$

The critical point in the cross-section is A, where the bending stress and torsional stress are maximum. Thus, obtain the principal stresses at A for estimating the safety of the structure.

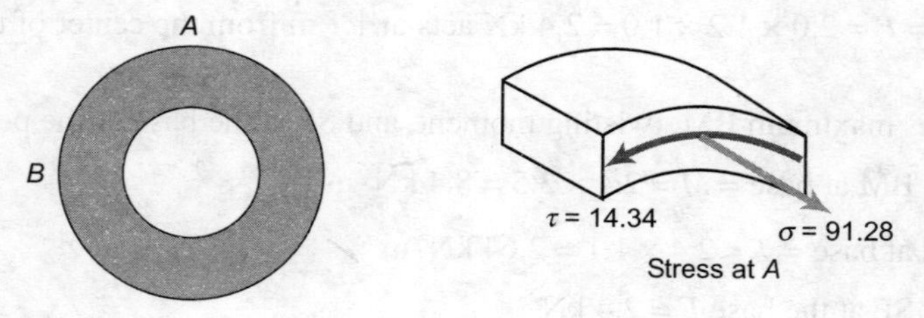

FIGURE 13.30

Maximum normal stress (maximum principal stress) at A due to combined bending and torsion is to be obtained.

$$\sigma_{1,2} = \left\{\frac{\sigma_x + \sigma_y}{2}\right\} \pm \sqrt{\left\{\frac{\sigma_x - \sigma_y}{2}\right\}^2 + \{\tau\}^2}$$

$$\sigma_x = 91.28\text{ MPa};\ \sigma_y = 0;\text{ and }\ \tau = 14.34\text{ MPa}$$

$$\sigma_1 = \left\{\frac{91.28}{2}\right\} + \sqrt{\left\{\frac{91.28}{2}\right\}^2 + \{14.34\}^2} = 93.48\text{ MPa}$$

$$\sigma_2 = \left\{\frac{91.28}{2}\right\} - \sqrt{\left\{\frac{91.28}{2}\right\}^2 + \{14.34\}^2} = -2.20\text{ MPa}$$

In the present case, the third principal stress is zero.

Let '*f*' be the FOS.

Then stress in the member is to be limited to $\sigma_{all} = \frac{\sigma_y}{f} = \frac{150}{f}$

The principal stresses in the strained body are given as

$$\sigma_1 = 93.48 \text{ MPa}, \ \sigma_2 = 0 \text{ MPa, and } \sigma_3 = -2.20 \text{ MPa}$$

(a) Rankine's theory:

The element fails when maximum principal stress reaches $\sigma_{all} = \frac{\sigma_y}{f} = \frac{150}{f}$

Maximum principal stress = σ_1 = 93.48 MPa

$$\sigma_{all} = \sigma_1$$

$$\frac{150}{f} = 93.48$$

$$f = 1.605$$

(b) Maximum shear stress theory:

Maximum shear stress is given by

$$\tau_{max} = \frac{\sigma_{max} - \sigma_{min}}{2}$$

in which σ_{max} and σ_{min} are the maximum and minimum values of three principal stresses.

In case of uniaxial tension case

$$\sigma_1 = \frac{\sigma_y}{f}; \ \sigma_2 = \sigma_3 = 0$$

$$\sigma_{max} = \frac{\sigma_y}{f}; \sigma_{min} = 0$$

Maximum shear stress in case of uniaxial tension after allowing for safety $\tau_{max\,f} = \frac{\frac{\sigma_y}{f}}{2} = \frac{\sigma_y}{2f}$

As per maximum shear stress theory, the element fails when maximum shear stress reaches $\tau_{max\,f} = \frac{\sigma_y}{2f} = \frac{150}{2f}$

Maximum shear stress in case of the strained pole = $\tau_{max} = \frac{\sigma_{max} - \sigma_{min}}{2}$

$$\sigma_1 = 93.48 \text{ MPa}, \ \sigma_2 = 0 \text{ MPa, and } \sigma_3 = -2.20 \text{ MPa}$$

$$\tau_{max} = \frac{\{93.48 + 2.20)}{2} = \frac{95.68}{2}$$

For assessing the safety in the element equate τ_{max} to $\tau_{max\,f}$

$$\frac{150}{2} = \frac{95.68}{f \times 2}$$

$$f = 1.568$$

(c) Shear strain energy density theory:

Shear strain energy density is given by

$$U_* = \frac{1}{12G}\{(\sigma_1 - \sigma_2)^2 + (\sigma_2 - \sigma_3)^2 + (\sigma_3 - \sigma_1)^2\}$$

In case of uniaxial tension

$$\sigma_1 = \frac{\sigma_y}{f};\ \sigma_2 = \sigma_3 = 0$$

Shear strain energy density in case of uniaxial tension after allowing for safety

$$U_{*f} = \frac{2\sigma_f^2}{(12G)f^2} = \frac{45000}{(12G)f^2}$$

As per shear strain energy theory, the element fails when shear strain energy density reaches $U_{*f} = \frac{45000}{(12G)f^2}$.

Shear strain energy density in case of the strained body

$$\sigma_1 = 93.48 \text{ MPa},\ \sigma_2 = 0 \text{ MPa, and } \sigma_3 = -2.20 \text{ MPa}$$

$$U_* = \frac{1}{12G}\{(\sigma_1 - \sigma_2)^2 + (\sigma_2 - \sigma_3)^2 + (\sigma_3 - \sigma_1)^2\}$$

$$U_* = \frac{1}{12G}\{(93.48 - 0)^2 + (0 + 2.20)^2 + (-2.20 - 93.48)^2\} = \frac{17898}{12G}$$

For assessing the safety in the element equate U_* to U_{*f}

$$\frac{17898}{12G} = \frac{45000}{f^2 \times 12G}$$

$$f = 1.585$$

Based on the five theories of failure, the factors of safety in the element are tabled below.

Table 13.2

Failure Theory	Factor of Safety
Maximum normal stress theory	1.605
Shear stress theory	1.568
Shear strain energy density theory	1.585

SUMMARY

- **Stress at any point $S(x, y)$ in the cross-section of a member subjected to eccentric loading of P at (e_x, e_y) is given by $\sigma = \frac{P}{A} + \frac{Pe_x}{I_{yy}}x + \frac{Pe_y}{I_{xx}}y$.**
- **The region within which application of longitudinal load avoids the tensile stress in the entire cross-section is called kern or core of the cross-section. This region does not depend on the intensity of the load.**
- **Kern of circular cross-section with diameter d is a circle of radius $d/8$.**
- **Equivalent BM is the BM that alone produces a maximum normal stress equal to the maximum normal stress developed due to combined bending and torsion.**

$$M_{eq} = \frac{1}{2}\{M + \sqrt{\{M\}^2 + \{T\}^2}\}$$

- **Equivalent twisting moment is the twisting moment that alone produces a maximum shear stress equal to the maximum shear stress developed due to combined bending and torsion.**

$$T_{eq} = \sqrt{\{M\}^2 + \{T\}^2}$$

- **Total strain energy density =**

$$U = \frac{1}{2E}\{\sigma_1^2 + \sigma_2^2 + \sigma_3^2 - 2\mu\sigma_1\sigma_2 - 2\mu\sigma_2\sigma_3 - 2\mu\sigma_3\sigma_1\}$$

- **The shear strain energy density or distorsional strain energy density is given by**

$$U_s = \frac{1}{12G}\{(\sigma_1 - \sigma_2)^2 + (\sigma_2 - \sigma_3)^2 + (\sigma_3 - \sigma_1)^2\}$$

OBJECTIVE TYPE QUESTIONS

1. Which theory of failure will you use for aluminum components under steady loading? [GATE-1999]
(a) Principal stress theory (b) Principal strain theory
(c) Strain energy theory (d) Maximum shear stress theory

2. A small element at the critical section of a component is in a biaxial state of stress with the two principal stresses being 360 MPa and 140 MPa. The maximum working stress according to distortion energy theory is: [GATE-1997]
(a) 220 MPa (b) 110 MPa (c) 14 MPa (d) 330 MPa

3. Design of shafts made of brittle materials is based on [IES-1993]
(a) Guest's theory (b) Rankine's theory
(c) Venant's theory (d) Von-Mises theory

4. According to the maximum shear stress theory of failure, permissible twisting moment in a circular shaft is 'T'. The permissible twisting moment for the same shaft as per the maximum principal stress theory of failure will be: [IES-1998; ISRO-2008]

(a) $T/2$ (b) T (c) $\sqrt{2T}$ (d) $2T$

5. A rod having cross-sectional area 100×10^{-6} m^2 is subjected to a tensile load. Based on the Tresca failure criterion, if the uniaxial yield stress of the material is 200 MPa, the failure load is: [IES-2001]

(a) 10 kN (b) 20 kN (c) 100 kN (d) 200 kN

6. A cold roller steel shaft is designed on the basis of maximum shear stress theory. The principal stresses induced at its critical section are 60 MPa and –60 MPa, respectively. If the yield stress for the shaft material is 360 MPa, the FOS of the design is: [IES-2002]

(a) 2 (b) 3 (c) 4 (d) 6

7. Who postulated the maximum distortion energy theory? [IES-2008]

(a) Tresca (b) Rankine (c) St. Venant (d) Mises-Henky

8. If a shaft made from ductile material is subjected to combined bending and twisting moments, calculations based on which one of the following failure theories would give the most conservative value? [IES-1996]

(a) Maximum principal stress theory
(b) Maximum shear stress theory
(c) Maximum strain energy theory
(d) Maximum distortion energy theory

9. Consider the following statements: [IAS-2007

1. Experiments have shown that the distortion energy theory gives an accurate prediction about failure of a ductile component than any other theory of failure.
2. According to the distortion energy theory, the yield strength in shear is less than the yield strength in tension.

Which of the statements given above is/are correct?

(a) 1 only (b) 2 only (c) Both 1 and 2 (d) Neither 1 nor 2

10. Consider the following statements: [IAS-2003]

1. Distortion energy theory is in better agreement for predicting the failure of ductile materials.
2. Maximum normal stress theory gives good prediction for the failure of brittle materials.
3. Module of elasticity in tension and compression are assumed to be different stress analysis of curved beams.

Which of these statements is/are correct?

(a) 1, 2, and 3 (b) 1 and 2 (c) 3 only (d) 1 and 3

11 Under the combined action of BM M and torque T, the equivalent BM is

(a) $M + \sqrt{M^2 + T}$ (b) $M + T$

(c) $1/2\ [M + \sqrt{M^2 + T^2}\ ]$ (d) $\sqrt{M^2 + T^2}$

12 Under the combined action of BM M and torque T, the equivalent torque is

(a) $M + \sqrt{M^2 + T}$ (b) $M + T$

(c) $1/2\ [M + \sqrt{M^2 + T^2}\ ]$ (d) $\sqrt{M^2 + T^2}$

13. For a masonry dam of base width B, to avoid tension, the eccentricity of loading e should be

(a) More than $B/6$ (b) Less than $B/6$ (c) Equal to $B/6$ (d) Equal to zero

Solutions for Objective Questions

Sl. No.	1.	2.	3.	4.	5.	6.	7.	8.	9.	10.
Answer	(d)	(c)	(b)	(d)	(b)	(b)	(d)	(b)	(c)	(b)

Sl.No.	11.	12.	13.
Answer	(c)	(d)	(b)

EXERCISE PROBLEMS

1. A 30-m-high brick chimney is 2.5 m square at the base and tapers to 1 m square at the top. If the total weight of the brickwork above the base is 1000 kN, find for what uniform intensity of wind pressure on one face of the chimney the stress distribution across the base just ceases to be wholly compressive. What is then the maximum value of the compressive stress on the section?

2. A brick chimney weighs 2000 kN and has internal and external diameters at the base of 3 m and 4 m, respectively. The chimney leans by 6° with the vertical. Calculate the maximum stresses in the base. Assume that there is no wind pressure and C.G. of chimney is 15 m above the base.

3. A masonry dam of trapezoidal section is 8 m high, 1.8 m thick at top, and 5 m thick at base. Calculate (a) the height to which water may be stored so that no tension is induced at the base, (b) the base width required to avoid tension at the base if water is stored up to the top of the dam. Density of masonry is 2240 kg/m^3 and that of water is 1000 kg/m^3.

4. A concrete dam which retains water to its full height is 6 m wide at top, 8 m wide at bottom, and 12 m high, with its upstream face vertical. Estimate stress distribution at the base.

   Density of concrete = 2500 kg/m^3

   Density of water = 1000 kg/m^3

5. A masonry retaining wall of 10 m high is 1 m wide at the top and 4 m wide at the base. Its water face is vertical. Determine the maximum height of water level so that no tensile stress is developed in the base. Specific weight of masonry is 20 kN/m^3 and water weighs 10 kN/m^3.

6. The principal stresses at a point in an elastic material are 22 N/mm^2, 110 N/mm^2, and 55 N/mm^2 if the elastic limit in simple tension is 220 N/mm^2 and $\mu = 0.3$, then determine whether the failure of material will occur or not according to
   (i) Maximum principal stress theory
   (ii) Maximum principal strain theory
   (iii) Maximum shear stress theory
   (ix) Maximum strain energy theory
   (x) Maximum shear strain energy theory.

7. Determine the diameter of a bolt which is subjected to an axial pull of 12 kN together with a transverse SF of 6 kN, when the elastic limit in tension is 300 N/mm^2, FOS = 3 and Poisson's ratio = 0.3 using:
   (i) Maximum principal stress theory
   (ii) Maximum principal strain theory
   (iii) Maximum shear stress theory
   (ix) Maximum strain energy theory
   (x) Maximum shear strain energy theory.

8. A bolt is under an axial thrust of 7.2 kN together with a transverse SF of 3.6 kN. Calculate the diameter of the bolt according to:
   (i) Maximum principal stress theory

(ii) Maximum shear stress theory
(iii) Maximum strain energy theory

9. A steel shaft is subjected to an end thrust producing a stress of 90 MPa and the maximum shearing stress on the surface arising from torsion is 60 MPa. The yield point of the material in simple tension was found to be 300 MPa. Calculate the FOS of the shaft according to the following theories:
(i) Maximum shear stress theory
(ii) Maximum distortion energy theory

10. At a section of a mild steel shaft of diameter 180 mm, the maximum torque is 67.5 kN/m and maximum BM is 40.5 kN·m. The elastic limit in simple tension is 220 N/mm^2. Determine whether the failure of the material will occur or not according to maximum shear stress theory. If not then find the FOS.

CHAPTER 14

UNSYMMETRICAL BENDING AND SHEAR CENTER

UNIT OBJECTIVE

This presentation attempts to help the student learn about:

Objective 1: Principal axes and their directions with respect to centroidal axes of the section.

Objective 2: Moment of inertia (MI) of the section about the principal axes.

Objective 3: Stresses developed at various points of the section due to unsymmetrical bending.

Objective 4: Product of inertia of the section about any coordinate axes.

Objective 5: Position of neutral axis (NA) with respect to centroidal axes of the unsymmetrical sections.

Objective 6: Location of shear center of sections which are symmetrical about one centroidal axes and sections which are not symmetrical about any centroidal axes.

14.1 INTRODUCTION

The flexure formula derived in the theory of simple bending assumes that the loading plane should pass through the axis of symmetry. This condition is not satisfied in some cases where the loading is gravity loading and the member rests on an inclined support such as purlins of a truss. In some cases, the cross-section may not have axis of symmetry such as angular sections. This chapter provides the solution for such problems where bending is referred as unsymmetrical bending.

14.2 UNSYMMETRICAL BENDING

If the plane of loading of a beam does not coincide with one of the principal axes of the cross-section, then the resulting bending is termed as unsymmetrical bending. Unsymmetrical bending may occur in two ways.

1. The section is symmetrical about two axes like I section, rectangular section, and circular section but the loading plane is inclined to both the principal axes.
2. The section itself is unsymmetrical like angle section or a channel.

The following examples illustrate unsymmetrical bending.

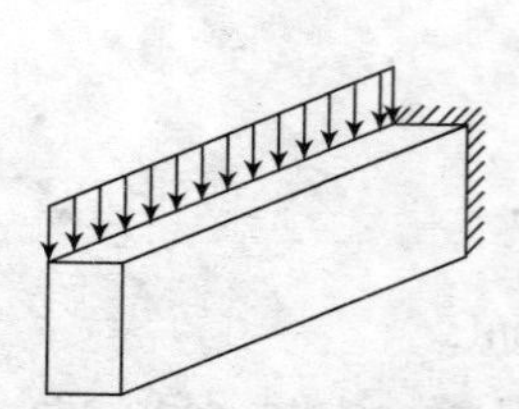

FIGURE 14.1 Plane of loading is not passing through the principal axis

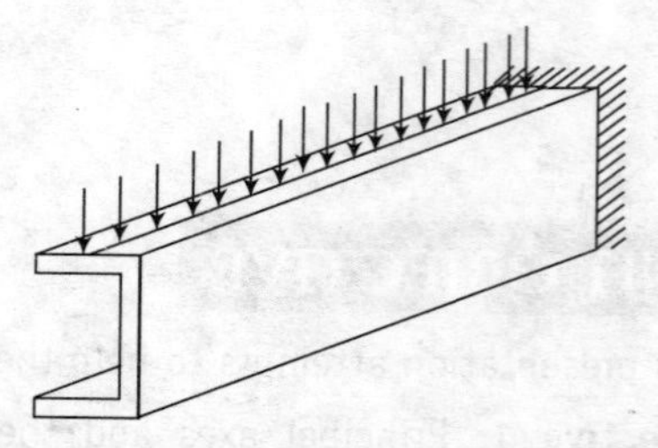

FIGURE 14.2 Plane of loading is not passing through the axis of symmetry.

14.3 PRINCIPAL MOMENT OF INERTIA

Let us consider an arbitrary cross-section. The centroid of the cross-section is taken as the origin of coordinate axes system. Consider an elemental area δA at point $P(x,y)$.

$P(x,y)$ are coordinates of the elemental area about axes X and Y passing through the centroid of the cross-section.

Let the coordinate axes X and Y be rotated and be defined by the angle θ.

Then, the coordinates of the elemental area δA is given by transformation matrix.

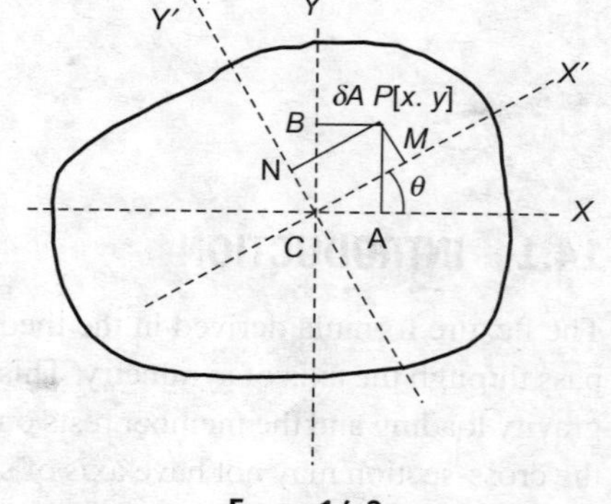

FIGURE 14.3

$$\text{CM} = x'$$

$$\text{PM} = \text{CN} = y'$$

$$\begin{bmatrix} x' \\ y' \end{bmatrix} = \begin{bmatrix} \cos\theta & \sin\theta \\ -\sin\theta & \cos\theta \end{bmatrix} \begin{bmatrix} x \\ y \end{bmatrix}$$

$$x' = x\cos\theta + y\sin\theta$$

$$y' = -x\sin\theta + y\cos\theta.$$

The MI

$$I_{x'x'} = \int (y')^2\, dA = \int (-x\sin\theta + y\cos\theta)^2\, dA$$

$$I_{x'x'} = \int (x^2\sin^2\theta + y^2\cos^2\theta - 2xy\cos\theta\sin\theta)\, dA$$

$$I_{x'x'} = \sin^2\theta \int x^2 dA + \cos^2\theta \int y^2 dA - 2\cos\theta\sin\theta \int xydA$$

$$I_{x'x'} = \cos^2\theta(I_{xx}) + \sin^2\theta(I_{yy}) - 2\cos\theta\sin\theta(I_{xy})$$

$$I_{x'x'} = I_{xx}\left(\frac{1+\cos 2\theta}{2}\right) + I_{yy}\left(\frac{1-\cos 2\theta}{2}\right) - I_{xy}\sin 2\theta$$

$$I_{x'x'} = \left(\frac{I_{xx}+I_{yy}}{2}\right) + \left(\frac{I_{xx}-I_{yy}}{2}\right)\cos 2\theta - I_{xy}\sin 2\theta$$

$$I_{x'y'} = \int x(y')dA = \int (x\cos\theta + y\sin\theta)(-x\sin\theta + y\cos\theta)dA$$

$$I_{x'y'} = \int (-x^2 \cos\theta\sin\theta + y^2\cos\theta\sin\theta - xy\sin^2\theta + xy\cos^2\theta)dA$$

$$I_{x'y'} = -\cos\theta\sin\theta \int x^2 dA + \cos\theta\sin\theta \int y^2 dA + (\cos^2\theta - \sin^2\theta)\int xydA$$

$$I_{x'y'} = -\cos\theta\,\sin\theta(I_{xx}) + -\cos\theta\,\sin\theta(I_{yy}) - (\cos^2\theta - \sin^2\theta)(I_{xy})$$

$$I_{x'y'} = \left(\frac{I_{xx}-I_{yy}}{2}\right)\sin 2\theta + I_{xy}\cos 2\theta$$

$I_{xx} = \int y^2 dA$ = MI about x axis

$I_{yy} = \int x^2 dA$ = MI about y axis

$I_{xy} = \int xydA$ = product of inertia.

Product of inertia may be positive or negative and it is zero if any one of the x or y axis is the axis of symmetry of the cross-section. A detailed calculation of the same is presented in the next coming topics of this chapter.

Principal axes are the axes about which the product of inertia of the cross-section is zero.

Hence for principal axes product of inertia $I_{x'y'}$ be equated to zero. Let α be the angle that defines the principal axes. Then

$$\left(\frac{I_{xx}-I_{yy}}{2}\right)\sin 2\alpha + I_{xy}\cos 2\alpha = 0.$$

This indicates that

$$\tan 2\alpha = -\frac{I_{xy}}{\left(\dfrac{I_{xx}-I_{yy}}{2}\right)}.$$

Substituting this value of tan 2α in the $I_{x'x'}$ expression, we get major principal MI and minor MI. The major principal MI is represented as I_{uu} and minor principal MI is represented as I_{vv}.

$$I_{uu} = \left(\frac{I_{xx} + I_{yy}}{2}\right) + \sqrt{\left\{\frac{I_{xx} - I_{yy}}{2}\right\}^2 + \{I_{xy}\}^2}$$

$$I_{vv} = \left(\frac{I_{xx} + I_{yy}}{2}\right) - \sqrt{\left\{\frac{I_{xx} - I_{yy}}{2}\right\}^2 + \{I_{xy}\}^2}.$$

These expressions are similar to the expressions of principal stresses. Thus, Mohr circle can be drawn for MIs of a cross-section. However, these things are not discussed here.

PROBLEM 14.1

Objective 2

Determine the product of inertia of the following cross-sections shown in Figure 14.4 about the axes mentioned.

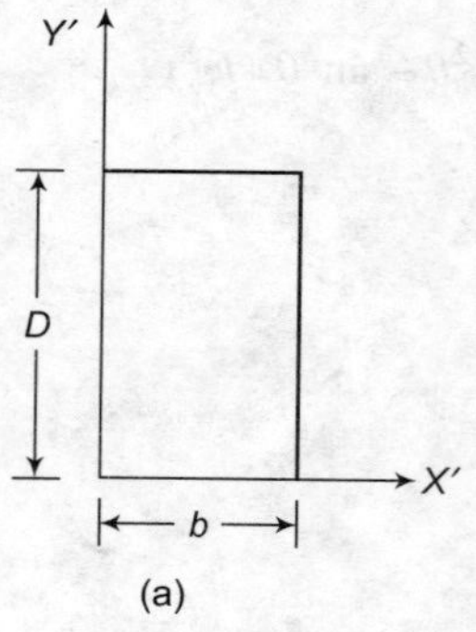

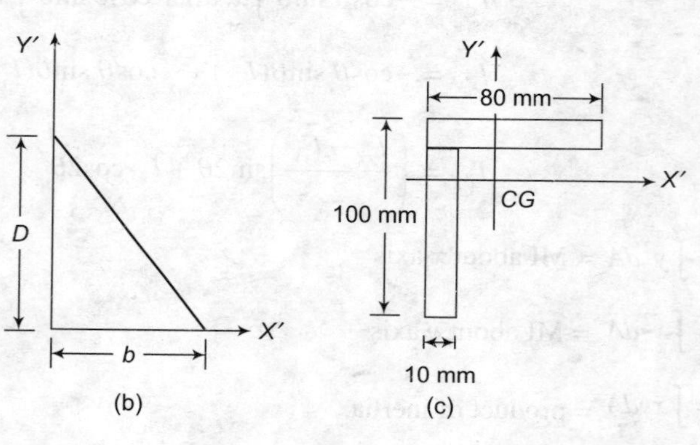

FIGURE 14.4

SOLUTION

(a) In Figure 14.4(a), the coordinate's axes given are passing through the corner of the rectangle. The cross-section is not symmetrical about any of the axes. Thus, the product of inertia is not equal to zero.

Parallel axis theorem for product of inertia, that is,

$$I_{x'y'} = I_{xy} + Axy$$

In which, x' axis should be parallel to x axis and y' axis be parallel to y axis. $I_{x'y'}$ is the product of inertia about an arbitrary axes. I_{xy} is the product of inertia about axes passing through centroid. x and y are the coordinates of the centroid of the cross-section about arbitrary axes x' and y'. A is the cross-sectional area.

Applying the parallel axes theorem, for rectangular cross-section I_{xy} about axes passing through centroid and parallel to x' and y', product of inertia is zero. $I_{xy} = 0$.

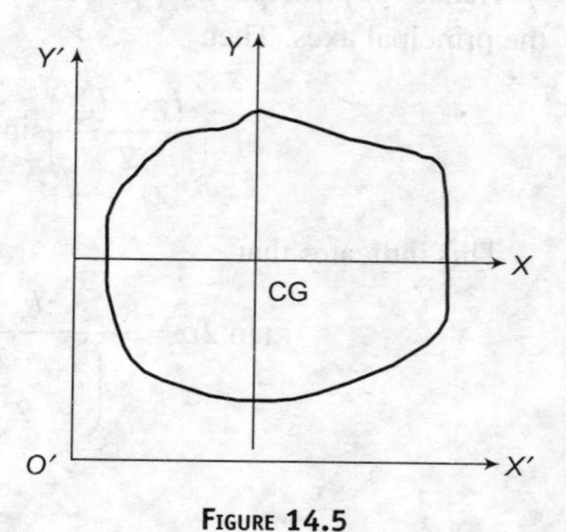

FIGURE 14.5

$$x = \frac{b}{2}\text{; and } y = \frac{D}{2}$$

Hence $I_{x'y'} = 0 + bD\left(\frac{b}{2} \times \frac{D}{2}\right) = \frac{b^2 D^2}{4}$.

(b) For the triangular section, there is no axis of symmetry passing through centroid parallel to X' and Y' axes like rectangular section. Hence the product of inertia is to be obtained from the fundamentals.

Consider an elemental strip of length x, and thickness δy located y distant from x axis.

This strip can be taken as rectangular strip of width x and thickness δy. For this elemental strip, product of inertia about axes passing through its centroid parallel to X and Y axes is zero.

So the product of inertia of this elemental strip with respect to X' and Y' axes can be taken as (applying parallel axes theorem)

$$\delta I_{x'y'} = \delta I_{xy} + \left(y \times \frac{x}{2}\right) \times x\,\delta y$$

$$\delta I_{xy} = 0$$

$$I_{x'y'} = \int_0^D \left(\frac{xy}{2}\right) x\,dy \quad \text{and} \quad x = \frac{D-y}{D}b$$

$$I_{x'y'} = \int_0^D \frac{y}{2}\left(\frac{D-y}{D}b\right)^2 dy$$

$$I_{x'y'} = \frac{b^2}{2D^2}\left\{\frac{D^4}{2} + \frac{D^4}{4} - \frac{2D^4}{3}\right\} = \frac{b^2 D^2}{24}.$$

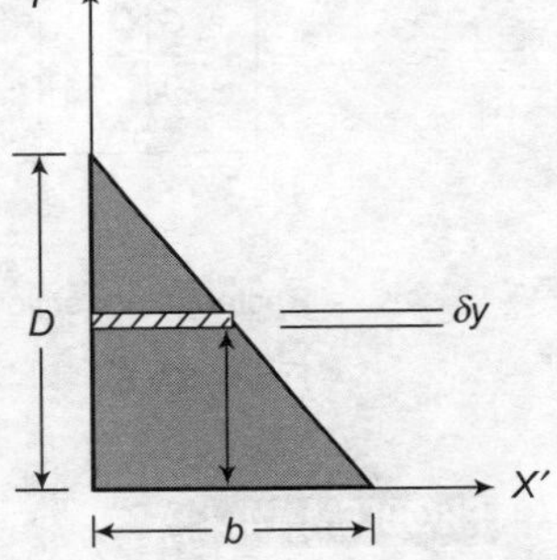

FIGURE 14.6

(c) Angle section is a compound section consisting of two rectangles. Thus, the products of inertia of the two rectangular sections need to be worked separately.

Find the centroid coordinates with respect to the X' and Y' reference axis.

$$\bar{x} = \frac{80 \times 10 \times 40 + 90 \times 10 \times 5}{80 \times 10 + 90 \times 10} = 21.47\,\text{mm}$$

$$\bar{y} = \frac{80 \times 10 \times 5 + 90 \times 10 \times (10 + 45)}{80 \times 10 + 90 \times 10} = 31.47\,\text{mm}$$

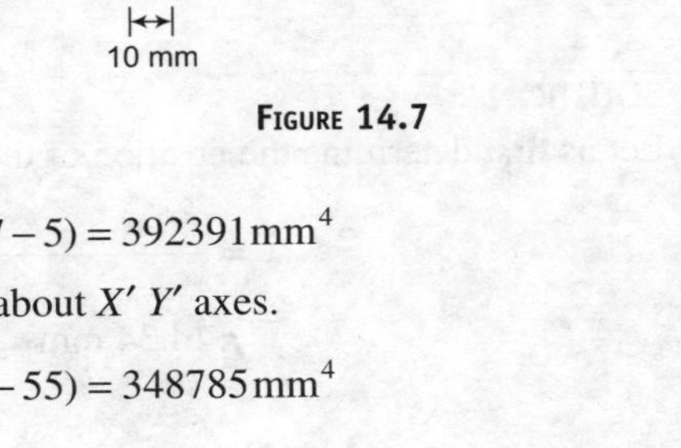

FIGURE 14.7

Product of inertia of top flange (80 mm × 10 mm) about X' Y' axes.

$$I_{1x'y'} = 80 \times 10 \times (40 - 21.47) \times (31.47 - 5) = 392391\,\text{mm}^4$$

Product of inertia of web portion (90 mm × 10 mm) about X' Y' axes.

$$I_{2x'y'} = 90 \times 10 \times (5 - 21.47) \times (31.47 - 55) = 348785\,\text{mm}^4$$

$$I_{x'y'} = 392391 + 348785 = 7.41 \times 10^5 \text{ mm}^4$$

14.4 STRESSES DUE TO UNSYMMETRICAL BENDING

When the loading plane of a beam does not coincide with one of the principal axes of the section, unsymmetrical bending takes place. Figure 14.8(a) shows a rectangular section, symmetrical about *XX* and *YY* axes or with *UU* and *VV* principal axes. Loading plane is inclined at an angle ϕ to the *Y* axis, and passing through *G* (centroid) or *C* (shear center) of the section.

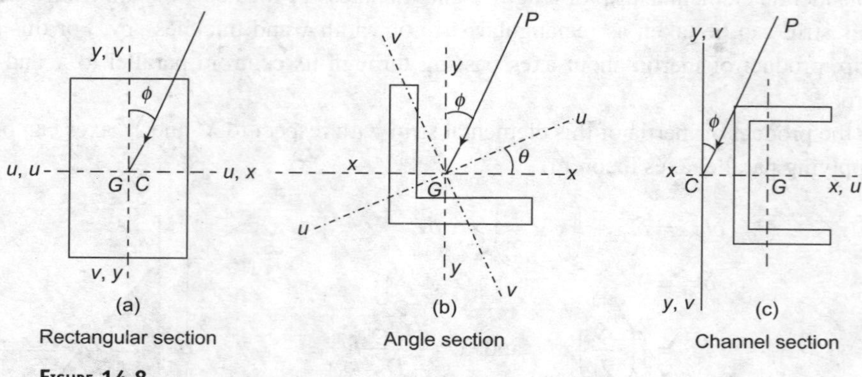

FIGURE 14.8

PROBLEM 14.2

Objective 3

A 50 mm × 50 mm × 5 mm angle section shown in Figure 14.9 is used for a cantilever beam over a span of 1.2 m. It carries a 0.1 kN load at free end, along the line *YG*, in which *G* is the centroid of the section. Determine the resultant bending stresses on points *A*, *B*, *C*, that is, outer corners of the section, at the free end of the beam.

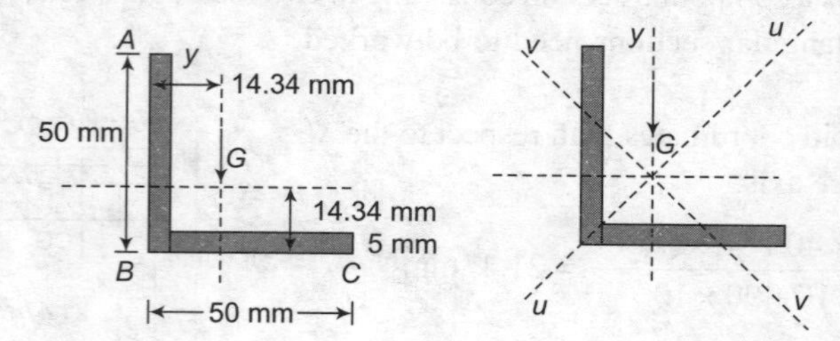

FIGURE 14.9

SOLUTION

Let us first determine the position of the centroid

$$\bar{x} = \frac{50 \times 5 \times 2.5 + 45 \times 5 \times (5 + 22.5)}{250 + 225} = 14.34 \text{ mm}$$

$$\bar{y} = 14.34 \text{ mm}$$

MI,

$$I_{xx} = \frac{50 \times (14.34)^3}{3} + \frac{50 \times (50 - 14.34)^3}{3} - \frac{45 \times (14.34)^3}{3} - \frac{45 \times (50 - 14.34)^3}{3}$$

$$I_{xx} = 100.5 \times 10^3 \text{ mm}^4$$

$$I_{yy} = 100.5 \times 10^3 \text{ mm}^4$$

$I_{xx} = I_{yy}$ as the cross-section is equal angle section.

Calculation of product of inertia:

Coordinates of the centroid of vertical leg 50 mm × 5 mm with reference to centroid of total cross-section.

x coordinate = –14.34 + 5 = –9.34 mm

y coordinate = 25 – 14.34 = 10.66 mm

Coordinates of the centroid of horizontal leg 45 mm × 5 mm with reference to centroid of total cross-section.

x coordinate = 5 + 22.5 – 14.34 = 13.16 mm

y coordinate = 5 – 14.34 = –9.34 mm

Product of inertia, $I_{xy} = 50 \times 5(-9.34) \times (10.66) + 45 \times 5(-9.34) \times (13.16)$

(Product of inertia about their own centroidal axes is zero because portions I and II are rectangular strips.)

$$I_{xy} = -24891.1 - 27655.74 = -52546.84 \text{ mm}^4$$

$$= -5.255 \times 10^4 \text{ mm}^4.$$

Principal axes angle,

$$\tan 2\theta = \frac{I_{xy}}{\frac{1}{2}(I_{yy} - I_{xx})} = \frac{-3.266 \times 10^4}{0} = \infty \text{ (infinity)}$$

$$= \tan 90°; \theta = 45°$$

Principal MIs

$$I_{uu} = 10^4 \left(\frac{10.05 + 10.05}{2} \right) + \sqrt{\left\{ \frac{10.05 - 10.05}{2} \right\}^2 + \{5.255\}^2}$$

$$I_{uu} = 15.275 \times 10^4 \text{ mm}^4$$

$$I_{vv} = 10^4 \left(\frac{10.05 + 10.05}{2} \right) - \sqrt{\left\{ \frac{10.05 - 10.05}{2} \right\}^2 + \{5.255\}^2}$$

$$I_{uu} = 4.825 \times 10^4 \text{ mm}^4$$

Bending moment (BM),

$$M = WL = 0.1 \times 1.2 = 0.12 \times 10^6 \text{ N·mm}$$

Components of BM,

$$M_1 = M \sin 45° = 0.12 \times 0.707 \times 10^6 = 84.84 \times 10^3 \text{ N·mm}$$

$$M_2 = M \cos 45° = 0.12 \times 0.707 \times 10^6 = 84.84 \times 10^3 \text{ N·mm}$$

u–v coordinates of the points

Point A, $x = -14.34$, $y = 50 - 14.34 = 35.66$ mm

$$u = x \cos \theta + y \sin \theta = -14.34 \times 0.707 + 35.66 \times 0.707 = 15.07 \text{ mm}$$

$$v = y \cos \theta - x \sin \theta = 35.66 \times 0.707 + 14.34 \times 0.707 = 35.35 \text{ mm}$$

Point B, $x = -14.34$, $y = -14.34$

$$u = -14.34 \times 0.707 - 14.34 \times 0.707 = -20.28 \text{ mm}$$

$$v = 0$$

Point C, $x = 50 - 14.34 = 35.66$, $y = -14.34$

$$u = 35.66 \times \cos 45° - 14.34 \sin 45°$$

$$= 35.66 \times 0.707 - 14.34 \times 0.707 = 15.07 \text{ mm}$$

$$v = -14.34 \times 0.707 - 35.66 \times 0.707 = -35.35 \text{ mm}$$

Resultant bending stresses at points A, B, and C

As the BM is hogging in nature, positive stress indicates tensile stress and negative stress indicates compressive stress.

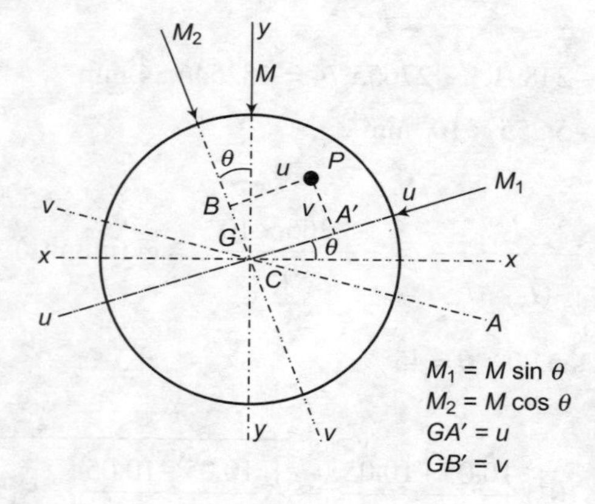

FIGURE 14.10

$$\sigma_A = \frac{M_1 u}{I_{vv}} + \frac{M_2 v}{I_{uu}}$$

$$\sigma_A = 84.84 \times 10^3 \left(\frac{15.07}{4.825 \times 10^4} + \frac{35.35}{15.275 \times 10^4} \right) = 46.13 \text{ N/mm}^2$$

$$\sigma_B = 84.84 \times 10^3 \left(\frac{-20.28}{4.825 \times 10^4} + \frac{0}{15.275 \times 10^4} \right) = -34.65 \text{ N/mm}^2$$

$$\sigma_C = 84.84 \times 10^3 \left(\frac{15.07}{4.825 \times 10^4} - \frac{35.35}{15.275 \times 10^4} \right) = 6.86 \text{ N/mm}^2$$

For NA

$$\sigma = \frac{M_1 u}{I_{vv}} + \frac{M_2 v}{I_{uu}} = 0$$

$$\frac{u}{I_{vv}} + \frac{v}{I_{uu}} = 0$$

$$15.275u + 4.825v = 0$$

$$15.275(x \cos 45 + y \sin 45) + 4.825(-x \sin 45 + y \cos 45) = 0$$

$$15.275(x \cos 45 + y \sin 45) + 4.825(-x \sin 45 + y \cos 45) = 0$$

$$7.388x + 14.21y = 0$$

Inclination of the NA with x axis is given by $\tan^{-1}\left(\frac{-7.388}{14.21}\right) = -27.47°$.

PROBLEM 14.3

Objective 3

A simply supported beam of a length of 2 m carries two central load of 20 kN inclined at 30° to the vertical and 10 kN making 60° as shown in Figure 14.11 and passing through the centroid of the section. Determine stresses at four corners. Top flange and bottom flange dimensions are 200 mm × 10 mm, overall depth is 200 mm, and web thickness is 12 mm.

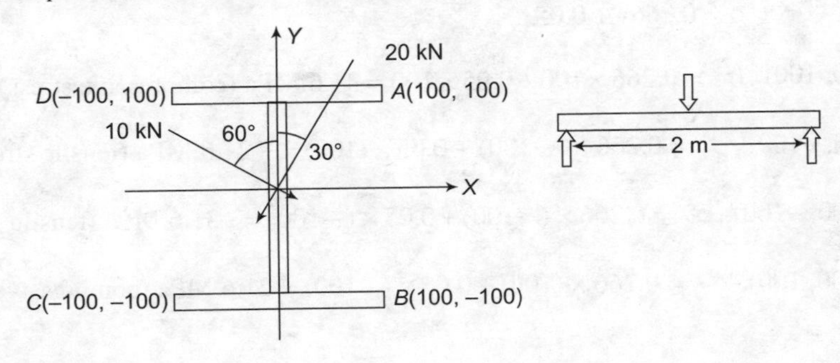

FIGURE 14.11

SOLUTION

In this case, the cross-section is symmetrical but plane of loading is inclined. Resolve the loading along Y axis and along X axis.

Resultant load along Y axis = $P_V = 20\cos 30° + 10\cos 60° = 22.32$ kN

Resultant load along X axis = $P_H = 20\sin 30° - 10\sin 60° = 1.34$ kN ($\leftarrow$)

P_V creates sagging BM at midspan about X axis = $M_x = \dfrac{22.32 \times 2}{4} = 11.16\,\text{kN·m}$

P_H creates sagging BM at midspan about Y axis = $M_y = \dfrac{1.34 \times 2}{4} = 0.67\,\text{kN·m}$

Both these moments create compression in the first quadrant of the cross-section with respect to the axes chosen.

MI,

$$I_{xx} = \frac{(200)^4}{12} - \frac{188 \times (180)^3}{12} = 41.965 \times 10^6 \text{ mm}^4$$

$$I_{yy} = \frac{180 \times (12)^3}{12} + 2 \times \frac{10 \times (200)^3}{12} = 13.359 \times 10^6 \text{ mm}^4$$

$$I_{xx} = 41.965 \times 10^6 \text{ mm}^4$$

$$I_{yy} = 13.359 \times 10^6 \text{ mm}^4.$$

Expression for stress at point defined by (x,y) due to this load at midspan section.

$$\sigma = \frac{M_x y}{I_{xx}} + \frac{M_y x}{I_{yy}}$$

$$\sigma = \frac{(11.16 \times 10^6)y}{41.965 \times 10^6} + \frac{(0.67 \times 10^6)x}{13.359 \times 10^6}$$

$$\sigma = 0.266y + 0.05x.$$

Stress at A(100, 100), $\sigma_A = 0.266 \times 100 + 0.05 \times 100 = 31.6$ MPa (compressive stress)

Stress at B(100, –100), $\sigma_B = 0.266 \times (-100) + 0.05 \times (100) = -21.6$ MPa (tensile stress)

Stress at C(–100, –100), $\sigma_C = 0.266 \times (-100) + 0.05 \times (-100) = -31.6$ MPa (tensile stress)

Stress at D(–100, 100), $\sigma_D = 0.266 \times (100) + 0.05 \times (-100) = 21.6$ MPa (compressive stress)

For NA, $\sigma = 0$

$$s = 0.266y + 0.05x = 0$$

This indicates that the equation of the NA is $y = -0.188x$

Slope of the NA is given by $\tan^{-1}(-0.188) = -10.65°$

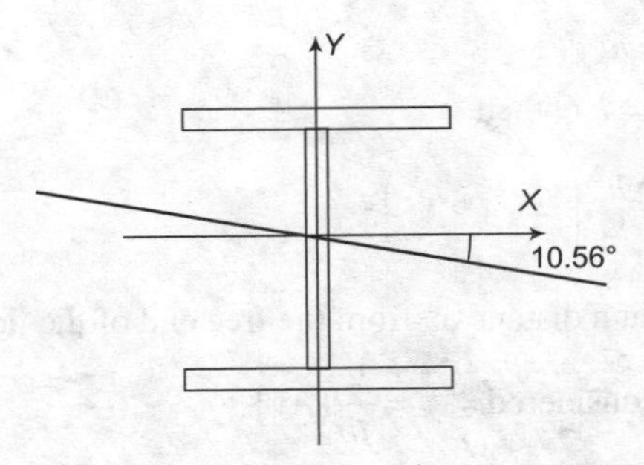

FIGURE 14.12

14.5 SHEAR CENTER

When a cross-section is subjected to shear due to flexure it results in shear stress. Summation of these shear stress components will be equal to the transverse shear force (SF). This resultant of shear stress components, in case of solid cross-sections, passes through the geometric centroid of the cross-section. In open cross-sections such as channel sections and angle sections, the resultant of shear stress components does not pass through the centroid of the cross-section. Thus, when the external load passes through the centroid and the resultant passes through a point other than centroid, results in additional torsional moment. If the external load also passes through the point where the resultant of shear stress components act, the cross-section would be free from additional torsional moment. *Shear center is the point where the resultant of all shear stress components of the cross-section acts.* A typical shear stress distribution and location of shear center is presented in Figure 14.13.

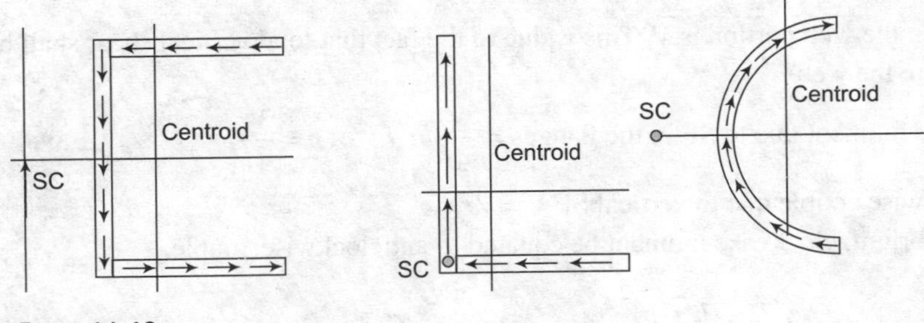

FIGURE 14.13

14.5.1 Determination of Shear Center of a Channel Section

Consider a channel section subjected to flexure shear.

Let b_f and t_f be the width and thickness of the flange, respectively.

Let h be the height and t_w be the thickness of the web.

MI about axis passing through centroid $= I_{xx} = I = \dfrac{t_w h^3}{12} + 2 \times b_f t_f \left(\dfrac{h}{2}\right)^2$

In above expression, the thickness of flange and thickness of web are very small compared to the breadth of the flange and depth of the channel section.

Taking area of web as $A_w = ht_w$

Taking area of flange as $A_f = b_f t_f$, then

$$I_{xx} = I = \frac{h^2}{4}\left(\frac{A_w}{3} + 2A_f\right).$$

Consider a section located at a distant 'x' from the free end of the flange.

Shear stress at the section considered = $\tau = \frac{V}{Ib}(A\bar{y})$

Here in the flange portion $b = t_f$.

$A\bar{y}$ is the moment of the area available above the section under consideration about centroidal axis.

$$A\bar{y} = xt_f \frac{h}{2}$$

Hence shear stress in the flange is given by

$$\tau = \frac{V}{It_f}\left(xt_f \frac{h}{2}\right) = \frac{V}{2I}hx$$

Total SF in the top flange = $\int_0^{b_f}\left(\frac{V}{2I}hx\right)t_f dx = \frac{V}{4I}ht_f b_f^{\,2}$

Total SF in the bottom flange owing to symmetry = $\int_0^{b_f}\left(\frac{V}{2I}hx\right)t_f dx = \frac{V}{4I}ht_f b_f^{\,2}$.

Total SF in the web portion is V. This is due to the fact that total vertical shear shall be balanced by the shear in the web.

Clockwise moment due to SF in the flanges = $\frac{V}{4I}ht_f b_f^{\,2} \times h = \frac{V}{4I}h^2 t_f b_f^{\,2}$

Anticlockwise couple due to vertical SF $V = Ve$

For equilibrium, clockwise moment be equated to anticlockwise couple.

$$Ve = \frac{V}{4I}h^2 t_f b_f^{\,2}$$

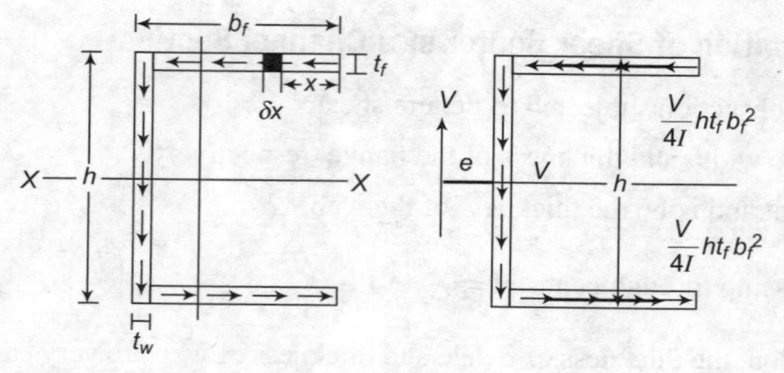

FIGURE 14.14

$$e = \frac{1}{4I} h^2 t_f b_f^{\,2}$$

'e' is the location of the shear center of the cross-section

$$e = \frac{1}{4\frac{h^2}{4}\left(\frac{A_w}{3} + 2A_f\right)} h^2 t_f b_f^{\,2} = \frac{b_f A_f}{\left(\frac{A_w}{3} + 2A_f\right)} = \frac{3b_f}{\left(\frac{A_w}{A_f} + 6\right)}.$$

Thus, the shear center of a channel section is given by

$$e = \frac{3b_f}{\left(\frac{A_w}{A_f} + 6\right)}.$$

PROBLEM 14.4

Objective 6

Determine the shear center and sketch the shear stress variation across for the cross-section shown in Figure 14.15. Thickness of the element is uniform equal to 8 mm. Section is symmetrical about X axis. SF acting on the section is 60 kN.

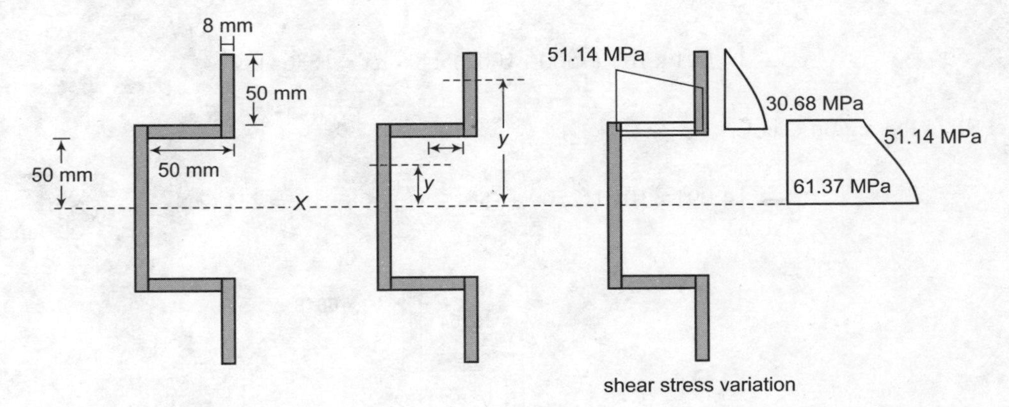

FIGURE 14.15

SOLUTION

Thickness of the elements is small compared to the other dimensions of the cross-section.

MI of the cross-section:

$$I_{xx} = 2\left\{\frac{8 \times 50^3}{12} + 8 \times 50 \times (25 + 50)^2 + 8 \times 50 \times (50)^2 + \frac{8 \times 50^3}{3}\right\}$$

$$I_{xx} = 7.333 \times 10^6 \text{ mm}^4$$

Shear stress variation in the portion A:

Consider a section located at a distant y from the NA (X–X axis)

$$\tau = \frac{V}{Ib} A\bar{y} = \frac{60 \times 1000}{7.333 \times 10^6 \times 8}\left(\frac{8}{2} \times (100^2 - y^2)\right) = 4.091 \times 10^{-3}(100^2 - y^2)$$

At $y = 100$, $\tau = 0$; and at $y = 50$, $\tau = 30.68$ MPa. The variation is parabolic.

In the region B, consider a point located x distant as shown in Figure 14.11.

$$\tau = \frac{V}{Ib} A\bar{y} = \frac{60 \times 1000}{7.333 \times 10^6 \times 8}\left(\frac{8}{2} \times (100^2 - 50^2) + 8 \times x \times 50\right)$$

$$\tau = 4.091 \times 10^{-3}\,(7500 + 100x)$$

At $x = 0$, $\tau = 30.68$ MPa and at $x = 50$, $\tau = 51.14$ MPa. The variation is linear.

In the region C, consider a point located y distant as shown in Figure 14.11.

$$\tau = \frac{V}{Ib} A\bar{y} = \frac{60 \times 1000}{7.333 \times 10^6 \times 8}\left(\frac{8}{2} \times (50^2 - y^2) + \frac{8}{2} \times (100^2 - 50^2) + 50 \times 8 \times 50\right)$$

$$\tau = 4.091 \times 10^{-3}(12500 + (50^2 - y^2))$$

At $y = 50$, $\tau = 51.14$ MPa; and at $y = 0$, $\tau = 61.34$ MPa. The variation is parabolic.

Total SF in the region A is V_1.

$$V_1 = \int_{50}^{100} 4.091 \times 10^{-3}(100^2 - y^2) \times 8 \times dy = 6818\,\text{N}.$$

Total SF in the region B is V_2.

$$V_2 = \int_{0}^{50} 4.091 \times 10^{-3}(7500 + 100x) \times 8 \times dx = 16364\,\text{kN}$$

Total SF in the region C is V_3

$$V_2 = \int_{0}^{50} 4.091 \times 10^{-3}(12500 + (50^2 - y^2)) \times 8 \times dx = 23182\,\text{N}$$

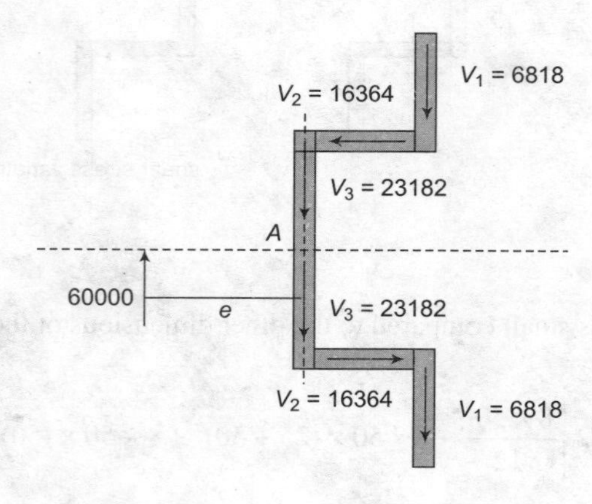

FIGURE 14.16

To find the shear center, take moments of all forces about A,

$$2V_1 \times 50 - V_2 \times 100 + 60000e = 0$$

$$e = \frac{(23182 - 6818) \times 100}{60000} = 15.91 \text{ mm}.$$

14.6 APPLICATIONS

The concept of shear center and unsymmetrical bending is illustrated in the following worked out example problems.

PROBLEM 14.5

Objective 6

Determine the shear center of the cross-section shown in Figure 14.17. Thickness of the cross-section is uniform and is equal to t. Mean radius of the cross-section is R. Included angle at the center is 2β.

SOLUTION

Let V be the SF acting on the cross-section.

Consider an elemental strip, which makes an angle $\delta\alpha$ at the center O. Elemental area is located and is defined by its angular measurement α as shown in Figure 14.17.

Elemental area = $R \times \delta\alpha$

$$\text{MI} = I_{xx} = I = 2\int_0^\beta y^2 dA = 2\int_0^\beta (R\cos\alpha)^2 Rtd\alpha = 2tR^3\int_0^\beta (\cos\alpha)^2 d\alpha$$

$$I = 2tR^3\left\{\frac{\beta}{2} + \frac{\sin 2\beta}{4}\right\}$$

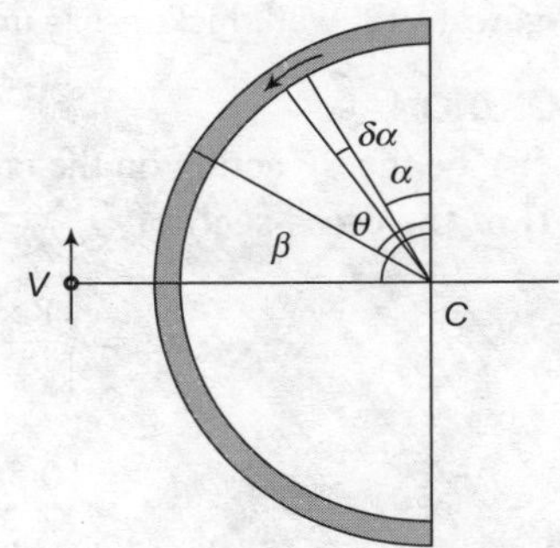

FIGURE 14.17

Shear stress at the section defined by θ is given by

$$\tau = \frac{V}{Ib} A\bar{y}$$

$$b = t \text{ and } A\bar{y} = \int_0^\theta y dA = \int_0^\theta (R\cos\alpha) Rtd\alpha = R^2 t\sin\theta$$

$$\tau = \frac{V}{2tR^3\left\{\frac{\beta}{2} + \frac{\sin 2\beta}{4}\right\} t} R^2 t\sin\theta$$

$$\tau = \frac{V}{2Rt} \times \frac{\sin\theta}{\left\{\frac{\beta}{2} + \frac{\sin 2\beta}{4}\right\}}.$$

Let e be the location of the shear center from center C.

Then, taking moments about C,

$$V \times e = \int_0^{2\beta} R\tau dA = \int_0^{2\beta} R\frac{V}{2Rt} \times \frac{\sin\theta}{\left\{\frac{\beta}{2} + \frac{\sin 2\beta}{4}\right\}} Rtd\theta$$

$$e = \frac{R}{2}\int_0^{2\beta} \frac{\sin\theta}{\left\{\frac{\beta}{2} + \frac{\sin 2\beta}{4}\right\}} d\theta$$

$$e = \frac{R\{1-\cos 2\beta\}}{\left\{\beta + \dfrac{\sin 2\beta}{2}\right\}} = \frac{2R\{\sin^2\beta\}}{\left\{\beta + \dfrac{\sin 2\beta}{2}\right\}}$$

For a semicircular arc of a section $\beta = \dfrac{\pi}{2}$

$$e = \frac{4R}{\pi}.$$

PROBLEM 14.6

Objective 6

Determine the shear center of thin rectangular cross-section with a slit or pin opening as shown in Figure 14.18. Wall thickness is uniform and is equal to 8 mm.

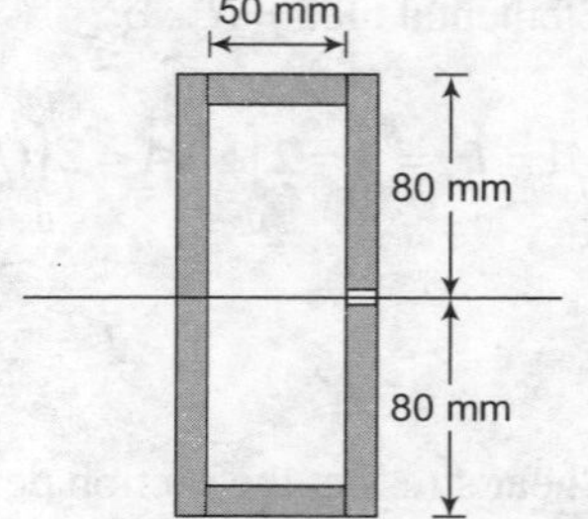

FIGURE 14.18

SOLUTION

Let V be the SF acting on the cross-section.

MI of the cross-section:

$$I_{xx} = \left\{\frac{8\times 80^3}{3}\times 4 + 2\times 8\times 50\times 80^2\right\}$$

$$I_{xx} = 10.581\times 10^6\ \text{mm}^4.$$

Shear stress variation in the portion A:

Consider a section located at a distant y from the NA (X–X axis)

$$\tau = \frac{V}{Ib}A\bar{y} = \frac{V}{10.581\times 10^6\times 8}\left(\frac{8}{2}\times y^2\right)$$

Total force in the region A:

$$V_1 = \int_0^{80} \frac{V}{10.581\times 10^6\times 8}\left(\frac{8}{2}\times y^2\right)\times 8\times dy = 0.0645V$$

In the region B, consider a point located x distant as shown in Figure 14.15.

$$\tau = \frac{V}{Ib}A\bar{y} = \frac{V}{10.581\times 10^6\times 8}\left(\frac{8\times 80\times 80}{2} + 8\times x\times 80\right)$$

Total force in the region B:

$$V_2 = \int_0^{50} \frac{V}{10.581\times 10^6\times 8}(25600 + 640x)\times 8\times dx$$

$$V_2 = \frac{V}{10.581\times 10^6}\left(25600\times 50 + 640\times \frac{50\times 50}{2}\right) = 0.1966V$$

In the region C, consider a point located y distant as shown in Figure 14.15.

$$\tau = \frac{V}{Ib}A\bar{y} = \frac{V}{10.581\times 10^6\times 8}\left(\frac{8\times 80\times 80}{2} + 8\times 50\times 80 + 8y(80 - \frac{y}{2}\right)$$

$$\tau = \frac{V}{10.581 \times 10^6 \times 8}(57600 + 640y - 4y^2)$$

Total SF in the region C is V_3

$$V_3 = \int_0^{80} \frac{V}{10.581 \times 10^6 \times 8}(57600 + 640y - 4y^2) \times 8dy$$

$$V_3 = \frac{V}{10.581 \times 10^6}\left(57600 \times 80 + 640 \times \frac{(80^2)}{2} - \frac{4 \times 80^3}{3}\right)$$

$$V_3 = 0.5645\ V$$

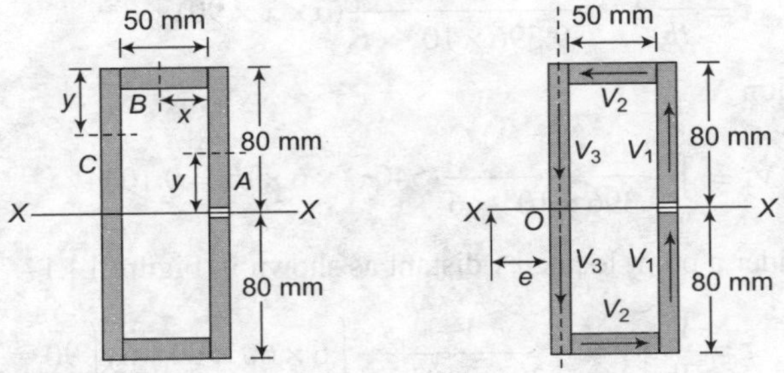

FIGURE 14.19

Referring to the free body diagram (FBD) in Figure 14.16, and taking moments about O, we get

$$Ve = 0.0645V \times 50 + 0.0645V \times 50 + 0.1966V \times 80 + 0.1966V \times 80$$

$$e = 37.91 \text{ mm.}$$

PROBLEM 14.7

Objective 6

Determine the shear center of the cross-section shown in Figure 14.20. Thickness of the elements is uniform and is equal to 6 mm.

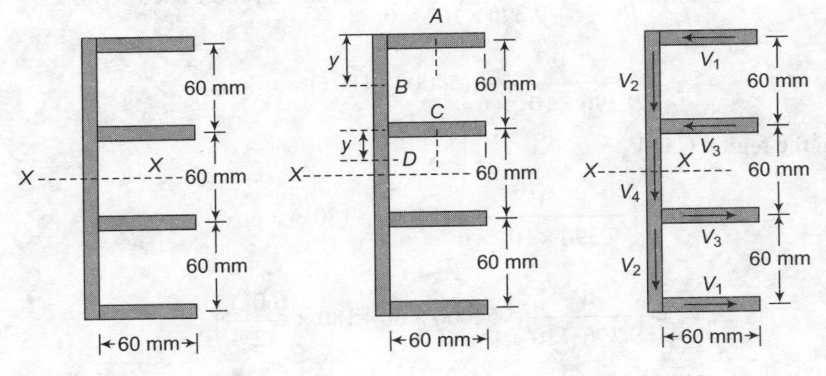

FIGURE 14.20

SOLUTION

Let V be the SF acting on the cross-section.

MI of the cross-section:

$$I_{xx} = \left\{ \frac{6 \times 180^3}{12} + 2 \times 6 \times 60 \times 30^2 + 2 \times 6 \times 60 \times 90^2 \right\}$$

$$I_{xx} = 9.396 \times 10^6 \text{ mm}^4$$

Shear stress variation in the portion A:

Consider a section located at a distant x as shown in Figure 14.20.

$$\tau = \frac{V}{Ib} A\bar{y} = \frac{V}{9.396 \times 10^6 \times 6} (6 \times x \times 90)$$

Total force in the region A:

$$V_1 = \int_0^{60} \frac{V}{9.396 \times 10^6 \times 6} (540x) \times 6 \times dx = 0.1034V$$

In the region B, consider a point located y distant as shown in Figure 14.17.

$$\tau = \frac{V}{Ib} A\bar{y} = \frac{V}{9.396 \times 10^6 \times 6} \left(6 \times 60 \times 90 + 6y\left(90 - \frac{y}{2}\right)\right)$$

Total force in the region B:

$$V_2 = \int_0^{60} \frac{V}{9.396 \times 10^6 \times 6} (32400 + 540y - 3y^2) \times 6 \times dy$$

$$V_2 = \frac{V}{9.396 \times 10^6} \left(32400 \times 60 + 540 \times \frac{60 \times 60}{2} - 3\frac{60^3}{3} \right) = 0.2874V$$

In the region C, consider a point located x distant as shown in Figure 14.20.

$$\tau = \frac{V}{Ib} A\bar{y} = \frac{V}{9.396 \times 10^6 \times 6} (32400 + 21600 + 6 \times x \times 30)$$

$$\tau = \frac{V}{9.396 \times 10^6 \times 6} (54000 + 180x)$$

Total SF in the region C is V_3

$$V_3 = \int_0^{60} \frac{V}{9.396 \times 10^6 \times 6} (54000 + 180x) \times 6dx$$

$$V_3 = \frac{V}{9.396 \times 10^6} \left(54000 \times 60 + 180 \times \frac{(60^2)}{2} \right)$$

$$V_3 = 0.379V$$

In the region D, consider a point located y distant as shown in Figure 14.20.

$$\tau = \frac{V}{Ib}A\bar{y} = \frac{V}{9.396\times10^6\times6}\left(32400+21600+10800+6y\left(30-\frac{y}{2}\right)\right)$$

$$\tau = \frac{V}{9.396\times10^6\times6}(64800+180y-3y^2)$$

Total SF in the region D is V_4

$$V_4 = \int_0^{30}\frac{V}{9.396\times10^6\times6}(64800+180y-3y^2)\times6dy$$

$$V_4 = \frac{V}{9.396\times10^6}\left(64800\times30+180\times\frac{(30^2)}{2}-3\frac{30^3}{3}\right)$$

$$V_4 = 0.2126V$$

For a check total vertical $(2(V_2 + V_4))$ shear must be equal to V.

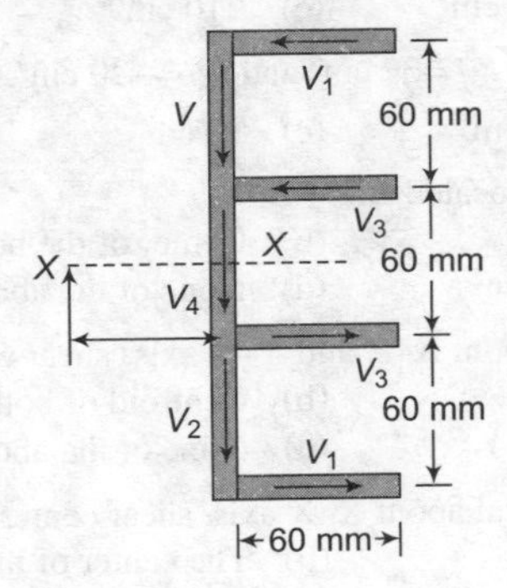

FIGURE 14.21

Referring to the FBD in Figure 14.21, and taking moments about O, we get

$$Ve = 0.1034V\times90+0.3793V\times30+0.3793V\times30+0.1034V\times90$$

$$e = 41.37 \text{ mm.}$$

SUMMARY

- **The expressions for principal moment of inertia are similar to that of principal stresses expressions.**
- **Shear center for channel section** $e = \dfrac{3bf}{(Aw/Af)+6}$
- **Shear center and unsymmetrical bending can be applicable to any type of section.**

OBJECTIVE TYPE QUESTIONS

1. The product of inertia of a rectangular section of a breadth of 4 cm and a depth of 6 cm about its centroidal axis is
   (a) 144 cm^4 (b) 72 cm^4 (c) 36 cm^4 (d) None of the above
2. The product of inertia of a rectangular section of a breadth of 3 cm and a depth of 6 cm about the coordinate axes passing at one corner of the section and parallel to the sides is
   (a) 81 cm^4 (b) 72 cm^4 (c) 54 cm^4 (d) None of these
3. For an equal angle section, coordinate axes XX and YY passing through centroid are parallel to its length. The principal axes are inclined to XY axes at an angle
   (a) 22.5° (b) 45.0° (c) 67.5° (d) None of the above
4. For an equal angle section, MI I_{xx} and I_{yy} are both equal to 120 cm^4. If one principal MI is 180 cm^4, the magnitude of other principal MI is
   (a) 180 cm^4 (b) 120 cm^4 (c) 60 cm^4 (d) 30 cm^4
5. For a section, principal MIs are I_{uu} = 360 cm^4 and I_{vv} = 160 cm^4. MI of the section about an axis inclined at 30° to the U–U axis is
   (a) 310 cm^4 (b) 260 cm^4 (c) 210 cm^4 (d) 120 cm^4
6. For an equal angle section, $I_{xx} = I_{yy}$ = 32 cm^4 and I_{xy} = –20 cm^4. The magnitude of one principal MI is
   (a) 52 cm^4 (b) 42 cm^4 (c) 32 cm^4 (d) 16 cm
7. For a T section, shear center is located at
   (a) Center of the vertical web (b) Center of the horizontal flange
   (c) At the centroid of the section (d) None of the above
8. For an I section (symmetrical about X–X and Y–Y axis) shear center lies at
   (a) Centroid of top flange (b) Centroid of bottom flange
   (c) Centroid of the web (d) None of the above
9. For a channel section symmetrical about X–X axis, shear center lies at
   (a) The centroid of the section (b) The center of the vertical web
   (c) The center of the top flange (d) None of the above
10. If the applied load passes through the shear center of the section of the beam, then there will be
    (a) No bending in the beam (b) No twisting in the beam
    (c) No deflection in the beam (d) None of these
11. The shear center of a section is defined as that point:
    (a) Through which load must be applied to produce zero twisting moment on the section
    (b) At which the SF is zero
    (c) At which SF is maximum
    (d) At which SF is minimum
12. Consider the following statements:
    (i) If a beam has two axes of symmetry even then shear center does not coincide with the centroid
    (ii) For a section having one axis of symmetry, the shear center does not coincide with the centroid but lies on the axis of symmetry
    (iii) If a load passes through the shear center, then there will be only bending in the cross-section and no twisting.

    Which of these statements are correct?
    (a) i, ii, and iii (b) i and ii (c) ii and iii (d) i and iii

13. A point load applied at shear center induces
(a) Zero SF (b) Zero bending (c) Pure twisting (d) Pure bending

14. The product of inertia of an area is about the axis of symmetry is
(a) Minimum (b) Maximum (c) Zero (d) Unpredictable

15. Unsymmetrical bending may be defined as the bending caused by loads that
(a) lie in or parallel to a plane containing the principal centroid axes of inertia of the cross-section
(b) do not lie in or parallel to a plane containing the principal centroid axes of inertia of the cross-section
(c) lie in a vertical plane
(d) lie in a horizontal plane

16. If the loads pass through the bending axis of a beam then there shall be
(a) no bending of the beam
(b) pure bending of the beam
(c) bending shall be accompanied by twisting
(d) twisting of the beam

17. The bending stress by unsymmetrical bending is given by

(a) $\dfrac{M_u \cdot u}{I_u} + \dfrac{M_v \cdot v}{I_v}$ (b) $\dfrac{M_u \cdot u}{I_v} + \dfrac{M_v \cdot v}{I_u}$

(c) $\dfrac{M_u \cdot v}{I_v} + \dfrac{M_v \cdot u}{I_v}$ (d) $\dfrac{M_u \cdot v}{I_v} + \dfrac{M_v \cdot u}{u}$

18. The resultant deflection of a beam under unsymmetrical bending is
(a) parallel to the NA (b) perpendicular to the NA
(c) parallel to the axis of symmetry (d) parallel to the axis of symmetry

19. If the load passes through the shear center then there shall be
(a) only bending in the beam (b) only twisting in the beam
(c) bending accompanied by twisting (d) no bending of the beam

20. Shear center of a semicircular arc of radius r is

(a) $\dfrac{r}{\pi}$ (b) $\dfrac{2r}{\pi}$ (c) $\dfrac{3r}{\pi}$ (d) $\dfrac{4r}{\pi}$ (E) $\dfrac{\pi}{r}$

21. Product of inertia of a beam about a set of axes may be
(a) positive or negative (b) always positive
(c) always negative (d) zero

22. Unsymmetrical bending of beam occur if
(a) the beam cross-section is doubly asymmetric
(b) the load line does not coincide with one of the axes of symmetry
(c) the load line does not pass through the shear center
(d) the load line is not perpendicular to the longitudinal axis of the beam

23. NA of a beam is
(a) an imaginary line whose location within the cross-section is fixed
(b) an axis of symmetry of the beam cross-section
(c) a line passing through the centroid of the cross-section
(d) the longitudinal axis of the beam

Solutions for Objective Questions

Sl. No.	1.	2.	3.	4.	5.	6.	7.	8.	9.	10.
Answer	(d)	(a)	(b)	(c)	(a)	(a)	(b)	(c)	(d)	(b)

Sl.No.	11.	12.	13.	14.	15.	16.	17.	18.	19.	20.
Answer	(a)	(c)	(c)	(c)	(b)	(b)	(b)	(b)	(a)	(d)

Sl.No.	21.	22.	23.
Answer	(a)	(d)	(c)

EXERCISE PROBLEMS

1. Figure 1 shows Z section of a beam simply appalled over a span of 2 m. A vertical load of 2 kN acts at the center of the beam and passes through the centroid of the section. Determine the resultant bending stress at points A and B. [Objective 3]

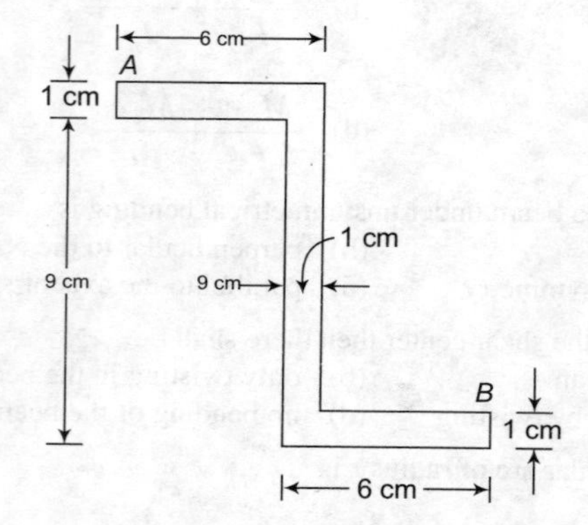

FIGURE 1

2. Figure 2 shows a section of a beam subjected to SF F. Locate the position of the shear center as defined by e. [Objective 6]

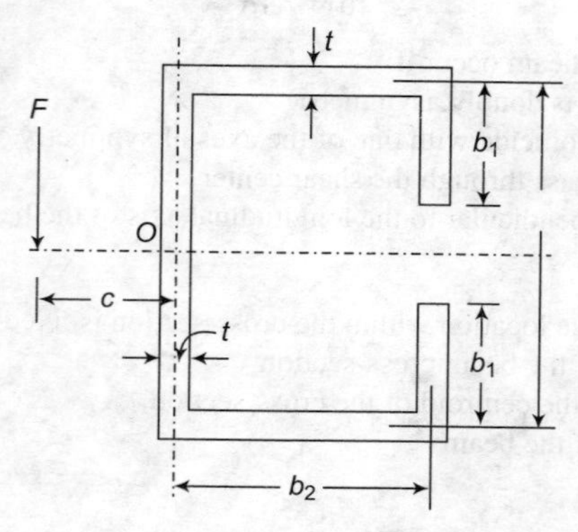

FIGURE 2

3. Determine the location e of the shear center point O for the thin-walled member having the cross-section shown in Figure 2, in which $b_2 > b_1$, the member segments have the same thickness t. [Objective 6]
4. For an extruded beam having the cross-section shown in Figure 3, determine (a) location of shear center and (b) distribution of shear stresses caused by vertical SF F = 12 kN. [Objectives 3 and 6]

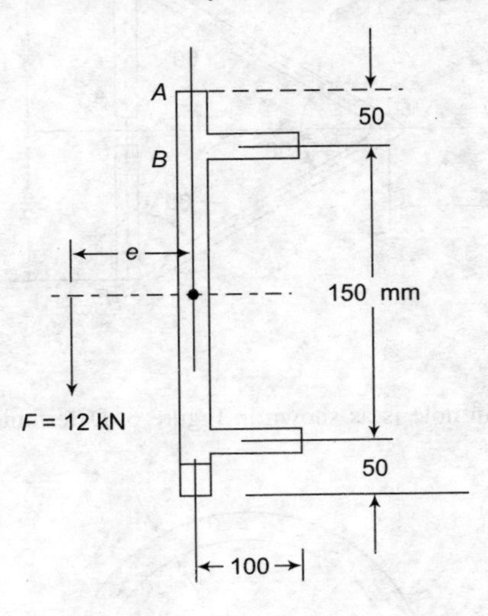

FIGURE 3

5. Determine the location e of the shear center for the thin-walled member having the cross-section shown in Figure 3. The member segments have the same thickness. [Objective 6]
6. Determine the location of shear center O of a thin-walled beam of uniform thickness having the equilateral triangular section as shown in Figure 4. [Objective 6]

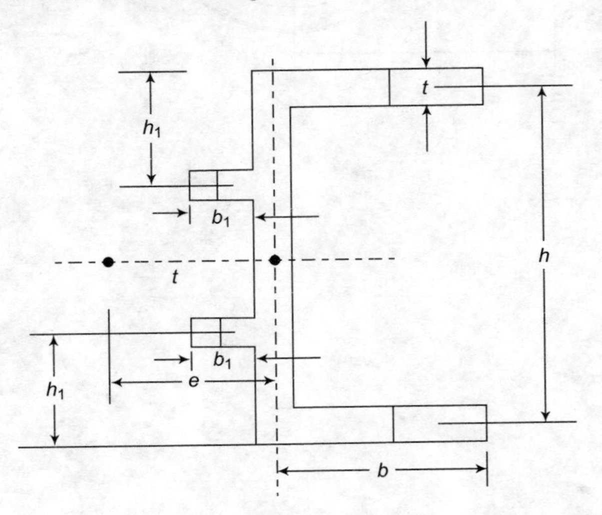

FIGURE 4

7. Determine the location of shear center of a thin-walled cross-section of a beam of uniform thickness having the cross-section shown in Figure 5. [Objective 6]

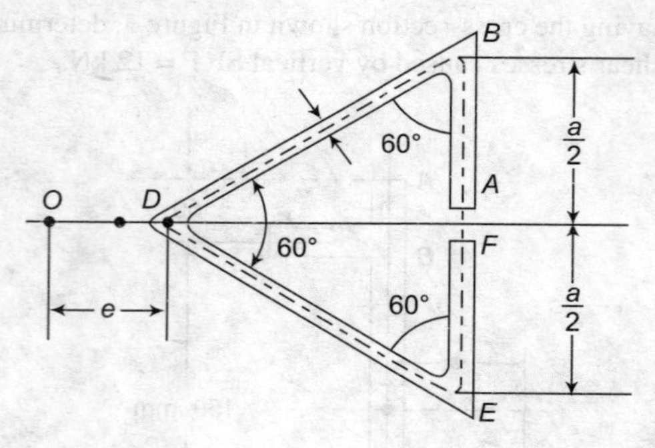

FIGURE 5

8. A tubular section with a pin hole is as shown in Figure 6. Determine the position of its shear center. [Objective 6]

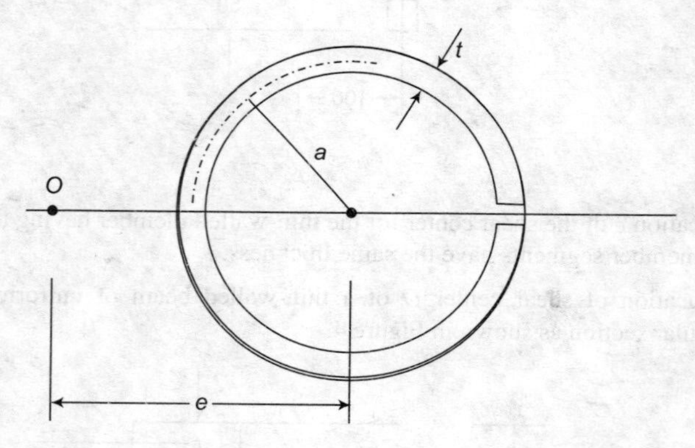

FIGURE 6

CHAPTER 15

ROTATING DISCS

UNIT OBJECTIVE

This chapter provides information on the stresses induced in the rotating discs (solid and hollow), rotating cylinders (solid and hollow), and stresses in spoked rims. The presentation attempts to help the student achieve the following:

Objective 1: Determine hoop and radial stresses in a disc and flywheel due to rotation.

Objective 2: Determine the maximum stresses and variation of radial and hoop stresses.

Objective 3: Determine the maximum allowable rotation speed for the disc and flywheel.

Objective 4: Evaluate the change in shrink fit pressure.

15.1 INTRODUCTION

Rotating structural elements are very common in many of the moving structures such as wheels of a cycle, bike, and shafts in transmitting the power. The stresses induced in the members owing to rotation are referred as centrifugal stresses. Rotation induces centrifugal forces within the body of the rotating object. These centrifugal forces are dependent on the mass of the objects. Thus, it is assumed that the mass of the structural element is uniform and mass density is constant throughout the body.

15.2 STRESSES IN ROTATING THIN DISC

Rotating discs are grouped in two categories such as thin rings and solid discs.

Analysis of Thin Rings

Consider a thin ring rotating about its center of gravity at O as shown in Figure 15.1.

Let ρ = Mass density of the ring in kg/m^3

r = Mean radius of ring in meters
ω = Angular speed of rotation in rad/s
t = Thickness of ring in meters

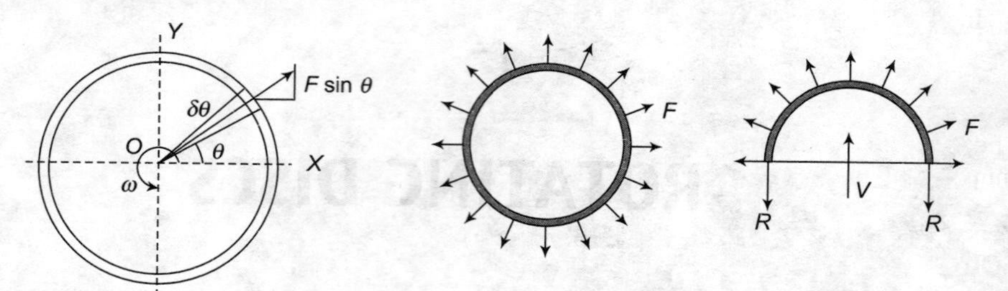

FIGURE 15.1

Consider an elemental strip, which makes an angle $\delta\theta$ at the center and its position is defined by θ as shown in Figure 15.1.

The radial acceleration on the elemental strip considered is $r\omega^2$.

The radial force acting on the element $F = \rho \times rd\theta \times t \times b \times r\omega^2 = \rho r^2\omega^2(tb \times d\theta)$

'b' is the dimension perpendicular to the width of the ring.

Component of the radial force F acting in the Y direction is

$$F\sin\theta = \rho r^2\omega^2(tb \times d\theta)\sin\theta$$

Net upward force in the ring = $V = \int_0^{\pi} F\sin\theta = \int_o^{\pi} \rho r^2\omega^2(tb \times d\theta)\sin\theta = 2\rho r^2\omega^2\, tb$

Taking the equilibrium equation that sum of vertical forces is equal to zero.

$$V = 2R$$

Hence $R = \rho r^2\omega^2\, tb$.

Hence the tensile stress in the ring = $\dfrac{R}{A} = \dfrac{\rho r^2\omega^2\, tb}{tb} = \rho r^2\omega^2$.

Thus, the ring is subjected to tensile stress called hoop stress = $\sigma_\theta = \rho r^2\omega^2$

If 'v' is the linear velocity of the point O, then $v = r\omega$.

Hence hoop stress = $\sigma_\theta = \rho v^2$.

PROBLEM 15.1

Objective 1

The linear velocity of the rim of a pulley is 20 m/s. Calculate hoop stress induced in the pulley rim. If the speed is increased by 30%, then find the increase in stress induced. Density of rim material = 7800 kg/m^3.

SOLUTION

Hoop stress,

$$\sigma_\theta = \rho v^2 = 7800 \times 20^2 = 3.12 \text{ MPa}$$

When the speed is increased by 30%, then

$$\sigma_\theta = \rho(1.3v)^2 = 1.69\rho v^2$$

Increase in stress $= (1.69 - 1)\rho v^2 = 0.69 \times 3.12 = 2.152$ MPa.

PROBLEM 15.2

Objective 1

The thin rim of a wheel is 100 cm diameter. Neglecting the effect of spokes, how many revolutions per minute (rpm) may it make without the hoop stress exceeding 100 MPa. The density is 7800 kg/m^3 and $E = 200$ GPa. Also find the change in diameter.

SOLUTION

$$\sigma_\theta = \rho\omega^2 r^2$$

$$100 \times 10^6 = 7800 \times \omega^2 \times 50^2 \times 10^{-4}$$

$$\omega = 226.45 \text{ rad/s}$$

$$\omega = \frac{2\pi N}{60}$$

$$N = \frac{226.45 \times 60}{2\pi} = 2163 \text{ rpm.}$$

Hoop strain

$$\varepsilon_\theta = \frac{\sigma_\theta}{E}$$

$$\frac{\delta d}{d} = \frac{100}{2 \times 10^5}$$

$$\delta d = \frac{100 \times 100}{2 \times 10^5} = 0.05 \text{ cm} = 0.50 \text{ mm.}$$

PROBLEM 15.3

Objective 1

A built-up ring consists of an inner copper ring and outer ring. The diameter of the surface of contact of the two rings is 80 cm. Determine the stresses set up in the steel and copper by rotation of the ring at 3000 rpm. Both the rings are of rectangular cross-section 15 mm in the radial direction and 20 mm in the direction perpendicular to the plane of the ring.

For steel	$E = 200$ GPa,	$\rho = 7800$ kg/m^3,
For copper	$E = 100$ GPa,	$\rho = 8900$ kg/m^3.

SOLUTION

Consider the built-up ring as shown in Figure 15.2.

Let p = contact pressure between the rings in MPa.

Hoop stress in steel ring,

$$(\sigma_\theta)_r = \frac{pd}{2t} = \frac{p \times 80}{2 \times 1.5} = 26.67p \text{ MPa (tensile).}$$

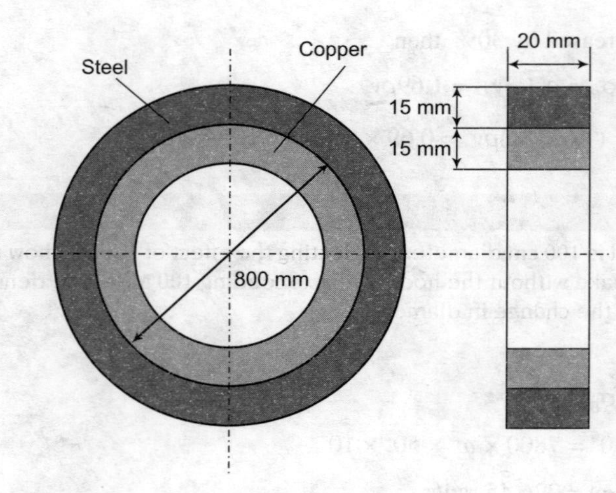

FIGURE 15.2

Hoop stress in copper ring,

$$(\sigma_\theta)_c = \frac{pd}{2t} = \frac{p \times 80}{2 \times 1.5} = 26.67\,p \text{ MPa (compressive).}$$

Because of rotation, the mean hoop stresses in rings are:

In steel ring, $(\sigma_\theta)'_s = \rho_s v^2$

$$= 7800 \times \left(\frac{2\pi \times 3000}{60} \times 40.75 \times 10^{-2}\right)^2 = 127.7 \text{ MPa (tensile)}$$

In copper ring, $(\sigma_\theta)'_c = \rho_c v^2$

$$= 8900 \times \left(\frac{2\pi \times 3000}{60} \times 39.25 \times 10^{-2}\right)^2 = 135.185 \text{ MPa (tensile)}$$

Total stress in steel = $(\sigma_\theta)_s + (\sigma_\theta)'_s = 26.67p + 127.7$

Total stress in copper = $(\sigma_\theta)'_c - (\sigma_\theta)_c = 135.185 - 26.67p.$

Now hoop strain in steel = Hoop strain in copper

$$\frac{1}{2 \times 10^5}(26.67p + 127.7) = \frac{1}{10^5}(135.185 - 26.67p)$$

$$13.335p + 63.85 = 135.185 - 26.67p$$

$$40p = 71.335$$

$$p = 1.783 \text{ MPa}$$

$\therefore$ Total stress in steel ring = 26.67 × 1.783 + 127.7 = 175.25 MPa

Total stress in copper ring = 135.185 – 26.67 × 1.783 = 87.63 MPa.

15.3 ROTATING DISC

Consider a flat rotating disc of uniform thickness t. Let r_1 and r_2 be the inner and outer radii of the disc as shown in Figure 15.3 rotating at speed ω.

An element of the disc ABCD at radius r is acted upon by stresses σ_r and $\sigma_r + d\sigma_r$ on faces AD and BC, respectively, and by stresses σ_θ on the faces AB and CD. On the flat faces of the disc, there is no normal stress, and hence there is free strain in the direction of the axis.

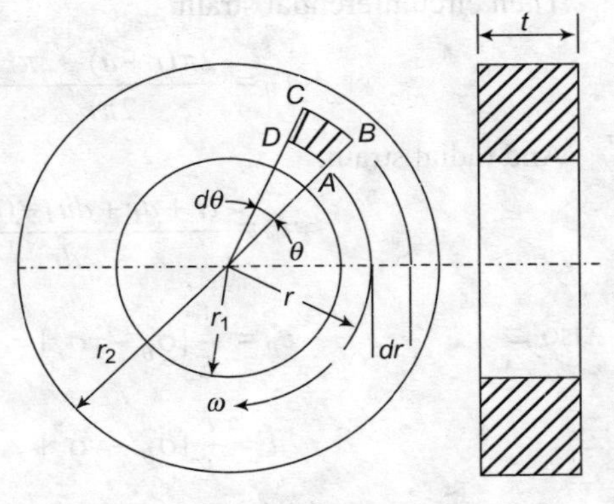

Rotating disc of uniform thickness

FIGURE 15.3

Volume of element ABCD $= rd\theta \cdot dr \cdot t$

Radial force on the element ABCD due to rotation

$$= \rho \times rd\theta \times dr \times t \times \omega^2 r$$

Outward force on face BC $= (r + dr)d\theta \times t(\sigma_r + d\sigma_r)$

Inward force on face AD $= rd\theta \times t\sigma_r$

Forces on faces AB and DC $= \sigma_\theta \times t \times dr$

These forces have been shown in Figure 15.4 (b).

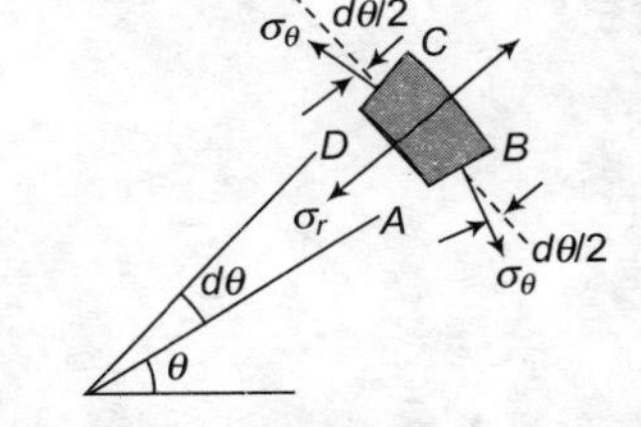

(a) Stresses acting on the element

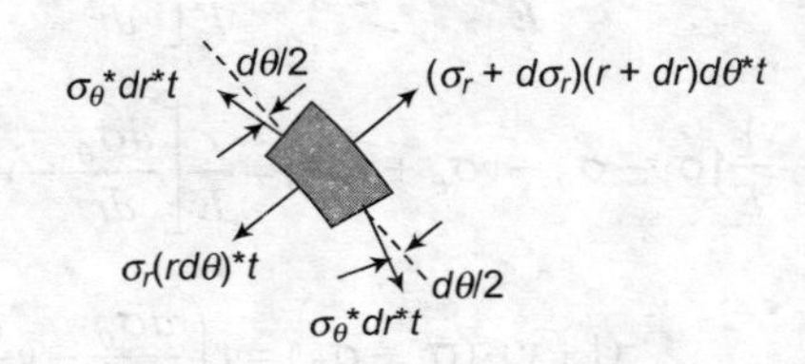

(b) Forces acting on hte element due to stresses

FIGURE 15.4

Resolving forces in the radial outward direction, we get

$$(\sigma_r + d\sigma_r) \times (r + dr)d\theta \times t - \sigma_r \times rd\theta \times t - \sigma_\theta \times tdr \times d\theta.$$

Simplifying and neglecting small quantities, we get $(\sigma_r - \sigma_\theta)dr \times d\theta \times t + d\sigma_r \times rd\theta \times t$.

For equilibrium of the element,

$$\rho \times rd\theta \times dr \times t \times \omega^2 r + (\sigma_r - \sigma_\theta)dr \times d\theta \times t + d\sigma_r \times rd\theta \times t = 0$$

$$\rho \cdot \omega^2 r^2 + (\sigma_r - \sigma_\theta) + r \times \frac{d\sigma_r}{dr} = 0$$

$$\sigma_r - \sigma_\theta = -r\frac{d\sigma_r}{dr} - \rho \cdot \omega^2 r^2 \tag{15.1}$$

On account of rotation let r becomes $r + u$ and $r + dr$ becomes $r + dr + du$.

Then circumferential strain,

$$\varepsilon_\theta = \frac{2\pi(r+u) - 2\pi r}{2\pi r} = \frac{u}{r}$$

And radial strain,

$$\varepsilon_r = \frac{(r+dr+du) - (r+dr)}{dr} = \frac{du}{dr}$$

Also $$\varepsilon_\theta = \frac{1}{E}[\sigma_\theta - v\sigma_r] = \frac{u}{r}$$

∴ $$u = \frac{r}{E}[\sigma_\theta - v\sigma_r] \quad (15.2)$$

and $$\varepsilon_r = \frac{1}{E}[\sigma_r - v\sigma_\theta] = \frac{du}{dr} \quad (15.3)$$

Differentiating Eq. (15.2) with respect to *r*, we get

$$\frac{du}{dr} = \frac{r}{E}\left[\frac{d\sigma_\theta}{dr} - v\frac{d\sigma_r}{dr}\right] + \frac{1}{E}[\sigma_\theta - v\sigma_r] \quad (15.4)$$

Comparing Eqs. (15.3) and (15.4), we get

$$\frac{1}{E}[\sigma_r - v\sigma_\theta] = \frac{r}{E}\left[\frac{d\sigma_\theta}{dr} - v\frac{d\sigma_r}{dr}\right] + \frac{1}{E}[\sigma_\theta - v\sigma_r]$$

$$\frac{1}{E}[\sigma_r - \sigma_\theta - v\sigma_\theta + v\sigma_r] = \frac{r}{E}\left[\frac{d\sigma_\theta}{dr} - v\frac{d\sigma_r}{dr}\right]$$

$$(1+v)\cdot(\sigma_r - \sigma_\theta) = r\left(\frac{d\sigma_\theta}{dr} - v\frac{d\sigma_r}{dr}\right) \quad (15.5)$$

Substituting Eq. (15.1) in Eq. (15.5), we get

$$(1+v)\left[-r\frac{d\sigma_r}{dr} - \rho\cdot\omega^2 r^2\right] = r\left(\frac{d\sigma_\theta}{dr} - v\frac{d\sigma_r}{dr}\right)$$

Simplifying, we get

$$\frac{d}{dr}(\sigma_r + \sigma_\theta) + (1+v)\rho\cdot\omega^2 r = 0 \quad (15.6)$$

Integrating, we get

$$\sigma_r + \sigma_\theta + \frac{1}{2}(1+v)\rho\omega^2 r^2 = C_1 \quad (15.7)$$

in which C_1 = constant of integration.

$$\sigma_\theta = C_1 - \sigma_r - \frac{1}{2}(1+v)\rho\omega^2 r^2.$$

Substituting in Eq. (15.1), we get

$$2\sigma_r + r\frac{d\sigma_r}{dr} = C_1 - \left(\frac{3+v}{2}\right)\rho\cdot\omega^2 r^2$$

Multiplying each side by r, we get

$$2r\sigma_r + r^2\frac{d\sigma_r}{dr} = C_1 r - \left(\frac{3+v}{2}\right)\rho\cdot\omega^2 r^3$$

or

$$\frac{d}{dr}(r^2\sigma_r) = C_1 r - \left(\frac{3+v}{2}\right)\rho\omega^2 r^3$$

Integrating, we get $r^2\sigma_r = \dfrac{C_1 r^2}{2} - \left(\dfrac{3+v}{2}\right)\rho\omega^2\dfrac{r^4}{4} + C_2$

$$\sigma_r = \frac{C_1}{2} - \left(\frac{3+v}{8}\right)\rho\cdot\omega^2 r^2 + \frac{C_2}{r^2} \tag{15.8}$$

in which C_2 is another constant of integration.

Substituting Eq. (15.8) in Eq. (15.7), we get

$$\sigma_\theta = \frac{C_1}{2} - \frac{C_2}{r^2} - \left(\frac{1+3v}{8}\right)\rho\cdot\omega^2 r^2 \tag{15.9}$$

The values of the constants C_1 and C_2 can be determined from the boundary conditions. Eq. (15.2) becomes,

$$u = \frac{r}{E}\left[(1-v)\frac{C_1}{2} - (1-v)\frac{C_2}{r^2} - \left(\frac{1-v^2}{8}\right)\rho\cdot\omega^2 r^2\right] \tag{15.10}$$

Eqs. (15.8) and (15.9) are the equations which give the stresses in the disc.

15.3.1 Solid Disc

From Eqs. (15.8) and (15.9), we have respectively

$$\sigma_r = \frac{C_1}{2} + \frac{C_2}{r^2} - \left(\frac{3+v}{8}\right)\rho\cdot\omega^2 r^2$$

$$\sigma_\theta = \frac{C_1}{2} - \frac{C_2}{r^2} - \left(\frac{1+3v}{8}\right)\rho\cdot\omega^2 r^2.$$

At $r = 0$, $u = 0$, and σ_r and σ_θ becomes infinite from above equations, which is not possible. Hence $C_2 = 0$.

$$\therefore \quad \sigma_r = \frac{C_1}{2} - \left(\frac{3+v}{8}\right)\rho\cdot\omega^2 r^2;\ \sigma_\theta = \frac{C_1}{2} - \left(\frac{1+3v}{8}\right)\rho\cdot\omega^2 r^2$$

At $r = r_2$, $\quad \sigma_r = 0$

$$\therefore \qquad C_1 = \left(\frac{3+v}{4}\right)\rho \cdot \omega^2 r_2^2$$

$$\therefore \qquad \sigma_r = \left(\frac{3+v}{8}\right)\rho \cdot \omega^2 (r_2^2 - r^2) \tag{15.11}$$

$$\sigma_\theta = \left(\frac{3+v}{8}\right)\rho \cdot \omega^2 r_2^2 - \left(\frac{1+3v}{8}\right)\rho \cdot \omega^2 r^2$$

Or

$$\sigma_\theta = \frac{\rho}{8}\omega^2[(3+v)r_2^2 - (1+3v)r^2] \tag{15.12}$$

At $r = r_2$

$$\sigma_\theta = \frac{(1-v)\rho}{4}\omega^2 r_2^2 \tag{15.13}$$

The values of σ_r and σ_θ are maximum at $r = 0$.

$$(\sigma_r)_{\max} = \left(\frac{3+v}{8}\right)\rho \cdot \omega^2 r_2^2 \tag{15.14}$$

$$(\sigma_\theta)_{\max} = \left(\frac{3+v}{8}\right)\rho \cdot \omega^2 r_2^2 \tag{15.15}$$

Radial displacement of the disc becomes

$$u = \frac{r}{E}[\sigma_\theta - v\sigma_r]$$

$$u = \frac{r}{E}\left[\left\{\left(\frac{3+v}{8}\right)\rho \cdot \omega^2 r_2^2 - \left(\frac{1+3v}{8}\right)\rho \cdot \omega^2 r^2\right\}\right] - v\left[\left(\frac{3+v}{8}\right)\rho \cdot \omega^2 (r_2^2 - r^2)\right]$$

$$= \frac{r(1-v)}{8E} \cdot \rho \cdot \omega^2[(3+v)r_2^2 - (1+v)r^2] \tag{15.16}$$

At $r = r_2$

$$u = \left(\frac{1-v}{4E}\right)\rho \cdot \omega^2 r_2^3 \tag{15.17}$$

15.3.2 Hollow Disc

From Eqs. (15.8) and (15.9), we have respectively

$$\sigma_r = \frac{C_1}{2} + \frac{C_2}{r^2} - \left(\frac{3+v}{8}\right)\rho \cdot \omega^2 r^2 \; ; \; \sigma_\theta = \frac{C_1}{2} - \frac{C_2}{r^2} - \left(\frac{1+3v}{8}\right)\rho \cdot \omega^2 r^2$$

At $r = r_1$, $\sigma_r = 0$ and at $r = r_2$, $\sigma_r = 0$

$$0 = \frac{C_1}{2} + \frac{C_2}{r_1^2} - \left(\frac{3+v}{8}\right)\rho \cdot \omega^2 r_1^2$$

$$0 = \frac{C_1}{2} + \frac{C_2}{r_2^2} - \left(\frac{3+v}{8}\right)\rho \cdot \omega^2 r_2^2$$

Solving for C_1 and C_2, we get

$$C_1 = \left(\frac{3+v}{4}\right)\rho \cdot \omega^2 (r_1^2 + r_2^2)\ ;\ C_2 = \left(\frac{3+v}{4}\right)\rho \cdot \omega^2 r_1^2 r_2^2$$

$$\sigma_r = \left(\frac{3+v}{8}\right)\rho \cdot \omega^2 \left[r_1^2 + r_2^2 - r^2 - \frac{r_1^2 r_2^2}{r^2}\right] \tag{15.18}$$

$$\sigma_\theta = \left(\frac{3+v}{8}\right)\rho \cdot \omega^2 \left[r_1^2 + r_2^2 + \frac{r_1^2 r_2^2}{r^2} - \left(\frac{1+3v}{3+v}\right) r^2\right] \tag{15.19}$$

σ_θ is maximum when $r = r_1$

$$(\sigma_\theta)_{\max} = \left(\frac{3+v}{8}\right)\rho \cdot \omega^2 \left[r_2^2 + \left(\frac{1-v}{3+v}\right) r_1^2\right] \tag{15.20}$$

For σ_r to be maximum

$$\frac{d\sigma_r}{dr} = 0$$

Using Eq. (15.8), we get

$$2r - \frac{2r_1^2 r_2^2}{r^3} = 0$$

$$r = \sqrt{r_1 r_2} \tag{15.21}$$

$$(\sigma_r)_{\max} = \left(\frac{3+v}{8}\right)\rho \cdot \omega^2 (r_2 - r_1)^2. \tag{15.22}$$

When $r_1 \to 0$, that is, a very small hole is drilled in the disc at the center, then Eq. (15.9) gives,

$$(\sigma_\theta)_{\max} = \left(\frac{3+v}{8}\right)\rho \cdot \omega^2 r_2^2. \tag{15.23}$$

Comparing with Eq. (15.15), it can be concluded that the maximum hoop stress in a rotating disc is twice as large, when there is a small hole at its axis of rotation, as that when the disc is solid. This is very important when solid discs are used as structural element. Due to corrosion or any other effect if at small opening (however small it may be) forms the stress at that pin hole doubles.

PROBLEM 15.4 **Objective 2**

A disc of 50 cm diameter and uniform thickness is rotating at 3000 rpm. Determine the maximum stress induced in the disc. If a hole of 20 cm diameter is drilled at the center of the disc, determine the maximum intensities of radial and hoop stresses induced. Take $v = 0.3$, and density of disc = 7800 kg/m^3.

SOLUTION

For the solid disc

$$(\sigma_r)_{\max} = (\sigma_\theta)_{\max} = \left(\frac{3+v}{8}\right)\rho\cdot\omega^2 r_2^2$$

$$= \left(\frac{3+0.3}{8}\right)\times 7800\times\left(\frac{2\pi\times 3000}{60}\right)^2\times 25^2\times 10^{-4}$$

$$= \frac{3.3}{8}\times 7800\times(100\pi)^2\times 625\times 10^{-4} = 19.827 \text{ MPa}$$

For the hollow disc

$$(\sigma_r)_{\max} = \left(\frac{3+v}{8}\right)\rho\cdot\omega^2(r_2-r_1)^2 = \left(\frac{3+0.3}{8}\right)\times 7800\times(100\pi)^2\times(25-10)^2\times 10^{-4}$$

$$= \left(\frac{3.3}{8}\right)\times 7800\times(100\pi)^2\times 225\times 10^{-4} = 7.137 \text{ MPa}$$

$$(\sigma_\theta)_{\max} = \left(\frac{3+v}{8}\right)\rho\cdot\omega^2\left[\left(\frac{1-v}{3+v}\right)r_1^2 + r_2^2\right]$$

$$= \left(\frac{3+0.3}{8}\right)\times 7800\times(100\pi)^2\times\left[\left(\frac{1-0.3}{3+0.3}\right)100+625\right]\times 10^{-4}$$

$$= \frac{3.3}{8}\times 7800\times(100\pi)^2\times 646.21\times 10^{-4} = 20.5 \text{ MPa.}$$

PROBLEM 15.5

Objective 2

A circular thin disc of outer radius 20 cm and inner radius 10 cm is rotating at 2500 rpm. Determine the variation of radial and hoop stresses in the disc. $\rho = 7800$ kg/m^3, $v = 0.28$.

SOLUTION

$$\sigma_r = \left(\frac{3+v}{8}\right)\rho\cdot\omega^2\left[r_1^2 + r_2^2 - r^2 - \frac{r_1^2 r_2^2}{r^2}\right]$$

$$= \left(\frac{3+0.28}{8}\right)\times 7800\times\left(\frac{2\pi\times 2500}{60}\right)^2\times\left[100+400-r^2-\frac{100\times 400}{r^2}\right]\times 10^{-4}$$

$$= 0.2189\times 10^{-4}\times\left[500-r^2-\frac{40000}{r^2}\right]$$

Also, σ_r is maximum at $r = \sqrt{r_1 r_2} = \sqrt{10\times 20} = 14.14$ cm.

r (cm)	10	14.14	15	20
σ_r (MPa)	0	2.189	2.128	0

$$\sigma_\theta = \left(\frac{3+v}{8}\right)\rho \cdot \omega^2\left[r_1^2 + r_2^2 + \frac{r_1^2 r_2^2}{r^2} - \left(\frac{1+3v}{3+v}\right)r^2\right]$$

$$= \left(\frac{3+0.28}{8}\right)\times 7800 \times \left(\frac{2\pi \times 2500}{60}\right)^2 \times \left[100 + 400 + \frac{100\times 400}{r^2} - \frac{1.84}{3.28}r^2\right]\times 10^{-4}$$

$$= \left(\frac{3.28}{8}\right)\times 7800 \times \left(\frac{2\pi \times 2500}{60}\right)^2 \times \left[500 + \frac{40000}{r^2} - 0.561r^2\right]\times 10^{-4}$$

$$= 0.2189\times 10^{-4}\times\left[500 + \frac{40000}{r^2} - 0.561r^2\right].$$

r (cm)	10	15	20
σ_θ (MPa)	18.473	12.073	8.221

The variation of stress distribution is shown in Figure 15.5.

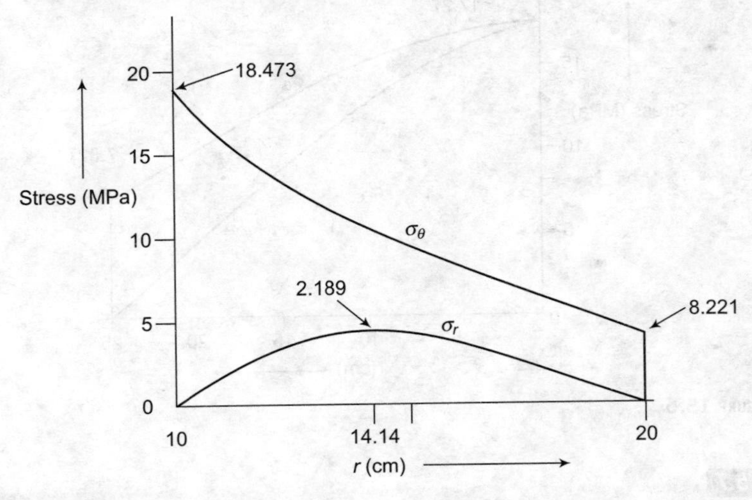

FIGURE 15.5

PROBLEM 15.6

Objective 2

A solid disc of diameter 50 cm is rotating at a speed of 2800 rpm. Determine the distribution of radial and hoop stresses in the disc. Density of disc material is 7800 kg/m^3 and $v = 0.3$.

SOLUTION

$$\sigma_r = \left(\frac{3+v}{8}\right)\rho\cdot\omega^2(r_2^2 - r^2)$$

$$= \left(\frac{3+0.3}{8}\right) \times 7800 \times \left(\frac{2\pi \times 2800}{60}\right)^2 (625 - r^2) \times 10^{-4}$$

$$= 0.2763 \times 10^{-4}(625 - r^2)$$

$$\sigma_\theta = \frac{\rho}{8}\omega^2[(3+v)r_2^2 - (1+3v)r^2]$$

$$= \frac{7800}{8} \times \left(\frac{2\pi \times 2800}{60}\right)^2 \times [3.3 \times 625 - 1.9r^2] \times 10^{-4}$$

$$= 0.08374 \times 10^{-4}[2062.5 - 1.9r^2]$$

r (cm)	0	5	10	15	20	25
σ_r (MPa)	17.27	16.578	14.505	11.052	6.216	0
σ_θ (MPa)	17.27	16.873	15.68	13.691	10.907	7.327

The variation of stresses has been plotted in Figure 15.6.

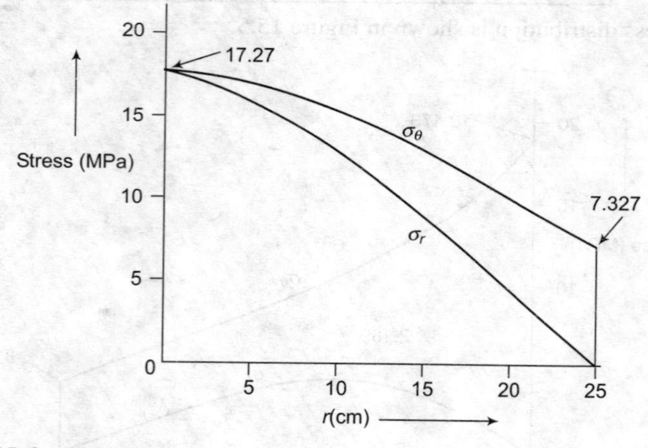

FIGURE 15.6

PROBLEM 15.7

Objective 3

A steel disc 30 cm outside diameter and 10 cm inside diameter is shrunk on a steel shaft so that the pressure between the shaft and the disc at standstill is 60 MPa. Assuming that the change in dimensions of the shaft is negligible, find at which speed the disc will loosen from the shaft. Density = 7800 kg/m^3 and $v = 0.3$.

SOLUTION

For hollow disc

$$\sigma_r = \frac{C_1}{2} + \frac{C_2}{r^2} - \left(\frac{3+v}{8}\right)\rho \cdot \omega^2 r^2$$

At standstill, $\omega = 0$

$$\therefore \quad \sigma_r = \frac{C_1}{2} + \frac{C_2}{r^2}$$

Similarly, $$\sigma_\theta = \frac{C_1}{2} - \frac{C_2}{r^2}.$$

Let p = shrinkage pressure between the disc and the shaft.

For the disc,

At $r = 5$ cm, $\sigma_r = -p$ and at $r = 15$ cm, $\sigma_r = 0$

$$-p = \frac{C_1}{2} + \frac{C_2}{25}$$

$$0 = \frac{C_1}{2} + \frac{C_2}{225}$$

$$\therefore \quad \frac{C_1}{2} = -\frac{C_2}{225}$$

$$-p = C_2\left(-\frac{1}{225} + \frac{1}{25}\right) = \frac{200C_2}{225 \times 25}$$

$$C_2 = C_2 = \frac{-225 \times 25}{200} p = -28.125p$$

$$\frac{C_1}{2} = 0.125p$$

$$\therefore \quad \sigma_r = 0.125p - \frac{28.125p}{r^2}; \; \sigma_\theta = 0.125p + \frac{28.125p}{r^2}$$

At $r = 5$ cm, the hoop stress becomes

$$\sigma_\theta = 0.125p + \frac{28.125}{25} p$$

$$= 0.125p + 1.125p = 1.25p = 1.25 \times 60 = 75 \text{ MPa.}$$

Hoop strain at inside of disc becomes,

$$\varepsilon_\theta = \frac{1}{E}[\sigma_\theta - v\sigma_r]$$

$$= \frac{1}{E}[75 + 0.3 \times 60] = \frac{93}{E}.$$

Because of rotation, hoop stress in the disc at inside radius is given by,

$$\sigma_\theta = \left(\frac{3+v}{8}\right)\rho \cdot \omega^2 \left[r_2^2 + \left(\frac{1-v}{3+v}\right) r_1^2\right]$$

$$= \left(\frac{3+0.3}{8}\right) \times 7800 \times \omega^2 \left[225 + \frac{0.7}{3.3} \times 25\right] \times 10^{-4}$$

$$= \frac{3.3 \times 7800 \times 230.303 \times 10^{-4}}{8} \times \omega^2 = 74.1\omega^2 \; ; \; \sigma_r = 0$$

$$\therefore \text{ Hoop strain } = \frac{\sigma_\theta}{E} = \frac{74.1\omega^2}{E}$$

For the disc to loosen from the shaft, the hoop strains should be equal.

$$\therefore \quad \frac{74.1\omega^2}{E} = \frac{93 \times 10^6}{E}$$

$$\therefore \quad \omega^2 = \frac{93 \times 10^6}{74.1}$$

$$\omega = 1120.294 \text{ rad/s}$$

$$\frac{2\pi N}{60} = 1120.294$$

$$N = \frac{1120.294 \times 60}{2\pi} = 10{,}703 \text{ rpm.}$$

PROBLEM 15.8

Objective 2, 3 and 4

A flat steel disc of 90 cm outside diameter with a 15 cm hole is shrunk around a solid steel shaft. The shrink allowance is 1 in 1000 parts. (i) At what rpm will the shrink fit loosen up as a result of rotation?, (ii) What are the maximum stresses when spinning at the speed calculated in part (i)?, (iii) What are the stresses at standstill?, and (iv) What are the stresses at half the speed calculated in part (i). $\rho = 7800$ kg/m^3, $v = 0.3$, $E = 200$ GPa.

SOLUTION

(i) When spinning at the required speed, there is no radial pressure between the disc and the shaft.

At standstill, the shrinkage allowance at 7.5 cm radius

$$= 0.001 \times 7.5 = 0.0075 \text{ cm}$$

$$u_{\text{disc}} = \frac{r_1}{E}[\sigma_\theta - v\sigma_r]$$

At $r = r_1 = 7.5$ cm

$$\sigma_\theta = \left(\frac{3+v}{8}\right)\rho \cdot \omega^2 \left[r_2^2 + \left(\frac{1-v}{3+v}\right) r_1^2\right]$$

$$\sigma_r = 0$$

$$u_{\text{disc}} = \frac{r_1}{200 \times 10^9}\left[\left(\frac{3+0.3}{4}\right) \times 7800 \times \omega^2 \left\{(45)^2 + \frac{0.7}{3.3} \times 56.25\right\}\right] \times 10^{-4}$$

$$= \frac{7.5 \times 10^{-2} \times 3.3 \times 7800}{200 \times 10^9 \times 4} \times 2154.93 \times \omega^2 \times 10^{-4}$$

$$= 520.013 \times 10^{-12} \times \omega^2$$

$$u_{\text{shaft}} = \frac{r_1}{E}[\sigma_\theta - v\sigma_r] = \left[\frac{1-v}{4E}\right]\rho\omega^2 r_1^3$$

$$= \frac{0.7}{4\times 200\times 10^9}\times 7800\times(7.5)^3\times\omega^2\times 10^{-6}$$

$$= 2.8793\times 10^{-12}\times\omega^2$$

$$u_{\text{disc}} - u_{\text{shaft}} = 0.0075$$

$$(520.013 - 2.8793)10^{-12}\times\omega^2 = 0.0075\times 10^{-2} = 0.0075\times 10^{-2}$$

$$517.1337\times 10^{-12}\times\omega^2 = 0.0075\times 10^{-2}$$

$$\omega^2 = 14.503\times 10^4$$

$$\omega = 380.828 \text{ rad/s}$$

$$\frac{2\pi N}{60} = 380.828$$

$$N = 3638.48 \text{ rpm.}$$

(ii) $$(\sigma_\theta)_{\max} = \left(\frac{3+v}{8}\right)\rho\cdot\omega^2\left[r_2^2 + \left(\frac{1-v}{3+v}\right)r_1^2\right]$$

$$= \frac{3.3}{4}\times 7800\times 14.503\times 10^4[2154.93]\times 10^{-4} = 201.112 \text{ MPa}$$

$$(\sigma_\theta)_{\max} = \left(\frac{3+v}{8}\right)\rho\cdot\omega^2(r_2 - r_1)^2$$

$$= \frac{3.3}{8}\times 7800\times 14.503\times 10^4(45-7.5)^2\times 10^{-4}$$

$$= \frac{3.3}{8}\times 7800\times 14.503\times 10^4\times 1406.25\times 10^{-4} = 65.62 \text{ MPa}$$

(iii) At standstill, $\sigma_r = \frac{C_1}{2} + \frac{C_2}{r^2}$

Let p = shrinkage pressure.
For the disc,
At $r = 7.5$ cm, $\sigma_r = -p$
At $r = 45$ cm, $\sigma_r = 0$

$$-p = \frac{C_1}{2} + \frac{C_2}{56.25}$$

$$0 = \frac{C_1}{2} + \frac{C_2}{2025}$$

$$\frac{C_1}{2} = -\frac{C_2}{2025}$$

$$-p = C_2\left[-\frac{1}{2025} + \frac{1}{56.25}\right]$$

$$C_2 = \frac{-2025 \times 56.25}{1968.75}p = -57.857p$$

$$\frac{C_1}{2} = 0.02857p$$

$$\therefore \quad \sigma_r = 0.02857p - \frac{57.857}{r^2}p\,;\ \sigma_\theta = 0.02857p + \frac{57.857}{r^2}p$$

At $r = 7.5$ cm, $\sigma_r = -p$

$$\sigma_\theta = 0.02857p + 1.02856p = 1.05713p$$

$$u_{\text{disc}} = \frac{r_1}{E}[\sigma_\theta - v\sigma_r]$$

$$= \frac{7.5 \times 10^{-2}\,p}{200 \times 10^9}[1.05713 + 0.3 \times 1]$$

$$= \frac{7.5 \times 10^{-2} \times 1.35713}{200 \times 10^9}p$$

$$= 5.0892 \times 10^{-13}p \text{ m}$$

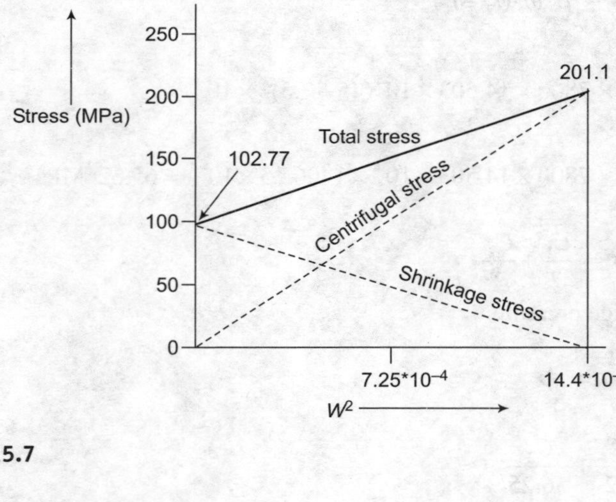

FIGURE 15.7

For the shaft, $\sigma_\theta = \sigma_r = -p$

$$\therefore \quad u_{\text{shaft}} = \frac{r_1}{E}[-p + vp]$$

$$= \frac{-p \times 7.5 \times 10^{-2} \times 0.7}{200 \times 10^3} = -2.625 \times 10^{-13}\,p$$

Now $u_{disc} - u_{shaft} = 0.0075 \times 10^{-2}$

$$\therefore \quad (5.0892 + 2.625)10^{-13}p = 0.0075 \times 10^{-2}$$

$$p = \frac{0.0075 \times 10^{11}}{7.7142}$$

$$= 97.223 \text{ MPa}$$

Hence in the disc,

$$\sigma_\theta = 1.05713 \times 97.223 = 102.77 \text{ MPa}$$

(iv) Stresses at half the speed:

$$\sigma_\theta = \left(\frac{1}{4} \times 201.112\right) + \left(\frac{3}{4} \times 102.77\right)$$

$$= 50.278 + 77.0775 = 127.355 \text{ MPa}$$

The variation of hoop stress with ω^3 is shown in Figure 15.7.

PROBLEM 15.9

Objective 2 and 3

A thin circular disc of external radius R is fitted tightly on to a rigid shaft of radius r. Show that when the shaft rotates with angular velocity ω, the reaction between the two is reduced by: $\dfrac{\gamma\omega^2(R^2 - r^2)\{(3+v)R^2 + (1+v)r^2\}}{4g(1+v)R^2 + (1-v)r^2}$.

SOLUTION

Let p be the shrinkage pressure.

At standstill, $\sigma_r = \dfrac{C_1}{2} + \dfrac{C_2}{r^2}$; $\sigma_\theta = \dfrac{C_1}{2} - \dfrac{C_2}{r^2}$

For the disc:

At $x = r$, $\sigma_r = -p$

At $x = R$, $\sigma_r = 0$

$$-p = \frac{C_1}{2} + \frac{C_2}{r^2}\,;\; 0 = \frac{C_1}{2} + \frac{C_2}{R^2}$$

$$\frac{C_1}{2} = -\frac{C_2}{R^2}$$

$$-p = C_2\left[-\frac{1}{R^2} + \frac{1}{r^2}\right]$$

$$C_2 = \frac{-r^2R^2p}{(R^2 - r^2)}$$

$$\frac{C_1}{2} = \frac{r^2p}{(R^2 - r^2)}$$

$$\therefore \qquad \sigma_r = \frac{r^2 p}{(R^2 - r^2)} - \frac{r^2 R^2 p}{(R^2 - r^2)x^2}\,;\ \sigma_\theta = \frac{r^2 p}{(R^2 - r^2)} + \frac{r^2 R^2 p}{(R^2 - r^2)x^2}$$

At $x = r$,

$$\sigma_\theta = \frac{r^2 p}{(R^2 - r^2)} + \frac{R^2 p}{(R^2 - r^2)} = p\left(\frac{R^2 + r^2}{R^2 - r^2}\right) \text{ (tensile)}$$

Hoop strain, $$\varepsilon_\theta = \frac{1}{E}[\sigma_\theta - v\sigma_r] = \frac{1}{E}\left[p^1\left(\frac{R^2 + r^2}{R^2 - r^2}\right) + vp\right] = \frac{p}{E}\left[\frac{R^2 + r^2}{R^2 - r^2} + v\right]$$

When the speed is ω, let p' be the pressure between the shaft and the disc. Therefore, hoop stress in disc due to pressure becomes,

$$\sigma'_\theta = p'\left(\frac{R^2 + r^2}{R^2 - r^2}\right) \text{ (tensile).}$$

Hoop stress in the disc due to rotation is,

$$\sigma''_\theta = \left(\frac{3 + v}{4}\right)\frac{\gamma}{g}\omega^2\left[R^2 + \left(\frac{1 - v}{3 + v}\right)r^2\right]$$

Total tensile hoop stress in disc becomes,

$$\sigma_\theta = \sigma'_\theta + \sigma''_\theta = p'\left(\frac{R^2 + r^2}{R^2 - r^2}\right) + \left(\frac{3 + v}{4}\right)\frac{\gamma}{g}\omega^2\left[R^2 + \left(\frac{1 - v}{3 + v}\right)r^2\right]$$

Hoop strain becomes, $$\varepsilon_\theta = \frac{1}{E}[\sigma_\theta - v\sigma_r] = \frac{1}{E}[\sigma_\theta + vp']$$

$$= \frac{1}{E}\left[p'\left(\frac{R^2 + r^2}{R^2 - r^2} + v\right) + \left(\frac{3 + v}{4}\right)\frac{\gamma}{g}\omega^2\left\{R^2 + \left(\frac{1 - v}{3 + v}\right)r^2\right\}\right]$$

Since hoop strains will be equal,

$$\therefore \qquad \frac{p}{E}\left[\frac{R^2 + r^2}{R^2 - r^2} + v\right] = \frac{1}{E}\left[p'\left(\frac{R^2 + r^2}{R^2 - r^2} + v\right) + \left(\frac{3 + v}{4}\right)\frac{\gamma}{g}\omega^2\left\{R^2 + \left(\frac{1 - v}{3 + v}\right)r^2\right\}\right]$$

$$(p - p')\left[\frac{R^2 + r^2}{R^2 - r^2} + v\right] = \left(\frac{3 + v}{4}\right)\frac{\gamma}{g}\omega^2\left\{R^2 + \left(\frac{1 - v}{3 + v}\right)r^2\right\}$$

$$(p - p') = \frac{\gamma\omega^2(R^2 - r^2)\{(3 + v)R^2 + (1 - v)r^2\}}{4g[(1 + v)R^2 + (1 - v)r^2]}$$

Hence proved.

PROBLEM 15.10

Objective 4

A steel disc of 10 cm internal and 25 cm external radius is shrunk on a cast iron (C.I.) disc of 5 cm internal radius. Determine the change in the shrink-fit pressure produced by inertia forces at 3000 rpm.

For steel $E = 200$ GPa, $\rho = 7800$ kg/m^3
For cast iron $E = 100$ GPa, $\rho = 7200$ kg/m^3
Poisson's ratio 0.3 for both.

SOLUTION

Let p = shrink fit pressure at standstill, MPa
For the steel disc,

$$\sigma_r = \frac{C_1}{2} + \frac{C_2}{r^2}$$

At $r = 10$ cm, $\sigma_r = -p$
At $r = 25$ cm, $\sigma_r = 0$

$$-p = \frac{C_1}{2} + \frac{C_2}{100}\ ;\ 0 = \frac{C_1}{2} + \frac{C_2}{625}$$

$$\frac{C_1}{2} = -\frac{C_2}{625}$$

$$-p = C_2\left[-\frac{1}{625} + \frac{1}{100}\right]$$

$$C_2 = \frac{-p \times 625 \times 100}{525} = -119.047p$$

$$\frac{C_1}{2} = 0.19p$$

$$\therefore \qquad \sigma_\theta = 0.19p + \frac{119.047p}{r^2}$$

At $r = 10$ cm, $\sigma_\theta = 1.38p$
For the inner cast iron disc

$$\sigma_\theta = \frac{C_1'}{2} + \frac{C_2'}{r^2}$$

At $r = 5$ cm, $\sigma_r = 0$
At $r = 10$ cm, $\sigma_r = -p$

$$0 = \frac{C_1'}{2} + \frac{C_2'}{25}$$

$$-p = \frac{C_1'}{2} + \frac{C_2'}{100}$$

$$\frac{C_1'}{2} = -\frac{C_2'}{25}$$

$$-p = C_2'\left[-\frac{1}{25} + \frac{1}{100}\right]$$

$$C_2' = \frac{25 \times 100p}{75} = 33.3p$$

$$\frac{C_1'}{2} = 1.39$$

$$\therefore \qquad \sigma_\theta = -\left(1.3p + \frac{33.3p}{r^2}\right)$$

At $r = 10$ cm, $\sigma_\theta = -1.633p$

Hoop strain at $r = 10$ cm in steel disc,

$$(\varepsilon_\theta)_S = \frac{1}{E_S}[\sigma_\theta - v\sigma_r] = \frac{1}{E_S}[\sigma_\theta + vp]$$

$$= \frac{1}{E_S}[1.38p + 0.3p] = \frac{1.68p}{E_S} = \frac{1.68p}{200 \times 10^3} = 0.84 \times 10^{-5}\,p$$

Hoop strain in cast iron disc at $r = 10$ cm;

$$(\varepsilon_\theta)_{\text{C.I.}} = \frac{1}{E_{\text{C.I.}}}[\sigma_\theta - v\sigma_r]$$

$$\frac{1}{100 \times 10^3}[-1.633p + 0.3p] = \frac{-1.333p}{10^5} = -1.333 \times 10^{-5}\,p$$

Total strain at standstill at $r = 8$ cm becomes,

$$\varepsilon_\theta = (\varepsilon_\theta)_S - (\varepsilon_\theta)_{\text{C.I.}} = (0.84 + 1.333) \times 10^{-5}\,p = 2.173 \times 10^{-5}\,p$$

At 3000 rpm, let p' be the pressure between the discs.

Hoop stress in steel disc is then,

$$(\sigma_\theta)'_S = 1.38p'$$

Hoop stress in cast iron disc at $r = 10$ cm,

$$(\sigma_\theta)'_{\text{C.I.}} = -1.633p'$$

Because of rotation, hoop stress in outer disc, at $r = 10$ cm is,

$$(\sigma_\theta)''_S = \left(\frac{3+v}{4}\right)\rho\omega^2\left[r_2^2 + \left(\frac{1-v}{3+v}\right)r_1^2\right]$$

$$= \left(\frac{3.3}{4}\right) \times 7800 \times \left(\frac{2\pi \times 3000}{60}\right)^2 \times \left[625 + \frac{0.7}{3.3} \times 100\right] \times 10^{-4}$$

$$= \frac{3.3}{4} \times 7800 \times (100\pi)^2 \times 646.21 \times 10^{-4} = 41 \text{ MPa}$$

Hoop stress in inner disc at $r = 10$ cm

$$(\sigma_\theta)''_{\text{C.I.}} = \left(\frac{3+v}{4}\right)\rho\omega^2\left[r_2^2 + \left(\frac{1-v}{3+v}\right)r_1^2\right]$$

$$= \left(\frac{3.3}{4}\right) \times 7200 \times \left(\frac{2\pi \times 3000}{60}\right)^2 \times \left[100 + \frac{0.7}{3.3} \times 25\right] \times 10^{-4}$$

$$= \frac{3.3}{4} \times 7200 \times (100\pi)^2 \times 65.91 \times 10^{-4} = 6.167 \text{ MPa}$$

Total hoop stress in steel disc becomes,

$$(\sigma_\theta)_S = (\sigma_\theta)'_S + (\sigma_\theta)''_S = 1.38p' + 41$$

Total hoop stress in cast iron disc becomes,

$$(\sigma_\theta)_{\text{C.I.}} = -1.633p' + 6.167$$

Hoop strain in steel disc,

$$(\sigma_\theta)_S = \frac{1}{E}[(\sigma_\theta)_S - v\sigma_r]$$

$$= \frac{1}{200 \times 10^3}[1.38p' + 41 + 0.3p]$$

$$= \frac{1}{200 \times 10^3}[1.68p' + 41] = (0.84p' + 20.5) \times 10^{-5}$$

Hoop strain in cast iron disc,

$$(\varepsilon_\theta)_{\text{C.I.}} = \frac{1}{E_{\text{C.I.}}}[(\sigma_\theta)_{\text{C.I.}} - v\sigma_r]$$

$$= \frac{1}{100 \times 10^3}[-1.633p' + 6.167 + 0.3p'] = (-1.333p' + 6.167) \times 10^{-5}$$

Total hoop strain at $r = 10$ cm during rotation,

$$= (\varepsilon_\theta)_S - (\varepsilon_\theta)_{\text{C.I.}}$$

$$= (0.84p' + 20.5 + 1.333p' - 6.167)10^{-5} = (2.173p' + 14.333) \times 10^{-5}$$

But total strains must be equal.

$$\therefore \quad 2.173 \times 10^{-5}\, p = 2.173\, p' + 14.333) \times 10^{-5}$$

$$2.173(p - p') = 14.333$$

$$p - p' = 6.596 \text{ MPa.}$$

15.4 DISC OF UNIFORM STRENGTH

From the earlier discussion, it is observed that the hoop stress is maximum at the inner face of the disc and decreases toward the outer faces of the disc. Where stress is low there if we decrease the material (thickness) the hoop stress along the radial would be constant. This way of doing the design makes the structural element economical. Thus, for the disc to have uniform strength the radial and hoop stresses must be equal at all points in the disc. Hence

$$\sigma_\theta = \sigma_r = \sigma = \text{ constant}$$

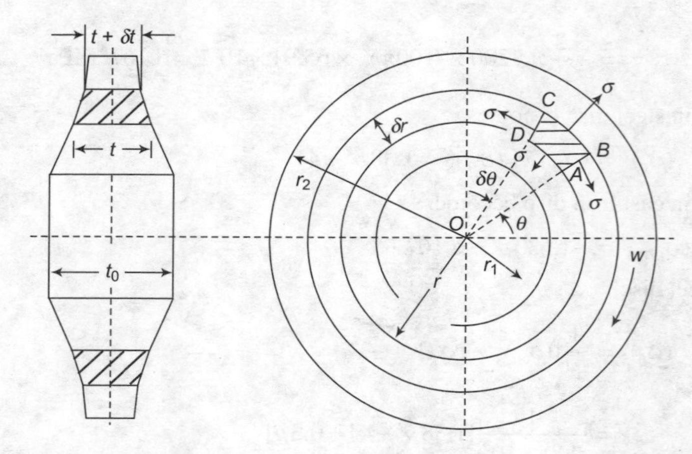

Figure 15.8

Consider the equilibrium of the element ABCD of the disc as shown in Figure 15.8. Let t be the thickness of the disc at radius r and $t + \delta t$ at radius $r + \delta r$.

Outward radial force acting on face BC

$$= \sigma(t+\delta t)(r+\delta r)\cdot\delta\theta \cong \sigma(tr + r\cdot\delta t + t\cdot\delta r)\delta\theta.$$

Centrifugal force acting on the element ABCD is

$$= \rho\cdot(r\delta\theta\cdot\delta r\cdot t)\cdot\omega^2 r$$

Inward radial force acting on face AD $= \sigma\cdot t\cdot r\cdot\delta\theta$

Inward radial force due to component of forces acting on faces AB and CD $= \sigma\cdot t\cdot\delta r\cdot\delta\theta$.

For equilibrium of the element,

Total inward radial force = Total outward radial force

$$\sigma\cdot t\cdot r\cdot\delta\theta + \sigma\cdot t\cdot\delta r\cdot\delta\theta = \sigma(tr + r\cdot\delta t + t\cdot\delta r)\delta\theta + \rho(r\delta\theta\cdot\delta r\cdot t)\omega^2 r$$

$\therefore$
$$\sigma\cdot r\delta t\cdot\delta\theta + \rho\cdot\delta\theta\cdot\delta r\cdot t\cdot\omega^2 r^2 = 0$$

$$\frac{\delta t}{t} = -p\frac{\omega^2}{\sigma}r\cdot\delta r$$

Or
$$\frac{dt}{t} = -p\frac{\omega^2}{\sigma}r\cdot dr$$

Integrating, we get $\ln t = -\rho\cdot\dfrac{\omega^2}{\sigma}\cdot\dfrac{r^2}{2} + \ln A$ in which ln A is a constant of integration.

$$\ln\frac{t}{A} = -\rho\cdot\left(\frac{\omega^2 r^2}{2\sigma}\right)$$

$$\therefore \qquad t = Ae^{-\rho\cdot\left(\frac{\omega^2 r^2}{2\sigma}\right)}$$

Let $t = t_0$ at $r = r_1$, then

$$t_0 = Ae^{-\rho\cdot\left(\frac{\omega^2 r_1^2}{2\sigma}\right)}$$

$$\therefore \qquad A = t_0 e^{\rho\cdot\left(\frac{\omega^2 r_1^2}{2\sigma}\right)}$$

$$\therefore \qquad t = t_0 e^{-\rho\cdot\frac{\omega^2}{2\sigma}(r^2 - r_1^2)} \tag{15.24}$$

Which gives the thickness of disc at any radius.

Let $\lambda = \sqrt{\dfrac{2\sigma}{\rho\omega^2}}$

$$\therefore \qquad t = t_0 e^{-\left(\frac{r^2 - r_1^2}{\lambda^2}\right)}$$

$$\frac{t}{t_0} = e^{-\left(\frac{r^2 - r_1^2}{\lambda^2}\right)}$$

$$\therefore \qquad \frac{d(t/t_0)}{d(r/\lambda)} = -\frac{2r}{\lambda} e^{-\left(\frac{r^2 - r_1^2}{\lambda^2}\right)} = -\frac{2r}{\lambda} \times \frac{t}{t_0}$$

Which is the slope of the curve. It is zero for $\dfrac{r}{\lambda} = 0$ and $\dfrac{t}{t_0} = 0$ occurring at $\dfrac{r}{\lambda} = \infty$.

The curvature is $\dfrac{d^2(t/t_0)}{d(r/\lambda)^2} = -\dfrac{2t}{t_0} - \dfrac{2r}{t}\dfrac{d(t/t_0)}{d(r/\lambda)} = -\dfrac{2t}{t_0} + 4\left(\dfrac{r}{\lambda}\right)^2 \times \dfrac{t}{t_0}$

Point of inflection is given by,

$$-2\frac{t}{t_0} + 4\left(\frac{r}{\lambda}\right)^2 \times \frac{t}{t_0} = 0$$

$$\left(\frac{r}{\lambda}\right)^2 = \frac{1}{2}$$

$$\text{Or} \qquad \frac{r}{\lambda} = 0.707. \tag{15.25}$$

PROBLEM 15.11

Objective 3

A steam turbine rotor is 25 cm diameter below the blade ring and 3 cm thick. The turbine is running at 45000 rpm. The allowable stress is 400 MPa. What is the thickness of the rotor at a radius of 8 cm and at the center. Assume uniform strength and take density of material = 7800 kg/m³.

SOLUTION

Here $t = t_0 e^{-\frac{\rho}{2\sigma}\omega^2(r^2 - r_1^2)}$

At $r = 8$ cm

$$t = 2e^{-\frac{7800}{2\times 400\times 10^6}\times\left(\frac{2\pi\times 45000}{60}\right)^2(64-156.25)\times 10^{-4}}$$

$$= 2e^{1.995} = 2\times 7.352 = 14.704 \text{ cm}$$

At $r = 0$

$$t = 2e^{-\frac{7800}{2\times 400\times 10^6}\times\left(\frac{2\pi\times 45000}{60}\right)^2(-156.25)\times 10^{-4}}$$

$$= 2e^{3.379} = 2\times 29.341 = 58.682 \text{ cm.}$$

15.5 ROTATING CYLINDER

Rotating cylinders are commonly encountered in shaft used for power transmission. Consider a circular cylinder of inside radius r_1 and outside radius r_2 rotating at speed ω as shown in Figure 15.9. Assume that plane sections of the cylinder remain plane during rotation; then the axial strain along the z axis will be independent of the radius r of the cylinder and will be constant.

Radial strain, $$\varepsilon_r = \frac{1}{E}[\sigma_r - v(\sigma_\theta + \sigma_z)] = \frac{du}{dr} \quad (15.26)$$

Hoop strain, $$\varepsilon_\theta = \frac{1}{E}[\sigma_\theta - v(\sigma_r + \sigma_z)] = \frac{u}{r} \quad (15.27)$$

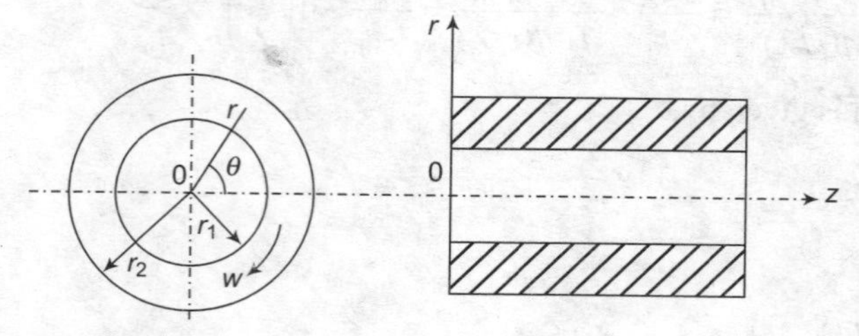

FIGURE 15.9 Rotating Cylinder

Axial strain, $$\varepsilon_z = \frac{1}{E}[\sigma_z - v(\sigma_r + \sigma_\theta)] \quad (15.28)$$

From Eq. (15.27), we have

$$Eu = r[\sigma_\theta - v(\sigma_r + \sigma_z)]$$

Differentiating with respect to r, we get

$$E\frac{du}{dr} = \sigma_\theta - \nu(\sigma_r + \sigma_z) + r\left[\frac{d\sigma_\theta}{dr} - \nu\left(\frac{d\sigma_r}{dr} + \frac{d\sigma_z}{dr}\right)\right] = \sigma_r - \nu(\sigma_\theta + \sigma_z)$$

Using Eq. (15.26),

$$\therefore \quad (\sigma_r - \sigma_\theta)(1 + \nu) = r\left[\frac{d\sigma_\theta}{dr} - \nu\left(\frac{d\sigma_r}{dr} + \frac{d\sigma_z}{dr}\right)\right] \tag{15.29}$$

From Eq. (15.28), we get

$$E\varepsilon_z = \sigma_z - \nu(\sigma_r - \sigma_\theta) = \text{constant} = C_1$$

$$\therefore \quad \sigma_z = C_1 + \nu(\sigma_r - \sigma_\theta)$$

Differentiating with respect to r, we get

$$\frac{d\sigma_z}{dr} = \nu\left(\frac{d\sigma_r}{dr} + \frac{d\sigma_\theta}{dr}\right)$$

Substituting in Eq. (15.29), we get

$$(\sigma_r - \sigma_\theta)(1 + \nu) = r\left[\frac{d\sigma_\theta}{dr} - \nu\left\{\frac{d\sigma_r}{dr} + \nu\left(\frac{d\sigma_r}{dr} + \frac{d\sigma_\theta}{dr}\right)\right\}\right] = r\left[(1-\nu^2)\frac{d\sigma_\theta}{dr} - \nu(1+\nu)\frac{d\sigma_r}{dr}\right]$$

$$\sigma_r - \sigma_\theta = r\left[(1-\nu)\frac{d\sigma_\theta}{dr} - \nu\frac{d\sigma_r}{dr}\right] \tag{15.30}$$

Also considering the equilibrium of an element of the cylinder between angular positions θ and $\theta + d\theta$ and radii r and $r + dr$, we can get as in the case of a rotating disc,

$$\sigma_r - \sigma_\theta = -\left(r\frac{d\sigma_r}{dr} + \rho\omega^2 r^2\right) \tag{15.31}$$

Comparing Eqs. (15.30) and (15.31), we get

$$r\left[(1-\nu)\frac{d\sigma_\theta}{dr} - \nu\frac{d\sigma_r}{dr}\right] = -r\left(\frac{d\sigma_r}{dr} + \rho\omega^2 r\right)$$

$$(1-\nu)\frac{d\sigma_\theta}{dr} - \nu\frac{d\sigma_r}{dr} = -\left(\frac{d\sigma_r}{dr} + \rho\omega^2 r\right)$$

$$(1-\nu)\frac{d\sigma_r}{dr} + (1-\nu)\frac{d\sigma_\theta}{dr} = -\rho\omega^2 r$$

$$(1-\nu)\left[\frac{d\sigma_r}{dr} + \frac{d\sigma_\theta}{dr}\right] = -\rho\omega^2 r$$

$$\frac{d\sigma_r}{dr} + \frac{d\sigma_\theta}{dr} = -\frac{\rho}{(1-\nu)}\omega^2 r$$

$$\frac{d}{dr}(\sigma_r + \sigma_\theta) = -\frac{\rho}{(1-v)}\omega^2 r$$

Integrating, we get $\sigma_r + \sigma_\theta = -\frac{\rho}{(1-v)} \cdot \frac{\omega^2 r^2}{2} + C_2$ (15.32)

in which C_2 is a constant of integration.

Adding Eqs. (15.31) and (15.32), we get,

$$2\sigma_r = -\left[r\frac{d\sigma_r}{dr} + \rho\omega^2 r^2 \right] - \frac{\rho}{(1-v)} \cdot \frac{\omega^2 r^2}{2} + C_2$$

$$2\sigma_r + r\frac{d\sigma_r}{dr} = -\rho\omega^2 r^2 \left(\frac{3-2v}{2(1-v)} \right) + C_2$$

Multiplying both sides by r, we get

$$2r\sigma_r + r^2\frac{d\sigma_r}{dr} = -\rho\omega^2 r^3 \left(\frac{3-2v}{2(1-v)} \right) + rC_2$$

or
$$\frac{d}{dr}(r^2\sigma_r) = -\rho \cdot \frac{\omega^2 r^3}{2} \left(\frac{3-2v}{1-v} \right) + rC_2$$

Integrating, we get

$$r^2\sigma_r = -\rho \cdot \frac{\omega^2 r^4}{8} \left(\frac{3-2v}{1-v} \right) + \frac{r^2}{2}C_2 + C_3$$

in which C_3 is another constant of integration

$$\sigma_r = -\rho \cdot \frac{\omega^2 r^2}{8} \left(\frac{3-2v}{1-v} \right) + \frac{C_2}{2} + \frac{C_3}{r^2} \quad (15.33)$$

Substituting in Eq. (32), we get,

$$\sigma_\theta = -\rho \cdot \frac{\omega^2 r^2}{8} \left(\frac{1+2v}{1-v} \right) + \frac{C_2}{2} - \frac{C_3}{r^2} \quad (15.34)$$

Eqs. (15.33) and (15.34) are the governing equations for a rotating cylinder.

15.5.1 Solid Cylinder

From Eqs. (15.21) and (15.22), we have respectively

$$\sigma_r = \frac{C_2}{2} + \frac{C_3}{r^2} - \rho \cdot \frac{\omega^2 r^2}{8} \left(\frac{3-2v}{1-v} \right)$$

$$\sigma_\theta = \frac{C_2}{2} - \frac{C_3}{r^2} - \rho \cdot \frac{\omega^2 r^2}{8} \left(\frac{1+2v}{1-v} \right)$$

Constant C_3 must be zero, because the stress remains finite at $r = 0$.

$$\therefore \quad \sigma_r = \frac{C_2}{2} - \frac{1}{8}\left(\frac{3-2v}{1-v}\right)\rho\omega^2 r^2$$

$$\sigma_\theta = \frac{C_2}{2} - \frac{1}{8}\left(\frac{1+2v}{1-v}\right)\rho\omega^2 r^2$$

For a solid cylinder with a free surface, at $r = r_2$, $\sigma_r = 0$

$$\therefore \quad \frac{C_2}{2} = \frac{1}{8}\left(\frac{3-2v}{1-v}\right)\rho\omega^2 r_2^2$$

$$\therefore \quad \sigma_r = \frac{1}{8}\left(\frac{3-2v}{1-v}\right)\rho\omega^2 (r_2^2 - r^2) \tag{15.35}$$

$$\sigma_\theta = \frac{1}{8}\left(\frac{3-2v}{1-v}\right)\rho\omega^2 \left[r_2^2 - \left(\frac{1+2v}{3-2v}\right)r^2\right] \tag{15.36}$$

The maximum stresses occur at the center of the cylinder, in which $r = 0$.

$$\therefore \quad (\sigma_r)_{\max} = (\sigma_\theta)_{\max} = \frac{1}{8}\left(\frac{3-2v}{1-v}\right)\rho\omega^2 r_2^2 \tag{15.37}$$

15.5.2 Hollow Cylinder

From Eq. (15.33), we have

$$\sigma_r = \frac{C_2}{2} + \frac{C_3}{r^2} - \frac{1}{8}\left(\frac{3-2v}{1-v}\right)\rho\omega^2 r^2$$

$\sigma_r = 0$ at $r = r_1$ and $r = r_2$

$$0 = \frac{C_2}{2} + \frac{C_3}{r_1^2} - \frac{1}{8}\left(\frac{3-2v}{1-v}\right)\rho\omega^2 r_1^2$$

$$0 = \frac{C_2}{2} + \frac{C_3}{r_2^2} - \frac{1}{8}\left(\frac{3-2v}{1-v}\right)\rho\omega^2 r_2^2$$

Solving for C_2 and C_3, we get

$$\frac{C_2}{2} = \frac{1}{8}\left(\frac{3-2v}{1-v}\right)\rho\omega^2 (r_1^2 + r_2^2)$$

$$C_3 = -\frac{1}{8}\left(\frac{3-2v}{1-v}\right)\rho\omega^2 r_1^2 r_2^2$$

$$\sigma_r = \frac{1}{8}\left(\frac{3-2v}{1-v}\right)\rho\omega^2 \left(r_1^2 + r_2^2 - \frac{r_1^2 r_2^2}{r^2} - r^2\right) \tag{15.38}$$

$$\sigma_\theta = \frac{1}{8}\left(\frac{3-2v}{1-v}\right)\rho\omega^2 \left[r_1^2 + r_2^2 + \frac{r_1^2 r_2^2}{r^2} - \left(\frac{1+2v}{3-2v}\right)r^2\right] \tag{15.39}$$

σ_θ is maximum at $r = r_1$

$$(\sigma_\theta)_{\max} = \frac{1}{4}\left(\frac{3-2v}{1-v}\right)\rho\omega^2 r_2^2\left[1+\left(\frac{1+2v}{3-2v}\right)\frac{r_1^2}{r_2^2}\right] \tag{15.40}$$

If $\frac{r_1}{r_2} = 0$, then

$$(\sigma_\theta)_{\max} = \frac{1}{4}\left(\frac{3-2v}{1-v}\right)\rho\omega^2 r_2^2 \tag{15.41}$$

Comparing Eqs. (15.37) and (15.41), we find that the maximum hoop stress in a cylinder with a small hole at the center is twice that of in a solid cylinder.

For σ_r to be maximum, $\frac{d\sigma_r}{dr} = 0$

$$-2r + \frac{2r_1^2 r_2^2}{r^2} = 0$$

$$r = \sqrt{r_1 r_2} \tag{15.42}$$

$$(\sigma_r)_{\max} = \frac{1}{8}\left(\frac{3-2v}{1-v}\right)\rho\omega^2 (r_2 - r_1)^2 \tag{15.43}$$

PROBLEM 15.12

Objective 2

A solid cylinder of 30 cm diameter is rotating at 2500 rpm. Determine the maximum hoop stress induced in the cylinder if its material density is 7800 kg/m^3. Poisson's ratio is 0.3. Also draw the variation of radial and hoop stresses in the cylinder.

SOLUTION

$$\sigma_r = \frac{1}{8}\left(\frac{3-2v}{1-v}\right)\rho\omega^2(r_2^2 - r^2)$$

$$= \frac{1}{8}\left(\frac{3-0.6}{1-0.3}\right)\times 7800 \times \left(\frac{2\pi\times 2500}{60}\right)^2 \times (225 - r^2)\times 10^{-4}$$

$$= 0.2288\times 10^{-4}\times(225 - r^2)$$

r (cm)	0	2.5	5	7.5	10	12.5	15
σ_r (MPa)	5.148	5.005	4.576	3.861	2.86	1.573	0

$$\sigma_\theta = \frac{1}{8}\left(\frac{3-2v}{1-v}\right)\rho\omega^2\left[r_2^2 - \left(\frac{1+2v}{3-2v}\right)r^2\right]$$

$$= \frac{1}{8}\left(\frac{3-0.6}{1-0.3}\right)\times 7800\times\left(\frac{2\pi\times 2500}{60}\right)^2\left[225 - \frac{1.6}{2.4}r^2\right]$$

$$= 0.2288 \times 10^{-4}[225 - 0.6r^2]$$

r (cm)	0	2.5	5	7.5	10	12.5	15
σ_θ (MPa)	5.148	5.062	4.804	4.375	3.775	3.003	2.059

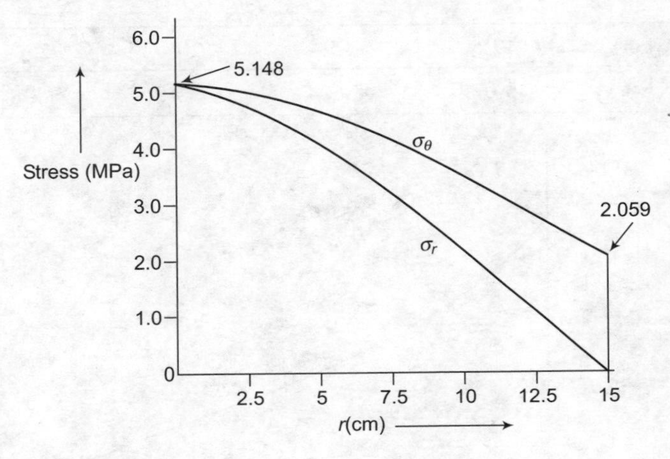

FIGURE 15.10

The variation of stresses is shown in Figure 15.10.

PROBLEM 15.13

Objective 2

A hollow cylinder of 50 cm external diameter and 25 cm internal diameter is rotating at 2000 rpm. Determine the distribution of radial and hoop stresses in the cylinder. Density of cylinder material is 7800 kg/m^3, $v = 0.3$.

SOLUTION

$$\sigma_r = \frac{1}{8}\left(\frac{3-2v}{1-v}\right)\rho\omega^2\left(r_1^2 + r_2^2 - \frac{r_1^2 r_2^2}{r^2} - r^2\right)$$

$$= \frac{1}{8}\left(\frac{3-0.6}{1-0.3}\right) \times 7800 \times \left(\frac{2\pi \times 2000}{60}\right)^2 \left(156.25 + 625 - \frac{156.25 \times 625}{r^2} - r^2\right) \times 10^{-4}$$

$$= 0.14648 \times 10^{-4} \times \left(781.25 - \frac{97656.25}{r^2} - r^2\right)$$

r (cm)	12.5	15	17.677	20	25
σ_r (MPa)	0	1.790	2.288	2.008	0

σ_r is maximum at

$$r = \sqrt{r_1 r_2} = \sqrt{12.5 \times 25} = 17.677 \text{ cm}$$

$$\sigma_\theta = \frac{1}{8}\left(\frac{3-2v}{1-v}\right)\rho\omega^2\left[r_1^2 + r_2^2 + \frac{r_1^2 r_2^2}{r^2} - \left(\frac{1+2v}{3-2v}\right)r^2\right]$$

$$= 0.14648 \times 10^{-4}\left[781.25 + \frac{97656.25}{r^2} - 0.6667r^2\right]$$

r (cm)	12.5	15	20	25
σ_θ (MPa)	19.072	15.604	11.113	7.628

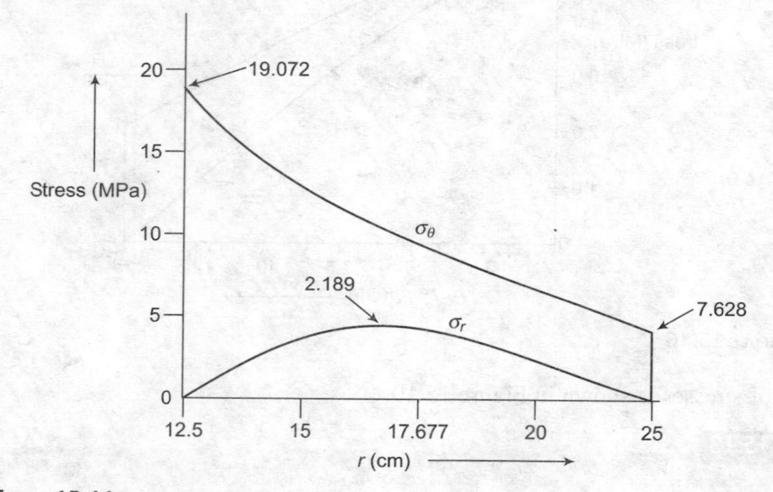

FIGURE 15.11

The variation of stresses is shown in Figure 15.11.

15.6 STRESSES IN A SPOKED RIM

Consider a spoked rim as shown in Figure 15.12 (a). Let

r = Mean radius of the rim, A_r = Cross-sectional area of rim = bd

A_S = Cross-sectional area of spokes, 2α = Angle between two adjacent spokes

I_r = Moment of inertia (MI) of the cross-section of the rim, ω = Angular speed of rotation

ρ_r = Mass per unit length of the rim, ρ_s = Mass per unit length of the spokes.

The rim is subjected to extension and bending due to spokes. In the section AB between two spokes, there is a longitudinal force P_0 and bending moment (BM) M_0 as shown in Figure 15.12 (b). If F is the force exerted by the spoke on the rim, then for the equilibrium of AB, we have

$$2P_0 \sin\alpha + F - 2\rho_r\omega^2 r^2 \sin\alpha = 0$$

$$P_0 = \rho_r\omega^2 r^2 \sin\alpha - \frac{F}{2\sin\alpha} \tag{15.44}$$

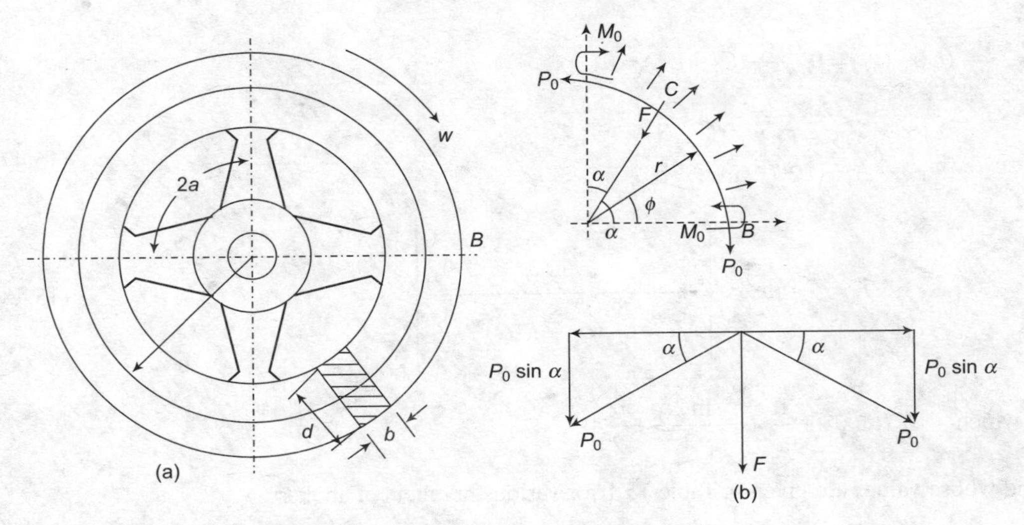

FIGURE 15.12 Stresses in a spoked rim.

The longitudinal force P at any cross-section x–x is,

$$P = P_0 \cos\alpha + \rho_r \omega^2 r^2 \cdot 2r \sin^2 \frac{\phi}{2} \tag{15.45}$$

$$P = \rho_r \omega^2 r^2 + \frac{F \cos\phi}{2 \sin\alpha}$$

BM at x–x is,

$$M = M_0 - P_0 r(1 - \cos\phi) + \rho_r \omega^2 r^3 \cdot 2 \sin^2 \frac{\phi}{2}$$

$$M = M_0 + \frac{Fr}{\sin\alpha} \cdot \sin^2 \frac{\phi}{2} \tag{15.46}$$

The strain energy of the portion AB of the rim is,

$$U_r = 2\int_0^\alpha \frac{M^2 r}{E_r I_r} d\phi + 2\int_0^\alpha \frac{P^2 r}{2A_r E_r} d\phi$$

The tensile force P_1 at any cross-section of the spoke at distance ρ from the center of the wheel is (taking length of spoke equal to r).

$$P_1 = F + \frac{\rho_S \omega^2}{2}(r^2 - \rho^2)$$

Strain energy of spoke,

$$U_S = \int_0^r \frac{P_1^2}{2A_S E_S} d\rho$$

Using Castigliano's theorem,

$$\frac{\partial}{\partial M_0}(U_r + U_S) = 0\,;\ \frac{\partial}{\partial F}(U_r + U_S) = 0$$

$$M_0 = \frac{-Fr}{2}\left(\frac{1}{\sin\alpha} - \frac{1}{\alpha}\right) \tag{15.47}$$

$$F = \frac{2}{3}\rho_r \omega^2 r^2 \left\{\frac{A_r}{\frac{A_r \cdot r^2}{I_r} f_2(\alpha) + f_1(\alpha) + \frac{A_r}{A_S}}\right\} \tag{15.48}$$

in which $f_1(\alpha) = \dfrac{1}{2\sin^2\alpha}\left(\dfrac{\sin 2\alpha}{4} + \dfrac{\alpha}{2}\right)$

And whose values are given in Table 15.1 for various numbers of spokes.

Table 15.1 Value of $f_1(\alpha)$ and $f_2(\alpha)$

No. of Spokes	4	6	8	10
$f_1(\alpha)$	0.6427	0.9566	1.2740	1.5917
$f_2(\alpha)$	0.00608	0.00169	0.00076	0.00015

PROBLEM 15.14

Objective 1

A flywheel of mean radius 125 cm is rotating at 700 rpm. The flywheel has six spokes. The rim of the flywheel is having cross-section of 40 cm × 40 cm and the spokes have a uniform area of cross-section of 200 cm^2. Determine the stresses in the flywheel when (i) the effect of spokes is ignored, and (ii) the effect of spokes is considered. ρ = 7800 kg/m^3.

SOLUTION

(i) Neglecting the effect of spokes,

$$\sigma_\theta = \rho v^2 = 7800 \times \left(\frac{2\pi \times 700}{60} \times 125\right)^2 \times 10^{-4}$$

$$= 65.422 \text{ MPa}$$

(ii) For $n = 6$, $\alpha = \dfrac{360}{12} = 30°$ and from Table 15.1,

$$f_1(\alpha) = 0.9566; f_2(\alpha) = 0.00169$$

Force in each spoke

$$F = \frac{2}{3}\rho_r \omega^2 r^2 \left\{\frac{A_r}{\frac{A_r \cdot r^2}{I_r} f_2(\alpha) + f_1(\alpha) + \frac{A_r}{A_S}}\right\}$$

$$= \frac{2}{3} \times 7800 \times \left(\frac{2\pi \times 700}{60} \times 125 \right)^2 \times 10^{-4} \left\{ \frac{1600 \times 10^{-4}}{\frac{1600 \times 15625}{\frac{40 \times 40^3}{12}} \times 0.00169 + 0.9566 + \frac{1600}{200}} \right\}$$

$$= 4361.5036 \left\{ \frac{1600}{0.198 + 0.9566 + 8} \right\} = 4361.5036 \times \frac{1600}{9.1546} = 762.284 \text{ kN}$$

Longitudinal force in the rim,

$$P_0 = A_r \times \rho_r \omega^2 r^2 - \frac{F}{2 \sin \alpha}$$

$$= 1600 \times 7800 \times \left(\frac{2\pi \times 700}{60} \times 125 \right)^2 \times 10^{-8} - \frac{762.284}{2 \times \sin 30}$$

$$= 10467.608 - 762.284 = 9705.324 \text{ kN.}$$

BM in the rim is,

$$M_0 = -\frac{Fr}{2} \left(\frac{1}{\sin \alpha} - \frac{1}{\alpha} \right) = \frac{-762.284 \times 125}{2} \left(\frac{1}{0.5} - \frac{1}{0.5236} \right) \times 10^{-2}$$

$$= -\frac{762.284 \times 125}{2} \times 0.0901 \times 10^{-2} = -42.926 \text{ kN·m}$$

$$\sigma_{max} = \frac{P_0}{A_r} - \frac{M_0}{Z}$$

$$= \frac{9705.324}{1600 \times 10^{-4}} + \frac{42.926}{\frac{40 \times 40^2}{6} \times 10^{-6}} = 60.658 + 4.024 = 64.682 \text{ MPa}$$

At the axis of the spoke, $\phi = \alpha$

$$P = A_r \rho_r \omega^2 r^2 - \frac{F \cos \alpha}{2 \sin \alpha}$$

$$= 7800 \times 1600 \times \left(\frac{2\pi \times 700}{60} \times 125 \right)^2 \times 10^{-8} - \frac{762.284}{2} \times \frac{1}{0.5774}$$

$$= 10467.608 - 660.1 = 9807.507 \text{ kN}$$

$$M = M_0 + \frac{F_r}{\sin \alpha} \times \sin^2 \frac{\alpha}{2}.$$

PROBLEM 15.15

Objective 3

A steel rotor disc which is part of a turbine assembly has a uniform thickness of 50 mm. The disc has an outer diameter of 500 mm and a central hole of 100 mm diameter. If there are 250 blades each of mass 0.2 kg pitched evenly around the periphery of the disc at an effective radius of 250 mm, determine the rotational speed if the maximum shear stress is limited to 200 MN/m^2.
For steel, $E = 200$ GN/m^2, $v = 0.3$, $\rho = 7500$ kg/m^3.

SOLUTION

Total mass of blades, $m = 250 \times 0.2 = 50$ kg

Centrifugal force on the blades

$$= m\omega^2 r = 50 \times \omega^2 \times 0.25 = 12.5\omega^2 \text{ N}$$

Area of disc rim = πdt

$$= \pi \times 0.5 \times 0.05 = 0.025\pi \text{ m}^2$$

As the blades are evenly pitched around the periphery, they may be assumed to produce a uniform radial stress at the outside surface of the disc.

$$\sigma_r = \frac{12.5\omega^2}{0.025\pi} = 159.09\omega^2 \text{ N/m}^2$$

Now $\sigma_r = \dfrac{C_1}{2} + \dfrac{C_2}{r^2} - \left(\dfrac{3+v}{8}\right)\rho\omega^2 r^2$; $\sigma_\theta = \dfrac{C_1}{2} - \dfrac{C_2}{r^2} - \left(\dfrac{1+3v}{8}\right)\rho\omega^2 r^2$

When $r = 0.05$ m, $\sigma_r = 0$

$$\therefore \quad 0 = \frac{C_1}{2} + \frac{C_2}{0.05^2} - \frac{3.3}{8} \times 7500 \times \omega^2 \times 0.05^2$$

Or $\quad 0.5C_1 + 400C_2 = 7.734\omega^2 \qquad (15.49)$

When $r = 0.25$ m, $\sigma_r = 159.09\omega^2$

$$159.0 \times \omega^2 = \frac{C_1}{2} + \frac{C_2}{0.25^2} - \frac{3.3}{8} \times 7500 \times \omega^2 \times 0.25^2$$

$$= 0.5C_1 + 16\,C_2 - 193.359\omega^2$$

Or $\quad 0.5C_1 + 16C_2 = 352.449\omega^2 \qquad (15.50)$

Subtracting Eq. (15.50) from Eq. (15.49), we get

$$384C_2 = -344.715\omega^2$$

$$C_2 = -0.897\omega^2$$

$$\therefore \quad 0.5C_1 - 14.363\omega^2 = 352.449\omega^2$$

$$C_1 = \frac{366.812\omega^2}{0.5} = 733.624\omega^2$$

Hence at $r = 0.05$ m

$$\sigma_\theta = \frac{733.624\omega^2}{2} + \frac{0.897\omega^2}{0.05^2} - \left(\frac{1+0.9}{8}\right) \times 7500 \times \omega^2 \times 0.05^2$$

$$= 366.812\omega^2 + 358.8\omega^2 - 4.453\omega^2 = 721.159\omega^2$$

and is the maximum stress.

Maximum shear stress occurs at the inside radius. Hence

$$\tau_{max} = \frac{\sigma_\theta - \sigma_r}{2} = \frac{721.159\omega^2 - 0}{2} = 360.58\omega^2 = 200 \times 10^6$$

$$\therefore \quad \omega = \sqrt{\frac{200 \times 10^6}{360.58}} = 235.512 \text{ rad/s} = 2250 \text{ rpm.}$$

PROBLEM 15.16

The flywheel of an engine is 8 m in diameter. The maximum allowable stress in the material of the flywheel is 12 MPa. Calculate the maximum speed at which the flywheel can be run if the density of the material of the flywheel is 7800 kg/m^3.

SOLUTION

$$\sigma_\theta = \rho\omega^2 r^2$$

$$\omega = \frac{1}{r}\sqrt{\frac{\sigma_\theta}{\rho}} = \frac{1}{4}\sqrt{\frac{12 \times 10^6}{7800}} = 9.805 \text{ rad/s}$$

$$N = \frac{60\omega}{2\pi} = \frac{60 \times 9.805}{2\pi} = 93.686 \text{ rpm.}$$

PROBLEM 15.17

A hollow thin disc of 80 cm outer diameter and 30 cm inner diameter is running at 2000 rpm. Calculate the maximum intensities of radial and hoop stresses. At what radius the radial stress is maximum? Density of disc material = 7800 kg/m^3. Poisson's ratio = 0.30.

SOLUTION

$$r = \sqrt{r_1 r_2} = \sqrt{15 \times 40} = 24.495 \text{ cm}$$

$$(\sigma_r)_{max} = \left(\frac{3+\nu}{8}\right)\rho\omega^2 (r_2 - r_1)^2$$

$$= \frac{3.3}{8} \times 7800 \times \left(\frac{2\pi \times 2000}{60}\right)^2 (40-15)^2 \times 10^{-4} = 8.812 \text{ MPa}$$

$$(\sigma_\theta)_{max} = \left(\frac{3+\nu}{4}\right)\rho\omega^2 \left[r_2^2 + \left(\frac{1-\nu}{3+\nu}\right) r_1^2\right]$$

$$= \frac{3.3}{4} \times 7800 \times \left(\frac{2\pi \times 2000}{60}\right)^2 \left(1600 + \frac{0.7}{3.3} \times 225\right) \times 10^{-4} = 46.463 \text{ MPa.}$$

PROBLEM 15.18

A thin uniform disc of 30 cm diameter with a central hole of 10 cm diameter runs at 5000 rpm. Calculate the maximum principal stresses and the maximum shearing stress in the disc. Poisson's ratio = 0.3 and density of material = 7500 kg/m^3.

SOLUTION

$$(\sigma_r)_{\max} = \frac{3.3}{4} \times 7500 \times \left(\frac{2\pi \times 5000}{60}\right)^{22} (15-5) \times 10^{-4} = 16.946 \text{ MPa}$$

$$(\sigma_\theta)_{\max} = \frac{3.3}{4} \times 7500 \times \left(\frac{2\pi \times 5000}{60}\right)^2 \left(225 + \frac{0.7}{3.3} \times 25\right) \times 10^{-4} = 39.027 \text{ MPa}$$

$$\tau_{\max} = \frac{\sigma_\theta - \sigma_r}{2} = 11.04 \text{ MPa.}$$

PROBLEM 15.19

The rotor of a stream turbine is a solid disc of uniform strength and is 30 cm diameter at the blade ring and 5 cm thick at the center. It is running at a constant speed of 20,000 rpm. Calculate the thickness of the rotor at a radius of 10 cm. The material density is 7800 kg/m^3 and the maximum allowable stress in the rotor is 200 MPa.

SOLUTION

$$t = t_0 \cdot \exp\left[-\frac{\rho\omega^2}{2\sigma}(r^2 - r_1^2)\right]$$

$$= 5 \times \exp\left[-\frac{7800 \times \left(\frac{2\pi \times 20{,}000}{60}\right)^2}{2 \times 200 \times 10^6}(100 - 225) \times 10^{-4}\right] = 14.547 \text{ cm.}$$

PROBLEM 15.20

A long hollow cylinder is of 30 cm external diameter and is 10 cm thick. It is revolving at a constant speed of 3000 rpm. Calculate the maximum radial and hoop stresses induced in the cylinder. The density of cylinder material is 7800 kg/m^3 and Poisson's ratio = 0.30.

SOLUTION

$$(\sigma_r)_{\max} = \frac{1}{8}\left(\frac{3-2\nu}{1-\nu}\right)\rho\omega^2(r_2 - r_1)^2$$

$$= \frac{1}{8}\times\left(\frac{2.4}{0.7}\right)\times 7800\times\left(\frac{2\pi\times 3000}{60}\right)^2 (15-5)^2\times 10^{-4} = 3.295 \text{ MPa}$$

$$(\sigma_\theta)_{\max} = \frac{1}{4}\left(\frac{3-2v}{1-v}\right)\rho\omega^2 r_2^2\left[1+\left(\frac{1-2v}{3-2v}\right)\left(\frac{r_1}{r_2}\right)^2\right]$$

$$= \frac{1}{4}\times\frac{2.4}{0.7}\times 7800\times\left(\frac{2\pi\times 3000}{60}\right)^2\times 225\times 10^{-4}\times\left(1+\frac{0.4}{2.4}\times 0.111\right)$$

$$= 15.106 \text{ MPa.}$$

PROBLEM 15.21

A thin circular disc of uniform thickness and of radius b is built up of two concentric portions, the surface of separation having a radius a. Find the minimum value of the radial pressure over the surface of separation when the disc is at rest, in order that the outer portion of the disc may not become loose upon the inner portion at an angular velocity ω.

SOLUTION

Outer disc:

At $r = a$, $\sigma_r = 0$ and

$$\sigma_r = \left(\frac{3+v}{4}\right)\rho\omega^2\left(b^2+\left(\frac{1-v}{3+v}\right)a^2\right)$$

Inner disc:

At $r = a$, $\sigma_r = 0$ and

$$\sigma_\theta = \left(\frac{3+v}{4}\right)\rho\omega^2\left(a^2-\left(\frac{1-v}{3+v}\right)a^2\right) = \left(\frac{1-v}{4}\right)\rho\omega^2 a^2$$

$$(\varepsilon_\theta)_0 = \frac{1}{E}\left[\left(\frac{3+v}{4}\right)\rho\omega^2 b^2+\left(\frac{1-v}{4}\right)\rho\omega^2 a^2\right]$$

$$(\varepsilon_\theta)_i = -\frac{1}{E}\left(\frac{1-v}{4}\right)\rho\omega^2 a^2$$

$$(\varepsilon_\theta)_0 - (\varepsilon_\theta)_i = \frac{1}{E}\left(\frac{3+v}{4}\right)\rho\omega^2 b^2$$

Due to pressure:

$$\frac{\delta a}{a} = \frac{1}{E}\left[(\sigma_\theta)_h + p\right] = \frac{p}{E}\left[\frac{a^2+b^2}{b^2-a^2}+1\right] = \frac{p}{E}\left[\frac{2b^2}{b^2-a^2}\right] = \frac{1}{E}\left(\frac{3+v}{4}\right)\rho\omega^2 b^2$$

$$p = \left(\frac{3+v}{8}\right)\rho\omega^2(b^2-a^2).$$

PROBLEM 15.22

A circular disc of outside and inside radii r_1 and r_2 is made up in two parts with the common radius being r_0. The outer portion is shrunk on so as to exert a pressure on the inner disc. Prove that the hoop tension at the inside and outside of the disc will be equal for an angular velocity ω, if the shrinkage pressure at the common surface, when the disc is stationary, has the value $\rho\omega^2\left\{\frac{(1+\nu)(r_1^2-r_0^2)(r_0^2-r_2^2)}{4r_0^2}\right\}$.

SOLUTION

Stationary Disc:

Inner disc: $(\sigma_\theta)_1' = \frac{-2p_S r_0^2}{r_0^2 - r_1^2}$

Outer disc: $(\sigma_\theta)_2' = \frac{-2p_S r_0^2}{r_2^2 - r_0^2}$

Rotating Disc:

Inner disc: $(\sigma_\theta)_1'' = \left(\frac{3+\nu}{4}\right)\rho\omega^2\left[r_2^2 + \left(\frac{1-\nu}{3+\nu}\right)r_1^2\right]$

Outer disc: $(\sigma_\theta)_2'' = \left(\frac{3+\nu}{4}\right)\rho\omega^2\left[r_1^2 + \left(\frac{1-\nu}{3+\nu}\right)r_2^2\right]$

Resultant stresses:

$$(\sigma_\theta)_1 = (\sigma_\theta)_1' + (\sigma_\theta)_1''\ ;\ (\sigma_\theta)_2 = (\sigma_\theta)_2' + (\sigma_\theta)_2''$$

$$(\sigma_\theta)_1 = (\sigma_\theta)_2'$$

$$\left(\frac{3+\nu}{4}\right)\rho\omega^2\left(r_2^2 + \left(\frac{1-\nu}{3+\nu}\right)r_1^2\right) - \frac{2p_S r_0^2}{r_0^2 - r_1^2} = \left(\frac{3+\nu}{4}\right)\rho\omega^2\left(r_1^2 + \left(\frac{1-\nu}{3+\nu}\right)r_2^2\right) + \frac{2p_S r_0^2}{r_2^2 - r_0^2}$$

$$2p_S r_0^2\left[\frac{1}{r_2^2 - r_0^2} + \frac{1}{r_0^2 - r_1^2}\right] = \left(\frac{3+\nu}{4}\right)\rho\omega^2\left[(r_2^2 - r_1^2)\left(-\frac{1-\nu}{3+\nu} + 1\right)\right]$$

$$\frac{2p_S r_0^2 (r_2^2 - r_1^2)}{(r_2^2 - r_0^2)(r_0^2 - r_1^2)} = \left(\frac{3+\nu}{4}\right)\rho\omega^2 (r_2^2 - r_1^2)\left[\frac{2(1+\nu)}{3+\nu}\right]$$

$$\frac{2p_S r_0^2}{(r_2^2 - r_0^2)(r_0^2 - r_1^2)} = \left(\frac{1+\nu}{2}\right)\rho\omega^2$$

$$p_S = \left(\frac{1+\nu}{4}\right)\frac{\rho\omega^2}{r_0^2}(r_2^2 - r_0^2)(r_0^2 - r_1^2).$$

PROBLEM 15.23

A disc of thickness t and outside diameter $2r_2$ is shrunk on to a shaft of diameter $2r_1$ producing a radial interference pressure p in the stand-still condition. It is then rotated with an angular speed ω rad/s. If μ is the coefficient of friction between disc and shaft and ω_0 is the value of angular speed for which the interference pressure falls to zero. Show that (i) the maximum horse power (HP) is transmitted when $\omega = \frac{1}{\sqrt{3}}\omega_0$, and (ii) the maximum HP is equal to $kr_1^2 t\mu p\omega_0$, in which k is a constant.

SOLUTION

Stationary:

Shaft:

$$\sigma_r = \sigma_\theta = -p_S$$

Disc:

At $r = r_1$, $\sigma_r = -p_S$ and

$$\sigma_\theta = p_S\left(\frac{r_1^2 + r_2^2}{r_2^2 - r_1^2}\right)$$

$$(\varepsilon_\theta)_{\text{shaft}} = \frac{1}{E}[\sigma_\theta - v\sigma_r] = -\frac{p_S}{E}(1-v)$$

$$(\varepsilon_\theta)\text{disc} = \frac{p_S}{E}\left[\frac{r_1^2 + r_2^2}{r_2^2 - r_1^2} + v\right]$$

$$\text{Total strain} = \frac{p_S}{E}\left[\frac{r_1^2 + r_2^2}{r_2^2 - r_1^2} + v + 1 - v\right] = \frac{2p_S r_2^2}{E(r_2^2 - r_1^2)}.$$

Rotating Disc:

$$(\sigma_r)_1 = 0$$

$$(\sigma_\theta)_1 = \left(\frac{3+v}{4}\right)\rho\omega^2\left(r_2^2 + \left(\frac{1-v}{3+v}\right)r_1^2\right)$$

$$\varepsilon_\theta = \frac{1}{E}[(\sigma_\theta)_1 - v(\sigma_r)_1] = \left(\frac{3+v}{4}\right)\times\frac{\rho\omega^2}{E}\times\left[r_2^2 + \left(\frac{1-v}{3+v}\right)r_1^2\right] = \frac{2p_S r_2^2}{E(r_2^2 - r_1^2)}$$

$$p_S = \left(\frac{r_2^2 - r_1^2}{2r_2^2}\right)\left(\frac{3+v}{4}\right)\rho\omega^2\left[r_2^2 + \left(\frac{1-v}{3+v}\right)r_1^2\right]$$

If p'_S = pressure at speed ω.

$$\text{Hoop strain} = \frac{2p'_S r_2^2}{E(r_2^2 - r_1^2)}$$

Total strain at speed ω,

$$\left(\frac{3+v}{4}\right)\frac{\rho\omega^2}{E}\left[r_2^2+\left(\frac{1-v}{3+v}\right)r_1^2\right]+\frac{2p_S'r_2^2}{E(r_2^2-r_1^2)}=\frac{2p_Sr_2^2}{E(r_2^2-r_1^2)}$$

Tangential force between disc and shaft, $F = 2\pi r_1 t\mu p_S$

Torque $T = 2\pi r_1^2 t\mu p_S$

$$\text{HP}=\frac{T\omega}{75}=\frac{2\pi r_1^2 t\mu p_S\omega}{75}$$

$$(p_S-p_S')\frac{2r_2^2}{E(r_2^2-r_1^2)}=\left(\frac{3+v}{4}\right)\frac{\rho\omega^2}{E}\left[r_2^2+\left(\frac{1-v}{3+v}\right)r_1^2\right]$$

$$p_S'=p_S-\left(\frac{3+v}{4}\right)\rho\omega^2\left[r_2^2+\left(\frac{1-v}{3+v}\right)r_1^2\right]\left(\frac{r_2^2-r_1^2}{2r_2^2}\right)$$

$$0=p_S-\left(\frac{3+v}{4}\right)\rho\omega_0^2\left[r_2^2+\left(\frac{1-v}{3+v}\right)r_1^2\right]\left(\frac{r_2^2-r_1^2}{2r_2^2}\right)$$

Now HP $\propto p_S\omega$ and $p_S \propto \omega^3$

$\therefore$ HP $\propto \omega^3$

$$\frac{d(\text{HP})}{d\omega}\propto 3\omega^3=\omega_0^2 \text{ or } \omega=\frac{\omega_0}{\sqrt{3}}$$

$$(\text{HP})_{\max}=\left[\frac{1}{\sqrt{3}}\left(\frac{2\pi}{75}\right)\right]r_1^2t\mu p_S\omega_0=kr_1^2tp_S\omega_0$$

in which k is a constant.

PROBLEM 15.24

Solid steel disc 400 mm diameter and of small constant thickness has a steel ring of outer diameter 600 mm and the same thickness shrunk onto it. If the interference pressure is reduced to zero at a rotational speed of 5000 rpm, calculate:

The radial pressure at the interference when stationary.

The difference in diameter of the mating surfaces of the disc and the ring before assembly. Take $E = 200$ GPa, $v = 0.3$, and $\rho = 7800$ kg/m^3.

SOLUTION

Stationary:

Shaft:

$$\sigma_r=\sigma_\theta=-p_S$$

$$(\varepsilon_\theta)_{\text{shaft}}=\frac{1}{E}[-p_S+vp_S]=-\frac{p_S}{E}(1-v)=-\frac{0.7p_S}{E}$$

Disc:

At $r = r_1$, $\sigma_r = -p_S$ and

$$\sigma_\theta = p_S\left(\frac{r_1^2 + r_2^2}{r_2^2 - r_1^2}\right) = p_S\left(\frac{300^2 + 200^2}{300^2 - 200^2}\right) = 2.6 p_S$$

$$(\varepsilon_\theta)_{\text{disc}} = \frac{1}{E}[2.6 p_S + 0.3 p_S] = \frac{2.9 p_S}{E}$$

$$\text{Total hoop strain } \varepsilon_\theta = 2.9\frac{p_S}{E} + 0.7\frac{p_S}{E} = 3.6\frac{p_S}{E}$$

Rotating:

At $r = r_1$, $\sigma_r = 0$

$$\sigma_\theta = \left(\frac{3+v}{4}\right)\rho\omega^2\left(r_2^2 + \left(\frac{1-v}{3+v}\right)r_1^2\right)$$

$$= \frac{3.3}{4} \times 7800 \times \left(\frac{2\pi \times 5000}{60}\right)^2 \left(900 + \frac{0.7}{3.3} \times 400\right) \times 10^{-4} = 173.57 \text{ MPa}$$

$$\varepsilon_\theta = \frac{\sigma_\theta}{E} = \frac{173.57 \times 10^6}{E} = \frac{3.6 p_S}{E}$$

$$p_S = 48.214 \text{ MPa}$$

$$\varepsilon_\theta = \frac{\delta d_1}{d_1} = \frac{3.6 \times 48.214 \times 10^6}{200 \times 10^9} = 0.8678 \times 10^{-3}$$

$$\delta d_1 = 400 \times \varepsilon_\theta = 347.14 \times 10^{-3}.$$

PROBLEM 15.25

A cast iron of length 2.1 m rotates with a constant angular speed about a vertical axis through its mid length. Determine the limiting number of revolutions for this rod if the density of cast iron is 7400 kg/m^3 and the allowable tensile stress is 50 MN/m^2. Determine the elongation of the rod at 2000 rpm if $E = 165$ GPa.

SOLUTION

$$dF_C = \rho A \cdot dx \cdot \omega^2 x$$

$$\sigma = \int \frac{dF_C}{A} = \rho\omega^2 \int_x^{l/2} x \cdot dx = \frac{\rho\omega^2}{2}\left(\frac{l^2}{4} - x^2\right)$$

$$\sigma_{\max}\Big|_{x=0} = \frac{1}{8}\rho\omega^2 l^2 = \sigma_t$$

$$\frac{1}{8} \times 7400 \times \left(\frac{2\pi N}{60}\right)^2 \times 2.1^2 = 50 \times 10^6$$

$$N = 1057.7 \text{ rpm}.$$

$$d(\Delta l) = \frac{\sigma}{E}dx = \frac{\rho\omega^2}{2E}\left(\frac{l^2}{4} - x^2\right)dx$$

$$\Delta l = 2\int_0^{l/2} \frac{\rho\omega^2}{2E}\left(\frac{l^2}{4} - x^2\right)dx = \frac{\rho\omega^2}{E}\int \frac{l^2}{4} - \frac{x^3}{3}\int_0^{l/2} = \frac{\rho\omega^2 l^3}{12E}$$

$$= 7400 \times \left(\frac{2\pi \times 2000}{60}\right)^2 \times \frac{2.1^3}{12 \times 165 \times 10^9} = 1.516 \text{ mm}$$

PROBLEM 15.26

The maximum safe peripheral speed for a cast iron flywheel is 50 m/s. Neglecting the spokes and taking the density as 7400 kg/m^3, determine the maximum tensile stress in the rim of the flywheel at this speed.

SOLUTION

$$\sigma_\theta = \rho v^2 = 7400 \times 50^2 = 18.5 \text{ MPa}$$

PROBLEM 15.27

A steel rotor of an electrical machine having outer diameter 2 m and inner diameter 0.5 m is provided with 0.2-m-deep trapezoidal slots axially around its outer periphery for the windings. It rotates at 1000 rpm. Assuming the weight of the windings in the slots to be the same as that of the material removed, calculate the maximum stress induced in the rotor. Take density of rotor 7400 kg/m^3 and $v = 0.30$.

SOLUTION

$r_1 = 0.25$ m, $r_2 = 1 - 0.2 = 0.8$ m

$N = 1000$ rpm., $\rho = 7400$ kg/m^3, $v = 0.3$

Centrifugal force due to winding

$$F_C = \int_{0.8}^{1} (2\pi r \cdot dr \cdot t)\rho r\omega^2 = 2\pi t\rho\omega^2 \int_{0.8}^{1} r^2 dr$$

$$= \frac{2\pi t\rho\omega^2}{3}\left|r^3\right|_{0.8}^{1} = \frac{2\pi t\rho\omega^2}{3}[(1)^3 - (0.8)^3] = \frac{2\pi}{3} * 0.488 * t\rho\omega^2$$

$$\sigma_r = \frac{F_C}{A} = \frac{F_C}{2\pi \times 0.8 \times t} = 0.2033\rho\omega^2$$

$$= 0.2033 \times 7400 \times \left(\frac{2\pi \times 1000}{60}\right)^2 = 16.48 \text{ MPa}$$

At $r = 0.25$ m,

$$\sigma_\theta = \frac{2 \times 0.8^2 \times 16.48}{\{(0.8)^2 - (0.25)^2\}} = 36.527 \text{ MPa.}$$

Due to rotation:

At $r = 0.25$ m

$$\sigma_\theta = \left(\frac{3+v}{4}\right)\rho\omega^2\left(r_2^2 + \left(\frac{1-v}{3+v}\right)r_1^2\right)$$

$$= \frac{3.3}{4} \times 7400 \times \left(\frac{2\pi \times 1000}{60}\right)^2\left(0.8^2 + \frac{0.7}{3.3} \times 0.25^2\right) = 43.69 \text{ MPa}$$

Total hoop stress = 36.527 + 43.69 = 80.217 MPa.

SUMMARY

1. For a rotating solid disc, the stresses at any radius r, are:

$$\sigma_\theta = \frac{\rho \cdot \omega^2}{8}\left[(3+v)r_2^2 - (1+3v)r^2\right]$$

$$\sigma_r = \left(\frac{3+v}{8}\right)\rho.\omega^2(r_2 - r)^2$$

2. At the center of rotating solid disc, the radial and circumferential stress are maximum and are equal.

$$(\sigma_r)_{max} = (\sigma_\theta)_{max} = \left(\frac{3+v}{8}\right)\rho.\omega^2 r_2^2$$

3. For a rotating hollow disc, the stresses at any radius r, are:

$$\sigma_r = \left(\frac{3+v}{8}\right)\rho.\omega^2\left[r_1^2 + r_2^2 - r^2 - \frac{r_1^2 r_2^2}{r^2}\right]$$

$$\sigma_\theta = \left(\frac{3+v}{8}\right)\rho.\omega^2\left[r_1^2 + r_2^2 + \frac{r_1^2 r_2^2}{r^2} - \left(\frac{1+3v}{3+v}\right)r^2\right]$$

4. A disc which has equal values of σ_r and σ_θ at all radii, is known as a disc of uniform strength.

$\therefore$ **$\sigma_\theta = \sigma_r = \sigma$ for all radii.**

5. The thickness of a disc of uniform strength is given by:

$$t = t_0 e^{\frac{-\rho.\omega^2.r^2}{2\sigma}}$$

OBJECTIVE TYPE QUESTIONS

1. The hoop stress in a thin flat ring of density ρ due to rotation at speed v is given by

(a) $\frac{1}{2}\rho v^2$ (b) $\frac{1}{2}\rho v$ (c) ρv (d) ρv^2

2. In a solid rotating circular disc, the radial stress is maximum at
(a) the center (b) the outer radius
(c) the mean radius (d) square root of the radius

3. In a solid rotating circular disc, the hoop stress is maximum at
(a) the center (b) the outer radius
(c) the mean radius (d) square root of the radius

4. In a hollow circular rotating disc, the hoop stress is maximum at
(a) the outer radius (b) the inner radius
(c) the mean radius (d) the geometric mean radius

5. In a hollow circular rotating disc, the radial stress is maximum at
(a) the outer radius (b) the inner radius
(c) the mean radius (d) the geometric mean radius

6. The ratio of the maximum hoop stress in a circular rotating disc having a very small hole at the center to the maximum hoop stress in a circular rotating disc is
(a) 1 (b) 1.5 (c) 2 (d) 2.5

7. In a rotating disc of uniform strength, the following statements are true
(a) radial stress is constant
(B) hoop stress is constant
(c) radial stress is equal to the hoop stress
(d) radial and hoop stresses are equal to each other and are constant

8. In a rotating solid circular cylinder, the hoop stress is maximum at
(a) the center (b) the outer radius
(c) the mean radius (d) the square root of the outer radius

9. In a hollow rotating circular cylinder, the radial stress is maximum at
(a) the inner radius (b) the outer radius
(c) the mean radius (d) the geometric mean radius

10. The ratio of the maximum hoop stress in a circular rotating cylinder having a small hole at the center to the maximum hoop stress in a solid circular rotating cylinder is
(a) 1 (b) 1.5 (c) 2 (D) 2.5

11. Thin rotating rings develop
(a) tensile hoop stress (b) compressive hoop stress
(c) radial stress (d) both hoop and radial stresses

12. An ice skater is spinning fast with her arms tight against her body. When she extends her arms, which of the following statements is not true?
(a) She increases her MI. (b) She decreases her angular velocity.
(c) Her MI remains constant. (d) Her total angular momentum will remain constant.
(E) She will spin slower.

13. In a hollow rotating disc, radial stress is maximum at
(a) $r = \sqrt{R_i R_0}$ (b) $r = \sqrt{R_i^2 R_0}$
(c) $r = \sqrt{R_i R_o^2}$ (d) $r = \sqrt{R_i / R_0}$

14. In a solid rotating disc, the hoop stress at the outer surface is
(a) $\dfrac{1-v}{4}\rho\omega^2 r^2$ (b) $\dfrac{3+v}{4}\rho\omega^2 r^2$
(c) $\dfrac{3-v}{4}\rho\omega^2 r^2$ (D) zero

15. In a solid rotating disc, the radial stress at the outer surface is

(a) $\frac{1-v}{4}\rho\omega^2 r^2$ (b) $\frac{3+v}{4}\rho\omega^2 r^2$

(c) $\frac{3-v}{4}\rho\omega^2 r^2$ (d) zero

16. In a solid rotating disc, the radial stress at the center is

(a) $\frac{1-v}{4}\rho\omega^2 r^2$ (b) $\frac{3+v}{4}\rho\omega^2 r^2$

(c) $\frac{3-v}{4}\rho\omega^2 r^2$ (d) zero

17. In a hollow rotating disc, hoop stress is maximum at

(a) inner surface (b) outer surface

(c) $r=\sqrt{R_i R_o^2}$ (d) none of these

18. In a long rotating solid cylinder, the radial stress at the center is given by

(a) $\frac{(3-2v)}{2(1-v)}\rho\omega^2 r^2$ (b) $\frac{(3-2v)}{4(1-v)}\rho\omega^2 r^2$

(c) $\frac{(3-2v)}{8(1-v)}\rho\omega^2 r^2$ (d) $\frac{(3-2v)}{(1-v)}\rho\omega^2 r^2$

Solutions for Objective Questions

Sl. No.	1.	2.	3.	4.	5.	6.	7.	8.	9.	10.
Answer	(d)	(a)	(a)	(b)	(d)	(c)	(d)	(a)	(d)	(c)

Sl.No.	11.	12.	13.	14.	15.	16.	17.	18.
Answer	(a)	(c)	(a)	(a)	(d)	(b)	(a)	(c)

EXERCISE PROBLEMS

1. A thin circular disc of 500 mm outside diameter having a central hole rotates at a uniform speed about its axis through its center. The diameter of hole is such that the maximum stress due to rotation is 80% of that in a thin ring whose mean diameter is 500 mm. If both are of the same material and rotates at the same speed, find out the diameter of the center hole and the speed of rotation if the maximum allowable stress in the disc is 150 MPa. Take $\rho = 7800$ kg/m^3 and $v = 0.28$.

2. A hollow disc of uniform thickness has an external diameter of 400 mm and internal diameter of 80 mm. Find the radial and hoop stresses at 100 mm if the disc rotates at 2000 rpm. Find also the change in external diameter of the disc. Take $\rho = 7500$ kg/m^3, $v = 0.28$, and $E = 200$ GPa.

3. The rotor of a turbine is to be designed for uniform strength throughout so that the stress does not exceed 350 MPa. The maximum operational speed is 5000 rpm. If the thickness of the rotor at the center is 60 mm and outside diameter is 600 mm, calculate the thickness at the top. Take $\rho = 7500$ kg/m^3.

4. A long hollow cylinder of 150 mm external diameter and 30 mm thickness is rotating at a constant speed of 2000 rpm. Calculate the maximum radial and hoop stresses developed in the cylinder. Take $\rho = 7500\ \text{kg/m}^3$ and $\nu = 0.3$.
5. A solid steel shaft 250 mm in diameter is rotating at a speed of 280 rpm. If the shaft is constrained at its ends so that it cannot expand or contract longitudinally, calculate the total longitudinal thrust over a cross-section due to rotational stresses. Take $\rho = 7500\ \text{kg/m}^3$, $\nu = 0.3$.

CHAPTER 16

STRESSES IN CURVED BARS

UNIT OBJECTIVE

This chapter provides information on the stresses in bars with curvature. The presentation attempts to help the student achieve the following:

Objective 1: Determine stresses at inner and outer faces of curved bar.

Objective 2: Determine tensile and compressive stresses of curved bar in crane hook and other structural members.

16.1 INTRODUCTION

Bernoulli's bending theory is applicable only for straight beams. In case the beams are curved then the simple bending theory cannot be used. Curved structural elements are common (a crane hook, ring of chain, etc.).

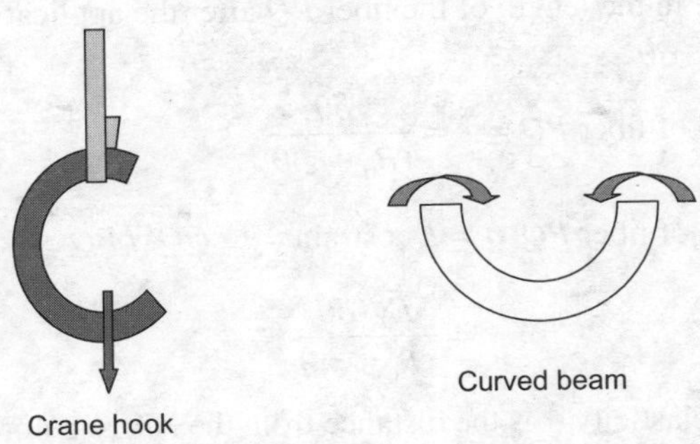

FIGURE 16.1 Examples of curved bars

16.2 ASSUMPTIONS

In a curved beam, plane sections before bending remains plane even after bending. This means that the strain variation along the depth of the cross-section of the beam is linear. Consider a curved beam shown in Figure 16.2.

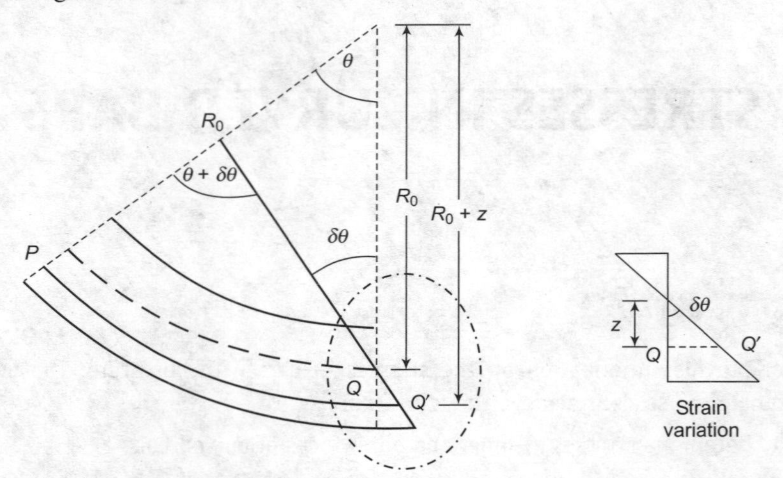

FIGURE 16.2 Examples of curved bars

16.3 ANALYSIS OF DIFFERENT CROSS-SECTIONS

The stress analysis of curved beams is developed by Winker Bach. The important aspect to be remembered in the analysis of curved beams is that the neutral axis (NA) and centroidal axis of the cross-section do not coincide.

Referring to Figure 16.2 and writing, PQ is the length of the fibers located y distant from NA of the cross-section before application of moment.

Let R_0 be the radius of the fiber of the cross-section at NA.

$$PQ = (R_0 + z)\theta$$

Let QQ' be the increase in the length of the fiber PQ after the application of moment.

$$QQ' = z\delta\theta$$

Then, strain at the level of fiber $PQ = \varepsilon = \dfrac{z\delta\theta}{(R_0 + z)\theta}$

Then, stress at the level of fiber PQ $\sigma = E \times \text{strain} = E \cdot QQ' / PQ$,

$$= \frac{Ey \cdot \delta\theta}{(R_0 + z)\theta} \tag{16.1}$$

in which, E is modulus of elasticity, y is the distance from the NA as before, and R_0 the initial radius of the neutral surface.

The total force acting along the axis of the beam is zero.

Total normal force on cross-section = 0 for pure bending, that is,

$$\int \sigma \cdot dA = \frac{E\delta\theta}{\theta}\int \frac{zdA}{R_0 + z} = 0 \tag{16.2}$$

Moment of resistance $M = \int \sigma \cdot z \cdot dA$

$$= \frac{E\delta\theta}{\theta}\int \frac{z^2 dA}{R_0 + z} \text{ from (16.1)} \tag{16.3}$$

But $\int \frac{z^2 dA}{R_0 + z} = \int \frac{z^2 + R_0 z - R_0 z}{R_0 + z} dA$

Hence $\int \frac{z^2 dA}{R_0 + z} = \int \frac{[z(z + R_0) - R_0 z]}{R_0 + z} \cdot dA$

$$= \int zdA - R_0 \int z \cdot dA / (R_0 + z)$$

$\int ydA$ = Moment of the area from reference axis, that is, axis from which 'y' is measured. In the present case, it is $A \times e$.

in which, 'e' is the location of NA from centroid of the cross-section.

Also from Eq. (16.2), it is known that $\int \frac{zdA}{R_0 + z} = 0$

Hence

$$\int \frac{z^2 dA}{R_0 + z} = Ae, \tag{16.4}$$

in which, e is the distance between the NA and the principal axis through the centroid (i.e., being positive for NA to be on the same side of the centroid as the center of curvature).

Substituting Eq. (16.4) in Eq. (16.3) gives

$$M = \left(\frac{E\delta\theta}{\theta}\right) Ae$$

$$= [\sigma(R_0 + z)/z]Ae \text{ from (16.1)}$$

Rearranging, $\sigma = Mz / Ae(R_0 + z)$. (16.5)

In this equation, 'z' is positive measured outward, a positive bending moment (BM) being one which tends to increase the curvature.

In the above formulation of equation for stress in a curved beam, the term 'e', position of centroid from NA is to be determined. R_0 is the radius to the NA of the cross-section. Generally, all engineering calculations are with respect to centroid of the cross-section. Thus, the reference axis needs to be shifted from NA to the centroidal axis.

This leads to the following changes as shown in Figure 16.3.

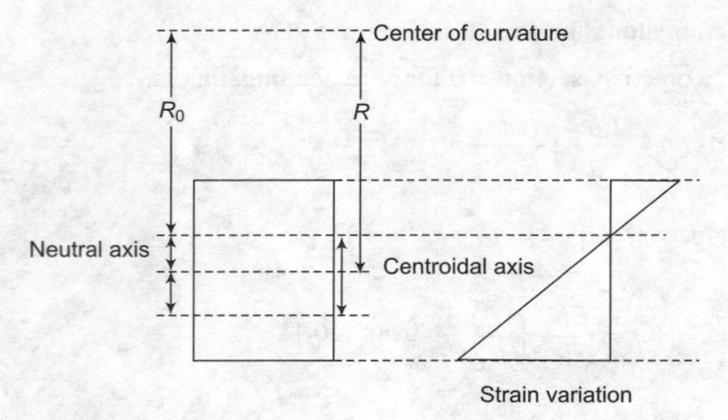

FIGURE 16.3

If reference axis is taken as centroidal axis then

$$R_0 = R - e$$

and

$$z = y + e$$

Thus, Eq. (16.5) can be rewritten as

$$\sigma = \frac{M}{Ae}\left\{\frac{y+e}{R+y}\right\}, \tag{16.6}$$

in which 'y' is the location of the fiber where the stress is to be determined.

And the expression for 'e' can be obtained from Eq. (16.2) which can be rewritten as

$$\int \frac{z}{R_0 + z} dA = 0$$

or

$$\int \frac{y+e}{R+y} dA = 0 \tag{16.7}$$

Eqs. (16.6) and (16.7) shall be used for determining the stresses in curved beams.

Stress at the centroidal axis is given by $y = 0$ in Eq. (16.6).

$$\sigma_c = \frac{M}{AR}$$

Strain at the centroid level

$$\varepsilon_c = \frac{M}{EAR}$$

Sign convention

The BM that tends to decrease the radius of curvature is positive otherwise negative.

Distance of fiber from centroidal axis (y) toward the extrados is positive otherwise negative.

Tensile stress (σ) is positive and compressive stress is negative.

Location of NA for different cross-sections:

NA position depends on the cross-section and radius of curvature of the beam (R). Consider a rectangular cross-section. Let 'b' be the breadth of the beam and depth, that is, dimension along the radial of the curved beam be 'D'. Let 'R' be the radius of curvature of the beam.

$$\int \frac{y+e}{R+y} dA = 0.$$

The above integral is taken as I_*.

$$I_* = \int \frac{y+e+R-R}{R+y} dA = A + (e-R)\int \frac{1}{R+y} dA = 0.$$

Consider an elemental strip of thickness 'dy' located 'y' distant from the centroid of the cross-section as shown in Figure 16.4.

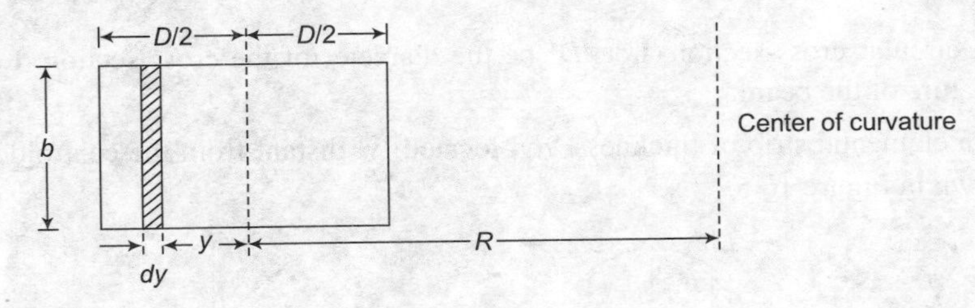

FIGURE 16.4

$$dA = bdy$$

Consider $$I_* = \int \frac{y+e}{R+y} dA = A + (e-R)\int \frac{1}{R+y} dA$$

$$I_* = A + (e-R)\int_{-\frac{D}{2}}^{\frac{D}{2}} \left\{\frac{1}{R+y}\right\} bdy$$

$$= A + b(e-R)\int_{-\frac{D}{2}}^{\frac{D}{2}} \left\{\frac{1}{R+y}\right\} dy$$

$$= A + b(e-R)\ln\left\{\frac{R+\frac{D}{2}}{R-\frac{D}{2}}\right\}$$

$$= A + b(e-R)\ln\left\{\frac{2R+D}{2R-D}\right\}$$

As $$I_* = 0,\ A + b(e-R)\ln\left\{\frac{2R+D}{2R-D}\right\} = 0.$$

Hence $$A = -b(e-R)\ln\left\{\frac{2R+D}{2R-D}\right\}$$

$$bD = -b(e-R)\ln\left\{\frac{2R+D}{2R-D}\right\}$$

Rearranging the above equation

$$e = R - \frac{D}{\ln\left\{\dfrac{2R+D}{2R-D}\right\}}.$$

Consider a circular cross-section. Let 'D' be the diameter of the cross-section. Let 'R' be the radius of curvature of the beam.

Consider an elemental strip of thickness 'dy' located 'y' distant from the centroid of the cross-section as shown in Figure 16.5.

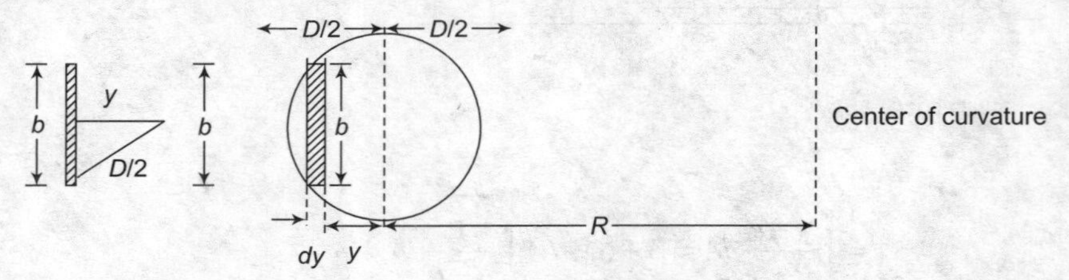

FIGURE 16.5

The integral I_* is given by

$$I_* = A + (e-R)\int_{-\frac{D}{2}}^{\frac{D}{2}}\left\{\frac{1}{R+y}\right\}dA$$

$$dA = bdy$$

$$b = 2\sqrt{\left(\frac{D}{2}\right)^2 - y^2}$$

$$I_* = A - (e-R)\int_{-\frac{D}{2}}^{\frac{D}{2}}\left\{\frac{1}{R+y}\right\}bdy$$

$$= A + 2(e - R)\int_{-\frac{D}{2}}^{\frac{D}{2}} \left\{\frac{1}{R+y}\right\}\sqrt{\left(\frac{D}{2}\right)^2 - y^2}\, dy$$

From the calculus principles

$$2\int_{-\frac{D}{2}}^{\frac{D}{2}} \left\{\frac{1}{R+y}\right\}\sqrt{\left(\frac{D}{2}\right)^2 - y^2}\, dy = 2\pi\left[R - \sqrt{R^2 - \left(\frac{D}{2}\right)^2}\right]$$

As $I_* = 0$

$$A + (e - R)2\pi\left[R - \sqrt{R^2 - \left(\frac{D}{2}\right)^2}\right] = 0$$

$$e = R - \frac{A}{2\pi\left[R - \sqrt{R^2 - \left(\frac{D}{2}\right)^2}\right]}.$$

Consider a trapezium cross-section:

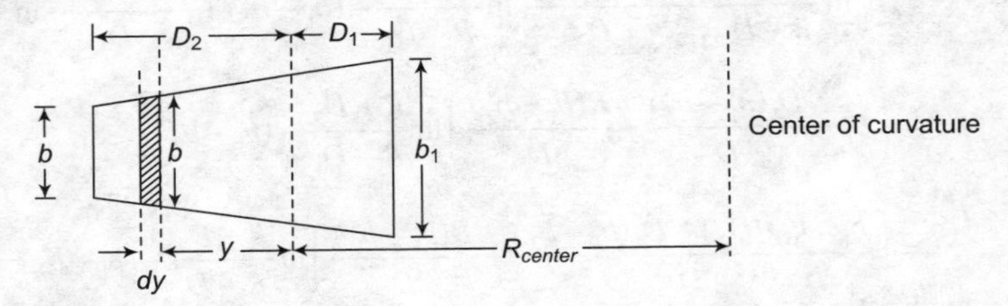

FIGURE 16.6

Consider an elemental strip of thickness 'dy' located 'y' distant from the centroid of the cross-section as shown in Figure 16.6.

The integral I_* is given by

$$I_* = A + (e - R)\int_{-D_1}^{D_2} \left\{\frac{1}{R+y}\right\} dA$$

From Figure 16.6,

$$D_1 + D_2 = D$$

$$dA = b dy$$

$$b = b_1 - \frac{b_1 - b_2}{D}(D_1 + y)$$

$$\int_{-D_1}^{D_2}\left\{\frac{1}{R+y}\right\}dA = \int_{-D_1}^{D_2}\left\{\frac{1}{R+y}\right\}bdy = \int_{-D_1}^{D_2}\left\{\frac{1}{R+y}\right\}\left(b_1 - \frac{b_1-b_2}{D}(D_1+y)\right)dy$$

$$= \int_{-D_1}^{D_2}\left\{\frac{b_1}{R+y}\right\}dy - \int_{-D_1}^{D_2}\left(\frac{b_1-b_2}{D}(D_1+y)\right)dy$$

$$= \int_{-D_1}^{D_2}\left\{\frac{b_1}{R+y}\right\}dy - \left(\frac{b_1-b_2}{D}\right)\int_{-D_1}^{D_2}\left\{\frac{D_1}{R+y}\right\}dy - \left(\frac{b_1-b_2}{D}\right)\int_{-D_1}^{D_2}\left\{\frac{y}{R+y}\right\}dy$$

$$= \int_{-D_1}^{D_2}\left\{\frac{b_1}{R+y}\right\}dy - \left(\frac{b_1-b_2}{D}\right)\int_{-D_1}^{D_2}\left\{\frac{D_1}{R+y}\right\}dy - \left(\frac{b_1-b_2}{D}\right)\int_{-D_1}^{D_2}\left\{\frac{y+R-R}{R+y}\right\}dy$$

$$= \int_{-D_1}^{D_2}\left\{\frac{b_1}{R+y}\right\}dy - \left(\frac{b_1-b_2}{D}\right)\int_{-D_1}^{D_2}\left\{\frac{D_1}{R+y}\right\}dy - \left(\frac{b_1-b_2}{D}\right)\left\{\int_{-D_1}^{D_2}\left\{1-\frac{R}{R+y}\right\}dy\right\}$$

$$= \int_{-D_1}^{D_2}\left\{\frac{b_1}{R+y}\right\}dy - \left(\frac{b_1-b_2}{D}\right)\int_{-D_1}^{D_2}\left\{\frac{D_1}{R+y}\right\}dy - \left(\frac{b_1-b_2}{D}\right)\left\{D - R\int_{-D_1}^{D_2}\left\{\frac{1}{R+y}\right\}dy\right\}$$

$$= b_1\ln\frac{R+D_2}{R-D_1} - \frac{D_1(b_1-b_2)}{D}\ln\frac{R+D_2}{R-D_1} - (b_1-b_2) + \frac{R(b_1-b_2)}{D}\ln\frac{R+D_2}{R-D_1}$$

$$= \left(b_1 - \frac{D_1(b_1-b_2)}{D} + \frac{R(b_1-b_2)}{D}\right)\ln\frac{R+D_2}{R-D_1} - (b_1-b_2)$$

$$= \left(\frac{b_1(D_1+D_2) - D_1(b_1-b_2)}{D} + \frac{R(b_1-b_2)}{D}\right)\ln\frac{R+D_2}{R-D_1} - (b_1-b_2)$$

$$= \left(\frac{b_1D_2 + D_1b_2}{D} + \frac{R(b_1-b_2)}{D}\right)\ln\frac{R+D_2}{R-D_1} - (b_1-b_2)$$

$$= \left(\frac{b_1(R+D_2) - b_2(R-D_1)}{D}\right)\ln\frac{R+D_2}{R-D_1} - (b_1-b_2)$$

We have the condition that

$$A + (e-R)\int\frac{1}{R+y}dA = 0$$

This gives that

$$A + (e-R)\left\{\left(\frac{b_1(R+D_2) - b_2(R-D_1)}{D}\right)\ln\frac{R+D_2}{R-D_1} - (b_1-b_2)\right\} = 0.$$

Simplifying the above expression, 'e' can be estimated.

PROBLEM 16.1

Objective 1

Determine the intrados to extrados stress ratio, in a curved bar of width 50 mm, depth 100 mm subjected to a BM of 10 kN·m. Center of curvature at the intrados is 150 mm. The BM decreases the radius of curvature.

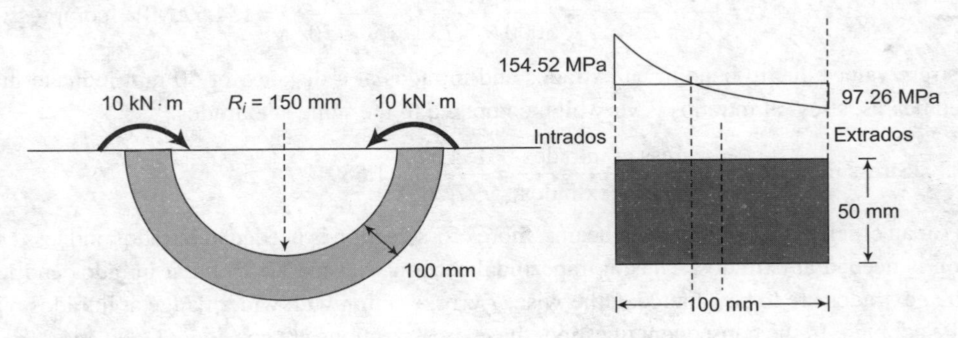

FIGURE 16.7

SOLUTION

Radius of curvature of the curved bar to the intrados $R_i = 100$ mm.
Cross-section is rectangular, width = 50 mm and depth = 100 mm.
Radius of the bar to the centroid of the cross-section = 100 + 50 = 150 mm.
BM acting on the bar

$M = 10$ kN·m (positive as the moment tries to reduce the radius of curvature)

A = Cross-sectional area of the bar = 50 × 1000 = 5000 mm^2
To find NA position 'e' from centroidal axis:

$$I_* = \int \frac{y+e}{R+y} dA = A + (e-R)\int \frac{1}{R+y} dA = 0$$

For a rectangular section of depth 'D'

$$e = R - \frac{D}{\ln\left\{\dfrac{2R+D}{2R-D}\right\}}$$

$$e = 150 - \frac{100}{\ln\left\{\dfrac{2\times150+100}{2\times150-100}\right\}}$$

$$e = 150 - \frac{100}{\ln\{2\}} = 5.73\,\text{mm}$$

Stress due to bending $\sigma = \dfrac{M}{Ae}\left\{\dfrac{y+e}{R+y}\right\}$

Stress due to bending at extrados is (y = 50 mm)

$$\sigma = \frac{M}{Ae}\left\{\frac{y+e}{R+y}\right\} = \frac{10\times10^6}{5000\times5.73}\left\{\frac{50+5.73}{150+50}\right\} = 97.26 \text{ MPa (tension)}$$

Stress due to bending at intrados is ($y = -50$ mm)

$$\sigma = \frac{M}{Ae}\left\{\frac{y+e}{R+y}\right\} = \frac{10\times10^6}{5000\times5.73}\left\{\frac{-50+5.73}{150-50}\right\} = 154.52 \text{ MPa (compression)}$$

The stress values due to bending at extrados and intrados at a distance of 50 mm indicate that in curved beams, stress at intrados is very high compared to the same at extrados.

$$\text{Stress ratio} = S_R = \frac{\text{Stress at intrados}}{\text{Stress at extrados}} = \frac{154.52}{97.26} = 1.589.$$

Thus for an efficient design in curved beams, more cross-section is needed at intrados and less cross-section is needed at extrados. Thus, a trapezoidal section with higher width at intrados and lesser width at extrados are to be adopted. Otherwise, *T* cross-section with wider flange at intrados region is to be adopted. In the consequent question, those cross-sections are considered.

PROBLEM 16.2

Objective 1

A curved bar of *T* cross-section shown in Figure 16.8 is subjected to a moment of 2 kN·m, which tends to increase the radius of curvature of the member. The radius of curvature to the intrados of the curved beam is 100 mm. Find the stresses at the intrados and extrados. Also find the stress at the centroid. Sketch the bending stresses. If $E = 200$ GPa, what is the strain at the centroid of the cross-section.

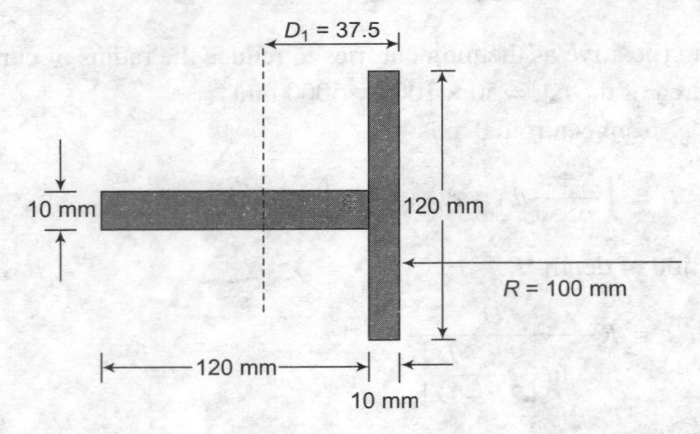

FIGURE 16.8

SOLUTION

Radius of curvature of the curved bar to the intrados $R = 100$ mm.

BM acting on the bar

$M = -2$ kN·m (negative as the moment tries to increase the radius of curvature)

A = Cross-sectional area of the beam = 120 × 10 + 120 × 10 = 2400 mm^2

Centroid of the cross-section from intrados

$$D_1 = \frac{120\times10\times5+120\times10\times(10+60)}{120\times10+120\times10} = 37.50 \text{ mm}.$$

Radius of curvature of the beam to the centroid of the cross-section = R = 100 + 37.5 = 137.5 mm.
To find NA position from centroidal axis:
As the cross-section is composed of two rectangular sections, consider a section in the flange and web separately.

$$I_* = \int \frac{y+e+R-R}{R+y} dA = A + (e-R)\int \frac{1}{R+y} dA = 0$$

For flange portion,

$$\int \frac{1}{R+y} dA = \int_{-37.5+10}^{-37.5} \frac{1}{R+y} b_f dy$$

$$\int_{-37.5}^{-37.5+10} \frac{1}{R+y} b_f dy = b_f \ln\left\{\frac{R-37.5}{R-27.5}\right\}$$

$$= 120 \times \ln\left\{\frac{137.5-37.5+10}{137.5-37.5}\right\} = 11.44$$

For web portion,

$$\int \frac{1}{R+y} dA = \int_{-37.5+10}^{130-37.5} \frac{1}{R+y} b_w dy$$

$$\int_{-37.5+10}^{130-37.5} \frac{1}{R+y} b_w dy = b_w \ln\left\{\frac{R+130-37.5}{R-27.5}\right\} = 7.376$$

$$= 10 \times \ln\left\{\frac{137.5+130-37.5}{137.5-27.5}\right\} = 7.376$$

$$A + (e-R)\int \frac{1}{R+y} dA = 0$$

$$2400 + (e - 137.5)(7.376 + 11.44) = 0$$

$$e = 9.95 \text{ mm.}$$

Stress due to bending $\sigma = \dfrac{M}{Ae}\left\{\dfrac{y+e}{R+y}\right\}$

Stress due to bending at extrados is (y = 130 – 37.5 = 92.5 mm)

$$\sigma = \frac{M}{Ae}\left\{\frac{y+e}{R+y}\right\} = \frac{-2000000}{2400 \times 9.95}\left\{\frac{92.5+9.95}{137.5+92.5}\right\} = -37.31 \text{ MPa (compression)}$$

Stress due to bending at intrados is $y = -37.5$ mm

$$\sigma = \frac{M}{Ae}\left\{\frac{y+e}{R+y}\right\} = \frac{-2000000}{2400 \times 9.95}\left\{\frac{-37.5+9.95}{137.5-37.5}\right\} = 23.07 \text{ MPa (tension)}$$

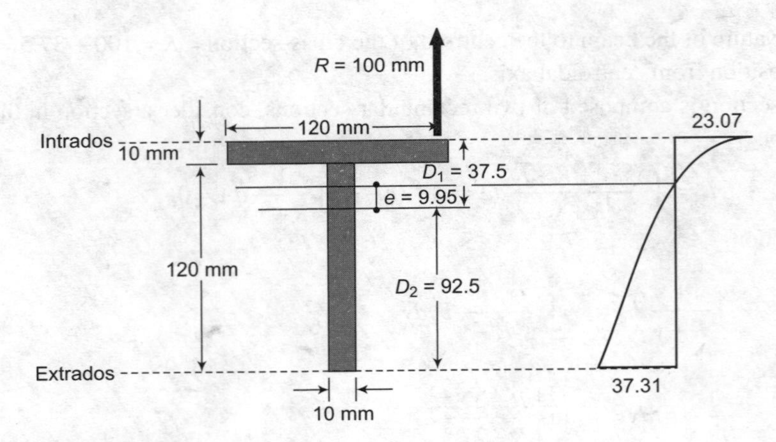

FIGURE 16.9

Stress due to bending at centroid is ($y = 0$)

$$\sigma = \frac{M}{Ae}\left\{\frac{y+e}{R+y}\right\} = \frac{-2000000}{2400 \times 9.95}\left\{\frac{0+9.95}{137.5-0}\right\} = -6.06 \text{ MPa (compression)}$$

$$\text{Strain at centroid} = \varepsilon_c = \frac{M}{EAR} = \frac{6.06}{2 \times 10^5} = 3.03 \times 10^{-5}.$$

PROBLEM 16.3

Objective 2

A crane hook whose horizontal cross-section is trapezoidal, 50 mm wide at the inside and 80 mm wide at the outside, thickness 60 mm, carries a vertical load of 10 kN whose line of action is 80 mm from the inside edge of this section. The center of curvature is 50 mm from the inside edge. Calculate the maximum tensile and compressive stress set up.

SOLUTION

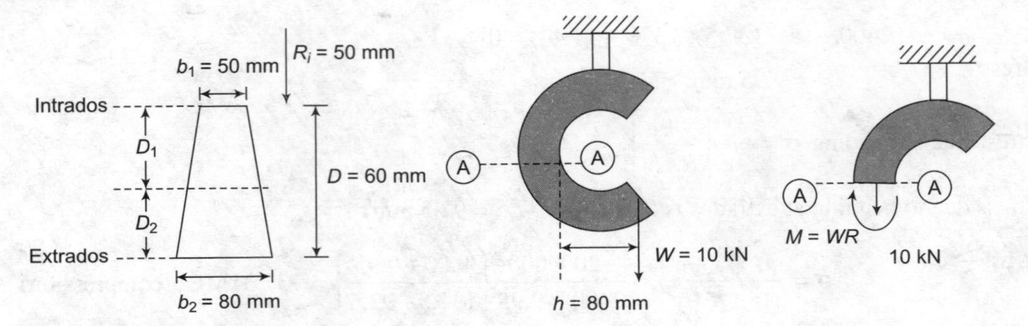

FIGURE 16.10

For determining the maximum stresses, maximum eccentricity is to be considered.
Maximum eccentricity occurs at section A-A is $h = 80$ mm.

Centroid of the cross-section from intrados

$$D_1 = \frac{2\times 80+50}{80+50}\times\frac{60}{3} = 32.31 \text{ mm}$$

$$D_2 = 60 - 32.31 = 27.69 \text{ mm}.$$

Hence the eccentricity of the load to the centroid at section AA is 50 + 32.31 = 82.31 mm.

At section A-A:

Axial load = P = 10,000 N

$$\text{BM} = M = -\frac{10\times 82.31}{1000} = -0.8231\,\text{kN}\cdot\text{m}.$$

The BM increases the radius of curvature. Thus, BM is negative.

$$A = \frac{b_1+b_2}{2}D = \frac{80+50}{2}\times 60 = 3900 \text{ mm}^2$$

Radius of crane to its centroid = R = 82.31 mm.

To find the position of NA

$$A+(e-R)\left\{\left(\frac{b_1(R+D_2)-b_2(R-D_1)}{D}\right)\ln\frac{R+D_2}{R-D_1}-(b_1-b_2)\right\}=0$$

$$3900+(e-82.31)\left\{\left(\frac{50(110)-80(50)}{60}\right)\ln\frac{110}{50}-(50-80)\right\}=0$$

$$3900+(e-82.31)\{19.71+30\}=0$$

$$(e-82.31) = -\frac{3900}{49.71}$$

$$e = 3.85 \text{ mm}$$

Axial stress $\sigma_a = \frac{P}{A} = \frac{10000}{3900} = 2.56\,\text{MPa (Tension)}$.

Stress due to bending $\sigma = \frac{M}{Ae}\left\{\frac{y+e}{R+y}\right\}$

Stress due to bending at extrados is

$$\sigma = \frac{M}{Ae}\left\{\frac{y+e}{R+y}\right\} = \frac{-823100}{3900\times 3.85}\left\{\frac{27.69+3.85}{82.31+27.69}\right\} = -15.57\,\text{MPa (compression)}$$

Stress due to bending at intrados is

$$\sigma = \frac{M}{Ae}\left\{\frac{y+e}{R+y}\right\} = \frac{-823100}{3900\times 3.85}\left\{\frac{-32.31+3.85}{82.31-32.31}\right\} = 32.20 \text{ MPa (tension)}$$

Final stresses:

Stress at intrados = 32.20 + 2.56 = 34.76 MPa (tensile)

Stress at extrados = −15.57 + 2.56 = −13.01 MPa (compressive).

PROBLEM 16.4 **Objective 1**

A curved bar of circular cross-section with 50 mm diameter and mean radius of curvature 80 mm subjected to a BM of 500 N·m is applied to the bar tending to shorten it; find the stresses at the intrados and extrados. Also find the stress at the centroid.

SOLUTION

Radius of curvature of the curved bar $R = 80$ mm.

BM acting on the bar

$M = 500$ N·m (positive as the moment tries to reduce the radius of curvature)

Diameter of the cross-section of the beam = $D = 50$ mm

A = Cross-sectional area of the bar

To find NA position from centroidal axis:

$$e = R - \frac{A}{2\pi\left[R - \sqrt{R^2 - \left(\frac{D}{2}\right)^2}\right]}$$

$$e = 80 - \frac{\frac{\pi(50)^2}{4}}{2\pi\left[80 - \sqrt{80^2 - \left(\frac{50}{2}\right)^2}\right]}$$

$$e = 80 - \frac{1963.5}{2\pi[80 - 75.99]} = 2.00 \text{ mm}$$

Stress due to bending $\sigma = \dfrac{M}{Ae}\left\{\dfrac{y+e}{R+y}\right\}$

Stress due to bending at extrados is ($y = 25$ mm)

$$\sigma = \frac{M}{Ae}\left\{\frac{y+e}{R+y}\right\} = \frac{500000}{1963.5 \times 2.00}\left\{\frac{25+2.00}{80+25}\right\} = 32.74 \text{ MPa (tension)}$$

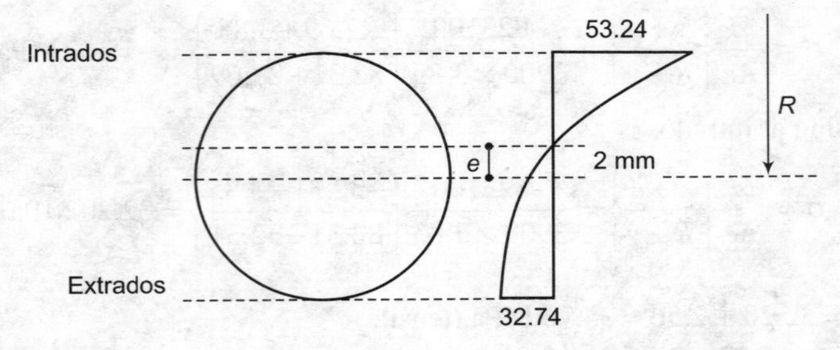

FIGURE 16.11

Stress due to bending at intrados is ($y = -25$ mm)

$$\sigma = \frac{M}{Ae}\left\{\frac{y+e}{R+y}\right\} = \frac{500000}{1963.5 \times 2.00}\left\{\frac{-25+2.00}{80-25}\right\} = 53.24 \text{ MPa (compression)}$$

PROBLEM 16.5 **Objective 2**

A *U* shaped frame is subjected to force of 40 kN as shown in Figure 16.12. Determine the stresses at the horizontal section A-A. The cross-section is an I section with depth 150 mm along the radius of curvature and flanges 50 mm × 10 mm at intrados and extrados. Web thickness of the cross-section is 8 mm.

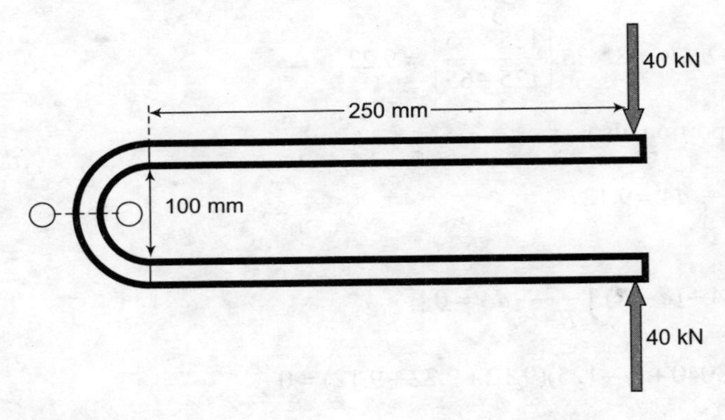

FIGURE 16.12

SOLUTION

Radius of curvature of the curved bar to the intrados $R_i = 50$ mm

Centroid of the cross-section from intrados = 75 mm

At section A-A, axial compressive load = 40 kN

Hence, $R = 50 + 75 = 125$ mm

BM acting on the bar at section A-A, $M = 40 \times (0.25 + 0.05 + 0.075)$

= 15 kN·m (positive as the moment tries to decrease the radius of curvature)

A = Cross-sectional area of the beam = 2 × 100 × 10 + (150 – 20) × 8 = 3040 mm^2

To find NA position from centroidal axis:

As the cross-section is composed of three rectangular sections, consider a section in the flanges and web separately.

$$I_* = \int \frac{y+e+R-R}{R+y}dA = A + (e-R)\int \frac{1}{R+y}dA = 0$$

For flange portion,

$$\int \frac{1}{R+y}dA = \int_{-75}^{-65} \frac{1}{R+y}b_f dy$$

$$\int_{-75}^{-65} \frac{1}{R+y} b_f dy = b_f \ln\left\{\frac{R-65}{R-75}\right\}$$

$$= 50 \times \ln\left\{\frac{125-65}{125-75}\right\} = 9.12$$

For web portion,

$$\int \frac{1}{R+y} dA = \int_{-65}^{65} \frac{1}{R+y} b_w dy$$

$$\int_{-65}^{65} \frac{1}{R+y} b_w dy = 8 \times \ln\left\{\frac{125+65}{125-65}\right\} = 9.22$$

For bottom flange portion also

$$\int \frac{1}{R+y} dA = 9.12$$

$$A + (e-R)\int \frac{1}{R+y} dA = 0$$

$$3040 + (e-125)(9.12+9.22+9.12) = 0$$

$$e = 14.29 \text{ mm}$$

Stress due to bending $\sigma = \dfrac{M}{Ae}\left\{\dfrac{y+e}{R+y}\right\}$

Stress due to bending at extrados is (y = 75 mm)

$$\sigma = \frac{M}{Ae}\left\{\frac{y+e}{R+y}\right\} = \frac{15\times10^6}{3040\times14.29}\left\{\frac{75+14.29}{125+75}\right\} = 154.16 \text{ MPa (tension)}$$

Stress due to bending at intrados is (y = –75 mm)

$$\sigma = \frac{M}{Ae}\left\{\frac{y+e}{R+y}\right\} = \frac{15\times10^6}{3040\times14.29}\left\{\frac{-75+14.29}{125-75}\right\} = -419.25 \text{ MPa (compression)}$$

Stress due to axial tension = $\sigma_a = \dfrac{40000}{3040} = 13.16$ (tension)

Final stresses

At intrados: $\sigma_i = \sigma_{ai} + \sigma_{bi} = 13.16 - 419.25 = -406.09$ MPa (compression)

At extrados: $\sigma_e = \sigma_{ae} + \sigma_{be} = 13.16 + 154.16 = 167.34$ MPa (tension)

SUMMARY

- **The normal stress in a curved beam subjected to moment M is given by $\sigma = \frac{M}{Ae}\left\{\frac{y+e}{R+y}\right\}$, in which e is the location of NA from centroid and R is the radius of curvature of the beam to the centroid.**
- **NA always lies toward the intrados of the cross-section from the centroidal axis.**
- **The stress at the centroidal axis of the curved beam is $\sigma = \frac{M}{AR}$.**

OBJECTIVE TYPE QUESTIONS

1. In a curved bar, the stress distribution is
   (a) Parabolic with maximum at centroidal axis
   (b) Linear with maximum at outermost fibers
   (c) Nonlinear with maximum either at extrados or intrados
   (d) Nonlinear with minimum either at extrados or intrados
2. In a curved beam
   (a) Strain variation is linear and stress variation is linear
   (b) Strain variation is linear and stress variation is nonlinear
   (c) Strain variation is nonlinear and stress variation is linear
   (d) Strain variation is nonlinear and stress variation is nonlinear
3. In curved beam of rectangular cross-section, the ratio of stress at intrados to stress at extrados is
   (a) Always more than 1
   (b) Always equal to 1
   (c) Always less than 1
   (d) Depends on the cross-section and radius of curvature
4. In a curved bar
   (a) Stress at centroid is zero
   (b) Nature of stress at the centroid is same as that of intrados
   (c) Nature of stress at the centroid is same as that of extrados
   (d) Nature of stress at the centroid is always tensile
5. The strain at centroid in a curved bars depends on
   (a) Moment and modulus of elasticity
   (b) Moment and radius of curvature
   (c) Moment, modulus of elasticity, and radius of curvature
   (d) Moment, modulus of elasticity, area of cross-section, and radius of curvature
6. Which one of the following cross-section can be adopted for the economical design of a curved bead
   (a) Rectangular (b) Circular (c) Square (d) Trapezoidal
7. A crane hook carries at gravity load at its end. A horizontal cross-section of the crane cross-section is subjected to
   (a) BM and shear force (b) BM and axial tension
   (c) BM and axial compression (d) Only BM

8. In a curved beam of circular cross-section subjected to bending, failure occurs at
   (a) Intrados
   (b) Extrados
   (c) Intrados and extrados simultaneously
   (d) Centroid
9. The deviation in straight beams and curved beams is that the
   (a) Strain variation
   (b) Stress variation
   (c) Both strain variation and stress variation
   (d) Stress at NA

Solutions for Objective Questions

Sl. No.	1.	2.	3.	4.	5.	6.	7.	8.	9.
Answer	(c)	(b)	(a)	(c)	(d)	(d)	(b)	(a)	(b)

EXERCISE PROBLEMS

1. A crane hook whose horizontal cross-section is trapezoidal, 40 mm wide at the inside, 20 mm wide at the outside, and thickness 40 mm, carries a vertical load of 800 kg whose line of action is 28 mm from the inside edge of this section. The center of curvature is 40 mm from the inside edge. Calculate the maximum tensile and compressive stress set up.
2. A curved bar of square section, 4-cm sides and mean radius of curvature 6 cm is initially unstressed. If a BM of 450 N·m is applied to the bar tending to straighten it, find the stresses at the inner and outer faces.
3. A curved bar of rectangular section 38 mm wide by 50 mm deep and of mean radius of curvature 100 mm is subjected to a BM of 1.5 kN·m tending to straighten the bar. Find the position of the NA and the magnitudes of the greatest bending stresses. Draw a diagram to show the variation of stress across the section. $E = 2{,}06{,}000$ N/mm^2.
4. A bar 5 cm diameter, curved to a mean radius of 5 cm, is subjected to a BM of 760 N·m tending to open out the bend. Plot the stress distribution across the section.
5. Obtain the stress at intrados and extrados of a triangular cross-section of base width 50 mm and depth 60 mm, subjected to a BM of 10 kN·m, which tries to increase the radius of curvature of the beam. The radius of curvature to the intrados of the cross-section is 75 mm.

CHAPTER 17

DEFLECTION OF TRUSSES

UNIT OBJECTIVE

This chapter provides information on the deflection at joints and also deflection due to temperature variation in trusses by unit load method and Castigliano's theorem. The presentation attempts to help the student achieve the following:

Objective 1: To determine the vertical horizontal displacement of a truss.

Objective 2: To determine the safe load for allowable deflection of a truss.

Objective 3: To determine the effective span for given deflection.

Objective 4: To determine the relative moment and relative displacement.

17.1 INTRODUCTION

An articulated structure or a truss is composed of a number of bars or members connected by frictionless pins, forming geometrical figures which are usually triangles. A truss is said to be statically determinate internally if the number of members

$$m = 2j - r$$

in which m is the total number of members, j is the total number of joints, and r is the total number of equilibrium equations.

In the above equation, the value of r is usually 3 but if there is an additional hinge separating the structure in two parts, $r = 3 + 1 = 4$, and if there is a link somewhere in the structure, $r = 3 + 2 = 5$. The frame is said to be perfect if the number m is equal to right side of the equation.

When external loads are applied on the truss, at the joints, the members carry internal forces, usually called the stresses, which may be either tensile or compressive. According to Hook's law, if any axial member of length L carries a force P, it will be deformed by an amount PL/AE, in which A

is the area of cross section of the members. In the truss, therefore, the members carrying tension will be elongated, whereas those carrying compression will be shortened. Thus, all the joints will move from their initial position and will occupy their final equilibrium position. The axial deformation in the members of the truss may also be due to temperature changes or due to errors in fabrication or lack of fit of some members, and the joints may move from their original position. This movement of each joint is defined as the deflection of the joint in the direction of movement. The resolved part of the movement in the vertical direction is called vertical deflection of the joint, and that in the horizontal direction is known as the horizontal deflection of the joint. In general, each joint has both vertical as well as horizontal deflection, unless constrained to move in a given direction.

There are various methods of computing the joint deflection of a perfect frame. We shall, however, discuss the following methods:

1. The unit load method
2. Deflection by Castigliano's first theorem
3. Graphical method: Williot–Mohr diagram.

The first two methods are analytical and require the full knowledge of the methods of finding the stresses in plane frames under static loading.

17.2 THE UNIT LOAD METHOD

To develop the method, let us consider a perfect frame as shown in Figure 17.1, in which W_1, W_2, W_3, …, W_n, etc., are external loads.

Let P_1, P_2, P_3, …, P_n, etc., be the forces or stresses in the members 1, 2, 3, …, n, etc., due to external loading.

Let us find the vertical displacement (δ_V) of the joint F due to the external system of loads shown.

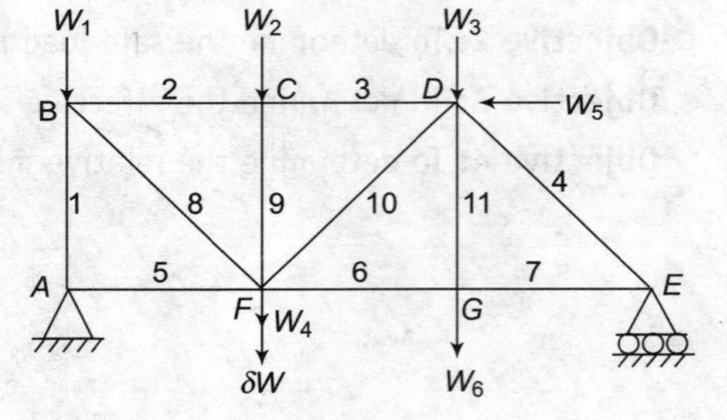

FIGURE 17.1

Apply gradually an infinitely small load δW at the joint F in the vertical direction, and let δV be the deflection of the joint. The work done by δW will be

$$\frac{1}{2}\delta w \cdot \delta v. \tag{17.1}$$

If u_1, u_2, u_3, …, u_n are the forces in the various members due to unit vertical load at the joint F, the forces in the members due to a load δW at F will be $u_1 \cdot \delta W$, $u_2 \cdot \delta W$, $u_3 \cdot \delta W$, …, $u_n \cdot \delta W$, respectively.

If any member of length L has the force $u \cdot \delta W$, its extra deformation due to δW at F will be

$$= \frac{(u \cdot \delta w)L}{AE}$$

$$\text{Work stored in the member} = \frac{1}{2}\text{(force) (deformation)}$$

$$= \frac{1}{2}(P + u \cdot \delta w) \times \frac{u \cdot \delta w L}{AE} = \frac{1}{2}\frac{P u L \delta w}{AE}$$

= (neglecting the product of small quantities)

Total work stored in all the members

$$= \sum_{i=1}^{n} \frac{1}{2} \frac{Pu_i L \delta w}{AE} = \frac{1}{2} \delta w \sum_{i=1}^{n} \frac{Pu_i L}{AE}. \tag{17.2}$$

Equating the work supplied to the work stored, we get

$$\frac{1}{2} \delta w \delta v = \frac{1}{2} \delta w \sum_{i=1}^{n} \frac{Pu_i L}{AE}$$

$$\delta_V = \sum_{i=1}^{n} \frac{Pu_i L}{AE} = \sum_{i=1}^{n} \frac{pu_i L}{E}, \tag{17.3}$$

in which n is the total number of members.

If, however, horizontal deflection is required, it can similarly be proved that

$$\delta_H = \sum_{i=1}^{n} \frac{Pu_i L}{AE} = \sum_{i=1}^{n} \frac{pu_i L}{E}, \tag{17.4}$$

in which P is the force in any member due to central loads, p is the intensity of stress in any member due to central loads, u is the force in any member due to unit vertical load applied at the joint where deflection is required, and u^1 is the force in any member due to unit horizontal load applied at the joint where deflection is required.

Steps: The method of computing the deflection of a joint can be summarized below:

1. Find the forces $P_1, P_2, \ldots\ldots\ldots, P_n$ in all the members due to external loads.
2. Remove the external loads and apply the unit vertical load at the joint if the vertical deflection of the joint is required, and the stresses $u_1, u_2, \ldots, u_7$ in all the members. (If horizontal deflection is required, apply unit horizontal load there and find $u_1^1, u_2^1, \ldots\ldots\ldots, u_n^1$.)
3. Apply Eq. (17.3) for vertical deflection and Eq. (17.4) for horizontal deflection of the point.

17.3 JOINT DEFLECTION IF LINEAR DEFORMATION OF ALL THE MEMBERS IS KNOWN

If, in the place of external loads, the deformations $\Delta_1, \Delta_2, \ldots, \Delta_n$, etc., of all the members are known, the deflection δ can be calculated as follows:

Eq. (17.3) can be rewritten as

$$\delta_V = \sum_{i=1}^{n} u_i \frac{PL}{AE}. \tag{17.5}$$

But $\frac{PL}{AE}$ = deformation of the member = Δ (according to Hook's law).

$$\delta_V = \sum_{i=1}^{n} u_i \Delta_i \tag{17.6}$$

Similarly,

$$\delta_H = \sum_{i=1}^{n} u_i \Delta \tag{17.5}$$

Hence, to find the deflection δ in such cases, apply unit load at the joint, in the direction where the deflection is required, and apply Eq. (17.6) or Eq. (17.7).

17.3.1 Deflection of a Joint due to Temperature Variation

Let $\Delta_1, \Delta_2, \ldots, \Delta_n$ be the changes in the lengths of various members of a perfect frame due to temperature variation. To find the deflection of an unloaded frame due to temperature variations, apply the unit load at the joint and calculate u_1, u_2, ... and apply Eq. (17.6) or Eq. (17.7) for the joint deflection.

Thus,

$$\delta = \sum_{i=1}^{n} u_i \Delta_i = u_1 \Delta_1 + u_2 \Delta_2 + \cdots + u_n \Delta_n. \tag{17.8}$$

If the change in length (Δ) of certain members is zero, the product $u.\,\Delta$ for those members will be substituted as zero in the above equation. If, for example, there is only one member in which there is change in length Δ_1, the deflection of a particular joint will be equal to $u_1\Delta_1$ in which u_1 is the stress in that member due to the unit load at the joint under consideration.

17.3.2 Deflection of a Joint due to Lack of Fit of Certain Members

Let $\Delta_1, \Delta_2, \ldots, \Delta_n$ be the lack of fit in the members. The joint deflection can be found by Eq. (17.6) or Eq. (17.7) that is,

$$\delta = \sum_{i=1}^{n} u_i \Delta_i = u_1 \Delta_1 + u_2 \Delta_2 + \cdots + u_n \Delta_n.$$

If there is only one member having lack of fit Δ_1, the deflection of a particular joint will be equal to $u_1\Delta_1$, in which u_1 is the stress in that member due to unit load at the joint under consideration.

PROBLEM 17.1 **Objective 1**

Determine the vertical and horizontal displacements of the point of the pin-jointed frame shown in Figure 17.2 (a). The cross-sectional area of AB is 150 mm^2 and of AC and BC 200 mm^2 each. $E = 2 \times 10^5$ N/mm^2.

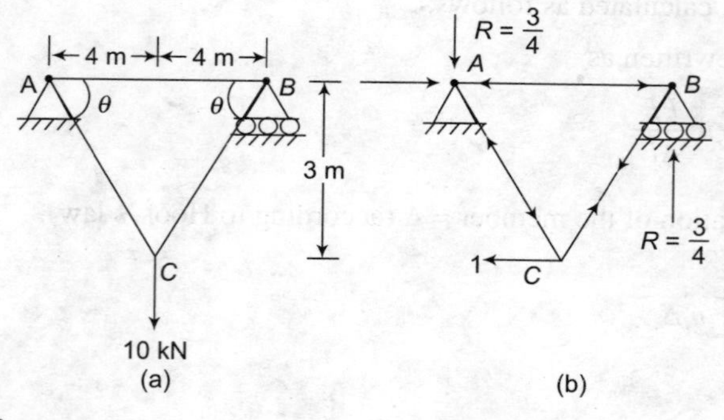

FIGURE 17.2

SOLUTION

The vertical and horizontal deflections of the joint C are given by,

$$\delta_V = \sum \frac{PuL}{AE} \tag{17.9}$$

$$\delta_H = \sum \frac{Pu^1L}{AE}. \tag{17.10}$$

Let us now find P, u, and u^1 in each member.

(a) Stresses due to external loading

$$\text{AC} = \sqrt{3^2 + 4^2} = 5 \text{ m}$$

$$\sin\theta = \frac{3}{5} = 0.6; \cos\theta = \frac{4}{5} = 0.8$$

Resolving at the joint C, we get

$$10 = P_{\text{AC}} \sin\theta + P_{\text{BC}} \cos\theta$$

Resolving horizontally, $P_{\text{AC}} = P_{\text{BC}}$

$$2P_{\text{AC}} \sin\theta = 10.$$

From which

$$P_{\text{AC}} = P_{\text{BC}} = \frac{10}{2\sin\theta} = \frac{10}{2\times 0.6} = +8.33 \text{ kN (tension)}$$

(Use + sign for tension and – sign for compression)

Resolving horizontally at A,

$$P_{\text{AB}} = P_{\text{AC}} \cos\theta = 8.33 \times 0.8 = 6.66 \text{ kN (compression)} = -6.66 \text{ kN}.$$

(b) Stresses due to unit vertical load at C

Apply unit vertical load at C. The stresses in each member will be 1/10 of those obtained above.

Thus, $u_{\text{AC}} = u_{\text{BC}} = +\dfrac{8.33}{10}$ and $u_{AB} = -\dfrac{6.66}{10}$.

(c) Stresses due to unit horizontal load at C

Assuming that the horizontal movement of joint C is to the left, apply a unit horizontal load at C as shown in Figure 17.2 (b), along with the reactions.

Resolving vertically at joint C, we get

$$u^1_{\text{CA}} = u^1_{\text{CB}} \text{ (numerically)}$$

Resolving horizontally,

$$u^1_{\text{CB}} \cos\theta + u^1_{\text{CA}} \cos\theta = 1$$

Or $\quad u^1_{\text{CB}} = u^1_{\text{CA}} = \dfrac{1}{2\cos\theta} = \dfrac{1}{2\times 0.8} = \dfrac{5}{8}$

$$u^1_{CB} = +\frac{5}{8}; u^1_{CA} = -\frac{5}{8}$$

Resolving horizontally at B, we get

$$u^1_{AB} = u^1_{BC}\cos\theta = \frac{5}{8}\times 0.8 = 0.5 \text{ (compression).}$$

To calculate $\frac{PuL}{A}$ and $\frac{Pu'L}{A}$, the results are tabulated below:

Table 17.1

Member	L (mm)	Area (mm²)	P (kN)	u	u^1	$\frac{PuL}{A}$	$\frac{Pu^1L}{A}$
AB	8000	150	−6.66	−0.66	−0.5	234.43	177.6
BC	5000	200	8.33	0.83	$+\frac{5}{8}$	173.47	130.15
CA	5000	200	8.33	0.83	$-\frac{5}{8}$	173.47	−130.15
					Sum	+581.37	177.6

$$E = 2\times 10^5 \text{ N/mm}^2 = 200 \text{ kN/mm}^2$$

$$\delta_V = \sum\frac{PuL}{AE} = +\frac{581.37}{200} = +2.906\text{ mm and } \delta_H = \sum\frac{Pu^1L}{AE} = +\frac{177.6}{200} = +0.888\text{ mm}$$

(The signs indicate that the assumed directions are correct).

Check: Since the end B is supported on roller, the movement of B is horizontal only, and is equal to the deformation of the bar AB. Since the structure is symmetrical about C and loading is also central, it is evident that horizontal movement of $C = \frac{1}{2}$ of deformation of AB.

$$\delta_H = \sum\frac{1}{2}\frac{P_{AB}\times L_{AB}}{A_{AB}\times E} = \frac{1}{2}\frac{6.66\times 8000}{150\times 200} = 0.888\text{ mm.}$$

PROBLEM 17.2

Objective 2

A frame ABCD consists of two equilateral triangles and is hinged at E and supported on rollers at H as shown in Figure 17.3. Determine the vertical deflection of G and horizontal movement of H due to a load W applied vertically at G. All the members are of length L. All the tension members are of area *a* and compression members of area 2*a*.

Solution

The vertical deflection of G is given by

$$\delta_{GV} = \sum_1^n \frac{PuL}{AE}$$

in which u is the stress due to unit vertical load at G.

The horizontal deflection of H is given by $\delta_{HH} = \sum_{1}^{n} \frac{Pu^1L}{AE}$

in which u^1 is the stress due to unit horizontal load at H.

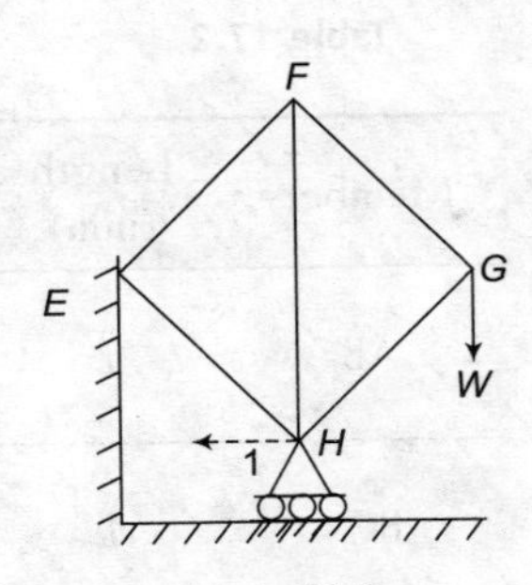

FIGURE 17.3

(a) Stresses due to external loading Figure 17.3

$$\cos 60° = 1/2; \ \cos 30° = \frac{\sqrt{3}}{2}$$

Resolving horizontally at joint G,

$$P_{GF} = P_{GH}$$

Resolving vertically at joint G,

$$2P_{GF} \cos 60° = W$$

$$P_{FG} = \text{W (tension)}$$

$$P_{GH} = W \text{ (compression)}$$

And resolving horizontally at F,

$$PFE = P_{FG} = W \text{ (tension)}$$

Resolving vertically at F,

$$P_{FH} = 2P_{EF} \cos 60° = W \text{ (compression)}$$

Resolving horizontally at F,

$$P_{HE} = 2P_{HG} = W \text{ (compression).}$$

(b) Stresses due to vertical load at G

To find u in all members, put unit $W = 1$ in the expressions found above.

(c) Stresses due to horizontal unit load at H

The find the horizontal deflection of H, apply unit horizontal load at H, in the directions shown in Figure 17.3.

By resolution at joint G,

$$u^1_{FG} = u^1_{GH} = 0 \text{ (since there is no load at B)}$$

Resolving horizontally at F,

$$u^1_{FE} = u^1_{FG} = 0.$$

Hence, $u^1_{FH} = 0$.

Resolving horizontally at H,

$$u^1_{HE} \cos 30° = 1$$

Or $u^1_{HE} = \dfrac{1}{\cos 30°} = \dfrac{2}{\sqrt{3}}$ (compression).

The results are tabulated below:

Table 17.2

Member	Length (mm)	Area (mm^2)	P (kN)	u	u^1	$\frac{PuL}{A}$	$\frac{Pu^1L}{A}$
AB	L	a	$+W$	$+1$	0	$+\frac{WL}{aE}$	0
BC	L	a	$+W$	$+1$	0	$+\frac{WL}{aE}$	0
CD	L	$2a$	$-W$	$+1$	0	$+\frac{WL}{2aE}$	0
AD	L	$2a$	$-W$	-1	$-\frac{2}{\sqrt{3}}$	$+\frac{WL}{2aE}$	$+\frac{WL}{\sqrt{3}aE}$
BD	L	$2a$	$-W$	-1	0	$+\frac{WL}{2aE}$	0
		Sum				$+\frac{7WL}{2aE}$	$+\frac{WL}{\sqrt{3}aE}$

Hence, $$\delta_{GV} = \sum \frac{PuL}{AE} = \frac{7WL}{2aE} (\downarrow)$$

And $$\delta_{HH} = \sum \frac{Pu^1L}{AE} = \frac{WL}{\sqrt{3}aE} (\leftarrow)$$

PROBLEM 17.3

Objective 1

Figure 17.4 represents a crane structure attached to a vertical wall and carrying a vertical load of 30 kN at C. All tension members are stressed to 100 N/mm^2 and all compression members to 60 N/mm^2. Determine the horizontal and vertical deflection of the end C. Take $E = 2 \times 10^5$ N/mm^2. All members except CD have a length of 2 m. AE = 2 m.

SOLUTION

The horizontal and vertical deflection at C are given by

$$\delta_V = \sum_1^n \frac{PuL}{AE} = \sum_1^n \frac{puL}{E} \qquad (17.11)$$

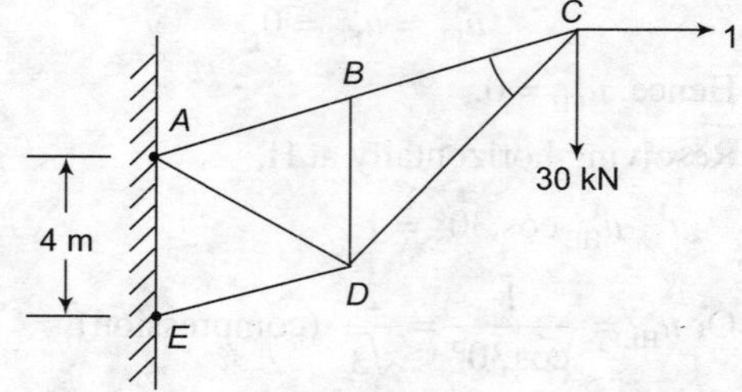

FIGURE 17.4

and $$\delta_H = \sum_1^n \frac{Pu^1L}{AE} = \sum_1^n \frac{pu^1L}{E} \tag{17.12}$$

in which p = intensity of stress in each member and is known.

(a) Calculation of stresses due to unit vertical load at C

To calculate u in all members, apply a unit vertical load at C,

Resolving perpendicular to BC at C,

$$u_{CD} \sin 30° = 1 \times \sin 60°$$

$$u_{CD} = \frac{\sin 60°}{\sin 30°} = \sqrt{3} \text{ (compression)}$$

$$u_{CB} = u_{CD} \cos 30° - 1 \cos 60°$$

$$= \sqrt{3} \times \frac{\sqrt{3}}{2} - \frac{1}{2} = 1.0 \text{ (tension)}$$

$$u_{AB} = u_{BC} = 1.0 \text{ (tension)}$$

$$u_{BD} = 0 \text{ (by resolving perpendicular to AB at B)}$$

Resolving perpendicular to ED at D,

$$u_{AD} \sin 60° = u_{DC} \sin 30°$$

$$u_{AD} = 1.0 \text{ (tension)}$$

Resolving along ED at C,

$$u_{ED} = u_{AD} \cos 60° + u_{DC} \cos 30° = 2.0 \text{ (compression).}$$

(b) Calculation of stresses due to unit horizontal load at C

Apply unit load at C, as shown.

Resolving perpendicular to BC at C,

$$u^1_{CD} = \sin 30° = 1.0 \sin 30° \; u^1_{CD} = 1 \text{ (compression)}$$

Resolving along BC at C,

$$u^1_{CB} = u^1_{CD} \cos 30° + 1.0 \cos 30° = \left(1.0 \times \frac{\sqrt{3}}{2}\right) + \left(1.0 \times \frac{\sqrt{3}}{2}\right) = \sqrt{3} = \text{(tension)}$$

$$u^1_{BA} = u^1_{CB} = \sqrt{3} \text{ (tension) and } u^1_{BD} = 0.$$

Resolving perpendicular to ED at D,

$$u^1_{AD} \sin 60° = u^1_{CD} \sin 30°$$

$$u^1_{AD} = \frac{1}{2} \times \frac{2}{\sqrt{3}} = \frac{1}{\sqrt{3}} \text{ (tension).}$$

Resolving along ED at D,

$$u^1_{ED} = u^1_{AD} \cos 60° + u^1_{DC} \cos 30°$$

$$= \left(\frac{1}{\sqrt{3}} \times \frac{1}{2}\right) + \left(1 \times \frac{\sqrt{3}}{2}\right) = \frac{2}{\sqrt{3}} \text{ (compression).}$$

The results are tabulated below:

Table 17.3

Member	L (mm)	p (N/mm^2)	u	u^1	puL	pu^1L
AB	2000	+100	+1.0	$+\sqrt{3}$	$+20 \times 10^4$	$+20 \times \sqrt{3} \times 10^4$
BC	2000	+100	+1.0	$+\sqrt{3}$	$+20 \times 10^4$	$+20 \times \sqrt{3} \times 10^4$
AD	2000	+100	+1.0	$\frac{1}{\sqrt{3}}$	$+20 \times 10^4$	$+\frac{20}{\sqrt{3}} \times 10^4$
BD	2000	0	0	0	0	0
ED	2000	– 60	–2.0	$-\frac{2}{\sqrt{3}}$	$+24 \times 10^4$	$+\frac{24}{\sqrt{3}} \times 10^4$
DC	$2000\sqrt{3}$	– 60	$-\sqrt{3}$	–1.0	$+36 \times 10^4$	$+12 \times \sqrt{3} \times 10^4$
				Sum	$+120 \times 10^4$	$+115.47 \times 10^4$

$$\delta_V = \sum_1^n \frac{puL}{E} = \frac{120 \times 10^4}{2 \times 10^5} = 6\,\text{mm}\,(\downarrow)$$

and
$$\delta_H = \sum_1^n \frac{pu^1L}{E} = \frac{115.47 \times 10^4}{2 \times 10^5} = 5.77\,\text{cm}\,(\rightarrow).$$

PROBLEM 17.4

Objective 1

The steel truss shown in Figure 17.5 is anchored at A and supported on roller at B. If the truss is so designed that, under the given loading, all tension members are stressed to 150 N/mm^2 and all compression members to 100 N/mm^2. Find also the lateral displacement of the end B.

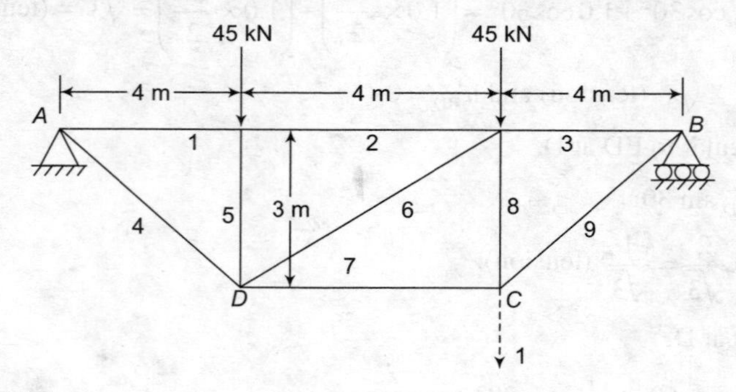

FIGURE 17.5

SOLUTION

(a) Vertical deflection:

The vertical deflection is given by

$$\delta_{CV} = \sum_{1}^{u} \frac{puL}{E}$$

in which p is the stress due to external loading.

To find the values of u, apply a unit vertical load at C and analyze the frame. The results are tabulated below (The students are advised to work out the stresses themselves.).

Table 17.4

(+ for tension; – for compression)

Member	L (mm)	p (N/mm^2)	u	uL
1	4000	-100	$-\frac{4}{9}$	$+\frac{160}{9}\times10^4$
2	4000	-100	$-\frac{4}{9}$	$+\frac{160}{9}\times10^4$
3	4000	-100	$-\frac{8}{9}$	$+\frac{320}{9}\times10^4$
4	5000	$+150$	$+\frac{5}{9}$	$+\frac{375}{9}\times10^4$
5	3000	-100	0	0
6	5000	0	$-\frac{5}{9}$	0
7	4000	-150	$+\frac{8}{9}$	$+\frac{480}{9}\times10^4$
8	3000	-100	$+\frac{1}{3}$	-10×10^4
9	5000	$+150$	$+\frac{10}{9}$	$+\frac{750}{9}\times10^4$
			Sum	-259.4×10^4

$$\delta_{CV} = \sum_{1}^{n} \frac{puL}{E} = \frac{259.44\times10^4}{2\times10^5} = 12.97\,\text{mm}\,(\downarrow).$$

(b) Horizontal deflection of B

Since the roller at B moves in the horizontal direction only, its movement will evidently be equal to the axial shortening of the members 1, 2, and 3.

Thus, $\delta_{BH} = \Delta_1 + \Delta_2 + \Delta_3$

$$= \sum_{1}^{3} \frac{pL}{E} = \frac{3(100 \times 4000)}{2 \times 10^5} = 6\,\text{mm}\,(\leftarrow).$$

PROBLEM 17.5

Objective 2

The frame shown in Figure 17.6 consists of four panels each 3 m wide, and the cross-sectional areas of the member are such that, when the frame carries equal loads at the panel points of the lower chord, the stress in all the tension members is f N/mm^2, and the stress in all the compression members of 0.75 f N/m^2. Determine the value of f if the ratio of the maximum deflection to span is $\frac{1}{750}$. Take $E = 2 \times 10^5$ N/mm^2.

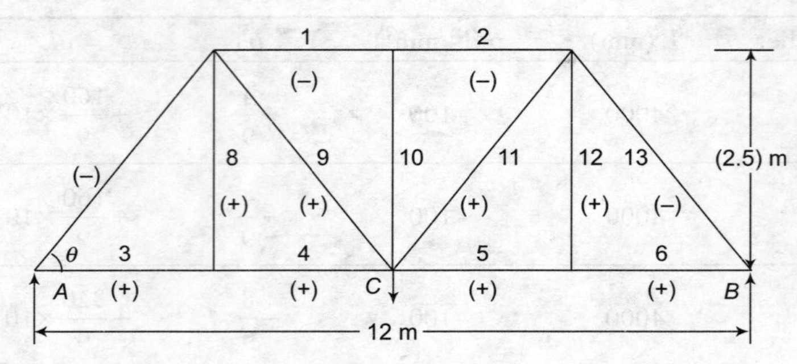

FIGURE 17.6

SOLUTION

By inspection, it can be seen that for the loads at the panel points of the lower panel, the top chord members will be in compression, and the bottom chord members, verticals, and diagonals will be in tension.

Because of symmetrical loading, the maximum deflection occurs at C. Apply unit load at C to find u in all the members. All the members have seen numbered 1, 2, …, etc. in Figure 17.6. By inspection

$$u_8 = 0;\ u_{10} = 0;\ u_{12} = 0$$

Reaction $R_A = R_B = \frac{1}{2}$;

$$\theta = 45°;\ \cos\theta = \sin\theta = \frac{1}{\sqrt{2}}$$

$$\therefore \quad u_7 = \frac{R_A}{\sin\theta} = \frac{\sqrt{2}}{2} \text{ (compression)}$$

$$u_3 = u_7 \cos\theta = \frac{\sqrt{2}}{2}\frac{1}{\sqrt{2}} = \frac{1}{2} = u_4 \text{ (tension)}$$

$$u_9 = \frac{u_4}{\cos\theta} = \frac{\sqrt{2}}{2} \text{ (tension)}$$

Also, $u_7 \cos\theta + u_9 \cos\theta = u_1$

$$u_1 = \frac{\sqrt{2}}{2} \times \frac{1}{\sqrt{2}} + \frac{\sqrt{2}}{2} \times \frac{1}{\sqrt{2}} = 1.0 \text{ (compression)}$$

The results for half the truss are tabulated below.

Table 17.5

Member	Length (mm)	p (N/mm^2)	u	puL
1	3000	$-0.75f$	-1.0	$+2250f$
3	3000	$+f$	$+\frac{1}{2}$	$+1500f$
4	3000	$+f$	$+\frac{1}{2}$	$+1500f$
7	$3000\sqrt{2}$	$-0.75f$	$-\frac{\sqrt{2}}{2}$	$+2250f$
8	3000	$+f$	0	0
9	$3000\sqrt{2}$	$+f$	$+\frac{\sqrt{2}}{2}$	$+3000f$
			Sum	$+10,500f$

$$\delta_C = \sum_1^n \frac{puL}{E} = \frac{(10,500 f \times 2)}{2 \times 10^5} = 0.105 f \text{ mm}$$

As per given condition,

$$\delta_C = \frac{1}{750} \times \text{span} = \frac{1}{750} \times 12000 = 16 \text{ mm}$$

Hence, $0.105f = 16$ or $f = \frac{16}{0.105} = 152.38 \text{ N/mm}^2$.

PROBLEM 17.6

Objective 1

Figure 17.7 (a) shows the outline of truss used for lifting a load which is distributed as 4 kN on each of the four points Q, R, S, and T. The members PQ, PR, PS, and PT, each have an area of 90 mm^2 and the members QR, RS and ST each have of 150 mm^2. Determine the vertical deflection of *Q* and *R* relative to support *P*. Take $E = 2 \times 10^5$ N/mm^2.

SOLUTION

Since the loading and the truss are symmetrical, consider half truss only. For the equilibrium of the half truss, the horizontal reaction (H) will be supplied by the other half of the truss as shown in Figure 17.7 (b). Thus, the half truss of Figure 17.7 (b) is obtained by assuming the truss having cut by a vertical section through P, and the basic system of Figure 17.7 (a) does not change. Figure 17.7 (c) shows the half truss under vertical load at Q, and Figure 17.7 (d) shows the half truss under vertical load at R. The vertical deflection of Q and R are given by

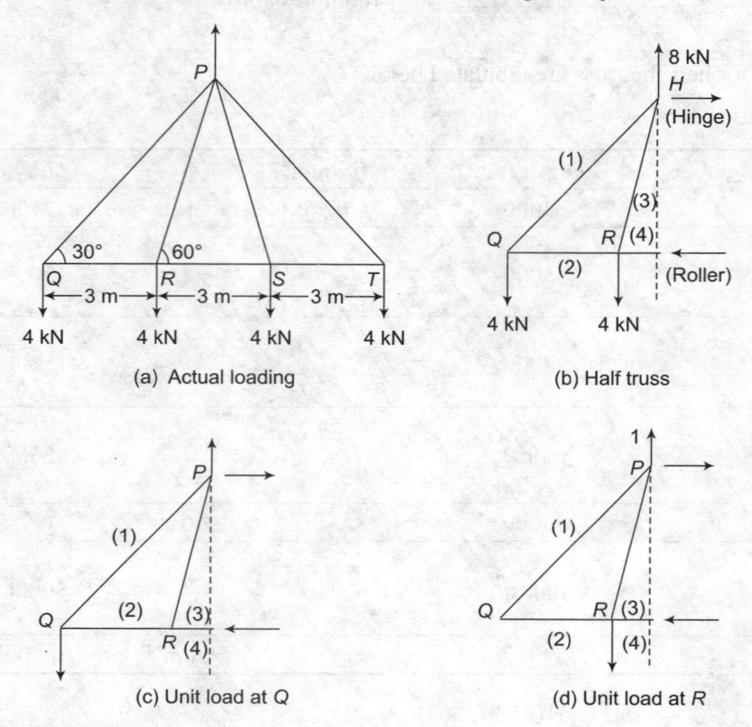

FIGURE 17.7

$$\delta_{QV} = \sum \frac{PuL}{AE} \text{ and } \delta_{RV} = \sum \frac{Pu^1L}{AE}$$

in which u is the stress in any member due to unit vertical load at Q, and u^1 is the stress due to unit vertical load at R. The members have been numbered 1, 2, 3, etc. The summation is made for half the truss only.

(a) Calculation of stresses due to external loading

$P_1 \sin 30° = 4$; $P_1 = 8$ (compression)

$$P_2 = P_1 \cos 30° = \frac{8\sqrt{3}}{2} = 4\sqrt{3} \text{ (compression)}$$

$$P_3 \sin 60° = 4; P_3 = \frac{4 \times 2}{\sqrt{3}} = \frac{8\sqrt{3}}{3} \text{ (tension)}$$

$$P_4 = P_3 \cos 60° + P_2 = \frac{8\sqrt{3}}{3} \times \frac{1}{2} + 4\sqrt{3} = \frac{16\sqrt{3}}{3} \text{ (compression).}$$

(b) Calculation of stresses due to unit vertical load at Q

By inspection, $u_3 = 0$

$$u_1 \sin 30° = 1;\ u_1 = 2 \text{ (tension)}$$

$$u_2 = u_1 \cos 30° = 2\times\frac{\sqrt{3}}{2} = \sqrt{3} \text{ (compression)};\quad u_4 = u_2 = \sqrt{3} \text{ (compression)}.$$

(c) Calculation of stresses due to unit vertical load at R

Since there is no load at Q, $u_1^1 = 0;\ u_2^1 = 0$

$$u_3^1 \sin 60° = 1;\ u_3^1 = \frac{2}{\sqrt{3}} = \frac{2\sqrt{3}}{3} \text{ (tension)}$$

$$u_4^1 \cos 60° = \frac{2\sqrt{3}}{3}\times\frac{1}{2} = \frac{\sqrt{3}}{3} \text{ (compression)}.$$

The results are tabulated below.

Table 17.6

(+ for tension; – for compression)

Member	L (mm)	A (mm^2)	p (kN)	u	u^1	$\frac{PuL}{A}$	$\frac{Pu^1L}{A}$
1	$3000\sqrt{3}$	90	+8	+2	0	+923.76	0
2	3000	150	$-4\sqrt{3}$	$-\sqrt{3}$	0	+240	0
3	3000	90	$+\frac{8\sqrt{3}}{3}$	0	$+\frac{2\sqrt{3}}{3}$	0	+177.77
4	1500	150	$-\frac{16\sqrt{3}}{3}$	$-\sqrt{3}$	$-\frac{\sqrt{3}}{3}$	+160	+53.33
					Sum	+1323.76	+231.1

$$E = 2\times 10^5 \text{ N/mm}^2 = 200 \text{ kN/mm}^2$$

$$\delta_{QV} = \frac{1323.76}{200} = 6.62 \text{ mm};\ \delta_{RV} = \frac{231.1}{200} = 1.155 \text{ mm}.$$

PROBLEM 17.7 **Objective 3**

The roof truss shown in Figure 17.8 has members with cross-sectional areas such that when the loading is as shown, all members are subjected to the same intensity of stress, either tensile or compressive. If the vertical deflection of joint C is 15 mm, determine the change in the span of the truss.

SOLUTION

To find the vertical deflection at point C, apply a unit vertical load at C as shown in Figure 17.8. Let $\pm P$ be the intensity of stress in the members due to external loading. The members have been numbered 1, 2, 3, etc.

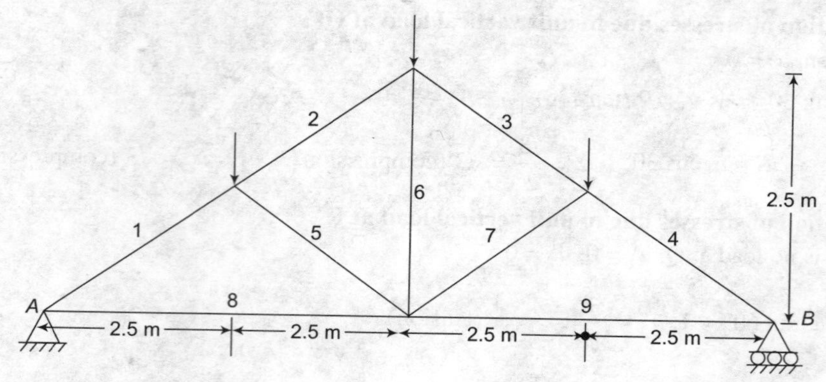

FIGURE 17.8

$$L_1 = L_2 = L_3 = L_4 = L_5 = L_7 = \sqrt{(2.5)^2 + \left(\frac{2.5}{2}\right)^2} = \frac{2.5}{2}\sqrt{5}\text{ m} = 1250\sqrt{5}\text{ mm}$$

$$L_6 = 2.5\text{ m};\ L_8 = L_9 = 5\text{ m and } \sin\theta = \frac{2.5}{2.5\sqrt{5}} = \frac{1}{\sqrt{5}};\ \cos\theta = \frac{2.5}{2.5\sqrt{5}} = \frac{2}{\sqrt{5}}$$

$$u_1 \sin\theta = \frac{1}{2} \text{ or } u_1 = \frac{\sqrt{5}}{2} = u_4 \text{ (compression)}$$

$$u_8 = u_1 \cos\theta = \frac{\sqrt{5}}{2} \times \frac{2}{\sqrt{5}} = 1 = u_9 \text{ (tension)};\ u_5 = u_7 = 0$$

$$\Rightarrow \quad u_6 = 1 \text{ (tension) and } u_2 = u_3 = u_1 = \frac{\sqrt{5}}{2} \text{ (compression)}$$

$$\delta_c = \sum \frac{PUL}{E} = \frac{25000P}{E}\text{ mm}$$

But $\delta_c = 15$ mm (given)

$$\Rightarrow \quad \frac{25000P}{E} = 15$$

$$\Rightarrow \quad \frac{P}{E} = \frac{3}{5000}$$

Now, change in the span AB $= \Delta_{AC} + \Delta_{CB} = 2\left\{\frac{5000P}{E}\right\} = 10000\frac{P}{E}$

Substituting the value of $\frac{P}{E}$, we get

Change in the span AB $= \frac{10000 \times 3}{5000} = \frac{30}{5} = 6\text{ mm}$

The results are tabulated below, giving correct signs to $\pm P$ obtained by inspection.

Table 17.7

(+ for tension, – for tension)

Member	Length L (mm)	Stress P	U	PUL
1	$1250\sqrt{5}$	$-P$	$-\frac{\sqrt{5}}{2}$	$+3125P$
2	$1250\sqrt{5}$	$-P$	$-\frac{\sqrt{5}}{2}$	$+3125P$
3	$1250\sqrt{5}$	$-P$	$-\frac{\sqrt{5}}{2}$	$+3125P$
4	$1250\sqrt{5}$	$-P$	$-\frac{\sqrt{5}}{2}$	$+3125P$
5	$1250\sqrt{5}$	$-P$	0	0
6	2500	$+P$	$+1$	$+2500P$
7	$1250\sqrt{5}$	$-P$	0	0
8	5000	$+P$	$+1$	$+5000P$
9	5000	$+P$	$+1$	$+5000P$
			Sum	$+25000P$

PROBLEM 17.8 **Objective 4**

The frame shown in Figure 17.9 consists of four panels each 2.5 m wide and cross-sectional area of the members are such that when the frame carries equal loads at the panel points of the lower chord the stress in all tension members is 100 N/mm^2 and the stress in all the compression members is 80 N/mm^2. Determine the relative moment between the joints C and K in the direction CK. Take $E = 2 \times 10^5$ N/mm^2.

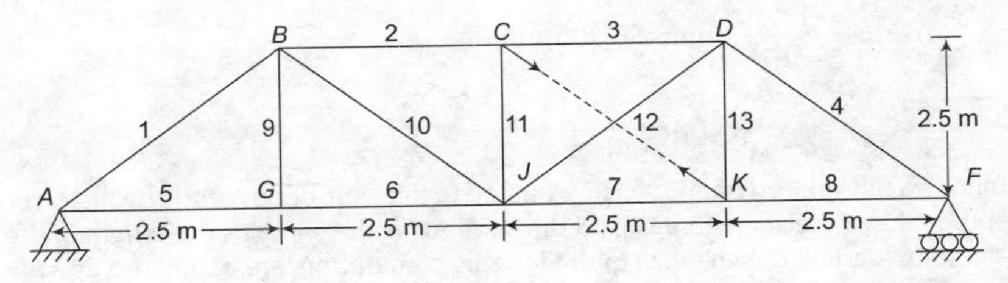

FIGURE 17.9

SOLUTION

To find the relative movement between joints C and K, apply unit loads at C and K in the direction CK. The movement δ of the joints C and K toward each other is then given by $\delta = \sum \frac{PUL}{E}$, and there will be no reactions at A and F due to the unit loads. Hence, $u_1 = 0$; $u_5 = 0$; $u_6 = 0$; $u_9 = 0$; $u_{10} = 0$; $u_2 = 0$; $u_4 = 0$; and $u_3 = 0$.

Resolving at joint C, $u_{11} = 1 \sin\theta = \frac{1}{\sqrt{2}}$ (compression); $u_3 = 1 \cos\theta = \frac{1}{\sqrt{2}}$ (compression).

Resolving at joint K, $u_7 = 1 \cos\theta = \frac{1}{\sqrt{2}}$ (compression); $u_{13} = 1 \sin\theta = \frac{1}{\sqrt{2}}$ (compression).

Resolving at D

$$u_{12} \cos\theta = u_{13} \Rightarrow u_{12} = 1 \text{ (tension)}$$

By inspection, it can be seen that for the loads at the panel points of the lower panel, the top chord members will be in compression and the bottom chord members, verticals, and diagonals will be in tension. Member CJ does not carry any stress.

Hence, $P_1 = P_2 = P_3 = P_4 = 80\ \text{N/mm}^2$ (Compression)

$P_5 = P_9 = P_6 = P_{10} = 100\ \text{N/mm}^2$ (tension.)

$P_1 = P_8 = P_{12} = P_{13} = 100\ \text{N/mm}^2$ and $P_{11} = 0$.

The results are tabulated below. However, since δ is the function of the product of P and U, those members having $U = 0$ or $P = 0$ have not been included in the table.

Table 17.8

Member	Length L (mm)	Stress P	U	PUL
U_3	2500	−80	$-\frac{1}{\sqrt{2}}$	$+14.14 \times 10^4$
U_7	2500	+100	$-\frac{1}{\sqrt{2}}$	-17.66×10^4
U_{12}	$2500\sqrt{2}$	+100	+1	$+35.32 \times 10^4$
U_{13}	2500	+100	$-\frac{1}{2}$	-17.66×10^4
			Sum	$+14.14 \times 10^4$

$$\delta = \frac{14.14 \times 10^5}{2 \times 10^5}\ \text{mm} = 0.706\ \text{mm}.$$

PROBLEM 17.9

Objective 1

Figure 17.10 shows a pin-jointed frame, which is hinged to rigid supports A and D, which are at the same level. All members have the same length and the span AD is the same as the length of the members. Because of a certain loading the changes in the lengths of members are estimated as AB = +0.185 in.; BC = +0.240 in.; BD = −0.200 in.; CD = −0.365 in. Find the horizontal and vertical movements of C.

SOLUTION

The horizontal and vertical deflections of C are given by

$$\delta_V = \sum \frac{PUL}{E} = \Sigma U\Delta \text{ and } \delta_H = \sum \frac{PUL}{E} = \Sigma U' \cdot \Delta,$$

in which U is the force in any member due to unit vertical load at C and U' is the force due to horizontal load at C.

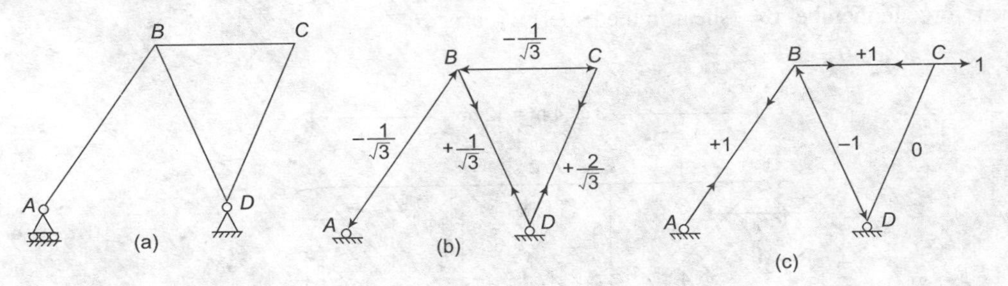

FIGURE 17.10

To find the values of C, apply unit vertical load at C. Figure 17.10 shows the induced stresses in the members. Similarly, Figure 2 shows the stresses due to unit horizontal load at C. The results are tabulated below.

Table 17.9

Member	Deformation Δ (in.)	U	U'	$U\Delta$	$U'\Delta$
AB	+0.185	$-\frac{1}{\sqrt{3}}$	+1	−0.107	+0.185
BC	+0.240	$-\frac{1}{\sqrt{3}}$	+1	−0.139	+0.240
CD	−0.365	$+\frac{2}{\sqrt{3}}$	0	−0.422	0
BD	−0.200	$+\frac{1}{\sqrt{3}}$	−1	−0.116	+0.200
			Sum	−0.784	+0.625

$$\delta_V = \sum U\Delta = -0.748 \text{ in. and } \delta_H = \sum U'\Delta = +0.625 \text{ in.}$$

PROBLEM 17.10 **Objective 1**

Determine the horizontal and vertical deflection of the joint C of the frame as shown in Figure 17.11, if member DF has a lack of fit of 1 cm (long).

SOLUTION

Let u be the stress in member DF due to unit vertical load at C [Fig. 17.11 (a)] and 'u' be the stress in due to unit horizontal load at C [Fig. 17.11 (b)]. Since only one member has lack of fit, we have

$$\delta_{CV} = u_6\Delta_6 = u_6 \times (+1) = u_6 \quad (\text{since } u_6 = +1 \text{ cm})$$

and

$$\delta_{CH} = u_6'\Delta_6 = u_6' \times (+1) = u_6'.$$

Refer Figure 17.11 (a), reaction at A and B will be 1/3 and 2/3, respectively. Pass a section to cut the members 2, 6, and 7.

Then, force in member 6 = (shear in the panel) × (cosec θ)

cosec θ = 5/3

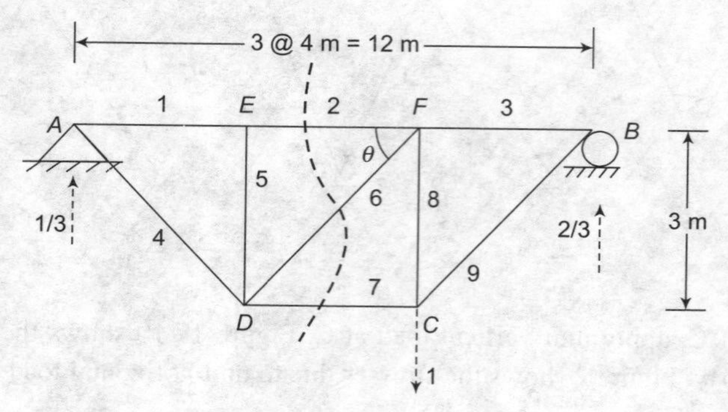

FIGURE 17.11 (A)

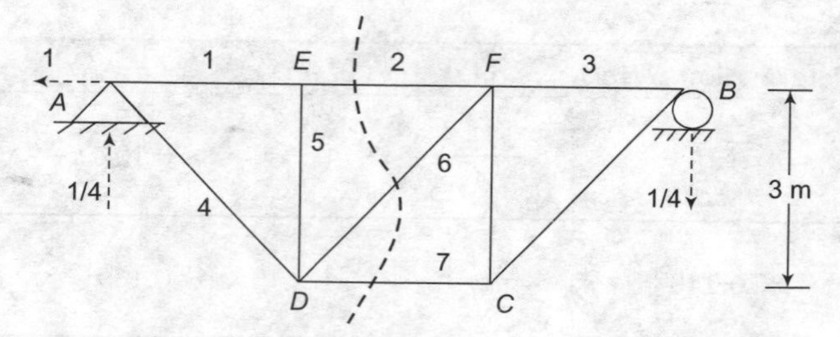

FIGURE 17.11 (B)

$$u_6 = \frac{1}{3} \times \frac{5}{3} = \frac{5}{9} \text{ (compression)} = -\frac{5}{9} \text{ (since compression has negative sign)}$$

$$\delta_{CV} = u_6 = -\frac{5}{9} \text{cm, i.e., } \frac{5}{9} \text{cm } (\uparrow)$$

Similarly, when a unit horizontal load is applied at C [Fig. 17.11(b)], the horizontal reaction at A = 1(←). Since the unit horizontal load at C and the reaction at A produce a couple in the anticlockwise direction, equal and opposite vertical reactions of magnitude $\frac{1 \times 3}{12} = \frac{1}{4}$ will be induced at A and B as shown.

To find the force u'_6 in member FD, pass a section to cut the member 2, 6, and 7. Then

$$u'_6 = \text{(shear in the panel)} \times (\text{cosec } \theta) = \frac{1}{4} \times \frac{5}{3} = \frac{5}{12} \text{ (compression)} = -\frac{5}{12}.$$

17.4 DEFLECTION BY CASTIGLIANO'S FIRST THEOREM

In previous chapters, it has been proved that the partial derivative of the total strain energy with respect to a force gives the deflection in the direction of the force. This is Castigliano's first theorem, and its application for finding out the deflection of the beams, etc., has already been studied earlier. In articulated structures, the loads are applied at panel points only, and hence the members carry axial forces (either tension or compression) only. Thus, the strain energy will be due to direct forces and is given by

$$U = \sum_{1}^{n} \frac{P^2 L}{2AE},$$

in which P is force in any member due to external loading. If W is an external load acting at a joint and it is required to find the deflection of the joint in the direction of the application of the load, we have, according to Castigliano's first theorem.

$$\frac{\partial U}{\partial W} = \sum_{1}^{n} \frac{PL}{AE} \frac{\partial P}{\partial W}. \tag{17.13}$$

This is the expression for the deflection of a joint. If, however, it is required to find the deflection of a joint where W is not acting, a fictitious load W is applied there in the required direction. The fictitious load W is then equated to zero. A similar method may be employed for the joint where external load is acting, but the deflection is required to be found in some other direction.

Procedure for Computing Deflection

1. Apply a fictitious load W at the joint in the direction in which the deflection is required if no such external load is acting.
2. Find the force P in all members. The force P will be a function of W and the external load. Thus, in general, $P = a + bW$
   in which a and b are the constants depending on the geometry of the truss, position of the load, the position of the member, and the system of external loading. In some cases, either 'a' may be zero, or b may be zero, or both a and b may be zero.
3. Find the value $\frac{\partial P}{\partial W}(= b)$.
4. Calculate $\frac{PL}{AE} \cdot \frac{\partial P}{\partial W}$ for each number. If W is a fictitious load, equate it to zero.
5. $\sum \frac{PL}{AE} \cdot \frac{\partial P}{\partial W}$ gives the required deflection.

Comparison with unit load method

The deflection by the unit load method is given by

$$\delta = \sum_{1}^{n} \frac{PUL}{AE} = \sum_{1}^{n} \frac{PL}{AE} \cdot U = \sum_{1}^{n} \Delta \cdot U$$

The deflection by Castigliano's theorem is given by

$$\mathrm{d} = \sum_{1}^{n} \frac{PL}{AE} \cdot \frac{\partial P}{\partial W}.$$

If expressions (1) and (2) are compared, it is evident that $\frac{\partial P}{\partial W} = U$. Thus, both the expressions are the same. However, because of different form, the procedure for computation is different. In the unit load method, one has to analyze the frame twice for finding P and U in each member, whereas in the latter method, only one analysis is needed. However, the expression for P, by the Castigliano's method, is sometimes long and cumbersome.

PROBLEM 17.11

Objective 1

Solve Problem 17.2 by Castigliano's first theorem.

SOLUTION

(a) Vertical deflection of C

An external load W is already acting vertically C. Hence

$$\delta_{CV} = \sum_{1}^{n} \frac{PL}{AE} \cdot \frac{\partial P}{\partial W}$$

The values of P in various members have already been calculated in example 17.2. The values of P, $\frac{\partial P}{\partial W}$, etc., have been entered, and computations done in a tabular form below.

Table 17.10

(+ for tension; – for compression)

Member	Length L	Area A	P	$\frac{\partial P}{\partial W}$	$\frac{PL}{A} \cdot \frac{\partial P}{\partial W}$
AB	L	a	$+W$	$+1$	$+\frac{WL}{a}$
BC	L	a	$+W$	$+1$	$+\frac{WL}{a}$
AD	L	$2a$	$-W$	-1	$+\frac{WL}{2a}$
CD	L	$2a$	$-W$	-1	$+\frac{WL}{2a}$
BD	L	$2a$	$-W$	-1	$+\frac{WL}{2a}$
					sum = $\frac{7WL}{2a}$

Hence $\delta_{CV} = \sum_{1}^{n} \frac{PL}{AE} \cdot \frac{\partial P}{\partial W} = \frac{7WL}{2a}$.

(b) Horizontal deflection of C

Apply a horizontal load H at the joint D (in the dotted direction shown in Fig. 17.3), with the external load W still acting on the frame. The horizontal deflection of C is given by,

$$\delta_{DH} = \sum_{1}^{n} \frac{PL}{AE} \cdot \frac{\partial P}{\partial H}.$$

It will be seen that the forces in all members except AD will be the same as in the previous case (i.e., when only W is acting and $H = 0$). The force in the AD can be found by resolving horizontally at A. As the horizontal reaction at A is H, we have

$$P_{AD} \cos 30° = H + P_{AB} \cos 30° = W \cos 30°$$

$$P_{AD} = H\frac{2}{\sqrt{3}} + W \text{ and } \frac{\partial P_{AD}}{\partial H} = \frac{2}{\sqrt{3}}.$$

The forces in the four members are as below:

$$P_{AB} = +W; \frac{\partial P_{AB}}{\partial H} = 0; P_{BC} = +W; \frac{\partial P_{BC}}{\partial H} = 0;$$

$$P_{CD} = -W; \frac{\partial P_{CD}}{\partial H} = 0; P_{BD} = -W; \frac{\partial P_{BD}}{\partial H} = 0;$$

Hence $\delta_{DH} = \sum_{1}^{n} \frac{PL}{AE} \cdot \frac{\partial P}{\partial H} = \left(\frac{2}{\sqrt{3}}H + W\right)\frac{2}{\sqrt{3}} \cdot \frac{L}{2AE} = \frac{WL}{\sqrt{3}AE}$ by substituting $H = 0$.

PROBLEM 17.12 **Objective 1**

A pin-jointed frame shown in Figure 17.12 is hinged to a rigid wall at A and is free to slide vertically at E. The frame carries a vertical load 'W' at B. The area of each tension member is 'a' and of each compression member '$2a$' and the length AE is 'L'. Obtain an expression for the vertical displacement of C.

SOLUTION

Because of external loading, members AB, AD, and AE will carry tension, and the area of each these members is therefore a. Members BD and DE carry compression and hence their area is $2a$. Members BC and CD carry zero stresses.

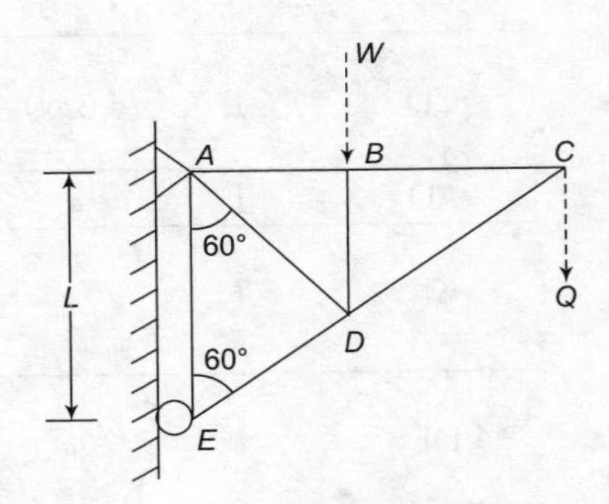

FIGURE 17.12

To find the vertical deflection of C, apply a fictitious vertical load Q there. Then

$$\delta_{CV} = \sum_{1}^{n} \frac{PL}{AE} \cdot \frac{\partial P}{\partial Q},$$

in which P = stress in any member due to both external and fictitious load.

Vertical reaction at E = 0 (roller)

Vertical reaction at A = $(W + Q)$

The stresses in BC, CD, and AB will be in terms of Q only and hence $P \cdot \frac{\partial P}{\partial Q}$ will be zero when the value of 'Q' is put zero. Similarly, stresses in BD and AD will be the function of W only and hence $\frac{\partial P}{\partial Q}$ and $P \cdot \frac{\partial P}{\partial Q}$ will be zero for these two members. Thus, $\sum \frac{PL}{AE} \cdot \frac{\partial P}{\partial Q}$ is required for members AE and DE only.

Pass a section to cut AB, AD, and ED. Taking moments about A, we get

$$P_{DE} = \frac{(W \times AB) + (Q + AC)}{L \sin 60°} = \frac{WL \cos 30° + 2QL \cos 30°}{L \sin 60°} = (W + 2Q) \text{ (compression)}$$

Resolving vertically at E, $P_{AE} = P_{DE} \cos 60° = \frac{1}{2}(W + 2Q)$ tension.

To make it more clear now $P \cdot \frac{\partial P}{\partial Q}$ is zero for the five members, value of P and $\frac{\partial P}{\partial Q}$ are tabulated below for all the members.

Table 17.11

Member	Length L	Area A	P	$\frac{\partial P}{\partial Q}$	$\frac{PL}{AE} \cdot \frac{\partial P}{\partial Q}$ after substituting Q = 0
AB	$\frac{\sqrt{3}}{2}L$	a	$+\frac{4Q}{\sqrt{3}}$	$+\frac{4}{\sqrt{3}}$	$\frac{16}{3} \times \frac{\sqrt{3}}{2} \frac{QL}{aE} = 0$
BD	$\frac{L}{2}$	$2a$	$-W$	0	0
BC	$\frac{\sqrt{3}}{2}L$	a (say)	$+\frac{4Q}{\sqrt{3}}$	$+\frac{4}{\sqrt{3}}$	$\frac{16}{3} \times \frac{\sqrt{3}}{2} \frac{QL}{aE} = 0$
CD	L	a (say)	$-2Q$	-2	$\frac{4L}{aE} Q = 0$
AD	L	a	$+W$	0	0
AE	L	a	$+\frac{1}{2}(W + 2Q)$	$+1$	$\frac{L(W + 2Q)}{2aE} = \frac{WL}{2aE}$
DE	L	$2a$	$-(W + 2Q)$	-2	$\frac{2L(W + 2Q)}{2aE} = \frac{WL}{aE}$
		Sum			$\frac{3WL}{2aE}$

$$\delta_{CV} = \sum_{1}^{n} \frac{PL}{AE} \cdot \frac{\partial P}{\partial Q} = \frac{3WL}{2aE}$$

17.5 MAXWELL'S RECIPROCAL THEOREM APPLIED TO FRAMES

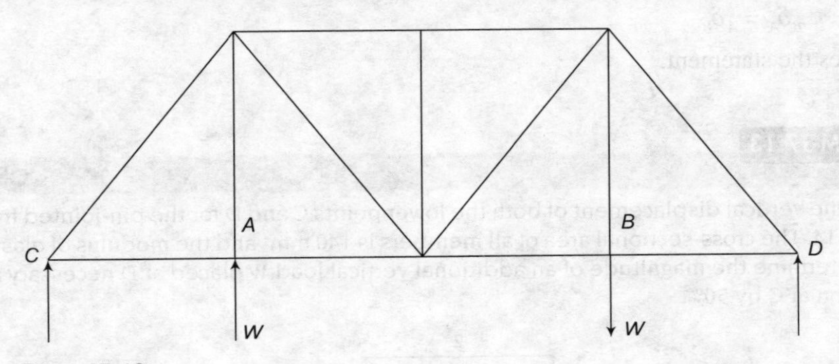

FIGURE 17.13

As applied to the deflection of articulated structures, Maxwell's theorem of reciprocal deflection has the following statement:

'In a perfect frame under equilibrium, the deflection of any joint A due to a load at the joint B is equal to the deflection of the joint B due to same load at the joint A.'

Expressed mathematically, ${}_{B}\delta_{A} = {}_{A}\delta_{B}$

in which ${}_{B}\delta_{A}$ = deflection of A due to load at B

${}_{A}\delta_{B}$ = deflection of B due to load at A

Thus, in Figure.17.13 let W be the load at B. According to unit load method, the deflection of the joint A is given by

$$ {}_{B}\delta_{A} = \sum_{1}^{n} \frac{PuL}{AE} \tag{17.14}$$

in which P = stresses in any member due to W at B

u = stress in any member due to unit load A.

Now remove the load from B and apply it at the joint A. Then, the deflection at the joint B is given by

$$ {}_{A}\delta_{B} = \sum_{1}^{n} \frac{P'u'L}{AE} \tag{17.15}$$

in which P' = stresses in any member due to W at $A = uW$

u' = stresses in any member due to unit load at $B = \dfrac{P}{W}$

Substituting these values of P' and u' in Eq. (2), we get

$$_A\delta_B = \sum_1^n \frac{L}{AE}(uW)\left(\frac{P}{W}\right) = \sum_1^n \frac{PuL}{AE} \tag{17.16}$$

Comparing Eqs. (17.13) and (17.15), we get

$$_B\delta_A = {}_A\delta_B \tag{17.17}$$

which proves the statement.

PROBLEM 17.13

Objective 3

Determine the vertical displacement of both the lower points C and D for the pin-jointed frame shown in Figure 17.14. The cross-sectional area of all members is 140 mm^2 and the modulus of elasticity is 200 kN/mm^2. Determine the magnitude of an additional vertical load W placed at D necessary to increase the deflection at C by 50%.

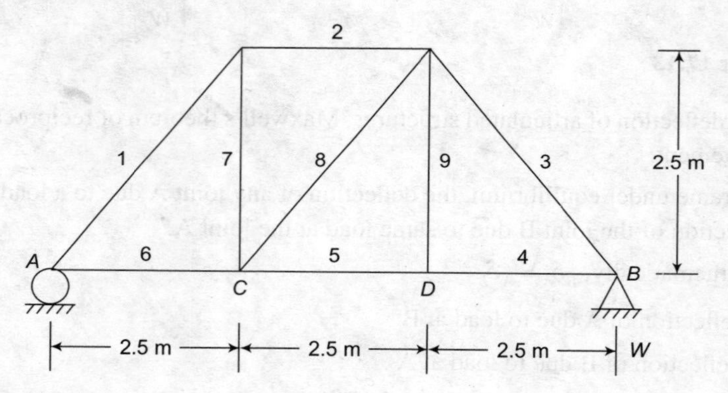

FIGURE 17.14

SOLUTION

From Castigliano's first theorem, the deflections of C and D due to external load is given by

$$\delta_C = \sum_1^n \frac{Pu_1L}{AE} \text{ and } \delta_D = \sum_1^n \frac{Pu_2L}{AE}$$

in which P = force in any member due to the load of 9 kN acting at C

u_1 = force in any member due to unit load at C

u_2 = force in any member due to unit load at D.

The calculations of P, u_1, and u_2 are presented in the tabular form below:

Table 17.12

(+ for tension; – for compression)

Member	L (mm)	P	u_1	u_2	Pu_1L	Pu_2L
1	$2500\sqrt{2}$	$-6\sqrt{2}$	$-\frac{2}{3}\sqrt{2}$	$-\frac{\sqrt{2}}{3}$	$+2.828 \times 10^4$	$+1.414 \times 10^4$
2	2500	-6	$-\frac{2}{3}$	$-\frac{1}{3}$	$+1.0 \times 10^4$	$+0.5 \times 10^4$
3	$2500\sqrt{2}$	$-3\sqrt{2}$	$-\frac{1}{3}\sqrt{2}$	$-\frac{2\sqrt{2}}{3}$	$+0.707 \times 10^4$	$+1.414 \times 10^4$
4	2500	$+3$	$+\frac{1}{3}$	$+\frac{2}{3}$	$+0.250 \times 10^4$	$+0.5 \times 10^4$
5	2500	$+3$	$+\frac{1}{3}$	$+\frac{2}{3}$	$+0.250 \times 10^4$	$+0.5 \times 10^4$
6	2500	$+6$	$+\frac{2}{3}$	$+\frac{1}{3}$	$+1.0 \times 10^4$	$+0.5 \times 10^4$
7	2500	$+6$	$+\frac{2}{3}$	$+\frac{1}{3}$	$+1.0 \times 10^4$	$+0.5 \times 10^4$
8	$2500\sqrt{2}$	$+3\sqrt{2}$	$+\frac{1}{3}\sqrt{2}$	$-\frac{\sqrt{2}}{3}$	$+0.707 \times 10^4$	-0.707×10^4
9	2500	0	0	$+1$	0	0
				Sum	7.742×10^4	4.621×10^4

$$\delta_C = \sum_1^n \frac{Pu_1L}{AE} = \frac{7.742 \times 10^4}{140 \times 200} = 2.76\,\text{mm}\ (\downarrow) \tag{17.19}$$

$$\delta_D = \sum_1^n \frac{Pu_2L}{AE} = \frac{4.621 \times 10^4}{140 \times 200} = 1.65\,\text{mm}\ (\downarrow) \tag{17.20}$$

Let the additional load at D be W. We want ${}_D\delta_C = \frac{2.76}{2} = 1.38\,\text{mm}$

According to Maxwell's reciprocal theorem

$${}_C\delta_D = {}_D\delta_C\text{; Hence } {}_C\delta_D = 1.38\,\text{mm} \tag{17.21}$$

When a load of W is at C, ${}_C\delta_D = \frac{1.65}{9}W$ (17.22)

Equating Eqs. (3) and (4), we get

$$\frac{1.65}{9}W = 1.38\text{ ; or } W = \frac{1.38 \times 9}{1.65} = 7.52\text{ kN.}$$

SUMMARY

- **The horizontal and vertical deflection at C are given by**

$$\delta_V = \sum_1^n \frac{PuL}{AE} = \sum_1^n \frac{puL}{E}, \quad \delta_H = \sum_1^n \frac{Pu^1L}{AE} = \sum_1^n \frac{pu^1L}{E}$$

- **Deflection of a joint due to temperature variation**
Let Δ_1, Δ_2 ... Δ_n be the changes in the lengths of various members of a perfect frame due to temperature variation. Thus,

$$\delta = \sum_1^n u^1\Delta = u_1\Delta_1 + u_2\Delta_2....u_n\Delta_n$$

- **The deflection by Castigiliano's theorem is given by**

$$\delta = \sum_{i=1}^n \frac{pL}{AE}\cdot\frac{\partial P}{\partial W}$$

- **Maxwell's reciprocal theorem applied to frames**

$${}_A\delta_B = \sum_1^n \frac{L}{AE(uW)\left(\dfrac{P}{W}\right)} = \sum_1^n \frac{PuL}{AE}$$

OBJECTIVE TYPE QUESTIONS

1. The deflection at any point of a perfect frame can be obtained by applying a unit load at the joint in
   (a) Vertical direction (b) Horizontal direction
   (c) Inclined direction (d) The direction in which the deflection is required
2. Castigliano's first theorem is applicable
   (a) For statically determinate structures only
   (b) When the system behaves elastically
   (c) Only when principle of superposition is valid
   (d) None of the above
3. The graphical method of determining the forces in the members of a truss is based on
   (a) Method of joint (b) Method of section
   (c) Either method (d) None of the two methods
4. The member forces in a statically in determinate truss
   (a) Can be obtained by graphic statics (b) Cannot be obtained by graphic statics
   (c) May be obtained by graphic statics (d) Can be obtained by graphic statics by trial and error

5. Independent displacement components at each joint of a rigid-jointed plane frame are
(a) Three linear movements
(b) Two linear movements and one rotation
(c) One linear movement and two rotations
(d) Three rotations

6. The number of independent displacement components at each joint of a rigid-jointed space frame is
(a) 1 (b) 2 (c) 3 (d) 6

7. The Castigliano's second theorem can be used to compute deflections
A) In statically determinate structures only
(b) For any type of structure
(c) At the point under the load only
(d) For beams and frames only

8. Bending moment (BM) at any section in a conjugate beam gives in the actual beam
(a) Slope (b) Curvature (c) Deflection (d) BM

9. Which of the following is not the displacement method?
(a) Equilibrium method (b) Column analogy method
(c) Moment distribution method (d) Kani's method

10. Select the correct statement
(a) Flexibility matrix is a square symmetrical matrix
(b) Stiffness matrix is a square symmetrical matrix
(c) Both (a) and (b)
(d) None of the above

11. While using three moments equation, a fixed end of a continuous beam is replaced by an additional span of
(a) Zero length (b) Infinite length
(c) Zero moment of inertia (d) None of the above

12. The three moments equation is applicable only when
(a) The beam is prismatic
(b) There is no settlement of supports
(c) There is no discontinuity such as hinges within the span
(d) The spans are equal

13. The degree of static indeterminacy up to which column analogy method can be used is
(a) 2 (b) 3 (c) 4 (d) Unrestricted

14. Principle of superposition is applicable when
(a) Deflections are linear functions of applied forces
(b) Material obeys Hooke's law
(c) The action of applied forces will be affected by small deformations of the structure
(d) None of the above

15. The principle of virtual work can be applied to elastic system by considering the virtual work of
(a) Internal forces only (b) External forces only
(c) Internal as well as external forces (d) None of the above

16. If in a rigid-jointed space frame $(6m + r) < 6j$, then the frame is
(a) Unstable (b) Stable and statically determinate
(c) Stable and statically indeterminate (d) None of the above

Solutions for Objective Questions

Sl. No.	1.	2.	3.	4.	5.	6.	7.	8.	9.	10.
Answer	(b)	(c)	(a)	(b)	(b)	(d)	(b)	(c)	(b)	(c)

Sl.No.	11.	12.	13.	14.	15.	16
Answer	(a)	(c)	(b)	(a)	(c)	(a)

EXERCISE PROBLEMS

1. Determine the horizontal deflection at C of the truss shown in Figure 1, if the temperature is raised to 40°. Take $E = 200$ GPa. The load P is applied in such a way that ties are stressed to 80 MPa and struts are stressed to 60 MPa. The coefficient of thermal expansion of the material of the truss α is $12 \times 10^{-6}/°$ raise of temperature.

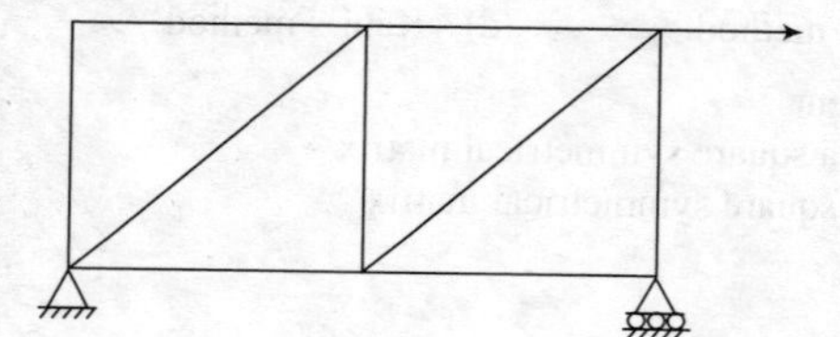

FIGURE 1

2. Determine the vertical deflection and horizontal deflection at C of the truss shown in Figure 4. AE = 10,000 kN. Height of the truss is 3 m. Each vertical load is 100 kN.

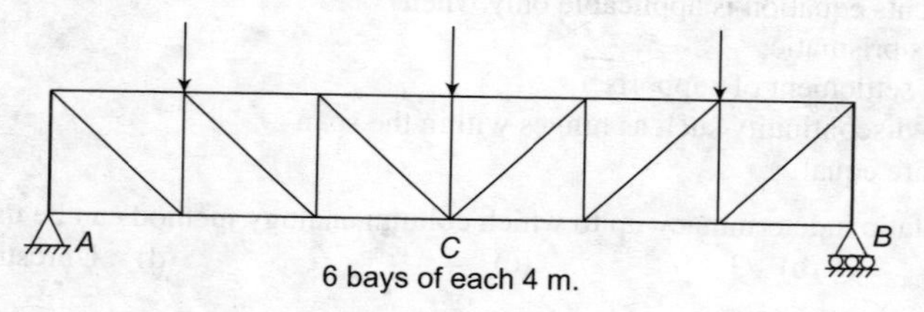

FIGURE 4

3. Determine the vertical deflection at C of the truss shown in Figure 3. Take $E = 200$ GPa. The load P is applied in such a way that ties are stressed to 80 MPa and struts are stressed to 60 MPa.

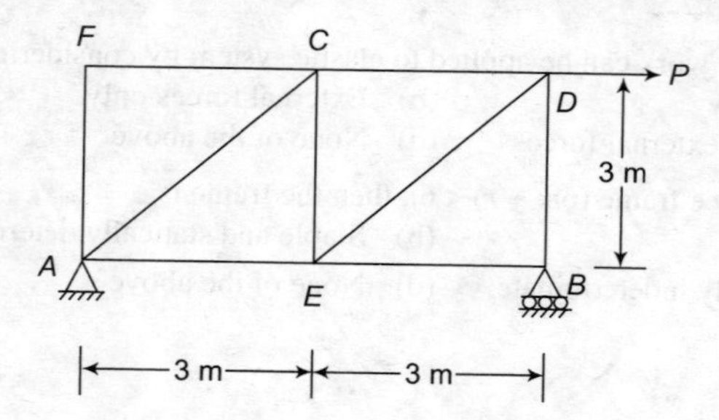

FIGURE 3

4. Determine the vertical and horizontal deflection at E of the truss shown in Figure 4. AE of all members is the same, and also find relative displacement between the points A and E. Objective 1

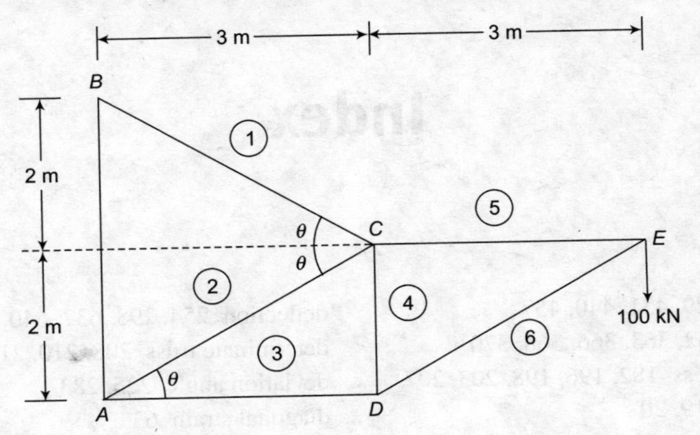

FIGURE 4

Index